AF181365

Statistics for Industry and Technology

Series Editor

N. Balakrishnan
McMaster University
Department of Mathematics and Statistics
1280 Main Street West
Hamilton, Ontario L8S 4K1
Canada

Editorial Advisory Board

Max Engelhardt
EG&G Idaho, Inc.
Idaho Falls, ID 83415

Harry F. Martz
Group A-1 MS F600
Los Alamos National Laboratory
Los Alamos, NM 87545

Gary C. McDonald
NAO Research & Development Center
30500 Mound Road
Box 9055
Warren, MI 48090-9055

Peter R. Nelson
Department of Mathematcal Sciences
Clemson University
Martin Hall
Box 341907
Clemson, SC 29634-1907

Kazuyuki Suzuki
Communication & Systems Engineering Department
University of Electro Communications
1-5-1 Chofugaoka
Chofu-shi
Tokyo 182
Japan

Parametric and Semiparametric Models with Applications to Reliability, Survival Analysis, and Quality of Life

M.S. Nikulin
N. Balakrishnan
M. Mesbah
N. Limnios
Editors

Springer Science+Business Media, LLC

M. S. Nikulin
Laboratoire Statistique Mathématique
Université Bordeaux 2
33076 Bordeaux Cedex
France
and
Laboratory of Statistical Methods
V. Steklov Mathematical Institute
191011 St. Petersburg
Russia

M. Mesbah
UFR SSI
Université de Bretagne-Sud
556017 Vannes Cedex
France

N. Balakrishnan
Department of Mathematics and Statistics
McMaster University
Hamilton, Ontario L8S 4K1
Canada

N. Limnios
Division Mathématiques Appliquées
Université de Technologie de Compiègne
Compiègne Cedex 60205
France

Library of Congress Cataloging-in-Publication Data
Parametric and semiparametric models with applications to reliability, survival analysis, and quality of life / M. Nikulin ... [et al.], editors.
p. cm. – (Statistics for industry and technology)
Includes bibliographical references and index.

1. Nonparametric statistics. 2. Nonparametric statistics–Industrial applications. I. Nikulin, M. S. (Mikhail Stepanovich) II. Series.

QA278.8.P37 2004
519.5–dc22 2004043766
CIP

AMS Subject Classifications: 62F03, 62F05, 62F10, 62F12, 62G05, 62G10, 62G20

ISBN 978-1-4612-6491-0 ISBN 978-0-8176-8206-4 (eBook)
DOI 10.1007/978-0-8176-8206-4

Printed on acid-free paper.

© 2004 Springer Science+Business Media New York
Originally published by Birkhäuser Boston in 2004
Softcover reprint of the hardcover 1st edition 2004

All rights reserved. This work may not be translated or copied in whole or in part without the written permission of the publisher (Birkhäuser Boston, c/o Springer, New York, LLC, 175 Fifth Avenue, New York, NY 10010, USA), except for brief excerpts in connection with reviews or scholarly analysis. Use in connection with any form of information storage and retrieval, electronic adaptation, computer software, or by similar or dissimilar methodology now known or hereafter developed is forbidden.
The use in this publication of trade names, trademarks, service marks and similar terms, even if they are not identified as such, is not to be taken as an expression of opinion as to whether or not they are subject to property rights.

9 8 7 6 5 4 3 2 1 SPIN 10925263

Birkhäuser is a part of *Springer Science+Business Media*

www.birkhauser.com

In Honor of Catherine Huber-Carol

Catherine Huber-Carol

Contents

Preface

Catherine Huber-Carol has made several significant contributions to both theoretical and applied statistics. This is clearly evident from her publications list and the numerous citations her publications have received over the past 25 years or so. She has also been instrumental in developing interest in applied statistical problems among the statistical community in France.

Our association with her has been at both the professional and personal levels. While we have all benefitted from her statistical knowledge and insight, we have also had the pleasure of having her as a friend over the years. We have enjoyed numerous hours of discussion on such varied topics as international politics, world peace, religious practice and music.

Reliability, Survival Analysis and Quality of Life have all become very active and fertile areas of research in the recent years. Since Catherine's recent research has focused on these areas, we decided to put tother a volume in order to (i) review some of the recent developments in these areas, (ii) highlight some of the new results and illustrate their applications, and (iii) point out directions for possible future research. With these aims, a number of authors actively involved in these areas of research were invited to write articles for this volume. The articles so collected have been carefully organized into this volume in the form of 33 chapters. For the convenience of the readers, the volume has been divided into the following 13 parts:

- Cox Models and Analyses
- Degradation Models and Analyses
- Accelerated Failure Time Models and Analyses
- Aging Properties and Analyses
- Analyses of Censored and Truncated Data
- Regression Methods and Applications
- Time Series Analyses
- Inferential Methods
- Multicentre Studies
- Quality of Life Studies

- Breast Cancer Studies
- Inference for Processes
- Probability Theory and Applications

From this list, it should be clear to the readers that both theory and applications have received due consideration in this volume. It should be mentioned that a number of authors were present at a Workshop that we organized at Mont Saint Michel, France, during May 15–17, 2003, in honor of Catherine. It is, therefore, with great pleasure that we dedicate this volume to **Catherine Huber-Carol**.

Acknowledgements. Special thanks are due to Mrs. Debbie Iscoe for the excellent typesetting of the entire volume. Finally, we thank Mr. Thomas Grasso and Mr. Seth Barnes for their assistance in the production of the volume.

M. Nikulin
N. Balakrishnan
M. Mesbah
N. Limnios

December 2003

Catherine Huber-Carol: Career and Accomplishments

Catherine Huber-Carol was born in Paris (France) on May 1st 1943. Her parents, who both originated from the South of France, moved back to the south this same year just after her father graduated from the "Ecole Normale Supérieure d'Ulm" located in Paris. They lived first in Carcassonne, then in Montpellier until Catherine turned fifteen. At that time, they moved back again to Paris.

When Catherine Huber-Carol entered the "Ecole Normale Supérieure de Sèvres" in 1962, she began a cursus both in Physics and Mathematics and finally chose Mathematics. She graduated in 1964 both in Pure and Applied Mathematics and received the "agrégation" in mathematics in 1965.

At that time she married André Huber who, as a Swiss, lived in Zurich whereto she moved after one year. That year, 1965, she spent in France, working for the French Ministry of Education, doing a statistical analysis of the flows of teachers in and out of the French Educational System, and struggling to obtain a position for the next year in the university closest to the Swiss border: Mulhouse in Alsace. During the five years she lived in Zurich, she worked in Mulhouse. She intended at first to do a thesis in Probability at the University of Strasbourg in Alsace, but as this meant a lot of traveling, she tried to do it in Switzerland in spite of the fact that the language spoken in this region of Switzerland was German, a language that she did not know. So, after some difficulties due to the non-equivalence of diploma between France and Switzerland, she began a PhD in Robust Statistics under the direction of Professor Peter J. Huber in Zurich at the Federal Institute of Technology (ETH: Eidgenössische Technische Hochschule), which she obtained in 1970 on "Asymptotic Behaviour of Robust Tests".

André and Catherine Huber returned to Paris in 1971, where both received positions. Catherine was Assistant Professor at the University of Paris 13 (Saint-Denis), and again the non-equivalence of Diploma made her PhD unrecognised in France. She went again for a French PhD called "Thèse d'Etat" which was more like an habilitation, with Professor Didier Dacunha-Castelle as advisor at University Paris 11 (Orsay). The subject was "Non Parametric Statistics and Minimax Properties of Functional Estimation". She obtained the "Thèse d'Etat" on this subject in 1979. Her daughter Hélène was born in July 1980.

Catherine decided then to turn from purely theoretical statistics to Biomedical Applications and joined the team of Professor J. Lellouch at the INSERM (Institut National d'Etudes et de Recherches Médicales) in Villejuif. The first problem she was interested in was "Computer Aided Diagnosis" using log-linear and mixed models. She visited then the major centers for those developments which were then Salt Lake City, Washington and Boston. Then she became interested in "Survival data analysis involving censoring and truncation". She collaborated on this topic mainly with Shulamith Gross (New York University) as well as Jean Bretagnolle (University of Orsay). With the latter, she had also joint work on density estimation. A group and a seminar on "Mathematical Methods for Survival Analysis, Reliability and Quality of Life" involving several colleagues of European countries, at first Germany and Italy, was founded jointly with N. Limnios (UTC of Compiègne), M. Mesbah (University of Vannes) and M. Nikulin (University of Bordeaux), in order to combine both approaches for problems showing similarities
(http://www.dma.utc.fr/ nlimnios/SEMINAIRE/).

Since 1981, Catherine has been a Professor of Mathematics and Statistics at the University of René Descartes (Paris-5), where she led for some time the laboratory of Biomedical Statistics. During her time as a Professor, Catherine Huber-Carol has participated actively in the international statistical activity. She was invited for short and longer visits to several foreign countries; among them Algeria (Constantine), Australia (Canberra), Canada (Montréal), Cyprus (Nicosia), United States (Salt Lake City, Berkeley), Germany (Berlin and Potsdam), and Russia (St. Petersbourg).

She was the advisor – jointly with Joseph Lellouch – of Mounir Mesbah for his Science Doctorate thesis (Thèse d'Etat, Habilitation) and she directed the PhD work of many students among whom are to be cited Adjabi Smain (University of Bejaia), Stephane Robin (Agronomy Institute, Paris), Andrew Kramar (Cancer Center, Montpellier), and Natacha Heutte (University of Caen).

She was the president of many juries and committees in France and abroad. She is an Associate Editor of *Lifetime Data Analysis* and of *Communications in Statistics.* She is also a member of the European Committee of the Bernoulli Society, elected member of the ISI since 1989 and a member of its Council since 1999.

In France, as a scientific researcher and teacher in the field of application of mathematics and statistics to medicine, Catherine Huber is probably the most famous academic. It is well known that France is the country of mathematics. In the French University, mathematics is very strong as a discipline, but applications of mathematics are somewhat overlooked. While theoretical Probability is viewed with some consideration, Statistics is often neglected. In this spirit, the "Ecole Normale Supérieure", where Catherine started her career, is considered as the Mecca of Pure Mathematics. It is greatly to her credit for choosing the way of Applied Statistics. Around the world, a lot of colleagues have a great deal of respect for her. Moreover, this professional respect goes often with a natural friendship that Catherine always inspires.

Publications

Books

1. *Théorie de la Robustesse*, Lecture Notes in Mathematics 1215, Springer-Verlag, New York, 128 pages, 1986.

2. *Statistique au PCEM*, Masson, Paris, 1989.

3. *Analyse Statistique des Dures de Vie, Modélisation des Données Censurées* (co-edited with M. Carbon, B. Lecoutre, P. Tassi and Gourieroux), Economica, Série ASU, 1989.

4. *Goodness of Fit Tests and Validity of Models* (co-edited with N. Balakrishnan, M. Nikouline and M. Mesbah), Birkhäuser, Boston, 500 pages, 2002.

Articles

1974

1. Estimation dans les tableaux de contingence à un grand nombre d'entrées (with J. Lellouch), *International Statistical Review*, **42**, 193–203.

2. Original approach of the hyperthiroidism diagnosis with computer (with J. Lellouch, P. Fragu, and E. Patois), *Medinfo*, **74**, 559–564.

1977

3. Exemples d'instabilité et voisinages d'un modèle, *Société Mathématique de France, Astérisque*, **43-44**, 11–21.

4. L-estimateurs: Combinaisons linéaires de statistiques d'ordre, *Société Mathématique de France, Astérisque*, **43-44**, 65–86.

5. Robustesse des L-estimateurs, *Société Mathématique de France, Astérisque*, **43-44**, 203–207.

1979

6. Estimation des densités: Risque minimax (with J. Bretagnolle), *Zeitschrift für Wahrscheinlichkeitstheorie und verwandte Gebiete*, **47**, 119–137.

1987

7. Matched pair experiments: Cox and maximum likelihood estimation (with S. Gross), *Scandinavian Journal of Statistics*, **14**, 27–41.

1988

8. Effects of omitting covariates in Cox's model for survival data (with J. Bretagnolle), *Scandinavian Journal of Statistics*, **15**, 125–138.

1990

9. Aniline blue staining as a marker of sperm chromatin defects associated with different semen characteristics discriminates between proven fertile and suspected infertile men (with J. Auger, M. Mesbah and J.P. Danoune), *International Journal of Andrology*, **13**, 452–462.

10. Discussion of D. Ewards and N. Wermuth's papers, "Hierarchical interaction models" (with M. Mesbah), *Journal of the Royal Statistical Society, Series B*, **52**.

1992

11. The choice of loglinear models in contingency tables when the variables of interest are not jointly observed (with M. Mesbah and J. Lellouch), *Biometrics*, **48**, 259–265.

12. Regression models for truncated survival data (with S. Gross), *Scandinavian Journal of Statistics*, **19**, 193–213.

13. Weighted low rank statistics for comparing two distributions (with T. Moreau, J. Maccario and J. Lellouch), *Biometrika*, **79**, 195–198.

1994

14. Durées de survie tronquées et censurées, *Journal de la Société Statistique de Paris*, **135**.

15. Asymptotically optimal balloon density estimates (with P. Hall, A. Owen, and A. Coventry), *Journal of Multivariate Analysis*, **51**, 352–371.

1995

16. Overexpression of Annexin V in cystic fibrosis epithelial vells from fetal trachea (with B. Della Gaspera, S. Weinman, M. Lemnaouar, A. Paul, J. Picard and D. C. Gruenert), *Experimental Cell Research*, **219**, 379–383.

1997

17. Lower bounds for function estimations, In *Festschrift für Lucien Le Cam, Research Papers in Probability and Statistics*, pp. 245–258, Springer-Verlag, New York.

18. Covariate-matched one-sided tests for the difference between functional means (with P. Hall and P. L. Speckman), *Journal of the American Statistical Association*, **92**, 1074–1083.

19. Remarques sur le maximum de vraisemblance (with M. Nikulin), *Qüestio*, **21**, 37–58.

2000

20. Hierarchical dependency models for multivariate survival data with censoring (with S. Gross), *Lifetime Data Analysis*, **6**, 299–320.

21. Censored and truncated lifetime data, In *Recent Advances in Reliability Theory*, pp. 291–300, Birkhäuser, Boston.

2002

22. Two competitors among semi-markov models for multi-state survival data analysis, *Mathematical Methods in Reliability*, MMR 2002.

23. Semi-Markovian models applied to AIDS with censoring (with N. Heutte and O. Pons), In *Third International Conference on Mathematical Methods in Reliability*, MMR 2002, Trondheim, Norway.

24. A new family of multivariate distributions for survival data (with S. Gross), In *Goodness of Fit Tests and Model Validity*, pp. 255–266, Birkhäuser, Boston.

25. Semi-Markov models for quality of life data with censoring (with N. Heutte), In *Statistical Methods for Quality of Life Studies*, pp. 207–218, Kluwer Academic Publishers.

2003

26. Genomic studies in AIDS: Problems and answers. Development of a statistical model integrating both longitudinal cohort studies and transversal observations of extreme cases (with O. Pons, H. Hendel, Laurent Jacquemin, S. Tamim and J. F. Zagury), *Biomed. Pharmacoth.*, **57**, 25–33.

27. Lower bounds for estimating a hazard (with B. Mac Gibbon), *Communications in Statistics* (to appear)

Contributors

Afchain, Anne-Laure Université Paris-Sud XI, France
anne_laure_afchain@yahoo.fr

Anisimov, Vladimir, V. GlaxoSmithKline, New Frontiers Science Park (South), Third Avenue, Harlow, Essex CM19 5AW, U.K.
Vladimir.V.Anisimov@gsk.com

Asselain, Bernard Département de Biostatistiques, Institut Curie, 75005 Paris, France

Bagdonavicius, V. Université de Bordeaux-2, UFR MI2S, B.P. 69, F33076 Bordeaux Cedex, France

Balakrishnan, N. Department of Mathematics & Statistics, McMaster University, 1280 Main Street West, Hamilton, Ontario, Canada L8S 4K1
bala@univmail.cis.mcmaster.ca

Barbu, Vlad Université de Technologie de Compiègne, Compiègne, France

Bertail, Patrice CREST, Laboratoire de Statistiques, Timbre J340, 3, Ave Pierre Larousse, 92245 Malakoff Cedex, France
Patrice.Bertail@ensae.fr

Boucher, Kenneth M. Huntsman Cancer Institute, University of Utah, Salt Lake City, Utah 84108 U.S.A.

Bretagnolle, Jean Laboratoire de Probabilité et Statistique, Université Paris-Sud, UMR 8628, Batiment 425, 91405 Orsay Cedex, France
Jean.Bretagnolle@wanadoo.fr

Caroni, C. Department of Mathematics, National Technical University of Athens, GR-15780 Athens, Greece
ccar@math.ntua.gr

Couallier, Vincent UFR Sciences et Modélisation, Université Victor Segalen Bordeaux 2, 31076 Bordeaux Cedex, France
vincent.couallier@sm.u-bordeaus2.fr

Cuadras, Carles M. Department of Statistics, University of Barcelona, 08023 Barcelona, Spain
cuadras@ub.edu

Deheuvels, Paul L.S.T.A., Université Paris VI, 7 avenue du Château, 92340 Bourg-la-Reine, France
pd@ccr.jussieu.fr

Dupuy, Jean-François CEREMADE, Université Paris IX-Dauphine, Place du Maréchal de Lattre de Tassigny, 75775 Paris Cedex 16, France

Feddag, Mohand-Larbi University of South Brittany, IUP de Vannes Tohannic, 5600 Vannes, France
Mohand-Larbi.Feddag@univ-ubs.fr

Fedorov, Valerii V. Research Statistics Unit, GlaxoSmithKline, Upper Providence, Philadelphia, U.S.A.

Fortiana, Josep Department of Statistics, University of Barcelona, 08007 Barcelona, Spain
fortiana@cerber.mat.ub.es

Gerville-Reache, Léo Laboratoire de statistique mathématique et ses applications, Université Victor Segalen Bordeaux 2, 33000 Bordeaux Cedex, France
gerville@u-bordeaux2.fr

Greenwood, Priscilla E. Department of Mathematics and Statistics, Arizona State University, Tempe, AZ 85287-1804 U.S.A.
pgreenw@math.asu.edu

Hoàng, Thu M. Laboratoire de Statistique Médicale, Université René Descartes, F75006 Paris, France

Huber-Carol, Catherine Université René Descartes, Laboratoire de Statistique Médicale, 45 rue des Saints Pères, F75006 Paris, France
huber@biomedicale.uni-paris5.fr

Ibragimov, I. St. Petersburg Branch of Steklov Mathematical Institute Russian Academy of Sciences, St. Petersburg State University

Jin, Zhezhen Department of Biostatistics, Columbia University, New York, NY 10032 U.S.A.

Kahle, Waltraud Faculty of Mathematics, Otto-von-Guericke-University, D-39016 Magdeburg, Germany
wkahle@imst.math.uni-magdeburg.de

Kim, J.-A. Department of Mathematics & Statistics, McMaster University, Hamilton, Ontario, Canada L8S 4K1

Kvaløy, Jan Terje Department of Mathematics and Natural Science, Stavanger University College, School of Science and Technology, P.O. Box 2557, Ullandhaug, N-4091 Stavanger, Norway
jan.t.kvaloy@tn.his.no

Kutoyants, Yu. A. Laboratoire de Statistique & Processus, Université du Maine, 72085 Le Mans, France
kutoyants@univ-lemans.fr

Läuter, Henning Institute of Mathematics, University of Potsdam, D-14469 Potsdam, Germany
laeuter@rz.uni-potsdam.de

Lehmann, Axel Institute für Mathematische Stochastik, Otto-von-Guericke-University, D-39016 Magdeburg, Germany
alehmann@saturn.Math.Uni-Magdeburg.de

Liero, Hannelore Institut für Mathematik, Universität Potsdam, Am Neuen Palais 10, D-14415 Potsdam Geb. 8, Germany
liero@rz.uni-potsdam.de

Limnios, Nikolaos Université de Technologie de Compiègne, Division Mathématiques Appliquées, Centre de Recherches de Royallieu, BP 20529, 60205 Compiègne Cedex, France
Nikolaos.Limnios@utc.fr

Lindqvist, Bo Henry Department of Mathematical Sciences, Norwegian University of Science and Technology, 7491 Trondheim, Norway
Bo.Lindqvist@math.ntnu.no

Mesbah, Mounir University of South Brittany, IUP de Vannes Tohannic, F56 000 Vannes, France
mounir.mesbah@univ.ubs.fr

Müller, Ursula U. Fachbereich Mathematik und Informatik, Universität Bremen, Bibliothekstrasse 1, 28359 Bremen, Germany

Nikulin, Mikhail Laboratoire de statistique mathématieque et ses applications (EA 2961) Université Victor Segalen Bordeaux 2, 33000 Bordeaux, France
nikou@sm.u-bordeaux2.fr

Parsons, Van L. National Center for Health Statistics, Hyattsville, MD U.S.A.

Sen, Pranab Kumar Departments of Biostatistics and Statistics, University of North Carolina, Chapel Hill, NC 27599-7420, U.S.A.
pksen@bios.unc.edu

Solev, V. Steklov Mathematical Institute, Saint Petersburg, Russia
solev@pdmi.ras.ru

Spizzichino, Fabio Department of Mathematics, University "La Sapienza", 00185 Rome, Italy
spizzichino@axrma.uniroma1.it

Tikhov, M.S. Nizhny Novgorod State University, Russia

Tsodikov, Alexander D. Huntsman Cancer Institute, University of Utah, Salt Lake City, Utah, U.S.A.

Viertl, Reinhard Department of Statistics, Probability Theory, and Actuarial Mathematics, Vienna University of Technology, 1040 Wien, Austria
viertl@statistik.tuwien.ac.at

Vonta, Filia Department of Mathematics and Statistics, University of Cyprus, 1678 Nicosia, Cyprus
vonta@ucy.ac.cy

Wefelmeyer, Wolfgang Fachbereich Mathematik, Universität Siegen, 57068 Siegen, Germany

Wendt, Heide Faculty of Mathematics, Otto-von-Guericke-University, D-39016 Magdeburg, Germany

Wu, Hong-Dar Isaac School of Public Health, China Medical College, 91 Hsueh-Shih Rd., Taichung 40443, Taiwan
honda@mail.cmc.edu.tw

Yakovlev, Andrei Y. Department of Biostatistics and Computational Biology, University of Rochester, Rochester, NY, U.S.A.

Yashin, Anatoli I. Max Planck Institute for Demographic Research, Doberaner Strasse 114, Rostock D18057, Germany
Yashin@demogr.mpg.de

Ying, Zhiliang Department of Biostatistics, Columbia University, New York, NY 10032 U.S.A.
`zying@stat.columbia.edu`

List of Tables

List of Figures

Figures [illegible]

PART I
Cox Models and Analyses

1

Estimation in Partly Parametric Additive Cox Models

Henning Läuter

Institute of Mathematics, University of Potsdam, Germany

Abstract: The dependence between survival times and covariates is described e.g., by proportional hazard models. We consider partly parametric Cox models and discuss here the estimation of interesting parameters. We represent the maximum likelihood approach and extend the results of Huang (1999) from linear to nonlinear parameters. Then we investigate the least squares estimation and formulate conditions for the a.s. boundedness and consistency of these estimators.

Keywords and phrases: Survival models with covariates, estimation of regression, maximum likelihood estimator, least squares estimator, boundedness, consistency

1.1 Introduction

The relationship between the survival time and covariates very often is expressed through the proportional hazards model proposed originally by Cox (1972). In several papers afterwards this model was investigated in several directions [Kalbfleisch and Prentice (1980), Andersen et al. (1993), Bagdonavicius and Nikulin (2001)]. For the estimation of parameters of interest many approaches exist. In this paper we consider the maximum likelihood and a least norm estimator. Both approaches have some advantages. With the maximum likelihood method we can obtain consistent and efficient estimators. With a least norm estimator it is possible to construct strongly consistent estimators. Survival models describe the probabilistic properties of a nonnegative failure time Y. We assume that Y depends on covariates Z and we model the hazard rate

$$\lambda(t \mid z) = \lim_{h \to 0} \frac{\mathsf{P}(t < Y \leq t + h \mid Y \geq t, z)}{h}.$$

We investigate parametric and semiparametric multiplicative proportional hazard models

$$\lambda(t \mid z) = \lambda_0(t)\, r(\beta, z) \tag{1.1}$$

for a completely unspecified baseline hazard function λ_0 and the relative risk function r. The function r is assumed to be a partly parametric function. The unknown parameter β is to be estimated. The deterministic or random covariate variable Z is constant in time.

We assume that $Z = (X, V) \in \mathbb{R}^k \times \mathbb{R}^m$ and

$$\log r(\beta, z) = g(x, \beta) + l(v) \tag{1.2}$$

for a known regression function g and an unknown function l. The covariate X is controlled, but V is an auxiliary covariate. The aim is to estimate β and so to find the influence of $X = (X^{(1)}, ..., X^{(k)})$ on the survival time. Such models for the hazard rate were considered by several authors. The starting point was Cox (1972) with his proportional hazard rate model

$$\lambda(t) = \lambda_0(t) \exp(\beta' x), \tag{1.3}$$

later on e.g., Andersen and Gill (1982) proved for the regression model (1.3) the asymptotic normality of the maximum partial likelihood estimator (mple) for β. The special case of (1.2) is the semiparametric partial linear model

$$\log r(\beta, z) = x'\beta + l(v) \tag{1.4}$$

which was considered by Sasieni (1992). He found an information bound for the mple for β in (1.4). In Andersen et al. (1993) a broad class of multiplicative and additive models for the hazard rate and related estimates was studied. Bagdonavicius and Nikulin (2001) formulated a lot of models for uncensored and censored variables and found properties for the estimates in these models. Huang (1999) gave an approach for the estimation of β in (1.4) and was able to prove the asymptotic efficiency of the approximated mple. We will extend the model (1.4) to nonlinear functions g and discuss the properties of estimates in these models. In Section 1.5 we discuss properties of least squares estimators where the basis is a nonparametric estimation of the hazard function. Here we use the Nelson–Aalen type estimator considered by several authors, for example Van Keilegom and Veraverbeke (2001) and Liero (2003). Especially we formulate conditions for the boundedness and consistency of the estimator for β.

1.2 Types of Regression Functions

Under a precise knowledge of the considered or observed process one has some special models for the hazard rate. The baseline function is unrestricted. The model specification consists in (1.1) of the function r or the function g. We consider different choices of g.

1.2.1 Linear models

The linear models play an important role in all analyses. In most cases one understands these models as approximations for g or λ. In the model

$$g(x, \beta) = x'\beta = x^{(1)}\beta_1 + \cdots + x^{(k)}\beta_k \tag{1.5}$$

with $x \in \mathbb{R}^k, \beta \in \mathbb{R}^k$ the dependence is linear in the controlled covariates and the parameters. If the covariates have a nonlinear influence, then

$$g(x, \beta) = w(x)'\beta = \beta_1 + \beta_2\, x + \cdots + \beta_k\, x^{k-1} \tag{1.6}$$

is a polynomial model in one covariate x. Here $w(x) \in \mathbb{R}^k$, $\beta \in \mathbb{R}^k$. Polynomial models in higher dimensional covariates are also possible.

1.2.2 Nonlinear models

The simplest type of functions in this class is

$$g(x, \beta) = \beta_1 \exp(\beta_2\, x) \tag{1.7}$$

for $x \in \mathbb{R}^1, \beta \in \mathbb{R}^2$. These are convex or concave curves. Other examples are the logistic curves which one can describe in a different form. The solution $g(x, \beta) := \nu(x)$ of the differential equation

$$\dot{\nu}(x) = \beta_1\, \nu(x)(\beta_2 - \nu(x)) \tag{1.8}$$

is used very often. These curves have a saturation point β_2. There is a huge set of possible other regression curves. In difference to linear regression curves the statistical properties of estimates in nonlinear models depend strongly on the space of parameters. In general the estimation problem leads to ill-posed problems and one needs additional restrictions for β. We are interested in estimating β. There are different possibilities. One approach is based on the likelihood principle. Here one expects that one gets consistent and asymptotically efficient estimates. Another approach is based on a least squares principle. Here one hopes that the corresponding estimates are strongly consistent. We will discuss both possibilities.

1.3 Random Covariates and Partial Likelihood

We consider the nonnegative failure time Y and the censoring time C. Y depends on the covariate Z and we assume that Y and C are conditionally independent given z. $\Delta = \mathbf{1}_{Y \leq C}$ is the censoring indicator and we set $T = \min\{Y, C\}$. Our observable variable is

$$(T, \Delta, Z) \in \mathbb{R}^+ \times \{0,1\} \times \mathbb{R}^{k+m}. \tag{1.9}$$

There is given a random sample

$$(T_i, \Delta_i, Z_i), \quad i = 1, ..., n$$

from the distribution of (T, Δ, Z). Here we have

$$Z_i = (X_i, V_i)$$

with $X_i = (X_i^{(1)}, ..., X_i^{(k)})^t, \quad V_i = (V_i^{(1)}, ..., V_i^{(m)})^t$.

The idea of Sasieni (1992) and Huang (1999) consists in approximating the nonparametric term $l(\cdot)$ in (1.1) and (1.2) by a smooth function. They proposed a spline approximation. Assuming

$$l(v) = \sum_{j=1}^{m} l_j(v^{(j)}) \tag{1.10}$$

then

$$l(v) \approx \sum_{j=1}^{m} s_j(v^{(j)}), \tag{1.11}$$

for splines s_j of order $q = q(n) \geq 1$. We assume that $a \leq V^{(j)} \leq b$ and choose knots $\xi_1, ..., \xi_p$ with $a \leq \xi_1 \leq \cdots \leq \xi_p \leq b$ and $p = p(n)$. Furthermore

$$\max_s |\xi_s - \xi_{s-1}| = O(n^{-\alpha})$$

is assumed. In the space $\tilde{S}_n$ of polynomial splines of order $q \geq 1$ there exists a local basis $B_1, ..., B_{\nu_n}$ such that the representation

$$s_j(\cdot) = \sum_{t=1}^{\nu_n} b_{jt} B_t(\cdot), \quad 1 \leq j \leq m \tag{1.12}$$

holds. Consequently we have with (1.11) the approximation

$$l(v) \approx \tilde{s}_n(v, b) := \sum_{j=1}^{m} \sum_{t=1}^{\nu_n} b_{jt} B_t(v^{(j)}) \tag{1.13}$$

where b is the vector of unknown coefficients $b_{11}, ..., b_{m\nu_n}$.

Assuming the model for the hazard rate function in the form

$$\lambda(t|x, v) = \lambda_0(t) \exp(g(x, \beta) + l(v)),$$

then with

$$f_n(x, v) = g(x, \beta) + \tilde{s}_n(v, b)$$

the partial log-likelihood function is given by

$$l_n(\beta, b) = \frac{1}{n} \sum_{i=1}^{n} \Delta_i \{f_n(X_i, V_i) - \log \sum_{j:T_j \geq T_i} \exp[f_n(X_j, V_j)]\}. \tag{1.14}$$

Any $(\hat{\beta}_n, \hat{b}_n)$ which maximizes $l_n(\beta, b)$ is called a maximum partial likelihood estimator (mple) for (β, b). The additive constant in (1.14) is free, so we determine such a $\hat{b}_n$ that

$$\sum_{i=1}^{n} \Delta_i \sum_{t=1}^{\nu_n} \hat{b}_n^{(j,t)} B_t(V_i^{(j)}) = 0, \quad j = 1, ..., m$$

with

$$\hat{b}_n^T = (\hat{b}_n^{(1,1)}, ..., \hat{b}_n^{(m,\nu_n)})^T.$$

The limit distribution of $\hat{\beta}_n$ is given by the next theorem. For this we formulate at first some assumptions.

Assumption 1 *g is continuously differentiable w.r.t. β.*

Assumption 2 *β is an inner point of a bounded set in $\mathbb{R}^k$ or for any $\tilde{\beta}$ and any sequence $x_1, x_2, ...$ of realizations of $X_1, X_2, ...$ we have*

$$\sum_{i=1}^{n} [g(x_i, \tilde{\beta}) - g(x_i, \tilde{\tilde{\beta}})]^2 \to \infty \quad \textit{for} \quad \|\tilde{\tilde{\beta}}\| \to \infty.$$

Assumption 3

$$\mathsf{E} g(X, \beta)\Delta = 0, \quad \mathsf{E} l_j(V^{(j)})\Delta = 0, \quad j = 1, ..., m.$$

Assumption 4

$$\|X\| \leq K \quad \textit{a.s. for a finite constant } K,$$

$$V \in [a, b]^m.$$

Theorem 1.3.1 *Under Assumptions 1 to 4 we have*

$$\sqrt{n}(\hat{\beta}_n - \beta) \xrightarrow{\mathrm{d}} \mathsf{N}(0, \Sigma), \tag{1.15}$$

where

$$\Sigma^{-1} = \mathsf{E}\Big[\Delta\Big(G(X,\beta) - a^*(T) - h^*(V)\Big)^{\otimes 2}\Big]$$

with $x^{\otimes 2} := xx'$ *and*

$$G(x,\beta) = \frac{\partial}{\partial\beta}\, g(x,\beta) \in \mathbb{R}^k.$$

Here a^*, h^* *are* k*-dimensional functions with*

$$h^*(V) = \sum_{j=1}^{m} h_j^*(V^{(j)})$$

and

$$\mathsf{E}\Delta\|G(X,\beta) - a(T) - h(V)\|^2$$

is minimized by a^*, h^*.

PROOF. The proof combines results from nonlinear regression estimation and results of Huang (1999) for estimation in partly linear models. ■

This theorem formulates an expected result. Important are the assumptions. We mention especially the finite supports for the covariates. The assumption **A2** ensures the estimability of the parameters. This assumption is important for infinite parameter spaces.

1.3.1 Linear case

The linear case

$$g(x,\beta) = x'\beta$$

is of special interest. For such models Sasieni (1992) found a lower information bound for the asymptotic variance of an estimator for β. Huang (1999) proved that the maximum partial likelihood estimator $\hat{\beta}_n$ achieves this lower bound. The result of theorem 1.3.1 can be used for choosing an appropriate experimental design. We have to find such a random variable X that Σ is in some sense minimal. This is then a general problem of nonlinear estimation.

1.4 Fix Covariates and Partial Likelihood

For fix covariates $z_1, ..., z_n$ the random variables (T_i, Δ_i, z_i) are independently distributed but not identically. The partial log-likelihood is given by

$$l_n(\beta, b) = \sum_{i=1}^{n} \Delta_i \{f_n(x_i, v_i) - \log \sum_{j:T_j \geq T_i} \exp[f_n(x_j, v_j)]\}. \tag{1.16}$$

Then a maximum partial likelihood estimator $\hat{\beta}_n$ is given as before by such a vector which maximizes l_n. The asymptotic properties of $\hat{\beta}_n$ depends on the sequence of z_i.

1.5 Least Squares Estimation

We consider a proportional hazard model (1.1) with $\log r(\beta, z) = g(x, \beta)$. The estimation of the parameter β can be constructed as before by a maximum likelihood principle but also by a least squares criterion. For this we point out that we are interested in the hazard rate

$$\lambda(t \mid z) = \lambda_0(t)\, \dot{r}(\beta, z) \tag{1.17}$$

or in the cumulative hazard rate

$$\Lambda(t \mid z) = \int_0^t \lambda_0(s)\, \mathrm{d}s \; r(\beta, z). \tag{1.18}$$

The term which includes β is the same in (1.17) and in (1.18). Therefore for the estimation of β there is no statistical difference in both equations.

In Van Keilegom and Veraverbeke (2001) one finds a nonparametric estimator for the cumulative hazard function Λ. This is a Nelson–Aalen type estimator and is constructed in the following way. We choose a kernel function W and a sequence of bandwidths $\{a_n\}$ and define for any value of the covariates $z_1, ..., z_n$ the weights

$$w_{nj}(z) = \frac{W(\frac{z - z_j}{a_n})}{\sum_{k=1}^{n} W(\frac{z - z_k}{a_n})}. \tag{1.19}$$

Then we have $\sum_{j=1}^{n} w_{nj}(z) = 1$ for any z. An estimator for Λ is then

$$\hat{\Lambda}_n(t|z) = \sum_{i=1}^{n} \frac{\mathbf{1}(T_i \leq t)\Delta_i w_{ni}(z)}{\sum_{j:T_j \geq T_i} w_{nj}(z)}. \tag{1.20}$$

The parameter β is estimated by comparing $\hat{\Lambda}_n$ with Λ under an appropriate distance measure. We formulate this for the quadratic norm. We call $\hat{\beta}_n$ a least squares estimator if

$$arg \min_{\beta,\lambda} \sum_{j=1}^{n} \int \Big(\hat{\Lambda}_n(t|z_j) - \int_0^t \lambda(s)\, ds \cdot r(\beta, z_j)\Big)^2 b_n(t)\, dt = (\hat{\beta}_n, \hat{\lambda}_n). \quad (1.21)$$

Here $\{b_n\}$ is a sequence of nonnegative weight functions. With (β_0, λ_0) we denote the true parameter in the hazard rate function and set

$$\Lambda_0(t) = \int_0^t \lambda_0(s)\, ds.$$

We use the notation

$$W_n^2(\beta_0, \Lambda_0, \beta, \Lambda) = \frac{1}{n} \sum_{j=1}^{n} \int \Big(\Lambda_0(t)\, r(\beta_0, z_j) - \Lambda(t)\, r(\beta, z_j)\Big)^2 b_n(t)\, dt,$$

$$\begin{aligned} Z_n(\beta_0, \Lambda_0, \beta, \Lambda) \;=\; & \frac{1}{n\, W_n(\beta_0, \Lambda_0, \beta, \Lambda)} \\ & \times \sum_{j=1}^{n} \int \Big(\hat{\Lambda}_n(t|z_j) - \Lambda_0(t)\, r(\beta_0, z_j)\Big) \\ & \times \Big(\Lambda_0(t)\, r(\beta_0, z_j) - \Lambda(t)\, r(\beta, z_j)\Big) b_n(t)\, dt \end{aligned}$$

if $W_n(\beta_0, \Lambda_0, \beta, \Lambda) \neq 0$ and put

$$\varepsilon_n^2 = \frac{1}{n} \sum_{j=1}^{n} \int \Big(\hat{\Lambda}_n(t|z_j) - \Lambda_0(t)\, r(\beta_0, z_j)\Big)^2 b_n(t)\, dt.$$

According to (1.21) the functional that we need to minimize is

$$Q_n(\beta, \Lambda) = \frac{1}{n} \sum_{j=1}^{n} \int \Big(\hat{\Lambda}_n(t|z_j) - \Lambda(t)\, r(\beta, z_j)\Big)^2 b_n(t)\, dt \quad (1.22)$$

and with the introduced notation we have the representation

$$\begin{aligned} Q_n(\beta, \Lambda) \;=\; & W_n^2(\beta_0, \Lambda_0, \beta, \Lambda) \\ & + 2\, W_n(\beta_0, \Lambda_0, \beta, \Lambda)\, Z_n(\beta_0, \Lambda_0, \beta, \Lambda) + \varepsilon_n^2. \end{aligned} \quad (1.23)$$

Assumption 5 *The support of X is a subset $\mathcal{X}$ in $\mathbb{R}^k$. The parameter β lies in a set $\mathcal{B}$. Both sets are not necessarily bounded.*

Assumption 6 *ε_n^2 tends a.s. to a finite value σ^2.*

With (1.23) we obtain immediately a result on the boundedness of the least squares estimator.

Theorem 1.5.1 *If* **A**5 *and* **A**6 *and*

$$\inf_{|\beta|\geq\eta} W_n(\beta_0, \Lambda_0, \beta, \Lambda) \to \infty \text{ for } \eta \to \infty$$

hold, then $\hat{\beta}_n$ is bounded a.s.

Sometimes one can check the behavior of Z_n. Then the following result is useful.

Theorem 1.5.2 *If for any β_0 there exist constants $c(\beta_0)$, $d(\beta_0)$ such that for $\|\beta\| \geq d(\beta_0)$ for all $n \geq n_0$ the condition*

$$W_n(\beta_0, \Lambda_0, \beta, \Lambda) \geq c(\beta_0),$$

and if for some positive η and a constant κ the inequality

$$\inf_{\|\beta\|\geq\eta} \frac{Z_n(\beta_0, \Lambda_0, \beta, \Lambda)}{W_n(\beta_0, \Lambda_0, \beta, \Lambda)} \leq \kappa < 0.5 \quad a.s.$$

holds, then $\hat{\beta}_n$ is bounded a.s.

The boundedness is necessary to prove consistency of $\hat{\beta}_n$. In general one has two possibilities to prove the consistency. Either one assumes that the parameter lies in a compact set or one finds such conditions that one can restrict it to compact sets. Our formulated conditions in the previous theorems are such assumptions that we can choose compact subsets which contain the true value β_0 and the estimates $\hat{\beta}_n$ lie a.s. in these subsets.

The representation (1.23) leads to

$$\begin{aligned} & n\, Q_n(\beta, \Lambda) - n\, Q_n(\beta_0, \Lambda_0) \\ = \;& n\, W_n^2(\beta_0, \Lambda_0, \beta, \Lambda) \left[1 + 2\, \frac{Z_n(\beta_0, \Lambda_0, \beta, \Lambda)}{W_n(\beta_0, \Lambda_0, \beta, \Lambda)}\right]. \end{aligned} \tag{1.24}$$

This representation shows that the difference of the values of the functional Q_n for different parameter values is expressed by W_n and $\frac{Z_n}{W_n}$. These terms played already a role in the theorems about the boundedness of the l.s.e.

Theorem 1.5.3 *If there is a compact subset $\tilde{\mathcal{B}}$ of the parameter space $\mathcal{B}$ such that $\hat{\beta}_n \in \tilde{\mathcal{B}}$ a.s. and if $\inf_{|\beta|\geq\eta} W_n(\beta_0, \Lambda_0, \beta, \Lambda) \to \infty$ for $\eta \to \infty$, then under* **A**5 *and* **A**6*, $\hat{\beta}_n$ is strongly consistent.*

References

1. Andersen, P. K., Borgan, O., Gill, R. D. and Keiding, N. (1993). *Statistical Models Based on Counting Processes*, Springer-Verlag, New York.

2. Andersen, P. K. and Gill, R. D. (1982). Cox's regression model for counting processes: A large sample study, *Annals of Statistics*, **10**, 1100–1120.

3. Bagdonavicius, V. and Nikulin, M. (2001). *Accelerated Life Models: Modeling and Statistical Analysis*, Chapman & Hall, Boca Raton, Florida.

4. Cox, D. R. (1972). Regression models and life-tables, *Journal of the Royal Statistical Society, Series B*, **34**, 187–220.

5. Huang, J. (1999). Efficient estimation of the partly linear additive Cox model, *Annals of Statistics*, **27**, 1536–1563.

6. Kalbfleisch, J. and Prentice, R. (1980). *The Statistical Analysis of Failure Time Data*, John Wiley & Sons, New York.

7. Liero, H. (2003). Goodness of fit tests of L_2-type, *In this Volume.*

8. Sasieni, P. (1992). Information bounds for the conditional hazard ratio in a nested family of regression models, *Journal of the Royal Statistical Society, Series B*, **54**, 617–635.

9. Van Keilegom, I. and Veraverbeke, N. (2001). Hazard rate estimation in nonparametric regression with censored data, *Annals of the Institute of Statistical Mathematics*, **53**, 730–745.

2

Nonparametric Maximum Likelihood Estimation in the Proportional Hazards Model with Covariate Measurement Error

Jean-François Dupuy

CEREMADE, Université Paris IX-Dauphine, Paris, France

Abstract: The proportional hazards (PH) regression model is commonly used to characterize the relationship between some time-to-event and covariates. Covariates values are frequently subject to measurement error. Substituting mismeasured values for the true covariates in the PH model leads to biased estimation. Hu, Tsiatis and Davidian (1998) have proposed to base estimation for the PH model with covariate measurement error on a joint likelihood for survival and the covariate. The authors have used nonparametric maximum likelihood estimation (NPMLE) and have conducted simulations to assess the asymptotic validity of this approach. In this paper, we derive rigorous proofs of existence and consistency of the NPML estimators.

Keywords and phrases: Proportional hazards model, covariate measurement error, joint likelihood, nonparametric maximum likelihood estimators, existence, consistency

2.1 Introduction

The PH model [Cox (1972)] assumes that the hazard rate for the survival of an individual with covariate X has the proportional form

$$\lambda(t|X) = \lambda(t)\exp(\beta^T X), \quad t \geq 0,$$

where β is a vector of unknown regression coefficients, and $\lambda(t)$ is an unknown and unspecified nonnegative function called baseline hazard function. Let T be the failure time and C be a right-censoring time. Define $V = \min(T, C)$ and the censoring indicator $\Delta = 1_{\{T \leq C\}}$, where $1_{\{.\}}$ is the indicator function. The statistical problem is that of estimating β and the cumulative baseline hazard function $\Lambda(t) = \int_0^t \lambda(s)\,ds$ on the basis of n independent replicates (V_i, Δ_i, X_i) $(i = 1, \ldots, n)$ of (V, Δ, X). By maximizing the partial likelihood

$$L(\beta) = \prod_{i=1}^{n} \left[\frac{e^{\beta^T X_i}}{\sum_{j=1}^{n} e^{\beta^T X_j} 1_{\{V_i \le V_j\}}} \right]^{\Delta_i}$$

proposed by Cox (1972, 1975), one can obtain a consistent and asymptotically normal estimator of β [see Tsiatis (1981) and Andersen and Gill (1982)]. Letting $\hat{\beta}$ be the maximum partial likelihood estimator of β, that is the solution to the score equation

$$\sum_{i=1}^{n} \Delta_i \left[X_i - \tilde{X}(\beta, V_i) \right] = 0$$

where

$$\tilde{X}(\beta, t) = \frac{\sum_{j=1}^{n} X_j e^{\beta^T X_j} 1_{\{t \le V_j\}}}{\sum_{j=1}^{n} e^{\beta^T X_j} 1_{\{t \le V_j\}}},$$

Breslow (1972, 1974) suggested estimating $\Lambda(t)$ by

$$\hat{\Lambda}(t) = \sum_{V_i \le t} \frac{\Delta_i}{\sum_{j=1}^{n} e^{\hat{\beta}^T X_j} 1_{\{V_i \le V_j\}}}.$$

Convergence in distribution of the standardised Breslow estimator to a Gaussian process was proved by Tsiatis (1981) and Andersen and Gill (1982).

In many applications, the covariate X_i $(i = 1, \ldots, n)$ is not observed directly but is measured through an additive error model $W_i = X_i + U_i$, where the distribution of the measurement error U_i is known [see Carroll et al. (1995)].

The presence of random error in a measured covariate in the PH model was shown by Prentice (1982) to yield biased estimation. Several approaches have been proposed to take account of covariate measurement error in the PH model. References include Tsiatis et al. (1995), Zhou and Pepe (1995), Buzas (1998), Dafni and Tsiatis (1998). Recent interest has focused on joint models for the survival and covariate data. This was developed in the case of a time-dependent covariate by Faucett and Thomas (1996), Wulfsohn and Tsiatis (1997), Henderson et al. (2000), Song et al. (2002a,b), Dupuy and Mesbah (2000, 2002), Dupuy et al. (2003). These models have the disadvantage of relying heavily on the parametric assumptions made on the distribution of the time-dependent covariate.

For a time-independent covariate however, Hu et al. (1998) were able to develop a more flexible joint model by avoiding specifying a particular parametric distribution for the true covariate X. The authors assumed an additive error model for X and derived a joint likelihood for the survival and covariate. NPMLE was used to estimate β and $\Lambda(t)$. An EM algorithm was applied to solve the maximization problem. Some simulations were conducted to investigate the asymptotic properties of the NPML estimators. In this paper, we give formal proofs of existence and consistency of these estimators.

The paper is organized as follows. Section 2.2 briefly describes the joint model developed by Hu et al. (1998) and explains the NPML method. In Section 2.3, we prove existence of the NPML estimators. In Section 2.4, we prove that the NPML estimators are consistent. Discussion follows in Section 2.5.

2.2 Nonparametric Maximum Likelihood Estimation

Notation and model

In the following, a capital letter Z will denote a random variable, while z will denote a realization of Z.

Assume that n subjects are observed. For each subject i ($i = 1, \ldots, n$), the observable random variables are $Y_i = (V_i, \Delta_i, W_i)$, where $V_i = \min(T_i, C_i)$, Δ_i is the censoring indicator, and W_i is the really observed covariate. An additive error model $W_i = X_i + U_i$ is assumed for the covariate, where the measurement errors U_i ($i = 1, \ldots, n$) are normally distributed with mean zero and variance $\sigma^2 > 0$, and are independent of each other and of all other random variables. As Hu et al. (1998), we will restrict attention, for simplicity of formulas, to the case where X_i is scalar, and we will assume that the data follow a PH model with $\lambda(t|X) = \lambda(t)\exp(\beta X)$. The arguments below can however easily be generalized to a vector of the covariate.

If replicate measurements on the covariate are available, one can consider the more general case where σ^2 is unknown and has to be estimated. We will consider this case in the following. Let $\mathbf{W}_i = (W_{i1}, \ldots, W_{ir_i})^T$ denote the r_i ($r_i \geq 1$) replicate measurements for the i-th subject. Assume that given X_i, W_{il} ($l = 1, \ldots, r_i$) are independent. Each $W_{il}|X_i$ is normally distributed with mean X_i and variance σ^2. The joint density $f_{W,X}$ of the random variables W and X may be written as the product of the conditional density $f_{W|X}$ and the marginal density h of X. The key aspect of the approach of Hu et al. (1998) is that no particular parametric assumption is made on the distribution of X. It will only be assumed that h can be parameterized by a vector $\nu \in \mathbf{R}^p$ ($p \geq 1$). We give some additional technical assumptions later in this paper.

The joint likelihood

Suppose that given (W_i, X_i), T_i and C_i are independent. Suppose that the distribution of C_i does not depend on $\theta = (\beta, \Lambda, \nu, \sigma^2)$ [Nielsen et al. (1992) refer to this hypothesis as noninformative censoring].

Under these assumptions, Hu et al. (1998) derive the following joint likelihood for θ from n independent realizations y_i $(i = 1, \dots, n)$:

$$\begin{aligned} L(\theta) &= \prod_{i=1}^{n} \left[\int \left\{ \lambda(v_i)\, e^{\beta x} \right\}^{\delta_i} \exp\left\{ -\Lambda(v_i) e^{\beta x} \right\} f_{W,X}(\mathbf{w}_i, x)\, dx, \right] \\ &= \prod_{i=1}^{n} \left[\int \left\{ \lambda(v_i)\, e^{\beta x} \right\}^{\delta_i} \exp\left\{ -\Lambda(v_i) e^{\beta x} \right\} \right. \\ &\qquad \left. \times (2\pi\sigma^2)^{-\frac{r_i}{2}} e^{-\frac{1}{2\sigma^2}(\mathbf{w}_i - x)^T(\mathbf{w}_i - x)} h(x;\nu)\, dx \right]. \end{aligned} \tag{2.1}$$

In the following, we shall denote by $l(y_i, x; \theta)$ the integrand

$$\left\{ \lambda(v_i)\, e^{\beta x} \right\}^{\delta_i} \exp\left\{ -\Lambda(v_i) e^{\beta x} \right\} \times (2\pi\sigma^2)^{-\frac{r_i}{2}} e^{-\frac{1}{2\sigma^2}(\mathbf{w}_i - x)^T(\mathbf{w}_i - x)} h(x;\nu)$$

of each term in the product (2.1).

The maximum of the likelihood $L(\theta)$ over the parameter space:

$$\Big\{ (\beta, \Lambda, \nu, \sigma^2) : \beta \in \mathbf{R}, \Lambda(t) \in \{\text{absolutely continuous cumulative hazards}\}, \\ \nu \in \mathbf{R}^p, \sigma^2 \in \mathbf{R} \Big\}$$

does not exist. Hu et al. (1998) then restrict the estimator of Λ to be a step function with jumps at the observed uncensored event times. For simplicity of notation, we reorder the indices of the data such that $V_1 < \cdots < V_m$ represent the increasingly ordered event times and $V_{m+1} \leq \cdots \leq V_n$ represent the nondecreasingly ordered censoring times. Instead of $\Lambda(t)$, we consider in (2.1) a stepwise version $\Lambda_n(t)$ $(t \geq 0)$ with the unknown deterministic values $\Lambda_n(V_i) = \Lambda_{n,i}$ in the points V_i, $i = 1, \dots, m$. The resulting function, often termed nonparametric (NP) likelihood can be maximized over the parameter space Θ_n:

$$\Theta_n = \left\{ \theta = (\beta, \Lambda_n, \nu, \sigma^2) : \beta \in \mathbf{R}, \Lambda_{n,1} \leq \cdots \leq \Lambda_{n,m}, \nu \in \mathbf{R}^p, \sigma^2 \in \mathbf{R} \right\}.$$

In the case of no covariate measurement error, this approach results in the usual maximum partial likelihood estimator of β and the Breslow estimator of $\Lambda(t)$ [see Johansen (1983) and Bagdonavičius and Nikulin (2001)].

We shall denote the NP likelihood by $L^n(\theta)$. It is obtained by multiplying over uncensored subjects i $(i = 1, \dots, m)$ the following individual contributions

$L_i(\theta)$:

$$\int \Delta\Lambda_{n,i} \exp\left\{\beta x - e^{\beta x} \sum_{k=1}^{m} \Delta_{n,k} 1_{\{v_k \le v_i\}}\right\} (2\pi\sigma^2)^{-\frac{r_i}{2}} \cdot \times e^{-\frac{1}{2\sigma^2}(\mathbf{w}_i - x)^T(\mathbf{w}_i - x)} h(x;\nu)\,dx,$$

and over censored subjects i ($i = m+1, \ldots, n$) the contributions $L_i(\theta)$:

$$\int \exp\left\{-e^{\beta x} \sum_{k=1}^{m} \Delta_{n,k} 1_{\{v_k \le v_i\}}\right\} (2\pi\sigma^2)^{-\frac{r_i}{2}} e^{-\frac{1}{2\sigma^2}(\mathbf{w}_i - x)^T(\mathbf{w}_i - x)} h(x;\nu)\,dx,$$

where $\Delta\Lambda_{n,k} = \Lambda_{n,k} - \Lambda_{n,k-1}$ ($k = 2, \ldots, m$) and $\Delta\Lambda_{n,1} = \Lambda_{n,1}$ are the jumps $\Delta\Lambda_n(V_i)$ of Λ_n at the points V_i, $i = 1, \ldots, m$.

In the following, for any random variable X with density function $f(x;\theta)$, we shall denote by $E_\theta[g(X)]$ the expected value of $g(X)$ that is, $E_\theta[g(X)]$ is the integral of $g(x)$ with respect to $f(x;\theta)$. Moreover, if X and Y are random variables, we shall denote by $E_\theta[g(X)|y]$ the expectation of $g(X)$ taken with respect to the conditional density function $f(x|y;\theta)$ of X given $Y = y$, that is $E_\theta[g(X)|y] = \int g(x) f(x|y;\theta)\,dx$.

2.3 Existence of the NPML Estimators

Let $\theta_0 = (\beta_0, \Lambda_0, \nu_0, \sigma_0^2)$ denote the true value of θ. Let P_{θ_0} denote the distribution of a single observation and let E_{θ_0} be the expectation taken under P_{θ_0}. We will assume throughout that the following conditions 1–5 are satisfied.

1. Let τ be a finite time point at which the study ends. Any individual still under study at τ is censored. Assume that $P_{\theta_0}(\sum_{i=1}^n 1_{\{u \le V_i\}} \ge 1$ for all $u \in [0,\tau]) > 0$.

2. The parameters β, ν, σ^2 belong to known compact sets $B \subset \mathbf{R}, N \subset \mathbf{R}^p$ and $S \subset \mathbf{R}$ respectively. The cumulative baseline Λ is a continuous and increasing function on $[0,\tau]$.

3. If $ag(W_i) = 0$ ($i = 1, \ldots, n$), where g is any nondegenerate function, then $a = 0$.

4. Assume that X is bounded in $\mathbf{R}$ by a constant c. Assume that $h(x;\nu)$ is uniformly bounded, continuous with respect to ν, and has continuous second order derivatives with respect to ν. Assume finally that $h(x;\nu)$ is identifiable that is, $h(x;\nu) = h(x;\nu')$ implies that $\nu = \nu'$.

5. Suppose that for all $u \in [0,\tau]$, $E_{\theta_0}\left[e^{\beta_0 X} 1_{\{u \leq V\}}\right] > 0$.

Condition 1 ensures that we can observe failures on the entire time interval and therefore can estimate $\Lambda_0(t)$ on the entire interval. Condition 3 excludes the case where the observed covariate is a constant. Condition 4 ensures existence of the NPML estimators. The last assumption in condition 4 ensures that we can identify the model parameters.

The proof of identifiability is similar to the proof of identifiability in the frailty model for clustered survival data with covariate measurement error by Li and Lin (2000), and is therefore omitted.

We first prove existence of the NPML estimator of θ as a maximizer of the NP likelihood.

Theorem 2.3.1 *A maximizer $\hat{\theta}_n = (\hat{\beta}_n, \hat{\Lambda}_n, \hat{\nu}_n, \hat{\sigma}_n^2)$ of the NP likelihood $L^n(\theta)$ over Θ_n exists and is achieved.*

PROOF. Suppose first that $\Delta\Lambda_{n,i} \leq U$ $(i = 1, \ldots, m)$ for some finite U. Since L^n is a continuous function of β, of the jump sizes $\Delta\Lambda_{n,i}$ $(i = 1, \ldots, m)$, of ν and σ^2 on the compact set $B \times [0,U]^m \times N \times S$, L^n achieves its maximum on this set.

To show that a maximum of L^n exists on $B \times [0,\infty)^m \times N \times S$, we show that there exists a finite U such that for all $\theta_U = (\beta_U, \Delta\Lambda_{n,1,U}, \ldots, \Delta\Lambda_{n,m,U}, \nu_U, \sigma_U^2) \in \{B \times [0,\infty)^m \times N \times S\} \setminus \{B \times [0,U]^m \times N \times S\}$, there exists an element $\theta = (\beta, \Delta\Lambda_{n,1}, \ldots, \Delta\Lambda_{n,m}, \nu, \sigma^2) \in \{B \times [0,U]^m \times N \times S\}$ such that $L^n(\theta) > L^n(\theta_U)$.

A proof by contradiction is adopted for this purpose. Assume that for all U, there exists $\theta_U = (\beta_U, \Delta\Lambda_{n,1,U}, \ldots, \Delta\Lambda_{n,m,U}, \nu_U, \sigma_U^2) \in \{B \times [0,\infty)^m \times N \times S\} \setminus \{B \times [0,U]^m \times N \times S\}$ such that for all $\theta = (\beta, \Delta\Lambda_{n,1}, \ldots, \Delta\Lambda_{n,m}, \nu, \sigma^2) \in \{B \times [0,U]^m \times N \times S\}$, $L^n(\theta) < L^n(\theta_U)$.

Let $\alpha = \min_{\beta \in B, |x| \leq c} e^{\beta x}$ and $\eta = \max_{\beta \in B, |x| \leq c} e^{\beta x}$. Then,

$$e^{\beta x} \sum_{k=1}^{m} \Delta\Lambda_{n,k} 1_{\{v_k \leq v\}} \geq \alpha \sum_{k=1}^{m} \Delta\Lambda_{n,k} 1_{\{v_k \leq v\}},$$

and the likelihood $L^n(\theta)$ is bounded above by

$$\prod_{i=1}^{n} \left[(\eta \Delta\Lambda_n(v_i))^{\delta_i} \exp\left(-\alpha \sum_{k=1}^{m} \Delta\Lambda_{n,k} 1_{\{v_k \leq v_i\}}\right) \int f_{W,X}(\mathbf{w}_i, x)\, dx \right],$$

where $\int f_{W,X}(\mathbf{w}_i, x)\, dx$ is bounded (the continuity of $h(x;\nu)$ with respect to ν implies the continuity of $\int f_{W,X}(\mathbf{w}_i, x)\, dx$ with respect to ν and σ^2).

If $\theta_U = (\beta_U, \Delta\Lambda_{n,1,U}, \ldots, \Delta\Lambda_{n,m,U}, \nu_U, \sigma_U^2) \in \{B \times [0,\infty)^m \times N \times S\} \setminus \{B \times [0,U]^m \times N \times S\}$, then there exists j $(j \in \{1, \ldots, m\})$ such that $\Delta\Lambda_{n,j,U} > U$. Hence there exists at least one $i_U \in \{1, \ldots, n\}$ such that $\sum_{k=1}^{m} \Delta\Lambda_{n,k,U} 1_{\{v_k \leq v_{i_U}\}}$

is larger than U. Hence $\sum_{k=1}^{m} \Delta\Lambda_{n,k,U} 1_{\{v_k \le v_{i_U}\}} \longrightarrow +\infty$ as $U \longrightarrow +\infty$. It follows that the upper bound of $L^n(\theta_U)$ (and hence $L^n(\theta_U)$) can be made as close to 0 as desired by increasing U. This is the desired contradiction. ■

Hu et al. (1998) maximize the logarithm of $L^n(\theta)$ via the EM algorithm [see Dempster et al. (1972) for further details on EM] that is, instead of $\ln L^n(\theta)$, an EM loglikelihood

$$n^{-1} \sum_{i=1}^{n} E_{\theta^{(p)}} \left[\ln l(Y, X; \theta) | y_i \right]$$

is maximized iteratively on Θ_n, where $\theta^{(p)}$ denotes the current parameter estimate. Let $H_n(u) = n^{-1} \sum_{i=1}^{n} \Delta_i 1_{\{V_i \le u\}}$. The following equation holds for $\hat{\Lambda}_n$:

$$\hat{\Lambda}_n(t) = \int_0^t \left(n^{-1} \sum_{i=1}^{n} E_{\hat{\theta}_n} \left[e^{\hat{\beta}_n X} 1_{\{u \le V\}} | y_i \right] \right)^{-1} dH_n(u). \tag{2.2}$$

This results from the M step of the EM algorithm (see (1998) for details). This expression will be central in our proof of consistency of the NPML estimators.

Note that there exists no closed form for the solution $\hat{\beta}_n$. It was approximated by Hu et al. (1998) by a Newton–Raphson algorithm. In the maximization step of the $(p+1)$th iteration of EM, one would thus obtain, at the $(d+1)$th iteration of the Newton–Raphson algorithm:

$$\hat{\beta}^{(d+1)} = \hat{\beta}^{(d)} + (I^{(p)}(\hat{\beta}^{(d)}))^{-1} \cdot U^{(p)}(\hat{\beta}^{(d)}),$$

where

$$U^{(p)}(\beta) = \sum_{i=1}^{n} \Delta_i \left[E_{\theta^{(p)}}[X | y_i] - \frac{\sum_{j=1}^{n} E_{\theta^{(p)}}[X e^{\beta X} | y_j] 1_{\{V_i \le V_j\}}}{\sum_{j=1}^{n} E_{\theta^{(p)}}[e^{\beta X} | y_j] 1_{\{V_i \le V_j\}}} \right]$$

and

$$\begin{aligned} I^{(p)}(\beta) &= \sum_{i=1}^{n} \Delta_i \left[\frac{\sum_{j=1}^{n} E_{\theta^{(p)}}[X^2 e^{\beta X} | y_j] 1_{\{V_i \le V_j\}}}{\sum_{j=1}^{n} E_{\theta^{(p)}}[e^{\beta X} | y_j] 1_{\{V_i \le V_j\}}} \right. \\ &\quad \left. - \left\{ \frac{\sum_{j=1}^{n} E_{\theta^{(p)}}[X e^{\beta X} | y_j] 1_{\{V_i \le V_j\}}}{\sum_{j=1}^{n} E_{\theta^{(p)}}[e^{\beta X} | y_j] 1_{\{V_i \le V_j\}}} \right\}^2 \right]. \end{aligned}$$

2.4 Consistency of the NPML Estimators

Let $\|\cdot\|_\infty$ denote the supremum norm on $[0,\tau]$ and $|\cdot|$ denote the Euclidean norm. Consistency of the NPML estimators is described in the following theorem.

Theorem 2.4.1 *Under conditions 1–5, the NPML estimator*

$$\hat{\theta}_n = (\hat{\beta}_n, \hat{\Lambda}_n, \hat{\nu}_n, \hat{\sigma}_n^2)$$

is consistent, that is,

$$|\hat{\beta}_n - \beta_0|,\ \|\hat{\Lambda}_n - \Lambda_0\|_\infty,\ |\hat{\nu}_n - \nu_0| \text{ and } |\hat{\sigma}_n^2 - \sigma_0^2|$$

all converge to 0 almost surely.

PROOF. In the following, it will be convenient to denote (β, ν, σ^2) by γ and to use the following notation: $W_n(u,\theta) = n^{-1}\sum_{i=1}^n E_{\hat{\theta}_n}\left[e^{\hat{\beta}_n X} 1_{\{u \le V\}} | y_i\right]$.

Our proof follows Murphy's proof (1994) of a.s. consistency in the frailty model. The scheme of the proof is a very general one, it was also successfully used in various settings by Murphy et al. (1997), Parner (1998), Scharfstein et al. (1998) and Dupuy et al. (2002). The plan for proving consistency is as follows. We first show that the set $\{\hat{\theta}_n = (\hat{\gamma}_n, \hat{\Lambda}_n), n \in \mathbf{N}\}$ is relatively compact. Using identifiability of the model, we then show that its closure reduces to the single element $\theta_0 = (\gamma_0, \Lambda_0)$.

Since we are interested in almost sure (a.s.) consistency, we work with fixed realizations of the data which are assumed to lie in a set of probability 1. The first step is to show that $(\hat{\Lambda}_n)_{n\in\mathbf{N}}$ stays bounded as $n \longrightarrow \infty$. We note from (2.2) that $\hat{\Lambda}_n(t) \ge 0$ for all $t \in [0,\tau]$ and that

$$\hat{\Lambda}_n(\tau) \le \frac{\sum_{i=1}^n \Delta_i 1_{\{X_i \le \tau\}}}{\alpha \sum_{i=1}^n 1_{\{\tau \le X_i\}}}.$$

Noting that there exists a positive constant c_1 such that $1/[\frac{1}{n}\sum_{i=1}^n 1_{\{\tau \le X_i\}}] \le 1/P_{\theta_0}(V \ge \tau) + c_1$ as $n \longrightarrow \infty$ (with $P_{\theta_0}(V \ge \tau) > 0$ by condition 1), it follows that $\hat{\Lambda}_n(\tau)$ is bounded above by $\frac{1}{\alpha}(1/P_{\theta_0}(V \ge \tau) + c_1)$. Hence $\hat{\Lambda}_n(\tau)$ does not diverge to infinity as $n \longrightarrow \infty$.

Let now $\phi(n)$ be an arbitrary subsequence of (n). From the Bolzanno–Weierstrass theorem, $(\hat{\gamma}_{\phi(n)})_{n\in\mathbf{N}}$ being a bounded sequence of $\mathbf{R}^{p+2}$ has a convergent subsequence $(\hat{\gamma}_{\varphi(\phi(n))})_{n\in\mathbf{N}}$ which converges to some γ^*. Since $\hat{\Lambda}_n$ is not allowed to diverge, Helly's theorem (see Ash (1972)) can be used to prove the existence of a subsequence $(\hat{\Lambda}_{\eta(\varphi(\phi(n)))})_{n\in\mathbf{N}}$ of $(\hat{\Lambda}_{\varphi(\phi(n))})_{n\in\mathbf{N}}$ which converges pointwise to some Λ^*. Since every subsequence of a convergent sequence in $\mathbf{R}^{p+2}$ must converge to the same limit, $(\hat{\gamma}_{\eta(\varphi(\phi(n)))})_{n\in\mathbf{N}}$ must converge to γ^*.

Hence, for any given subsequence $\hat{\theta}_{\phi(n)} = (\hat{\gamma}_{\phi(n)}, \hat{\Lambda}_{\phi(n)})_{n\in\mathbf{N}}$, we can find a further subsequence $\hat{\theta}_{\eta(\varphi(\phi(n)))} = (\hat{\gamma}_{\eta(\varphi(\phi(n)))}, \hat{\Lambda}_{\eta(\varphi(\phi(n)))})_{n\in\mathbf{N}}$ which converges to some $\theta^* = (\gamma^*, \Lambda^*)$. We now show that $\hat{\Lambda}_{\eta(\varphi(\phi(n)))}$ converges uniformly to Λ^*. In the following, for the sake of clarity of formulas, we shall use the notation $g(n) = \eta(\varphi(\phi(n)))$.

We shall use in the sequel the Helly–Bray lemma [see Korsholm (1998)] and the result that the class of all functions $f : [0, \tau] \longrightarrow \mathbf{R}$ that are uniformly bounded and of variation bounded is Glivenko–Cantelli [see Van Der Vaart and Wellner (1996)].

We first define the intermediate quantity $\bar{\Lambda}_n$ by $\bar{\Lambda}_n(t) = \int_0^t \frac{dH_n(u)}{W_n(u,\theta_0)}$, which will help mediate between $\hat{\Lambda}_n$ and Λ_0. By the Glivenko–Cantelli theorem, $(H_n(u))_{n\in\mathbf{N}}$ converges uniformly on $[0, \tau]$ to $H(u) = E_{\theta_0}\left[\Delta 1_{\{V\le u\}}\right]$. Note that $\Lambda_0(t) = \int_0^t \frac{dH(u)}{W(u,\theta_0)}$, where $W(u, \theta_0) = E_{\theta_0}\left[e^{\beta_0 X} 1_{\{u\le V\}}\right]$.

The functions $u \longmapsto E_{\theta_0}[e^{\beta_0 X} 1_{\{u\le V\}}|y]$ are uniformly bounded and of bounded variation. Hence $(W_n(u, \theta_0))_{n\in\mathbf{N}}$ converges uniformly to $E_{\theta_0}[e^{\beta_0 X} 1_{\{u\le V\}}]$, which is bounded away from 0 by condition 5. Hence $(1/W_n(u, \theta_0))_{n\in\mathbf{N}}$ converges to $1/W(u, \theta_0)$ uniformly on $[0, \tau]$.

Applying the Helly–Bray lemma gives

$$\sup_{t\in[0,\tau]} \left| \int_0^t \frac{dH_n(u)}{W_n(u,\theta_0)} - \int_0^t \frac{dH(u)}{W(u,\theta_0)} \right| \longrightarrow 0 \text{ a.s. as } n \longrightarrow \infty.$$

Similarly

$$\sup_{t\in[0,\tau]} \left| \int_0^t \frac{dH_{g(n)}(u)}{W_{g(n)}(u,\hat{\theta}_{g(n)})} - \int_0^t \frac{dH(u)}{W(u,\theta^*)} \right| \longrightarrow 0 \text{ a.s. as } n \longrightarrow \infty.$$

Hence $\|\bar{\Lambda}_n - \Lambda_0\|_\infty$ and $\|\hat{\Lambda}_{g(n)} - \Lambda^*\|_\infty$ converge almost surely to 0. This establishes the relative compactness of $\{\hat{\theta}_n, n \in \mathbf{N}\}$.

We now show that every subsequence $(\hat{\theta}_{g(n)})_{n\in\mathbf{N}}$ must converge to the true value $\theta_0 = (\gamma_0, \Lambda_0)$. Since $\hat{\theta}_{g(n)}$ maximizes the loglikelihood,

$$\frac{1}{g(n)} \sum_{i=1}^{g(n)} \left[\ln L_i(\hat{\theta}_{g(n)}) - \ln L_i(\gamma_0, \bar{\Lambda}_{g(n)})\right] \ge 0.$$

Note that for all $g(n)$, as $m \longrightarrow \infty$,

$$\frac{1}{m} \sum_{i=1}^{m} \left[\ln L_i(\hat{\theta}_{g(n)}) - \ln L_i(\gamma_0, \bar{\Lambda}_{g(n)})\right] \longrightarrow E_{\theta_0}\left[\ln L(\hat{\theta}_{g(n)}) - \ln L(\gamma_0, \bar{\Lambda}_{g(n)})\right] \text{ a.s.}$$

It follows that

$$E_{\theta_0}\left[\ln L(\hat{\theta}_{g(n)}) - \ln L(\gamma_0, \bar{\Lambda}_{g(n)})\right] \ge -o(1). \tag{2.3}$$

We have

$$\ln L(\hat{\theta}_{g(n)}) - \ln L(\gamma_0, \bar{\Lambda}_{g(n)}) \longrightarrow \ln L(\theta^*) - \ln L(\theta_0) \text{ a.s.}$$

By Lebesgue's theorem,

$$E_{\theta_0}\left[\ln \frac{L(\hat{\theta}_{g(n)})}{L(\gamma_0, \bar{\Lambda}_{g(n)})}\right] \longrightarrow E_{\theta_0}\left[\ln \frac{L(\theta^*)}{L(\theta_0)}\right] \text{ a.s.}$$

From (2.3), $E_{\theta_0}\left[\ln \frac{L(\theta^*)}{L(\theta_0)}\right] \geq 0$. Since $E_{\theta_0}\left[\ln \frac{L(\theta^*)}{L(\theta_0)}\right]$ is minus the Kullback-Leibler distance, which is positive, the Kullback–Leibler distance is equal to 0. From identifiability, it follows that $\theta^* = \theta_0$. ■

2.5 Discussion

In this paper, we have proved adequacy of the joint modeling approach and of the nonparametric maximum likelihood estimation method in the proportional hazards model with covariate measurement error. Using theory developed by Murphy (1994), we have shown that the NPML estimators obtained from a joint model assuming an additive error model for the covariate exist and are consistent. Asymptotic normality of these estimators is a subject of current work.

References

1. Andersen, P. K. and Gill, R. D. (1982). Cox's regression model for counting processes: A large sample study, *Annals of Statistics*, **10**, 1100–1120.

2. Ash, R. B. (1972). *Real Analysis and Probability*, Academic Press, New York.

3. Bagdonavičius, V. and Nikulin, M. S. (2001). *Accelerated Life Models: Modeling and Statistical Analysis*, Chapman & Hall, Boca Raton, Florida.

4. Breslow, N. E. (1972). Contribution to the discussion on the paper by D.R. Cox, *Journal of the Royal Statistical Society, Series B*, **34**, 187–220.

5. Breslow, N. E. (1974). Covariance analysis of censored survival data, *Biometrics*, **30**, 89–99.

6. Buzas, J. S. (1998). Unbiased scores in proportional hazards regression with covariate measurement error, *Journal of Statistical Planning and Inference*, **67**, 247–257.

7. Carroll, R. J., Ruppert, D. and Stefanski, L. A. (1995). *Measurement Error in Nonlinear Models*, Chapman & Hall, London, England.

8. Cox, D. R. (1972). Regression models and life-tables (with discussion), *Journal of the Royal Statistical Society, Series B*, **34**, 187–220.

9. Cox, D. R. (1975). Partial likelihood, *Biometrika*, **62**, 269–276.

10. Dafni, U. G. and Tsiatis, A. A. (1998). Evaluating surrogate markers of clinical outcome when measured with error, *Biometrics*, **54**, 1445–1462.

11. Dempster, A. P., Laird, N. M. and Rubin, D. B. (1972). Maximum likelihood from incomplete data via the EM algorithm (with discussion), *Journal of the Royal Statistical Society, Series B*, **39**, 1–38.

12. Dupuy, J.-F., Grama, I. and Mesbah, M. (2003). Normalité asymptotique des estimateurs semiparamétriques dans le modèle de Cox avec covariable manquante nonignorable, *Comptes Rendus de l'Académie des Sciences de Paris, Série I*, **336**, 81–84.

13. Dupuy, J.-F., Grama, I. and Mesbah, M. (2004). Asymptotic theory for the Cox model with missing time-dependent covariate, *Communications in Statistics—Theory and Methods*, **33** (to appear).

14. Dupuy, J.-F. and Mesbah, M. (2000). Joint modeling of survival and informatively missing continuous longitudinal data, In *Proceedings of the Second International Conference on Mathematical Methods in Reliability*, pp. 357–360.

15. Dupuy, J.-F. and Mesbah, M. (2002). Joint modeling of event time and nonignorable missing longitudinal data, *Lifetime Data Analysis*, **8**, 99–115.

16. Faucett, C. L. and Thomas, D. C. (1996). Simultaneously modeling censored survival data and repeatedly measured covariates: a Gibbs sampling approach, *Statistics in Medicine*, **15**, 1663–1685.

17. Henderson, R., Diggle, P. and Dobson, A. (2000). Joint modelling of longitudinal measurements and event time data, *Biostatistics*, **1**, 465–480.

18. Hu, P., Tsiatis, A. A. and Davidian, M. (1998). A semiparametric estimator for the proportional hazards model with longitudinal covariates measured with error, *Biometrics*, **88**, 447–458.

19. Johansen, S. (1983). An extension of Cox's regression model, *International Statistical Review*, **51**, 258–262.

20. Korsholm, L. (1998). Likelihood ration test in the correlated gamma-frailty model. Research Report 98–11, University of Aarhus and the Aahrus School of Busness.

21. Li, Y. and Lin, X. (2000). Covariate measurement errors in frailty models for clustered survival data, *Biometrika*, **81**, 61–71.

22. Murphy, S. A. (1994). Consistency in a proportional hazards model incorporating a random effect, *Annals of Statistics*, **22**, 712–731.

23. Murphy, S. A., Rossini, A. J. and Van Der Vaart, A. W. (1997). Maximum likelihood estimation in the proportional odds model, *Journal of the American Statistical Association*, **92**, 968–976.

24. Nielsen, G. G., Gill, R. D., Andersen, P. K. and Sørensen, T. I. A. (1992). A counting process approach to maximum likelihood estimation in frailty models, *Scandinavian Journal of Statistics*, **19**, 25–43.

25. Parner, E. (1998). Asymptotic theory for the correlated gamma-frailty model, *Annals of Statistics*, **26**, 183–214.

26. Prentice, R. L. (1982). Covariate measurement errors and parameter estimation in a failure time regression model, *Biometrika*, **69**, 331–342.

27. Scharfstein, D. O., Tsiatis, A. A. and Gilbert, P. B. (1998). Semiparametric efficient estimation in the generalized odds-rate class of regression models for right-censored time-to-event data, *Lifetime Data Analysis*, **4**, 355–391.

28. Song, X., Davidian, D. and Tsiatis, A. A. (2002a). A semiparametric likelihood approach to joint modeling of longitudinal and time-to-event data, *Biometrics*, **58**, 742–753.

29. Song, X., Davidian, D. and Tsiatis, A. A. (2002b). An estimator for the proportional hazards model with multiple longitudinal covariates measured with error, *Biostatistics*, **3**, 511–528.

30. Tsiatis, A. A. (1981). A large sample study of Cox's regression model, *Annals of Statistics*, **9**, 93–108.

31. Tsiatis, A. A., De Gruttola, V. and Wulfsohn, M. S. (1995). Modeling the relationship of survival to longitudinal data measured with error. Applications to survival and CD4 counts in patients with AIDS, *Journal of the American Statistical Association*, **90**, 27–37.

32. Van Der Vaart, A. W. and Wellner, J. A. (1996). *Weak Convergence and Empirical Processes*, Springer-Verlag, New York.

33. Wulfsohn, M. S. and Tsiatis, A. A. (1997). A joint model for survival and longitudinal data measured with error, *Biometrics*, **53**, 330–339.

34. Zhou, H. and Pepe, M. S. (1995). Auxiliary covariate data in failure time regression, *Biometrika*, **82**, 139–149.

3

Diagnostics for Cox's Proportional Hazards Model

C. Caroni

National Technical University of Athens, Athens, Greece

Abstract: Although Cox's proportional hazards model is applied so often, diagnostic methods are not well known to non-expert users. Latest results suggest that individually scaled Schoenfeld residuals, dfbetas and added variable plots should be provided for assessing the model. Most of these methods are absent from widely used packages.

Keywords and phrases: Lifetime data, Cox's proportional hazards model, diagnostic methods, residuals, influence, added variable plots

3.1 Introduction

A large number of diagnostic methods for model checking have been proposed over the years for Cox's proportional hazards (PH) regression model, but they are still not used as widely as they should be. The purpose of this paper is to present an up-to-date review of the model checking procedures that seem to have the greatest practical importance and deserve to receive the most attention. The PH model [Cox (1972)] specifies

$$\lambda(t; \mathbf{x}) = \lambda_0(t) e^{\beta' \mathbf{x}}$$

for the hazard function for the failure time T associated with a p-dimensional vector of covariates $\mathbf{x}$, given an unknown underlying hazard function λ_0 and a vector of coefficients β which must be estimated. Articles dealing with this model often start with the observation that in many fields it has become the standard way of analyzing lifetime data with covariates. This makes Cox's model one of the big success stories of statistics. At the same time, once a model has reached the status of being the standard, there is a great danger that people will apply it because everyone else does, and that referees of scientific papers will accept it for the same reason. Yet it is a statistical model, therefore it contains assumptions that might not be suitable for the problem in question. The fact that it is a semi-parametric model does not change this. It is therefore important that diagnostic methods become widely known.

Previous work along the same lines includes a review by Hess (1995) of graphical methods for examining the PH assumption and an empirical comparison of statistical tests for PH by Ng'andu (1997). The more modern books, such as Klein and Moeschberger (1997), Therneau and Grambsch (2000) and Harrell (2001) also contain much relevant material, although heavily linked in some cases to particular packages such as S-Plus.

From the hazard function of the PH model, we obtain the survivor function

$$S(t;\mathbf{x}) = exp\{-\Lambda_0(t)e^{\beta'\mathbf{x}}\}$$

where Λ_0 is the cumulative hazard corresponding to λ_0. Hence

$$ln\{-lnS(t;\mathbf{x})\} = ln\Lambda_0 + \beta'\mathbf{x}.$$

This means that any survivor function $S(t;\mathbf{x})$, when plotted on the complementary log-log scale, differs from $ln\Lambda_0$ by the constant amount $\beta'\mathbf{x}$ throughout its duration. Hence any two such functions, $S(t;\mathbf{x_1})$ and $S(t;\mathbf{x_2})$ for different values of the covariate vector $\mathbf{x}$, will be parallel. This gives the basic means of inspecting the PH assumption, maybe the only one suggested in the more elementary books such as Parmar and Machin (1995):

- obtain Kaplan-Meier estimates $\hat{S}(t;\mathbf{x})$ for various selected $\mathbf{x}$
- examine visually whether the curves $\ln\{-\ln\hat{S}(t;\mathbf{x})\}$ against t (or a function of t) are parallel for different $\mathbf{x}$.

The point of using a function of t is to make it easier to assess the graph. If $\ln t$ is used, then plots of $\ln\{-\ln\hat{S}(t;\mathbf{x})\}$ will be straight lines when lifetimes follow the Weibull distribution. It is probably easier for the eye to assess straight lines than curves.

The main drawback of this procedure is obvious. We need to have reliable estimates $\hat{S}$ and this means having a large amount of data available for each chosen $\mathbf{x}$. This will only happen if there are few covariates and few distinct values (although continuous variates can be grouped), for example if there is just one covariate with a small number of levels. Therefore, we will not often be able to use this plot although, when we can, it is essential to look at it before doing the work of fitting the PH model.

We now move on to the more important diagnostic methods based on residuals and other information obtained from the fit of the PH model to the data.

3.2 Schoenfeld Residuals

Residuals from a fitted model are most easily understood when they express in some way the difference between observed data values and the corresponding fitted (predicted) values. This can usually be done in various ways, as shown by the fact that SPSS offers five residuals for ordinary linear regression. It is not surprising that the more complex PH model also gives rise to several alternative forms of residual. One of the most important was introduced by Schoenfeld (1982).

At the time t_i of death or failure of unit or individual i, there is a risk set R_i consisting of the individuals who have not yet failed (or been censored) so were at risk of failure at this time. The conditional probability that i failed, given that there was a failure at t_i is given by

$$\begin{aligned} p_i &= \frac{\lambda_i(t_i)}{\sum_{j\in R_i}\lambda_j(t_i)} \\ &= \frac{\lambda_0(t_i)e^{\beta'\mathbf{x_i}}}{\sum_{j\in R_i}\lambda_0(t_i)e^{\beta'\mathbf{x_j}}} \\ &= \frac{e^{\beta'\mathbf{x_i}}}{\sum_{j\in R_i}e^{\beta'\mathbf{x_j}}}. \end{aligned}$$

Now consider $E(\mathbf{X}|R_i)$, the expected value (before we know which unit failed) of the covariate vector $\mathbf{X}$ of the failure at time t_i, given the risk set.

$$\begin{aligned} E(\mathbf{X}|R_i) &= \sum_{k\in R_i}\mathbf{x_k}p_k \\ &= \frac{\sum_{k\in R_i}\mathbf{x_k}e^{\beta'\mathbf{x_k}}}{\sum_{j\in R_i}e^{\beta'\mathbf{x_j}}}. \end{aligned}$$

A residual, observed minus expected, can thus be defined as the vector

$$\mathbf{r_i} = \mathbf{x_i} - E(\mathbf{X}|R_i)$$

giving the Schoenfeld residual

$$\hat{\mathbf{r}}_\mathbf{i} = \mathbf{x_i} - \hat{E}(\mathbf{X}|R_i)$$

where $\hat{E}$ denotes that $\hat{\beta}$ has been substituted for β in E. Note that these residuals are defined only for the units with an observed failure time, and not for censored times. Unlike the residuals that are familiar from linear and generalized linear models, the Schoenfeld residual is not based on the dependent variable (which could be time, or the event of failure) but on the covariate. It represents the deviation between the covariate of the unit failing at t_i and the weighted average of all covariates in the risk set, so a large residual shows that the unit failing at t_i was extreme in its **x**-values *at that time.* (Since $\mathbf{r}_i$ is a vector, it might be large in one or more components, not necessarily in all). Hence this kind of residual appears to be strongly connected to the idea of influence.

The Schoenfeld residual is closely related to the estimation procedure of Cox's PH model. The product of the above conditional probabilities p_i over the set D of all individuals who failed is, for the case of distinct failure times, the partial likelihood

$$L = \prod_{i \in D} p_i$$

whence

$$l = \ln L = \sum_{i \in D} \{\beta' \mathbf{x_i} - \ln \sum_{j \in R_i} e^{\beta' \mathbf{x_j}}\}.$$

Hence the score vector

$$\begin{aligned} \mathbf{U} = \frac{\partial l}{\partial \beta} &= \sum_{i \in D} \{\mathbf{x_i} - \sum_{k \in R_i} \mathbf{x_k} e^{\beta' \mathbf{x_k}} / \sum_{j \in R_i} e^{\beta' \mathbf{x_j}}\} \\ &= \sum_{i \in D} \mathbf{r_i}. \end{aligned}$$

Since the estimation procedure solves for $\beta = \hat{\beta}$ by setting $\mathbf{U}(\hat{\beta}) = \mathbf{0}$, we see immediately one of the desirable properties of residuals, namely

$$\sum_{i \in D} \hat{\mathbf{r}}_\mathbf{i} = \mathbf{0}.$$

3.3 Martingale Residuals

Cox's PH model can be generalized in various ways. One very important development was to fit it into the framework of counting processes [Andersen and Gill (1982)]. Let

$$N_i(t), \quad t \geq 0$$

for subject i indicate the number of events observed over time t. Assume that $N_i(t)$ has the intensity function

$$Y_i(t) e^{\beta' \mathbf{x}} d\Lambda_0(t)$$

where $Y_i(t)$ is a 0-1 process indicating whether the ith subject is at risk at time t, and $X_i(t)$ is the vector of covariate processes. For Cox's PH model, $\Lambda_0(t)$ is completely unspecified, and $Y_i(t)$ takes the value 1 until the first event or censoring and 0 afterwards.

Residuals for Cox's PH model in this framework were developed by Therneau et al. (1990), following Barlow and Prentice (1988) who introduced residuals defined from the difference between the counting process and the integrated intensity function

$$M_i(t) = N_i(t) - \int_0^t Y_i(s) e^{\beta' \mathbf{x}_i(s)} d\Lambda_0(s), \quad i = 1, ..., n.$$

Under standard conditions, $M_i(t)$ is a martingale. The corresponding estimate

$$\hat{M}_i(t) = N_i(t) - \int_0^t Y_i(s) e^{\hat{\beta}' \mathbf{x}_i(s)} d\hat{\Lambda}_0(s), \quad i = 1, ..., n$$

where $\hat{\Lambda}_0$ is Breslow's estimator of the baseline cumulative hazard (reducing to the Nelson–Aalen estimator when there are no covariates), is therefore termed the martingale residual. It has the interpretation of the observed minus expected number of events over [0,t]. For the special case of Cox's PH model with no time-dependent covariates, it takes the simpler form

$$\hat{M}_i = \delta_i - \hat{\Lambda}_0(t_i) e^{\hat{\beta}' \mathbf{x}_i}$$

where δ_i is the final status (1 for the event and 0 for censored) and t_i the observation time of the ith unit. The second term is an estimate of

$$\Lambda_0(t_i) e^{\beta' \mathbf{x}_i}$$

which is distributed as unit exponential. Therefore, the corresponding estimates can themselves be regarded as residuals of the general type introduced by Cox and Snell (1968), which had earlier been considered by Kay (1977) and Crowley and Hu (1977).

Outliers may be detectable in the plot of martingale residuals against the risk score $\hat{\beta}'\mathbf{x}$. A difficulty here is that, as can be seen immediately, the martingale residuals have an extremely skew distribution, since their range is $(-\infty, +1)$. Therefore it is only possible to see the unusually large negative values, corresponding to individuals who survived for an unexpectedly long time. Individuals who died unexpectedly early will not stand out, because large positive values are impossible. It seems a good idea to avoid this problem by transforming the martingale residuals in some way, for example, transforming to approximate normality, like the familiar residuals from simpler statistical models. For this purpose, Therneau et al. (1990) introduced deviance residuals which have a motivation similar to the deviance residuals in generalized linear models. In the special case of Cox's PH model, the ith deviance residual is

$$d_i = sgn(\hat{M}_i)[-2\{\hat{M}_i + \delta_i log(\delta_i - \hat{M}_i)\}]^{1/2}.$$

However, according to Therneau and Grambsch (2000), these residuals have not turned out to be useful so far. The individuals who died much earlier than expected might show up in the deviance residuals, but not always.

Another suggested use for the martingale residuals is to indicate the correct functional form for a covariate. The idea here is that, if the covariate j should enter the model as $exp(f(x_j)\beta_j)$, then a smoothed plot of the martingale residuals from the model with no covariates against x_j will show the curve $f(x_j)$ [Therneau and Grambsch (2000)]. This idea is only expected to work well if the covariates are weakly correlated. Added variable plots (see below) do handle correlation between covariates, although they are designed to show only linear relationships.

3.4 Using Residuals to Examine the PH Assumption

The PH assumption requires that the ratio $\lambda(t;\mathbf{x})/\lambda_0(t)$ is constant over time (when the covariates do not vary with time). Therefore, one way of expressing the failure of the PH assumption is to include dependence on time by rewriting the model as

$$\lambda(t;\mathbf{x}) = \lambda_0(t)e^{\beta(t)'\mathbf{x}}$$

where at least one component of the vector of coefficients $\beta(t)$ is not constant. A test for $H_0 : \beta(t) = \beta,\ \forall\, t$ is therefore a test for PH. As shown by Grambsch and Therneau (1994), appropriate tests are based on the Schoenfeld residuals and many tests published before that date fit into this category. They justify the test that now seems to have become dominant as follows.

Write $\beta(t) = \beta + \mathbf{g}(t)$ and let $\hat{\beta}$ be the usual estimate of β from the ordinary PH model. It can be shown that

$$E(\hat{\mathbf{r}}_i) \simeq V_i\mathbf{g}(t_i)$$

where V_i is the covariance matrix of the Schoenfeld residual $\mathbf{r_i}$. Inserting estimates, this suggests that

$$\beta(t) \simeq \beta + \hat{V}_i^{-1}\hat{r}_i = \beta + \hat{r}_i^*$$

where $\hat{\mathbf{r}}_i^*$ are the scaled Schoenfeld residuals (referred to as the individually-scaled residuals from here on). Hence a plot of $\hat{r}_{ij}^* + \hat{\beta}_j$ against time (or some function of time) will indicate departures from PH and suggest the correct functional form of time dependence. In fact, it is essential to superimpose a scatterplot smoother on the diagram in order to be able to see what is going on.

Grambsch and Therneau (1994) suggested replacing the separate $\hat{V}_i$ in this procedure by their average

$$\bar{V} = I(\hat{\beta})/d$$

that is, the observed information matrix $I(\hat{\beta})$ divided by the number of uncensored survival times d. This avoids inverting all the separate matrices $\hat{V}_i$, but this is not an important gain with modern computing power. Actually, Grambsch and Therneau regarded this substitution as an improvement, since the later $\hat{V}_i$ will be based on small numbers of subjects and may be poor estimates. This proposal has been examined recently in detail by Winnett and Sasieni (2001). They found that in many cases it makes very little difference, but in other cases it can lead to misleading estimates of the time-varying coefficients. Therefore they concluded that it should not be used. They prefer to use the individually-scaled residuals, although they left open the possibility that the V_i could be modified another way, perhaps by smoothing.

The separate $\hat{V}_i$ are available in the output from S-Plus and users of that language would have no difficulty in manipulating them. But many non-specialists use SPSS. What facilities are available there? The situation in Version 11, is that the unscaled Schoenfeld residuals are provided (called simply partial residuals). Since the estimated correlation matrix of the parameter estimates can be requested — through the command syntax, not the windows — and their standard errors are on the usual output, one can calculate the estimated covariance of the estimates. This is I^{-1}, so $\bar{V}$ is available and Grambsch and Therneau's modified version of the scaled residuals can be obtained. But the unskilled user is not very likely to do this work. Therefore, it has to be said that, at the moment, SPSS does not enable the ordinary user to use these residuals in the accepted way. Furthermore, there is no way at all to obtain the individually-scaled residuals. In any case, SPSS does not provide smoothers, without which it is hard to assess what these residuals show anyway.

3.5 Influence

The influence of one data point on the fitting of a statistical model is defined most directly as the change in parameter estimates when this point is omitted. In the PH model, the effect of omitting point j and thus moving the parameter estimates from $\hat{\beta}$ to $\hat{\beta}_{(j)}$, can only be computed exactly by refitting the model. To avoid this, an approximation is generally used. This can be derived very simply by taking the first term of a Taylor series expansion of the score function, giving

$$\hat{\beta} - \hat{\beta}_{(j)} \simeq I^{-1}\mathbf{d}_j$$

where

$$\mathbf{d}_j = \delta_j \hat{\mathbf{r}}_j - \sum_{i \in D_j} [e^{\beta' \mathbf{x}_j}\{\mathbf{x}_j - \hat{E}(\mathbf{x}|R_i)\}/\sum_{k \in R_i} e^{\beta' \mathbf{x}_k}].$$

[Cain and Lange (1984)]. D_j denotes the set of individuals who died at or before t_j, the death or censoring time of the jth individual. The vectors $\mathbf{d}_j$ are called score residuals. The $n \times p$ matrix whose rows are the $I^{-1}\hat{\mathbf{d}}_j$ has become known as the matrix of *dfbeta* values. They are available in SPSS as well as in the more specialized packages.

The first term of $\mathbf{d}_j$ is just the Schoenfeld residual and is present only if this individual died. The second term, which is present for all individuals, represents the contributions from all the risk sets R_i that included this individual. If an individual dies early in the study, the first term will be more important than the second. The longer the individual survives, the more important becomes the second term. For an individual who is censored early in the study, the first term is zero and the second term is small, so the influence is minimal, as expected.

3.6 Added Variable Plots

In linear regression, the added variable plot is a very useful tool for investigating whether a particular covariate should be added to the model, also showing whether any observations are particularly influential in this respect (Atkinson, 1985). Because of its usefulness, it has been extended to other models. Thus O'Hara Hines and Carter (1993) developed an added variable plot for generalized linear models. The same idea was subsequently adapted to Cox's PH model by Hall et al. (1996). It is very simple to give the idea of their procedure. Firstly, the added variable plot for a covariate x_j in a linear regression consists of a scatterplot of the residuals from two regressions: the regression of the dependent variable y on the covariates except for x_j, against the residuals from the regression of x_j on the other covariates. This means plotting

$$(I - H_{(j)})\mathbf{y} \quad \text{against} \quad (I - H_{(j)})\mathbf{x}_j$$

where

$$H_{(j)} = I - X_{(j)}(X'_{(j)}X_{(j)})^{-1}X_{(j)}$$

and $X_{(j)}$ denotes the matrix of covariates with x_j omitted. The slope of a regression line through the origin fitted to these points equals the estimated regression coefficient that will be obtained for x_j if it is included in the model.

Secondly, observe that generalized linear models can be fitted by iterative weighted least squares [McNullagh and Nelder (1989)]. In this representation, maximum likelihood estimates of the parameters β are obtained by solving the equations

$$(X'WX)\hat{\beta} = X'W\mathbf{z}.$$

These are identical in appearance to the equations appearing in weighted linear regression, but they require iteration with recalculation of the weights W and the dependent variable Z at each step. This resemblance to ordinary linear regression suggests producing the added variable plot for generalized linear models by replacing $\mathbf{y}$ by $W^{1/2}\mathbf{z}$ and $\mathbf{x}_j$ by $W^{1/2}\mathbf{x}_j$ in the above plots, where W and $\mathbf{z}$ are the final values from the iteration. It is recommended that W should be obtained from the full model including x_j, not from the reduced model omitting x_j [O'Hara Hines and Carter (1993)]. Finally, observe that an iterative procedure for fitting Cox's PH model can also be written in the iterative weighted least squares form [Hall et al. (1996) and Lindkvist (2000)]. Hence the version of the added variable plot produced for generalized linear models can also be applied to Cox's PH model. The plot improves an earlier version by Chen and Wang (1991).

The difference between the two models is that for the PH model, this least squares representation is not applied directly to the n cases, but to the "pseudo-observations", which represent a case every time it appears in a risk set. [Rewriting the data in terms of pseudo-observations is a device that has been used to enable the PH model to be fitted in the GLIM program for generalized linear models: Whitehead (1980), Aitkin et al. (1989).] Therefore Hall et al.'s first version of the added variable plot, called the *risk level added variable plot* (RLAV), is not necessarily very useful because it contains a very large number of points. They therefore developed two other versions. In one, the *subject level* plot (SLAV), each point represents a case. In the other, the *event level* plot (ELAV), each point represents a failure time (that is, a risk set). The plotting positions in these new versions (x^*, y^*) are obtained from the positions (x, y) in the RLAV by

$$x^* = (\sum x_i^2)^{1/2}, \quad y^* = \sum x_i y_i / (\sum x_i^2)^{1/2}$$

where each summation is over the pseudo-observations in a particular risk set, for the ELAV, or over the pseudo-observations corresponding to one case, for the SLAV. This choice ensures that the slopes of the SLAV and ELAV are the same as in the RLAV, and equal the estimated parameter β_j for x_j if it is added to the model [Hall et al. (1996)]. Further details of the properties of these added variable plots, and of their relation to residuals and the influence measure, are given by Lindkvist (2000). The correspondence with model diagnostics is one of the reasons why the added variable plot is so helpful.

These plots are apparently not available yet in any package. They seem to be worth adding.

3.7 Conclusions

It has not been easy to extend all the familiar linear model diagnostics to the PH model. Many methods that have been proposed have already been replaced by better versions and, as the above literature shows, improvements are still being made. The current state of the art seems to be that Schoenfeld residuals, but with individual scaling, and dfbetas must be provided. Martingale residuals are less important but may be useful too. The new added variable plots should become available. Unfortunately, one of the packages most widely used by non-statisticians for fitting the PH model, SPSS, offers nothing from this list except the values of dfbeta.

References

1. Aitkin, M., Anderson, D., Francis, B. and Hinde, J. (1989). *Statistical Modelling in GLIM,* Oxford University Press, Oxford, England.

2. Andersen, P. K. and Gill, R. D. (1982). Cox's regression model for counting processes: A large sample study, *Annals of Statistics*, **10**, 1100–1120.

3. Atkinson, A. C. (1985). *Plots, Transformations and Regression: An Introduction to Graphical Methods of Diagnostic Regression Analysis,* Oxford University Press, Oxford, England.

4. Barlow, W. E. and Prentice, R. L. (1988). Residuals for relative risk regression, *Biometrika*, **75**, 65–74.

5. Cain, K. C. and Lange, N. T. (1984). Approximate case influence for the proprtional hazards regression model with censored data, *Biometrics*, **40**, 493–499.

6. Chen, C.-H. and Wang, P. C. (1991). Diagnostic plots in Cox's regression model, *Biometrics*, **47**, 841–850.

7. Cox, D. R. (1972). Regression models and life tables (with discussion), *Journal of the Royal Statistical Society, Series B*, **34**, 187–220.

8. Cox, D. R. and Snell, E. J. (1968). A general definition of residuals (with discussion), *Journal of the Royal Statistical Society, Series B*, **30**, 248–275.

9. Crowley, J. and Hu, M. (1977). Covariance analysis of heart transplant survival data, *Journal of the American Statistical Association*, **72**, 27–36.

10. Grambsch, P. M. and Therneau, T. M. (1994). Proportional hazards tests and diagnostics based on weighted residuals, *Biometrika*, **81**, 515–526.

11. Hall, C. B., Zeger, S. L. and Bandeen-Roche, K. J. (1996). Adjusted variable plots for Cox's proportional hazards regression model, *Lifetime Data Analysis*, **2**, 73–90.

12. Harrell, F. E., Jr. (2001). *Regression Modeling Strategies, with Applications to Linear Models, Logistic Regression, and Survival Analysis*, Springer-Verlag, New York.

13. Hess, K. R. (1995). Graphical methods for assessing violations of the proportional hazards assumption in Cox regression, *Statistics in Medicine*, **14**, 1707–1723.

14. Kay, R. (1977). Proportional hazard regression models and the analysis of censored survival data, *Applied Statistics*, **26**, 227–237.

15. Klein, J. P. and Moeschberger, M. L. (1997). *Survival Analysis: Techniques for Censored and Truncated Data*, Springer-Verlag, New York.

16. Lindkvist, M. (2000). Properties of added variable plots in Cox's regression model, *Lifetime Data Analysis*, **6**, 23–38.

17. McCullagh, P. and Nelder, J. A. (1989). *Generalized Linear Models*, Second edition, Chapman & Hall, London, England.

18. Ng'andu, N. H. (1997). An empirical comparison of statistical tests for assessing the proportional hazards assumption of Cox's model, *Statistics in Medicine*, **16**, 611–626.

19. O' Hara Hines, R. J. and Carter, E. M. (1993). Improved added variable and partial residual plots for the detection of influential observations in generalized linear models, *Applied Statistics*, **42**, 3–20.

20. Parmar, M. K. B. and Machin, D. (1995). *Survival Analysis: A Practical Approach*, John Wiley & Sons, Chichester, England.

21. Schoenfeld, D. (1982). Partial residuals for the proportional hazards regression model, *Biometrika*, **69**, 239–241.

22. Therneau, T. M. and Grambsch, P. M. (2000). *Modeling Survival Data: Extending the Cox Model*, Springer-Verlag, New York.

23. Therneau, T. M., Grambsch, P. M. and Fleming, T. R. (1990). Martingale-based residuals for survival models, *Biometrika*, **77**, 147–160.

24. Whitehead, J. (1980). Fitting Cox's regression model to survival data using GLIM, *Applied Statistics*, **29**, 268–275.

25. Winnett, A. and Sasieni, P. (2001). A note on scaled Schoenfeld residuals for the proportional hazards model, *Biometrika*, **88**, 565–571.

PART II
DEGRADATION MODELS AND ANALYSES

4

Semiparametric Analysis of Degradation and Failure Time Data with Covariates

V. Bagdonavičius[1] **and M. Nikulin**[2]

[1]*Department of Mathematical Statistics, University of Vilnius, Vilnius, Lithuania*
[2] *Victor Segalen University Bordeaux 2, Bordeaux, France & Laboratory of Statistical Methods, Steklov Mathematical Institute, Saint Petersburg, Russia*

Abstract: The paper considers the semiparametric analysis of several new degradation and failure time regression models without and with time depending covariates. These joint models for survival and longitudinal data measured with errors can be applied in studies of longevity, aging and degradation in survival analysis, biostatistics, epidemiology, demography, oncology, biology and reliability.

Keywords and phrases: Aging, censored data, conjoint model, degradation process, disability, failure, longitudinal data, measurement error, Nelson–Aalen type estimator, noise, nonparametric estimation, path model, regression parameters, reliability, semiparametric estimation, survival analysis, traumatic event, Wulfsohn–Tsiatis model

4.1 Introduction

Functioning of a unit is characterized by its degradation process and by the random moments of its potential failures. We call a failure of a unit natural if the degradation attains some critical level. Other failures are called traumatic. These can be related with production defects, caused by mechanical damages or by fatigue of components etc.

The intensities of the traumatic failures depend on degradation. As a rule these intensities are increasing functions of degradation values.

Suppose that the lifetime of a unit is determined by the degradation process $Z(t)$ and the moment of its potential traumatic failure T. For example, $Z(t)$ may be the value of tire wear at the moment when a tire has run t km (in this case "time" is the tire run), the size of fatigue crack, the size of failure-causing conducting filament of chlorine-copper compound in a printed-circuit board, luminosity of light emitting diode at the moment t [Meeker and Escobar (1998), Bagdonavičius and Nikulin (2002), Zdorova-Cheminade (2003)], etc.

Denote by T^0 the moment of non-traumatic failure, i.e., the moment when the degradation attains some critical value z_0. The moment of the unit's failure is

$$\tau = T^0 \wedge T.$$

4.2 Modeling the Degradation-Failure Time Process

We suppose that the real degradation process is modeled by the general path model [Meeker and Escobar (1998)]

$$Z_r(t) = g(t, A); \tag{4.1}$$

here $A = (A_1, \ldots, A_r)$ is a random vector with positive components and the distribution function F_A, and g is a specified continuously differentiable increasing in t function. The typical form of the degradation curves is

$$g(t, a) = e^{a_1}(1 + t)^{a_2}. \tag{4.2}$$

In the particular case of linear degradation $a_2 = 0$.

Denote by h the function inverse to g with respect to the first argument. Evidently, it is continuously differentiable and increasing in t. Moreover,

$$T^0 = h(z_0, A). \tag{4.3}$$

The observed degradation process $Z = Z(t), t \geq 0$ may be slightly different from the real degradation process.

Suppose that the values of the real degradation process are measured at time moments $t_1, \ldots, t_m$.

Model 1 (Degradation with measurement errors). *The observed degradation values are*

$$Z(t_j) = Z_r(t_j)\, U(t_j), \tag{4.4}$$

where $e_j = \ln U(t_j)$ *are i.i.d. random variables,* $e_j \sim N(0, \sigma^2)$.

Model 2 (Degradation with noise). *The observed degradation process is*

$$Z(t) = Z_r(t)\, U(t), \tag{4.5}$$

where

$$V(t) = \ln U(t) = \sigma W(c(t)), \tag{4.6}$$

W is the standard Wiener process independent on A, and $c : [0, \infty \to [0, \infty)$, $c(0) = 0$, is a specified continuous and increasing time function, $c(0) = 0$.

For any $t > 0$ the median of the random variable $U(t)$is 1.

The cumulative distribution function (c.d.f.) of the real degradation at the moment t is

$$F_{r0}(z \mid t) = P(g(t, A) \leq z) = \int_{g(t,a) \leq z} dF_A(a), \tag{4.7}$$

the mean real degradation attained at the moment t is

$$m_{r0}(t) = Eg(t, A) = \int_{\mathcal{A}} g(t, a) dF_A(a), \tag{4.8}$$

where $\mathcal{A}$ is the set of possible values of the random vector A.

Note that the characteristics (4.7) and (4.8) have sense only in the ideal situation when the probability of failure until the moment t is zero. So we shall modify them defining the c.d.f. and other characteristics of the degradation at any moment t as conditional given survival up to this moment.

Let T be the time to traumatic failure.

Model 3. *The conditional survival of T given the real degradation process has the form:*

$$S_T(t \mid A) = P\{T > t \mid g(s, A), 0 \leq s \leq t\} = \exp\Big\{-\int_0^t \lambda_0(s, \alpha)\lambda(g(s, A))ds\Big\}; \tag{4.9}$$

here λ is the unknown intensity function, $\lambda_0(s, \alpha)$ being from a parametric family of hazard functions.

Note that the function λ is defined on the set of degradation values, not on the time scale.

The model states that the conditional hazard rate $\lambda_T(t \mid A)$ at the moment t given the degradation $g(s, A), 0 \leq s \leq t$, has the multiplicative form as in the famous Cox model:

$$\lambda_T(t \mid A) = \lambda_0(t, \alpha)\lambda(g(t, A)). \tag{4.10}$$

In this model the degradation is considered as a random covariable, and the second factor $\lambda(g(t, A))$ shows the influence of degradation on the hazard rate. The first factor $\lambda_0(t, \alpha)$ shows the influence of time on the hazard rate not explained by degradation. If, for example, $\lambda_0(t, \alpha) = (1 + t)^\alpha, e^{\alpha t}$, then $\alpha = 0$ corresponds to the case when the hazard rate at any moment t is a function of the degradation level at this moment.

Wulfsohn and Tsiatis (1997) considered a joint model for survival and longitudinal data measured with error, given by

$$\lambda_T(t \mid A) = \lambda_0(t)e^{\beta(A_1 + A_2 t)}$$

with bivariate normal distribution of (A_1, A_2). The difference: in our model the function λ, characterizing the influence of degradation on the hazard rate, is non-parametric, in the Wulfsohn–Tsiatis model this function is parametric. On the other hand, the baseline hazard rate λ_0 (it is proportional to the hazard rate which should be observed if the degradation would be absent) is parametric in our model and non-parametric in the Wulfsohn–Tsiatis model. We consider the case when the distribution of A is not specified.

Let us consider the model (4.9) and set

$$\Lambda(z) = \int_0^z \lambda(y)dy. \tag{4.11}$$

The conditional survival function (4.5) can be rewritten in the form

$$S_T(t \mid A) = \exp\Big\{-\int_{g(0,A)}^{g(t,A)} \lambda_0(h(z,A),\alpha)h'(z,A)\,d\Lambda(z)\Big\}, \tag{4.12}$$

where h is the function inverse to g with respect to the first argument.

4.3 Survival and Degradation Characteristics

The survival function and the mean of the time-to-failure τ are

$$S_\tau(t) = P(\tau > t) = \int_{g(t,a)\leq z_0} S_T(t \mid a)dF_A(a), \tag{4.13}$$

and

$$\begin{aligned} e_\tau &= E(\tau) = \int_A \{h(z_0,a)S_T(h(z_0,a) \mid a) \\ &\quad + \int_{g(0,a)\wedge z_0}^{z_0} \lambda_0(h(z,a))h(z,a)h'(z,a)S_T(h(z,a) \mid a)d\Lambda(z)\}dF_A(a), \end{aligned} \tag{4.14}$$

respectively.

Set

$$\delta = \begin{cases} 0, & \text{if} \quad \tau = T^{(0)}, \\ 1, & \text{if} \quad \tau = T. \end{cases} \tag{4.15}$$

Important reliability characteristics are the probability of traumatic failure in the interval $[0;t]$:

$$P^{(tr)}(t) = P\{\tau \leq t,\ \delta = 1\} = 1 - \int_A S_T(t \wedge h(z_0,a)) \mid a)dF_A(a) \tag{4.16}$$

and the probability of non-traumatic failure in the interval $[0;t]$:

$$P^{(0)}(t) = \{\tau \leq t,\ \delta = 0\} = \int_{g(t,a)\geq z_0} S_T(h(z_0,a)) \mid a)dF_A(a). \tag{4.17}$$

Now we are able to modify the characteristics (4.7) and (4.8) to the case of the real situation.

The c.d.f. of the real degradation at the moment t given the survival until this moment is:

$$F_\tau(z \mid t) = P(g(t, A) \leq z \mid \tau > t) = \frac{1}{S_\tau(t)} \int_{g(t,a) \leq z} S_T(t \mid a) dF_A(a), \qquad (4.18)$$

if $z \leq z_0$, and $F_\tau(z \mid t) = 1$, if $z > z_0$. The mean real degradation attained at the moment t given survival until this moment is

$$\begin{aligned} m_\tau(t) &= E(g(t, A) \mid \tau > t) \\ &= \frac{1}{S_\tau(t)} \int_A \{g(t, a) \wedge z_0 - z_*\} S_T(t \mid a) dF_A(a), \end{aligned} \qquad (4.19)$$

where z_* is the minimal possible value of degradation.

For any $B \subset \mathbf{R}, B_0 \subset \mathbf{R}$, $t_1 \leq s < t$ the probability that at the moment t the value of the real degradation will be in the set B, given that at the moment s it is in the set B_0 and given the survival until t, is

$$\begin{aligned} &P(g(t, A) \in B \mid g(s, A) \in B_0, T > t) \\ &= \int_{g(t,a) \in B, g(s,a) \in B_0} S_T(t \mid a) dF_A(a \mid t) / \int_{g(s,a) \in B_0} S_T(t \mid a) dF_A(a \mid t). \end{aligned} \qquad (4.20)$$

Denote by

$$T^0(z) = h(z, A)$$

the time needed to attain the level z ($z \leq z_0$) of the real degradation. The conditional survival function and the conditional mean of $T^0(z)$ given survival until the degradation attains the level z and given survival until the moment $T^0(z)$ are

$$P(T^0(z) > t \mid T > T^0(z)) = \frac{\int_{g(t,a) < z} S_T(h(z, a) \mid a) dF_A(a)}{\int_A S_T(h(z, a) \mid a) dF_A(a)}, \qquad (4.21)$$

and

$$E(T^0(z) \mid T > T^0(z)) = \frac{\int_A h(z_0, a) S_T(h(z, a) \mid a) dF_A(a)}{\int_A S_T(h(z, a) \mid a) dF_A(a)}, \qquad (4.22)$$

respectively.

4.4 Modeling Covariate Effects on Degradation and Survival

Suppose units are observed under the covariate $x = x(t) = (x_1(t), \ldots, x_m(t))$, $t \geq 0$, and the values of the real degradation process are measured at time moments $t_1, \ldots, t_m$.

In this case, following Bagdonavičius and Nikulin (2001), the real degradation process is modeled by

$$Z_r(t \mid x) = g(\varphi(t, \beta, x), A), \tag{4.23}$$

where

$$\varphi(t, \beta, x) = \int_0^t e^{\beta x^T(u)} du, \tag{4.24}$$

is the link function, $\beta = (\beta_1, \ldots, \beta_m)$ is the vector of the regression parameters.

If the covariate is constant over time, then

$$\varphi(t, \beta, x) = e^{\beta x^T} t. \tag{4.25}$$

Model 4 (Degradation with measurement errors). *The observed degradation values are*

$$Z(t_j \mid x) = Z_r(t_j \mid x)\, U(t_j), \tag{4.26}$$

where $e_j = \ln U(t_j)$ are i.i.d. random variables, $e_j \sim N(0, \sigma^2)$.

Model 5 (Degradation with noise). *The observed degradation process is*

$$Z(t \mid x) = Z_r(t \mid x)\, U(t),$$

where

$$V(t) = \ln U(t) = \sigma W(c(t)), \tag{4.27}$$

W is the standard Wiener process independent on A, and $c : [0, \infty \to [0, \infty)$, is a specified increasing and continuous time function, $c(0) = 0$.

The c.d.f. of the real degradation at the moment t given the covariate $x(s), 0 \leq s \leq t$, is

$$F_{r0}(z \mid t, x) = \int_{g(\varphi(t,\beta,x),a) \leq z} dF_A(a), \tag{4.28}$$

the mean real degradation attained at the moment t is

$$m_{r0}(t, x) = \int_A g(\varphi(t, \beta, x), a) dF_A(a). \tag{4.29}$$

As in the case without covariates these characteristics have sense only in the ideal situation when the probability of failure until the moment t is zero.

Similarly as in (4.9), we model the survival distribution:

Model 6 *The probability of survival until the moment t given the real degradation process $Z_r(s \mid x), 0 \leq s \leq t$ is*

$$\begin{aligned} S_T(t \mid a, x) &= P\{T > t \mid A = a, x(s), 0 \leq s \leq t\} \\ &= \exp\Big\{-\int_0^t \lambda_0(\varphi(s,\beta,x),\alpha)\lambda(g(\varphi(t,\beta,x),a))ds\Big\}, \quad (4.30) \end{aligned}$$

where λ is an unknown intensity function, $\lambda_0(s,\alpha)$ being from a parametric family of hazard functions.

The probability (4.30) can be rewritten in the form

$$S_T(t \mid a, x) = \exp\Big\{-\int_{g(\varphi(0,\beta,x),a)}^{g(\varphi(t,\beta,x),a)} \lambda_0(h(z,a),\alpha)\psi'(h(z,a),\beta,x)h'(z,a)\,d\Lambda(z)\Big\}, \quad (4.31)$$

where $\psi(s,\beta,x)$ is the inverse function of $\varphi(t,\beta,x)$ with respect to the first argument, ψ_1' is the derivative of ψ with respect to the first argument, and Λ is defined by the formula (4.11).

In the case of constant covariates,

$$\psi(t,\beta,x) = e^{-\beta x^T} t, \quad \psi'(t,\beta,x) = e^{-\beta x^T}.$$

The given model implies that the survival function and the mean of the random variable τ given the covariate x are

$$S_\tau(t \mid x) = P(\tau > t \mid x(s), 0 \leq s \leq t) = \int_{g(\varphi(t,\beta,x),a) < z_0} S_T(t \mid a, x) dF_A(a). \quad (4.32)$$

The probability of traumatic failure in the interval $[0;t]$under the covariate x is:

$$\begin{aligned} P^{(tr)}(t \mid x) &= P\{\tau \leq t, \delta = 1 \mid x(s), 0 \leq s \leq t\} \\ &= 1 - \int_A S_T(t \wedge \psi(h(z_0,a),\beta,x)) \mid a, x) dF_A(a) \quad (4.33) \end{aligned}$$

and the probability of non-traumatic failure in the interval $[0;t]$ under the covariate x is

$$\begin{aligned} P^{(0)}(t \mid x) &= \{\tau \leq t, \delta = 0 \mid x(s), 0 \leq s \leq t\} \\ &= \int_{g(\varphi(t,\beta,x)),a) \geq z_0} S_T(\psi(h(z_0,a),\beta,x)) \mid a, x) dF_A(a), \quad (4.34) \end{aligned}$$

where δ is given by (4.15).

The c.d.f. of the real degradation at the moment t given survival until this moment is

$$\begin{aligned} F_r(z \mid t, x) &= P(Z_r(t \mid x) \le z \mid x(s), 0 \le s \le t, \tau > t) \\ &= \frac{1}{S_\tau(t \mid x)} \int_{g(\varphi(t,\beta,x),a) \le z} S_T(t \mid a, x) dF_A(a), \end{aligned} \tag{4.35}$$

and the mean real degradation attained at the moment t given survival until this moment is

$$m_r(t,x) = EZ_r(t \mid x) = \frac{1}{S_T(t \mid x)} \int_A (g(\varphi(t,\beta,x),a) \wedge z_0 - z_*) S_T(t \mid a, x) dF_A(a). \tag{4.36}$$

For any $B \subset \mathbf{R}, B_0 \subset \mathbf{R}$, the probability that at the moment t the value of the real degradation will be in the set B, given that at the moment s it is in the set B_0 and given $x(u), 0 \le u \le t$, is

$$\begin{aligned} &P(Z_r(t \mid x) \in B \mid Z_r(s \mid x) \in B_0, \tau > t, x(u), 0 \le u \le t) \\ &\quad = \frac{\int_{g(\varphi(t,\beta,x),a) \in B, g(\varphi(s,\beta,x),a) \in B_0} S_T(t \mid a, x) dF_A(a)}{\int_{g(\varphi(s,\beta,x),a) \in B_0} S_T(t \mid a, x) dF_A(a)}. \end{aligned} \tag{4.37}$$

Denote by

$$T^0(z,x) = \psi(h(z,A), \beta, x) \tag{4.38}$$

the time needed to attain the level z of the real degradation under the covariate x. The conditional survival function and the conditional mean of $T^0(z,x)$ given survival until $T^0(z,x)$, and given $x(u), 0 \le u \le t$, are

$$\begin{aligned} &P(T^0(z,x) > t \mid \tau > T^0(z,x), x(u), 0 \le u \le t) \\ &\quad = \frac{\int_{g(\varphi(t,\beta,x),a) < z} S_T(\psi(h(z,a),\beta,x) \mid a, x) dF_A(a)}{\int_A S_T(\psi(h(z,a),\beta,x) \mid a, x) dF_A(a)}, \end{aligned} \tag{4.39}$$

and

$$\begin{aligned} &E(T^0(z) \mid \tau > T^0(z), x(u), 0 \le u \le t) \\ &\quad = \frac{\int_A \psi(h(z,a),\beta,x) S_T(\psi(h(z,a),\beta,x) \mid a, x) dF_A(a)}{\int_A S_T(\psi(h(z,a),\beta,x) \mid a, x) dF_A(a)}, \end{aligned} \tag{4.40}$$

respectively.

4.5 Semiparametric Estimation of Degradation and Survival Characteristics: Models Without Covariates

Assume that the c.d.f. F_A and the intensities λ are completely unknown.

Fix the degradation measurement moments $t_{i,1}, \ldots, t_{i,m_i}$ of the ith unit ($i = 1, \ldots, n$). If the failure time τ_i of this unit occurs in the interval $[t_{i,j_i}, t_{i,j_i+1})$ ($j_i = 1, \ldots, m_i; t_{m_i+1} = \infty$) then the values $Z_{i1}, \ldots, Z_{i,j_i}$ of the degradation process Z_i of the ith unit are observed at the time moments $t_{i1}, \ldots, t_{i,j_i}$. Set

$$Y_{ij} = \ln Z_{ij}, \quad Y_i = (Y_{i1}, \ldots, Y_{ij_i})^T.$$

Then given $A_i = a_i$, j_i

$$Y_i \sim N(\mu_i, \sigma^2 \Sigma_i),$$

where

$$\mu_i = (\mu_{i1}, \ldots, \mu_{i,j_i})^T, \quad \mu_{ij} = \mu_{ij}(a_i) = \ln g(t_{ij}, a_i), \quad \Sigma_i = \| s_{ikl} \|_{j_i \times j_i},$$

$$s_{ikl} = \mathbf{1}_{\{k=l\}} \quad \text{(first model)},$$

$$s_{ikl} = c_{ik} \wedge c_{il}, \quad \text{(second model)}, \quad c_{ij} = c(t_{ij}). \tag{4.41}$$

Denote by b_{ikl} the elements of the inverse matrix Σ_i^{-1}, and by N the number of units such that $j_i \geq r$ (usually $r = 2$).

The predictors $\hat{A}_i$ of the random vectors A_i are found by minimizing with respect to $a_1, \ldots, a_N$ the quadratic form

$$\sum_{i=1}^{N} (Y_i - \mu_i(a_i))^T \, \Sigma_i^{-1} \, (Y_i - \mu_i(a_i)). \tag{4.42}$$

Suppose that the function $g(t, a)$ is differentiable with respect to a and set $m = \sum_{i=1}^{N} j_i$. The estimator $\hat{\sigma}^2$ of the parameter σ^2 is found by maximizing with respect to σ^2 the conditional likelihood function

$$\begin{aligned} &L(\sigma^2 \mid \hat{A}_1, \ldots, \hat{A}_N) \\ &= \frac{1}{(2\pi)^{m/2} \sigma^m} \prod_{i=1}^{N} | \Sigma_i |^{-1/2} \exp\Big\{ -\frac{1}{2\sigma^2} \sum_{i=1}^{N} (Y_i - \mu_i(\hat{A}_i))^T \\ &\qquad \times \Sigma_i^{-1} \, (Y_i - \mu_i(\hat{A}_i)) \Big\}. \end{aligned} \tag{4.43}$$

The minimization of the quadratic form (4.42) and the maximization of the conditional likelihood function gives the following equations for computing $\hat{\sigma}^2$ and $\hat{A}_i$:

$$\sum_{k=1}^{j_i} \frac{\partial}{\partial \hat{A}_i} \ln g(t_{ik}, \hat{A}_i) \sum_{l=1}^{j_i} \{Y_{il} - \ln g(t_{il}, \hat{A}_i)\} b_{ikl} = 0; \tag{4.44}$$

$$\hat{\sigma}^2 = \frac{\hat{c}}{m} \tag{4.45}$$

where

$$\hat{c} = \sum_{i=1}^{N} (Y_i - \mu_i(\hat{A}_i))^T \, \Sigma_i^{-1} \, (Y_i - \mu_i(\hat{A}_i)). \tag{4.46}$$

Let us consider the following model for the paths $g(t, A_i)$:

$$g(t, A_i) = e^{A_{i1}} (1+t)^{A_{i2}}. \tag{4.47}$$

Set $\mathbf{1} = (1, \ldots, 1)^T_{j_i}$, $C_i = (c_{i1}, \ldots, c_{i,j_i})^T$. In such a case we have

$$\hat{A}_{i1} = \frac{c_i d_i - e_i f_i}{c_i^2 - b_i e_i}, \quad \hat{A}_{i2} = \frac{c_i f_i - b_i d_i}{c_i^2 - b_i e_i}, \tag{4.48}$$

and

$$\hat{\sigma}^2 = \frac{\hat{c}}{m}, \tag{4.49}$$

where

$$\hat{c} = \sum_{i=1}^{N} (g_i + b_i \hat{A}_{1i}^2 + e_i \hat{A}_{2i}^2 + 2 c_i \hat{A}_{1i} \hat{A}_{2i} - 2 f_i \hat{A}_{1i} - 2 d_i \hat{A}_{2i}),$$

$$b_i = \mathbf{1}^T \Sigma_i^{-1} \mathbf{1}, \quad c_i = C_i^T \Sigma_i^{-1} \mathbf{1}, \quad d_i = Y_i^T \Sigma_i^{-1} C_i,$$
$$e_i = C_i^T \Sigma_i^{-1} C_i, \quad f_i = Y_i^T \Sigma_i^{-1} \mathbf{1}, \quad g_i = Y_i^T \Sigma_i^{-1} Y_i. \tag{4.50}$$

The conditional mean of the estimator $\hat{\sigma}^2$ given N and j_i is:

$$E(\hat{\sigma}^2 \mid j_i, i = 1, \ldots, N) = \frac{m+2}{m} \sigma^2,$$

so $\hat{\sigma}^2$ is a consistent estimator of σ^2.

Set

$$V_{ij} = e_{ij} \quad \text{(first model)}, \quad V_{ij} = \sigma W_i(c_{ij}) \quad \text{(second model)},$$

$$V_i = (V_{i1}, \ldots, V_{i,j_i})^T.$$

Replacing Y_{ij} by their expressions $Y_{ij} = A_{i1} + A_{i2} c_{ij} + V_{ij}$ in the formulas (4.48), we obtain that

$$\hat{A}_i = A_i + \varepsilon_i,$$

where

$$\varepsilon_i = (\varepsilon_{i1}, \varepsilon_{i1}), \quad \varepsilon_{i1} = \frac{c_i v_i - e_i u_i}{c_i^2 - b_i e_i}, \quad \varepsilon_{i2} = \frac{c_i u_i - b_i v_i}{c_i^2 - b_i e_i},$$

$$u_i = V_i^T \Sigma_i^{-1} \mathbf{1}, \quad v_i = V_i^T \Sigma_i^{-1} C_i.$$

Given j_i the vector $V_i \sim N(0, \sigma^2 \Sigma_i)$, so the bi-dimensional vector ε_i is normally distributed with zero mean and the density

$$\varphi_i(x;\sigma^2) = \frac{1}{2\pi\sigma_{i1}\sigma_{i2}\sqrt{1-\rho_i^2}} \exp\left\{-\frac{1}{2(1-\rho_i^2)}\left[\frac{x_1^2}{\sigma_{i1}^2} - 2\rho_i\frac{x_1 x_2}{\sigma_{i1}\sigma_{i2}} + \frac{x_2^2}{\sigma_{i2}^2}\right]\right\},$$

which depends only on σ^2, because

$$\sigma_{i1}^2 = Var(\varepsilon_{i1}) = \frac{e_i}{b_i e_i - c_i^2}\sigma^2, \quad \sigma_{i2}^2 = Var(\varepsilon_{i2}) = \frac{b_i}{b_i e_i - c_i^2}\sigma^2,$$

$$cov(\varepsilon_{i1}, \varepsilon_{i2}) = -\frac{c_i}{b_i e_i - c_i^2}\sigma^2,$$

and the correlation coefficient

$$\rho_i = -\frac{c_i}{\sqrt{b_i e_i}}.$$

Note that the means of the random variables A_i and $\hat{A}_i$ are equal:

$$E(\hat{A}_i) = E(A_i),$$

and

$$Eg(t, \hat{A}_i) = m_{r0}(t)\exp\{\frac{1}{2}[\sigma_{i1}^2 + 2\sigma_{i1}\sigma_{i2}\rho_i \ln(1+t) + \sigma_{i2}^2 \ln^2(1+t)]\}.$$

So a consistent estimator of the mean real degradation $m_{r0}(t)$ is

$$\hat{m}_{r0}(t) = \frac{\sum_{i=1}^N g(t, \hat{A}_i)}{\sum_{i=1}^N \exp\{\frac{1}{2}[\hat{\sigma}_{i1}^2 + 2\hat{\sigma}_{i1}\hat{\sigma}_{i2}\hat{\rho}_i \ln(1+t) + \hat{\sigma}_{i2}^2 \ln^2(1+t)]\}},$$

and a consistent estimator of the mean observed degradation m_{o0} (for Model 2) is

$$\hat{m}_{o0}(t) = e^{\frac{1}{2}\hat{\sigma}^2 c^2(t)}\hat{m}_{r0}(t),$$

where

$$\hat{\sigma}_{i1}^2 = \frac{e_i}{b_i e_i - c_i^2}\hat{\sigma}^2, \quad \hat{\sigma}_{i2}^2 = \frac{b_i}{b_i e_i - c_i^2}\sigma^2, \quad \hat{\rho}_i = -\frac{c_i}{\sqrt{b_i e_i}}.$$

Lemma 4.5.1 *If $t_{ij} = q_i(j/m_i)$, $j = 1, \ldots, m_i$, where q_i is a continuous increasing on $[0,1]$ function, $q_i(0) = 0, q_i(1) \le c < \infty$. Under Model 1 almost surely*

$$\lim_{m_i \to \infty} \sigma_{i1}^2 = \lim_{m_i \to \infty} \sigma_{i2}^2 = 0.$$

PROOF. It is sufficient to prove that almost surely

$$b_i - \frac{c_i^2}{e_i} = \sum_{j=1}^{j_i} c_{ij}^2 - \frac{1}{n}(\sum_{j=1}^{j_i} c_{ij})^2 \overset{a.s.}{\to} \infty, \quad e_i - \frac{c_i^2}{b_i} = j_i - \frac{(\sum_{j=1}^{j_i} c_{ij})^2}{\sum_{j=1}^{j_i} c_{ij}^2} \overset{a.s.}{\to} \infty.$$

The function $r_i(x) = c(q_i(x))$ is continuous and increasing on $(0,1)$, $r_i(0) = 0$. We have

$$\frac{1}{j_i}(b_i - \frac{c_i^2}{e_i}) = \frac{1}{j_i}\sum_{j=1}^{j_i} r_i^2(j/j_i) - (\frac{1}{j_i^2}\sum_{j=1}^{j_i} r_i(j/j_i))^2 \overset{a.s.}{\to}$$

$$\int_0^{a_i} r_i^2(x)dx - (\int_0^{a_i} r_i(x)dx)^2 > 0,$$

where

$$a_i = q_i^{-1}(E(\tau_i \wedge q_i(1)).$$

So $b_i - c_i^2/e_i \overset{a.s.}{\to} \infty$.

Analogously

$$\frac{1}{j_i}(e_i - \frac{c_i^2}{b_i}) \overset{a.s.}{\to} 1 - \frac{(\int_0^{a_i} r_i(x)dx)^2}{\int_0^{a_i} r_i^2(x)dx} \in (0,1),$$

so $e_i - c_i^2/b_i \overset{a.s.}{\to} \infty$.

The proof is complete. ∎

Remark 4.5.1 The lemma is true in the particular case $q_i(x) = x$, i.e., when $t_{ij} = j/m_i$ are equidistant.

Set

$$\tilde{F}_A(a) = \frac{1}{n}\sum_{i=1}^{n} \mathbf{1}_{\{A_i \le a\}} = \frac{1}{n}\sum_{i=1}^{n} \mathbf{1}_{\{A_{i1} \le a_1, \ldots, A_{ir} \le a_r\}} \tag{4.51}$$

and let us consider the statistic

$$\hat{F}_A(a) = \frac{1}{n}\sum_{i=1}^{n} \mathbf{1}_{\{\hat{A}_i \le a\}} \tag{4.52}$$

as the estimator of F_A. If the conditions of Lemma 4.5.1 are verified, then under Model 1 the difference $\hat{F}_A - \tilde{F}_A$ converges in probability to zero uniformly on $\mathcal{A}$ as $\min m_i \to \infty$.

Simulation results show that if the noise is not very large (i.e., the variance σ^2 is not large) then the estimators $\hat{F}_A$ and $\tilde{F}_A$ are close for Model 2.

The estimator $\tilde{F}_A$ is the best estimator of the c.d.f. F_A. If m_i are not very small (Model 1) or σ^2 is not large (Models 1 and 2), then $\hat{F}_A$ is a good estimator of F_A and the c.d.f. of the real degradation at the moment t may be estimated by

$$\hat{F}_{r0}(z \mid t) = \frac{1}{n}\sum_{i=1}^{n} \mathbf{1}_{\{g(t,\hat{A}_i)\le z\}}.$$

Let us consider estimation of Λ. Suppose at first that $\sigma = \alpha = 0$.

Denote by $\tau_i = T_i \wedge T_i^0$ the failure time of the ith individual, T_i, T_i^0 being the traumatic and non-traumatic failure moments. Set

$$X_i = \min(\tau_i, t_{i,m_i}), \quad \delta_i = \begin{cases} 1, & \text{if } X_i = T_i, \\ 0, & \text{otherwise.} \end{cases}$$

If $\sigma = 0$ and $j_i \ge r$, then the values of A_i can be found and the degradation value $g(T_i, A_i)$ just prior to the failure also can be found. Set

$$Z_i = g(X_i, A_i),$$

and on $[z_*, z_0]$ define the following counting processes:

$$N_i(z) = \mathbf{1}_{\{Z_i \le z, \delta_i = 1\}}, \quad N(z) = \sum_{i=1}^{n} N_i(z).$$

$N(z)$ is the number of units having a failure before or at the moment when the degradation attains the level z.

Let $\mathcal{F}_z$ be the σ-algebra generated by the random variables

$$A_1, \ldots, A_n \quad \text{and} \quad N_1(y), \ldots, N_n(y) \quad \text{with} \quad y \le z.$$

Then [see Bagdonavičius et al. (2002, 2003)] the process $N_i(z)$ can be written as the sum

$$N_i(z) = \int_{z_*}^{z} \lambda(y) Y_i(y)dy + M_i(z), \tag{4.53}$$

where

$$Y_i(z) = \mathbf{1}_{\{Z_i \ge z\}} h'(z, A_i), \tag{4.54}$$

and M_i is a martingale with respect to the filtration $(\mathcal{F}_z \mid z \ge 0)$.

This implies that the optimal estimator of the cumulative intensity $\Lambda(z)$ is of Nelson–Aalen type [see Andersen et al. (1993, pp. 177–178)]:

$$\tilde{\Lambda}(z) = \int_{z_*}^{z} Y^{-1}(y)\, dN(y) = \sum_{Z_i \le z, \delta_i = 1} Y^{-1}(Z_i) = \sum_{Z_i \le z, \delta_i = 1} \left(\sum_{Z_j \ge Z_i} h'(z, A_i) \right)^{-1}, \tag{4.55}$$

where $Y(z) = \sum_{j=1}^{n} Y_j(z)$.

The estimator is correctly defined if $Z_i \leq z$ and $\delta_i = 1$ for some i. If such i do not exist, then the estimator is defined as 0.

To get asymptotic properties of $\hat{\Lambda}$, assumptions about the choice of t_{i,m_i} are needed. We assume that t_{i,m_i} are the realizations of (censoring) random variables C_i which are mutually independent, identically distributed and conditionally (given $A_i = a$) independent of (τ_i, δ_i). Let $S_c(t \mid a)$ denote the conditional survival function of C_i:

$$S_c(t \mid a) = P(C_i > t \mid A_i = a).$$

Two additional assumptions are needed. Set

$$c_0(a) = \sup_{z \leq z_0} h'(z, a), \quad c_1(a) = \sup_{z_1 < z_2 \leq z_0} \frac{\mid h'(z_2, a) - h'(z_1, a) \mid}{z_2 - z_1}$$

and

$$b(z) = E[h'(z, A_i)\mathbf{1}_{\{Z_i \geq z\}}] = \int h'(z, a) S_c(h(z, a) \mid a)\, S_T(h(z, a) \mid a)\, dF_A(a).$$

Assumption 1 $E[c_0(A) + c_1(A)] < \infty$.

Assumption 2 $\inf_{z \leq z_0} b(z) > 0$.

Similarly as in Bagdonavicius et al. (2002, 2003) we obtain:

Theorem 4.5.1 *Under Assumptions 1–2, the estimator $\hat{\Lambda}$ is uniformly consistent, i.e.,*

$$\sup_{z \leq z_0} \mid \hat{\Lambda}(z) - \Lambda(z) \mid \xrightarrow{P} 0,$$

as $n \to \infty$.

Theorem 4.5.2 *If Assumptions 1–2 hold, then the random vector function $\sqrt{n}(\hat{\Lambda} - \Lambda)$ tends in distribution in the space $D[z_*; z_0]$ to a zero mean Gaussian process W with the covariance function*

$$E[W(z)W(z')] = \sigma^2(z \wedge z'),$$

where

$$\sigma^2(z) = \int_{z_*}^{z} \frac{\lambda(y)}{b(y)} dy.$$

The c.d.f F_A is estimated by $\tilde{F}_A$ given by (4.51). It is well known that almost surely

$$\sup_a \mid \tilde{F}_A(a) - F_A(a) \mid \to 0,$$

as $n \to \infty$. Moreover, if the function F_A is absolutely continuous, then the random function $\sqrt{n}(\hat{F}_A - F_A)$ tends in distribution in the Skorokhod space $D^r[0;\infty]$ to a zero mean Gaussian field W_0 with the covariance function

$$E[W_0(a)W_0(a')] = F_A(a \wedge a') - F_A(a)\,F_A(a'),$$

where $a \wedge a' = (a_1 \wedge a_1', \ldots, a_r \wedge a_r')$.

Theorem 4.5.3 *If Assumptions 1–2 hold and the function F_A is absolutely continuous, then the random vector function $\sqrt{n}(\hat{F}_A - F_A,\, \hat{\Lambda} - \Lambda)$ tends in distribution in the space $D^r[0;\infty] \times D[z_*; z_0]$ to a random vector function with independent components.*

Estimating the survival and degradation characteristics we replace the functions F_A and Λ in the expressions of these characteristics by $\tilde{F}_A$ and $\hat{\Lambda}$. Asymptotic properties of such estimators can be easily found because they are smooth functionals of the pair $(\tilde{F}_A, \hat{\Lambda})$.

If $\alpha \neq 0$, then two-step estimation of Λ is possible. If α would be known then similarly as in the case $\alpha \neq 0$ the following estimator of Λ could be used:

$$\Lambda^*(z,\alpha) = \int_{z_*}^{z} Y^{-1}(y,\alpha)\,dN(y) = \sum_{Z_i \le z, \delta_i = 1} Y^{-1}(Z_i,\alpha),$$

where $Y(z) = \sum_{j=1}^{n} Y_j(z)$, and

$$Y_j(z) = \mathbf{1}_{\{Z_j \ge z\}} \lambda_0(h(z,A_j),\alpha) h'(z,A_j).$$

If α is unknown, the estimator $\hat{\alpha}$ is obtained by maximizing the loglikelihood function

$$\begin{aligned}\ln L(\alpha) \;\asymp\; & \sum_{i=1}^{n} \delta_i \ln \lambda_0(X_i,\alpha) + \sum_{i=1}^{n} \delta_i \ln \lambda^*(Z_i,\alpha) \\ & - \sum_{i=1}^{n} \int_{g(0,A_i)}^{g(X_i,A_i)} \lambda_0(h(z,A_i))h'(z,A_i) \frac{dN(z)}{Y(z,\alpha)},\end{aligned}$$

where

$$\lambda^*(z,\alpha) = \sum_{j=1}^{n} \frac{\delta_j \mathbf{1}_{\{z = Z_j\}}}{Y(z,\alpha)}.$$

The final estimator of Λ is

$$\tilde{\Lambda}(z) = \Lambda^*(z,\hat{\alpha}).$$

If $\sigma \neq 0$ then the values of stochastic processes $Z_i(t)$ instead of $g(t,A_i)$ are observed.

Using the values of $Z_i(t)$ the predictors $\hat{A}_i$ of the random variables A_i are found using the equations (48). Replacing A_i by their predictors $\hat{A}_i$ we obtain the following estimator of the cumulative intensity Λ:

$$\hat{\Lambda}(z) = \int_{z_*}^{z} \frac{d\hat{N}(y)}{\hat{Y}(y)} = \sum_{\hat{Z}_i \le z, \delta_i = 1} \frac{1}{\hat{Y}(\hat{Z}_i)}, \tag{4.56}$$

where

$$\hat{Z}_i = g(X_i, \hat{A}_i), \quad \hat{N}(z) = \sum_{i=1}^{n} \hat{N}_i(z), \quad \hat{N}_i(z) = \mathbf{1}_{\{\hat{Z}_i \le z, \delta_i = 1\}},$$

$$\hat{Y}(z) = \tilde{Y}(z, \hat{\alpha}), \quad \tilde{Y}(z, \alpha) = \sum_{i=1}^{n} \tilde{Y}_i(z, \alpha),$$

$$\tilde{Y}_i(z, \alpha) = \mathbf{1}_{\{\hat{Z}_i \ge z\}} \lambda_0(h(z, \hat{A}_i), \alpha) h'(z, \hat{A}_i),$$

and $\hat{\alpha}$ is obtained by maximizing the loglikelihood function

$$\begin{aligned}
\ln L(\alpha) \asymp\ & \sum_{i=1}^{n} \delta_i \ln \lambda_0(X_i, \alpha) + \sum_{i=1}^{n} \delta_i \ln \tilde{\lambda}(\hat{Z}_i, \alpha) \\
& - \sum_{i=1}^{n} \int_{g(0,\hat{A}_i)}^{g(X_i,\hat{A}_i)} \lambda_0(h(z, \hat{A}_i), \alpha) h'(z, \hat{A}_i) \frac{d\hat{N}(z)}{\tilde{Y}(z, \alpha)} \\
=\ & \sum_{i=1}^{n} \delta_i \ln \lambda_0(X_i, \alpha) + \sum_{i=1}^{n} \delta_i \ln \tilde{\lambda}(\hat{Z}_i, \alpha) \\
& - \sum_{i=1}^{n} \sum_{\hat{Z}_j \le g(X_i, \hat{A}_i), \delta_j = 1} \frac{\lambda_0(h(\hat{Z}_j, \hat{A}_i), \alpha) h'(\hat{Z}_j, \hat{A}_i)}{\tilde{Y}(\hat{Z}_j, \alpha)},
\end{aligned}$$

where

$$\tilde{\lambda}(z, \alpha) = \sum_{j=1}^{n} \frac{\delta_j \mathbf{1}_{\{z = \hat{Z}_j\}}}{\tilde{Y}(z, \alpha)}.$$

Note that

$$h(z, A_j) = e^{-A_{1j}/A_{2j}} z^{1/A_{2j}} - 1, \quad h'(z, A_j) = \frac{1}{A_{2j}} e^{-A_{1j}/A_{2j}} z^{1/A_{2j} - 1}. \tag{4.57}$$

The formula (4.13) implies the following estimator of the probability $S_\tau(t)$ to survival time t:

$$\hat{S}_\tau(t) = \int_{g(t,a) \le z_0} \hat{S}_T(t \mid a) d\hat{F}_A(a) = \frac{1}{n} \sum_{i: g(t, \hat{A}_i) \le z_0} \hat{S}_T(t \mid \hat{A}_i),$$

where

$$\hat{S}_T(t \mid \hat{A}_i) = \exp\Big\{-\int_{g(0,\hat{A}_i)}^{g(t,\hat{A}_i)} h'(z,\hat{A}_i)\,d\hat{\Lambda}(z)\Big\}$$
$$= \exp\{-\sum_{j=1}^{n} \mathbf{1}_{\{\delta_j=1, g(0,\hat{A}_i)\le \hat{Z}_j < g(t,\hat{A}_i)\}} \frac{h'(\hat{Z}_j,\hat{A}_i)}{\hat{Y}(\hat{Z}_j)}\}.$$

The probabilities $P^{(tr)}(t)$ and $P^{(0)}(t)$ are estimated by

$$\hat{P}^{(tr)}(t) = 1 - \frac{1}{n}\sum_{i=1}^{n} \hat{S}_T(t \wedge h(z_0,\hat{A}_i) \mid \hat{A}_i)) \tag{4.58}$$

and

$$\hat{P}^{(0)}(t) = \frac{1}{n} \sum_{g(t,\hat{A}_i)\ge z_0} \hat{S}_T(h(z_0,\hat{A}_i) \mid \hat{A}_i). \tag{4.59}$$

The c.d.f. of the real degradation at the moment t given survival until this moment is estimated by

$$\hat{F}_\tau(z \mid t) = \frac{1}{\hat{S}_\tau(t)} \int_{g(t,a)\le z} \hat{S}_T(t \mid a)d\hat{F}_A(a) = \frac{1}{n} \sum_{i:g(t,\hat{A}_i)\le z} \hat{S}_T(t \mid \hat{A}_i)/\hat{S}_\tau(t),$$

and the mean real degradation attained at the moment t given survival until this moment is estimated by

$$\hat{m}_\tau(t) = \frac{1}{n}\sum_{i=1}^{n}(g(t,\hat{A}_i) \wedge z_0 - z_*)\hat{S}_T(t \mid \hat{A}_i)/\hat{S}_T(t).$$

The probability that at the moment t the value of the real degradation will be in the set B, given that at the moment s it is in the set B_0, is estimated by

$$\hat{P}(g(t,A) \in B \mid g(s,A) \in B_0, \tau > t)$$
$$= \int_{g(t,a)\in B, g(s,a)\in B_0} \hat{S}_T(t \mid a)d\hat{F}_A(a) \Big/ \int_{g(s,a)\in B_0} \hat{S}_T(t \mid a)d\hat{F}_A(a)$$
$$= \frac{\sum_{i:g(t,\hat{A}_i)\in B, g(s,\hat{A}_i)\in B_0} \hat{S}_T(t \mid \hat{A}_i)}{\sum_{i:g(s,\hat{A}_i)\in B_0} \hat{S}_T(t \mid \hat{A}_i)}.$$

The conditional survival function of $T(z_0)$ and the conditional mean given survival until the degradation attains the level z_0 are estimated by

$$\hat{P}(T^0(z) > t \mid \tau > T^0(z)) = \frac{\sum_{i:g(t,\hat{A}_i)<z} \hat{S}_T(h(z,\hat{A}_i) \mid \hat{A}_i)}{\sum_{i=1}^{n} \hat{S}_T(h(z,\hat{A}_i) \mid \hat{A}_i)},$$

and

$$\hat{E}(T^0(z) \mid \tau > T^0(z)) = \frac{\sum_{i=1}^n h(z, \hat{A}_i)\hat{S}_T(h(z, \hat{A}_i) \mid \hat{A}_i)}{\sum_{i=1}^n \hat{S}_T(h(z, \hat{A}_i) \mid \hat{A}_i)},$$

respectively.

Zdorova–Cheminade (2003) studied by simulation the considered model as a statistical degradation model of disablement in the elderly to verify that a hierarchical relationship exists between the concepts of Activities Daily Living, Instrumental Activities of Daily Living and mobility and to use this model to study the evolution of disability. The cumulative disability scale was used to describe the degradation process in time. In longitudinal analysis an additional level was considered to the disability index to take death into account. It is evident that this approach can be used in many other medical studies where a degradation is observed, especially in oncology.

4.6 Semiparametric Estimation of Degradation and Survival Characteristics: Models with Covariates

Assume that the c.d.f. F_A and the intensities λ are completely unknown.

Suppose that n units are tested , and the ith of them is observed under the covariate $x^{(i)}$. Fix the moments of degradation measurements $t_{i1}, \ldots, t_{i,m_i}$ of this unit. If the failure time τ_i of the ith unit occurs in the interval $[t_{i,j_i}, t_{i,j_i+1})$ $(j_i = 1, \ldots, m_i; t_{m_i+1} = \infty)$, then the values $Z_{i1}, \ldots, Z_{i,j_i}$ of the degradation process Z_i of the ith unit are observed at the time moments $t_{i1}, \ldots, t_{i,j_i}$.

Set

$$Y_{ij} = \ln Z_{ij}, \quad Y_i = (Y_{i1}, \ldots, Y_{i,j_i})^T.$$

Then given $A_i = a_i$,

$$Y_i \sim N(\mu_i, \sigma^2\Sigma_i),$$

where

$$\mu_i = (\mu_{i1}, \ldots, \mu_{i,j_i})^T, \quad \mu_{ij} = \mu_{ij}(a_i, \beta) = \ln g(\varphi(t_{ij}, \beta, x^{(i)}), a_i),$$

Σ_i is given in (4.41). Denote by N the number of units such that $j_i \geq r$.

The estimators $\hat{\sigma}^2$, $\hat{\beta}$ of the parameters σ^2, β and the predictors $\hat{A}_i$ of the random vectors A_i are found by maximizing with respect to $\sigma^2, \beta, a_1, \ldots, a_n$ the conditional likelihood function

$$L(\sigma^2, \beta, a_1, \dots, a_n)$$
$$= \frac{1}{(2\pi)^{m/2}\sigma^m} \prod_{i=1}^{N} |\Sigma_i|^{-1/2} \exp\Big\{ -\frac{1}{2\sigma^2} \sum_{i=1}^{N} (Y_i - \mu_i(a_i, \beta))^T$$
$$\times \Sigma_i^{-1} (Y_i - \mu_i(a_i, \beta)) \Big\}. \quad (4.60)$$

Suppose that the function $g(\varphi(t, \beta, x), a)$ is differentiable with respect to β and a. The maximization of the likelihood function gives the following equations for computing $\hat{\sigma}^2$, $\hat{\beta}$, and $\hat{A}_i$:

$$\sum_{k=1}^{j_i} \frac{\partial}{\partial A_i} \ln g(\varphi(t_{ik}, \hat{\beta}, x^{(i)}), \hat{A}_i) \sum_{l=1}^{j_i} \{Y_{il} - \ln g(\varphi(t_{ij}, \hat{\beta}, x^{(i)}), \hat{A}_i)\} b_{ikl} = 0,$$

$$\sum_{i=1}^{N} \sum_{k=1}^{j_i} \frac{\partial}{\partial \beta} \ln g(\varphi(t_{ik}, \hat{\beta}, x^{(i)}), \hat{A}_i) \sum_{l=1}^{j_i} \{Y_{il} - \ln g(\varphi(t_{ij}, \hat{\beta}, x^{(i)}), \hat{A}_i)\} b_{ikl},$$

$$\hat{\sigma}^2 = \frac{c(\hat{\beta})}{m} \quad (4.61)$$

where

$$c(\hat{\beta}) = \sum_{i=1}^{N} (Y_i - \mu_i(\hat{A}_i, \hat{\beta}))\ \Sigma_i^{-1}\ (Y_i - \mu_i(\hat{A}_i, \hat{\beta}))^T.$$

Set

$$c_{ij}(\beta) = c(\varphi(t_{ij}, \beta, x^{(i)})).$$

For the model (4.47) we have

$$\hat{A}_{i1} = \frac{c_i(\hat{\beta}) d_i(\hat{\beta}) - e_i(\hat{\beta}) f_i}{c_i^2(\hat{\beta}) - b_i e_i(\hat{\beta})}, \quad \hat{A}_{i2} = \frac{c_i(\hat{\beta}) f_i - b_i d_i(\hat{\beta})}{c_i^2(\hat{\beta}) - b_i e_i(\hat{\beta})}, \quad (4.62)$$

$$\hat{\sigma}^2 = \frac{c(\hat{\beta})}{m}, \quad (4.63)$$

and $\hat{\beta}$ verifies the system of equations

$$\sum_{i=1}^{N} \hat{A}_{2i} \sum_{k=1}^{j_i} \psi_{iks}(\hat{\beta}) \sum_{l=1}^{j_i} \{Y_{il} - \hat{A}_{1i} - \hat{A}_{2i} c_{il}(\hat{\beta})\} b_{ikl} = 0,$$

where

$$c(\hat{\beta}) = \sum_{i=1}^{n} (g_i + b_i \hat{A}_{1i}^2 + e_i(\hat{\beta}) \hat{A}_{2i}^2 + 2c_i(\hat{\beta}) \hat{A}_{1i} \hat{A}_{2i} - 2 f_i \hat{A}_{1i} - 2 d_i(\hat{\beta}) \hat{A}_{2i}),$$

$$c_i(\hat\beta) = C_i^T(\hat\beta)\Sigma_i^{-1}\mathbf{1}, \quad d_i(\hat\beta) = Y_i^T\Sigma_i^{-1}C_i(\hat\beta),$$

$$e_i(\hat\beta) = C_i^T(\hat\beta)\Sigma_i^{-1}C_i(\hat\beta), \quad \psi_{iks}(\hat\beta) = \frac{\int_0^{t_{ij}} x_s^{(i)}(u)e^{x^{(i)}(u)\hat\beta^T}du}{1+\int_0^{t_{ij}} e^{x^{(i)}(u)\hat\beta^T}du};$$

$$\mathbf{1} = (1,\dots,1)_{j_i}^T, \quad C_i(\hat\beta) = (c_{i1}(\hat\beta),\dots,c_{i,j_i}(\hat\beta))^T, \tag{4.64}$$

and b_i, f_i, g_i are given by (4.50).

If the vector of covariates $x^{(i)}$ is a step-function , and has the form

$$x^{(i)}(t) = x^{(ik)}, \quad \text{if} \quad t\in[t_{i,k-1}^*, t_{ik}^*) \quad (k=1,\dots l_i,\ t_{i0}^*=0),$$

then

$$\psi_{iks}(\hat\beta) = \frac{\sum_{l=1}^{\nu_{ik}-1} x_s^{(il)}e^{x^{(ik)}\hat\beta^T}(t_{il}^*-t_{i,l-1}^*) + x_s^{(i,\nu_{ik})}e^{x^{(i\nu_{ik})}\hat\beta^T}(t_{i,\nu_{ik}}^*-t_{i,\nu_{ik}-1}^*)}{1+\sum_{k=1}^{\nu_{ij}-1} e^{x^{(ik)}\hat\beta^T}(t_{ik}^*-t_{i,k-1}^*) + e^{x^{(i\nu_{ik})}\hat\beta^T}(t_{i,\nu_{ik}}^*-t_{i,\nu_{ik}-1}^*)},$$

where $\nu_{ik} = \{l : t_{ik}\in[t_{i,l-1}^*, t_{il}^*)\}$.

In particular, if $x^{(i)}$ has only one jump ($l_i = 2$), then

$$\psi_{iks}(\hat\beta) = \frac{x_s^{(i1)}e^{x^{(i1)}\beta^T}\min(\tau_i, t_{ik}^*) + x_s^{(i2)}e^{x^{(i2)}\beta^T}\max(t_{ik}^*-\tau_i, 0)}{1+e^{x^{(i1)}\beta^T}\min(\tau_i, t_{ik}^*) + e^{x^{(i2)}\beta^T}\max(t_{ik}^*-\tau_i, 0)}.$$

The estimator of the c.d.f. $F_A(a)$ has the form (4.52).

The estimators of the c.d.f. and the mean of the real degradation at the moment t given the covariate $x = x(s), 0\le s\le t$ is

$$\hat F_{r0}(z \mid t, x) = \frac{1}{n}\sum_{i=1}^n \mathbf{1}_{\{g(\varphi(t,\hat\beta,x),\hat A_i)\le z\}};$$

the mean real degradation attained at the moment t is estimated by

$$\hat m_{r0}(t,x) = \frac{1}{n}\sum_{i=1}^n g(\varphi(t,\hat\beta,x),\hat A_i).$$

Similarly as in the case without covariates we define the estimators

$$\hat\Lambda(z) = \tilde\Lambda(z,\hat\alpha),$$

where

$$\begin{aligned}
\tilde\Lambda(z,\alpha) &= \int_0^z \hat Y^{-1}(y,\alpha)\,d\hat N(y) = \sum_{\hat Z_i\le z,\delta_i=1} \tilde Y^{-1}(\hat Z_i,\alpha),\\
\hat Z_i &= g(\varphi(X_i,\hat\beta,x^{(i)}),\hat A_i),\\
\hat N(z) &= \sum_{i=1}^N \hat N_i(z), \quad \hat N_i(z) = \mathbf{1}_{\{\hat Z_i\le z,\delta_i=1\}},\\
\tilde Y(z,\alpha) &= \sum_{i=1}^N \tilde Y_i(z,\alpha),\\
\tilde Y_i(z,\alpha) &= \mathbf{1}_{\{\hat Z_i\ge z\}}\lambda_0(h(z,\hat A_i),\alpha)\psi'(h(z,\hat A_i),\hat\beta,x^{(i)})h'(z,\hat A_i),
\end{aligned}$$

and $\hat{\alpha}$ is obtained by maximizing the loglikelihood function

$$\begin{aligned}
&\ln L(\alpha)\\
&\asymp \sum_{i=1}^{N} \delta_i \ln \lambda_0(\varphi(X_i, \hat{\beta}, x^{(i)}), \alpha) + \sum_{i=1}^{n} \delta_i \ln \tilde{\lambda}(\hat{Z}_i, \alpha)\\
&\quad - \sum_{i=1}^{N} \int_{g(\varphi(0,\hat{\beta},x^{(i)}),\hat{A}_i)}^{g(\varphi(X_i,\hat{\beta},x^{(i)}),\hat{A}_i)} \lambda_0(h(z, \hat{A}_i), \alpha)\psi'(h(z, \hat{A}_i), \hat{\beta}, x^{(i)})h'(z, \hat{A}_i)\frac{d\hat{N}(z)}{\tilde{Y}(z, \alpha)}\\
&= \sum_{i=1}^{N} \delta_i \ln \lambda_0(\varphi(X_i, \hat{\beta}, x^{(i)}), \alpha) + \sum_{i=1}^{n} \delta_i \ln \tilde{\lambda}(\hat{Z}_i, \alpha)\\
&\quad - \sum_{i=1}^{N} \sum_{g(\varphi(0,\hat{\beta},x^{(i)}),\hat{A}_i) \le \hat{Z}_j \le g(\varphi(X_i,\hat{\beta},x^{(i)}),\hat{A}_i), \delta_j=1}\\
&\quad\quad \times \frac{\lambda_0(h(\hat{Z}_j, \hat{A}_i), \alpha)\psi'(h(\hat{Z}_j, \hat{A}_i), \hat{\beta}, x^{(i)})h'(\hat{Z}_j, \hat{A}_i)}{\tilde{Y}(\hat{Z}_j, \alpha)},
\end{aligned} \tag{4.65}$$

where

$$\tilde{\lambda}(z, \alpha) = \sum_{j=1}^{N} \frac{\delta_j \mathbf{1}_{\{z=\hat{Z}_j\}}}{\tilde{Y}(z, \alpha)}.$$

The estimators of the survival function

$$\hat{S}_r(t \mid x) = \frac{1}{n} \sum_{ig(\varphi(t,\hat{\beta},x)\hat{A}_i) < z_0} \hat{S}_T(t \mid \hat{A}_i, x), \tag{4.66}$$

where

$$\begin{aligned}
&\hat{S}_T(t \mid \hat{A}_i, x)\\
&= \exp\Bigg\{- \sum_{\delta_j=1, g(\varphi(0,\hat{\beta},x),\hat{A}_i) \le \hat{Z}_j < g(\varphi(t,\hat{\beta},x),\hat{A}_i)} \lambda_0(h(\hat{Z}_j, \hat{A}_i), \hat{\alpha})\, \psi'(h(\hat{Z}_j, \hat{A}_i), \hat{\beta}, x)\\
&\qquad \times \frac{h'(\hat{Z}_j, \hat{A}_i)}{\hat{Y}(\hat{Z}_j)}\Bigg\}.
\end{aligned}$$

The c.d.f. and the mean of the real degradation at the moment t given survival until this moment are estimated by

$$\hat{F}_r(z \mid t, x) = \frac{1}{N} \sum_{i: g(\varphi(t,\hat{\beta},x),\hat{A}_i) \le z} \hat{S}_T(t \mid \hat{A}_i, x)/\hat{S}_T(t \mid x), \tag{4.67}$$

$$\hat{m}_r(t, x) = \frac{1}{N} \sum_{i=1}^{n} g(\varphi(t, \hat{\beta}, x), \hat{A}_i)\hat{S}_T(t \mid \hat{A}_i, x)/\hat{S}_T(t \mid x). \tag{4.68}$$

The probabilities $P^{(tr)}(t \mid x)$ and $P^{(0)}(t \mid x)$ are estimated by

$$\hat{P}^{(tr)}(t \mid x) = 1 - \frac{1}{N}\sum_{i=1}^{N} \hat{S}_T(t \wedge \psi(h(z_0, \hat{A}_i), \hat{\beta}, x)) \mid \hat{A}_i, x), \tag{4.69}$$

$$\hat{P}^{(0)}(t \mid x) = \frac{1}{N} \sum_{g(\varphi(t,\hat{\beta},x)),\hat{A}_i) \geq z_0} S_T(\psi(h(z_0, \hat{A}_i), \hat{\beta}, x)) \mid \hat{A}_i, x). \tag{4.70}$$

The probability that at the moment t the value of the real degradation will be in the set B, given that at the moment s it is in the set B_0 and given $x(u), 0 \leq u \leq t$, is estimated by

$$\hat{P}(g(\varphi(t, \beta, x), a) \in B \mid g(\varphi(s, \beta, x), a) \in B_0, \tau > t, x(u), 0 \leq u \leq t)$$
$$= \frac{\sum_{i:g(\varphi(t,\hat{\beta},x),\hat{A}_i)\in B, g(\varphi(s,\hat{\beta},x),\hat{A}_i)\in B_0} \hat{S}_T(t \mid \hat{A}_i, x)}{\sum_{i:g(\varphi(s,\hat{\beta},x),\hat{A}_i)\in B_0} \hat{S}_T(t \mid \hat{A}_i, x)}. \tag{4.71}$$

The conditional survival function of $T^0(z, x)$ given survival until the degradation attains the level z and survival until the moment t_1, and given $x(u), 0 \leq u \leq t$, is estimated by

$$\hat{P}(T^0(z) > t \mid \tau > T^0(z), x(u), 0 \leq u \leq t)$$
$$= \frac{\sum_{i:g(\varphi(t,\hat{\beta},x),\hat{A}_i)<z} \hat{S}_T(\psi(h(z, \hat{A}_i), \hat{\beta}, x) \mid \hat{A}_i, x)}{\sum_{i=1}^{n} \hat{S}_T(\psi(h(z, \hat{A}_i), \hat{\beta}, x) \mid \hat{A}_i, x)},$$

and

$$\hat{E}(T^0(z) \mid \tau > T(z_0), x(u), 0 \leq u \leq t)$$
$$= \frac{\sum_{i=1}^{n} \psi(h(z, \hat{A}_i), \hat{\beta}, x)\hat{S}_T(\psi(h(z, \hat{A}_i), \hat{\beta}, x) \mid \hat{A}_i, x)}{\sum_{i=1}^{n} \hat{S}_T(\psi(h(z, \hat{A}_i), \hat{\beta}, x) \mid \hat{A}_i, x)},$$

respectively.

References

1. Andersen, P. K., Borgan, O., Gill, R. and Keiding, N. (1993). *Statistical Models Based on the Counting Processes*, Springer-Verlag, New York.

2. Bagdonavičius, V. and Nikulin, M. (2001). Estimation in degradation models with explanatory variables, *Lifetime Data Analysis*, **7**, 85-103.

3. Bagdonavičius, V. and Nikulin, M. (2002). *Accelerated Life Models: Modeling and Statistical Analysis*, Chapman & Hall/CRC, Boca Raton, Florida.

4. Bagdonavičius, V., Bikelis, A., Kazakevičius, A. and Nikulin, M. (2002). Non-parametric estimation from simultaneous degradation and failure data, *Comptes Rendus, Academie des Sciences de Paris*, **335**, 183–188.

5. Bagdonavičius, V., Bikelis, A., Kazakevičius, A. and Nikulin, M. (2003). Estimation from simultaneous degradation and failure time data, In *Mathematical and Statistical Methods in Reliability* (Eds., B. Lindqvist and K. A. Doksum), pp. 301–318, World Scientific Publishing, Singapore.

6. Zdorova-Cheminade, O. (2003). *Modélisation du processus d'évolution de l'incapacité chez les personnes agées*, Mémoire de DEA "Epidémiologie et Intervention en Santé Publique", Université Bordeaux 2, Juin 2003, 77 pages.

7. Meeker, W. Q. and Escobar, L. (1998). *Statistical Analysis for Reliability Data*, John Wiley & Sons, New York.

8. Wulfsohn, M. and Tsiatis, A. (1997). A joint model for survival and longitudinal data measured with error, *Biometrics*, **53**, 330–339.

5

On a Degradation-Failure Model for Repairable Items

Axel Lehmann

Otto-von-Guericke-University, Magdeburg, Germany

Abstract: We present a marked point process reliability model that is based on failure and degradation data of repairable items operating in a dynamic environment. An item is regarded as failed when degradation reaches a critical threshold level or when a censoring traumatic event occurs. Degradation is modelled by a process with stationary independent increments and is related to external covariates through a random time scale. Traumatic events occur due to a point process with an intensity that may depend on degradation and external covariates. Parametric and semiparametric estimation is considered.

Keywords and phrases: Covariate, degradation process, doubly stochastic Poisson process, DTS-model, marked point process, maximum likelihood estimation, semiparametric estimation, time scale, traumatic event

5.1 Introduction

In most practical applications items or systems operate in heterogeneous environments in which loads, environmental stresses, and other factors may vary over time. So, their reliability depends on dynamically changing environmental factors, which induce different internal stresses in an item that may influence its failure rate. Referring to these explanatory factors as covariates, Kalbfleisch and Prentice (1980) distinguish between external and internal covariates, which may be time-varying. External covariates characterize the operating environment of an item in a way that they may influence, but are not influenced by, the failure experience of a trial. Examples are, for instance, experimental factors like temperature, pressure or voltage, different loads and stress levels, age, and usage measures. Internal covariates characterize the deterioration of an item. They are time measurements taken on the item and can be observed only as long as the item survives and is uncensored. Such covariates like wear, fatigue-crack-growth, corrosion, material fatigue, general performance measures, or, in medical settings, biomarker processes represent the level of deterioration of an item and provide additional information to failure time observations.

In this paper, we present a reliability model that is based on failure and degradation data. An item is regarded as failed when degradation reaches a critical threshold level or when a censoring traumatic event occurs. To model the influence on failure of an item's dynamic operating environment, the degradation process and the intensity of traumatic events may depend on possibly time-varying covariates, for instance on different stress levels. The degradation process is modeled by a univariate process with stationary independent increments with a time scale transformation to model the influence on failure of a dynamic operating environment of the item and to cover non-linear degradation behavior. Degradation is related to external covariates through a random time scale describing slowing or accelerating degradation in real time.

To extend the threshold model to repairable items we use a marked point process approach. An item will be repaired in negligible time upon each failure and possibly preventively at regular inspection times depending on the degradation level. Each repair action sets back the degradation level to a value between its previous level and the level of a new item. The failure-repair process is described by a marked point process with failure and inspection times as events and degradation level, repair level and censoring information as marks. Statistical inference for the model is based on event times and marker measurements. The covariate process is assumed to be continuously observable whereas degradation can be observed only at event times.

Several processes with stationary independent increments have been used in degradation models. Wiener diffusion processes with a nonrandom time scale transformation have found application in Whitmore and Schenkelberg (1997) but not related to a threshold level and additional failure time data. Bagdonavičius and Nikulin (2000) use the gamma-process with a random time scale depending on covariates to model degradation. An excellent overview on probabilistic modelling was given by Singpurwalla (1995).

5.2 Types of Covariates

The characterization of an external covariate Z as influencing, but not being influenced by, the failure experience, means that its history is determined independently of the failure time. To be more precise, let $Z = \{Z(t) : t \in \mathbb{R}_+\}$ be a stochastic process on a fixed probability space $(\Omega, \mathcal{F}, \mathrm{P})$ and set $Z_t = \{Z(s) : 0 \leq s \leq t\}$ and $Z_{(t,\infty)} = \{Z(s) : t < s < \infty\}$. We define the histories

$\mathcal{F}_t^Z = \sigma(Z_t)$ and $\mathcal{F}_{(t,\infty)}^Z = \sigma(Z_{(t,\infty)})$ which refer to the past and the future of Z at time t. Let $\Phi = (T_n)_{n\geq 1}$ denote a point process of successive failure times, defined on $(\Omega, \mathcal{F}, \mathrm{P})$ and adapted to a filtration $\{\mathcal{F}_t\}$ of observable information, and let $\Phi(t) = \mathrm{card}\{n \geq 1 : T_n \leq t\}$ denote its counting process. Again, $\mathcal{F}_t^\Phi = \sigma(\Phi_t) = \sigma(\{\Phi(s) : 0 \leq s \leq t\})$ refers to the past of Φ at time t.

Z being external should mean, that at time t, information on the past of Φ additional to information on the past of Z is irrelevant with regard to events of the future of Z, i.e.,

$$\begin{aligned} & \mathrm{P}(A \mid \mathcal{F}_t^Z \vee \mathcal{F}_t^\Phi) = \mathrm{P}(A \mid \mathcal{F}_t^Z) && \forall\, A \in \mathcal{F}_{(t,\infty)}^Z \\ \iff & \mathrm{P}(B \mid \mathcal{F}_t^Z \vee \mathcal{F}_{(t,\infty)}^Z) = \mathrm{P}(B \mid \mathcal{F}_t^Z) && \forall\, B \in \mathcal{F}_t^\Phi \\ \iff & \mathcal{F}_t^\Phi \perp \mathcal{F}_{(t,\infty)}^Z \text{ given } \mathcal{F}_t^Z \end{aligned}$$

where $\perp$ denotes conditional independence.

Definition 5.2.1 Z is called an *external* covariate with respect to Φ, if $\mathcal{F}_t^\Phi$ and $\mathcal{F}_{(t,\infty)}^Z$ are for all $t \geq 0$ conditionally independent given $\mathcal{F}_t^Z$, i.e.,

$$\mathrm{P}(\{\Phi_t \in \cdot\} \cap \{Z_{(t,\infty)} \in \cdot\} \mid \mathcal{F}_t^Z) = \mathrm{P}(\Phi_t \in \cdot \mid \mathcal{F}_t^Z)\, \mathrm{P}(Z_{(t,\infty)} \in \cdot \mid \mathcal{F}_t^Z) \tag{5.1}$$

for all $t \in \mathbb{R}_+$. Otherwise, Z is called *internal* with respect to Φ.

This definition extends that given by Kalbfleisch and Prentice (1980) with respect to point processes. Hence, by (5.1) we have for the conditional distribution function of a failure time T_n,

$$F_{T_n}(t \mid Z_t) := \mathrm{P}(T_n \leq t \mid \mathcal{F}_t^Z) = \mathrm{P}(T_n \leq t \mid \mathcal{F}_\infty^Z).$$

That means, at each time t, knowledge of only the past of Z or the whole path of Z is equivalent with regard to events of the past of Φ. Thus, the whole path $z(\cdot)$ can always be assumed to be known in advance, i.e., $\mathcal{F}_\infty^Z \subseteq \mathcal{F}_0$. Hence, for an external covariate Z, the process $\{F_{T_n}(t \mid Z_t)\}$ is nondecreasing whereas it is only a submartingale in general. Trivial examples of external covariates are nonrandom functions $f(t, Z_0)$ of a random variable Z_0 which is fixed in time and stochastic processes Z such that each T_n is an $\{\mathcal{F}_t^Z\}$-stopping time.

Further we consider a simple threshold model. Let the covariate Z characterize the state of the item and set for $n \geq 1$

$$T_n = \inf\{t > T_{n-1} : Z(t) \geq h\}$$

with $T_0 = 0$. The type of Z depends on the available information about Z. If Z is continuously observable, then Z is external with respect to Φ, since $\mathcal{F}_t^\Phi \subseteq \mathcal{F}_t^Z$ for all $t \geq 0$, i.e., each T_n is an $\{\mathcal{F}_t^Z\}$-stopping time. If Z is observable with a time delay $\Delta > 0$, then

$$Z^\Delta = \{Z((t-\Delta)^+) : t \in \mathbb{R}_+\}$$

is internal with respect to Φ, since

$$\mathrm{P}(T_n \le t \,|\, \mathcal{F}_t^{Z^\Delta}) = \mathrm{P}(T_n \le t \,|\, \mathcal{F}_{(t-\Delta)+}^Z) \ne \mathbf{1}\{T_n \le t\} = \mathrm{P}(T_n \le t \,|\, \mathcal{F}_t^Z).$$

If Z can be observed only at discrete times t_n with $0 = t_0 < t_1 < \cdots$, the resulting covariate process

$$Z^* = \left\{\sum\nolimits_{n\ge 0} \mathbf{1}\{t_n \le t < t_{n+1}\} Z(t_n) : t \in \mathbb{R}_+\right\}$$

is internal, as well, since for $t_n < t < t_{n+1}$,

$$\mathrm{P}(T_n \le t \,|\, \mathcal{F}_t^{Z^*}) = \mathrm{P}(T_n \le t \,|\, \mathcal{F}_{t_n}^Z) \ne \mathbf{1}\{T_n \le t\} = \mathrm{P}(T_n \le t \,|\, \mathcal{F}_t^Z).$$

Frequently, the point process Φ is modelled as an $\{\mathcal{F}_t\}$-doubly stochastic Poisson process (or Cox process) with a stochastic intensity $\kappa(t, Z(t))$ depending on a covariate process Z, i.e.,

$$\mathrm{P}(\Phi(s,t] = m \,|\, \mathcal{F}_s) = \frac{(K(s,t])^m}{m!}\, \mathrm{e}^{-K(s,t]}$$

for $m \in \mathbb{N}$, where $\Phi(s,t] = \mathrm{card}\{n \ge 1 : T_n \in (s,t]\}$ and K is a $\mathcal{F}_0$-measurable random measure with $K(s,t] = \int_s^t \kappa(u, Z(u))\, du$. The $\mathcal{F}_0$-measurability of K, which is equivalent to $\mathcal{F}_\infty^Z \subseteq \mathcal{F}_0$, implies that Φ has $\{\mathcal{F}_t\}$-conditional independent increments, i.e.,

$$\mathrm{P}(\Phi(s,t] = \cdot \,|\, \mathcal{F}_s) = \mathrm{P}(\Phi(s,t] = \cdot \,|\, \mathcal{F}_0)$$

for $s \le t$, and that Φ is a minimal repair process, which means that none of the random variables T_n is measurable with respect to the history generated by the stochastic intensity κ [see Aven and Jensen (1999)]. The following proposition says that such a modelling is only possible if Z is an external covariate.

Proposition 5.2.1 *If Φ has $\{\mathcal{F}_t\}$-conditional independent increments and a stochastic intensity $\kappa(t, Z(t))$, then Z is external with respect to Φ.*

PROOF. If Φ has $\{\mathcal{F}_t\}$-conditional independent increments, then $\mathrm{P}(\Phi_t = \cdot \,|\, \mathcal{F}_0) = f(t, K(0,t])$ for some nonrandom function f [see Last and Brandt (1995, Theorem 5.1.10)]. Hence, $\mathrm{P}(\Phi_t = \cdot \,|\, \mathcal{F}_0)$ and

$$\mathrm{P}(\Phi_t \in \cdot \,|\, \mathcal{F}_\infty^Z) = \mathrm{E}[\mathrm{E}[\mathbf{1}\{\Phi_t \in \cdot\} \,|\, \mathcal{F}_0] \,|\, \mathcal{F}_\infty^Z] = \mathrm{E}[\mathbf{1}\{\Phi_t \in \cdot\} \,|\, \mathcal{F}_0] = \mathrm{P}(\Phi_t \in \cdot \,|\, \mathcal{F}_0)$$

is $\mathcal{F}_t^Z$-measurable. Finally, by

$$\begin{aligned} \mathrm{P}(\Phi_t \in \cdot \,|\, \mathcal{F}_\infty^Z) &= \mathrm{E}[\mathbf{1}\{\Phi_t \in \cdot\} \,|\, \mathcal{F}_\infty^Z] \\ &= \mathrm{E}[\mathrm{E}[\mathbf{1}\{\Phi_t \in \cdot\} \,|\, \mathcal{F}_\infty^Z] \,|\, \mathcal{F}_t^Z] \\ &= \mathrm{E}[\mathbf{1}\{\Phi_t \in \cdot\} \,|\, \mathcal{F}_t^Z] \\ &= \mathrm{P}(\Phi_t \in \cdot \,|\, \mathcal{F}_t^Z) \end{aligned}$$

we conclude that Z is external with respect to Φ. ∎

5.3 Degradation-Failure Model for Non-Repairable Items

Let Z denote a covariate process in $(\mathbb{R}^q, \mathcal{B}^q)$ describing the dynamic environment and X a real-valued covariate process describing a key degradation measure. First, assume that $X = Y$, where Y is a process with stationary independent increments, e.g., a Wiener process, a gamma process or a Poisson process. Then, Y has linear mean and variance functions $\mathrm{E}[Y(t)] = \mu t$ and $\mathrm{Var}[Y(t)] = \sigma^2 t$. To cover nonlinear degradation behavior we use a time scale $\tau = \tau(t)$, i.e., a positive, increasing, and continuously differentiable function of real time t with $\tau(0) = 0$. It describes slowing or accelerating degradation in real time and is referred to as operational time. For instance, Bagdonavičius and Nikulin (2000) consider

$$\tau(t) = \gamma_1 t + \gamma_2(1 - \exp(-\gamma_3 t)) \,, \quad \gamma_i > 0 \tag{5.2}$$

for tyre protector wear and Whitmore and Schenkelberg (1997)

$$\tau(t) = 1 - \exp(-\gamma_1 t^{\gamma_2}) \qquad \text{and} \qquad \tau(t) = t^{\gamma_1} \,, \quad \gamma_i > 0 \tag{5.3}$$

in the context of self-regulating heating cables. The choice of τ depends on whether degradation is unbounded or approaches a saturation point.

To model the influence on failure of an item's dynamic operating environment, the time scale may depend on the covariate process Z, for instance on different stress levels, i.e., $\tau_Z(t) = \tau(t, Z_t)$. An example is the model of additive accumulation of damage (AAD-model) (see Bagdonavičius (1978)):

$$\tau_Z(t) = \tau\left(\int_0^t \exp(\beta^T Z(s))\, ds\right). \tag{5.4}$$

Our degradation model is given by

$$X(t) = X_0 + Y(\tau_Z(t)) =: X_0 + Y^Z(t),$$

where Y^Z is the time scaled process Y and X_0 is some possibly random initial degradation level independent of Y. Conditioned on $\{Z = z(\cdot)\}$, $Y^{z(\cdot)}$ is a process with independent but not stationary increments. Obviously,

$$\mathrm{E}[X(t) \,|\, \mathcal{F}_t^Z] = \mathrm{E}[X_0] + \mu\tau_Z(t) \quad \text{and} \quad \mathrm{Var}[X(t) \,|\, \mathcal{F}_t^Z] = \mathrm{Var}[X_0] + \sigma^2\tau_Z(t).$$

An item is regarded as failed when the degradation process reaches a critical threshold level h or when a traumatic event like a shock of large magnitude occurs. The traumatic event time is modelled as the first point of an $\{\mathcal{H}_t\}$-doubly stochastic Poisson process $\Psi = (C'_n)_{n\geq 1}$ with $\mathcal{H}_t = \mathcal{F}_\infty^Z \vee \mathcal{F}_\infty^X \vee \mathcal{F}_t^\Psi$

and with an intensity $\kappa(t, Z(t), X(t))$ that depends on the environment and on the degradation level. That means, the intensity κ of a traumatic event at time t depends on t, on the covariate value $Z(t)$, and on the degradation level $X(t) = Y(\tau(t, Z_t))$. Of course, Ψ can be an $\{\mathcal{F}_t\}$-doubly stochastic Poisson process only if Z and X are external with respect to Φ. This is not the case, if X is observable only at discrete times.

Hence, the failure time of an item is defined by

$$T = \min(D, C), \tag{5.5}$$

where

$$D = \inf\{t \geq 0 : X(t) \geq h\}, \quad C = \inf\{t \geq 0 : \Psi(t) = 1\}$$

are the nontraumatic failure time caused by degradation and the traumatic failure time caused by a traumatic event, respectively. We call this model *degradation-threshold-shock-model* (DTS-model). To find the survival function of T we use a theorem given by Yashin and Manton (1997):

Theorem 5.3.1 (Yashin, Manton) *Let X and Z be stochastic processes influencing a failure rate $\alpha(t, Z, X)$ and satisfying measurability conditions such that, for $t \geq 0$,*

$$\mathrm{E} \int_0^t \alpha(u, Z, X)\, du < \infty,$$

and let T be related to X_t and Z_t by

$$\mathrm{P}(T > t \mid X_t, Z_t) = \exp\left(-\int_0^t \alpha(u, Z, X)\, du\right). \tag{5.6}$$

If the trajectories of Z are observed up to t, then

$$\mathrm{P}(T > t \mid Z_t) = \exp\left(-\int_0^t \bar{\alpha}(u, Z_u)\, du\right),$$

where

$$\bar{\alpha}(t, Z_t) = \mathrm{E}[\alpha(t, Z, X) \mid Z_t, T > t].$$

The random failure rate $\alpha(t, Z, X)$ may depend on either the current values $X(t)$ and $Z(t)$ or on the trajectories X_t and Z_t up to t. Since

$$\mathrm{P}(C > t \mid X_t, Z_t) = \exp\left(-\int_0^t \kappa(s, Z(s), X(s))\, ds\right), \tag{5.7}$$

by Theorem 5.3.1, the conditional survival function of C is

$$\mathrm{P}(C > t \mid Z_t) = \exp\left(-\int_0^t \widetilde{\kappa}(s, Z_s)\, ds\right), \tag{5.8}$$

with $\tilde{\kappa}(t, Z_t) = \mathrm{E}[\kappa(t, Z(t), X(t)) \,|\, Z_t, C > t]$.

To compute the conditional survival function of D let $T_Y(h) = \inf\{t \geq 0 : Y(t) \geq h\}$ be the first passage time of Y to the threshold h with a survival function $\bar{F}_0(t, h)$. Assume that $T_Y(h)$ has a density $\mathrm{P}(T_Y(h) \in dt) = f_0(t, h)dt$ and a failure rate $\alpha_0(t, h) = f_0(t, h)/\bar{F}_0(t, h)$ for $h > 0$ and that the density at t of those trajectories of Y that have not exceeded h is $\mathrm{P}(Y(t) \in dy, T_Y(h) > t) = g_0(t, y, h)dt$. Further, assume that Y is independent of (Z, X_0) and that $\mathrm{P}(X_0 < h) = 1$. Then we have

$$\begin{aligned}
\mathrm{P}(D > t \,|\, Z_t, X_0) &= \mathrm{P}\left(\sup_{0 \leq s \leq \tau_Z(t)} Y(s) < h - X_0 \,\middle|\, Z_t, X_0\right) \\
&= \bar{F}_0(\tau_Z(t), h - X_0) \\
&= \exp\left(-\int_0^t \lambda(s, Z_s, X_0)\, ds\right)
\end{aligned}$$

where $\lambda(t, Z_t, X_0) = \tau_Z'(t)\, \alpha_0(\tau_Z(t), h - X_0)$. Analogously, by Theorem 5.3.1

$$\mathrm{P}(D > t \,|\, Z_t) = \exp\left(-\int_0^t \bar{\lambda}(s, Z_s)\, ds\right), \tag{5.9}$$

where $\bar{\lambda}(t, Z_t) = \tau_Z'(t)\, \bar{\alpha}_0(\tau_Z(t), h) = \tau_Z'(t)\, \mathrm{E}[\alpha_0(\tau_Z(t), h - X_0) \,|\, Z_t, D > t]$.

Although we observe that

$$\begin{aligned}
\mathrm{P}(T > t \,|\, Z_t, X_t) &= \mathrm{E}[\mathbf{1}\{D > t\}\mathbf{1}\{C > t\} \,|\, Z_t, X_t] \\
&= \mathbf{1}\{D > t\}\mathrm{P}(C > t \,|\, Z_t, X_t)
\end{aligned}$$

is not of the form (5.6), Theorem 5.3.1 can be used to compute $\mathrm{P}(T > t \,|\, Z_t)$.

Theorem 5.3.2 *Let Y be independent of (Z, X_0) with $\mathrm{P}(X_0 < h) = 1$. Conditioned on Z, the failure time T defined by (5.5) has for $t \geq 0$ a survival function given by*

$$\bar{F}_T(t \,|\, Z_t) = \mathrm{P}(T > t \,|\, Z_t) = \exp\left(-\int_0^t (\bar{\kappa}(s, Z_s) + \bar{\lambda}(s, Z_s))\, ds\right)$$

and a failure rate given by

$$\alpha(s, Z_s) = \bar{\kappa}(s, Z_s) + \bar{\lambda}(s, Z_s), \tag{5.10}$$

with $\bar{\kappa}(t, Z_t) = \mathrm{E}[\kappa(t, Z(t), X(t)) \,|\, Z_t, T > t]$ and $\bar{\lambda}(t, Z_t) = \tau_Z'(t)\, \mathrm{E}[\alpha_0(\tau_Z(t), h - X_0) \,|\, Z_t, D > t]$.

PROOF. Let $z_t = \{z(s) : 0 \le s \le t\}$ be a realisation of Z_t. By successive conditioning we have

$$\begin{aligned}\mathrm{P}(T > t \mid Z_t = z_t) &= \mathrm{P}(\mathbf{1}\{D > t\} = 1, \mathbf{1}\{C > t\} = 1 \mid Z_t = z_t) \\ &= \mathrm{P}(\mathbf{1}\{C > t\} = 1 \mid Z_t = z_t, \mathbf{1}\{D > t\} = 1)\, \mathrm{P}(\mathbf{1}\{D > t\} = 1 \mid Z_t = z_t) \\ &= \mathrm{P}(C > t \mid Z_t = z_t, D > t)\, \mathrm{P}(D > t \mid Z_t = z_t)\end{aligned}$$

and consequently,

$$\mathrm{P}(T > t \mid Z_t) = \mathrm{P}(C > t \mid Z_t, D > t)\, \mathrm{P}(D > t \mid Z_t). \tag{5.11}$$

Because of $\mathrm{P}(C > t \mid Z_t, X_t) = \mathrm{P}(C > t \mid Z_t, X_t, \mathbf{1}\{D > t\})$, we obtain by an application of Theorem 5.3.1 to (5.7)

$$\mathrm{P}(C > t \mid Z_t, \mathbf{1}\{D > t\}) = \exp\left(-\int_0^t \kappa^*(s, Z_s)\, ds\right)$$

where

$$\kappa^*(t, Z_t) = \mathrm{E}[\kappa(t, Z(t), X(t)) \mid Z_t, \mathbf{1}\{D > t\}, C > t].$$

Obviously, on $\{D > t\}$,

$$\kappa^*(t, Z_t) = \mathrm{E}[\kappa(t, Z(t), X(t)) \mid Z_t, T > t] = \bar{\kappa}(t, Z_t).$$

Hence, combining (5.11) and (5.9) we conclude

$$\mathrm{P}(T > t \mid Z_t) = \exp\left(-\int_0^t (\bar{\kappa}(s, Z_s) + \bar{\lambda}(s, Z_s))\, ds\right).$$

■

In particular, the failure rate α simplifies to

$$\alpha(t, Z_t) = \kappa(t, Z_t) + \lambda(t, Z_t, h - x_0)$$

if the intensity κ depends only on Z and not on X and if the initial degradation level x_0 is known.

There are two important subclasses of DTS-models. If no threshold exists, i.e., if $h = \infty$, we have a pure degradation-shock-model (DS-model) with $T = C$ and $\alpha(t, Z_t) = \bar{\kappa}(t, Z_t)$. If traumatic events cannot occur, i.e., if $\kappa \equiv 0$, the model is a pure degradation-threshold-model (DT-model) with $T = D$ and $\alpha(t, Z_t) = \bar{\lambda}(t, Z_t)$.

Looking at the proof of Theorem 5.3.2 and the formulae of $\bar{\kappa}$ and $\bar{\lambda}$ we can easily see that $\mathrm{P}(T > t \mid Z_t) = \mathrm{P}(T > t \mid Z_\infty)$, which implies, that Z is external with respect to $\Phi_1(t) = \mathbf{1}\{T \le t\}$.

Proposition 5.3.1 *Under the conditions of Theorem 5.3.2, the covariate Z in the DTS-model is external with respect to the one-point process Φ_1.*

The condition of independence of Y and Z means that the degradational evolution process Y^Z depends on an external covariate Z only through the time scale τ_Z.

5.4 Degradation-Failure Model for Repairable Items

To extend the threshold model to repairable items we use a marked point process approach. Upon each failure and possibly at regular inspection times $0 = b_0 < b_1 < \cdots$, an item will be repaired in negligible time depending on the current degradation level. Each repair action reduces the degradation level from the current level $X(T_n)$ to $(1 - R_n)X(T_n)$, where $(R_n)_{n\geq 1}$ is a sequence of $[0,1]$-valued random repair degrees. A repair degree zero denotes a minimal repair without reduction of the degradation level, or no repair, if a repair action is not necessary at an inspection time. A repair degree 1 is equivalent to a perfect repair which sets back the degradation level to 0, which is assumed to be the level of a new item. Any repair degree between 0 and 1 corresponds to an imperfect repair.

The degradation-failure marked point process $\Phi = ((T_n, V_n))_{n\geq 1}$ consists of event times T_n and marks V_n that contain additional information. The event times T_n and the process $X(t)$ on $(T_n, T_{n+1}]$ are defined recursively in the following way. First set $B_0 = C_0 = D_0 = T_0 = R_0 = 0$ and $X(0) = X_0$, where X_0 is some random initial degradation level. If X is already defined on $(0, T_n]$ for $n \geq 0$, set

$$X(t) = (1 - R_n)X(T_n) + Y^Z(t) - Y^Z(T_n)$$

on $(T_n, T_{n+1}]$ where $T_{n+1} = B_{n+1} \wedge C_{n+1} \wedge D_{n+1}$ with

$$\begin{aligned}
B_{n+1} &= \min\{b_l : b_l > T_n\} && \text{(next inspection time)},\\
C_{n+1} &= C'_{T_n,1} && \text{(next traumatic event time after } T_n),\\
D_{n+1} &= \inf\{t > T_n : X(t) \geq h\} && \text{(next nontraumatic failure time)}.
\end{aligned}$$

The time $C'_{T_n,1}$ denotes the first point after T_n of the traumatic event point process $\Psi = (C'_n)_{n\geq 1}$. Ψ is modelled as an $\{\mathcal{H}_t\}$-doubly stochastic Poisson process with $\mathcal{H}_t = \mathcal{F}^Z_\infty \vee \mathcal{F}^X_\infty \vee \mathcal{F}^\Psi_t$ and with an intensity $\kappa(t, Z(t), X(t))$ depending on the environment and on the degradation level. Thus, the resulting degradation process X is given as

$$X(t) = \pi_{1,n}X_0 + \sum_{i=1}^{n} \pi_{i,n}\Delta Y_i + Y^Z(t) - Y^Z(T_n), \quad T_n < t \leq T_{n+1},$$

where $\pi_{i,n} = \prod_{j=i}^{n}(1 - R_j)$ and $\Delta Y_i = Y^Z(T_i) - Y^Z(T_{i-1})$.

The marks $V_n = (V_n^1, V_n^2, V_n^3)^\top$ contain additional information on the event modes, the degradational increments between event times, and the results of repair actions. They are defined by $V_0 = (0, 0, X_0)^\top$ and

$$V_n^1 = \delta_n = \begin{cases} 0, & \text{if } B_n < C_n \wedge D_n \quad \text{(regular inspection at } T_n\text{)},\\ 1, & \text{if } D_n < C_n, D_n \le B_n \quad \text{(nontraumatic failure at } T_n\text{)},\\ -1, & \text{if } C_n \le B_n \wedge D_n \quad \text{(traumatic failure at } T_n\text{)}, \end{cases}$$

$$V_n^2 = \Delta Y_n = Y^Z(T_n) - Y^Z(T_{n-1}),$$

and

$$V_n^3 = X(T_n+) = (1 - R_n)X(T_n)$$

for $n \ge 1$. We suppose that the repair degrees $(R_n)_{n\ge1}$ are conditionally independent given $(T_n, X(T_n), V_n^1)_{n\ge1}$ and that $R_n > 0$ if $V_n^1 = 1$, i.e., in case of a nontraumatic failure. The point process Φ is adapted to the global history $\{\mathcal{F}_t\}$ where $\mathcal{F}_t = \mathcal{F}_t^Z \vee \mathcal{F}_t^\Phi$ contains all information on Z and Φ up to time t. If Z is external with respect to Φ, we can assume $\mathcal{F}_\infty^Z \subseteq \mathcal{F}_0$, i.e., $\mathcal{F}_t = \mathcal{F}_\infty^Z \vee \mathcal{F}_t^\Phi$.

The next step is computing the conditional survival function of T_{n+1} given $\mathcal{F}_{T_n}$, i.e., given all observable information up to T_n. Let $T_n < t \le T_{n+1}$. Analogously to the equations (5.7) and (5.8),

$$\mathrm{P}(C_{n+1} > t \,|\, \mathcal{H}_{T_n}) = \mathrm{P}(C_{n+1} > t \,|\, \mathcal{F}_{T_n}, X_t) = \exp\left(-\int_{T_n}^{t} \kappa(s, Z(s), X(s))\, ds\right)$$

and, by Theorem 5.3.1,

$$\mathrm{P}(C_{n+1} > t \,|\, \mathcal{F}_{T_n}) = \exp\left(-\int_{T_n}^{t} \widetilde{\kappa}(s \,|\, \mathcal{F}_{T_n})\, ds\right),$$

with $\widetilde{\kappa}(t \,|\, \mathcal{F}_{T_n}) = \mathrm{E}[\kappa(t, Z(t), X(t)) \,|\, \mathcal{F}_{T_n}, C_{n+1} > t]$.

Further, assume that Y is independent of (Z, X_0, R) and that $\mathrm{P}(X_0 < h) = 1$. As Y is a process with stationary independent increments, we find

$$\begin{aligned} \mathrm{P}(D_{n+1} > t \,|\, \mathcal{F}_{T_n}) &= \mathrm{P}\left(\sup_{\tau_Z(T_n)\le s\le\tau_Z(t)} Y(s) - Y(\tau_Z(T_n)) < h - X(T_n+) \,\Big|\, \mathcal{F}_{T_n}\right)\\ &= \mathrm{P}\left(\sup_{0\le s\le\tau_Z(t)-\tau_Z(T_n)} Y(s) < h - X(T_n+)\right)\\ &= \bar{F}_0(\tau_Z(t) - \tau_Z(T_n), h - X(T_n+))\\ &= \exp\left(-\int_{T_n}^{t} \lambda(s \,|\, \mathcal{F}_{T_n})\, ds\right), \end{aligned}$$

where $\lambda(t \,|\, \mathcal{F}_{T_n}) = \tau_Z'(t)\, \alpha_0(\tau_Z(t) - \tau_Z(T_n), h - X(T_n+))$.

Note that the values of the initial degradation level X_0 and all further starting levels after repair action are given by the marks V_n^3 and contained in the filtration $\{\mathcal{F}_t\}$. The respective intensity $\bar{\lambda}$ in (5.9) would be the $\{\mathcal{F}'_t\}$-intensity of the points (D_n) of the reduced marked point process $\Phi' = ((T_n, (V_n^1, V_n^2)))_{n\geq 1}$. On $(T_n, T_{n+1}]$, it is given by $\bar{\lambda}(t \mid \mathcal{F}_{T_n}) = \mathrm{E}[\lambda(t \mid \mathcal{F}_{T_n}) \mid \mathcal{F}'_{T_n}, D_{n+1} > t]$.

The following theorem provides expressions for the conditional survival function and the $\{\mathcal{F}_t\}$-intensity of the univariate point process $\bar{\Phi} = (T_n)_{n\geq 1}$. Its proof follows the lines of the proof of Theorem 5.3.2 and is therefore not given here. Let $\boldsymbol{\lambda}^1$ denote the one-dimensional Lebesgue measure and ϵ_t the Dirac measure at t.

Theorem 5.4.1 *Let $\Phi = ((T_n, V_n))_{n\geq 1}$ be the marked point process in the DTS-model and $\mathcal{F}_t = \mathcal{F}_t^Z \vee \mathcal{F}_t^\Phi$ the history of observable information. Assume that Y is independent of (Z, X_0, R) and $\mathrm{P}(X_0 < h) = 1$. Then Z is external with respect to $\bar{\Phi} = (T_n)_{n\geq 1}$ and, on (T_n, ∞), the conditional survival function of T_{n+1} is given by*

$$\mathrm{P}(T_{n+1} > t \mid \mathcal{F}_{T_n}) = \mathbf{1}\{B_{n+1} \geq t\} \exp\left(-\int_{T_n}^t (\bar{\kappa}(s \mid \mathcal{F}_{T_n}) + \lambda(s \mid \mathcal{F}_{T_n})\, ds\right)$$

with

$$\bar{\kappa}(t \mid \mathcal{F}_{T_n}) = \mathrm{E}[\kappa(t, Z(t), X(t)) \mid \mathcal{F}_{T_n}, T_{n+1} > t]$$

and

$$\lambda(t \mid \mathcal{F}_{T_n}) = \tau'_Z(t)\, \alpha_0(\tau_Z(t) - \tau_Z(T_n), h - X(T_n+)).$$

The univariate point process $\bar{\Phi} = (T_n)_{n\geq 1}$ has a stochastic $\{\mathcal{F}_t\}$-intensity

$$\alpha(t \mid \mathcal{F}_t) = \sum_{n=1}^{\infty} \mathbf{1}\{t = b_n\} + \bar{\kappa}(t \mid \mathcal{F}_{T_n}) + \lambda(t \mid \mathcal{F}_{T_n}) \tag{5.12}$$

with respect to the measure $\boldsymbol{\lambda}^1 + \sum_{n=1}^{\infty} \epsilon_{b_n}$.

5.5 Maximum Likelihood Estimation

Suppose that n independent items are observed in $[0, t]$. For the ith item we observe the covariate trajectory $(z_i)_t$, the event times t_{ij}, the indicators $\delta_{ij} = v_{ij}^1$ describing the event modes, the degradational increments $\Delta y_{ij} = Y_i^{z_i}(t_{ij}) - Y_i^{z_i}(t_{i,j-1})$, and the repair results $x_{ij}^+ = x(t_{ij}+)$ for $j = 1, \ldots, m_i$, where $m_i = m_i(t) = \max\{j \geq 0 : t_{ij} \leq t\}$. The likelihood function of observable information can be found by computing the $\{\mathcal{F}_t\}$-compensator ν of the marked point process Φ. It is given as

$$\nu(d(t,v)) = \sum_{n\geq 0} I\{T_n < t \leq T_{n+1}\} \frac{\mathrm{P}((T_{n+1}, V_{n+1}) \in d(t,v) \,|\, \mathcal{F}_{T_n})}{\mathrm{P}(T_{n+1} \geq t \,|\, \mathcal{F}_{T_n})}$$

(see Last and Brandt (1995)). Based on the compensator the likelihood function of one item can be computed by

$$L_t = \prod_{T_n \leq t} \nu(d(T_n, V_n)) \prod_{s \leq t} (1 - \bar{\nu}(\{s\})) \exp(-\bar{\nu}^c(t))$$

where $\bar{\nu}^c$ denotes the continuous part of the compensator $\bar{\nu}$ of $\bar{\Phi} = (T_n)_{n\geq 1}$.

Provided that the distribution of the n-th repair degree R_n depends only on the degradation level $X(T_n)$ at T_n and on the indicator δ_n with a common distribution function $F_R(dr, x, v) = \mathrm{P}(R_n \in dr \,|\, X(T_n) = x, \delta_n = v)$, the likelihood function in the DTS-model for n independent items is, conditioned on the covariate trajectories $(z_i)_t$, given by

$$\begin{aligned} L_t = \prod_{i=1}^{n} \Bigg\{ \prod_{t_{ij} \leq t} & \left(g_0(\Delta s_{ij}, y_{ij}, h - x^+_{i,j-1}) \right)^{I\{\delta_{ij} < 1\}} \\ & \times \left(\tau'_{z_i}(t_{ij})\, f_0(\Delta s_{ij}, h - x^+_{i,j-1}) \right)^{I\{\delta_{ij} = 1\}} \\ & \times \bar{\kappa}(t_{ij} \,|\, \mathcal{F}_{t_{i,j-1}})^{I\{\delta_{ij} = -1\}} \exp\left(- \int_{t_{i,j-1}}^{t_{ij}} \bar{\kappa}(s \,|\, \mathcal{F}_{t_{i,j-1}})\, ds \right) \\ & \times F_R\left(d\left(1 - \frac{x^+_{ij}}{x_{ij}}\right), x_{ij}, \delta_{ij} \right) \Bigg\} \end{aligned} \tag{5.13}$$

where $s_{ij} = \tau_{z_i}(t_{ij})$ and $\Delta s_{ij} = s_{ij} - s_{i,j-1}$.

Suppose that the functions $\tau, Q, g_0, f_0, \kappa$, and F_R are parameterized in some specific way:

$$\begin{aligned} Q(t, Z_t) &= \int_0^t q(Z(s); \beta)\, ds, & \beta &= (\beta_0, \ldots, \beta_l)^\top, \\ \tau(t) &= \tau(t; \gamma), & \gamma &= (\mu, \gamma_1, \ldots, \gamma_m)^\top, \\ g_0(t, y, h) &= g_0(t, y, h; \zeta), & \zeta &= (\sigma^2, \zeta_1, \ldots, \zeta_p)^\top, \\ f_0(t, h) &= f_0(t, h; \zeta), & & \\ \kappa(t, z, x) &= \kappa(t, z, x; \eta), & \eta &= (\eta_0, \ldots, \eta_q)^\top, \\ F_R(r, x, v) &= F_R(r, x, v; \rho), & \rho &= (\rho_0, \ldots, \rho_s)^\top. \end{aligned}$$

We set $\tau_Z(t) = \tau(Q(t, Z_t))$. For instance, the function τ may be chosen as one of the functions (5.2) or (5.3) and Q according to the AAD-model (5.4). A possible parameterization of the repair degree distribution would be

$$F_R(dr, v; \rho) = \sum_{i=0}^{s} p_i(v)\epsilon_{\rho_i}(dr)$$

with known probabilities $p_i(v) = \mathrm{P}(R_n = \rho_i \mid \delta_n = v)$ that a repair of degree ρ_i is carried out if the event mode is v.

Set $\theta = (\beta^\top, \gamma^\top, \zeta^\top, \eta^\top, \rho^\top)^\top$. Having specified the model, in most cases a maximum likelihood estimator $\hat{\theta}$ has to be found numerically. Explicit estimators of the degradation parameters in a DT-model based on a Wiener process are given in Lehmann (2001).

5.6 Semiparametric Estimation

Suppose that the mean degradation function τ in the above parameterized model is completely unknown and that the covariate process Z is constant in time or a step function. In this case we can use an approach proposed by Bagdonavičius and Nikulin (2000) which is presented here in adapted form.

Denote by $P(t, Z_t; \beta)$ the inverse of the increasing function

$$Q(t, Z_t; \beta) = \int_0^t q(Z(s); \beta)\, ds.$$

Then, the process $Y(\tau_Z(P(t, Z_t; \beta)))$ is a process with independent increments and

$$\mathrm{E}[Y(\tau_Z(P(t, Z_t; \beta)))] = \mu\,\tau(t), \qquad \mathrm{Var}[Y(\tau_Z(P(t, Z_t; \beta)))] = \sigma^2\tau(t). \quad (5.14)$$

Now, we drop the parameter μ as only $\mu\tau(t)$ can be estimated by this approach. If $Z(t) \equiv z$ is constant in time, then

$$Q(t, Z_t; \beta) = q(z; \beta)\, t, \qquad P(t, Z_t; \beta) = (q(z; \beta))^{-1}\, t =: p(z; \beta)\, t.$$

If $Z(t)$ is a step function of the form

$$Z(t) = z_j, \qquad t \in [s_{j-1}, s_j), \quad j = 1, \ldots, J$$

with $0 = s_0 < s_1 < \cdots < s_J$, then

$$Q(t, Z_t; \beta) = s_{j-1}^*(\beta) + q(z_j; \beta)\,(t - s_{j-1}), \qquad t \in [s_{j-1}, s_j],$$

$$P(t, Z_t; \beta) = s_{j-1} + p(z_j; \beta)\,(t - s_{j-1}^*(\beta)), \qquad t \in [s_{j-1}^*(\beta), s_j^*(\beta)],$$

where

$$s_j^*(\beta) = \sum_{k=1}^{j} q(z_k;\beta)(s_k - s_{k-1}), \qquad j = 1,\dots,J, \quad s_0^*(\beta) = 0.$$

Consider the following piecewise-linear approximation $\tilde{Y}_i(t)$ of the process $Y_i^{z_i(\cdot)}(t) = Y_i(\tau(Q(t, z_i(\cdot))))$, both defined on $[0, t_{im_i}]$:

$$\tilde{Y}_i(t) = \sum_{j=1}^{m_i} \left[Y_i^{z_i(\cdot)}(t_{i,j-1}) + \frac{t - t_{i,j-1}}{t_{ij} - t_{i,j-1}} \Delta y_{ij} \right] \mathbf{1}_{[t_{i,j-1}, t_{ij}]}(t).$$

Setting $P_i(t;\beta) = P(t, z_i(\cdot);\beta)$ we conclude from (5.14) that for all $i = 1,\dots,n$,

$$\mathrm{E}\left[\tilde{Y}_i(P_i(t;\beta))\right] \approx \tau(t), \qquad \mathrm{Var}\left[\tilde{Y}_i(P_i(t;\beta))\right] \approx \sigma^2\tau(t),$$

if the differences Δt_{ij} are small. In this case the mean $\tilde{\tau}(t;\beta)$ of the values $\tilde{Y}_i(P_i(t;\beta))$, which, of course, still depends on β, should be a reasonable estimator of the unknown function $\tau(t)$ at t . Note that the processes $\tilde{Y}_i(P_i(t;\beta))$ are censored at the points $t_i^*(\beta) = P_i(t_{im_i};\beta)$. Consider the ordered sequence

$$t_{(1)}^*(\beta) < \cdots < t_{(n)}^*(\beta)$$

of these points. Then the estimator $\tilde{\tau}(t;\beta)$ is given by

$$\tilde{\tau}(t;\beta) = \frac{1}{n}\sum_{i=1}^{n} \tilde{Y}_i(P_i(t;\beta))$$

for $t \in [0, t_{(1)}^*(\beta))$ and

$$\begin{aligned}\tilde{\tau}(t;\beta) &= \tilde{\tau}(t_{(j-1)}^*(\beta);\beta) \\ &\quad + \frac{1}{n-j+1} \sum_{i:\, t_i^*(\beta) > t_{(j-1)}^*(\beta)} \left(\tilde{Y}_i(P_i(t;\beta)) - \tilde{Y}_i(P_i(t_{(j-1)}^*(\beta);\beta)) \right)\end{aligned}$$

for $t \in [t_{(j-1)}^*(\beta), t_{(j)}^*(\beta))$.

Plugging $\tilde{\tau}(t;\beta)$ in the likelihood function (5.13) and computing the maximum likelihood estimator $\widehat{\theta^*}$ of $\theta^* = (\beta^\top, \zeta^\top, \eta^\top, \rho^\top)^\top$ we can estimate $\tau(t)$ by $\tilde{\tau}(t;\hat{\beta})$.

5.7 Concluding Remarks

The DTS-model can be easily extended to the case that the inspection times b_i are random following a point process Ψ_0 and that m different modes of traumatic events are considered such that traumatic failures of mode i occur due to a point process Ψ_i. If all these point processes are conditionally independent doubly stochastic Poisson processes adapted to appropriate filtrations with intensities $\kappa_i(t, Z(t), X(t))$, then the Theorems 5.3.2 and 5.4.1 remain valid if we replace $\bar{\kappa}(s, Z_s)$ by $\sum_{i=0}^{m} \bar{\kappa}_i(s, Z_s)$ in the equations (5.10) and (5.12).

References

1. Aven, T. and Jensen, U. (1999). *Stochastic Models in Reliability*, Springer-Verlag, New York.

2. Bagdonavičius, V. (1978). Testing the hypothesis of the additive accumulation of damage, *Probability Theory and its Applications*, **23**, 403–408.

3. Bagdonavičius, V. and Nikulin, M. S. (2000). Estimation in degradation models with explanatory variables, *Lifetime Data Analysis*, **7**, 85–103.

4. Kalbfleisch, J. D. and Prentice, R. S. (1980). *The Statistical Analysis of Failure Time Data*, John Wiley & Sons, New York.

5. Last, G. and Brandt, A. (1995). *Marked Point Processes on the Real Line*, Springer-Verlag, New York.

6. Lehmann, A. (2001). A Wiener process based model for failure and degradation data in dynamic environments, *Dresdner Schriften zur Mathemat. Stochastik*, **4/2001**, 35–40.

7. Singpurwalla, N. D. (1995). Survival in dynamic environments, *Statistical Science*, **10**, 86–103.

8. Whitmore, G. A. and Schenkelberg, F. (1997). Modelling accelerated degradation data using Wiener diffusion with a time scale transformation, *Lifetime Data Analysis*, **3**, 27–45.

9. Yashin, A. I. and Manton, G. M. (1997). Effects of unobserved and partially observed covariate processes on system failure: A review of models and estimation strategies, *Statistical Science*, **12**, 20–34.

6

Comparison of Parametric and Semiparametric Estimates in a Degradation Model with Covariates and Traumatic Censoring

Vincent Couallier

Victor Segalen University Bordeaux 2, Bordeaux, France

Abstract: There exists numerous ways in survival analysis and reliability to construct a statistical model for the evolution of the degradation of a subject. The most important models include finite state Markov chains or semi-Markov processes [for which some transitions mean deteriorating or improving and where the states reflect the levels of degradation, see Limnios and Oprisan (2001)] or models describing the degradation by the evolution (often increasing) of a random function of time [Meeker and Escobar (1998), Bagdonavicius and Nikulin (2001) and references therein]. In this last case, the individual trajectories can themselves be modelled by some known parametric function with random coefficients or by the paths of a general stochastic process. This process has to reflect the way degradation evolves; in some cases, it increases slowly and gradually, in other cases it results from shocks, each of them leading to an increment of degradation.

The main aim of this work is to analyse and compare the estimation in a specific statistical degradation model studied in Bagdonavicius and Nikulin, (2000 and 2001), with parametric or semi-parametric assumptions. Computational aspects of the maximum likelihood or pseudo maximum likelihood estimates are discussed. A second aim of the paper is to address the computational issues which arise in such complex degradation models with covariates and censored data. A new degradation model based on a path model with error measurement is also presented [Bagdonavicius and Nikulin (2003) and Zdorova-Cheminade (2003)].

Keywords and phrases: BFGS and Nelder–Mead methods, degradation process, failure, gamma process, likelihood, numerical optimization, semiparametric models, survival analysis, reliability, traumatic event

6.1 The Degradation Model

As in Bagdonavicius and Nikulin (2001), Singpurwalla (1995) and others we choose here a model in which the space of degradation levels is not finite nor countable but an interval of $\mathbb{R}$. The degradation process is modelled by an increasing stochastic process [hence Wiener diffusion processes can not be considered as in Whitmore and Schenkelberg (1997)]. This process is supposed to verify:

- **H1** $EZ(t) = m(t)$ where m is an unknown smooth function,
- **H2** Z is an increasing process with independent increments.

We will consider the nonparametric and parametric cases to model the mean degradation function m. In this last case we shall have

$$EZ(t) = m(t, \theta), \quad t > 0, \tag{6.1}$$

where m is a known function and $\theta \in \mathbb{R}^p$ is a parameter to be estimated.

Following Bagdonavicius and Nikulin (2000), the individual degradation process Z is supposed to lie in the family of Gamma processes with right-continuous trajectories. It ensures the growth of the paths and the fact that, denoting

$$Z(t) = \sigma^2\gamma(t) \quad \text{with } \gamma(t) \sim G(1, \nu(t)) = G(1, \frac{m(t)}{\sigma^2}),$$

the increment from t to $t+\Delta t$ for $\Delta t > 0$ lies in the same family of distribution,

$$\Delta\gamma(t) = \gamma(t+\Delta t) - \gamma(t) \sim G(1, \frac{\Delta m(t)}{\sigma^2}).$$

A failure caused by degradation occurs when $Z(t)$ reaches a fixed threshold z_0. The moment of failure is

$$T = \sup\{\ t \mid \ Z(t) < z_0\ \} = \inf\{\ t \mid \ Z(t) \geq z_0\ \}.$$

6.2 Some Traumatic Events Censure the Degradation Paths

In addition to the breakdown due to wear, it is natural to consider breakdowns whose origin is not directly related to ageing. The probability of these traumatic events can however depend on degradation. Thus, we model the censoring times C due to a traumatic event as the time of first jump of a nonhomogeneous Poisson process $(N(t), t \geq 0)$ whose intensity function can depend on the level of degradation reached at the time t.

Given the degradation path until time t, we assume that the intensity function depends on $Z(t)$,

$$\lambda(t) = \lambda(Z(t)),$$

and the conditional distribution of the random variable $N(t)$ which is the number of traumatic events until time t has the Poisson distribution with mean $\int_0^t \lambda(Z(s))ds$.

Thus, the conditional survival function of C is

$$S(t) = \mathbf{P}\{C > t | Z(s), 0 \leq s \leq t\} = \exp(-\int_0^t \lambda(Z(s))ds), \tag{6.2}$$

and its density function is

$$S(t) = \mathbf{P}\{C > t | Z(s), 0 \leq s \leq t\} = \lambda(Z(t)) \exp(-\int_0^t \lambda(Z(s))ds). \tag{6.3}$$

Note that an obvious and instantaneous generalization is to relax the assumption that the point process is poissonian. In order to do this, it is possible to take every positive real-valued function G decreasing from 1 to 0 instead of $\exp^{-x}$ in S.

Whatever the cause of failure is, let us denote $U = \min(T, C)$ the time of failure and G its survival function. We have

$$G(t) = \mathbf{P}\{U > t\} = \mathbf{E}\Big\{\exp(-\int_0^t \lambda(Z(s))ds)\mathbf{1}_{\{Z(t)<z_0\}}\Big\}. \tag{6.4}$$

In the following, we will assume that the conditional (on Z) intensity function is a parametric function of $Z(t)$.

6.3 The Covariates

We model the influence of the environment on the degradation process by covariates in the following way: we have for each item the value $\mathbf{x} = (x_1, ..., x_s)$ of the explanatory variable $\mathbf{X} = (X_1, ..., X_s) \in \mathbb{R}^s$ assumed to be constant in time. [see Bagdonavicius and Nikulin (2000), Singpurwalla (1995) and Lehmann (2004) for dynamic time-varying explanatory variables].

The classical framework of accelerated failure time model can be applied to the degradation model by considering that acceleration of time is due to a scaling by means of a function $f_x(t)$ depending on the value of $\mathbf{X}$. We will consider in the simulations and the implementation of the model that f is linear, $f_x(t) = e^{\beta^T \mathbf{x}} t$, and $\mathbf{X}$ is a categorical multivariate random variable. In this case, we will have, conditionally on $\mathbf{X}$=$\mathbf{x}$,

$$Z_{\mathbf{x}}(t) \overset{\mathcal{D}}{\sim} \sigma^2 \gamma(e^{\beta^T \mathbf{x}} t), \quad t \geq 0.$$

The parameter $\beta = (\beta_1, .., \beta_s)$ measures the influence of the different covariates. Given $\mathbf{X}$=$\mathbf{x}$, The real coefficient $e^{\beta^T \mathbf{x}}$ is a relative measurement of the (constant) stress of the environment and we have $E\ Z_{\mathbf{x}}(t) = m(e^{\beta^T \mathbf{x}} t), \quad t \geq 0$.

The issue is thus to estimate the following reliability characteristics: the mean degradation function m, τ the mean time needed to attain the threshold z_0, the survival functions Q, S and G of failures due to degradation only, traumatic event only, or an unspecified cause respectively.

6.4 The Data and the Estimation

Suppose that n independent items are observed. The ith item is observed under the vector of covariates $\mathbf{x}^{(i)}$ at m_i ordered moments $t_{i0} = 0 < t_{i1} < \cdots < t_{im_i}$. The values of the degradation levels $Z(t_{i0}) = 0, Z(t_{i1}), \ldots, Z(t_{im_i})$ are supposed to be measured without errors and some of them can be censured by a threshold failure or a traumatic failure. We have thus for the ith item the successive values of increments of degradation until failure, the value $\mathbf{x}^{(i)}$ describing the environment, the indicator of censorship $\delta_i = 1_{\{T_i \leq C_i\}}$.

The estimates are based on maximum likelihood or pseudo-maximum likelihood estimates. Numerical optimization is needed to achieve this estimation. We want to compare through simulations the efficiency of the method for two different assumptions on the mean degradation function m. Also we discuss the choice of numerical algorithms for the optimization of the likelihood for the two models. Let us consider

- **Model 1** $EZ(t) = m(t, \theta)$, $t > 0$, where m is a known function and $\theta \in \mathbb{R}^p$ is a parameter to be estimated.

- **Model 2** $EZ(t) = m(t)$, $t > 0$, where m is the unknown degradation smooth (at least continuous) function to be estimated.

In both models, we supposed that

H3: The intensity of the point process modelling the traumatic event satisfies the following parametric form :

$$\lambda(t) = \lambda(Z(t)) = \alpha Z(t)^p, \quad \alpha > 0, \quad p \geq 1. \tag{6.5}$$

α and p are two unknown coefficients to be estimated. Furthermore our model assume that

H4: X is a discrete random variable describing p different environments.

We have chosen $p = 3$ in the simulations. Different subsets of simulations

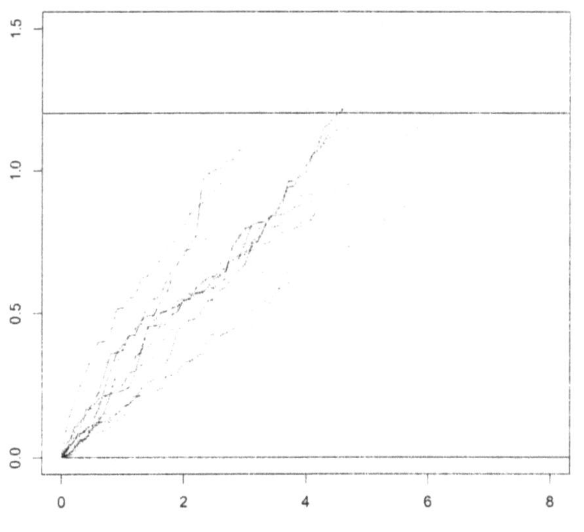

Figure 6.1: 18 paths under 3 different stresses.

have been carried out in order to measure the influence of the number of units, the number of degradation data for each unit, the type of assumption (parametric versus nonparametric) and the issue of lack-of-fit in the parametric model.

6.4.1 Estimation in the parametric framework

Given $x_1, \ldots, x_n$ the values of the n stresses, the likelihood function is [Bagdonavicius and Nikulin (2000)]

$$L(\theta^*) = L(\theta, \beta, \alpha, p, \sigma^2) \quad = \tag{6.6}$$

$$\prod_{n \text{ paths}} \Big(\prod_{n_i \text{ increments}} p_{ij}(\Delta Z_{ij}) \Big) \times \begin{cases} f(C_i) & \text{if traumatic,} \\ S(C_i) & \text{if not,} \end{cases} \tag{6.7}$$

where p_{ij} is the d.f. of a $\Gamma\big(1, \delta_{ij}\big)$ with $\delta_{ij} = \sigma^{-2}\big(m(t_{i(j+1)}, \theta) - m(t_{ij}, \theta)\big)$ and f and S are the conditional (on X and Z) density function and survival function of C written in (6.3) and (6.2). $\hat{\theta}^*$ is defined as the maximum likelihood estimate.

For the purely parametric Poisson-Gamma degradation model, L is a regular function of $\theta^* = (\theta, \beta, \alpha, p, \sigma^2)^T \in \mathbb{R}^p$ but $\hat{\theta}, \hat{\beta}, \hat{\alpha}, \hat{p}, \hat{\sigma^2}$ are not explicit (see Figure 6.2).

Nevertheless, standard likelihood theory gives asymptotic normality of the

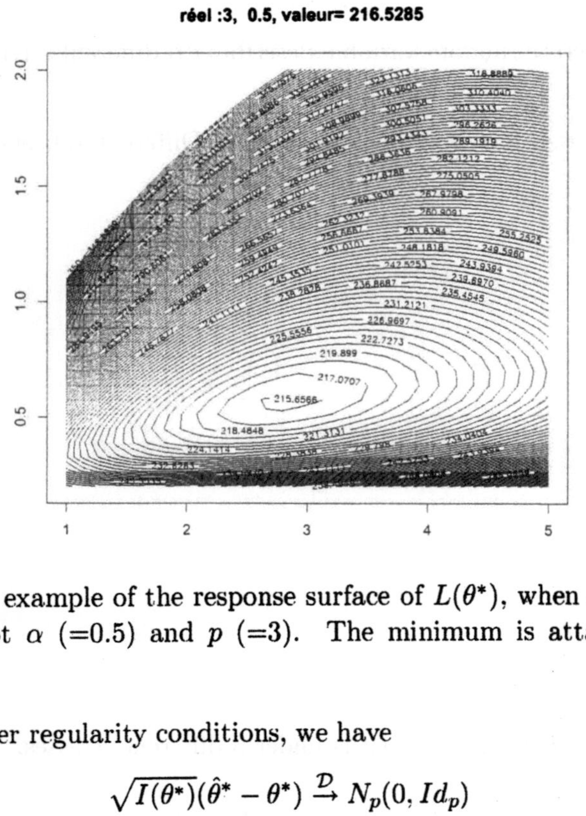

Figure 6.2: An example of the response surface of $L(\theta^*)$, when all parameters are fixed except α (=0.5) and p (=3). The minimum is attained at point (2.89;0.58).

estimates. Under regularity conditions, we have

$$\sqrt{I(\theta^*)}(\hat{\theta}^* - \theta^*) \xrightarrow{\mathcal{D}} N_p(0, Id_p)$$

where Id_p is the identity matrix in $\mathbb{R}^p$ and $I(\theta^*) = -\frac{\delta^2 \ln L(\theta^*)}{\delta\theta^{*2}}$ is the Fisher information matrix for the parameter θ^*. The calculation of the asymptotic variance is impossible in this general case hence, to use this asymptotic result, we need to approximate the Fisher matrix. For methods of optimization of Newton–Raphson or quasi-Newton type, we propose to minimize the negative log likelihood and to estimate the information matrix by a numerical version of the Hessian matrix evaluated at $\hat{\theta}^*$. This can be done with any statistics software having optimization procedures like *R*. In the simulations below we have chosen a quasi-Newton or *variable metric method* using the *Broyden–Fletcher–Goldfarb–Shanno* (B-F-G-S) algorithm which is well suited for a function which can be locally approximated by a quadratic form. With this estimate $\hat{H}(\hat{\theta}^*)$ of the fisher information matrix of θ^*, we can derive an approximated confidence region for θ at level $(1-\alpha)$:

$$\{\theta^* \mid (\hat{\theta}^* - \theta^*)^T \ \hat{H}(\hat{\theta}^*) \ (\hat{\theta}^* - \theta^*) \leq \mathcal{X}_p^2(1-\alpha)\}.$$

Note also that it can provide a confidence interval for every one-dimensional coordinate of θ^*. The mean degradation function is then estimated by plugging the estimation of θ in m,

$$\hat{m}(t) = m(t, \hat{\theta})$$

and in a similar manner for the unconditional and conditional survival functions S and G, for instance

$$\hat{G}(t) = \exp\left\{-\int_0^t \hat{\alpha}\ m^{\hat{p}}(s, \hat{\theta})\ ds\right\}\ \mathbf{1}_{\{\hat{m}(t) < z_0\}}.$$

6.4.2 Estimation in the semiparametric framework

The likelihood function has the same form as above,

$$\begin{aligned} &L(\beta, \alpha, p, \sigma^2) \\ &= \prod_{n \text{ paths}} \Big(\prod_{n_i \text{ increments}} p_{ij}(\Delta Z_{ij}) \Big) \times \begin{cases} f(C_i) & \text{if traumatic,} \\ S(C_i) & \text{if not,} \end{cases} \end{aligned} \tag{6.8}$$

except that the density p_{ij} which is the d.f. of a $\Gamma\big(1, \sigma^{-2}(m(t_{i(j+1)}) - m(t_{ij})\big)$ depends on the unknown infinite dimensional parameter m. The idea is as follows if we know β, given $X = x$, the inversion of the time-transformation f_x gives reference stress value x_0 for all items that is:

$$\forall i, Z_x(f_x^{-1}(t)) \overset{\mathcal{D}}{\sim} \sigma^2 \gamma(t), \quad t \geq 0.$$

In this case, all the scaled paths lie in the same basic environment and thus a piecewise-linear approximation of m can be calculated by taking for the estimation at point t the mean of all the scaled paths which are not censored at point t. This estimate of m denoted by $\tilde{m}_\beta$ can be plugged into the pseudo-likelihood function $\tilde{L}_\beta(\beta, \alpha, p, \sigma^2)$ at each iteration of the numerical optimization, leading to the pseudo-MLE $(\hat{\beta}, \hat{\alpha}, \hat{p}, \hat{\sigma^2})$.

Once this estimation is achieved, the nonparametric estimation m is given by $\hat{m} = \tilde{m}_{\hat{\beta}}$ and its support is $[0, \hat{W}]$ where

$$\hat{W} = \max_{i=1,\ldots,n} (e^{-\hat{\beta} X_i} U_i). \tag{6.9}$$

Unlike in the parametric framework, we do not have a regular enough function of $\beta, \alpha, p, \sigma^2$ (see Figure 6.3). Hence, a numerical optimization can only be possible with some robust method adapted to nondifferentiable functions such as the Nelder–Mead method which is a *Downhill simplex method* in multidimensions. This method requires only function evaluations, not derivatives. Thus in this case no estimate of the asymptotic variance-covariance matrix is given. Note that it could be possible to provide an approximated confidence region for the unknown parameters by using resampling methods but it is out of the scope of this paper.

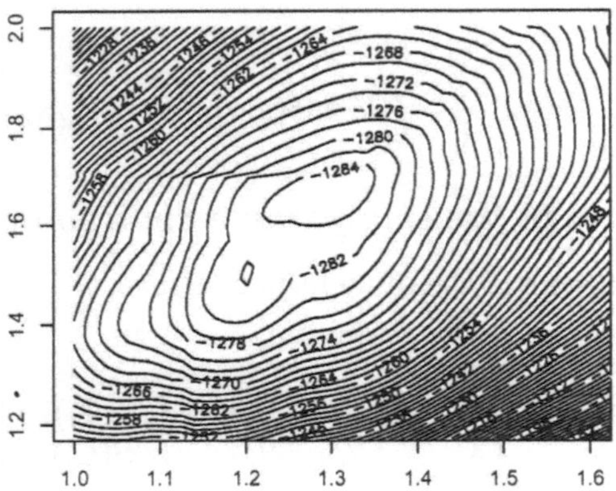

Figure 6.3: Optimization under semiparametric assumptions: response surface of the likelihood function restricted to 2 parameters (β_1 and β_2), the others being fixed to their real values. The minimum is attained at (1.28;1.63), the true parameter is (1.2;1.5).

6.5 Comparison and Analysis by Monte Carlo Simulations

6.5.1 Computational aspects

The comparison of the estimations in Model 1 and Model 2 is based on Monte Carlo simulations. Several issues arise in such a study. The first one is the way of handling continuous-type objects such as integral evaluations or continuous-time processes. The integrals have been calculated by a piecewise linear approximation of the functions. The continuous-time gamma processes have been simulated in the following manner: a very fine grid of time with time-step 2E-02 is the support of the sequençe of independent increments each of them having a gamma distribution with given parameters. These cumulated increments give a "continuous aspect" of the path after piecewise linear interpolation even if the path is only right-continuous. Then for the i-th unit, the only evaluation at the m_i points $t_{i1}, \ldots, t_{im_i}$ of the values $Z_{i1}, \ldots, Z_{im_i}$ is kept.

The second one is the way to compare the different results. A first answer concerns every subset of parameters which are common to the two models. This is the case of σ^2, β, α and p. For these quantities, numerical approximation of the mean squared error $MSE(\theta) = E(\hat{\theta} - \theta)^2$ can be calculated through Monte Carlo simulations. It is also possible to compare the estimates by the mean of their standard error. But these numerical values are not able to highlight differences in the shape of the distributions of the estimates. To answer this question, histograms have also been provided.

Remark. In order to compare the estimations of m in the two models, we could compute a Monte Carlo approximation of the integrated mean squared error

$$IMSE(\hat{m}) = \int (\hat{m}(x) - m(x))^2 w(x) dx,$$

(where w is a weight function) given by

$$IMSE_{app} = \frac{1}{N_{rep}} \sum_{j=1}^{n_{rep}} \int (\hat{m}_j(x) - m(x))^2 w_j(x) dx,$$

where N_{rep} is the number of Monte Carlo replications and $\hat{m}_j$ is the j-th estimation of m. The issue is the way we handle the difference in the support of the functions $\hat{m}_j$. A solution is to use (6.9) and to take the random weight $w_j(x) = \frac{1}{\hat{W}_j} 1_{[0, \hat{W}_j]}(x)$.

6.5.2 The simulations

The simulations have been implemented on *R* Software. The algorithms of optimization were already included in the software and available with the *optim* function. The choice for the parametric quantities is the following:

- The function m is linear $m(t) = at$ and we took $a = 0.2$.
- The intensity function is a power function $\lambda(t) = \alpha t^p$ where $\alpha = 0.5$ and $p = 3$.
- The covariate reflects three different environments in which the stress is constant in time : $X = (1, 1_{\{Env2\}}, 1_{\{Env3\}})$. We have chosen

  $$\beta = (1, \beta_2, \beta_3)^T = (1, 1.2, 1.5)^T$$

  and thus $e^{\beta^T \mathbf{x}} = 1$ or $= 1.2$ or $= 1.5$ according to the environment. The first one is the basic environment, the second is 20% more stressing and the third 50% more stressing for the evolution of the degradation.
- The value of the threshold z_0 is equal to 1.2 which gives reasonable ratios of censored paths in the three environments.

In order to evaluate the influence of the number of data on the results, three cases for the simulations have been carried out : the first case includes $n_1 = n_2 = n_3 = 30$ units for each level of stress, and a time-step for data equal to 0.5 (that is approximately at most 12 points by trajectories). The second case has the same number of units ($n_1 = n_2 = n_3 = 30$) and a time-step for data equal to 0.1 (that is approximately at most 60 points by trajectories). The third case requires $n_1 = n_2 = n_3 = 100$ and a time step for data equal to 0.2 (at most 30 points by trajectories).

For the three cases, n_1, n_2 and n_3 (censored) paths have been simulated and the estimations in the two models have been carried out. The number of replications for Monte Carlo simulation is $N_{rep} = 100$. Hence, the histograms of the empirical distributions of real-valued estimates (such that a, β_2, σ^2 ...) are based on an i.i.d. sample of size 100. Figure 6.6 shows the good behavior of the estimates of β_2 in the two first cases in both models. Note that the distributions seem to be heavy-right-tailed in the parametric model. This remark is also valid for the other parameters. The calculation of the standard errors and approximated mean squared errors of the estimates shows for all the parameters some correct values (see the table) but note that these standard errors are systematically more important in the parametric framework than in the semiparametric one. This remark holds also for the quadratic risk $MSE(\hat{\theta}) = E(\hat{\theta} - \theta)^2$. These surprising results can be explained by the the heavy-tailed distribution of the estimates in the parametric framework. In fact

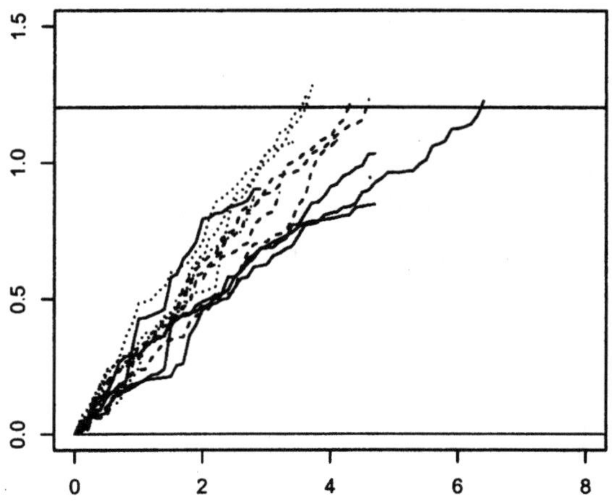

Figure 6.4: 4 real paths for each value of covariate X : $e^{\beta^T \mathbf{x}} = 1$ (solid) or $= 1.2$ (dashed) or $= 1.5$ (dotted).

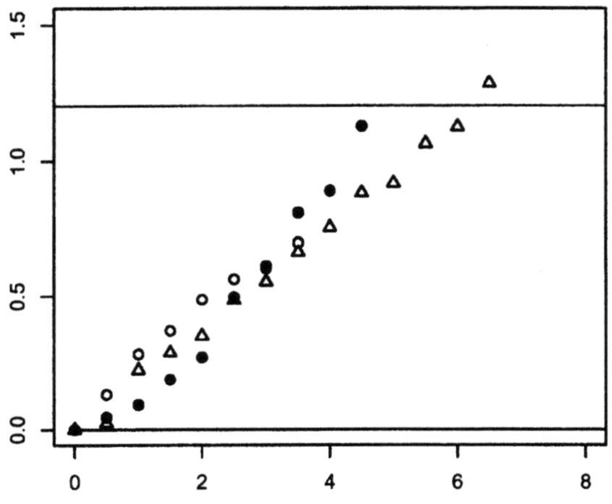

Figure 6.5: 3 paths measured without error with time-step 0.5: only the "triangle" one reaches the threshold z_0.

a few failures in numerical optimization lead to increasing the empirical MSE which does not seem to be a good criterion for comparison. This can be seen in Figures 6.7 and 6.8 where the few replications which did not converge to the true value of β give an important bias and thus an important contribution to the calculation of the empirical mean squared error (the index is the replication number). Nevertheless, a zooming shows how smaller the bias is in the parametric framework in the majority of cases. Instead of L2-type error, we could prefer L1-type error $R_1(\theta) = E|\hat{\theta} - \theta|$ and its empirical version. However the numerical results show the same phenomenon (see the table of $L1$-type errors).

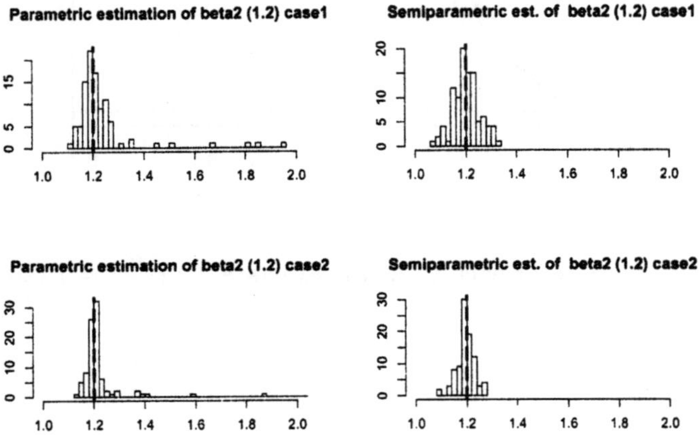

Figure 6.6: Example of empirical distribution of estimates of β_2 in both models for two cases. The vertical dashed line shows the true value of $\beta_2 = 1.2$.

Standard Error	a	β_2	β_3	σ^2
Case1 Para / (NP)	.017	.173 / .082	.200 / .078	.0010 / .0004
Case 2 Para / (NP)	.012	.135 / .052	.183 / .062	.0020 / .0011
Case 3 Para / (NP)	.012	.147 / .034	.160 / .042	.0014 / .0003

MSE	β_2	β_3	σ^2
Case1 Para / (NP)	6.1e−02 / (1.9e−02)	8.2e−02 / (3.3e−02)	2e−06 / (4e−07)
Case2 Para / (NP)	3.7e−02 / (5.49e−03)	6.9e−02 / (7.8e−03)	8.1e−06 / (2.9e−06)
Case 3 Para / (NP)	4.4e−02 / (2.4e−03)	5.2e−02 / (3.6e−03)	3.9e−06 / (3.4e−07)

L1-type error	β_2	β_3	σ^2
Case 1 Para / (NP)	$\frac{0.069}{(0.099)}$	$\frac{0.087}{(0.149)}$	$\frac{5.8e-04}{(4.2e-04)}$
Case 2 Para / (NP)	$\frac{0.061}{(0.039)}$	$\frac{0.080}{(0.051)}$	$\frac{12e-04}{(10e-04)}$
Case 3 Para / (NP)	$\frac{0.049}{(0.026)}$	$\frac{0.065}{(0.033)}$	$\frac{6.6e-04}{(3.6e-04)}$

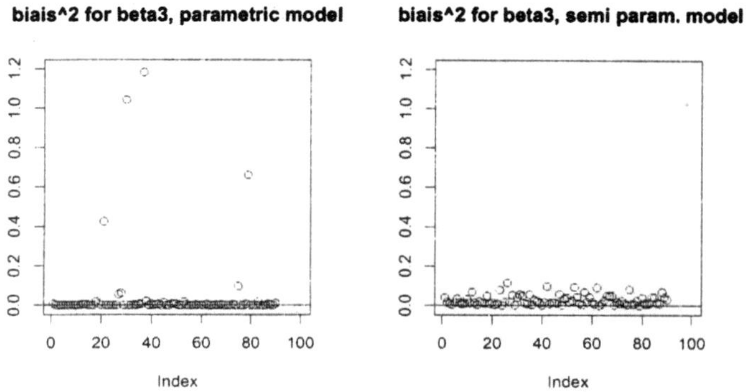

Figure 6.7: Squared bias of the N_{rep} replications in both models; a few cases seem to have failed in the BFGS optimization for parametric model.

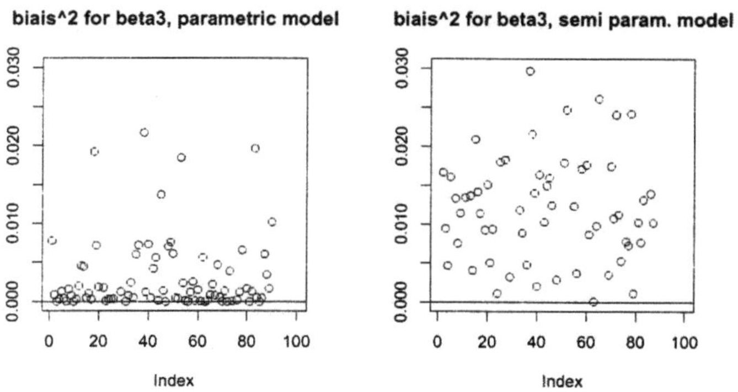

Figure 6.8: Squared bias of the N_{rep} replications in both models with modified scale of the vertical axis.

Thus, the computation and the optimization of such a likelihood structure is possible, numerical optimization in regular cases can give approximated confidence interval of parameters, and thus some procedures of test but the optimization has to be supervised in order to avoid some false convergence result, especially for methods of optimization using approximations of Hessian matrices.

6.6 The Path Model with Noise

The path model with noise is an alternative to the preceding model. Following Bagdonavicius and Nikulin (2003) and according to Meeker and Escobar (1998), we suppose that the real degradation process is modeled by the general *path model*

$$Z_r(t) = g(t, A);$$

here $A = (A_1, \ldots, A_r)$ is a random vector with positive components and with the distribution function F_A. g is a specified continuously differentiable increasing function in t. The typical form of the degradation curves is

$$g(t, a) = e^{a_1}(1+t)^{a_2}. \tag{6.10}$$

The *observed degradation process* $Z = Z(t), t \geq 0$ may be slightly different from the *real degradation process.* For this reason the next model for *the observed degradation process* can be considered. According to this model *the observed degradation process* is

$$Z(t) = Z_r(t)\; U(t),$$

where

$$V(t) = \ln U(t) = \sigma W(c(t)),$$

W is the standard Wiener process independent of A, and $c : [0, \infty) \to [0, \infty)$, $c(0) = 0$, is a specified continuous and increasing time transformation function, $c(0) = 0$. For any $t > 0$ the median of the random variable $U(t)$is 1.

Denote by h the function inverse to g with respect to the first argument. Evidently, it is continuously differentiable and increasing in t. Moreover,

$$T = h(z_0, A),$$

is a random variable which verifies $Z_r(T) = z_0$ and thus is called the *real non-traumatic failure time.* As in the Poisson-Gamma degradation model, let C be the time to traumatic failure. We have [Bagdonavicius and Nikulin (2003)]
Model 3. *The conditional survival of C given the real degradation process has the form*

$$S_C(t \mid A) = P\{C > t \mid g(s, A), 0 \leq s \leq t\} = \exp\Big\{-\int_0^t \lambda_0(s, \alpha)\lambda\big(g(s, A)\big)ds\Big\};$$

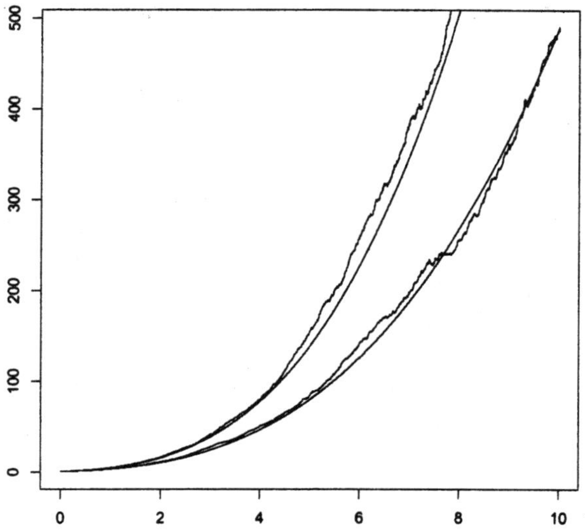

Figure 6.9: Two *observed degradation processes* for model (6.10) with individual realization of A given by $(a_1, a_2) = (-1, 3)$ and $(-0.8, 3.2)$. Brownian motion W simulated with $\sigma = 0.1$ and $c(t) = t$.

here λ is the unknown intensity function, $\lambda_0(s, \alpha)$ being from a parametric family of hazard functions.

The degradation process plays here the role of longitudinal explanatory covariate. This model has been studied in [Bagdonavicius and Nikulin (2003)] where survival and degradation characteristics are derived. Also, some covariate effects are modelled. The numerical implementation has been studied by Zdorova-Cheminade (2003) who considered this model for a model of disablement in the elderly to verify that a hierarchical relationship exists between the concepts of Activities Daily Living, Instrumental Activities of Daily Living and mobility.

References

1. Bagdonavicius, V. and Nikulin, M. (2000). Estimation in degradation models with explanatory variables, *Lifetime Data Analysis*, **7**, 85–103.

2. Bagdonavicius, V. and Nikulin, M. (2001). *Accelerated Life Models: Mod-*

eling and Statistical Analysis, Chapman & Hall/CRC, Boca Raton, Florida.

3. Lehmann, A. (2004). On a degradation-failure model for repairable items, *In this volume.*

4. Limnios, N. and Oprisan, G. (2001). *Semi-Markov Processes and Reliability*, Birkhäuser, Boston.

5. Meeker, W. Q. and Escobar, L. (1998). *Statistical Analysis for Reliability Data*, John Wiley & Sons, New York.

6. Singpurwalla , N. D. (1995). Survival in dynamic environnements, *Statistical Science*, **1**, 86–103.

7. Whitmore, G. A. and Schenkelberg, F. (1997). Modelling accelerated degradation data using Wiener diffusion with a time scale transformation, *Lifetime Data Analysis*, **3**, 27–45.

8. Zdorova-Cheminade, O. (2003). *Modélisation du processus d'évolution de l'incapacité chez les personnes agées*, Mémoire de DEA "Epidémiologie et Intervention en Santé Publique", Université Bordeaux 2, Juin 2003.

PART III

Accelerated Failure Time Models and Analyses

7

Accelerated Life Testing, Fuzzy Information and Generalized Probability

Reinhard Viertl

Department of Statistics, Probability Theory and Actuarial Mathematics, Technische Universität Wien, Wien, Austria

Abstract: Accelerated life testing (ALT) often is the only realistic method to obtain information on life time distributions at usual stress. Real continuous life time data are always more or less non-precise. This kind of data uncertainty is called fuzziness. The best up to date mathematical description of non-precise life times is by so-called fuzzy numbers. In ALT the fuzziness is increased by the models used to transform life times under high stress levels in order to estimate life time distribution under usual stress. Therefore data fuzziness has to be quantitatively described to obtain realistic results.

Another kind of fuzzy information is a priori information in Bayesian methods for ALT. A generalization of Bayes' theorem is necessary to take care of fuzzy data as well as imprecision of a priori distributions. For this a more general concept of probability is necessary, called fuzzy probability distributions.

Keywords and phrases: Accelerated life testing, lifetime distributions, fuzzy numbers, fuzziness, Bayesian methods, a-priori distributions, Bayes theorem, fuzzy probability distributions

7.1 Introduction

Nonparametric accelerated life testing is modelled in the following way: Let S denote the stress variable and S_u the usual stress. For two stress levels S_1 and S_2 the relationship between the c.d.f. $F(\cdot|S_2)$ of the life time distribution under stress S_2 and the c.d.f. $F(\cdot|S_1)$ is given by a so-called *acceleration function* $a_{1,2}(\cdot)$ by

$$F(t|S_2) = F(a_{1,2}(t)|S_1) \qquad \forall \quad t \geq 0.$$

In case of linear acceleration functions, i.e.,

$$a_{1,2}(t) = \alpha_{1,2} \cdot t \qquad \forall \quad t \geq 0,$$

there exists a method to estimate the acceleration constants $\alpha_{1,2}$ [see Bagdonavicius and Nikulin (2002) and Viertl (1988)].

The mentioned models allow us to calculate a predictive distribution for the life time under usual stress S_u by the relationship

$$F(t|S_u) = F\left(\frac{t}{\alpha(S)}\middle| S\right) \qquad \forall \quad t \geq 0,$$

where $\alpha(S)$ is the acceleration constant between usual stress S_u and stress S.

Estimation methods based on the above relationship show that small imprecisions of observed life times under stress S are drastically increased in estimating the c.d.f. of the life time under usual stress S_u. Therefore imprecisions of data, which are best modelled by fuzzy numbers (see the next section), cannot be neglected in order to obtain realistic results of the analysis.

7.2 Fuzzy Numbers, Fuzzy Vectors and Fuzzy Valued Functions

In order to model non-precise life time data, so-called fuzzy numbers are the most suitable model for their description.

A fuzzy number $x^\star$ is defined by its characterizing function $\xi(\cdot)$ which is a real function of one real variable x obeying the following:

(1) $0 \leq \xi(x) \leq 1 \quad \forall \quad x \in I\!R$.

(2) $\exists x_0 \in I\!R : \xi(x_0) = 1$.

(3) $\forall\, \delta \in (0,1]$ the so-called δ-cut $C_\delta\big(\xi(\cdot)\big) := \big\{x \in I\!R\colon \xi(x) \geq \delta\big\}$ is a closed finite interval $[\underline{x}_\delta, \overline{x}_\delta]$.

Remark. Characterizing functions are generalizations of indicator functions from set theory.

For multivariate data and for the combination of fuzzy data the following concept of *fuzzy vectors* is important.

A k-dimensional fuzzy vector $\underline{x}^\star$ is defined by its vector-characterizing function $\zeta(\cdot, \cdots, \cdot)$ which is a real function of k real variables $(x_1, \ldots, x_k) = \underline{x}$ obeying the following:

(1) $0 \leq \zeta(x_1, \ldots, x_k) \leq 1 \quad \forall \quad (x_1, \ldots, x_k) \in \mathbb{R}^k$.

(2) $\exists(x_1^0, \ldots, x_k^0) \in \mathbb{R}^k$: $\zeta(x_1^0, \ldots, x_k^0) = 1$.

(3) $\forall \delta \in (0, 1]$ the so-called δ-cut
$C_\delta\left[\zeta(\cdot, \cdots, \cdot)\right] := \left\{(x_1, \ldots, x_k) : \zeta(x_1, \ldots, x_k) \geq \delta\right\}$
is a star-shaped compact subset of $\mathbb{R}^k$.

Remark. A fuzzy vector $\underline{x}^\star = (x_1, \ldots, x_k)^\star$ is different from a vector $(x_1^\star, \ldots, x_k^\star)$ of fuzzy numbers. Based on the concept of fuzzy vectors and the so-called extension principle from fuzzy set theory, the arithmetic operations for real numbers can be extended to fuzzy numbers.

In order to adapt Bayesian inference to the situation of fuzzy information the following concept of fuzzy valued functions is necessary.

A *fuzzy valued function* $g^\star(\cdot)$ of one real variable x is a function whose values $g^\star(x)$ are fuzzy numbers, i.e., $y^\star = g^\star(x)$ for all $x \in \text{Def}\left[g^\star(\cdot)\right]$.

Let $\psi_x(\cdot)$ be the characterizing function of the fuzzy number $g^\star(x)$, then the fuzzy valued function $g^\star(\cdot)$ is given by the family $\left(\psi_x(\cdot);\ x \in \text{Def}\ [g^\star(\cdot)]\right)$ of characterizing functions $\psi_x(\cdot)$ of the fuzzy values $g^\star(x)$.

Using the δ-cuts of $\psi_x(\cdot)$ so-called *δ-level curves* of $g^\star(\cdot)$ are defined in the following way: For $\delta \in (0, 1]$ let $C_\delta\left[\psi_x(\cdot)\right] = \left[\underline{g}_\delta(x), \overline{g}_\delta(x)\right]$ denote the δ-cut of $g^\star(x)$. For variable $x \in \text{Def}\left[g^\star(\cdot)\right]$ two classical real functions $\underline{g}_\delta(\cdot)$ and $\overline{g}_\delta(\cdot)$ are obtained, called lower and upper δ-level curves respectively.

Assuming all δ-level curves are integrable, the generalized integral

$$⨍_a^b g^\star(x)dx = I^\star \qquad \text{for} \quad \text{Def}\left[g^\star(\cdot)\right] = [a, b]$$

is defined as the fuzzy number $I^\star$ whose characterizing function is determined by the family $\left[\underline{I}_\delta, \overline{I}_\delta\right], \delta \in (0, 1]$ of its δ-cuts, where the boundaries $\underline{I}_\delta$ and $\overline{I}_\delta$ are given by the classical integrals

$$\underline{I}_\delta = \int_a^b \underline{g}_\delta(x)dx \quad \text{and} \quad \overline{I}_\delta = \int_a^b \overline{g}_\delta(x)dx.$$

Remark. By a classical result the characterizing function $\eta(\cdot)$ of $I^\star$ is given by its values

$$\eta(y) = \max_{\delta \in (0,1]} \delta \cdot I_{[\underline{I}_\delta, \overline{I}_\delta]}(y) \qquad \forall \quad y \in \mathbb{R},$$

where $I_{[\underline{I}_\delta, \overline{I}_\delta]}(\cdot)$ denotes the indicator function of the interval $[\underline{I}_\delta, \overline{I}_\delta]$.

For examples related to statistical data compare Viertl (1996).

7.3 Semiparametric ALT and Bayes' Theorem

Under accelerating stress levels, usually enough life time data are obtained in order to justify parametric stochastic models, i.e.,

$$X \sim F\Big(\cdot\,|\theta(S)\Big), \qquad \theta \in \Theta$$

for the life time distribution under stress S. But for extrapolation to usual stress S_u the assumption of a parametric model is usually not testable and therefore not justified. Therefore the nonparametric model from Section 7.1 is necessary. Now a semiparametric approach is useful. Compare the reference of Viertl (1987).

Semiparametric methods are important in ALT because parametric assumptions are hardly justified at stress levels near the usual stress S_u. This is because under usual stress often no observed life times are available.

At accelerating stress levels by the parametric model Bayesian inference is possible, which is explained in Viertl (1987).

Again the problem of non-precise life time data has to be addressed. In Bayesian analysis it turns out that for fuzzy life time data a generalized form of Bayes' theorem is necessary [see Viertl and Hareter (2002)]. By the iterative nature of Bayesian inference Bayes' theorem has to be generalized also to the situation of fuzzy a priori densities. This results in a generalized probability concept, called *fuzzy probability distributions* [compare Viertl and Hareter (2002)].

7.4 Fuzzy Data and Generalized Bayes' Theorem

Since many life time data are not precise values but more or less non-precise, this imprecision (which should not be confused with errors) is best described by so-called fuzzy numbers from Section 7.2. Therefore Bayes' theorem has to be generalized to the situation of fuzzy data and fuzzy a priori densities.

Let $\pi^\star(\cdot)$ be a fuzzy probability density on the parameter space $\Theta \subseteq \mathbb{R}^k$, i.e., $\pi^\star(\cdot)$ is a function whose values are fuzzy numbers (compare Section 7.2). In case of continuous parameter θ the fuzzy-valued function has to obey

$$⨍_{\Theta} \pi^\star(\theta)d\theta = 1^\star$$

where $1^\star$ is a fuzzy number fulfilling $1 \in C_1(1^\star)$.

Using the set $\mathcal{D}_\delta$ of all classical probability densities $f(\cdot)$ on Θ fulfilling

$$\underline{\pi}_\delta(\theta) \le f(\theta) \le \overline{\pi}_\delta(\theta) \qquad \forall \quad \theta \in \Theta,$$

where $\underline{\pi}_\delta(\cdot)$ and $\overline{\pi}_\delta(\cdot)$ are the δ-level functions of the fuzzy-valued function $\pi^\star(\cdot)$, the generalized probability $P^\star(A)$ of a subset $A \subseteq \Theta$ is a fuzzy number, whose δ-cuts $\left[\underline{P}_\delta(A), \overline{P}_\delta(A)\right]$ are defined by

$$\overline{P}_\delta(A) = \sup_{f \in \mathcal{D}_\delta} \int_A f(\theta) d\theta, \tag{7.1}$$

$$\underline{P}_\delta(A) = \inf_{f \in \mathcal{D}_\delta} \int_A f(\theta) d\theta. \tag{7.2}$$

In order to keep the sequential updating procedure from classical Bayesian inference, Bayes' theorem has to be adapted to the situation of fuzzy a-priori densities and fuzzy data. This is possible in the following way:

In reasonable stochastic models, for fuzzy a-priori density $\pi^\star(\cdot)$ with δ-level functions $\underline{\pi}_\delta(\cdot)$ and $\overline{\pi}_\delta(\cdot)$ as well as fuzzy data $x_1^\star, \ldots, x_n^\star$ with characterizing functions $\xi_1(\cdot), \cdots, \xi_n(\cdot)$, and combined fuzzy sample $\underline{x}^\star$ whose vector characterizing function $\zeta(\cdot, \cdots, \cdot)$ is defined by

$$\zeta(x_1, \ldots, x_n) = \min_{i=1(1)n} \xi_i(x_i) \qquad \forall \quad (x_1, \ldots, x_n) \in I\!R^n,$$

the δ-level functions $\underline{\pi}_\delta(\cdot|\underline{x}^\star)$ and $\overline{\pi}_\delta(\cdot|\underline{x}^\star)$ of the fuzzy a posteriori density $\pi^\star(\cdot|\underline{x}^\star)$ are calculated by using the so-called *fuzzy valued likelihood function* $\ell^\star(\cdot; \underline{x}^\star)$. The fuzzy values $\ell^\star(\theta; \underline{x}^\star)$ of $\ell^\star(\cdot; \underline{x}^\star)$ have characterizing functions $\eta_\theta(\cdot)$ for all $\theta \in \Theta$, given by their values

$$\eta_\theta(y) = \left\{ \begin{array}{lll} \sup\{\zeta(\underline{x}) \colon \ell(\theta; \underline{x}) = y\} & \text{if} & \ell_\theta^{-1}(\{y\}) \neq \emptyset \\ 0 & \text{if} & \ell_\theta^{-1}(\{y\}) = \emptyset \end{array} \right\} \qquad \forall \quad y \in I\!R.$$

Denoting the δ-level functions of $\ell^\star(\cdot; \underline{x}^\star)$ by $\underline{\ell}_\delta(\cdot; \underline{x}^\star)$ and $\overline{\ell}_\delta(\cdot; \underline{x}^\star)$, the δ-level functions of the fuzzy a posteriori density are defined by

$$\overline{\pi}_\delta(\theta|\underline{x}^\star) = \frac{\overline{\pi}_\delta(\theta) \cdot \overline{\ell}_\delta(\theta; \underline{x}^\star)}{\int\limits_\Theta \frac{1}{2}\left[\underline{\pi}_\delta(\theta) \cdot \underline{\ell}_\delta(\theta; \underline{x}^\star) + \overline{\pi}_\delta(\theta) \cdot \overline{\ell}_\delta(\theta; \underline{x}^\star)\right] d\theta}$$

and

$$\underline{\pi}_\delta(\theta|\underline{x}^\star) = \frac{\underline{\pi}_\delta(\theta) \cdot \underline{\ell}_\delta(\theta; \underline{x}^\star)}{\int\limits_\Theta \frac{1}{2}\left[\underline{\pi}_\delta(\theta) \cdot \underline{\ell}_\delta(\theta; \underline{x}^\star) + \overline{\pi}_\delta(\theta) \cdot \overline{\ell}_\delta(\theta; \underline{x}^\star)\right] d\theta}.$$

By this definition two major requirements are fulfilled: In case of classical a priori densities and precise data $x_1, \ldots, x_n$ the above construction yields the classical Bayes' theorem. Moreover the sequential nature of Bayesian updating is kept by the above definition in case of fuzzy information.

7.5 Conclusion

By the definition of the fuzzy probabilities $P^\star(A)$ in Section 7.4, so-called *fuzzy probability distributions* $P^\star(\cdot)$ on measurable spaces $(M, \mathcal{A})$ are defined, having the following characteristics:

(P1) $P^\star(\varnothing) = 0$ and $P^\star(M) = 1$.

(P2) $P^\star(A)$ is a fuzzy number for every $A \in \mathcal{A}$.

(P3) For two disjoint events A and B from $\mathcal{A}$ the following holds for all $\delta \in (0,1]$:

$$\begin{aligned} \overline{P}_\delta(A \cup B) &\leq \overline{P}_\delta(A) + \overline{P}_\delta(B). \\ \underline{P}_\delta(A \cup B) &\geq \underline{P}_\delta(A) + \underline{P}_\delta(B). \end{aligned}$$

Remark. The foregoing explanations show that the concept of probability spaces is not sufficient to model fuzzy information. Therefore a more general probability concept is necessary. The concept of fuzzy probability distributions provides a solution to this.

Moreover probabilities as abstraction of relative frequencies in case of fuzzy data support the requirements (P1) to (P3) for fuzzy probability distributions.

References

1. Bagdonavicius, V. and Nikulin, M. (2002). *Accelerated Life Models: Modeling and Statistical Analysis*, Chapman & Hall/CRC, Boca Raton, Florida.

2. Ross, T., Booker, J. and Parkinson, W. (Eds.) (2002). *Fuzzy Logic and Probability Applications - Bridging the Gap*, American Statistical Association and SIAM, Philadelphia, Pennsylvania.

3. Viertl, R. (1987). Bayesian inference in accelerated life testing, In *Proceedings of the 46th Session of the ISI, Invited Papers, Book 4*, Internationtal Statistical Institute, Tokyo, Japan.

4. Viertl, R. (1988). *Statistical Methods in Accelerated Life Testing*, Vandenhoeck & Ruprecht, Göttingen, Germany.

5. Viertl, R. (1996). *Statistical Methods for Non-Precise Data*, CRC Press, Boca Raton, Florida.

6. Viertl, R. and Hareter, D. (2002). Fuzzy Information and Imprecise Probability, *Research Report RIS-2002-4*, Department of Statistics and Probability Theory, Vienna University of Technology, Wien, Germany.

7. Viertl, R. and Hule, H. (1991). On Bayes' theorem for fuzzy data, *Statistical Papers*, **32**, 115–122.

8

Asymptotic Theory in Rank Estimation for AFT Model Under Fixed Censorship

Zhezhen Jin and Zhiliang Ying

Department of Biostatistics, Columbia University, New York, NY U.S.A.

Abstract: Rank based parameter estimation in the accelerated failure time model with right-censoring has been studied rigorously by many authors [Tsiatis (1990), Lai and Ying (1991) and Ying (1993)]. A key assumption in establishing the asymptotic normality for the rank estimator is that, conditional on covariates, the censoring time has a density function. This assumption may be violated in many situations. It may also be non-intrinsic in that the limiting variance formula does not involve such a density. We show in this paper that the usual asymptotic theory for rank estimation can indeed be established without this stringent and unnatural assumption.

Keywords and phrases: Accelerated failure time model, asymptotic normality, exponential inequality, fixed censoring, linear rank statistics, survival data

8.1 Introduction

The accelerated failure time (AFT) model relates the logarithm of the failure time linearly to the covariates,

$$Y_i = X_i^{\mathrm{T}}\beta + \epsilon_i, \quad i = 1, \dots, n, \tag{8.1}$$

where $Y_i = \log T_i$ with T_i being the failure times, β is a p-vector of unknown parameters and ϵ_i are independent and identically distributed (iid) error terms; cf. Cox and Oakes (1984), Kalbfleisch and Prentice (2002) and Miller (1981). It arises naturally from applications in biomedicine and many other disciplines where a subject is followed until an event of interest occurs. In such applications, the event time is often subject to a right censorship. Thus, model (8.1) with right censoring is also called a censored linear regression model.

When the common error distribution is parametrized, the maximum likelihood estimation may be used to obtain straightforward inference methods. Without such parametrization, the model is semiparametric in nature and much less amenable. For hypothesis testing, the weighted log-rank tests, which include the Gehan, the Peto–Prentice and the long-rank tests, have been widely used; cf. Mantel and Haenszel (1959), Gehan (1965), Peto and Peto (1972) and Prentice (1978). Asymptotic distributions for such test statistics can easily be derived by the martingale theory for counting processes; cf. Aalen (1975), Fleming and Harrington (1991) and Andersen et al. (1993).

For estimating the regression parameter vector β under the semiparametric AFT model, the problem becomes technically more challenging. The weighted log-rank statistics can be used as estimating functions and solutions of which are the so-called rank estimators. However, these estimating functions are discontinuous, so that the Taylor series expansion, which is typically used in establishing asymptotic normality, may not be applicable. In addition, the discontinuity makes the roots numerically difficult to compute.

One way to rigorously establish asymptotic normality for rank estimators is to show the so-called asymptotic linearity for the rank estimating functions. This is essentially the approach adopted in Tsiatis (1990), Lai and Ying (1991) and Ying (1993), where asymptotic normality for the rank estimators are proved rigorously under suitable regularity conditions. Because the expansions of weighted empirical processes of the censoring times involving β are used, results in these paper requires, among others, a regularity condition which is essentially the existence of conditional densities of the censoring times given covariates. A closer examination of this condition reveals that it is not only stringent, as in many situations the censoring times easily have point masses due to study designs, but also artificial because these conditional densities show up neither in the slope in asymptotic linearity nor in the asymptotic variance of the resulting estimator. In fact, Ritov (1990) in his study of a closely related problem, established the asymptotic expansion in the $O(n^{-1/2})$ neighborhood without this condition.

The main purpose of the present paper is to establish rigorously an asymptotic theory for the rank estimation with minimum assumptions on the censoring distribution. Like Ying (1993), there is no tail modification and expansions are for the asymptotic behavior of the estimating function in any shrinking neighborhood. It makes use of modern empirical process theory as elucidated in Pollard (1990) and van der Vaart and Wellner (1996). The paper is organized as follows. The main results are presented in the next section. Section 8.3 contains major steps in proving these results. Some extensions and remarks are given in Section 8.4.

8.2 Main Results

The main results of the present paper are given in this section. They include the crucial asymptotic linearity for the log-rank estimating functions under the right (fixed) censoring. By the fixed censorship, we mean that the linearity holds conditionally on the covariates and censoring variables. In other words, the censoring time distributions can have continuous, discrete, or even a single point mass distributions. Under an additional nonsingularity assumption, the consistency and asymptotic normality of the log-rank estimator are established.

8.2.1 Notation and assumptions

Recall that T_i is the survival time for the ith subject with $Y_i = \log T_i$ and X_i denoting its associated covariates. We will consider the AFT model as specified by (1), with T_i subject to right censoring. Let C_i denote the corresponding (log) censoring time so that observations consist of $(\tilde{Y}_i, \delta_i, X_i)$ $(i = 1, \ldots, n)$, where $\tilde{Y}_i = Y_i \wedge C_i$ and $\delta_i = 1\{Y_i \leq C_i\}$. Here, $a \wedge b = \min(a, b)$ and $1\{.\}$ is the indicator function.

Throughout the paper, we shall derive asymptotic results which are valid conditional on the X_i and C_i. Therefore, we can equivalently view the X_i and C_i as fixed. The cumulative distribution of errors, ϵ_i, will be denoted by F. In addition, f, λ and Λ denote its density, hazard rate and cumulative hazard functions, respectively. We shall use the following notation:

$$N_i(b, \tilde{b}; t) = 1\{Y_i - X_i^{\mathrm{T}} b \leq t, Y_i - X_i^{\mathrm{T}} b \leq C_i - X_i^{\mathrm{T}} \tilde{b}\},$$

$$\tilde{N}_i(b, \tilde{b}; t) = 1\{Y_i - X_i^{\mathrm{T}} b \geq t, Y_i - X_i^{\mathrm{T}} b \leq C_i - X_i^{\mathrm{T}} \tilde{b}\}$$

and

$$Z_i(b, \tilde{b}; t) = 1\{Y_i - X_i^{\mathrm{T}} b \geq t, C_i - X_i^{\mathrm{T}} \tilde{b} \geq t\}.$$

Define weighted empirical processes

$$\begin{aligned}
N(b, \tilde{b}; t) &= \sum_{i=1}^{n} N_i(b, \tilde{b}; t), \\
N^x(b, \tilde{b}; t) &= \sum_{i=1}^{n} X_i N_i(b, \tilde{b}; t), \\
\tilde{N}(b, \tilde{b}; t) &= \sum_{i=1}^{n} \tilde{N}_i(b, \tilde{b}; t), \\
\tilde{N}^x(b, \tilde{b}; t) &= \sum_{i=1}^{n} X_i N_i(b, \tilde{b}; t),
\end{aligned}$$

$$Z(b, \tilde{b}; t) = \sum_{i=1}^{n} Z_i(b, \tilde{b}; t),$$

and

$$Z^x(b, \tilde{b}; t) = \sum_{i=1}^{n} X_i Z_i(b, \tilde{b}; t).$$

With the above notation, we define

$$\xi(b, \tilde{b}) = \sum_{i=1}^{n} \int_{-\infty}^{\infty} \left(X_i - \frac{Z^x(b, \tilde{b}; t)}{Z(b, \tilde{b}; t)} \right) dN_i(b, \tilde{b}; t).$$

Note that $\xi(b, b)$ is the original log-rank estimating function and the log-rank estimator is defined as, roughly, a root to $\xi(b, b) = 0$; cf. Ying (1993). It is "roughly" because $\xi(b, b)$ as a function of b is a step function and, consequently, the term root in its strict sense may not always be meaningful.

Comparing with Ying (1993), the key new development here is the inclusion of $\tilde{b}$ in those empirical processes and thus also in the estimating function ξ. Of course only $\xi(b, b)$ not $\xi(b, \tilde{b})$ can be regarded as a legitimate estimating function since the latter involves latent (unobserved) quantities. Our idea is to expand ξ with respect to b only, while holding $\tilde{b}$ fixed. In doing so, we shall demonstrate that the density assumption on C_i can be avoided.

The following two conditions will be used.

(C1) The covariates are uniformly bounded, without loss of generality, it is assumed that $\sup_i \|X_i\| \leq 1$.

(C2) The error density f and its derivative f' are bounded and

$$\int_{-\infty}^{\infty} \left(\frac{f'(t)}{f(t)} \right)^2 f(t) dt < \infty.$$

The conditions (C1) and (C2) are identical to the first two conditions in Ying (1993). As mentioned there, (C1) may be relaxed to that $\sup_{1 \leq i \leq n} \|X_i\| = O(n^\eta)$ for every $\eta > 0$. Introduce additional notation

$$\Gamma_{n,k}(t, \tilde{b}) = \frac{1}{n} \sum_{i=1}^{n} X_i^k 1\{C_i - X_i^{\mathrm{T}} \tilde{b} \geq t\}, \qquad k = 0, 1, 2$$

where $X_i^0 = 1$, $X_i^1 = X_i$ and $X_i^2 = X_i X_i^{\mathrm{T}}$,

$$R_n(t, \tilde{b}) = \Gamma_{n,2}(t, \tilde{b}) - \frac{\Gamma_{n,1}(t, \tilde{b}) \Gamma_{n,1}^{\mathrm{T}}(t, \tilde{b})}{\Gamma_{n,0}(t, \tilde{b})},$$

$$A_n(\tilde{b}) = \int_{-\infty}^{\infty} R_n(t, \tilde{b}) \frac{\lambda'(t)}{\lambda(t)} dF(t)$$

and

$$\Sigma_n(\tilde{b}) = \int_{-\infty}^{\infty} R_n(t, \tilde{b}) dF(t).$$

For fixed $\tilde{b}$, similar to the arguments in Ying (1993), it can be shown that the matrices $A_n(\tilde{b})$, $\Sigma_n(\tilde{b})$ and $R_n(t, \tilde{b})$ have the following properties:

(P1) The $R_n(t, \tilde{b})$ is nonnegative definite, and nonincreasing in t.

(P2) The $A_n(\tilde{b})$ is nonnegative definite.

(P3) The eigenvalues of $A_n(\tilde{b})$ and $\Sigma_n(\tilde{b})$ are bounded away from 0 for all large n if the eigenvalues of $R_n(t, \tilde{b})$ are bounded away from 0 for all large n and some t in the support of f.

(P4) The eigenvalues of $A_n(\tilde{b})$ are bounded away from 0 if and only if the eigenvalues of $\Sigma_n(\tilde{b})$ are bounded away from 0.

8.2.2 Main results

We now state the main results. The first theorem establishes the asymptotic linearity of $\xi(b, \tilde{b})$ in b near β, uniformly in $\tilde{b}$, while the second gives consistency and asymptotic normality.

Theorem 8.2.1 *Under conditions (C1) and (C2), $\xi(b, \tilde{b})$ is asymptotically linear in b at β in the sense that, for every sequence $d_n > 0$ with $d_n \to 0$ a.s. and constant $B > 0$,*

$$\sup_{\|b-\beta\| \le d_n, \|\tilde{b}\| \le B} \frac{\|\xi(b, \tilde{b}) - \xi(\beta, \tilde{b}) - A_n(\tilde{b}) n(b - \beta)\|}{\sqrt{n} + n\|b - \beta\|} = o(1) \quad a.s.$$

In addition, if $d_n \to 0$ in probability, then

$$\sup_{\|b-\beta\| \le d_n, \|\tilde{b}\| \le B} \frac{\|\xi(b, \tilde{b}) - \xi(\beta, \tilde{b}) - A_n(\tilde{b}) n(b - \beta)\|}{\sqrt{n} + n\|b - \beta\|} = o_p(1).$$

Theorem 8.2.1 can be used to characterize the local properties of the original log-rank estimator which is a root of $\xi(b, b) = 0$ or, more precisely, a minimizer of $\|\xi(b, b)\|$. This is summarized in the following theorem.

Theorem 8.2.2 *Under the conditions (C1) and (C2), if all the eigenvalues of $A_n = A_n(\beta)$ or, equivalently, $\Sigma_n = \Sigma_n(\beta)$, are bounded away from 0 for all large n, then the following two conclusions hold.*

(i) There exists a closed neighborhood $\mathcal{N}$ containing β as its interior point such that $\hat{\beta}$, defined as a minimizer of $\|\xi(b, b)\|$ in $\mathcal{N}$, is strongly consistent.

(ii) For any $\mathcal{N}$ containing β as its interior point and $\hat{\beta}$ defined as in (i), if $\hat{\beta}$ is consistent, then

$$\sqrt{n}\Sigma_n^{-\frac{1}{2}} A_n(\hat{\beta} - \beta) \to_{\mathcal{D}} N(0, I_p).$$

In particular, if $\Gamma_{n,k}(t, \beta) \to \Gamma_k(t)$ exist for all t and $k = 0, 1, 2$, then

$$A_n \to A = \int_{-\infty}^{\infty} R(t) \frac{\lambda'(t)}{\lambda(t)} dF(t), \qquad \Sigma_n \to \Sigma = \int_{-\infty}^{\infty} R(t) dF(t),$$

and

$$\sqrt{n}(\hat{\beta} - \beta) \to_{\mathcal{D}} N(0, A^{-1}\Sigma A^{-1})$$

where

$$R(t) = \Gamma_2(t) - \frac{\Gamma_1(t)\Gamma_1^{\mathrm{T}}(t)}{\Gamma_0(t)}.$$

8.3 Proof of Main Results

The focus of this section is to provide major steps in the proofs for the two theorems of the preceding section. The main technique is the modern empirical process theory [Pollard (1990) and van der Vaart and Wellner (1996)].

8.3.1 Preliminaries

We begin with a key lemma, which can produce the counterparts of Lemma 1 in Ying (1993) with $\tilde{b}$ separated from b. It is essentially an application of the maximal inequalities for general empirical processes.

Lemma 8.3.1 *For each $g \in \mathcal{G}$, let $u_i(g)$, $i = 1, \ldots, n$, be independent random variables. Suppose that $n^{-1}E\sum_{i=1}^{n}[u_i(g)]^2 \le d_n^2$ for all $g \in \mathcal{G}$ and that $\{(u_i(g); i = 1, \ldots, n) : g \in \mathcal{G}\}$ is Euclidean with a constant envelope [Nolan and Pollard (1987), Pakes and Pollard (1989) and Pollard (1990)]. Then, for any non-increasing sequence of positive constants α_n satisfying $n(\log n)^{-1}d_n^2\alpha_n^2 \to \infty$, we have*

$$\sup_{g \in \mathcal{G}_n} \left|\sum_{i=1}^{n}[u_i(g) - Eu_i(g)]\right| = o(nd_n^2\alpha_n) \quad a.s. \tag{8.2}$$

PROOF. By extending the Lemma 33 of Pollard (1984, p. 31) to the independent but not identically distributed case, we can follow exactly the same steps as those in the proof of Theorem 37 in Pollard (1984, p. 34) to prove the result. Thus the details are omitted. ■

To apply Lemma 8.3.1 to our setting, we notice that the empirical processes, N, $\tilde{N}$, etc. can be easily shown to be Euclidean as they can be decomposed into sums of monotone functions in each component. Thus we have the following approximations. Note that, without loss of generality, we assume $\beta = 0$ hereafter.

Proposition 8.3.1 *Let $\{w_i\}$ be a bounded sequence of constants. Under conditions (C1) and (C2), for every $\gamma \in [0,1)$, $B > 0$, $K > 0$ and $\theta > 0$, with probability 1,*

$$\sup_{\|b\|\le B, \|\tilde{b}\|\le B, Ev(b,\tilde{b};t)\le Kn^{1-\gamma}} \left| \sum_{i=1}^{n} w_i[v_i(b,\tilde{b};t) - Ev_i(b,\tilde{b};t)] \right| = o(n^{\frac{1-\gamma}{2}+\theta}), \quad (8.3)$$

$$\sup_{\|b_1-b_2\|\le Kn^{-\gamma},\ \|\tilde{b}\|\le B} \left| \sum_{i=1}^{n} w_i[v_i(b_1,\tilde{b};t) - Ev_i(b_1,\tilde{b};t) - v_i(b_2,\tilde{b};t) + Ev_i(b_2,\tilde{b};t)] \right|$$
$$= o(n^{\frac{1-\gamma}{2}+\theta}), \quad (8.4)$$

where v_i is any one of Z_i, N_i, $\tilde{N}_i$ and $v = \sum_{i=1}^{n} v_i$.

PROOF. We prove Proposition 8.3.1 with $v_i = N_i$. The proofs for $v_i = Z_i$ and $v_i = \tilde{N}_i$ are similar and thus omitted. Without loss of generality, assume $|w_i| \le 1$. Let $g_i(b,\tilde{b},t) = w_i 1\{Y_i - X_i^{\mathrm{T}}b \le t, Y_i - X_i^{\mathrm{T}}b \le C_i - X_i^{\mathrm{T}}\tilde{b}\}$ and $\mathcal{G}_n = \{\{g_i(b,\tilde{b},t), i = 1,\cdots,n\} : \|b\| \le B, \|\tilde{b}\| \le B, t \in R^1\}$. Then it can be shown that the class of function $\mathcal{G}_n$ is Euclidean with a constant envelope [Nolan and Pollard (1987), Pakes and Pollard (1989) and Pollard (1990)] and Lemma 8.3.1 can be used.

It is easy to see that if (8.3) holds for $\theta \in (0, \frac{1-\gamma}{2})$, then so is (8.4) for $\theta \ge \frac{1-\gamma}{2}$. Therefore, without loss of generality, we assume $0 < \theta < \frac{1-\gamma}{2}$ in the rest of the proof.

Take $d_n = n^{-\frac{\gamma}{2}}, \alpha_n = n^{-\frac{1}{2}+\frac{\gamma}{2}+\theta}$. Then $\{\alpha_n\}$ is a non-increasing sequence of positive numbers and $nd_n^2\alpha_n^2 = n^{2\theta}$, which satisfies $n(\log n)^{-1}d_n^2\alpha_n^2 \to \infty$ as $n \to \infty$.

Under (C1) and (C2), if $\|b\| \le B, \|\tilde{b}\| \le B$ and $E\sum_{i=1}^{n} N_i(b,\tilde{b};t) \le Kn^{1-\gamma}$, then

$$\frac{1}{n}E\sum_{i=1}^{n}\left[w_i N_i(b,\tilde{b};t)\right]^2 = \frac{1}{n}E\sum_{i=1}^{n} w_i^2 N_i(b,\tilde{b};t) \le \frac{1}{n}Kn^{1-\gamma} = K\delta_n^2.$$

Therefore, by Lemma 8.3.1,

$$\sup_{\|b\|\le B, \|\tilde{b}\|\le B, Ev(b,\tilde{b};t)\le Kn^{1-\gamma}} \left|\sum_{i=1}^{n} w_i[v_i(b,\tilde{b};t) - Ev_i(b,\tilde{b};t)]\right| = o(n\delta_n^2\alpha_n) = o(n^{\frac{1-\gamma}{2}+\theta}).$$

Now, we claim that under (C1) and (C2),

$$E\left[g_i(b_1,\tilde{b},t) - g_i(b_2,\tilde{b},t)\right]^2 \le \delta_n^2$$

for $\|b_1 - b_2\| \le Kn^{-\gamma}$. This is obvious since the ϵ_i have a bounded density. Hence, applying Lemma 8.3.1 again, we obtain (8.4). ■

Proposition 8.3.2 *Let $t_{b,\tilde{b}}(\alpha) = \inf\{t : EZ(b,\tilde{b};t) \le n^{1-\alpha}\}$ with $0 < \alpha < 1$. Then, under conditions (C1) and (C2), for every $\theta > 0$ and $B > 0$,*

$$\sup_{\|b\|\le B, \|\tilde{b}\|\le B, t\ge t_{b,\tilde{b}}(\alpha)} \left\| \int_{t_{b,\tilde{b}}(\alpha)}^{t} \sum_{i=1}^{n} \left[X_i - \frac{Z^x(b,\tilde{b};s)}{Z(b,\tilde{b};s)}\right] dN_i(b,\tilde{b};s) - \int_{t_{b,\tilde{b}}(\alpha)}^{t} \sum_{i=1}^{n} \left[X_i - \frac{EZ^x(b,\tilde{b};s)}{EZ(b,\tilde{b};s)}\right] dEN_i(b,\tilde{b};s)\right\| = o(n^{\frac{1-\alpha}{2}+\theta})$$

holds a.s. In particular, the left-hand side is of order $o(n^{\frac{1}{2}})$.

PROOF. Using Proposition 8.3.1 here instead of Lemma 1 in Ying (1993), we can follow exactly the same steps for the proof of Lemma 2 in Ying (1993) to prove Proposition 8.3.2. Thus, we omit the details. ■

To introduce the next result, we define

$$\xi(b,\tilde{b};t) = \sum_{i=1}^{n} \int_{-\infty}^{t} \left(X_i - \frac{Z^x(b,\tilde{b};s)}{Z(b,\tilde{b};s)}\right) dN_i(b,\tilde{b};s),$$

$$m(b,\tilde{b};t) = \sum_{i=1}^{n} \int_{-\infty}^{t} \left(X_i - \frac{EZ^x(b,\tilde{b};s)}{EZ(b,\tilde{b};s)}\right) dEN_i(b,\tilde{b};s).$$

The two quantities are analogues of those in Lemma 3 of Ying (1993), but contain an additional argument $\tilde{b}$.

Proposition 8.3.3 *Under (C1) and (C2), we have the following two approximations.*

(i) For any $B > 0$ and any $\theta > 0$,

$$\sup_{\|b\|\le B, \|\tilde{b}\|\le B, t\in R^1} \|\xi(b,\tilde{b};t) - m(b,\tilde{b};t)\| = o(n^{\frac{1}{2}+\theta}) \quad a.s.$$

(ii) There exists $\theta_0 > 0$ such that

$$\sup_{\|b\|\le n^{-\frac{1}{3}}, \|\tilde{b}\|\le B, t\in R^1} \|\xi(b,\tilde{b};t) - \xi(0,\tilde{b};t) - m(b,\tilde{b};t)\| = o(n^{\frac{1}{2}-\theta_0}) \quad a.s.$$

Proof. The proof is again identical to that of Lemma 3 of Ying (1993). In fact, we can follow exactly the same steps with $\tilde{b}$ being fixed, noting that all approximations will be uniform with respect to $\tilde{b}$. Thus we omit the details. ■

The next result shows that the deterministic function $m(b,\tilde{b};t)$ can be linearly expanded with respect to b at 0, uniformly in t and $\tilde{b}$. Recall that the X_i and C_i are regarded as fixed, not random and that $\beta = 0$.

Proposition 8.3.4 *Let*

$$A_n(t,\tilde{b}) = \int_{-\infty}^{t} R_n(s,\tilde{b}) \frac{\lambda'(s)}{\lambda(s)} dF(s).$$

Under (C1) and (C2),

$$\int_{-\infty}^{t} \left(X_i - \frac{EZ^x(b,\tilde{b};s)}{EZ(b,\tilde{b};s)} \right) dEN_i(b,\tilde{b};s) = n\left[A_n(t,\tilde{b})b + \eta_n(b,\tilde{b};t) \right],$$

where $\eta_n(b,\tilde{b};t) = o(b)$ uniformly in t and $\tilde{b}$ in the sense that $\sup\{\|\eta_n(b,\tilde{b};t)\|/\|b\| : 0 < \|b\| < d_n, t \in R^1, \|\tilde{b}\| \le B\} \to 0$ *for any $d_n \to 0$ and $B > 0$.*

Proof. By definition, $dEN_i(b,\tilde{b};t) = 1\{C_i - X_i^{\mathrm{T}}\tilde{b} \ge t\} f(t + X_i^{\mathrm{T}} b)dt$. Therefore,

$$\begin{aligned} m(b,\tilde{b};t) &= \int_{-\infty}^{t} \sum_{i=1}^{n} \left(X_i - \frac{EZ^x(b,\tilde{b};s)}{EZ(b,\tilde{b};s)} \right) 1\{C_i - X_i^{\mathrm{T}}\tilde{b} \ge s\} \\ &\quad \times \bar{F}(s + X_i^{\mathrm{T}} b)\lambda(s + X_i^{\mathrm{T}} b)dt \\ &= \int_{-\infty}^{t} \sum_{i=1}^{n} \left(X_i - \frac{EZ^x(b,\tilde{b};s)}{EZ(b,\tilde{b};s)} \right) 1\{C_i - X_i^{\mathrm{T}}\tilde{b} \ge s\} \\ &\quad \times \bar{F}(s + X_i^{\mathrm{T}} b) \left[\lambda(s + X_i^{\mathrm{T}} b) - \lambda(s) \right] dt. \end{aligned}$$

Define $\tilde{t}_{1,b,\tilde{b}} = \inf\{t : EZ(b,\tilde{b};t) \le n\|b\|^{\frac{4}{3}}\}$ and $\tilde{t}_{2,b,\tilde{b}} = \inf\{t : EZ(b,\tilde{b};t) \le n\|b\|^{\frac{1}{3}}\}$. It is easy to show that

$$\int_{\tilde{t}_{1,b,\tilde{b}}}^{\infty} \sum_{i=1}^{n} \left\| X_i - \frac{EZ^x(b,\tilde{b};s)}{EZ(b,\tilde{b};s)} \right\| dEN_i(b,\tilde{b};s) \le 2n\|b\|^{\frac{4}{3}}.$$

For $t \geq \tilde{t}_{2,b,\tilde{b}}$,

$$\begin{aligned} & m(b,\tilde{b};t) - m(b,\tilde{b};\tilde{t}_{2,b,\tilde{b}}) \\ = \; & nb \int_{\tilde{t}_{2,b,\tilde{b}}}^{t} \frac{1}{n} \sum_{i=1}^{n} \left[X_i - \frac{EZ^x(b,\tilde{b};s)}{EZ(b,\tilde{b};s)} \right] X_i^{\mathrm{T}} 1\{C_i - X_i^{\mathrm{T}}\tilde{b} \geq s\} \\ & \times \frac{\bar{F}(s+X_i^{\mathrm{T}}b)}{\bar{F}(s+X_i^{\mathrm{T}}b^*)} \frac{\lambda'(s+X_i^{\mathrm{T}}b^*)}{\lambda(s+X_i^{\mathrm{T}}b^*)} dF(s+X_i^{\mathrm{T}}b^*) \end{aligned}$$

where b^* lies between 0 and b. As shown in Ying (1993, eq. 2.25),

$$\sup_{|u|\leq\|b\|, \bar{F}(s)\geq\|b\|^{\frac{4}{3}}} \left| \frac{\bar{F}(s+u)}{\bar{F}(s)} - 1 \right| \rightarrow 0 \quad as \; \|b\| \rightarrow 0.$$

Therefore,

$$\begin{aligned} & \sup_{t\in(\tilde{t}_{2,b,\tilde{b}}, \tilde{t}_{1,b,\tilde{b}})} \frac{\|m(b,\tilde{b};t) - m(b,\tilde{b};\tilde{t}_{2,b,\tilde{b}})\|}{\|b\|} \\ \leq \; & 3\sum_{i=1}^{n} \int_{\tilde{t}_{2,b,\tilde{b}}}^{\tilde{t}_{1,b,\tilde{b}}} 1\{C_i - X_i^{\mathrm{T}}\tilde{b} \geq s\} \left| \frac{\lambda'(s+X_i^{\mathrm{T}}b^*)}{\lambda(s+X_i^{\mathrm{T}}b^*)} \right| dF(s+X_i^{\mathrm{T}}b^*) \\ \leq \; & 3\sum_{i=1}^{n} \left[\int_{\tilde{t}_{2,b,\tilde{b}}}^{\tilde{t}_{1,b,\tilde{b}}} 1\{C_i - X_i^{\mathrm{T}}\tilde{b} \geq s\} dF(s+X_i^{\mathrm{T}}b^*) \right. \\ & \left. \times \int_{\tilde{t}_{2,b,\tilde{b}}}^{\tilde{t}_{1,b,\tilde{b}}} \left(\frac{\lambda'(s+X_i^{\mathrm{T}}b^*)}{\lambda(s+X_i^{\mathrm{T}}b^*)} \right)^2 dF(s+X_i^{\mathrm{T}}b^*) \right]^{\frac{1}{2}} \\ \leq \; & 3\sum_{i=1}^{n} \left[\bar{F}(\tilde{t}_{2,b,\tilde{b}} - \|b\|) \int_{-\infty}^{\infty} \left(\frac{f'(t)}{f(t)} \right)^2 dF(t) \right]^{\frac{1}{2}} \\ = \; & o(n) \end{aligned}$$

as $\|b\| \rightarrow 0$. This implies that

$$\sup_{t\geq\tilde{t}_{2,b,\tilde{b}}} \|m(b,\tilde{b};t) - m(b,\tilde{b};\tilde{t}_{2,b,\tilde{b}})\|/(n\|b\|) \rightarrow 0$$

uniformly in $\|\tilde{b}\| \leq B$ and $\|b\| \rightarrow 0$. This together with the fact that

$$\sup_{n,t\geq\tilde{t}_{2,b,\tilde{b}}} \|A_n(t,\tilde{b}) - A_n(\tilde{t}_{2,b,\tilde{b}},\tilde{b})\| \rightarrow 0$$

as $\|b\| \rightarrow 0$, implies that we only need to prove the approximation for $t \leq \tilde{t}_{2,b,\tilde{b}}$, i.e.,

$$\sup_{t\leq\tilde{t}_{2,b,\tilde{b}}, \|\tilde{b}\|\leq B, 0<\|b\|\leq d_n} \{\|\eta_n(b,\tilde{b};t)\|/\|b\|\} \rightarrow 0.$$

By the definition of $\tilde{t}_{2,b,\tilde{b}}$, when b^* lies between 0 and b, one can show that uniformly in $\{X_i\}$ and $\|\tilde{b}\| \leq B$,

$$\sup_{s\leq \tilde{t}_{2,b,\tilde{b}}} \left\{ \left\| \frac{EZ^x(b,\tilde{b};s)}{EZ(b,\tilde{b};s)} - \frac{EZ^x(0,\tilde{b};s)}{EZ(0,\tilde{b};s)} \right\| + \left| \frac{\bar{F}(s+X_i^{\mathrm{T}}b)}{\bar{F}(s+X_i^{\mathrm{T}}b^*)} - 1 \right| \right\} \to 0.$$

Thus, when $t \leq \tilde{t}_{2,b,\tilde{b}}$,

$$\begin{aligned} m(b,\tilde{b};t) &= b\int_{-\infty}^{t} \sum_{i=1}^{n} \left(X_i - \frac{EZ^x(b,\tilde{b};s)}{EZ(b,\tilde{b};s)} \right) X_i^{\mathrm{T}} 1\{C_i - X_i^{\mathrm{T}}\tilde{b} \geq s\} \\ &\quad \times \frac{\bar{F}(s+X_i^{\mathrm{T}}b)}{\bar{F}(s+X_i^{\mathrm{T}}b^*)} \frac{\lambda'(s+X_i^{\mathrm{T}}b^*)}{\lambda(s+X_i^{\mathrm{T}}b^*)} dF(s+X_i^{\mathrm{T}}b^*) \\ &= (1+o(1))b\int_{-\infty}^{t} \sum_{i=1}^{n} \left(X_i - \frac{EZ^x(0,\tilde{b};s)}{EZ(0,\tilde{b};s)} \right) X_i^{\mathrm{T}} 1\{C_i - X_i^{\mathrm{T}}\tilde{b} \geq s\} \\ &\quad \times \frac{\lambda'(s+X_i^{\mathrm{T}}b^*)}{\lambda(s+X_i^{\mathrm{T}}b^*)} dF(s+X_i^{\mathrm{T}}b^*) \\ &= n\left[A_n(t,\tilde{b})b + \eta_n(b,\tilde{b};t) \right] \end{aligned}$$

with $\sup\{\|\eta_n(b,\tilde{b};t)\|/\|b\| : t \in R^1, \|\tilde{b}\| \leq B, 0 < \|b\| < d_n\} \to 0$. This completes the proof. ∎

8.3.2 Proofs of Theorems 8.2.1 and 8.2.2

Proof of Theorem 8.2.1. Theorem 8.2.1 follows directly from Propositions 8.3.3 and 8.3.4. ∎

Proof of Theorem 8.2.2. First, because the eigenvalues of A_n and thus Σ_n are bounded away from 0, the eigenvalues of $A_n(\tilde{b})$ and thus $\Sigma_n(\tilde{b})$ must also be bounded away from 0 for $\tilde{b}$ in some small neighborhood of β. In addition, from Proposition 8.3.3(i), $\sup_{\tilde{b}} \|\xi(\beta,\tilde{b})\| = o(n^{\frac{1}{2}+\theta})$ *a.s.* for any $\theta > 0$ since $m(\beta,\tilde{b}) = 0$. Consequently, Theorem 8.2.1 implies that $\hat{\beta}$ is strongly consistent.

By taking subsequences if necessary, we can assume without loss of generality that $n^{-1}\sum_{i=1}^{n} X_i^k X_i^l 1\{C_i - X_i^T\tilde{b} \geq t\}$ converge to their respective limiting functions for $k = 0,1$ and $l = 0,1$, since they can be expressed as sums of functions which are monotone in each argument. Thus, in view of Theorem 8.2.1, to prove part (ii) it suffices to show that $n^{-1/2}\xi(\beta,\hat{\beta})$ converges in distribution to $N(0,\Sigma)$.

Following the proof of Lemma 3 of Ying (1993), we can show that

$$\begin{aligned} n^{-\frac{1}{2}}\xi(\beta,\tilde{b}) &= n^{-\frac{1}{2}} \sum_{i=1}^{n} \int_0^{\infty} \left[X_i - \frac{EZ^x(\beta,\tilde{b};t)}{EZ(\beta,\tilde{b};t)} \right] 1\{C_i - X_i^T\tilde{b} \geq t\} \\ &\quad \times [d1\{\epsilon_i \leq t\} - 1\{\epsilon_i \geq t\} d\Lambda(t)] + o_p(1), \end{aligned}$$

where $o_p(1)$ is uniformly in $\|\tilde{b}\| \leq B$. Thus by the Functional Central Limit Theorem [Pollard (1990, p. 53)], the above process, indexed by $\tilde{b}$, converges weakly. In particular, $n^{-1/2}\xi(\beta, \hat{\beta})$ converges to the same limiting distribution as that of $n^{-1/2}\xi(\beta, \beta)$ since $\hat{\beta} \to \beta$.

8.4 Remarks

In this paper, we establish the asymptotic properties of the rank estimators from the usual log-rank estimating equations for the right-censored linear model. This is achieved without imposing the restrictive assumption on the density of the censoring variables. As in Ying (1993), no modification to the tail is needed. By following the same line of developments as that in Ying (1993), the asymptotic properties can be extended to the weighted log-rank estimators.

A key difficulty in the implementation of the (weighted) log-rank estimator lies in the numerical aspect, as it is often difficult to locate a root of a function which is discontinuous. Recently, Jin et al. (2003) overcomes this difficulty by modifying the estimating equations slightly so that the resulting estimators can be obtained via linear programming. Furthermore, they also developed a resampling approach to estimate the covariance matrices of the resulting estimators. We believe that the same techniques can be applied to establish the same asymptotic properties with the minimal assumption on the censoring times as we have done here.

By properly choosing the weight function, one can obtain estimators that achieve the semiparametric efficiency bound. Adaptive data-dependent construction of such efficient estimators is provided in Lai and Ying (1991). Again, this can be achieved with little assumption on the censoring times.

The same techniques can also be used in dealing with truncated data; cf. Klein and Moeschbeger (1997), Lai and Ying (1992), Huber (2000), Gross and Lai (1996), Gross and Huber-Carol (1992), among others. However, it is much more delicate to handle the tail behavior for the truncated data than for the censored data, so that modifications to the tails may be necessary.

Acknowledgements. This research was supported in part by the National Science Foundation, the National Institutes of Health, and the New York City Council Speaker's Fund for Public Health Research.

References

1. Aalen, O. O. (1975). *Statistical inference for a family of counting processes*, Ph.D. thesis, University of California, Berkeley.

2. Andersen, P. K., Borgan, O., Gill, R. D. and Keiding, N. (1993). *Statistical Models Based on Counting Processes*, Springer-Verlag, New York.

3. Cox, D. R. and Oakes, D. (1984). *Analysis of Survival Data*, Chapman & Hall, London, England.

4. Fleming, T. R. and Harrington, D. P. (1991). *Counting Processes and Survival Analysis*, John Wiley & Sons, New York.

5. Gehan, E. A. (1965). A generalized Wilcoxon test for comparing arbitrarily single-censored samples, *Biometrika*, **52**, 203–223.

6. Gross, S. T. and Huber-Carol, C. (1992). Regression models for truncated survival data, *Scandinavian Journal of Statistics*, **19**, 193–213.

7. Gross, S. T. and Lai, T. L. (1996). Nonparametric estimation and regression analysis with left-truncated and right-censored data, *Journal of the American Statistical Association*, **91**, 1166–1180.

8. Huber, C. (2000). Censored and truncated lifetime data, In *Recent Advances in Reliability Theory* (Eds., N. Limnios and M. Nikulin), pp. 291–305, Birkhäuser, Boston, Massachusetts.

9. Jin, Z., Lin, D.Y., Wei, L. J. and Ying, Z. (2003). Rank-based inference for the accelerated failure time models, *Biometrika*, **90**, 341–353.

10. Kalbfleisch, J. D. and Prentice, R. L. (2002). *The Statistical Analysis of Failure Time Data*, Second edition, John Wiley & Sons, New York.

11. Klein, J.P. and Moeschberger, M.L. (1997). *Survival Analysis. Techniques for censored and truncated data*, Springer-Verlag, New York.

12. Lai, T. L. and Ying, Z. (1991). Rank regression methods for left-truncated and right-censored data, *Annals of Statistics*, **19**, 531–556.

13. Lai, T. L. and Ying, Z. (1992). Linear rank statistics in regression analysis with censored or truncated data, *Journal of Multivariate Analaysis*, **40**, 13–45.

14. Mantel, N. and Haenszel, W. (1959). Statistical aspects of the analysis of data from retrospective studies of disease, *Journal of National Cancer Institute*, **22**, 719–748.

15. Miller, R. G. (1981). *Survival Analysis*, John Wiley & Sons, New York.

16. Nolan, D. and Pollard, D. (1987). U-processes: rates of convergence, *Annals of Statistics*, **15**, 780–799.

17. Pakes, A. and Pollard, D. (1989). Simulation and the asymptotics of optimization estimators, *Econometrica*, **57**, 1027–2057.

18. Peto, R. and Peto, J. (1972). Asymptotically efficient rank invariant test procedures (with discussion), *Journal of the Royal Statistical Society, Series A*, **135**, 185–206.

19. Pollard, D. (1990). *Empirical Processes: Theory and Application*, NSF-CBMS regional conference series in Probability and Statistics, Vol 2, Institute of Mathematical Statistics and American Statistical Association, Hayward, California.

20. Prentice, R. L. (1978). Linear rank tests with right censored data, *Biometrika*, **65**, 167–79.

21. Ritov, Y. (1990). Estimation in a linear regression model with censored data, *Annals of Statistics*, **18**, 303–28.

22. Tsiatis, A. A. (1990). Estimating regression parameters using linear rank tests for censored data, *Annals of Statistics*, **18**, 354–72.

23. van der Vaart, A. W. and Wellner, J. A. (1996). *Weak Convergence and Empirical Processes*, Springer-Verlag, New York.

24. Ying, Z. (1993). A large sample study of rank estimation for censored regression data, *Annals of Statistics*, **21**, 76–99.

9

An Example of Optimal Design in Accelerated Experiments

Léo Gerville-Reache

Université Victor Segalen, Bordeaux, France

Abstract: The AFT (Accelerated Failure Time) model is considered in order to study the opportunity of optimal design in accelerated experiment. Semiparametric estimation is proposed for a step stress experiment. Monte Carlo simulations are used to evaluate a strategy of optimal design for finite samples.

Keywords and phrases: AFT model, semiparametric estimators, Monte Carlo simulations, optimal design

9.1 Introduction

Accelerated experiments are useful as soon as the duration of an experiment is too short to observe enough failures for a good inference. A lot of statisticians deal with this problem [see for example Meeker and Escobar (1998), Nelson (1990), Singpurwalla (1973) or Bagdonavičius and Nikulin (2002)] from thirty years ago. As for all experiments, the question of optimal design is posed. We propose in this paper to present some ideas and results of simulations for the organization of a good accelerated experiment. The model used is the AFT model (Accelerated Failure Time model).

Suppose that $S_{x(\cdot)}$ is the reliability function of the time-to-failure $T_{x(\cdot)}$ under the time varying stress $x : [0, \infty[\to B \subset \mathbb{R}^m$.

Consider the classical accelerated failure time (AFT) model on $\mathcal{E}$ [see Bagdonavicius et al. (2000)]:

$$S_{x(\cdot)}(t) = S_{x_0}\{\int_0^t r[x(t)]d\tau\}, \qquad x(.) \in \mathcal{E},$$

where $\mathcal{E}$ is a set of all possible stresses, S_{x_0} is the time to failure under the usual constant in time stress x_0, r some positive function $r : B \to [0, \infty[$.

If $x(t) \equiv x = const$, then $S_x(t) = S_{x_0}\{r(x)t\}$.

If the classical parametric and nonparametric estimation procedures [see, Nelson (1990) and Viertl (1988)] the function r is parameterized in the following way:

$$r(x(t)) = e^{\boldsymbol{\beta}^T \boldsymbol{z}(\tau)}$$

where $\boldsymbol{\beta} = (\beta_0, ..., \beta_m)^T$ is the vector of unknown parameters, $\boldsymbol{z}(t) = (z_0(t), ..., z_m(t))^T$ is the vector of some known functions of stress, $r_0(x(t)) \equiv 1$.

We consider the case when the function r is completely unknown, and the experiment is as follows:

Suppose that two groups of items are tested:

the first group of n_1 items is tested under the constant in time accelerated stress x_1 and the complete sample $T_{11} \leq \cdots \leq T_{1n_1}$ is obtained;

the second group is tested under the stress

$$x_2(\tau) = \begin{cases} x_1, & \text{if} \quad 0 \leq \tau \leq t_1, \\ x_0, & \text{if} \quad t_1 < \tau \leq t \end{cases}$$

and the type I censored sample $T_{21} \leq \cdots \leq T_{2m_2}$ $(m_2 \leq n_2)$ is obtained.

Such an experiment is useful when the variation coefficient of time-to-failure under the usual stress x_0 is not large and most failures occur in some interval $[t_1, t_2]$, where t_1 is larger than the time t given for the experiment.

The items of the second group use much of their resources under the accelerated stress x_1 and after the moment t_1, even under the usual stress x_0, failures of items can be obtained.

9.2 Semiparametric Estimation

Suppose that $S_{\boldsymbol{x}_0}(t) = S_0\left((t/\theta)^\alpha\right)$, where S_0 is some specified function, for example, if $S_0(t) = e^{-t}, (1+t)^{-1}, 1 - \Phi(\ln(t))$, we obtain the class of Weibull, loglogistic and lognormal distribution. So the model (1) can be written

$$S_{\boldsymbol{x}(\cdot)}(t) = S_0\left(\left(\int_0^t r[\boldsymbol{x}(t)]d\tau/\theta\right)^\alpha\right).$$

Under the model of the first section we can write

$$\begin{aligned} S_{\boldsymbol{x}_1}(t) &= S_0\left(\left(\frac{rt}{\theta}\right)^\alpha\right), \\ S_{\boldsymbol{x}_2}(t) &= S_0\left(\left((r(t_1 \wedge t) + (t - t_1) \vee 0)/\theta\right)^\alpha\right), \end{aligned}$$

where $r = r(\boldsymbol{x}_1)$.

Denote $\rho = \ln r$, $\psi = \ln \theta$, $S(t) = S_0(e^t)$, $f(t) = -S'(t)$, $\lambda(t) = f(t)/S(t)$. Then

$$
\begin{aligned}
S_{\boldsymbol{x}_1}(t) &= S(\alpha(\ln t + \rho - \psi));\\
S_{\boldsymbol{x}_2}(t) &= \begin{cases} S(\alpha(\ln t + \rho - \psi)), & t \le t_1,\\ S(\alpha(\ln(e^{\rho}t_1 + t - t_1) - \psi)), & t > t_1; \end{cases}\\
f_{\boldsymbol{x}_1}(t) &= f(\alpha(\ln t + \rho - \psi))\frac{\alpha}{t};\\
f_{\boldsymbol{x}_2}(t) &= \begin{cases} f(\alpha(\ln t + \rho - \psi))\frac{\alpha}{t}, & t \le t_1,\\ f(\alpha(\ln(e^{\rho}t_1 + t - t_1) - \psi))\frac{\alpha}{e^{\rho}t_1 + t - t_1}, & t > t_1. \end{cases}
\end{aligned}
$$

Denote r_2 the random number of failures of items of the second group before time t_1. The likelihood function can be written as follows:

$$
\begin{aligned}
L &= \prod_{j=1}^{n_1} f(\alpha(\ln T_{1j} + \rho - \psi))\frac{\alpha}{T_{1j}} \prod_{j=1}^{r_2} f(\alpha(\ln T_{2j} + \rho - \psi))\frac{\alpha}{T_{2j}}\\
&\times \prod_{j=r_2+1}^{m_2} f(\alpha(\ln(e^{\rho}t_1 + T_{2j} - t_1) - \psi))\frac{\alpha}{e^{\rho}t_1 + T_{2j} - t_1}\\
&\times S^{n_2 - m_2}(\alpha(\ln(e^{\rho}t_1 + t_2 - t_1) - \psi)),
\end{aligned}
$$

so

$$
\begin{aligned}
U_1(\alpha,\rho,\psi) &= \frac{\partial \ln L}{\partial \alpha} = \sum_{j=1}^{n_1}(\ln f)'(c(T_{1j}))\frac{c(T_{1j})}{\alpha}\frac{n_1 + m_2}{\alpha}\\
&\quad + \sum_{j=1}^{r_2}(\ln f)'(c(T_{2j}))\frac{c(T_{2j})}{\alpha} + \sum_{j=r_2+1}^{m_2}(\ln f)'(d(T_{2j}))\frac{d(T_{2j})}{\alpha}\\
&\quad - (n_2 - m_2)\lambda(d(t_2))\frac{d(t_2)}{\alpha},
\end{aligned}
$$

$$
\begin{aligned}
U_2(\alpha,\rho,\psi) &= \frac{\partial \ln L}{\partial \rho} = \sum_{j=1}^{n_1}(\ln f)'(c(T_{1j}))\alpha + \sum_{j=1}^{r_2}(\ln f)'(c(T_{2j}))\alpha\\
&\quad + \sum_{j=r_2+1}^{m_2}(\ln f)'(d(T_{2j}))\frac{\alpha e^{\rho}t_1}{e^{\rho}t_1 + T_{2j} - t_1} - \sum_{j=r_2+1}^{m_2}\frac{e^{\rho}t_1}{e^{\rho}t_1 + T_{2j} - t_1}\\
&\quad - (n_2 - m_2)\lambda(d(t_2))\frac{\alpha e^{\rho}t_1}{e^{\rho}t_1 + t_2 - t_1},
\end{aligned}
$$

$$
\begin{aligned}
U_3(\alpha,\rho,\psi) &= \frac{\partial \ln L}{\partial \psi} = -\alpha\left[\sum_{j=1}^{n_1}(\ln f)'(c(T_{1j})) + \sum_{j=1}^{r_2}(\ln f)'(c(T_{2j}))\right.\\
&\quad \left. + \sum_{j=r_2+1}^{m_2}(\ln f)'(d(T_{2j})) - (n_2 - m_2)\lambda(d(t_2))\right],
\end{aligned}
$$

where $c(u) = \alpha(\ln u + \rho - \psi)$, $d(u) = \alpha(\ln(e^{\rho}t_1 + u - t_1) - \psi)$. In the case of Weibull, logistic and lognormal law,

$$(\ln f)'(t) = e^t; \quad \frac{1-e^t}{1+e^t}; \quad -t,$$

respectively, and

$$\lambda(t) = e^t; \quad (1+e^{-t})^{-1}; \quad \frac{\varphi(t)}{1-\Phi(t)},$$

respectively.

If maximum likelihood estimators $\hat{\alpha}$, $\hat{\rho}$, $\hat{\psi}$ are obtained, the survival function estimator $S_{\boldsymbol{x}_0}$ and the p-quantile estimator $t_p(\boldsymbol{x}_0)$ are

$$\hat{S}_{\boldsymbol{x}_0}(t) = S(\hat{\alpha}(\ln t - \hat{\psi})), \quad \hat{t}_p = \exp\left\{\hat{\psi} + \frac{1}{\hat{\alpha}}S^{-1}(1-p)\right\}.$$

In the case of the Weibull, loglogistic and lognormal law,

$$S^{-1}(p) = \ln(-\ln(1-p)); \quad -\ln\left(\frac{1}{p} - 1\right); \quad \Phi^{-1}(1-p).$$

Denote $\mathbf{I}(\alpha, \rho, \psi) = (I_{ij}(\alpha, \rho, \psi)$ a (3×4) symmetric matrix of the following elements:

$$\begin{aligned} I_{11} &= -\frac{\partial^2 \ln L}{\partial \alpha^2} = -\frac{1}{\alpha^2}\left\{\sum_{j=1}^{n_1}(\ln f)''(c(T_{1j}))[c(T_{1j})]^2 - n_1 - m_2\right. \\ &\quad + \sum_{j=1}^{r_2}(\ln f)''(c(T_{2j}))[c(T_{2j})]^2 + \sum_{j=r_2+1}^{m_2}(\ln f)''(d(T_{2j}))[d(T_{2j})]^2 \\ &\quad \left. - (n_2 - m_2)\lambda'(d(t_2))[d(t_2)]^2\right\}, \end{aligned}$$

$$\begin{aligned} I_{12} &= I_{21} = -\frac{\partial^2 \ln L}{\partial \alpha \partial \rho} = -\sum_{j=1}^{n_1}(\ln f)''(c(T_{1j}))c(T_{1j}) \\ &\quad - \sum_{j=1}^{r_2}(\ln f)''(c(T_{2j})) - \sum_{j=r_2+1}^{m_2}(\ln f)''(d(T_{2j}))d(T_{2j}) \\ &\quad + (n_2 - m_2)\lambda'(d(t_2))\frac{e^{\rho}t_1}{e^{\rho}t_1 + t_2 - t_1} \\ &\quad - \frac{1}{\alpha}U_2(\alpha, \rho, \psi) - \frac{1}{\alpha}\sum_{j=r_2+1}^{m_2}\frac{e^{\rho}t_1}{e^{\rho}t_1 + T_{2j} - t_1}, \end{aligned}$$

$$\begin{aligned} I_{13} &= I_{31} = -\frac{\partial^2 \ln L}{\partial \alpha \partial \psi} = \sum_{j=1}^{n_1}(\ln f)''(c(T_{1j}))c(T_{1j}) + \sum_{j=1}^{r_2}(\ln f)''(c(T_{2j}))c(T_{2j}) \\ &\quad + \sum_{j=r_2+1}^{m_2}(\ln f)''(d(T_{2j}))d(T_{2j}) - (n_2 - m_2)\lambda'(d(t_2))d(t_2) - \frac{1}{\alpha}U_3(\alpha, \rho, \psi), \end{aligned}$$

$$
\begin{aligned}
I_{22} &= -\frac{\partial^2 \ln L}{\partial \rho^2} = -\alpha^2 \sum_{j=1}^{n_1} (\ln f)''(c(T_{1j})) \\
&\quad -\alpha^2 \sum_{j=1}^{r_2} (\ln f)''(c(T_{2j})) - \alpha^2 \sum_{j=r_2+1}^{m_2} (\ln f)''(d(T_{2j})) \left(\frac{e^{\rho} t_1}{e^{\rho} t_1 + T_{2j} - t_1} \right)^2 \\
&\quad - \sum_{j=r_2+1}^{m_2} [\alpha (\ln f)'(d(T_{2j})) - 1] \frac{e^{\rho} t_1 (T_{2j} - t_1)}{(e^{\rho} t_1 + T_{2j} - t_1)^2} \\
&\quad + (n_2 - m_2) \lambda'(d(t_2)) \left(\frac{\alpha e^{\rho} t_1}{e^{\rho} t_1 + t_2 - t_1} \right)^2 \\
&\quad + (n_2 - m_2) \lambda(d(t_2)) \frac{\alpha e^{\rho} t_1 (t_2 - t_1)}{(e^{\rho} t_1 + t_2 - t_1)^2},
\end{aligned}
$$

$$
\begin{aligned}
I_{23} &= I_{32} = -\frac{\partial^2 \ln L}{\partial \rho \partial \psi} = \alpha^2 \left\{ \sum_{j=1}^{n_1} (\ln f)''(c(T_{1j})) + \sum_{j=1}^{r_2} (\ln f)''(c(T_{2j})) \right. \\
&\quad + \sum_{j=r_2+1}^{m_2} (\ln f)''(d(T_{2j})) \frac{e^{\rho} t_1}{e^{\rho} t_1 + T_{2j} - t_1} \\
&\quad \left. - (n_2 - m_2) \lambda'(d(t_2)) \frac{e^{\rho} t_1}{(e^{\rho} t_1 + t_2 - t_1)} \right\},
\end{aligned}
$$

$$
\begin{aligned}
I_{33} &= -\frac{\partial^2 \ln L}{\partial \psi^2} = -\alpha^2 \left[\sum_{j=1}^{n_1} (\ln f)''(c(T_{1j})) + \sum_{j=1}^{r_2} (\ln f)''(c(T_{2j})) \right. \\
&\quad \left. + \sum_{j=r_2+1}^{m_2} (\ln f)''(d(T_{2j})) - (n_2 - m_2) \lambda'(d(t_2)) \right].
\end{aligned}
$$

In the case of Weibull, loglogistic and lognormal law,

$$
(\ln f)''(t) = e^t; \ \frac{-2e^t}{(1+e^t)^2}; \ -1,
$$

respectively, and

$$
\lambda'(t) = e^t; \ \frac{e^t}{(1+e^t)^2}; \ -t\frac{\varphi(t)}{1-\Phi(t)} + \left(\frac{\varphi(t)}{1-\Phi(t)} \right)^2,
$$

respectively.

If n is great, the law of $(\hat{\alpha}, \hat{\rho}, \hat{\psi})$ can be approximated by a normal law with mean (α, ρ, ψ) and the covariance matrix estimated by

$$
\mathbf{I}^{-1} \left(\hat{\alpha}, \hat{\rho}, \hat{\psi} \right) = \left(I^{ls} (\hat{\alpha}, \hat{\rho}, \hat{\psi}) \right)_{3\times 3}.
$$

9.3 Numerical Simulations

We suppose that the time for experiment t_2 and the total number of units n are fixed. The problem of optimal design is to choose t_1, n_1 and x_1 such that estimations of parameters α and θ are "good" as possible. (r is not of interest). It is clear that optimal values of t_1, n_1 and x_1 depend on baseline function and on r. The asymptotic properties could be useful to find those "optimal values". Those estimators are asymptotically unbiased. Then, values of t_1, n_1 and x_1 which minimize the variances of those estimators could be found.

In our approach, we prefer using numerical simulations because in applications, the number of units is not large enough to neglect bias. For our simulation, we study the Weibull case:

$$S_0(t) = \exp\left(-\left(\frac{t}{\theta}\right)^{\alpha}\right),$$

with $\alpha = 5$, $\theta = 2000$, and we fix $n = 200$ and $t_2 = 500$.

Figure 9.1 presents the result of one simulation of the two groups of units (here, $n_1 = n_2 = 100$, $t_1 = 80$ and $r = 20$). We can see in the first group that a complete sample is obtained, for the second group, we have censored data.

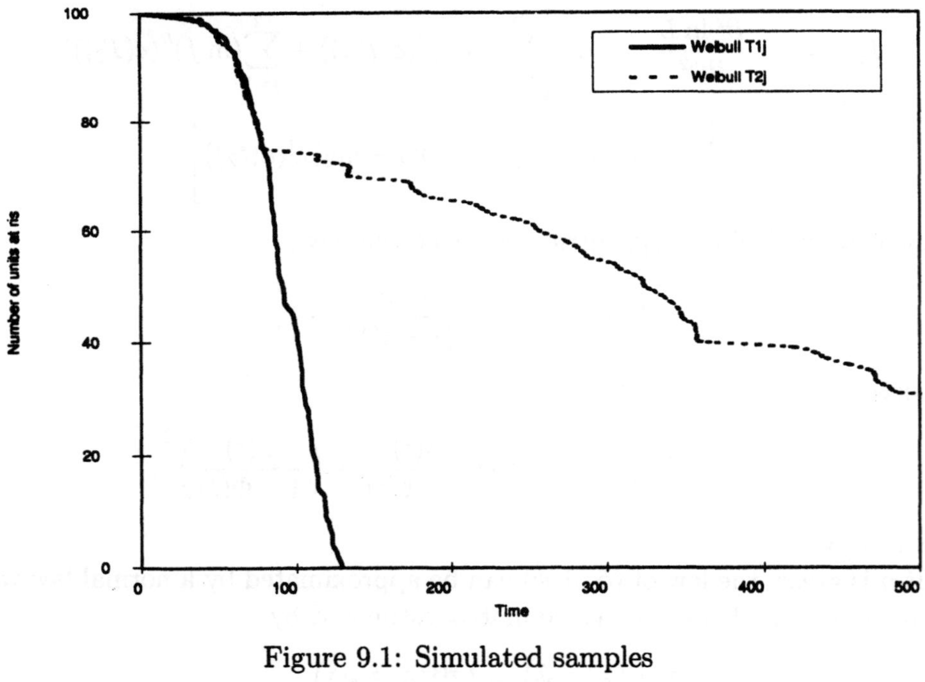

Figure 9.1: Simulated samples

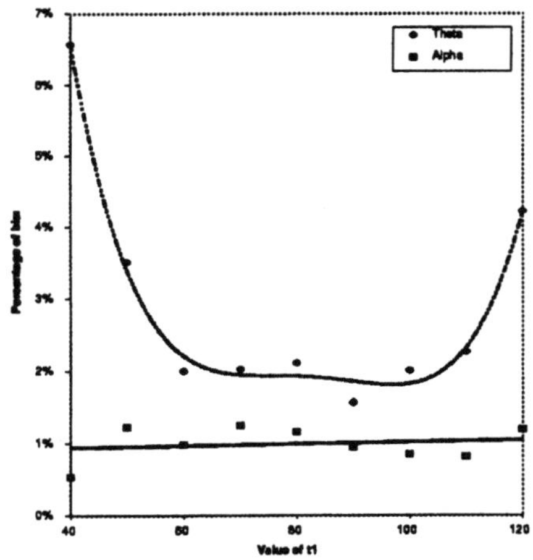

Figure 9.2: Bias depending on t_1

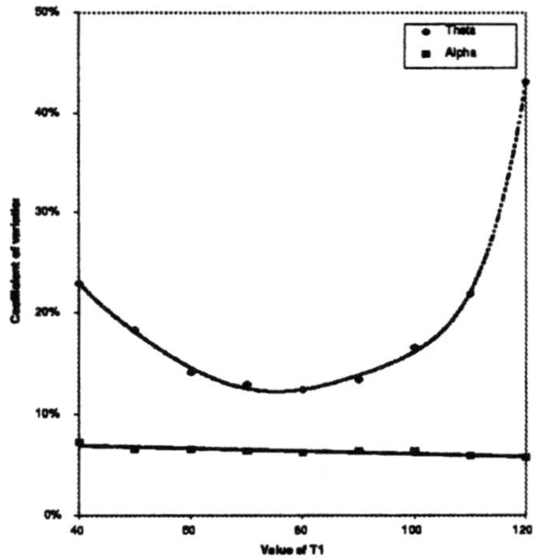

Figure 9.3: CV depending on t_1

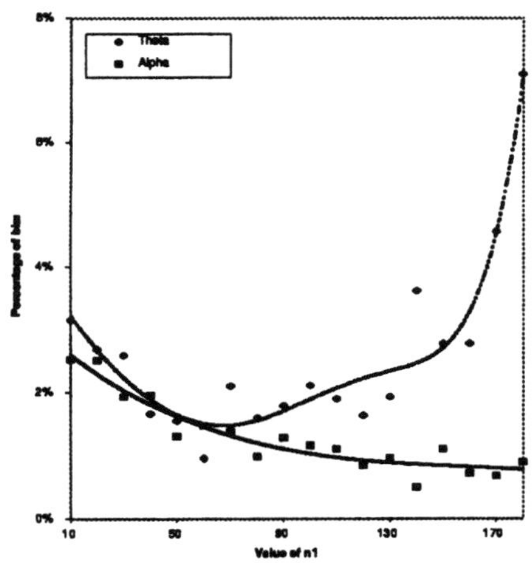

Figure 9.4: Bias depending on n_1

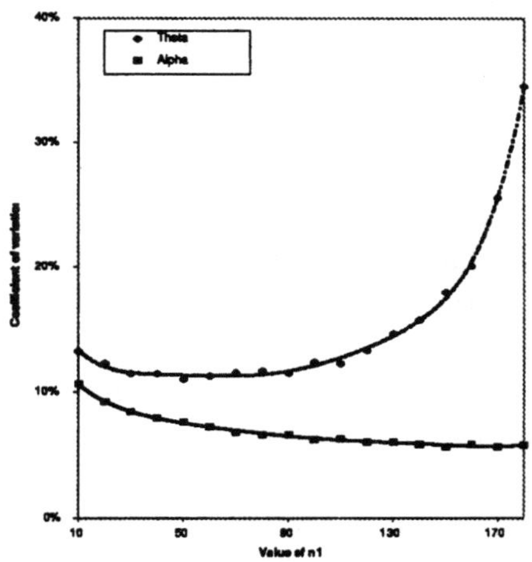

Figure 9.5: CV depending on n_1

By generation of 1000 samples for different values of t_1 (here, $n_1 = n_2$), we can compute the percentage of bias and the coefficient of variation. Results are presented in Figures 9.2 and 9.3. We can notice that the variations are important for θ with a minimum around $t_1 = 80$. For α, results show nonreal tendency.

In the same way, by generation of 1000 samples for different values of n_1 (here, $t_1 = 80$), we obtain Figures 9.4 and 9.5. We can notice that the variations are important for θ with a minimum around $n_1 = 50$. For α, the minimum is for a great value of n_2. At this stage of the study we can suppose that $n_1 \approx n_2$ is a reasonable choice.

Those first results show us that optimal design is possible. The problem is then to propose a procedure to help us to find values of t_1, n_1 and x_1 in

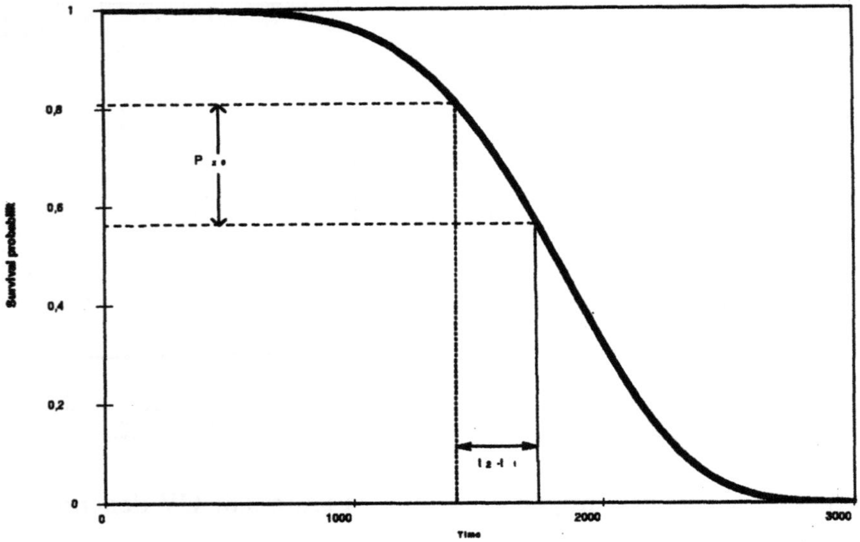

Figure 9.6: Idea for optimal design

real cases. The idea is resumed in Figure 9.6. It is well known that censored data have to be taken into account to obtain reasonable estimations. It is also well known that uncensored data give more information than censored data. The idea is then to maximize the number of failures in the second group. The problem is that for a good estimation of θ, the sizes of the groups have to be in equilibrium. Therefore we can only deal with the probability of failure in the second group. It is clear that to have a good estimation of θ, it is the number of failures under the stress x_0 which are important. Then our goal is to maximize the probability of failure under the stress x_0 (denote p_{x0}). We can notice that p_{x0} depends on t_1, x_1 and S_0.

First of all, we can calculate the function $p_{x0}(r)$ as follows:

$$p_{x0}(r) = \max_{t_1} \left[S_0(rt_1) - S_0(rt_1 + t_2 - t_1) \right].$$

Figure 9.7 represents this function. It shows us an increasing function of r which means that you must use a stress x_1 as great as possible. This propriety is due to the fact that a great value of $p_{x0}(r)$ corresponds to a small value of t_1. It is clear that if t_1 is small, the time of experiment under the stress x_0 is greater and then the probability of failure increases. Of course, the nature of the experiment, the cost of accelerated stress and the precision of failure time generate an unrealistic set for stresses and then, an unrealistic set for r. For our simulation we suppose that $r \leq 20$. Then we take $r_{opt} = 20$ for our experiment. As r is found, we can plot (see Figure 9.8) the following function:

$$p_{x0}(t_1) = S_0(r_{opt}t_1) - S_0(r_{opt}t_1 + t_2 - t_1).$$

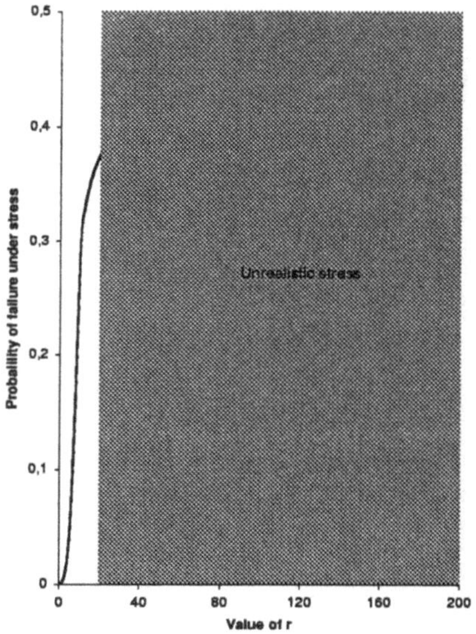

Figure 9.7: Optimal value of stress x_1

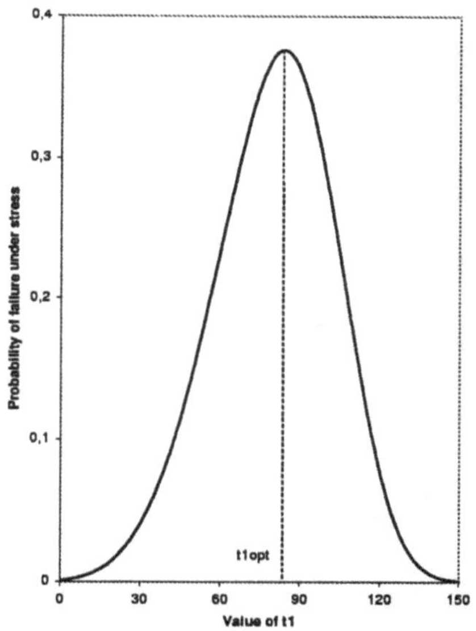

Figure 9.8: Optimal value of t_1

The really interesting result that we have to notice in Figure 9.8 is that the optimal value of t_1 is around 80. It means that good value of t_1 obtained by simulation coincides with the value of t_1 obtained by maximizing the probability of failure under the stress x_0.

9.4 Conclusion

In this example, we show that the problem of optimal design in an accelerated experiment is not simple. In classical models where optimal designs are used (ANOVA, Linear regression), estimators are unbiased and normal and optimal designs do not depend on parameters themselves. Simulations presented here show that good experimental designs exist. Our problem is to find it in real case. In our general planning of the experiment, we show that the idea, consisting of maximizing the probability of failure under the standard stress, is good. Our recommendations could be the following:

- Choose $n_1 \approx n_2$, it is reasonable.
- Choose x_0 as great as "possible".
- Choose t_1 such that the probability of failure under the standard stress is maximum.

Maybe chose t_1 after the result of the first group and do inference of the second group conditionally to the random value t_1.

References

1. Andersen, P. K., Borgan, D., Gill, R. D. and Keiding, N. (1993). *Stochastic Models Based on Counting Processes*, Springer-Verlag, New York.

2. Bagdonavicius, V., Gerville-Réache, L., Nikoulina, V. and Nikulin M. S. (2000). Analyse statistique du modèle standard de vie accélérée, *Rev. Statistiques Appliquées*, **XLVIII**, 5–38.

3. Bagdonavičius, V., Gerville-Réache, L. and Nikulin M. S. (2002). On parametric inference for step-stress models, *IEEE Transactions on Reliability*, **51**, 27–31.

4. Bagdonavičius, V. and Nikulin, M. (2001). *Accelerated Life Models: Modeling and Statistical Analysis*, Chapman & Hall/CRC, Boca Raton, Florida.

5. Ceci, C. and Mazliak, L. (2000). Optimal design in nonparametric lifetime testing, *Preprint du Laboratoire de Probabilité et Modèles Aléatoire*, Université Paris VI.

6. Cox, D. R. and Oakes, D. (1984). *Analysis of Survival Data*, Chapman & Hall, London, England.

7. Gerville-Réache, L. and Nikoulina, V. (1998). Analysis of reliability characteritics estimators in accelerated life testing, In *Statistical and Probabilistic Models in Reliability* (Eds., D. Ionescu and N. Limnios), pp. 91–99, Birkhäuser, Boston, Massachusetts.

8. Escobar, L. A. and Meeker, W. Q. (1995). Planning accelerated life tests with two or more experimental factors, *Technometrics*, **37**, 411–427.

9. Meeker, W. Q. and Escobar, L. A. (1998). Pitfalls of accelerated testing, *IEEE Transactions on Reliability*, **47**, 114–118.

10. Nelson, W. (1990). *Accelerated Testing: Statistical Models, Test Plan and Data Analysis*, John Wiley & Sons, New York.

11. Shaked, M. and Singpurwalla, N. D.(1983). Inference for step-stress accelerated life tests, *Journal of Statistical Planning and Inference*, **7**, 295–306.

12. Singpurwalla, N. D. (1973). Inference from accelerated life tests using Arrhenius type re-parametrisations, *Technometrics*, **15**, 289–299.

13. Viertl, R. (1988). *Statistical Methods in Accelerated Life Testing*, Vandenhoeck & Ruprecht, Göttingen, Germany.

PART IV
AGING PROPERTIES AND ANALYSES

10

Aspects of Multivariate Aging in Exchangeable Frailty Models

Fabio Spizzichino

Department of Mathematics, University "La Sapienza", Rome, Italy

Abstract: We consider some notions of multivariate aging, suitable to analyze vectors of exchangeable lifetimes; such notions are used to describe some basic aspects of a specific heterogeneity model, characterized by exchangeable frailties

Keywords and phrases: Mixture distributions, dependence, occupancy numbers, stochastic orderings

10.1 Introduction

The impact of heterogeneity on observed aging properties of a population is one of the classical issues of survival analysis. However most of the relevant literature has been focused on properties of univariate aging.

Here we aim, on the contrary, to present a brief discussion concerning the effects of heterogeneity on some multivariate aging properties of n dependent individuals.

Since in the frame of survival analysis the term "multivariate" appears under a number of different meanings, it is worthwhile specifying at once that we consider here the case where a single time-to-event is observable for each individual in the population; the term "multivariate" is then used only in order to stress that individuals can be dependent.

From a technical point of view, the main object of our interest is the joint survival function of the individuals' lifetimes $T_1, T_2, ..., T_n$:

$$\overline{F}^{(n)}(t_1, ..., t_n) \equiv P\{T_1 > t_1, ..., T_n > t_n\}.$$

In the by now classical literature on survival analysis, it has been pointed out several times, e.g., in Aalen (1988) and in Hougaard (1991), that heterogeneity due to the effect of observable covariates is to be eliminated whenever the interest is focused on the analysis of frailty aspects, i.e., of unexplained residual heterogeneity among individuals.

For this reason we consider individuals that are apparently similar; from a probabilistic point of view this may be described by the condition that $T_1, T_2, ..., T_n$ are exchangeable, i.e., it is

$$\overline{F}(t_1, ..., t_n) = \overline{F}(t_{\rho_1}, ..., t_{\rho_n})$$

where $\rho \equiv (\rho_1, ..., \rho_n)$ is any permutation of $\{1, 2, ..., n\}$.

Such condition turns out to be significantly simplifying in the analysis of the basic aspects of aging and dependence notions and of the relations existing among them.

In the present study we consider a suitable model of exchangeability, based on exchangeable frailties, that is specifically useful to analyze relations between frailty and multivariate aging; such a model will be briefly described in the next Section 10.2.

What do we mean by "multivariate aging"? This will be explained in Section 10.3; the latter will be devoted in fact to describing the type of multivariate aging that we want to consider and to point out some related aspects of main interest here.

For the special frailty models described in Section 10.2, we present in Section 10.4 a discussion concerning the role of the considered notions of multivariate aging in the analysis of the aging-distortion phenomena, that can be created by heterogeneity.

A few hints for further research in the field of exchangeable frailty models and multivariate aging are disseminated throughout the paper.

10.2 Exchangeable Frailty Models

In this section we describe the main features of a model of exchangeability, generated by exchangeable frailties; some basic aspects of this model were studied in Gerardi, Spizzichino, and Torti (2000).

By $I_1, ..., I_n$ we denote n different individuals whose lifetimes are $T_1, ..., T_n$ and consider the case where each I_i belongs to one of r *categories* $C_1, ..., C_r$. We think of the situations when categories are not observable; then we define the individuals' frailties $Z_1, ..., Z_n$ as, say, $\{v_1, ..., v_r\}$-valued random variables such that, for $1 \leq i \leq n$,

$$Z_i = v_k \Longleftrightarrow I_i \in C_k.$$

We assume that, given $Z_1, ..., Z_n$, $T_1, ..., T_n$ are conditionally independent and more precisely that a family of univariate survival functions $\{\overline{G}(\cdot|z), z = v_1, ..., v_r\}$ exists such that

$$\begin{aligned} P\{T_i > t | Z_1 &= z_1, ..., Z_n = z_n, T_1 = t_1, ..., T_{i-1} = t_{i-1}, T_{i+1} = t_{i+1}, T_n = t_n\} \\ &= \overline{G}(t|z_i). \end{aligned} \tag{10.1}$$

We want to give $Z_1, ..., Z_n$ the meaning of frailties, then we assume that categories are labelled in way such that $\overline{G}(\cdot|z)$ is decreasing in z, in some stochastic sense; one also usually assumes that $\mathbb{E}(Z_i) = 1$.

Since we assumed the existence of a finite number, r, of categories, for notational convenience we actually define $\{1, 2, ..., r\}$-valued variables, obtained by transforming the frailties as follows: we let, for $i = 1, 2, ..., n, k = 1, 2, ..., r$,

$$W_i = k \Longleftrightarrow Z_i = v_k \Longleftrightarrow I_i \in C_k.$$

We also put

$$\overline{G}_k(\cdot) \equiv \overline{G}(\cdot|v_k).$$

In conclusion we have for $T_1, ..., T_n$ a mixture model of the type

$$\overline{F}^{(n)}(t_1, ..., t_n) = \sum_{\mathbf{k}} \pi(\mathbf{k}) \left[\overline{G}_{k_1}(t_1) \cdot \cdots \cdot \overline{G}_{k_n}(t_n)\right], \tag{10.2}$$

where $\pi(\mathbf{k})$ denotes the joint discrete density of $(W_1, ..., W_n)$.

Furthermore we assume that $Z_1, ..., Z_n$, and then equivalently $W_1, ..., W_n$, are exchangeable, i.e.,

$$\pi(k_1, ..., k_n) = \pi(k_{\rho_1}, ..., k_{\rho_n})$$

for any permutation ρ.

Notice that the joint distribution of $T_1, ..., T_n$ is completely determined once we fixed the set $\{1, 2, ..., r\}$ of possible categories, the family of one-dimensional conditional survival function $\{\overline{G}_k(t), k \doteq 1, 2, ..., r\}$, and, finally, the joint density π; such a probabilistic model is shortly denoted by the symbol $\mathcal{H}(n; r; \{\overline{G}_k(\cdot)\}, \pi)$.

Let us denote by $\pi^{(1)}(\cdot)$ the (discrete) density of the one-dimensional marginal of π. As far as the marginal distribution of T_i is concerned, we notice that it has the mixture-type survival function given by

$$\overline{F}^{(1)}(t) = \sum_{k=1}^{r} \pi^{(1)}(k) \overline{G}_k(t). \tag{10.3}$$

It is important to remark that, for $\overline{G}_1(\cdot), ..., \overline{G}_r(\cdot)$ given, the marginal distribution of T_i only depends on $\pi^{(1)}(\cdot)$ and not on the type of correlation among $W_1, ..., W_n$.

It is easy furthermore to realize that, under the condition (10.1) and assuming that the frailties are exchangeable, $T_1, ..., T_n$ turn out to be exchangeable as well.

Remark 10.2.1 Typically, situations of exchangeability among random variables arise from the condition that the variables are independent, identically distributed, given a suitable parameter. It is interesting to notice that we described here a model of exchangeable variables that are conditionally independent, yet not identically distributed given the vector $(W_1, ..., W_n)$. Exchangeability among lifetimes here is rather created by exchangeability among $(W_1, ..., W_n)$.

Even though the assumption of exchangeability among the frailties is quite restrictive, yet it admits as special cases several situations that are of interest in survival analysis; for instance:

(a) $W_1, ..., W_n$ are independent and identically distributed.

(b) $W_1, ..., W_n$ are conditionally independent and identically distributed, given a random variable Θ.

(c) There is perfect dependence among the frailties, i.e.,

$$P\{W_1 = \cdots = W_n\} = 1.$$

(d) The cardinalities of the different subcohorts are known, i.e., we know how many individuals belong to each C_k, but we are not in a position to detect which are the individuals in I_k. This means that for some vector $\mathbf{k} \equiv (k_1, ..., k_n)$ it is

$$P\{\bigcup_{\rho} (W_1 = k_{\rho_1}, ..., W_n = k_{\rho_n})\} = 1.$$

In the case (a) we obtain that $T_1, ..., T_n$ are i.i.d. with the univariate mixture-type distribution in (10.3).

In the cases (b) and (c) $T_1, ..., T_n$ are conditionally i.i.d.

In the case (d) $T_1, ..., T_n$ can be negatively dependent.

Also in other cases of exchangeability among the frailties, one can directly obtain some results about stochastic dependence among $T_1, ..., T_n$. This can be done e.g., by applying some results contained in Shaked and Spizzichino (1998) and in Khaledi and Kochar (2001).

Remark 10.2.2 In view of the assumption that $W_1, ..., W_n$ are exchangeable, discrete, random variables, their joint distribution can be equivalently described by means of the joint distribution of the corresponding occupancy numbers $J_1, ..., J_r$, where

$$J_h \equiv \sum_{i=1}^{n} 1_{\{W_i = h\}}, h = 1, 2, ..., r.$$

In fact we can write

$$\pi(k_1, ..., k_n) = \frac{P\{J_1 = j_1, ..., J_r = j_r\}}{\binom{n}{j_1,...,j_r}},$$

where

$$j_h = \sum_{i=1}^{n} 1_{\{k_i=h\}}, h = 1, 2, ..., r.$$

It is then to be noticed that, in some cases, dependence properties of π can be usefully described in terms of some feature of the joint distribution of $J_1, ..., J_r$. Describing dependence in terms of the occupancy numbers can be in particular useful in the cases when r is small and n is big.

A further interesting property of the frailty models considered here is that, as it is easy to check, it is "conjugate" with respect to relevant operations such as

(a) *marginalization*

and

(b)*conditioning* with respect to the observation of longitudinal histories of the type

$$h_t \equiv \{T_{(1)} = t_1, ..., T_{(h)} = t_h, T_{(h+1)} > t\}$$

where $T_{(1)}, ..., T_{(n)}$ denote the order statistics of the individuals' lifetimes.

More precisely:

(a) Let $I_{j_1}, ..., I_{j_m} (j_1 \neq \cdots \neq j_m)$ be a random sample from the population $\{I_1, ..., I_n\}$ $(m < n)$.

The joint distribution of the lifetimes $T_{j_1}, ..., T_{j_m}$ is described by the model $\mathcal{H}(m; r; \{\overline{G}_k(\cdot)\}; \pi^{(m)})$, where $\pi^{(m)}$ is the m-dimensional marginal of π.

(b) Let us consider the observation of the longitudinal history h_t and let

$$T_{j_1} - t, ..., T_{j_{n-h}} - t$$

be the residual lifetimes of the surviving individuals. Then the conditional distribution of $T_{j_1} - t, ..., T_{j_{n-h}} - t$, given h_t is of the form

$$\mathcal{H}(n - h; r; \{\frac{\overline{G}_k(t + \cdot)}{\overline{G}_k(t)}\}; \pi_{h_t}^{(n-h)})$$

where $\pi_{h_t}^{(n-h)}$ denotes the conditional distribution of $W_{j_1}, ..., W_{j_m}$ given h_t.

In the field of survival analysis, there is a wide literature about specific probability models for the frailties [see e.g., Aalen (1992), Aalen and Hjort (2002), Commenges and Rondeau (2002), Hougaard (1986), Hougaard (1987), Hougaard (1991), and references cited therein]. For our purposes here however, we need not specify particular joint distributions for the frailties, apart from the qualitative assumption of exchangeability; nor have we to specify the relations among frailties and observable lifetimes (e.g., we are not assuming that the frailties act in a "multiplicative way" on a baseline hazard rate). The fact that we conceive situations of dependence among the frailties is rather the central aspect in our discussion.

We also notice that our model can be seen as a hierarchical model; actually this is in a sense that is quite natural from the Bayesian standpoint and it is different from other hierarchical models of interest considered in the literature [see e.g., Gross and Huber (2000)].

Let us conclude this section with a further remark, concerning the role of the distribution π in the analysis of aging properties of $T_1, ..., T_n$.

It is clear that aging properties of observed lifetimes are essentially influenced by the variability in the behavior of the frailties. It is important in this respect to realize that there are two different types of variability: the dispersion of the one-dimensional marginal distribution of each individual's frailty and the differences between different individuals' frailties.

In a more detailed analysis of our exchangeable frailty model, an interesting point could be the study of the separate effects that the distribution $\pi^{(1)}(\cdot)$ and the copula of the joint distribution π have on the aging properties of $T_1, ..., T_n$.

The concept of copula (and in particular of archimedean copula) has a direct interest in semi-parametric survival analysis [see e.g., the paper by Bagdonavicius, Malov and Nikulin (1999) and references therein]. Here we rather suggest the use of copulas in the specific analysis of multivariate aging properties of lifetimes in exchangeable frailty models [see also Bassan and Spizzichino (2002)].

10.3 Multivariate Aging For Exchangeable Lifetimes

Some special definitions of multivariate aging turn out to be useful in the analysis of the exchangeable frailty models described so far; this kind of definitions will be briefly described in the present section.

Let $\leq_*$ be a one-dimensional stochastic ordering; for instance $\leq_*$ can be the usual stochastic ordering (i.e., $* = \text{st}$), or the hazard rate ordering (i.e., $* = \text{hr}$), or the likelihood-ratio ordering (i.e., $* = \text{lr}$), or the mean residual life ordering (i.e., $* = \text{mrl}$), and so on ... ; see e.g., Shaked and Shantikhumar (1994) for definitions and basic properties of the cited stochastic orderings and for notions of one-dimensional aging to be cited below.

Let A be a subset of $[0, +\infty) \times [0, +\infty)$ such as

$$A \equiv \{(t', t'') | 0 \leq t' \leq t''\} \text{ or } A \equiv \{(t', t'') | 0 = t' \leq t''\}.$$

Several well-known (univariate) notions of positive aging can be obtained as special cases of the general definition that we give next. Let T be a non-negative random variable (T will have the meaning of the lifetime of a fixed individual or the waiting time up to the arrival of any specified event for the individual, as it is typical in the field of survival analysis). Here we essentially think of lifetimes as continuous variables, but a similar discussion can be developed also for discrete cases.

Let $\overline{F}$ denote the survival function of T and, for $s > 0$, let $\overline{F}_s$ be the conditional survival function of the residual lifetime $(T - s)$, given $\{T > s\}$, i.e.,

$$\overline{F}_s(t) \equiv \frac{\overline{F}(t+s)}{\overline{F}(s)}.$$

Definition 10.3.1 T has the $(\leq_*; A)$-positive aging property if

$$\overline{F}_{s''}(\cdot) \leq_* \overline{F}_{s'}(\cdot), \forall (s', s'') \in A.$$

Notice that one-dimensional positive aging notions such as IFR, NBU, DMRL, PF. $_2$, ... can easily be put into the above form by suitably specifying the pair $(\leq_*; A)$; discrete-time aging notions of interest can also be seen as special cases of the same general definition, as it will be detailed somewhere else.

Similarly we can define corresponding negative aging properties by letting

$$\overline{F}_{s'}(\cdot) \leq_* \overline{F}_{s''}(\cdot), \forall (s', s'') \in A.$$

Consider now, for (jointly absolutely continuous) lifetimes $T_1, ..., T_n$, events $D[n; \mathbf{s}]$ and $D[k; \mathbf{t}; \mathbf{s}]$, respectively of the form

$$D[n; \mathbf{s}] \equiv \{T_1 > s_1, ..., T_k > s_n\},$$

$$D[k; \mathbf{t}; \mathbf{s}] \equiv \{T_1 > s_1, ..., T_k > s_k, T_{k+1} = t_1, ..., T_n = t_{n-k}\},$$

where $k \leq n-2$; i.e., $D[n; \mathbf{s}]$ is a history containing all *survivals* and $D[k; \mathbf{t}; \mathbf{s}]$ is a history of *failures and survivals*.

We shall also put

$$\begin{aligned} \overline{F}_{s_1, D[n;\mathbf{s}]}(t) &\equiv P\{T_1 > s_1 + t | D[n; \mathbf{s}]\}, \\ \overline{F}_{s_1, D[k;\mathbf{t};\mathbf{s}]}(t) &\equiv P\{T_1 > s_1 + t | D[k; \mathbf{t}; \mathbf{s}]\}. \end{aligned}$$

Similarly we define $\overline{F}_{s_2, D[n;\mathbf{s}]}(t)$ and $\overline{F}_{s_2, D[k;\mathbf{t};\mathbf{s}]}(t)$.

Let us now consider, for the moment, the case when $T_1, ..., T_n$ are i.i.d.. It is immediate to check the validity of the following

Proposition 10.3.1 *Let $T_1, ..., T_n$ be i.i.d. with a survival function $\overline{F}$. Then the following conditions are equivalent:*

(i) $\overline{F}$ has the $(\leq_*; A)$-positive aging property;

(ii) for $(s_1, s_2) \in A$ *and any* $\mathbf{s} \equiv (s_1, s_2, s_3, ..., s_n)$ *it is*

$$\overline{F}_{s_2, D[n;\mathbf{s}]}(\cdot) \leq_* \overline{F}_{s_1, D[n;\mathbf{s}]}(\cdot); \tag{10.4}$$

(iii) for $(s_1, s_2) \in A$, *any* $k \leq n-2$, *and any* $\mathbf{t} \equiv (t_1, ..., t_{n-k})$, $\mathbf{s} \equiv (s_1, ..., s_k)$, *it is*

$$\overline{F}_{s_2, D[k;\mathbf{t};\mathbf{s}]}(t) \leq_* \overline{F}_{s_1, D[k;\mathbf{t};\mathbf{s}]}(t). \tag{10.5}$$

This trivial result can be the starting point for multivariate extensions, to be defined next. In fact the conditions (i), (ii), and (iii) are not generally equivalent anymore, when $T_1, ..., T_n$ are not i.i.d. .

In the case when $T_1, ..., T_n$ are exchangeable the conditions (ii) and (iii) thus inspire useful definitions, that can be thought of as different notions of multivariate aging. Such notions are multivariate in the sense that $T_1, ..., T_n$ are, generally, not independent.

Notice that such conditions establish a stochastic comparison between the residual lifetimes of two surviving individuals of different ages, given an observed history of survivals, and perhaps failures, of other individuals in the population. For brevity's sake, we fix attention here on the only condition (ii). Formally, we give the following

Definition 10.3.2 Let $\overline{F}^{(n)}$ be an exchangeable, n-dimensional, survival function. Then $\overline{F}^{(n)}$ has the n-dimensional $(\leq_*; A)$-positive aging property if the condition (ii) holds.

The idea of studying this general type of aging for exchangeable lifetimes was introduced by Bassan and Spizzichino (1999), where sufficient conditions for several such notions had be derived. This idea developed in a natural way, starting from a special case considered in Spizzichino (1992) and equivalent to Schur-concavity of the joint survival function; the latter condition in fact can also be seen as a multivariate condition of IFR, as discussed by Barlow and Mendel (1992) and by Barlow and Spizzichino (1993).

Summarizing, here we notice that a notion of positive multivariate aging for exchangeable lifetimes is singled out once we fix:

- the stochastic ordering $\leq_*$,
- the subset $A \subset [0, +\infty) \times [0, +\infty)$,
- the choice between the condition (ii) or (iii).

We obtain corresponding notions of negative aging by inverting the directions of the involved inequalities.

For instance, the positive aging notion mentioned above, which is equivalent to Schur-concavity of $\overline{F}^{(n)}$, is obtained by letting $\leq_*$ coincide with the usual stochastic ordering, $A \equiv \{(t', t'') | 0 \leq t' \leq t''\}$, and by requiring (ii); we obtain Schur-convexity of $\overline{F}^{(n)}$ by making the same choices and reverting the direction of the inequality (ii).

Concerning the discrepancies existing between the i.i.d. case and the more general case of exchangeability, the following difference is also to be noticed: different choices of the stochastic ordering $\leq_*$ may give rise to the same univariate aging notion. For instance, by fixing $A \equiv \{(t', t'')|0 \leq t' \leq t''\}$, the familiar one-dimensional notion of IFR (increasing failure rate) can be obtained by choosing both $\leq_* = \leq_{st}$ or $\leq_* = \leq_{hr}$; this is on the contrary not true generally in the case of exchangeability, as it turns out by looking at some aging notions analyzed by Bassan and Spizzichino (1999) and by Bassan, Kochar and Spizzichino (2002).

An important point concerning the aging notions above is that positive aging is compatible with negative dependence and, symmetrically, negative aging is compatible with positive dependence.

The latter aspect is of relevant interest in the field of frailty models.

Other aspects concerning the above notions of multivariate aging are discussed in Spizzichino (2001).

10.4 Exchangeable Frailties and Multivariate Aging

We now discuss the role that notions of multivariate aging (in the sense described so far) can have in analyzing some aspects of exchangeable frailty models of the type $\mathcal{H}(n; r; \{\overline{G}_k(\cdot)\}, \pi)$.

First we recall that any of the above notions of aging involves some type of stochastic comparison between $\overline{F}_{s_1, D[n;\mathbf{s}]}(\cdot)$ and $\overline{F}_{s_2, D[n;\mathbf{s}]}(\cdot)$, or between $\overline{F}_{s_1, D[k;\mathbf{t};\mathbf{s}]}(\cdot)$ and $\overline{F}_{s_2, D[k;\mathbf{t};\mathbf{s}]}(\cdot)$.

It is then of interest, from a technical point of view, to point out the form of such conditional probabilities; in this respect, we have

Proposition 10.4.1 *Let the joint distribution of* $T_1, ..., T_n$ *be described as an* $\mathcal{H}(n; r; \{\overline{G}_z(\cdot)\}, \pi)$ *model. Then it is*

$$\overline{F}_{s_1, D[n;\mathbf{s}]}(t) = \frac{\sum_{k=1}^{r} \sum_{k_2, ..., k_n} \pi(k, k_2, ..., k_n) \cdot \overline{G}_k(s_1 + t) \cdot \overline{G}_{k_2}(s_2) \cdot \dots \cdot \overline{G}_{k_n}(s_n)}{\sum_{k=1}^{r} \sum_{k_2, ..., k_n} \pi(k, k_2, ..., k_n) \cdot \overline{G}_k(s_1) \cdot \overline{G}_{k_2}(s_2) \cdot \dots \cdot \overline{G}_{k_n}(s_n)}. \tag{10.6}$$

PROOF. In view of the conditional independence assumption, we can write

$$\begin{aligned}\overline{F}_{s_1, D[n;\mathbf{s}]}(t) &= \sum_{k=1}^{r} P\{T_1 > s_1 + t | W_1 = k, T_1 > s_1\} \cdot P\{W_1 = k | D[n; \mathbf{s}]\} \\ &= \sum_{k=1}^{r} P\{W_1 = k | D[n; \mathbf{s}]\} \cdot \frac{\overline{G}_k(s_1 + t)}{\overline{G}_k(s_1)}.\end{aligned}$$

Furthermore

$$P\{W_1 = k|D[n;\mathbf{s}]\} = \sum_{k_2,...,k_n} P\{W_1 = k, W_2 = k_2, ..., W_n = k_n|D[n;\mathbf{s}]\}$$

and, by Bayes Formula,

$$P\{W_1 = k, W_2 = k_2, ..., W_n = k_n|D[n;\mathbf{s}]\}$$
$$= \frac{\pi(k, k_2, ..., k_n)\overline{G}_k(s_1)\cdot\overline{G}_{k_2}(s_2)\cdot\dots\cdot\overline{G}_{k_n}(s_n)}{\sum_{k,k_2,...,k_n}\pi(k, k_2, ..., k_n)\overline{G}_k(s_1)\cdot\overline{G}_{k_2}(s_2)\cdot\dots\cdot\overline{G}_{k_n}(s_n)}.$$

■

Similar representations can be given for $\overline{F}_{s_2,D[n;\mathbf{s}]}(t)$, $\overline{F}_{s_1,D[k;\mathbf{t};\mathbf{s}]}(t)$, and $\overline{F}_{s_2,D[k;\mathbf{t};\mathbf{s}]}(t)$.

A typical point of interest in survival analysis is the comparison between aging properties of the conditional survival functions $\overline{G}_k(\cdot)$, $k = 1, 2, ..., r$, and aging properties of the, mixture-type, unconditional survival function $\overline{F}^{(1)}(\cdot)$, given by (10.3). In this respect we first notice that it is

$$\begin{aligned}\overline{F}_s^{(1)}(t) &= \frac{\sum_{k=1}^r \pi^{(1)}(k)\cdot\overline{G}_k(t+s)}{\sum_{k=1}^r \pi^{(1)}(k)\cdot\overline{G}_k(s)} \\ &= \frac{\sum_{k=1}^r\sum_{k_2,...,k_n}\pi(k, k_2, ..., k_n)\cdot\overline{G}_k(t+s)}{\sum_{k=1}^r\sum_{k_2,...,k_n}\pi(k, k_2, ..., k_n)\cdot\overline{G}_k(s)}. \qquad (10.7)\end{aligned}$$

Fix a pair $(\leq_*; A)$. A comparison between (10.7) and (10.6) confirms that the n-dimensional $(\leq_*; A)$-positive (or negative) aging property of $\overline{F}^{(n)}$ and $(\leq_*; A)$-positive (or negative) aging property of $\overline{F}^{(1)}$ are two different conditions.

We now want to compare such two conditions and to point out some facts of interest in the fields of aging, frailty, and dependence.

We keep, just to fix ideas, $A \equiv \{(t', t'')|0 \leq t' \leq t''\}$ and $\leq_* = \leq_{st}$, so as to obtain the notions of IFR and DFR; but similar arguments can be developed for other choices, as well.

We concentrate attention on the fact that situations of hidden heterogeneity tend, roughly speaking, to "eliminate" positive aging and to "reinforce" negative aging:

We consider, for example, the assumptions
$\mathbf{C}_{IFR} \equiv \overline{G}_k(\cdot)$ is IFR for $k = 1, 2, ..., r$,
$\mathbf{C}_{DFR} \equiv \overline{G}_k(\cdot)$ is DFR for $k = 1, 2, ..., r$.

As is well known, if $\mathbf{C}_{DFR}$ holds, then $\overline{F}^{(1)}(\cdot)$ is DFR, whereas the assumption $\mathbf{C}_{IFR}$ does not generally imply that $\overline{F}^{(1)}(\cdot)$ is IFR.

The latter remark is really relevant in the case when the lifetimes are independent, identically distributed; in fact, in such a case, their aging properties are just described by properties of $\overline{F}^{(1)}(\cdot)$ and the latter also describe multivariate aging notions in the sense of Definition 10.3.2, in view of Proposition 10.3.1.

Let us comment on this point, in the perspective of our exchangeable frailty models, where the condition of i.i.d. lifetimes is interpreted as a condition of independence and identical distribution for the frailties.

Let the survival functions $\{\overline{G}_k(\cdot), k = 1, 2, ..., r\}$ be given and consider the assumption C_{IFR}.

The latter means that, for two individuals, say I_1 and I_2, their lifetimes T_1 and T_2 are such that, for $s_1 < s_2$,

$$P\{T_1 > s_1 + t | T_1 > s_1\} \geq P\{T_2 > s_2 + t | T_2 > s_2\}, \tag{10.8}$$

whenever it is known that $W_1 = W_2$, i.e., that I_1 and I_2 belong to the same category (or share a same frailty).

When I_1 and I_2 have been selected by random sampling from the whole population (in which case we ignore if $W_1 = W_2$ or $W_1 \neq W_2$), the assumption $\mathbf{C}_{IFR}$, on the contrary, does not imply the inequality (10.8) anymore.

Actually, in such a case, the validity of (10.8) depends on the mixing distribution $\pi^{(1)}$, whereas it is not affected by the structure of dependence among $W_1, ..., W_n$.

Under the assumption $\mathbf{C}_{IFR}$, what can be said about multivariate aging, when we drop the condition of stochastic independence among the frailties?

In what follows we briefly discuss this point with respect to the condition of multivariate $(\leq_*; A)$ positive aging introduced in Section 10.3, namely Schur-concavity of the joint survival function $\overline{F}^{(n)}$.

It can be argued that the validity of this condition depends both on $\pi^{(1)}$ and on the structure of dependence among $W_1, ..., W_n$.

Consider, for instance the extreme case when $P\{W_1 = \cdots = W_n\} = 1$, i.e. the case when $T_1, ..., T_n$ are conditionally independent given W_1; more precisely we let

$$P\{W_1 = \cdots = W_n = k\} = \pi^{(1)}(k), k = 1, 2, ..., r.$$

It is very easy to check that, under C_{IFR}, $\overline{F}^{(n)}$ is Schur-concave irrespectively of $\pi^{(1)}$, i.e., irrespectively of whether $\overline{F}^{(1)}$ is IFR or not.

If on the contrary we consider the other extreme case when $W_1, ..., W_n$ are stochastically independent, still with the conditions C_{IFR} and

$$P\{W_i = k\} = \pi^{(1)}(k), k = 1, 2, ..., r, i = 1, ..., n,$$

then, by Proposition 10.3.1, Schur-concavity of the joint survival function $\overline{F}^{(n)}$ is equivalent to the one-dimensional positive aging property (i.e., the IFR property of $\overline{F}^{(1)}$), which holds or not, depending only on the dispersion of the mixing distribution $\pi^{(1)}$.

Remark 10.4.1 In the cases when $\overline{F}^{(1)}$ is not IFR, the discrepancy between the case of independence and of perfect dependence ($P\{W_1 = \cdots = W_n\} = 1$) with respect to the validity of n-dimensional positive aging is fully explained by the fact that the latter situation does not really involve a case of heterogeneity; on the contrary, in the case when $W_1, ..., W_n$ are random, their independence actually means a condition of heterogeneity among the individuals.

One can argue more generally that negative dependence among $W_1, ..., W_n$ reinforces the tendency to *eliminate* the n-dimensional positive aging property, existing within each category.

On the other hand the latter tends to be maintained by assumptions of positive dependence among $W_1, ..., W_n$. In particular the validity of the one-dimensional marginal IFR property, combined with suitably strong conditions of positive dependence, guarantees the n-dimensional positive aging.

With respect to applications to survival analysis, we then remark that lack of positive aging in the failure rate function at the population level (i.e., lack of one-dimensional positive aging), combined with experimental evidence of better survival of younger than elder individuals (i.e., presence of n-dimensional positive aging) is an index of positive dependence among frailties (and among the individuals).

To the author's knowledge, the previous considerations have not been put into precise terms so far. As a hint for future research let us mention that such arguments can probably be made precise by using results along the line developed by Bassan and Spizzichino (2002).

We thus stressed so far the fact that, depending on the dependence structure among frailties, a one-dimensional positive aging property and the correspondent n-dimensional positive aging property may coincide or may be two different conditions.

What is the interest of properties of multivariate aging, however?

We remark that the latter are typically much more significant of univariate aging (excepting the cases of independence, where the two notions actually do coincide).

In fact it asserts that, for two randomly selected individuals respectively surviving for periods s_1 and s_2 , the inequality (10.4) holds and typically, when comparing the survival probabilities of two individuals of different ages, our state of information actually is of the type $D[n; \mathbf{s}]$ (or $D[k; \mathbf{t}; \mathbf{s}]$).

Notice on the contrary that we are seldom interested in the comparison (10.8).

Let us now turn to notions of negative aging; still we fix ideas on the case $A \equiv \{(t', t'')|0 \leq t' \leq t''\}$ and $\leq_* = \leq$.

Then we consider

- the assumption $\mathbf{C}_{DFR}$,

- the one-dimensional $(\leq_*, A)$-negative property of $T_1, ..., T_n$, namely the DFR property of $\overline{F}^{(1)}$,

- the multivariate $(\leq_*, A)$-negative property of $T_1, ..., T_n$, namely the Schur-convexity property of $\overline{F}^{(n)}$.

As mentioned, it is very well known that $\mathbf{C}_{DFR}$ implies the DFR property of $\overline{F}^{(1)}$, irrespectively of the form of the mixing distribution $\pi^{(1)}$.

Also with respect to multivariate aging we have a similar result and a similar difference with the case of positive aging: it is easy to show that the Schur-convexity property of $\overline{F}^{(n)}$ holds, irrespectively of the type of dependence among the frailties, as was proved by Spizzichino and Torrisi (2001).

References

1. Aalen, O. O. (1988). Heterogeneity in survival analysis, *Statistics in Medicine*, **7**, 1121–1137.

2. Aalen, O. O. (1992). Modelling heterogeneity in survival analysis by the compound Poisson distribution, *Annals of Applied Probability*, **2**, 951–972.

3. Aalen, O. O. and Hjort, N. (2002). Frailty models that yield proportional hazards, *Statistics & Probability Letters*, **58**, 335–342.

4. Barlow, R. E. and Mendel, M. B. (1992). de Finetti-type representations for life distributions, *Journal of the American Statistical Association*, **87**, 1116–1122.

5. Barlow, R.E. and Spizzichino, F. (1993). Schur-concave survival functions and survival analysis, *Journal of Computation and Applied Mathematics*, **46**, 437–447.

6. Bassan, B., Kochar, S. and Spizzichino F. (2002). Some bivariate notions of IFR and DMRL and related properties, *Journal of Applied Probability*, **39**, 533–544.

7. Bassan, B. and Spizzichino, F. (1999). Stochastic comparison for residual lifetimes and Bayesian notions of multivariate ageing, *Advances in Applied Probability*, **31**, 1078–1094.

8. Bassan, B. and Spizzichino, F. (2002). Relations among univariate aging, bivariate aging, and dependence for exchangeable lifetimes, *Submitted for publication.*

9. Bagdonavicius, V., Malov, S. and Nikulin, M. (1999). Characterizations and semiparametric regression estimation in Archimedean copulas, *Journal of Applied Statistical Science*, **8**, 137–153.

10. Commenges, D. and Rondeau, V. (2002). Modélisation de la fragilité en survie, *Preprint*.

11. Gerardi, A., Spizzichino, F. and Torti, B. (2000). Exchangeable mixture models for lifetimes: the role of "occupation numbers", *Statistics & Probability Letters*, **49**, 365–375.

12. Gross, S. and Huber, C. (2000). Hierarchical dependency models for multivariate survival data with censoring, *Lifetime Data Analysis*, **6**, 299–320.

13. Hougaard, P. (1986). Survival models for heterogeneous populations derived from stable distributions, *Biometrika*, **73**, 387–396.

14. Hougaard, P. (1987). Modelling multivariate survival, *Scandinavian Journal of Statistics*, **14**, 291–304.

15. Hougaard, P. (1991). Modelling heterogeneity in survival data, *Journal of Applied Probability*, **28**, 695–701.

16. Khaledi, S. and Kochar, S. (2001). Dependence properties of multivariate mixture distributions and their applications, *Annals of the Institute of Statistical Mathematics*, **53**, 620–630.

17. Shaked, M. and Shanthikumar, J. G. (1994). *Stochastic Orders and Their Applications*, Academic Press, San Diego, California.

18. Shaked, M. and Spizzichino, F. (1998). Positive dependence properties of conditionally independent random lifetimes, *Mathematics of Operations Research*, **23**, 944–959.

19. Spizzichino, F. (1992). Reliability decision problems under conditions of ageing, In *Bayesian Statistic 4* (Eds., J. Bernardo, J. Berger, A. P. Dawid, A. F. M. Smith), pp. 803–811, Clarendon Press, Oxford, England.

20. Spizzichino, F. (2001). *Subjective Probability Models for Lifetimes*, Chapman & Hall/CRC, Boca Raton, Florida.

21. Spizzichino, F. and Torrisi, G. (2001). Multivariate negative aging in an exchangeable model of heterogeneity, *Statistics & Probability Letters*, **55**, 71–82.

11

Semiparametric Models in the Studies of Aging and Longevity

Anatoli I. Yashin

Max Planck Institute for Demographic Research, Rostock, Germany

Abstract: Semiparametric models used in the studies of aging and longevity provide an important link between several research disciplines including demography, biostatistics, genetic epidemiology and survival analysis. In this paper we discuss semiparametric models which can be constructed using the notion of "frailty". We consider several classes of such models appropriate for the analysis of data on related individuals. These include shared and correlated frailty models. We discuss properties of these models, areas of their application, and further directions of research.

Keywords and phrases: Heterogeneity, bivariate survival data, twins, correlated frailty

11.1 Introduction

Many semiparametric models studied in survival analysis may be constructed using the notion of frailty. Frailty models are characterized by stochastic hazards, $\mu(Z, x) = Z\mu_0(\mathrm{x})$, or, $\mu(Z, x) = Z\mu_0(x) + \tilde{\mu}(x)$, where non-negative random variable, Z, is called frailty, $\mu_0(\mathrm{x})$ is an underlying hazard and $\tilde{\mu}(x)$ is a "background" hazard. Such models without observed covariates have been used in demographic studies of mortality in heterogeneous populations [Vaupel et al. (1979), Hougaard (2000), Manton et al. (1986), Vaupel and Yashin (1985)]. To make such models identifiable one has to fix the parametric structure of the underlying hazard (e.g., $\mu_0(x) = ae^{bx}$, and $\tilde{\mu}(x) = c$) and use parametric methods in the statistical analysis of survival data (e.g., the maximum likelihood method). Univariate frailty models were used in explanation of deviant age-trajectories of mortality rate at old ages [Vaupel et al. (1979)]. The model with $\mu(Z, x) = c + Zae^{bx}$ where Z is gamma distributed with mean 1 and variance σ^2 is called the "gamma-Makeham" model. It turns out that this model (logistic curve) fits mortality data better than the traditional Gompertz–Makeham (exponential) curve [Yashin and Iachine (1994)].

The assumption on proportionality of hazard makes frailty models a simple and convenient tool for analytical and computer calculations. Despite this simplicity these models turn out to be useful in providing insights and ideas for analyzing reality. They explain the deceleration of the mortality rate at older ages [Vaupel et al. (1998)]. They allow for the evaluation of bias in regression coefficients when calculating the effects of observed covariates on survival in the presence of heterogeneity [Chamberlain (1985), Bretagnolle and Huber-Carol (1988)]. They help to clarify the roles of genes and environment in life span, [Yashin and Iachine (1997)], etc.

In many epidemiological and biostatistical applications frailty models include vectors of observed covariates, U [Andersen et al. (1992)]. The individual hazard — i.e., the hazard conditioned on Z and U — is often used in a Cox-type proportional hazard form with random effect. Here we use the simplest form of such hazard

$$\mu(Z,U,x) = Z\mu_0(x)e^{\beta^* U}. \tag{11.1}$$

However, the results can easily be extended to a more general form of proportional hazard model. Note that the distribution of frailty, Z, in model (11.1) does not depend on observed covariates, U; the latter one is often used to account for omitted covariates. The univariate models are identifiable if $EZ < \infty$, i.e., in this case no assumptions about $\mu_0(x)$ are needed [Elbers and Ridder (1982)].

Bivariate biostatistical models have exploited the idea of shared frailty. They are used in studies of the effects of dependence between life spans (or other durations) on the estimates of regression coefficients in the Cox-type model [Clayton (1978), Clayton and Cuzick (1985) and Hougaard (2000)]. In all frailty models the distribution of frailty may, in principle, have an arbitrary form. For example, the gamma-frailty has been popular in demographic and biostatistical studies [Vaupel et al. (1979), Manton et al. (1986), Clayton (1978), Clayton and Cuzick (1985)]. The inverse-Gaussian, positive stable [Hougaard (2000)], lognormal [Xue and Brookmeyer (1996)] and other frailty distribution functions [Aalen (1994)] have also been used. Note that shared gamma-, inverse-Gaussian and positive stable frailty models are identifiable without observed covariates [Yashin and Iachine (1999a,b)].

In this paper we discuss several aspects of frailty modeling. In particular, we explain why properties of frailty models used in univariate and traditional bivariate analyses (i.e. analyses based on the idea of shared frailty) are radically different. We discuss why bivariate correlated frailty models are important in genetic studies of susceptibility to death, and how ideas from biostatistics and genetic epidemiology can be merged to analyze multivariate survival data.

11.2 The Shared Gamma-frailty Model

After its appearance in demography [Vaupel et al. (1979)], the notion of frailty reappeared in the biostatistical literature [Oakes (1989)] to denote shared relative risk in bivariate proportional hazard models [Clayton (1978) and Clayton and Cuzick (1985)]. In these models, the individual conditional hazard is also represented by (11.1), where the variable Z is also called "frailty". Despite this formal mathematical similarity in hazards, the two notions of frailty used in univariate and bivariate analyses are radically different.

Versions of the shared-frailty model have been used successfully in the analysis of bivariate survival data. Vaupel et al. (1998) uses the gamma-frailty model to prove that MZ twins share more longevity-related genetic material than DZ twins do. Hougaard (2000) comes to a similar conclusion using the shared-frailty model with both positive stable and gamma distributions of frailty. The effects of observed covariates on survival in the case of the shared-frailty model have also been considered. Estimates of the association parameter and its influence on regression coefficients in the Cox-type model with random effects are discussed by Clayton (1978), Clayton and Cuzick (1985).

In the absence of observed covariates the conditional bivariate survival function in the case of this model is

$$S(x_1, x_2 \,|Z) = e^{-Z(H_1(x_1)+H_2(x_2))} \tag{11.2}$$

where Z is a shared frailty, $H_i(x_i) = \int_0^{x_i} \mu_{0i}(u)du, i = 1, 2$, and $\mu_{0i}(x), i = 1, 2$ are the underlying hazards associated with two related individuals. Note that the assumption of conditional independence of life spans T_1 and T_2 given Z is used in (11.2). When frailty is gamma-distributed with the mean 1 and variance σ^2, the marginal bivariate survival function is [Cox and Oakes (1984)]

$$S(x_1, x_2) = \left(S_1(x_1)^{-\sigma^2} + S_2(x_2)^{-\sigma^2} - 1\right)^{-\frac{1}{\sigma^2}}. \tag{11.3}$$

Here $S_i(x), i = 1, 2$ are marginal univariate survival functions. One can see that the only parameter responsible for association between T_1 and T_2 is σ^2.

This model has several remarkable properties. It is identifiable without having to make assumptions about the parametric specification of the conditional survival function: this function and the parameters of other models can be identified from bivariate survival data [Iachine and Yashin (1998)]. When using this model, one can easily establish the difference in association parameters, σ^2 for populations of pairs of individuals with different levels of familial relationship [e.g., MZ and DZ twins, cf. Vaupel et al. (1998)]. However, the application of this approach to some problems of aging and survival meets certain limitations. If, for example, the true conditional hazards are

$$\mu_i(Z, U_i, x) = Z\,\mu_{0i}(x)\,e^{\beta\,U_i}, \quad i = 1, 2, \tag{11.4}$$

with U_i and β_i, $i = 1, 2$ as observed covariates and their respective regression coefficients, then the bivariate survival function given observed covariates is:

$$S(x_1, x_2 \,|\, U_1, U_2) = \left(S_1(x_1 \,|\, U_1)^{-\sigma^2} + S_2(x_2 \,|\, U_2)^{-\sigma^2} - 1\right)^{-\frac{1}{\sigma^2}} \tag{11.5}$$

where $S_i(x_i \,|\, U_i) = \left(1 + \sigma^2 H_i(x)\,e^{\beta_i\,U_i}\right)^{-\frac{1}{\sigma^2}}$. In this model the estimates of the association parameter σ_i^2 and the regression coefficients β in the Cox-like hazards with random effects are always biased if data on related individuals are used. This property has important methodological consequences for the statistical analysis of bivariate data.

11.3 Identifiability and Confounding

Elbers and Ridder (1982) show that if frailty distribution has a finite mean, then the presence of observed covariates (as in the case of hazard rate (11.1)) makes the univariate proportional hazard frailty model identifiable. This finding stimulated the use of random effect Cox-type regression models in the statistical analysis of univariate survival data [Andersen et al. (1992)]. Hazards with the same functional form (11.1) were used in bivariate survival modeling with shared random effect [Clayton (1978), Clayton and Cuzick (1985)], which later was also called "frailty" [Oakes (1989)]. However, "shared frailty" in bivariate survival models differs from " individual frailty" used in the univariate case. This difference in the notions of frailty has not been clearly understood. For example, starting from the discussion in Clayton's and Cuzick's (1985) paper where the shared gamma-frailty model (with hazard (11.1), the mean and the variance of frailty equal to 1 and σ^2, respectively) was considered, several authors repeatedly cited the following limitation of this model: "in the presence of observed covariates the association parameter σ^2 can be estimated from the univariate subset of bivariate data." They claim that this statement follows from the Elbers and Ridder (1982) identifiability criterion. It is worth noting, however, that the value of σ^2 estimated from the univariate data may, in fact, have nothing to do with association. Indeed, let us consider two hypothetical bivariate data sets with different associations between life spans of respective related individuals: survival data for a group of women and their daughters and for the same group of women and their granddaughters. In both situations the value of parameter σ^2 estimated from the data on women will be the same despite the fact that life span associations between related individuals in these two data sets (women and their daughters in one set, and women and their granddaughters in the other set) are different.

Finally, in the presence of observed Cox-type covariates, this model fails to satisfy some natural assumptions. For example, in the analysis of several bivariate data sets (e.g., life span data for MZ and DZ twins) the marginal univariate survival functions often turn out to be the same. The result of Elbers and Rider (1982) immediately yields that σ^2 and baseline hazards for survival functions of MZ and DZ twins are the same. Hence estimates obtained in the analysis of bivariate data will be biased because the same parameter (σ^2) is used to describe life spans' association and univariate fit. This property excludes the possibility of data integration for MZ and DZ twins. Such integration, however, plays a crucial role in the evaluation of genetic characteristics of longevity in genetic-epidemiological studies of aging and life span.

The limitations discussed above are inherent features of all shared-frailty models with a finite mean of frailty distribution. To avoid these problems, Hougaard (2000) suggests the positive stable distribution of frailty, which has an infinite mean. In this case the univariate model with observed covariates is not identifiable. So, one can expect more flexibility from the shared-frailty model with such frailty distribution than from the model with gamma-frailty.

11.4 The Positive-stable-frailty Models

The positive stable distribution with the parameter $0 < \alpha < 1$ is characterized by its Laplace transform $L(s) = e^{-\alpha s}$. So, if frailty Z is positive-stable distributed, the univariate marginal survival function is

$$S_i(x) = e^{-H_i^{\alpha}(x)} \tag{11.6}$$

where $H_i(x) = \int_0^x \mu_{0i}(u)du$. In the case of shared frailty the bivariate marginal survival function is [Hougaard (2000)]

$$S(x_1, x_2) = e^{-\left((-\ln S_1(x_1))^{\frac{1}{\alpha}} + (-\ln S_2(x_2))^{\frac{1}{\alpha}}\right)^{\alpha}}. \tag{11.7}$$

When covariates U_i, $i = 1, 2$ are observed, and the conditional hazard is (11.1), then the univariate survival functions conditional on observed covariates

U_i are:

$$S_i(x|U_i) = \exp\Big(- H_i^{\alpha}(x) e^{\alpha\beta_i U_i}\Big) = \exp\Big(- \tilde{H}_i(x) e^{\tilde{\beta}_i U_i}\Big). \tag{11.8}$$

Using a positive stable shared frailty model one can estimate both an association parameter α and regression coefficients β_i from the data on related individuals. One problem, however, remains unsolved: the interpretation of regression coefficients β_i. To illustrate this problem let us assume first that $\beta = \beta_1 = \beta_2$ is the same for members of MZ and DZ twin-pairs. Let us consider two hypothetical bivariate data sets: one for MZ twins, another for DZ twins. It is clear that the values of parameter α estimated for MZ and DZ twins from data sets large enough will be different since they characterize the association between life spans of MZ and DZ twins. Hence the values of parameter $\tilde{\beta} = \alpha\beta$ and $\tilde{H}\,(x)$in equation (11.7) are also different for the members of MZ and DZ pairs, which contradicts the natural assumption that the survival of these individuals conditioned on observed covariates follows the same (Cox-type) hazards. If we assume that parameters $\tilde{\beta} = \alpha\beta$ are the same for members of MZ and DZ pairs, then parameters β and the underlying hazards $\mu_0\,(x)$ should be different for these individuals, which creates a problem for the interpretation of the conditional hazard (11.1) for this model.

To avoid such methodological problems of a shared frailty model, both with finite and infinite mean of a frailty distribution, the idea of correlated frailty was suggested. In this model one more parameter — the correlation coefficient of frailty ρ_z — is introduced to describe the association between life spans of related individuals. In the presence of observed covariates such a model is internally consistent: if bivariate data correspond to the correlated frailty model, then large-sample parameters from univariate and bivariate data coincide. This model is convenient for the analysis of genetic aspects of susceptibility to death and longevity since the estimates of heritability in frailty have a one-to-one correspondence to the estimates of correlation coefficients of frailty for MZ and DZ twins.

11.5 The Correlated Gamma-frailty Model

Let $\mu_i(Z_i, x) = Z_i\,\mu_{0i}(x), i$ =1,2 be the two random hazards for two related individuals. Let us assume that frailties $Z_i, i = 1, 2$ are gamma-distributed with means 1, variances σ_1^2, σ_2^2 and the correlation coefficient ρ_z. Yashin and Iachine (1994) found that the respective bivariate survival function $S(x_1, x_2)$ for the latent failure times is

$$S(x_1, x_2) = S_1(x_1)^{1-\frac{\sigma_1}{\sigma_2}\rho_z} S_2(x_2)^{1-\frac{\sigma_2}{\sigma_1}\rho_z} \Big(S_1(x_1)^{-\sigma_1^2} + S_2(x_2)^{-\sigma_2^2} - 1\Big)^{-\frac{\rho_z}{\sigma_1\sigma_2}}. \tag{11.9}$$

Such a survival function may be convenient in the analysis of survival data for unlike-sex DZ twins, or in the case of a dependent competing risk problem. Note that the range of possible values of the correlation coefficient between frailties depends on the values of σ_1 and σ_2:

$$0 \le \rho_z \le min \cdot \left(\frac{\sigma_1}{\sigma_2}, \frac{\sigma_2}{\sigma_1}\right).$$

Hence if $\sigma_1 \neq \sigma_2$, it is always less than 1. This property can seriously limit the analysis when values of σ_1 and σ_2 differ substantially. Versions of the correlated gamma-frailty model with identical σ_1 and σ_2 were discussed by Pickles and Crouchley (1995). Applications to the analysis of genetic aspects of mortality by cause are discussed by Wienke et al. (2000, 2001, 2002), Zdravkovic et al.(2002). Log-normal bivariate frailty models are investigated by Xue and Brookmeyer (1996), Xue and Ding (1999). Locatelli et al. (2003) investigated the MCMC method in the analysis of bivariate survival data using a log-normal frailty model. A remarkable property of the correlated gamma-frailty model is that it can be identifiable even without observed covariates. A more precisely, let $S(x_1, x_2)$ be a known bivariate survival function given by (11.9) and let $\bar{\mu}_i(0) = -\frac{d \ln S_i(x)}{dx}|_{x=0} \neq 0, \quad i = 1, 2$. Then the parameters ρ, σ_1, σ_2 can be uniquely identified [Yashin and Iachine (1997)]. More general result on identifiability of bivariate correlated frailty models with arbitrary frailty distribution is proved by Iachine and Yashin (1998).

11.6 Other Correlated Frailty Models

The inverse Gaussian correlated frailty model is characterized by the marginal bivariate survival function

$$\begin{aligned} S(x_1, x_2) &= S(x_1)^{1-\rho_z} S(x_2)^{1-\rho_z} \\ &\times \exp\left\{\tfrac{\rho_z}{\sigma^2}\left(1 - \left((1-\sigma^2 \ln S(x_1))^2 + (1-\sigma^2 \ln S(x_2))^2 - 1\right)^{1/2}\right)\right\}. \end{aligned} \tag{11.10}$$

One can see that this function also has a semiparametric representation, i.e., it is a function of its marginals and parameters of bivariate frailty distribution.

One more correlated frailty model is based on bivariate extension of three parameter distribution of frailty introduced in the univariate survival analysis by Hougaard (2000). This model is characterized by the bivariate survival function of the form

$$S(x_1,x_2)=S(x_1)^{1-\rho_z}S(x_2)^{1-\rho_z}$$
$$\times\exp\left\{\tfrac{\rho_z(1-\alpha)}{\alpha\sigma^2}\left[1-\left(\left(1-\tfrac{\alpha\sigma^2}{1-\alpha}\ln S(x_1)\right)^{\frac{1}{\alpha}}+\left(1-\tfrac{\alpha\sigma^2}{1-\alpha}\ln S(x_2)\right)^{\frac{1}{\alpha}}-1\right)^{\alpha}\right]\right\}. \tag{11.11}$$

This model includes both gamma- and inverse Gaussian correlated frailty models as particular cases: the gamma model corresponds to $\alpha=0$; the inverse Gaussian model has α=0.5 [Yashin et al. (1999)].

11.7 The Underlying Hazard

Note that the estimation of σ_z^2 from bivariate data using (11.9) with $\sigma_1^2=\sigma_2^2=\sigma^2$ does not involve assumptions about μ_0(x). When the estimates of univariate survival function $S(x)$ and the bivariate estimate of σ_z^2 are known, the cumulative $H(x)$ and the instantaneous $\mu_0(x)$ underlying hazards can be calculated as

$$H(x)=\frac{S(x)^{-\sigma^2}-1}{\sigma^2} \tag{11.12}$$

with

$$\mu_0(x)=\overline{\mu}(x)S(x)^{-\sigma_z^2}, \tag{11.13}$$

for the gamma-frailty model, or

$$H(x)=\frac{1}{2\sigma^2}\left(\left(1-\sigma^2\ln S(x)\right)^2-1\right) \tag{11.14}$$

and

$$\mu_0(x)=\bar{\mu}(x)\left(1-\sigma^2\ln S(x)\right) \tag{11.15}$$

for the inverse Gaussian frailty model, or

$$H(x)=\frac{1-\alpha}{\sigma^2}\left(\left(1-\frac{\sigma^2\alpha}{1-\alpha}\ln S(x)\right)^{\frac{1}{\alpha}}-1\right) \tag{11.16}$$

and

$$\mu_0(x)=\bar{\mu}(x)\left(1-\sigma^2\frac{\alpha}{1-\alpha}\ln S(x)\right)^{\frac{\alpha}{1-\alpha}} \tag{11.17}$$

for the three parameter frailty model, where $\overline{\mu}(x)=-\frac{dS(x)/dx}{S(x)}$ is the marginal hazard.

To sum up, the information provided by bivariate data permits semiparametric estimation of the underlying hazard. Hence, in the case of a correlated frailty model one does not need any other assumptions for the conditional survival function except for the proportional hazards assumption. Note that function $S(x)$ may be estimated parametrically or non-parametrically (whatever is most convenient): the parametric description of $S(x)$ can be easily done since univariate subsets of survival data are available. In contrast, the assumptions about the parametric structure of $\mu_0(x)$ are more difficult to justify: there are no empirical data to support any conjecture about $\mu_0(x)$. The results of semiparametric analyses show that the slope of the underlying hazards is steeper than the slope of the observed mortality [Yashin and Iachine (1997)]. Note that the underlying hazards obtained in bivariate studies do not necessarily coincide with the respective hazards estimated in the univariate heterogeneity studies when parametric structure is assumed for $\mu_0(x)$. For example, if the marginal hazard $\overline{\mu}(x)$ corresponds to the Gompertz curve and the parametric structure of the underlying hazard $\mu_0(x)$ is assumed to be a Gompertz curve as well, then the univariate estimate of σ_z^2 will be zero, and $\mu_0(x) = \overline{\mu}(x)$. In contrast, a bivariate analysis will render a non-zero estimate of σ_z^2 to keep the life span correlation at the appropriate level. The underlying hazard will then be estimated by (11.12). Its slope will be steeper than for the Gompertz hazard [Yashin and Iachine (1997)]. This example shows that correction for unobserved heterogeneity in demographic life tables may be needed not only for the oldest old mortality but also for the traditional interval of aging (i.e., 35–85 years of age), where the marginal mortality rate is described by the Gompertz Law.

11.8 An Advantage for Genetic Studies

The variable characterizing susceptibility to disease and death would be an extremely important characteristic of aging and longevity if it could mediate all genetic influence on life span. In the univariate case such a requirement is difficult to meet: in cases where the univariate model is identifiable [Elbers and Ridder (1982)] the genetic contribution to frailty and to an underlying hazard remains unclear. However, this idea may, in fact, be obtainable in bivariate frailty models. The assumption of conditional independence of life spans of related individuals given their frailties guarantees that all genetic influence on life span is mediated by frailty. To show this, it is sufficient to apply this assumption to the data on identical twins. Because of this property the frailty variable in bivariate frailty models is expected to have higher heritability estimates than do the respective life spans.

The presence of correlation coefficients of frailty among model parameters (in the correlated frailty case) opens a new avenue for genetic research on susceptibility to death: the methods and approaches of demography and biostatistics may be combined with the ideas of genetic epidemiology and quantitative genetics to evaluate genetic parameters of susceptibility from incomplete data [Yashin and Iachine (1997)]. For example let A, C, E, refer to additive genetic effects, shared environmental effects and non-shared environmental effects, respectively, in the decomposition of individual frailty on three independent components:

$$Z = A + C + E. \tag{11.18}$$

Then

$$a^2 = 2(\rho_{MZ} - \rho_{DZ}) \tag{11.19}$$

and

$$c^2 = 2\rho_{MZ} - \rho_{DZ}. \tag{11.20}$$

We use small letters to denote respective proportions of variance. Proportion a^2 is called "narrow sense heritability". Note that these calculations are based on the assumption that frailty distribution for MZ and DZ twins have the same variances. The connection between heritability and evaluation of longevity limits is discussed by Yashin and Iachine (1997).

11.9 Dependence Between Observed and Hidden Covariates

The presence of dependence between frailty and observed covariates may make a univariate model non-identifiable. Attempts to use traditional methods of data analysis based on independency assumption in the presence of dependency may substantially bias parameter estimates and lead to erroneous conclusions.

Let us consider the case of the gamma-frailty model with mean 1 and variance $\sigma^2(U)$ (i.e., where frailty distribution depends on observed covariate). The hazard conditional on U is

$$\bar{\mu}(x;U) = \frac{\mu_0(x)e^{\beta U}}{1+\sigma^2(U)e^{\beta U}H(x)}. \tag{11.21}$$

Traditional biostatistical methods will approximate (11.17) with the hazard

$$\tilde{\mu}(x;U) = \frac{\tilde{\mu}_0(x)e^{\tilde{\beta}U}}{1+\tilde{\sigma}^2 e^{\tilde{\beta}U}\tilde{H}(x)} \tag{11.22}$$

following from (11.1), where $\tilde{\sigma}^2$ is the variance of gamma-frailty in case of the traditional model. As a result the estimates of $\mu_0(x), \beta, \sigma^2(U)$ will be biased.

For example, in case of $\beta \equiv 0$, a spurious influence of the covariate on survival ($\tilde{\beta} \neq 0$) may be detected when it does not exist. It turns out that in many important cases the use of bivariate frailty models makes the estimation of characteristics of observed and unobserved covariates possible in the dependent case as well.

Let us now assume that observed covariate U takes two values, 0 and 1, and that conditional on Z and U the hazard function is given by (11.1). Suppose that Z is gamma-distributed with mean 1 and variance $\sigma^2(U)$. We will use the notation σ_0^2 for $\sigma^2(0)$ and σ_1^2 for $\sigma^2(1)$. Thus, the distribution of Z depends on the observed covariate values. In this particular case the population can be stratified: one group is characterized by the hazard $\mu_0(x)$ and a frailty distribution with variance σ_0^2; and a second with the hazard $\mu_0(x)e^{\beta} = \mu_1(x)$ and the variance σ_1^2 for the frailty distribution. Note that neither model is identifiable from univariate data.

Assume now that these conditional hazards describe a population of pairs of dependent individuals with correlations of frailty,$\rho_z(U_1, U_2)$, where U_1 and U_2 are observed covariates associated with hazards of the respective individuals. Then in the case of gamma-frailty the bivariate survival function conditional on U_1,U_2 is

$$S(x_1, x_2 | U_1, U_2) = S(x_1 | U_1)^{1-\rho_z(U_1,U_2)} S(x_2 | U_2)^{1-\rho_z(U_1,U_2)}$$

$$\times \left(S(x_1, U_1)^{-\sigma^2(U_1)} + S(x_2 | U_2)^{-\sigma^2(U_2)} - 1\right)^{-\frac{\rho_z(U_1,U_2)}{\sigma_1(U_1)\sigma_2(U_2)}}. \tag{11.23}$$

Note that (11.23) represents three different bivariate survival functions: $S(x_1, x_2 | 0, 0)$, $S(x_1, x_2 | 0, 1)$, which is symmetric with $S(x_1, x_2 | 1, 0)$, and $S(x_1, x_2 | 1, 1)$. Cases with $U_1 = U_2 = 0$ and $U_1 = U_2 = 1$ correspond to the correlated frailty models described and analysed earlier by Yashin, Vaupel and Iachine (1995). The case with $U_1 = 0$, $U_2 = 1$ corresponds to the bivariate frailty model with different variances of frailty distribution.

Thus, survival models where observed covariates depend on unobserved gamma-distributed frailty may not be identifiable from univariate data. It turns out that they can be identified from bivariate data when they are imbedded into a bivariate correlated gamma frailty-model.

11.10 The Identifiability of the Dependent Competing Risk Model for Twins

Let us assume that cause-specific mortality data are available for the related individuals (e.g., twins). Let Z_i, $EZ_i = 1$, and $Var(Z_i) = \sigma_i^2$ be the gamma-frailty for the ith individual in a twin pair, $i = 1, 2$. Let us assume that for each individual the two competing risks operate in accordance with the shared-frailty model. If $\mu_{i1}(x_i), \mu_{i2}(x_i), i = 1, 2$ are the underlying hazards for the ith individual in a pair, $i = 1, 2$, then the individual survival function is

$$S_i(x_i) = \left(S_{i1}(x_i)^{-\sigma_i^2} + S_{i2}(x_i)^{-\sigma_i^2} - 1\right)^{-\frac{1}{\sigma_i^2}}, \qquad (11.24)$$

$i = 1, 2$. Here $S_{ij}(x_i), i = 1, 2;\ j = 1, 2$ are survival functions corresponding to the latent failure times T_{i1}, T_{i2}, which are associated with the net hazards for the ith individual in a twin pair. Note that these survival functions cannot be estimated from the data since only a sample of independent realizations of $X_i = T_{i1} \wedge T_{i2}, I_i\{T_{i1} < T_{i2}\}$, i=1,2 has been observed for each individual.

Suppose that life-span association between individuals in a pair corresponds to the correlated frailty model with the underlying hazards $\mu_{0i}(x) = \mu_{i1}(x_i) + \mu_{i2}(x_i)$, $i = 1, 2$, correlation coefficient between frailties ρ_z and variances of frailty σ_i^2, $i = 1, 2$. Then their bivariate survival function is:

$$S(x_1, x_2) = S_1(x_1)^{1-\frac{\sigma_1}{\sigma_2}\rho_z} S_2(x_2)^{1-\frac{\sigma_2}{\sigma_1}\rho_z} \left(S_1(x_1)^{-\sigma_1^2} + S_2(x_2)^{-\sigma_2^2} - 1\right)^{-\frac{\rho_z}{\sigma_1\sigma_2}} \qquad (11.25)$$

where x_1, x_2 are the survival times of related individuals and $S_i(x_i)$, $i = 1, 2$ are two marginal univariate survival functions given by (11.25). It turns out that the dependent competing risks model for related individuals described above is identifiable. This means that the parameters ρ_z, $\sigma_i^2, i = 1, 2$ and the underlying hazards $\mu_{0i}(x), \mu_{i1}(x), \mu_{i2}(x)$, $i = 1, 2$ can be estimated from cause-specific survival data for related individuals. The proof of this result is given by Yashin and Iachine (1999a).

11.11 Discussion

This brief review shows that semiparametric models aer powerful tools in survival studies. These models have important connections to the frailty modelling used in demography, epidemiology and biostatistics. Semiparametric representation of multivariate frailty models allows us to avoid unjustifiable technical assumptions about the parametric form of the underlying hazard. It also facilitates the estimation procedure and allows for proper interpretation of the results of data analysis. The structure of the correlated frailty models is appropriate for addressing questions about the genetic nature of individual susceptibility to disease and death. The bivariate models are identifiable even without observed covariates. The dependent competing risks problem (which is non-identifiable in the univariate case [Tsiatis (1975)]) becomes identifiable from bivariate data about cause-specific mortality for related individuals.

Despite clear advantage of such modeling, further work in improving the methods of analysis of bivariate and univariate survival data is needed. This is because the assumption of fixed frailty and the proportionality of hazard, used in these models, are not always realistic. The interpretation of frailty often depends on identifiability conditions. The characteristics of frailty identified by parametric demographic (e.g., gamma-Makeham) models differs from characteristics of frailty in biostatistical models with Elbers and Ridder (1982) identifiability conditions, which in turn differ from characteristics of frailty identified from bivariate data using a correlated gamma-frailty model. The assumption that frailty is independent of observed covariates is questionable especially in studies which address genetic questions. Shared-frailty models for related individuals should be used with care. Correlated frailty models may not be flexible enough in describing multiple correlations. Many useful semiparametric multivariate survival models may be constructed using copula structure. Methods developed by Bagdonavicius and Nikulin (2002) also look promising.

To overcome current problems with the genetic analysis of durations, extended versions of susceptibility models have to be developed. For example, dynamic models of frailty need to be introduced and analyzed. Possible dependence between frailty and observed covariates should be taken into account. Models for pedigree data with observed covariates need to be developed. Quadratic hazards models of liability should be tested. The dependent health-histories models for related individuals have to be developed.

References

1. Aalen, O. O. (1994). Effects of frailty in survival analysis, *Statistical Methods in Medical Research*, **3**, 227–243.

2. Andersen, P. K., Borgan, O., Gill, R. D. and, Keiding, N. (1992). *Statistical Methods Based on Counting Processes*, Springer-Verlag, New York.

3. Bagdonavicius, V. and Nikulin, M. (2002). *Accelerated Life Models: Modeling and Statistical Analysis*, Chapman & Hall/CRC, Boca Raton, Florida.

4. Bretagnolle, J. and Huber-Carol, C. (1988). Effects of omitting covariates in Cox's model with survival data, *Scandinavian Journal of Statistics*, **15**, 125–138.

5. Chamberlain, G. (1985). Heterogeneity, omitted variable bias, and duration dependence, In *Longitudinal Analysis of Labour Market Data* (Eds.,

J. Heckman and B. Singer), pp. 3–38, Cambridge University Press, London, England.

6. Clayton, D. G. (1978). A model for association in bivariate life tables and its application in epidemiological studies of a familial tendency in chronic disease incidence, *Biometrika*, **65**, 141–151.

7. Clayton, D. G. and Cuzick, J. (1985). Multivariate generalizations of the proportional hazards model (with discussion), *Journal of the Royal Statistical Society, Series A*, **148**, 82–117.

8. Cox, D. G. and Oakes, D. (1984). *Analysis of Survival Data*, Chapman & Hall, London, England.

9. Elbers, C. and Ridder, G. (1982). True and spurious duration dependence: The identifiability of proportitional hazard model, *Review of Economic Studies*, **XLIX**, 403–409.

10. Gail, M. H., Wie, S. and Piantadosi, S. (1984). Biased estimates of treatment effect in randomized experiments with nonlinear regressions and omitted covariates, *Biometrika*, **71**, 431–444.

11. Hougaard, P. (2000). *Analysis of Multivariate Survival Data*, Springer-Verlag, New York.

12. Iachine, I. A., Begun, A. Z., Iachina, M., Holm, N., Harris, R. H., Latinen, M., Kaprio, J. and Yashin, A. I. (1999). How heritable is individual susceptibility to death? The results of an analysis of survival data on Danish, Swedish and Finnish twins, *Twin Research*, **1**, 196–205.

13. Iachine, I. A. and Yashin, A. I. (1998). Identifiability of bivariate frailty models based on additive independent components, *Research Reports*, **8**, Department of Statistics and Demography, Odense University, Denmark.

14. Locatelli, I. (2003). A Bayesian approach to the correlated frailty model, Paper presented at the PAA Meeting, May 2003.

15. Manton, K. G., Stallard, E. and Vaupel, J. W.(1986). Alternative models of heterogeneity in mortality risks among the aged, *Journal of the American Statistical Association*, **81**, 635–644.

16. Oakes, D. (1989). Bivariate survival models induced by frailties, *Journal of the American Statistical Association*, **84**, 487–493.

17. Pickles, A. and Crouchley, R. (1995). A comparison of frailty models for multivariate survival data, *Statistics in Medicine*, **14**, 1447–1461.

18. Tsiatis, A. A. (1975). A nonidentifiability aspect of the problem of competing risks, *Proceedings of the Academy of Sciences, USA*, **72**, 20–22.

19. Vaupel, J. W., Carey, J. R., Christensen, K., Johnson, T. E., Yashin, A. I., Holm, N. V., Iachine, I. A., Kannisto, V., Khazaeli, A. A., Liedo, P., Longo, V. D., Zeng, Y., Manton, K. G. and Curtsinger, J. W. (1998). Biodemographic trajectories of longevity, *Science*, **280**, 855–860.

20. Vaupel, J. W., Manton, K. G. and Stallard, E. (1979). The impact of heterogeneity in individual frailty on the dynamics of mortality, *Demography*, **16**, 439–454.

21. Vaupel, J. W. and Yashin, A. I. (1985). Heterogeneity's ruses: Some surprising effects of selection on population dynamics, *The American Statistician*, **39**, 176–185.

22. Wienke, A., Christensen, K., Holm, N. V. and Yashin, A. I. (2000). Heritability of death from respiratory diseases: an analysis of Danish twin survival data using a correlated frailty model, In *Medical Infobahn for Europe* (Proceedings of MIE 2000 and GMDS 2000) (Eds., A, Hasman, B. Blobel, J. Dudeck, R. Engelbrecht, G. Gell, and H.-U. Prokosch), pp. 407-411, IOS Press, Amsterdam.

23. Wienke, A., Christensen, K., Skytte, A. and Yashin, A. I. (2002). Genetic analysis of cause of death in a mixture model of bivariate life time data, *Statistical Modelling*, **2**, 1–14.

24. Wienke, A., Holm, N. V., Skytthe, A. and Yashin, A. I. (2001). The heritability of mortality due to heart disease: A correlated frailty model applied to Danish twins, *Twin Research*, **4**, 266–274.

25. Yashin, A. I. and Iachine, I. A. (1994). Mortality models with application to twin survival data, In *CISS-First Joint Conference of International Simulation Societies Proceedings* (Eds., J. Halin, W. Karplus and R. Rimane), pp. 567–571, Zurich, Switzerland.

26. Yashin, A. I. and Iachine, I. A. (1997). How frailty models can be used for evaluating longevity limits: Taking advantage of an interdisciplinary approach, *Demography*, **34**, 31–48.

27. Yashin, A. I. and Iachine, I. A. (1999a). Dependent hazards in the problems of multivariate survival, *Journal of Multivariate Analysis*, **71** 241–261.

28. Yashin, A. I. and Iachine, I. A. (1999b). What difference does the dependence between durations make? Insights for the population studies of aging, *Lifetime Data Analysis*, **5**, 5–22.

29. Yashin, A. I., Vaupel, J. W. and Iachine, I. A. (1995). Correlated individual frailty: An advantageous approach to survival analysis of bivariate data, *Mathematical Population Studies*, **5**, 1–15.

30. Xue, X. and Brookmeyer, R. (1996). Bivariate frailty models for the analysis of multivariate survival time, *Lifetime Data Analysis*, **2**, 277–289.

31. Xue, X. and Ding, Y. (1999). Assessing heterogeneity and correlation of paired failure times with the bivariate frailty model, *Statistics in Medicine*, **18**, 907–918.

32. Zdravkovic, S., Wienke, A., Pedersen, N. L., Marenberg, M. E., Yashin, A. I. and de Faire, U. (2002). Heritability of death from coronary heart disease: 36 years follow-up of 20,966 Swedish twins, *Journal of Internal Medicine*, **252**, 1–8.

PART V
Analyses of Censored and Truncated Data

12

Semiparametric Transformation Models for Arbitrarily Censored and Truncated Data

Catherine Huber-Carol[1] **and Filia Vonta**[2]

[1] *University of Paris 5 and U 472 Inserm, France*
[2] *University of Cyprus, Nicosia, Cyprus*

Abstract: A regression analysis for arbitrarily censored and truncated data was proposed by Alioum and Commenges (1996), using Cox's proportional hazards model, and based on Turnbull's method (1976) for nonparametric estimation of a distribution function. We propose here a generalization of their method to the case where there is an unobserved heterogeneity in the data taken into account by a frailty model. Our methodology is applied to a set of real data on transfusion-related AIDS that has been used among others by Kalbfleisch and Lawless (1989).

Keywords and phrases: Frailty models, transformation models, censored data, truncated data, nonparametric maximum likelihood estimation, gamma frailty, inverse Gaussian frailty

12.1 Introduction

A common feature of many failure time data is that they are simultaneously truncated and interval-censored. According to Hosmer and Lemeshow (1999), a censored observation is one whose value is incomplete due to random factors for each subject. A truncated observation, on the other hand, is one which is incomplete due to a selection process in the design of the study. Thus, truncation changes the sample size while censoring does not. Examples of right truncated data occur in registers. An acquired immune deficiency syndrome (AIDS) register only contains AIDS cases which have already been reported, which generates right truncated samples of induction times. Censored data are commonly observed in various types of experiments such as for life testing and reliability improvement. A special type of censored data is the so-called interval censored data or inspection data. Interval censored data contain uncertainty as to when the units actually failed. The interval censoring usually comes from grouped or clustered data or from the fact that patients are examined at certain dates and the event of interest is only known to have occured between two specific checking times, one of which may be infinite in case of right censoring, when at the end of the study the event has not yet occured.

The most widely used model in survival analysis is the Cox proportional hazards model [Cox (1972)]. The Cox model is used in several fields of application in clinical, epidemiological and engineering research. The Cox model and its extensions have been applied to various settings such as the time-to-event, multiple events for each subject and competing events settings. Lifetime data of a component, product or system can be categorized as follows: complete data, right censored data, interval censored data, and left censored data. Each data type requires a modification to the analysis in order to correctly estimate the underlying lifetime distribution. The analysis of censored data presents a challenging problem to experimenters because many traditional methods become inadequate. Although the cases of right censored and/or left truncated data can be handled through the standard method of estimation in the Cox model, namely, the partial likelihood, the cases for example of interval censored or right truncated data should be treated differently. Turnbull (1976) and Frydman (1994) dealt with the nonparametric estimation of the distribution function F when the data are interval censored and truncated. Finkelstein (1986) fitted a proportional hazards model to interval censored data and Finkelstein et al. (1993) to right truncated data. Alioum and Commenges (1996) extended the existing results by fitting a proportional hazards model to arbitrarily censored and truncated data and concentrated on hypothesis testing. Pan and Chappell (2002) concentrated on estimation of parameters in the Cox model with left-truncated and interval censored data. In this paper we introduce frailty models for the case of arbitrarily censored and truncated data and focus on estimation of the parameters of interest as well as the nuisance parameter of our model.

Frailty models are random effects models designed to handle censored survival data. In other words, they are used to model survival times in the presence of overdispersion or group-specific random effects. Interest in the use of random effects in survival analysis settings has been increasing in recent years. Frailty models could be considered as the survival data analog to regression models, which account for heterogeneity and random effects. The random effect is a continuous variable which describes the frailty (excess risk) of an individual or family, group or cluster of individuals, over and above any measured covariates. The case where the standard assumption of independence seems unrealistic is yet another situation where the random effect is introduced to model dependence between survival times. The random effect is defined to act multiplicatively on the hazard rate $h(t|z)$ of an individual with covariate vector z. A frailty model therefore arises naturally from a Cox model with unobserved covariates which materialize the frailty parameter. Such models have been proposed for example by Vaupel et al. (1979) who introduced the concept of frailty and studied the model with gamma distributed frailties. Vaupel and Yashin (1983) proposed other distributions such as the uniform distribution, the Weibull and the log normal for the frailty parameter. A three-parameter family of distributions has been proposed by Hougaard (1986) which includes not only the Gamma but also the Inverse Gaussian and the Positive Stable distributions.

The paper is organised as follows. The general framework of arbitrarily censored and truncated data for i.i.d. random variables is presented in Section 12.2. In Section 12.3 a general class of semiparametric frailty or transformation models is defined, the likelihood is furnished and its maximization in order to produce the NPMLE's of β, the parameter of interest, and the baseline cumulative hazard function Λ, namely, the nuisance parameter, is examined. Furthermore, issues concerning the identifiability of both, the parameter of interest and the nuisance parameter are briefly discussed. The performance of the proposed model is illustrated in Section 4 through a set of real data on transfusion-related AIDS [Kalbfleisch and Lawless (1989)]. Note that the present manuscript provides a brief overview of Huber-Carol and Vonta (2003).

12.2 Nonparametric Estimation of the Survival Function

Let $X_1, X_2, \ldots, X_n$ be independent and identically distributed positive random variables with survival function $S(x)$ which are 'incomplete' in the sense that they are arbitrarily censored and truncated. More specifically, there exist sets $B_1, B_2, \ldots, B_n$, subsets of the real line, which are called truncating sets, such that each random variable X_i belongs to the sample, only if X_i falls into the set B_i. Thus the investigator is aware of the existence of the observation X_i only if $X_i \in B_i$. Moreover, X_i is not being observed exactly but rather it is only known that X_i lies in a set A_i, which is called the censoring set, where $A_i \subseteq B_i$. The observed data within this framework are therefore the n pairs $(A_1, B_1), (A_2, B_2), \ldots, (A_n, B_n)$. In the sequel we will follow the formulation of Turnbull (1976), Frydman (1994), and Alioum and Commenges (1996) to define a nonparametric maximum likelihood estimator of the survival function S, which decreases only in a finite number of disjoint intervals.

Let us assume that the sets A_i belong to a partition $\mathcal{P}_i$ of $[0, \infty)$ and also that B_i and $\mathcal{P}_i$ are independent of X_i and of the parameters of interest. We assume that the censoring sets A_i, $i = 1, \ldots, n$ can be expressed as a finite union of disjoint closed intervals, that is,

$$A_i = \cup_{j=1}^{k_i}[L_{ij}, R_{ij}]$$

where $0 \leq L_{i1} \leq R_{i1} < L_{i2} \leq R_{i2} < \cdots < L_{ik_i} \leq R_{ik_i} \leq \infty$ for $i = 1, \ldots, n$, $R_{i1} > 0$, and $L_{ik_i} < \infty$. Moreover, we assume that the truncating sets B_i can be expressed as a finite union of open intervals

$$B_i = \cup_{j=1}^{n_i}(\mathcal{L}_{ij}, \mathcal{R}_{ij})$$

where $0 \leq \mathcal{L}_{i1} < \mathcal{R}_{i1} < \mathcal{L}_{i2} < \mathcal{R}_{i2} < \cdots < \mathcal{L}_{in_i} < \mathcal{R}_{in_i} \leq \infty$ for $i = 1, \ldots, n$.

The likelihood of the n pairs of observations (A_i, B_i), $i = 1, 2, \ldots, n$ is proportional to

$$l(S) = \prod_{i=1}^{n} l_i(S) = \prod_{i=1}^{n} \frac{P_S(A_i)}{P_S(B_i)} = \prod_{i=1}^{n} \frac{\sum_{j=1}^{k_i}\{S(L_{ij}^-) - S(R_{ij}^+)\}}{\sum_{j=1}^{n_i}\{S(\mathcal{L}_{ij}^+) - S(\mathcal{R}_{ij}^-)\}}. \tag{12.1}$$

Let us define now the sets

$$\tilde{L} = \{L_{ij},\ 1 \leq j \leq k_i, 1 \leq i \leq n\} \cup \{\mathcal{R}_{ij},\ 1 \leq j \leq n_i, 1 \leq i \leq n\} \cup \{0\}$$

and

$$\tilde{R} = \{R_{ij},\ 1 \leq j \leq k_i, 1 \leq i \leq n\} \cup \{\mathcal{L}_{ij},\ 1 \leq j \leq n_i, 1 \leq i \leq n\} \cup \{\infty\}.$$

In order to find the function S that maximizes the above likelihood, one has to notice that it is maximized when the values of $S(x)$ are as large as possible for $x \in \tilde{L}$ and as small as possible for $x \in \tilde{R}$. Accordingly, a set Q is defined uniquely as the union of disjoint closed intervals whose left endpoints lie in the set $\tilde{L}$ and right endpoints in the set $\tilde{R}$ respectively, and which contain no other members of $\tilde{L}$ or $\tilde{R}$. Thus,

$$Q = \cup_{j=1}^{v}[q_j', p_j']$$

where $0 = q_1' \leq p_1' < q_2' \leq p_2' < \cdots < q_v' \leq p_v' = \infty$. Subsequently, we denote by C the union of intervals $[q_j', p_j']$ covered by at least one censoring set, W the union of intervals $[q_j', p_j']$ covered by at least one truncating set but not covered by any censoring set and $D = \overline{(\cup B_i)}$ the union of intervals $[q_j', p_j']$ not covered by any truncating set. D is actually included in the union of intervals $[q_j', p_j']$. Notice that the set Q can be written as $Q = C \cup W \cup D$. The following two lemmas [that appear in Turnbull (1976) and Alioum and Commenges (1996)] are essential for the maximization of (12.1) with respect to S and become apparent upon examination of (12.1).

Lemma 12.2.1 *Any survival distribution function which decreases outside the set $C \cup D$ cannot be the NPMLE of S.*

From the above lemma we deduce that a NPMLE of S lies among the functions that are constant outside the set $C \cup D$. Moreover notice that it is only possible to estimate the conditional survival function $S_{\overline{D}}(x) = P(X > x | X \in \overline{D})$ since we do not have any information from the observed data about the proportion of observations that belong to D. It is easy to see that $S_{\overline{D}}$ and S give rise to the same likelihood. Therefore one should concentrate his efforts into finding a NPMLE of $S_{\overline{D}}$ when $P_S(D)$ is unknown. $P_S(D)$ could be assumed equal to 0 as it was done in Turnbull (1976) and Frydman (1994). Let us denote the set C as

$$C = \cup_{i=1}^{m} [q_i, p_i]$$

where $q_1 \leq p_1 < q_2 \leq p_2 < \cdots < q_m \leq p_m$. Let also $s_j = S_{\overline{D}}(q_j{}^-) - S_{\overline{D}}(p_j{}^+)$. The problem of maximizing (12.1) actually reduces to the problem of maximizing the following likelihood written as a function of $s_1, s_2, \ldots, s_m$, that is,

$$l(s_1, \ldots, s_m) = \prod_{i=1}^{n} \frac{\sum_{j=1}^{m} \mu_{ij} s_j}{\sum_{j=1}^{m} \nu_{ij} s_j} \tag{12.2}$$

subject to $\sum s_j = 1$, $s_j \geq 0$ for $0 \leq j \leq m$, where $\mu_{ij} = I_{[\ [q_j, p_j] \subset A_i]}$ and $\nu_{ij} = I_{[\ [q_j, p_j] \subset B_i]}$, $i = 1, \ldots, n$ and $j = 1, \ldots, m$.

Lemma 12.2.2 *For fixed values of* $S_{\overline{D}}(q_j{}^-)$ *and* $S_{\overline{D}}(p_j{}^+)$*, for* $1 \leq j \leq m$*, the likelihood is independent of how the decrease actually occurs in the interval* $[q_j, p_j]$*, so that* $S_{\overline{D}}$ *is undefined within each interval* $[q_j, p_j]$*.*

The previous lemma implies that the NPMLE of $S_{\overline{D}}$ is unique only up to an equivalence class of NPMLE's that share the same values $s_1, s_2, \ldots, s_m$.

12.3 Semiparametric Transformation Models and the Likelihood Formulation

The hazard rate, that is, the probability of instantaneous failure of an individual with p-dimensional covariate vector z within the Cox proportional hazards model, is defined as

$$h(t|z) = e^{\beta^T z} h_0(t)$$

where $\beta \in R^p$ is the parameter of interest and $h_0(t)$ is the baseline hazard rate. This is a semiparametric model as the nuisance parameter, namely, the function $h_0(t)$ is infinite-dimensional. Although the Cox model is the model of choice for analyzing survival data, there are times when the assumption of proportional hazards seems unrealistic. A class of nonproportional hazards models is naturally generated from the Cox model when a positive random variable η, called frailty [Vaupel et al. (1979)], is introduced into the above hazard intensity function in the following way:

$$h(t|z,\eta) = \eta e^{\beta^T z} h_0(t).$$

Then

$$S(t|z,\eta) = e^{-\eta e^{\beta^T z}\Lambda(t)}$$

where $\Lambda(t)$ denotes the baseline cumulative hazard function. Thus,

$$S(t|z) = \int_0^\infty e^{-xe^{\beta^T z}\Lambda(t)} dF_\eta(x) \equiv e^{-G(e^{\beta^T z}\Lambda(t))} \quad (12.3)$$

where

$$G(y) = -\ln(\int_0^\infty e^{-xy} dF_\eta(x))$$

and F_η is the distribution function of the frailty parameter. The Cox model is of course included in model (12.3) when the function G is taken to be the identity function. A popular choice of a frailty model, due to its mathematical convenience, is the Clayton–Cuzick semiparametric model [Clayton–Cuzick (1986)] in which the frailty parameter is taken to be distributed as Gamma with mean 1 and variance c. The law of the Gamma frailty actually involves two parameters that reduce to one, when due to identifiability assumptions, it is imposed that $E(\eta) = 1$. The remaining parameter may be expressed simply in terms of the variance of the frailty so that choosing a parameter for the gamma frailty amounts to choosing its variance. Another choice for the frailty distribution is the Inverse Gaussian distribution [Hougaard (1984, 1986)] which also involves two parameters that reduce to one when $E(\eta) = 1$ is imposed. The remaining parameter appears in the variance of the frailty distribution as in this case $Var(\eta) = 1/2b$. The corresponding G functions for these models are given by

$$G(x,c) = \frac{1}{c}\ln(1+cx),\ c > 0$$

and

$$G(x,b) = 2(\sqrt{b(b+x)} - b),\ b > 0.$$

The class of semiparametric transformation models as was defined in Cheng et al. (1995) for right censored data, namely,

$$g(S(t|z)) = h(t) + \beta^T z \quad (12.4)$$

where g is known and h unknown, is equivalent to the class of models (12.3) through the relations

$$g(x) \equiv log(G^{-1}(-\log(x)), \quad h(t) \equiv \log(\Lambda(t)).$$

In the sequel, we will assume that the frailty distribution is completely known and we will focus on the joint estimation of β and Λ within the frailty model (12.3). Let $(X_1, Z_1), ..., (X_n, Z_n)$ be i.i.d. random pairs of variables with marginal survival function defined in (12.3) as in Vonta (1996) and Slud and Vonta (2004). The function $G \in C^3$ is assumed to be a known strictly increasing concave function with $G(0) = 0$ and $G(\infty) = \infty$. As in the previous section we assume that the random variables X_i are subject to arbitrary censoring and truncation. Within the framework of the frailty model (12.3), the likelihood (12.1) takes the form

$$l(\Lambda, \beta|z) = \prod_{i=1}^{n} l_i(\Lambda, \beta|z) = \prod_{i=1}^{n} \frac{\sum_{j=1}^{k_i} \{e^{-G(e^{\beta^T z}\Lambda(L_{ij}{}^-))} - e^{-G(e^{\beta^T z}\Lambda(R_{ij}{}^+))}\}}{\sum_{j=1}^{n_i} \{e^{-G(e^{\beta^T z}\Lambda(\mathcal{L}_{ij}{}^+))} - e^{-G(e^{\beta^T z}\Lambda(\mathcal{R}_{ij}{}^-))}\}}. \tag{12.5}$$

Our interest is to obtain the joint NPMLE's of the parameter of interest β and the nuisance parameter Λ. In the maximization of (12.5) with respect to the baseline cumulative hazard function Λ we employ Lemma 12.2.1 and Lemma 12.2.2 of the previous section, that continue to hold under the present generalization. The analogues of Lemmas 12.2.1 and 12.2.2 for the situation at hand are given below. The proof of Lemma 12.3.1 can be found in Huber-Carol and Vonta (2003).

Lemma 12.3.1 *Any cumulative hazard-type function Λ within model (12.3) which increases outside the set $C \cup D$ cannot be the NPMLE of Λ.*

Lemma 12.3.2 *For fixed values of $\Lambda(q_j{}^-)$ and $\Lambda(p_j{}^+)$, for $1 \leq j \leq m$, the likelihood is independent of how the increase actually occurs in the interval $[q_j, p_j]$, so that Λ is undefined within each interval $[q_j, p_j]$.*

Notice that we need not assume that $P_\Lambda(D)$ is zero as we have found that when $\beta \neq 0$ the parameters β and Λ are identifiable as long as we have at hand at least two covariates. In fact, one covariate with three levels would also suffice for identifiability of the parameters.

We need to rewrite the likelihood (12.5) in a more convenient form so that the maximization with respect to Λ and β will be possible. Since the set $C = \cup_{j=1}^{m}[q_j, p_j]$, the set D can be written as $D = \cup_{j=0}^{m} D_j$, where $D_j = D \cap (p_j, q_{j+1})$, where $p_0 = 0$ and $q_{m+1} = \infty$. Notice that D_j is either a closed interval or a union of disjoint closed intervals. Let $\delta_j = P_\Lambda(D_j)$ denote the mass of the baseline cumulative hazard function Λ on the set D_j. From Lemma 12.3.1 we have that $\Lambda(q_j^-) = \Lambda(p_{j-1}^+) + \delta_{j-1}$ for $1 \leq j \leq m+1$. The log-likelihood can then be expressed as

$$\log l(\Lambda, \beta|z) = \sum_{i=1}^{n} \Big\{ \log \Big(\sum_{j=1}^{m} \mu_{ij} (e^{-G(e^{\beta^T z}(\Lambda(p_{j-1}^+)+\delta_{j-1}))} - e^{-G(e^{\beta^T z}\Lambda(p_j+))}) \Big)$$
$$- \log \Big(\sum_{j=1}^{m} \nu_{ij} (e^{-G(e^{\beta^T z}(\Lambda(p_{j-1}^+)+\delta_{j-1}))} - e^{-G(e^{\beta^T z}\Lambda(p_j+))}) \Big) \Big\}. \quad (12.6)$$

In most real data problems, the set D consists of the union of two intervals, namely, D_0 and D_m. If we are in the situation of right truncated data, then the set $D = D_m$. If we are in the situation of left truncated data, then the set $D = D_0$. Therefore the case $D = D_0 \cup D_m$ covers most of the problems one would encounter in practice and therefore we will simplify our likelihood by setting $\delta_1 = \delta_2 = \cdots = \delta_{m-1} = 0$. Therefore the parameters that remain in (12.6) are $\beta, \delta_0, \Lambda(p_0), \ldots, \Lambda(p_m)$. Since $\Lambda(p_0) = 0$ we have to maximize the log-likelihood (12.6) with respect to the $p+m+1-$dimensional parameter $(\beta, \delta_0, \Lambda(p_1), \ldots, \Lambda(p_m))$. Similarly to Finkelstein et al. (1993) and Alioum and Commenges (1996) we proceed with the reparametrization $\tau_0 = \log(\delta_0)$, $\tau_1 = \log(\Lambda(p_1))$ and $\tau_j = \log(\log(\Lambda(p_j)) - \log(\Lambda(p_{j-1})))$ for $j = 2, \ldots, m$. The reparametrization is required not only for computational convenience but also for improving the speed of convergence. The maximization has been achieved with the use of Splus and Fortran 77 software.

12.4 Real Data Example

Kalbfleisch's data on AIDS acquired by transfusion [Kalbfleisch and Lawless (1989)] include $n = 295$ patients for whom induction time, age at infection and date of infection are reported. The induction time t_i and the date of infection x_i are reported in months, beginning on January 1st, 1978, and ending January 1st, 1987. As the earliest infection was reported in January 1978 and labeled as month 1, the maximum observable induction time is x_* equal to 102 months. The induction times are right-truncated because an individual i is only included in the sample if $t_i \leq x_* - x_i$. We consider three groups of individuals according to their age, namely children, adults and elderly (with corresponding age intervals [0,12], (12,60) and [60,80]). The covariate Z is assumed to be equal to 0,1 and 2 according to the three classes of age. The presence of this covariate allows our model to be identifiable. In our analysis we keep the month as the unit of time [as in Alioum and Commenges (1996)]. We considered two frailty models of the class defined in (12.3), namely, the Inverse Gaussian and the Gamma frailty model.

In both cases we maximize the log-likelihood over β and Λ but also with respect to the frailty parameter involved in the function G. For the Inverse Gaussian model the maximum of the log-likelihood with respect to the parameter b occurs at $b = 0.039$ giving an estimated value of the regression parameter $\hat{\beta} = -4.69$. This value of b corresponds to frailty variance equal to 12.82. On the other hand, for the Clayton–Cuzick model the log-likelihood values continue to increase as c tends to ∞ but they remain quite stable after $c = 10$. This value of the variance of the frailty, that is, c equal to 10 corresponds to an estimated $\hat{\beta} = -3.62$. From the above notice that a random effect is essential to be included in the hazard rate in order to describe the heterogeneity present in the transfusion related AIDS data. This conclusion is greatly supported by the fact that the log-likelihood values increase in the Clayton–Cuzick model as the parameter c, which is actually the variance of the frailty parameter, tends to ∞, implying that there exists heterogeneity in the population. The variance of the frailty parameter for the Inverse Gaussian model is equal to $1/2b$ and since the log-likelihood values in this case are maximized at a value of b close to 0 we arrive at the same conclusion.

Further details and an extended discussion on the proposed model for the general scheme of truncated and censored data can be found in Huber-Carol and Vonta (2003).

References

1. Alioum, A. and Commenges, D. (1996). A proportional hazards model for arbitrarily censored and truncated data, *Biometrics*, **52**, 512–524.

2. Cheng, S. C., Wei, L. J. and Ying, Z. (1995). Analysis of transformation models with censored data, *Biometrika*, **82**, 835–845.

3. Clayton, D. and Cuzick, J. (1986). The semiparametric Pareto model for regression analysis of survival times, In *Papers on Semiparametric Models*, MS-R8614, 19-31, Centrum voor Wiskunde en Informatica, Amsterdam.

4. Cox, D. R. (1972). Regression models and life tables (with discussion), *Journal of the Royal Statistical Society, Series B*, **34**, 187–220.

5. Finkelstein, D. M. (1986). A proportional hazards model for interval-censored failure time data, *Biometrics*, **42**, 845-854.

6. Finkelstein, D. M., Moore, D. F. and Schoenfeld , D. A. (1993). A proportional hazards model for truncated AIDS data, *Biometrics*, **49**, 731-740.

7. Frydman, H. (1994). A note on nonparametric estimation of the distribution function from interval-censored and truncated observations, *Journal of the Royal Statistical Society, Series B*, **56**, 71–74.

8. Hosmer, D. W. and Lemeshow, S. (1999). *Applied Survival Analysis: Regression Modeling of Time to Event Data*, John Wiley & Sons, New York.

9. Hougaard, P. (1984). Life table methods for heterogeneous populations: Distributions describing the heterogeneity, *Biometrika*, **71**, 75–83.

10. Hougaard, P. (1986). Survival models for heterogeneous populations derived from stable distributions, *Biometrika*, **73**, 387–396.

11. Huber-Carol, C. and Vonta, F. (2003). Frailty models for arbitrarily censored and truncated data, *Preprint*.

12. Kalbfleisch, J. D. and Lawless, J. F. (1989). Inference based on retrospective ascertainment: An analysis of data on transfusion-associated AIDS, *Journal of the American Statistical Association*, **84**, 360–372.

13. Pan, W. and Chappell, R. (2002). Estimation in the Cox proportional hazards model with left-truncated and interval-censored data, *Biometrics*, **58**, 64–70.

14. Slud, E. V. and Vonta, F. (2004). Consistency of the NPML estimator in the right-censored transformation model, *Scandinavian Journal of Statistics*, **31**, 21–41.

15. Turnbull, B. W. (1976). The empirical distribution function with arbitrarily grouped, censored and truncated data, *Journal of the Royal Statistical Society, Series B*, **38**, 290–295.

16. Vaupel, J. W., Manton, K. G. and Stallard, E. (1979). The impact of heterogeneity in individual frailty on the dynamics of mortality, *Demography*, **16**, 439–454.

17. Vaupel, J. W. and Yashin, A. I. (1983). The deviant dynamics of death in heterogeneous populations, *Research Report 83-1*, International Institute for Applied Systems Analysis, Laxenburg, Austria.

18. Vonta, F. (1996). Efficient estimation in a non-proportional hazards model in survival analysis, *Scandinavian Journal of Statistics*, **23**, 49–61.

13

EM Algorithm for Type-II Right Censored Bivariate Normal Data

N. Balakrishnan and J.-A. Kim

Department of Mathematics & Statistics, McMaster University, Hamilton, Ontario, Canada

Abstract: EM algorithm is used to find the maximum likelihood estimates based on Type-II right censored samples from a bivariate normal distribution. The asymptotic variances and covariances of the MLEs are derived, using the missing information principle, from the Fisher information matrix and the partially observed information matrix. Using the asymptotic normality of MLEs and the asymptotic variances and covariances derived, probability coverages of 90% and 95% confidence intervals for all the parameters are evaluated by means of Monte Carlo simulations.

Keywords and phrases: EM algorithm, maximum likelihood estimates, concomitants of order statistics, Type-II right censoring, asymptotic variances, missing information principle

13.1 Introduction

Let $(X_1, Y_1), ..., (X_n, Y_n)$ be a random sample from a bivariate distribution with cdf $F(x, y)$ and pdf $f(x, y)$. Let $X_1, X_2, ..., X_n$ be arranged in ascending order of magnitude, with $X_{1:n}$ denoting the smallest among X values, $X_{2:n}$ the second smallest, ..., and $X_{n:n}$ denoting the largest observation. We call $X_{i:n}$ the i-th order statistic, $i = 1, 2, ..., n$, and let $Y_{[i:n]} = Y_j$ if $X_{i:n} = X_j$ for $i, j = 1, ..., n$. Then, $Y_{[i:n]}$ is termed the concomitant of the i-th order statistic [David (1973)] or the induced order statistic [Bhattacharya (1974)]. These concomitants are of interest in selection and prediction problems based on the ranks of the X's. For example, when k $(< n)$ individuals having the highest X-scores are selected, we may wish to know the behavior of the corresponding concomitant Y-scores.

The exact and asymptotic distribution theory of $Y_{[i:n]}$ from some bivariate distributions were studied by Yang (1977) and the asymptotic distribution of the concomitants from a bivariate normal distribution were derived by David and Galambos (1974). David and Nagaraja (1998) have provided a review of the area of concomitants of order statistics including the role of concomitants in the estimation of regression and correlation coefficients and the analysis of censored bivariate data.

We denote a bivariate normal density by $\phi_\theta(x, y), \theta = (\mu_X, \sigma_X, \mu_Y, \sigma_Y, \rho)$, where (μ_X, μ_Y) are means, (σ_X, σ_Y) are the standard deviations, and ρ is the correlation coefficient between X_i and Y_i. Based on a Type-II right censored bivariate normal sample consisting of $(X_{1:n}, ..., X_{k:n})^T$ and $(Y_{[1:n]}, ..., Y_{[k:n]})^T$, Harrell and Sen (1979) derived the maximum likelihood estimators of the parameters and the associated large-sample variance-covariance matrix.

The Expectation-Maximization (EM) algorithm [Dempster et al. (1977) and McLachlan and Krishnan (1997)] is a useful tool to estimate the parameters of the distribution based on incomplete data, especially when the complete data problems are relatively easy. In the EM algorithm for incomplete data problems, the parameters are estimated after filling in initial values for the missing data. Then the initial values for missing data are updated by their expected values using the initially estimated parameters. The parameters are then re-estimated, and so on, proceeding iteratively until convergence. The censored data mentioned above can be viewed as an incomplete data and the EM algorithm can then be applied to obtain the maximum likelihood estimators. It should be mentioned that the EM algorithm for progressively Type-II right censored univariate data has been discussed recently by Ng et al. (2002). These authors also used the Missing Information Principle [Orchard and Woodbury (1970), Louis (1982) and Tanner (1993)] to derive the asymptotic variances and covariances of the MLEs when the EM algorithm is used for progressively Type-II right censored univariate data.

In Section 13.2, conditional distributions of censored data given observed data are determined for Type-II right censored samples from a bivariate distribution, using which conditional expectations of censored data given observed data are derived when the samples are from a bivariate normal distribution. In Section 13.3, EM algorithm for finding the MLEs of parameter θ of a bivariate normal distribution is discussed. In Section 13.4, asymptotic variances and covariances of the MLEs are derived using the missing information principle from the Fisher information matrix as well as from the partially observed information matrix. An illustrative example is presented in Section 13.5. A Monte Carlo simulation study was conducted for various sample sizes n, censoring proportions and the correlation coefficient ρ, and the corresponding results are presented in Section 13.6. Using the asymptotic normality of MLEs and the asymptotic variances and covariances of MLEs derived from the Fisher information matrix as well as from the partially observed information matrix, probability coverages of 90% and 95% confidence intervals are determined by Monte Carlo simulations and are presented in Section 13.7 with some comments.

13.2 Conditional Distributions and Expectations of Censored Data

For convenience, the following notation will be adopted throughout this article:

$$\begin{aligned} F_Z(z) &\text{ — cdf of the random variable Z,} \\ f_Z(z) &\text{ — pdf of the random variable Z,} \\ f_{Y|X}(y|x) &\text{ — conditional pdf of } Y\text{, given } X = x. \end{aligned}$$

Let $X_{obs} = (X_{1:n}, X_{2:n}, \ldots, X_{k:n})^T$ and $Y_{obs} = (Y_{[1:n]}, Y_{[2:n]}, \ldots, Y_{[k:n]})^T$ be the observed data, and $X_{cen} = (X_{k+1:n}, X_{k+2:n}, \ldots, X_{n:n})^T$ and $Y_{cen} = (Y_{[k+1:n]}, Y_{[k+2:n]}, \ldots, Y_{[n:n]})^T$ be the censored data. Combine (X_{obs}, Y_{obs}) and (X_{cen}, Y_{cen}) to form (X, Y) which is the complete data, where $X = (X_{obs}, X_{cen})^T$ and $Y = (Y_{obs}, Y_{cen})^T$. The joint density of (X, Y) is given by

$$\begin{aligned} f_{(X,Y)}(x, y) &= n! \prod_{i=1}^{n} f_{X,Y}(x_{i:n}, y_{[i:n]}) \\ &= n! \prod_{i=1}^{n} f_{Y|X}(y_{[i:n]}|x_{i:n}) f_X(x_{i:n}), \\ &\qquad\qquad x_{1:n} \leq \cdots \leq x_{n:n}. \end{aligned} \tag{13.1}$$

The joint density of (X_{obs}, Y_{obs}) is given by

$$\begin{aligned} &f_{(X_{obs},Y_{obs})}(x_{obs}, y_{obs}) \\ &= \frac{n!}{(n-k)!} \prod_{i=1}^{k} f_{Y|X}(y_{[i:n]}|x_{i:n}) f_X(x_{i:n}) [1 - F_X(x_{k:n})]^{n-k}, \\ &\qquad x_{1:n} \leq \cdots \leq x_{k:n}, \ -\infty < y_{[i:n]} < \infty. \end{aligned} \tag{13.2}$$

From (13.1) and (13.2), the conditional joint distribution of censored data, given observed data, can be written as

$$\begin{aligned} &f_{(X_{cen},Y_{cen})|(X_{obs},Y_{obs})}(x_{cen}, y_{cen}|x_{obs}, y_{obs}) \\ &= (n-k)! \prod_{i=k+1}^{n} f_{Y|X}(y_{[i:n]}|x_{i:n}) \frac{f_X(x_{i:n})}{[1 - F_X(x_{k:n})]}, \\ &\qquad x_{k:n} \leq x_{k+1:n} \leq \cdots \leq x_{n:n}, \ -\infty < y_{[i:n]} < +\infty, \\ &\qquad i = k+1, \ldots, n. \end{aligned} \tag{13.3}$$

Therefore, the density function of (X_j, Y_j) given (X_{obs}, Y_{obs}), where (X_j, Y_j) is one of the censored data, is

$$\begin{aligned} &f_{(X_j,Y_j)|(X_{obs},Y_{obs})}(x_j, y_j|x_{obs}, y_{obs}) \\ &\quad = f_{(X_j,Y_j)|X_{k:n}}(x_j, y_j|X_{k:n} = x_{k:n}) \\ &\quad = f_{Y|X}(y_j|x_j)\frac{f_X(x_j)}{[1 - F_X(x_{k:n})]}, \\ &\qquad x_{k:n} \le x_j < +\infty, \quad -\infty < y_j < +\infty. \end{aligned} \tag{13.4}$$

From (13.4), we can derive the first and second moments of Y_j and the product moment of X_j and Y_j, given (X_{obs}, Y_{obs}), when (X, Y) are from a bivariate normal distribution with parameter $\theta = (\mu_X, \sigma_X, \mu_Y, \sigma_Y, \rho)$, as

$$E[Y_j|X_{k:n} = x_{k:n}] = \mu_Y + \rho\sigma_Y Q, \tag{13.5}$$

$$E[Y_j^2|X_{k:n} = x_{k:n}] = \sigma_Y^2(1 + \rho^2\xi Q) + 2\rho\sigma_Y\mu_Y Q + \mu_Y^2, \tag{13.6}$$

$$E[X_jY_j|X_{k:n} = x_{k:n}] = \mu_Y(\mu_X + \sigma_X Q) + \rho\sigma_Y(\mu_X Q + \sigma_X + \sigma_X\xi Q), \tag{13.7}$$

where $\xi = \frac{x_{k:n} - \mu_X}{\sigma_X}$ and $Q = \frac{\phi(\xi)}{1-\Phi(\xi)}$ with $\phi(\cdot)$ and $\Phi(\cdot)$ denoting the standard normal pdf and cdf, respectively.

13.3 EM Algorithm

The Type-II right censored bivariate normal data can be viewed as an incomplete data and the EM algorithm can then be applied to determine the MLE of the parameter θ. The log-likelihood function $l(\theta; x, y)$ based on complete data (X, Y) is

$$\begin{aligned} &l(\theta; x, y) \\ &= \text{constant} - n\log\sigma_X - n\log\sigma_Y - \frac{n}{2}\log(1-\rho^2) \\ &\quad -\frac{1}{2(1-\rho^2)}\sum_{i=1}^{n}\left\{\left(\frac{x_{i:n}-\mu_X}{\sigma_X}\right)^2 - 2\rho\left(\frac{x_{i:n}-\mu_X}{\sigma_X}\right)\left(\frac{y_{[i:n]}-\mu_Y}{\sigma_Y}\right)\right. \\ &\quad \left. + \left(\frac{y_{[i:n]}-\mu_Y}{\sigma_Y}\right)^2\right\} \\ &= \text{constant} - n\log\sigma_X - n\log\sigma_Y - \frac{n}{2}\log(1-\rho^2) \\ &\quad -\frac{1}{2(1-\rho^2)}\sum_{i=1}^{k}\left\{\left(\frac{x_{i:n}-\mu_X}{\sigma_X}\right)^2 - 2\rho\left(\frac{x_{i:n}-\mu_X}{\sigma_X}\right)\left(\frac{y_{[i:n]}-\mu_Y}{\sigma_Y}\right)\right. \end{aligned}$$

$$+\left(\frac{y_{[i:n]}-\mu_Y}{\sigma_Y}\right)^2\Bigg\}$$

$$-\frac{1}{2(1-\rho^2)}\sum_{j=1}^{n-k}\left\{\left(\frac{x_j-\mu_X}{\sigma_X}\right)^2-2\rho\left(\frac{x_j-\mu_X}{\sigma_X}\right)\left(\frac{y_j-\mu_Y}{\sigma_Y}\right)\right.$$
$$\left.+\left(\frac{y_j-\mu_Y}{\sigma_Y}\right)^2\right\},$$

where (x_j, y_j), $j = 1, \ldots, n-k$, are the censored data. Based on the complete data, it is well-known that the MLE of $\theta = (\mu_X, \sigma_X, \mu_Y, \sigma_Y, \rho)$ is given by

$$\hat{\mu}_X = \frac{1}{n}\sum_{i=1}^{n} X_{i:n}, \tag{13.8}$$

$$\hat{\sigma}_X = \left[\frac{1}{n}\sum_{i=1}^{n}(X_{i:n}-\hat{\mu}_X)^2\right]^{\frac{1}{2}}, \tag{13.9}$$

$$\hat{\mu}_Y = \frac{1}{n}\sum_{i=1}^{n} Y_{[i:n]}, \tag{13.10}$$

$$\hat{\sigma}_Y = \left[\frac{1}{n}\sum_{i=1}^{n}(Y_{[i:n]}-\hat{\mu}_Y)^2\right]^{\frac{1}{2}}, \tag{13.11}$$

$$\hat{\rho} = \frac{\sum_{i=1}^{n}(X_{i:n}-\hat{\mu}_X)(Y_{[i:n]}-\hat{\mu}_Y)}{\sqrt{\sum_{i=1}^{n}(X_{i:n}-\hat{\mu}_X)^2}\sqrt{\sum_{i=1}^{n}(Y_{[i:n]}-\hat{\mu}_Y)^2}}. \tag{13.12}$$

Hence, in the $(h+1)$-th iteration of the EM algorithm, the values of $\hat{\mu}_X^{(h+1)}$, $\hat{\sigma}_X^{(h+1)}$, $\hat{\mu}_Y^{(h+1)}$, $\hat{\sigma}_Y^{(h+1)}$ and $\hat{\rho}^{(h+1)}$ are calculated as follows:

$$\hat{\mu}_X^{(h+1)} = \frac{1}{n}\left[\sum_{i=1}^{k} x_{i:n} + (n-k)E(X_j|X_j \ge x_{k:n}, \hat{\mu}_X^{(h)}, \hat{\sigma}_X^{(h)})\right], \tag{13.13}$$

$$\hat{\sigma}_X^{(h+1)} = \left[\frac{1}{n}\left\{\sum_{i=1}^{k} x_{i:n}^2 + (n-k)E(X_j^2|X_j \ge x_{k:n}, \hat{\mu}_X^{(h+1)}, \hat{\sigma}_X^{(h)})\right\}\right.$$
$$\left. -(\hat{\mu}_X^{(h+1)})^2\right]^{\frac{1}{2}}, \tag{13.14}$$

$$\hat{\mu}_Y^{(h+1)} = \frac{1}{n}\left[\sum_{i=1}^{k} y_{[i:n]} + (n-k)E(Y_j|(X_{obs}, Y_{obs}), \hat{\mu}_X^{(h+1)}, \hat{\sigma}_X^{(h+1)},\right.$$
$$\left.\hat{\mu}_Y^{(h)}, \hat{\sigma}_Y^{(h)}, \hat{\rho}^{(h)})\right], \tag{13.15}$$

$$\hat{\sigma}_Y^{(h+1)} = \left[\frac{1}{n}\left\{\sum_{i=1}^{k} y_{[i:n]}^2 + (n-k)E(Y_j^2|(X_{obs}, Y_{obs}), \hat{\mu}_X^{(h+1)}, \hat{\sigma}_X^{(h+1)}, \hat{\mu}_Y^{(h+1)}, \hat{\sigma}_Y^{(h)}, \hat{\rho}^{(h)})\right\} - (\hat{\mu}_Y^{(h+1)})^2\right]^{\frac{1}{2}}, \tag{13.16}$$

$$\hat{\rho}^{(h+1)} = \frac{A - n\hat{\mu}_X^{(h+1)}\hat{\mu}_Y^{(h+1)}}{BC}, \tag{13.17}$$

where

$$A = \sum_{i=1}^{k} x_{i:n} y_{[i:n]} + (n-k)E\left[X_j Y_j|(X_{obs}, Y_{obs}), \hat{\mu}_X^{(h+1)}, \hat{\sigma}_X^{(h+1)}, \hat{\mu}_Y^{(h+1)}, \hat{\sigma}_Y^{(h+1)}, \hat{\rho}^{(h)}\right],$$

$$B = \left\{\sum_{i=1}^{k} x_{i:n}^2 + (n-k)E[X_j^2|X_j \geq x_{k:n}, \hat{\mu}_X^{(h+1)}, \hat{\sigma}_X^{(h+1)}] - n(\hat{\mu}_X^{(h+1)})^2\right\}^{\frac{1}{2}},$$

and

$$C = \left\{\sum_{i=1}^{k} y_{[i:n]}^2 + (n-k)E[Y_j^2|(X_{obs}, Y_{obs}), \hat{\mu}_X^{(h+1)}, \hat{\sigma}_X^{(h+1)}, \hat{\mu}_Y^{(h+1)}, \hat{\sigma}_Y^{(h+1)}, \hat{\rho}^{(h)}] - n(\hat{\mu}_Y^{(h+1)})^2\right\}^{\frac{1}{2}}$$

The first and second moments of X_j, given $X_j \geq x_{k:n}$, are given by [see Cohen (1991)]

$$E(X_j|X_j \geq x_{k:n}, \hat{\mu}_X, \hat{\sigma}_X) = \hat{\sigma}_X Q + \hat{\mu}_X, \tag{13.18}$$

$$E(X_j^2|X_j \geq x_{k:n}, \hat{\mu}_X, \hat{\sigma}_X) = \hat{\sigma}_X^2(1 + \xi Q) + 2\hat{\sigma}_X \hat{\mu}_X Q + \hat{\mu}_X^2, \tag{13.19}$$

where ξ and Q are as defined in Section 13.2 (with μ_X and σ_X replaced by $\hat{\mu}_X$ and $\hat{\sigma}_X$). Now, by using Eqs. (13.5), (13.6), (13.7), (13.18) and (13.19) in Eqs. (13.13)–(13.17), we can find the MLE of θ via the EM algorithm.

By the fact that $(\hat{\mu}_X, \hat{\sigma}_X)$ is the MLE of (μ_X, σ_X) based on X_{obs} alone [see Harrell and Sen (1979)], the EM algorithm can be separated into two parts, with one iterating for $(\hat{\mu}_X, \hat{\sigma}_X)$ and the other for $(\hat{\mu}_Y, \hat{\sigma}_Y, \hat{\rho})$, which results in reducing the computational time.

13.4 Asymptotic Variances and Covariances

The Fisher information of complete data, $I_{comp}(\theta)$, for a bivariate normal distribution is known to be [see Kotz, Balakrishnan and Johnson (2000)]

$$
\begin{aligned}
I_{comp}(\theta) & \\
= & -E\left[\frac{\partial^2 l(\theta; X, Y)}{\partial \theta^2}\right] \\
= & \; n \begin{pmatrix}
\frac{1}{\sigma_X^2(1-\rho^2)} & 0 & -\frac{\rho}{\sigma_X\sigma_Y(1-\rho^2)} & 0 & 0 \\
 & \frac{2-\rho^2}{\sigma_X^2(1-\rho^2)} & 0 & -\frac{\rho^2}{\sigma_X\sigma_Y(1-\rho^2)} & -\frac{\rho}{\sigma_X(1-\rho^2)} \\
 & & \frac{1}{\sigma_Y^2(1-\rho^2)} & 0 & 0 \\
 & & & \frac{2-\rho^2}{\sigma_Y^2(1-\rho^2)} & -\frac{\rho}{\sigma_Y(1-\rho^2)} \\
 & & & & \frac{1+\rho^2}{(1-\rho^2)^2}
\end{pmatrix}.
\end{aligned}
\tag{13.20}
$$

The log-likelihood function for one of the missing data, given the observed data, can be derived from (13.4) to be

$$
\begin{aligned}
& l_{miss}(\theta, (X_j, Y_j)|X_{k:n} = x_{k:n}) \\
& \quad = \text{constant} - \log\sigma_Y - \frac{1}{2}\log(1-\rho^2) \\
& \quad\quad - \frac{1}{2\sigma_Y^2(1-\rho^2)}\left\{y_j - \mu_Y - \frac{\rho\sigma_Y}{\sigma_X}(x_j - \mu_X)\right\}^2 \\
& \quad\quad - \log\sigma_X - \frac{1}{2\sigma_X^2}(x_j - \mu_X)^2 - \log\{1 - \Phi(\xi)\}.
\end{aligned}
\tag{13.21}
$$

Then, the missing data information $I_{miss}(\theta)$ is given by

$$
\begin{aligned}
I_{miss}(\theta) &= -E_{X_{k:n}}\left[(n-k)E\left(\frac{\partial^2 l_{miss}(\theta, (X_j, Y_j)|X_{k:n} = x_{k:n})}{\partial\theta^2}\right)\right] \\
&= (n-k)E_{X_{k:n}}\begin{pmatrix}
I_{miss}^{*,11} & I_{miss}^{*,12} & I_{miss}^{*,13} & I_{miss}^{*,14} & I_{miss}^{*,15} \\
 & I_{miss}^{*,22} & I_{miss}^{*,23} & I_{miss}^{*,24} & I_{miss}^{*,25} \\
 & & I_{miss}^{*,33} & I_{miss}^{*,34} & I_{miss}^{*,35} \\
 & & & I_{miss}^{*,44} & I_{miss}^{*,45} \\
 & & & & I_{miss}^{*,55}
\end{pmatrix} \qquad (13.22) \\
&= E_{X_{k:n}}\left[I_{miss}^*\right], \qquad (13.23)
\end{aligned}
$$

where

$$
\begin{aligned}
I^{*,11}_{miss} &= \frac{\rho^2}{\sigma_X^2(1-\rho^2)} + \frac{1+\xi Q-Q^2}{\sigma_X^2}, & I^{*,12}_{miss} &= \frac{\rho^2 Q}{\sigma_X^2(1-\rho^2)} + \frac{Q(1+\xi^2-\xi Q)}{\sigma_X^2},\\
I^{*,13}_{miss} &= -\frac{\rho}{\sigma_X\sigma_Y(1-\rho^2)}, & I^{*,14}_{miss} &= -\frac{\rho^2 Q}{\sigma_X\sigma_Y(1-\rho^2)},\\
I^{*,15}_{miss} &= -\frac{\rho Q}{\sigma_X(1-\rho^2)}, & I^{*,22}_{miss} &= \frac{\rho^2(1+\xi Q)}{\sigma_X^2(1-\rho^2)} + \frac{2+\xi Q+\xi^3 Q-\xi^2 Q^2}{\sigma_X^2},\\
I^{*,23}_{miss} &= -\frac{\rho Q}{\sigma_X\sigma_Y(1-\rho^2)}, & I^{*,24}_{miss} &= -\frac{\rho^2(1+\xi Q)}{\sigma_X\sigma_Y(1-\rho^2)},\\
I^{*,25}_{miss} &= -\frac{\rho(1+\xi Q)}{\sigma_X(1-\rho^2)}, & I^{*,33}_{miss} &= \frac{1}{\sigma_Y^2(1-\rho^2)},\\
I^{*,34}_{miss} &= \frac{\rho Q}{\sigma_Y^2(1-\rho^2)}, & I^{*,35}_{miss} &= \frac{Q}{\sigma_Y(1-\rho^2)},\\
I^{*,44}_{miss} &= \frac{2}{\sigma_Y^2} + \frac{\rho^2(1+\xi Q)}{\sigma_Y^2(1-\rho^2)}, & I^{*,45}_{miss} &= -\frac{\rho(1-\xi Q)}{\sigma_Y(1-\rho^2)},\\
I^{*,55}_{miss} &= \frac{2\rho^2}{(1-\rho^2)^2} + \frac{1+\xi Q}{(1-\rho^2)}.
\end{aligned}
$$

Let ψ_{abcd} be defined by

$$\psi_{abcd} = E_{X_{k:n}}\left[\frac{\{\phi(\xi)\}^a \xi^d}{\{\Phi(\xi)\}^b\{1-\Phi(\xi)\}^c}\right].$$

By taking expectation of $I^{*,ij}_{miss}$ with respect to the k-th order statistic $X_{k:n}$, we can obtain the information for the missing data as

$$I_{miss}(\theta) \quad = \quad (n-k)\begin{pmatrix} I^{11}_{miss} & I^{12}_{miss} & I^{13}_{miss} & I^{14}_{miss} & I^{15}_{miss}\\ & I^{22}_{miss} & I^{23}_{miss} & I^{24}_{miss} & I^{25}_{miss}\\ & & I^{33}_{miss} & I^{34}_{miss} & I^{35}_{miss}\\ & & & I^{44}_{miss} & I^{45}_{miss}\\ & & & & I^{55}_{miss}\end{pmatrix}, \tag{13.24}$$

where

$$
\begin{aligned}
I^{11}_{miss} &= \frac{\rho^2}{\sigma_X^2(1-\rho^2)} + \frac{1+\psi_{1011}-\psi_{2020}}{\sigma_X^2}, & I^{12}_{miss} &= \frac{\rho^2\psi_{1010}}{\sigma_X^2(1-\rho^2)} + \frac{(\psi_{1010}+\psi_{1012}-\psi_{2021})}{\sigma_X^2},\\
I^{13}_{miss} &= -\frac{\rho}{\sigma_X\sigma_Y(1-\rho^2)}, & I^{14}_{miss} &= -\frac{\rho^2\psi_{1010}}{\sigma_X\sigma_Y(1-\rho^2)},\\
I^{15}_{miss} &= -\frac{\rho\psi_{1010}}{\sigma_X(1-\rho^2)}, & I^{22}_{miss} &= \frac{\rho^2(1+\psi_{1011})}{\sigma_X^2(1-\rho^2)} + \frac{2+\psi_{1011}+\psi_{1013}-\psi_{2022}}{\sigma_X^2},\\
I^{23}_{miss} &= -\frac{\rho\psi_{1010}}{\sigma_X\sigma_Y(1-\rho^2)}, & I^{24}_{miss} &= -\frac{\rho^2(1+\psi_{1011})}{\sigma_X\sigma_Y(1-\rho^2)},\\
I^{25}_{miss} &= -\frac{\rho(1+\psi_{1011})}{\sigma_X(1-\rho^2)}, & I^{33}_{miss} &= \frac{1}{\sigma_Y^2(1-\rho^2)},\\
I^{34}_{miss} &= \frac{\rho\psi_{1010}}{\sigma_Y^2(1-\rho^2)}, & I^{35}_{miss} &= \frac{\psi_{1010}}{\sigma_Y(1-\rho^2)},\\
I^{44}_{miss} &= \frac{2}{\sigma_Y^2} + \frac{\rho^2(1+\psi_{1011})}{\sigma_Y^2(1-\rho^2)}, & I^{45}_{miss} &= -\frac{\rho(1-\psi_{1011})}{\sigma_Y(1-\rho^2)},\\
I^{55}_{miss} &= \frac{2\rho^2}{(1-\rho^2)^2} + \frac{1+\psi_{1011}}{(1-\rho^2)}.
\end{aligned}
$$

From (13.20) and (13.24), we can obtain the Fisher information matrix for the observed data as

$$I = I_{comp} - I_{miss}. \tag{13.25}$$

Table 13.1: Simulated Type-II censored sample from a bivariate normal distribution

(-0.8813, -0.2167)	(2.6496,-0.0245)	(4.5950, 5.9566)
(0.2483, 2.6941)	(2.8902, 3.6322)	(4.8213, 5.2777)
(0.3563, 0.0337)	(3.1831, 2.8176)	(4.9187, 2.9420)
(0.4266, 1.6868)	(3.2617, 2.2995)	(5.2477, 6.1104)
(0.8208, 1.9015)	(3.6223, 0.8458)	(5.4001, 1.8456)
(1.0397, 2.0873)	(4.0311, 4.8035)	(5.4401, 3.4883)
(1.3031, 1.9768)	(4.4385, 1.1516)	(5.7363, 4.5110)
(1.3221, 0.3528)	(4.4954,-0.0340)	(5.9222, 5.3993)
(1.8833,-0.2700)	(4.5833, 2.7815)	(5.9358, 4.3120)
(2.3410, 3.1157)	(4.5875, 5.0524)	(6.0156, 3.9099)

Alternatively, the information matrix for the observed data can also be obtained directly from the likelihood function of the observed data in (13.2), and the corresponding expressions can be found in Harrell and Sen (1979). It can be shown that

$$\psi_{1010} = -\frac{1}{(n-k)} \sum_{i=1}^{k} E[Z_{i:n}], \qquad \psi_{1011} = \frac{1}{(n-k)} \left\{k - \sum_{j=1}^{k} E[Z_{j:n}^2]\right\}, \quad (13.26)$$

where $Z_{i:n} = \frac{X_{i:n} - \mu_X}{\sigma_X}$; see Appendix for a proof. Using the relationships in (13.26), it can be shown that the information matrix presented by Harrell and Sen (1979) and the information matrix derived in (13.25) by the use of the missing information principle are exactly the same. If I^*_{miss} is used in Eq. (13.24) instead of I_{miss}, then partially observed information matrix $I^* = I_{comp} - I^*_{miss}$ is obtained. Note that in this case no numerical integration needs to be performed [as required in the computation of I in (13.25)].

13.5 Illustrative Example

To illustrate the proposed EM algorithm for the computation of the MLE of θ, a Type-II censored bivariate normal data with $n = 50$, $k = 30$ and $\theta = (5, 3, 3, 2, 0.7)$ were generated. These are presented in Table 13.1.

A computer program has been written in FORTRAN to compute the MLE of θ via EM algorithm. Sample means, standard deviations and sample correlation coefficient are used as initial estimates for the parameters in the EM algorithm. After 17 EM-iterations, the estimates for (μ_X, σ_X) converged to $\mu_{X(\infty)} = 5.2836$ and $\sigma_{X(\infty)} = 3.0195$. The EM algorithm for (μ_Y, σ_Y, ρ), using $\mu_{X(\infty)}$ and $\sigma_{X(\infty)}$, yielded $\mu_{Y(\infty)} = 3.8240$, $\sigma_{Y(\infty)} = 2.3335$ and $\rho_{(\infty)} = 0.7664$

after 110 iterations. It should be mentioned that ψ_{abcd} values needed for calculating $I_{miss}(\hat{\theta})$ in this case were obtained by numerical integration using MAPLE. Based on the censored data in Table 13.1, the information matrices are computed to be

$$I_{comp}(\hat{\theta}) = \begin{pmatrix} 13.2915 & 0 & -13.1814 & 0 & 0 \\ & 18.7756 & 0 & -10.1025 & -30.7591 \\ & & 22.2543 & 0 & 0 \\ & & & 31.4365 & -39.8010 \\ & & & & 466.2273 \end{pmatrix},$$

$$I_{miss}(\hat{\theta}) = \begin{pmatrix} 3.8221 & 4.7016 & -5.2726 & -3.8411 & -11.6949 \\ & 8.6400 & -5.0117 & -4.9985 & -15.2190 \\ & & 8.9017 & 6.4849 & 19.7447 \\ & & & 13.8136 & -12.1480 \\ & & & & 197.9767 \end{pmatrix},$$

and

$$\begin{aligned} I(\hat{\theta}) &= I_{comp}(\hat{\theta}) - I_{miss}(\hat{\theta}) \\ &= \begin{pmatrix} 9.4694 & -4.7016 & -7.9088 & 3.8411 & 11.6949 \\ & 10.1355 & 5.0117 & -5.1040 & -15.5401 \\ & & 13.3526 & -6.4849 & -19.7447 \\ & & & 17.6229 & -27.6531 \\ & & & & 268.2504 \end{pmatrix}. \end{aligned}$$

From this, we obtain the asymptotic variance-covariance matrix of $\hat{\theta} = (\hat{\mu}_X, \hat{\sigma}_X, \hat{\mu}_Y, \hat{\sigma}_Y, \hat{\rho})$ as

$$I^{-1}(\hat{\theta}) = \begin{pmatrix} 0.2326 & 0.0651 & 0.1378 & 0.0296 & 0.0068 \\ & 0.1797 & 0.0386 & 0.0816 & 0.0188 \\ & & 0.2248 & 0.1001 & 0.0231 \\ & & & 0.1524 & 0.0265 \\ & & & & 0.0090 \end{pmatrix}.$$

On the other hand, we obtain the asymptotic variance-covariance matrix of $\hat{\theta}$ from the partially observed information matrix I^* as

$$(I^*)^{-1}(\hat{\theta}) = \begin{pmatrix} 0.2323 & 0.0655 & 0.1376 & 0.0297 & 0.0069 \\ & 0.1808 & 0.0388 & 0.0821 & 0.0189 \\ & & 0.2263 & 0.1015 & 0.0234 \\ & & & 0.1534 & 0.0267 \\ & & & & 0.0090 \end{pmatrix}.$$

Note that there are slight differences between $I^{-1}(\hat{\theta})$ and $(I^*)^{-1}(\hat{\theta})$, but the advantage of the use of $(I^*)^{-1}(\hat{\theta})$ is that it did not require any numerical integration.

13.6 Simulation Results

Type-II right censored samples from bivariate normal distribution for various n, k and ρ values with $(\mu_X, \sigma_X, \mu_Y, \sigma_Y) = (0, 1, 0, 1)$ were generated and MLE of θ was obtained using the EM algorithm. For sample size $n = 50$, the average values of the parameter estimates and their variances are presented in Table 13.2 and their covariances are presented in Table 13.3. The values of variances and covariances from the Fisher information matrix and the partially observed information matrix are averaged, and are presented in Tables 13.4–13.6, respectively. All these values are computed over 1000 Monte Carlo simulations. The average values of the estimates and their variances for sample sizes 20, 100 and 200 are presented in Tables 13.7–13.12.

As expected, we observe from Table 13.2 that variances of all the estimates increase as censoring proportion increases. σ_Y is estimated better than σ_X for small ρ with large censoring proportion. The variance of $\hat{\rho}$ decreases when the value of ρ increases, regardless of censoring proportion. Similar results are also observed for $n =$ 20, 100, and 200 from Tables 13.7–13.12. From Tables 13.4–13.6, we note that variances from I and I^* are comparable for all the cases. Difference between variance of $\hat{\rho}$ from simulation and averaged variance of $\hat{\rho}$ from the information matrix becomes noticeable in many situations, and this has an adverse effect on the coverage probabilities of asymptotic confidence intervals of ρ.

13.7 Probability Coverages

Using the asymptotic normality of MLEs, one can construct approximate 90% or 95% confidence intervals for the parameters. Monte Carlo simulations were used to evaluate the probability coverages of these confidence intervals. The values of variances obtained from the Fisher information matrix as well as from the partially observed information matrix were used for comparison, and the probability coverages of all the confidence intervals are presented in Tables 13.13–13.20 for $n =$ 20, 50, 100 and 200.

Probability coverages of 95% and 90% CIs for all the parameters are good for $n = 200$ even when the censoring proportion is as high as 0.5. For $n = 100$, probability coverages of the CIs for location and scale parameters of both X and Y are quite satisfactory when there is no censoring or moderate censoring. The probability coverage of the approximate confidence interval for ρ is very unsatisfactory when there is heavy censoring and n is less than 100, and it is therefore not advisable to use the approximate confidence interval for ρ when there is heavy censoring unless n is very large.

Appendix

Here, as presented earlier in Eq. (13.26), we will show that

$$\psi_{1010} = -\frac{1}{(n-k)}\sum_{i=1}^{k} E[Z_{i:n}] \quad \text{and} \quad \psi_{1011} = \frac{1}{(n-k)}\left\{k - \sum_{j=1}^{k} E[Z_{j:n}^2]\right\},$$

where $Z_{i:n} = \frac{X_{i:n}-\mu_X}{\sigma_X}$.

PROOF. First, let us write

$$\begin{aligned}
\psi_{1010} &= E\left[\frac{\phi(Z_{k:n})}{1-\Phi(Z_{k:n})}\right] \\
&= \frac{n!}{(k-1)!(n-k)!}\int_{-\infty}^{\infty}\frac{\phi(z_{k:n})}{1-\Phi(z_{k:n})}\{\Phi(z_{k:n})\}^{k-1} \\
&\quad \times\{1-\Phi(z_{k:n})\}^{n-k}\phi(z_{k:n})dz_{k:n} \\
&= \frac{n!}{(k-1)!(n-k)!}\int_{-\infty}^{\infty}\{\Phi(z_{k:n})\}^{k-1}\{1-\Phi(z_{k:n})\}^{n-k-1} \\
&\quad \times\{\phi(z_{k:n})\}^{2}dz_{k:n} \\
&= \frac{n!}{(k-1)!(n-k)!}\int_{-\infty}^{\infty}\{\Phi(z_{k:n})\}^{k-1}\{1-\Phi(z_{k:n})\}^{n-k-1}\phi(z_{k:n}) \\
&\quad \times\int_{z_{k:n}}^{\infty}-\phi'(t)dtdz_{k:n} \\
&= \frac{n!}{(k-1)!(n-k)!}\int_{-\infty}^{\infty}\{\Phi(z_{k:n})\}^{k-1}\{1-\Phi(z_{k:n})\}^{n-k-1}\phi(z_{k:n}) \\
&\quad \times\int_{z_{k:n}}^{\infty}t\phi(t)dtdz_{k:n} \\
&= \frac{n!}{(k-1)!(n-k)!}\int_{-\infty}^{\infty}\int_{z_{k:n}}^{\infty}\{\Phi(z_{k:n})\}^{k-1} \\
&\quad \times\left\{\sum_{i=0}^{n-k-1}\binom{n-k-1}{i}[\Phi(t)-\Phi(z_{k:n})]^{i}[1-\Phi(t)]^{n-k-1-i}\right\} \\
&\quad \times\phi(z_{k:n})t\phi(t)dtdz_{k:n}
\end{aligned}$$

$$
\begin{aligned}
&= \sum_{i=0}^{n-k-1} \frac{n!}{(k-1)!(n-k)!} \frac{(n-k-1)!}{i!(n-k-i-1)!} \\
&\quad \times \int_{-\infty}^{\infty} \int_{z_{k:n}}^{\infty} t\{\Phi(z_{k:n})\}^{k-1}\{\Phi(t)-\Phi(z_{k:n})\}^{i}[1-\Phi(t)]^{n-k-1-i} \\
&\quad \times \phi(t)\phi(z_{k:n})dt dz_{k:n} \\
&= \frac{1}{(n-k)} \sum_{i=0}^{n-k-1} E[Z_{k+i+1:n}] \\
&= \frac{1}{(n-k)} \sum_{i=1}^{n-k} E[Z_{k+i:n}] \\
&= -\frac{1}{(n-k)} \sum_{i=1}^{k} E[Z_{i:n}]
\end{aligned}
$$

using the fact that $\sum_{i=1}^{n} E[Z_{i:n}] = 0$.

Next, let us write

$$
\begin{aligned}
\psi_{1011} &= E\left[\frac{Z_{k:n}\phi(Z_{k:n})}{1-\Phi(Z_{k:n})}\right] \\
&= \frac{n!}{(k-1)!(n-k)!} \int_{-\infty}^{\infty} z_{k:n} \frac{\phi(z_{k:n})}{1-\Phi(z_{k:n})} \{\Phi(z_{k:n})\}^{k-1}\{1-\Phi(z_{k:n})\}^{n-k} \\
&\quad \times \phi(z_{k:n})dz_{k:n} \\
&= \frac{n!}{(k-1)!(n-k)!} \int_{-\infty}^{\infty} z_{k:n}\{\Phi(z_{k:n})\}^{k-1}\{1-\Phi(z_{k:n})\}^{n-k-1} \\
&\quad \times \{\phi(z_{k:n})\}^{2}dz_{k:n} \\
&= \frac{n!}{(k-1)!(n-k)!} \int_{-\infty}^{\infty} z_{k:n}\{\Phi(z_{k:n})\}^{k-1}\{1-\Phi(z_{k:n})\}^{n-k-1}\phi(z_{k:n}) \\
&\quad \times \int_{z_{k:n}}^{\infty} -\phi'(t)dt dz_{k:n} \\
&= \frac{n!}{(k-1)!(n-k)!} \int_{-\infty}^{\infty} z_{k:n}\{\Phi(z_{k:n})\}^{k-1}\{1-\Phi(z_{k:n})\}^{n-k-1}\phi(z_{k:n}) \\
&\quad \times \int_{z_{k:n}}^{\infty} t\phi(t)dt dz_{k:n} \\
&= \frac{n!}{(k-1)!(n-k)!} \int_{-\infty}^{\infty} \int_{z_{k:n}}^{\infty} z_{k:n}\{\Phi(z_{k:n})\}^{k-1} \\
&\quad \times \left\{\sum_{i=0}^{n-k-1} \binom{n-k-1}{i} [\Phi(t)-\Phi(z_{k:n})]^{i}[1-\Phi(t)]^{n-k-1-i}\right\} \\
&\quad \times \phi(z_{k:n})t\phi(t)dt dz_{k:n}
\end{aligned}
$$

$$
\begin{aligned}
&= \sum_{i=0}^{n-k-1} \frac{n!}{(k-1)!(n-k)!} \frac{(n-k-1)!}{i!(n-k-i-1)!} \\
&\quad \times \int_{-\infty}^{\infty} \int_{z_{k:n}}^{\infty} z_{k:n} t \{\Phi(z_{k:n})\}^{k-1} \{\Phi(t) - \Phi(z_{k:n})\}^{i} [1 - \Phi(t)]^{n-k-1-i} \\
&\quad \times \phi(t)\phi(z_{k:n}) dt dz_{k:n} \\
&= \frac{1}{(n-k)} \sum_{i=0}^{n-k-1} E[Z_{k:n} Z_{k+i+1:n}] \\
&= \frac{1}{(n-k)} \sum_{i=1}^{n-k} E[Z_{k:n} Z_{k+i:n}] \\
&= \frac{1}{(n-k)} \sum_{j=k+1}^{n} E[Z_{k:n} Z_{j:n}] \\
&= \frac{1}{(n-k)} \left\{ \sum_{j=k+1}^{n} E[Z_{j:n}^2] - (n-k) \right\}
\end{aligned}
$$

using an identity of Joshi and Balakrishnan (1981). From this expression, we readily obtain

$$
\begin{aligned}
\psi_{1011} &= \frac{1}{(n-k)} \left\{ n - \sum_{j=1}^{k} E[Z_{j:n}^2] - (n-k) \right\} \\
&= \frac{1}{(n-k)} \left\{ k - \sum_{j=1}^{k} E[Z_{j:n}^2] \right\},
\end{aligned}
$$

which completes the proof. ∎

References

1. Bhattacharya, P. K. (1974). Convergence of sample paths of normalized sums of induced order statistics, *Annals of Statistics,* **2**, 1034–1039.

2. Cohen, A. C. (1991). *Truncated and Censored Samples: Theory and Applications*, Marcel Dekker, New York.

3. David, H. A. (1973). Concomitants of order statistics. *Bull. Internat. Statist. Inst.* **45**, 295–300.

4. David, H. A. and Galambos, J. (1974). The asymptotic theory of concomitants of order statistics, *Journal of Applied Probability* **11**, 762–770.

5. David, H. A. and Nagaraja, H. N. (1998). Concomitants of order statistics, In *Handbook of Statistics - Vol.* **16** (Eds., N. Balakrishnan and C. R. Rao), pp. 487–513, North-Holland, Amsterdam.

6. Dempster, A. P., Laird, N. M. and Rubin, D. B. (1977). Maximum likelihood from incomplete data via the EM algorithm, *Journal of the Royal Statistical Society, Series B,* **39**, 1–38.

7. Harrell, F. E. and Sen, P. K. (1979). Statistical inference for censored bivariate normal distributions based on induced order statistics, *Biometrika,* **66**, 293–298.

8. Joshi, P. C. and Balakrishnan, N. (1981). An identity for the moments of normal order statistics with applications, *Scandinavian Actuarial Journal,* 203–213.

9. Kotz, S., Balakrishnan, N. and Johnson, N. L. (2000). *Continuous Multivariate Distributions - Vol. 1*, Second edition, John Wiley & Sons, New York.

10. Louis, T. A. (1982). Finding the observed information matrix when using the EM algorithm, *Journal of the Royal Statistical Society, Series B,* **44**, 226–233.

11. McLachlan, G. J. and Krishnan, T. (1997). *The EM Algorithm and Extensions*, John Wiley & Sons, New York.

12. Ng, H. K. T., Chan, P. S. and Balakrishnan, N. (2002). Estimation of parameters from progressively censored data using EM algorithm, *Computational Statistics & Data Analysis,* **39**, 371–386.

13. Orchard, T. and Woodbury, M. A. (1970). A missing information principle: Theory and applications, In *Proceedings of the Sixth Berkeley Symposium on Mathematical Statistics and Probability, Vol. 1*, 697–715.

14. Tanner, M. A. (1993). *Tools for Statistical Inference: Observed Data and Data Augmentation Methods*, Second edition, Springer-Verlag, New York.

15. Yang, S. S. (1977). General distribution theory of the concomitants of order statistics, *Annals of Statistics* **5**, 996–1002.

Table 13.2: Means and variances of the MLEs for $n = 50$

n	k	Proportion of Censoring	ρ	$\hat{\mu}_X$	$\hat{\sigma}_X$	$\hat{\mu}_Y$	$\hat{\sigma}_Y$	$\hat{\rho}$	Var $[\hat{\mu}_X]$	Var $[\hat{\sigma}_X]$	Var $[\hat{\mu}_Y]$	Var $[\hat{\sigma}_Y]$	Var $[\hat{\rho}]$
50	50	0	0	0.0025	0.9827	-0.0048	0.9817	-0.0093	0.0190	0.0091	0.0210	0.0095	0.0214
			0.1	-0.0028	0.9870	0.0009	0.9904	0.0912	0.0191	0.0102	0.0210	0.0098	0.0196
			0.2	0.0048	0.9874	0.0037	0.9869	0.2024	0.0196	0.0102	0.0203	0.0105	0.0191
			0.3	0.0036	0.9799	0.0035	0.9887	0.2982	0.0203	0.0093	0.0211	0.0098	0.0185
			0.4	0.0060	0.9821	0.0032	0.9841	0.3949	0.0204	0.0105	0.0207	0.0099	0.0150
			0.5	-0.0054	0.9868	0.0015	0.9869	0.4968	0.0208	0.0097	0.0204	0.0101	0.0121
			0.6	0.0056	0.9854	0.0061	0.9851	0.5968	0.0199	0.0099	0.0199	0.0100	0.0094
			0.7	-0.0020	0.9871	0.0009	0.9877	0.6971	0.0184	0.0099	0.0195	0.0095	0.0059
			0.8	-0.0063	0.9874	-0.0043	0.9871	0.7986	0.0208	0.0101	0.0201	0.0099	0.0027
			0.9	0.0052	0.9861	0.0046	0.9866	0.8986	0.0206	0.0099	0.0207	0.0094	0.0009
	45	0.1	0	0.0019	0.9773	-0.0062	0.9892	0.0004	0.0213	0.0122	0.0235	0.0107	0.0300
			0.1	-0.0033	0.9795	-0.0054	0.9886	0.0901	0.0207	0.0110	0.0243	0.0119	0.0306
			0.2	-0.0031	0.9820	-0.0035	0.9872	0.1982	0.0229	0.0123	0.0231	0.0110	0.0296
			0.3	-0.0036	0.9795	0.0008	0.9871	0.3034	0.0216	0.0120	0.0232	0.0111	0.0232
			0.4	0.0013	0.9844	0.0011	0.9885	0.3922	0.0191	0.0116	0.0241	0.0123	0.0201
			0.5	-0.0106	0.9884	-0.0041	0.9914	0.4947	0.0212	0.0126	0.0227	0.0128	0.0173
			0.6	-0.0034	0.9820	-0.0022	0.9892	0.5921	0.0194	0.0119	0.0226	0.0129	0.0119
			0.7	0.0022	0.9794	0.0000	0.9797	0.6906	0.0210	0.0111	0.0226	0.0134	0.0080
			0.8	0.0011	0.9816	0.0010	0.9833	0.7962	0.0203	0.0112	0.0212	0.0122	0.0035
			0.9	-0.0071	0.9849	-0.0030	0.9880	0.8963	0.0203	0.0115	0.0208	0.0127	0.0010
	40	0.2	0	-0.0021	0.9848	-0.0075	0.9894	-0.0023	0.0204	0.0134	0.0311	0.0132	0.0392
			0.1	0.0048	0.9822	0.0020	0.9906	0.1023	0.0229	0.0136	0.0314	0.0135	0.0380
			0.2	-0.0065	0.9877	-0.0025	0.9884	0.1893	0.0245	0.0142	0.0295	0.0141	0.0364
			0.3	-0.0032	0.9759	-0.0021	0.9870	0.2863	0.0222	0.0135	0.0301	0.0133	0.0339
			0.4	-0.0014	0.9770	0.0013	0.9844	0.3913	0.0206	0.0137	0.0279	0.0146	0.0284
			0.5	0.0028	0.9813	0.0083	0.9929	0.4910	0.0184	0.0147	0.0288	0.0168	0.0242
			0.6	-0.0041	0.9873	0.0003	0.9882	0.5915	0.0213	0.0133	0.0244	0.0169	0.0171
			0.7	-0.0047	0.9840	-0.0059	0.9784	0.6821	0.0209	0.0137	0.0270	0.0170	0.0115
			0.8	-0.0063	0.9852	-0.0054	0.9838	0.7959	0.0229	0.0126	0.0268	0.0159	0.0044
			0.9	-0.0035	0.9812	-0.0082	0.9804	0.8969	0.0209	0.0133	0.0230	0.0154	0.0012
	35	0.3	0	-0.0064	0.9790	0.0041	0.9881	-0.0009	0.0221	0.0167	0.0394	0.0150	0.0553
			0.1	-0.0069	0.9763	-0.0147	0.9835	0.0767	0.0236	0.0164	0.0406	0.0152	0.0523
			0.2	-0.0127	0.9764	0.0026	0.9973	0.1954	0.0236	0.0167	0.0429	0.0148	0.0535
			0.3	-0.0087	0.9785	-0.0076	0.9928	0.2764	0.0242	0.0158	0.0444	0.0169	0.0487
			0.4	-0.0127	0.9702	-0.0095	0.9841	0.3792	0.0254	0.0154	0.0374	0.0164	0.0372
			0.5	-0.0045	0.9773	-0.0017	0.9867	0.4762	0.0220	0.0173	0.0405	0.0187	0.0322
			0.6	-0.0191	0.9817	-0.0107	0.9830	0.5844	0.0223	0.0166	0.0364	0.0208	0.0234
			0.7	-0.0123	0.9777	-0.0025	0.9769	0.6816	0.0230	0.0155	0.0343	0.0222	0.0138
			0.8	-0.0139	0.9786	-0.0070	0.9887	0.7936	0.0242	0.0166	0.0292	0.0219	0.0064
			0.9	-0.0037	0.9776	0.0006	0.9847	0.8956	0.0226	0.0161	0.0265	0.0203	0.0017
	30	0.4	0	-0.0151	0.9726	-0.0032	0.9947	-0.0021	0.0275	0.0199	0.0662	0.0189	0.0696
			0.1	-0.0111	0.9749	-0.0086	0.9901	0.0877	0.0250	0.0205	0.0650	0.0185	0.0707
			0.2	-0.0038	0.9743	0.0036	0.9966	0.1964	0.0252	0.0195	0.0621	0.0187	0.0668
			0.3	-0.0037	0.9670	-0.0004	0.9896	0.2783	0.0262	0.0199	0.0642	0.0197	0.0618
			0.4	-0.0123	0.9650	-0.0145	0.9951	0.3829	0.0256	0.0195	0.0583	0.0227	0.0490
			0.5	-0.0106	0.9748	-0.0071	0.9949	0.4817	0.0269	0.0191	0.0581	0.0265	0.0416
			0.6	-0.0147	0.9620	-0.0052	0.9804	0.5750	0.0258	0.0206	0.0519	0.0267	0.0306
			0.7	-0.0088	0.9755	-0.0086	0.9783	0.6773	0.0259	0.0193	0.0490	0.0293	0.0207
			0.8	-0.0059	0.9766	-0.0111	0.9706	0.7834	0.0262	0.0197	0.0427	0.0278	0.0097
			0.9	-0.0093	0.9737	-0.0046	0.9785	0.8948	0.0270	0.0205	0.0328	0.0251	0.0021
	25	0.5	0	-0.0202	0.9673	-0.0084	1.0003	-0.0004	0.0295	0.0241	0.1146	0.0254	0.1008
			0.1	-0.0129	0.9654	0.0049	0.9985	0.0895	0.0321	0.0253	0.1042	0.0229	0.0917
			0.2	-0.0190	0.9620	-0.0121	0.9949	0.1774	0.0293	0.0252	0.1050	0.0245	0.0903
			0.3	-0.0197	0.9651	-0.0041	1.0068	0.2793	0.0306	0.0266	0.1028	0.0273	0.0827
			0.4	-0.0174	0.9678	0.0066	0.9996	0.3752	0.0282	0.0233	0.0944	0.0296	0.0730
			0.5	-0.0218	0.9638	-0.0168	0.9889	0.4545	0.0315	0.0253	0.0906	0.0309	0.0672
			0.6	-0.0199	0.9650	-0.0129	0.9841	0.5640	0.0298	0.0246	0.0785	0.0341	0.0436
			0.7	-0.0257	0.9674	-0.0144	0.9814	0.6692	0.0289	0.0241	0.0762	0.0380	0.0283
			0.8	-0.0236	0.9631	-0.0095	0.9746	0.7825	0.0313	0.0230	0.0588	0.0355	0.0132
			0.9	-0.0178	0.9675	-0.0251	0.9638	0.8897	0.0317	0.0257	0.0456	0.0329	0.0044

Table 13.3: Covariances of the MLEs for $n = 50$

n	k	Proportion of Censoring	ρ	Cov $[\hat{\mu}_X, \hat{\sigma}_X]$	Cov $[\hat{\mu}_X, \hat{\mu}_Y]$	Cov $[\hat{\mu}_X, \hat{\sigma}_Y]$	Cov $[\hat{\mu}_X, \hat{\rho}]$	Cov $[\hat{\sigma}_X, \hat{\mu}_Y]$	Cov $[\hat{\sigma}_X, \hat{\sigma}_Y]$	Cov $[\hat{\sigma}_X, \hat{\rho}]$	Cov $[\hat{\mu}_Y, \hat{\sigma}_Y]$	Cov $[\hat{\mu}_Y, \hat{\rho}]$	Cov $[\hat{\sigma}_Y, \hat{\rho}]$
50	50	0	0	0.0003	0.0003	-0.0001	0.0017	0.0000	-0.0003	0.0001	-0.0004	-0.0005	-0.0003
			0.1	0.0003	0.0012	-0.0008	-0.0001	0.0003	0.0002	0.0003	0.0001	0.0000	-0.0001
			0.2	-0.0002	0.0048	-0.0002	0.0003	0.0006	0.0008	0.0022	0.0006	0.0002	0.0023
			0.3	-0.0002	0.0052	-0.0007	-0.0006	-0.0007	0.0012	0.0029	0.0005	-0.0003	0.0029
			0.4	-0.0004	0.0085	0.0004	0.0007	-0.0008	0.0021	0.0034	0.0002	0.0003	0.0044
			0.5	-0.0003	0.0105	0.0001	0.0003	-0.0001	0.0029	0.0037	0.0000	0.0004	0.0041
			0.6	0.0000	0.0121	0.0006	0.0005	-0.0002	0.0033	0.0041	0.0006	0.0005	0.0040
			0.7	-0.0003	0.0132	0.0001	0.0000	-0.0006	0.0046	0.0034	0.0004	0.0001	0.0034
			0.8	0.0000	0.0161	-0.0005	0.0000	0.0001	0.0066	0.0030	-0.0003	0.0002	0.0029
			0.9	0.0002	0.0185	0.0003	0.0000	-0.0001	0.0076	0.0018	0.0002	-0.0001	0.0017
	45	0.1	0	0.0011	-0.0002	0.0005	-0.0007	0.0008	-0.0004	-0.0003	-0.0001	0.0058	0.0004
			0.1	0.0014	0.0011	-0.0007	0.0003	0.0008	-0.0003	0.0012	0.0002	0.0061	0.0019
			0.2	-0.0002	0.0042	-0.0002	-0.0002	0.0000	0.0007	0.0017	0.0016	0.0071	0.0038
			0.3	0.0021	0.0059	0.0002	0.0013	0.0012	0.0011	0.0033	0.0018	0.0056	0.0042
			0.4	0.0021	0.0072	0.0003	0.0013	0.0011	0.0019	0.0039	0.0024	0.0051	0.0057
			0.5	0.0003	0.0112	-0.0005	0.0002	0.0003	0.0027	0.0042	0.0020	0.0030	0.0072
			0.6	0.0008	0.0115	0.0003	-0.0005	0.0004	0.0050	0.0052	0.0014	0.0017	0.0064
			0.7	0.0002	0.0147	-0.0003	0.0000	0.0004	0.0057	0.0044	0.0014	0.0011	0.0059
			0.8	0.0003	0.0162	0.0006	0.0001	0.0005	0.0069	0.0032	0.0022	0.0010	0.0040
			0.9	0.0002	0.0182	-0.0002	-0.0001	0.0002	0.0094	0.0020	0.0009	0.0002	0.0023
	40	0.2	0	0.0019	0.0000	0.0003	-0.0002	0.0010	0.0000	-0.0004	-0.0004	0.0141	-0.0003
			0.1	0.0023	0.0025	0.0005	0.0006	-0.0001	0.0003	0.0017	0.0020	0.0160	0.0034
			0.2	0.0021	0.0046	0.0004	0.0002	0.0005	0.0005	0.0028	0.0027	0.0130	0.0053
			0.3	0.0021	0.0063	-0.0019	-0.0011	0.0008	0.0010	0.0035	0.0036	0.0142	0.0065
			0.4	0.0016	0.0073	-0.0004	-0.0001	-0.0004	0.0014	0.0040	0.0051	0.0103	0.0086
			0.5	0.0027	0.0097	0.0009	0.0010	0.0012	0.0032	0.0054	0.0077	0.0104	0.0102
			0.6	0.0029	0.0127	0.0013	0.0010	0.0020	0.0043	0.0049	0.0062	0.0071	0.0094
			0.7	0.0021	0.0154	0.0013	0.0009	0.0025	0.0071	0.0055	0.0063	0.0051	0.0091
			0.8	0.0028	0.0194	0.0026	0.0012	0.0024	0.0076	0.0037	0.0068	0.0029	0.0053
			0.9	0.0017	0.0188	0.0010	0.0003	0.0016	0.0107	0.0024	0.0037	0.0008	0.0029
	35	0.3	0	0.0042	-0.0010	0.0014	-0.0020	-0.0002	-0.0003	-0.0012	0.0009	0.0258	0.0011
			0.1	0.0047	0.0029	0.0004	0.0012	-0.0011	-0.0001	0.0007	0.0021	0.0252	0.0033
			0.2	0.0051	0.0056	-0.0002	0.0001	0.0003	0.0001	0.0027	0.0059	0.0270	0.0083
			0.3	0.0048	0.0080	0.0007	0.0028	0.0029	0.0010	0.0052	0.0083	0.0279	0.0105
			0.4	0.0047	0.0099	0.0002	0.0017	0.0027	0.0024	0.0055	0.0084	0.0188	0.0102
			0.5	0.0039	0.0114	0.0015	0.0014	0.0011	0.0044	0.0065	0.0115	0.0174	0.0131
			0.6	0.0041	0.0134	0.0021	0.0013	0.0027	0.0058	0.0064	0.0112	0.0139	0.0125
			0.7	0.0049	0.0167	0.0032	0.0023	0.0038	0.0081	0.0062	0.0130	0.0097	0.0118
			0.8	0.0041	0.0184	0.0023	0.0014	0.0035	0.0107	0.0053	0.0095	0.0050	0.0081
			0.9	0.0037	0.0203	0.0028	0.0006	0.0036	0.0131	0.0031	0.0079	0.0017	0.0041
	30	0.4	0	0.0089	-0.0011	-0.0001	-0.0007	-0.0001	0.0001	-0.0007	0.0001	0.0477	0.0004
			0.1	0.0064	0.0014	0.0016	-0.0012	-0.0008	0.0006	0.0005	0.0040	0.0478	0.0045
			0.2	0.0074	0.0051	0.0004	0.0002	0.0002	0.0020	0.0026	0.0082	0.0449	0.0109
			0.3	0.0067	0.0087	0.0007	0.0019	0.0027	0.0017	0.0053	0.0125	0.0418	0.0109
			0.4	0.0074	0.0105	0.0012	0.0021	0.0025	0.0024	0.0055	0.0163	0.0349	0.0143
			0.5	0.0078	0.0146	0.0017	0.0035	0.0049	0.0044	0.0077	0.0196	0.0313	0.0194
			0.6	0.0081	0.0145	0.0029	0.0020	0.0059	0.0086	0.0089	0.0206	0.0248	0.0181
			0.7	0.0068	0.0190	0.0037	0.0029	0.0055	0.0110	0.0088	0.0212	0.0178	0.0168
			0.8	0.0071	0.0210	0.0054	0.0020	0.0042	0.0112	0.0057	0.0196	0.0098	0.0112
			0.9	0.0086	0.0236	0.0058	0.0014	0.0077	0.0163	0.0038	0.0132	0.0033	0.0050
	25	0.5	0	0.0119	0.0011	-0.0008	0.0010	0.0021	0.0010	0.0010	-0.0023	0.0860	-0.0008
			0.1	0.0135	0.0037	0.0002	0.0026	0.0016	0.0003	0.0041	0.0097	0.0763	0.0081
			0.2	0.0124	0.0054	-0.0002	0.0015	0.0035	0.0002	0.0038	0.0115	0.0743	0.0101
			0.3	0.0128	0.0083	0.0011	0.0036	0.0057	0.0025	0.0091	0.0225	0.0711	0.0187
			0.4	0.0115	0.0087	0.0006	0.0034	0.0028	0.0016	0.0066	0.0286	0.0623	0.0227
			0.5	0.0126	0.0152	0.0026	0.0061	0.0093	0.0068	0.0124	0.0316	0.0590	0.0252
			0.6	0.0122	0.0180	0.0045	0.0042	0.0076	0.0086	0.0089	0.0340	0.0406	0.0237
			0.7	0.0109	0.0222	0.0057	0.0053	0.0078	0.0111	0.0096	0.0354	0.0317	0.0217
			0.8	0.0118	0.0253	0.0078	0.0038	0.0086	0.0136	0.0075	0.0305	0.0160	0.0150
			0.9	0.0125	0.0271	0.0096	0.0027	0.0101	0.0199	0.0055	0.0234	0.0070	0.0082

Table 13.4: Variances of the MLEs from the Fisher information matrix for $n = 50$

n	k	Proportion of Censoring	ρ	From I^* Var $[\hat{\mu}_X]$	Var $[\hat{\sigma}_X]$	Var $[\hat{\mu}_Y]$	Var $[\hat{\sigma}_Y]$	Var $[\hat{\rho}]$	From I Var $[\hat{\mu}_X]$	Var $[\hat{\sigma}_X]$	Var $[\hat{\mu}_Y]$	Var $[\hat{\sigma}_Y]$	Var $[\hat{\rho}]$
50	50	0	0	0.0195	0.0097	0.0195	0.0097	0.0192	0.0195	0.0097	0.0195	0.0097	0.0192
			0.1	0.0197	0.0098	0.0198	0.0099	0.0189	0.0197	0.0098	0.0198	0.0099	0.0189
			0.2	0.0197	0.0099	0.0197	0.0098	0.0177	0.0197	0.0099	0.0197	0.0098	0.0177
			0.3	0.0194	0.0097	0.0197	0.0099	0.0161	0.0194	0.0097	0.0197	0.0099	0.0161
			0.4	0.0195	0.0097	0.0196	0.0098	0.0139	0.0195	0.0097	0.0196	0.0098	0.0139
			0.5	0.0197	0.0098	0.0197	0.0098	0.0112	0.0197	0.0098	0.0197	0.0098	0.0112
			0.6	0.0196	0.0098	0.0196	0.0098	0.0083	0.0196	0.0098	0.0196	0.0098	0.0083
			0.7	0.0197	0.0098	0.0197	0.0098	0.0054	0.0197	0.0098	0.0197	0.0098	0.0054
			0.8	0.0197	0.0099	0.0197	0.0098	0.0027	0.0197	0.0099	0.0197	0.0098	0.0027
			0.9	0.0196	0.0098	0.0197	0.0098	0.0008	0.0196	0.0098	0.0197	0.0098	0.0008
	45	0.1	0	0.0197	0.0113	0.0230	0.0112	0.0288	0.0198	0.0113	0.0230	0.0112	0.0286
			0.1	0.0198	0.0114	0.0230	0.0113	0.0282	0.0198	0.0113	0.0230	0.0113	0.0280
			0.2	0.0199	0.0114	0.0228	0.0115	0.0265	0.0199	0.0114	0.0228	0.0115	0.0262
			0.3	0.0198	0.0114	0.0227	0.0118	0.0237	0.0198	0.0114	0.0226	0.0118	0.0235
			0.4	0.0200	0.0115	0.0226	0.0122	0.0205	0.0200	0.0115	0.0225	0.0122	0.0204
			0.5	0.0202	0.0116	0.0225	0.0127	0.0162	0.0202	0.0116	0.0224	0.0126	0.0161
			0.6	0.0199	0.0114	0.0221	0.0130	0.0119	0.0199	0.0114	0.0220	0.0129	0.0118
			0.7	0.0198	0.0114	0.0213	0.0129	0.0077	0.0198	0.0113	0.0213	0.0129	0.0076
			0.8	0.0199	0.0114	0.0210	0.0130	0.0036	0.0199	0.0114	0.0210	0.0129	0.0036
			0.9	0.0200	0.0115	0.0207	0.0126	0.0010	0.0200	0.0115	0.0207	0.0126	0.0010
	40	0.2	0	0.0209	0.0135	0.0295	0.0130	0.0387	0.0209	0.0135	0.0293	0.0130	0.0381
			0.1	0.0208	0.0135	0.0295	0.0132	0.0379	0.0208	0.0134	0.0293	0.0132	0.0373
			0.2	0.0210	0.0136	0.0292	0.0135	0.0361	0.0211	0.0136	0.0290	0.0135	0.0353
			0.3	0.0205	0.0133	0.0287	0.0140	0.0323	0.0206	0.0132	0.0285	0.0140	0.0319
			0.4	0.0206	0.0133	0.0279	0.0148	0.0274	0.0206	0.0133	0.0278	0.0147	0.0270
			0.5	0.0208	0.0134	0.0276	0.0158	0.0217	0.0208	0.0134	0.0275	0.0157	0.0215
			0.6	0.0210	0.0136	0.0265	0.0164	0.0158	0.0210	0.0136	0.0264	0.0162	0.0156
			0.7	0.0209	0.0135	0.0251	0.0165	0.0105	0.0209	0.0135	0.0249	0.0163	0.0103
			0.8	0.0209	0.0135	0.0239	0.0166	0.0047	0.0209	0.0135	0.0239	0.0164	0.0046
			0.9	0.0207	0.0134	0.0224	0.0155	0.0012	0.0208	0.0134	0.0223	0.0154	0.0012
	35	0.3	0	0.0222	0.0160	0.0412	0.0156	0.0504	0.0222	0.0159	0.0405	0.0155	0.0491
			0.1	0.0221	0.0159	0.0406	0.0155	0.0500	0.0221	0.0158	0.0400	0.0154	0.0489
			0.2	0.0221	0.0159	0.0411	0.0168	0.0466	0.0221	0.0158	0.0404	0.0166	0.0453
			0.3	0.0222	0.0159	0.0400	0.0174	0.0428	0.0222	0.0159	0.0394	0.0172	0.0418
			0.4	0.0218	0.0157	0.0382	0.0181	0.0371	0.0218	0.0156	0.0376	0.0179	0.0362
			0.5	0.0221	0.0159	0.0368	0.0195	0.0300	0.0222	0.0158	0.0362	0.0192	0.0292
			0.6	0.0223	0.0160	0.0344	0.0205	0.0213	0.0224	0.0160	0.0340	0.0202	0.0208
			0.7	0.0221	0.0159	0.0318	0.0210	0.0138	0.0222	0.0158	0.0314	0.0207	0.0135
			0.8	0.0222	0.0159	0.0295	0.0214	0.0063	0.0222	0.0159	0.0293	0.0210	0.0061
			0.9	0.0221	0.0159	0.0262	0.0197	0.0016	0.0222	0.0158	0.0260	0.0194	0.0016
	30	0.4	0	0.0246	0.0192	0.0633	0.0198	0.0664	0.0246	0.0190	0.0613	0.0195	0.0642
			0.1	0.0247	0.0193	0.0625	0.0198	0.0653	0.0248	0.0191	0.0604	0.0195	0.0630
			0.2	0.0247	0.0192	0.0617	0.0210	0.0611	0.0247	0.0191	0.0599	0.0207	0.0591
			0.3	0.0243	0.0190	0.0595	0.0217	0.0563	0.0244	0.0188	0.0578	0.0214	0.0544
			0.4	0.0242	0.0189	0.0579	0.0238	0.0489	0.0243	0.0187	0.0561	0.0233	0.0471
			0.5	0.0247	0.0192	0.0540	0.0257	0.0389	0.0247	0.0191	0.0528	0.0252	0.0378
			0.6	0.0241	0.0188	0.0491	0.0263	0.0298	0.0241	0.0187	0.0478	0.0257	0.0287
			0.7	0.0247	0.0193	0.0442	0.0273	0.0191	0.0248	0.0191	0.0432	0.0266	0.0184
			0.8	0.0248	0.0193	0.0381	0.0267	0.0092	0.0248	0.0192	0.0375	0.0261	0.0089
			0.9	0.0247	0.0192	0.0322	0.0249	0.0022	0.0247	0.0191	0.0318	0.0243	0.0021
	25	0.5	0	0.0291	0.0238	0.1032	0.0268	0.0857	0.0292	0.0235	0.0986	0.0262	0.0822
			0.1	0.0291	0.0237	0.1025	0.0265	0.0859	0.0291	0.0235	0.0981	0.0260	0.0824
			0.2	0.0288	0.0236	0.1005	0.0276	0.0823	0.0289	0.0233	0.0954	0.0268	0.0784
			0.3	0.0291	0.0237	0.0986	0.0300	0.0745	0.0291	0.0235	0.0944	0.0294	0.0713
			0.4	0.0291	0.0238	0.0928	0.0318	0.0654	0.0292	0.0235	0.0886	0.0311	0.0622
			0.5	0.0290	0.0236	0.0856	0.0334	0.0553	0.0290	0.0234	0.0817	0.0324	0.0527
			0.6	0.0290	0.0237	0.0769	0.0352	0.0414	0.0290	0.0234	0.0742	0.0341	0.0399
			0.7	0.0291	0.0238	0.0682	0.0369	0.0274	0.0292	0.0235	0.0654	0.0353	0.0262
			0.8	0.0289	0.0235	0.0561	0.0359	0.0130	0.0289	0.0233	0.0542	0.0346	0.0124
			0.9	0.0292	0.0238	0.0428	0.0317	0.0036	0.0292	0.0236	0.0419	0.0308	0.0034

Table 13.5: Covariances of the MLEs from the Fisher information matrix for $n = 50$

n	k	Proportion of Censoring	ρ	Cov $[\hat{\mu}_X, \hat{\sigma}_X]$	Cov $[\hat{\mu}_X, \hat{\mu}_Y]$	Cov $[\hat{\mu}_X, \hat{\sigma}_Y]$	Cov $[\hat{\mu}_X, \hat{\rho}]$	Cov $[\hat{\sigma}_X, \hat{\mu}_Y]$	Cov $[\hat{\sigma}_X, \hat{\sigma}_Y]$	Cov $[\hat{\sigma}_X, \hat{\rho}]$	Cov $[\hat{\mu}_Y, \hat{\sigma}_Y]$	Cov $[\hat{\mu}_Y, \hat{\rho}]$	Cov $[\hat{\sigma}_Y, \hat{\rho}]$
50	50	0	0	0.0000	-0.0002	0.0000	0.0000	0.0000	0.0002	-0.0001	0.0000	0.0000	-0.0001
			0.1	0.0000	0.0018	0.0000	0.0000	0.0000	0.0003	0.0008	0.0000	0.0000	0.0008
			0.2	0.0000	0.0040	0.0000	0.0000	0.0000	0.0006	0.0018	0.0000	0.0000	0.0018
			0.3	0.0000	0.0059	0.0000	0.0000	0.0000	0.0011	0.0025	0.0000	0.0000	0.0026
			0.4	0.0000	0.0078	0.0000	0.0000	0.0000	0.0017	0.0031	0.0000	0.0000	0.0031
			0.5	0.0000	0.0099	0.0000	0.0000	0.0000	0.0026	0.0035	0.0000	0.0000	0.0035
			0.6	0.0000	0.0118	0.0000	0.0000	0.0000	0.0037	0.0036	0.0000	0.0000	0.0036
			0.7	0.0000	0.0138	0.0000	0.0000	0.0000	0.0049	0.0034	0.0000	0.0000	0.0034
			0.8	0.0000	0.0158	0.0000	0.0000	0.0000	0.0064	0.0028	0.0000	0.0000	0.0028
			0.9	0.0000	0.0177	0.0000	0.0000	0.0000	0.0080	0.0017	0.0000	0.0000	0.0017
	45	0.1	0	0.0008	0.0000	0.0000	0.0000	0.0000	0.0003	0.0000	0.0000	0.0054	0.0000
			0.1	0.0008	0.0018	0.0000	0.0001	0.0001	0.0004	0.0010	0.0005	0.0053	0.0015
			0.2	0.0008	0.0040	0.0001	0.0001	0.0002	0.0008	0.0020	0.0010	0.0050	0.0032
			0.3	0.0008	0.0061	0.0001	0.0002	0.0002	0.0014	0.0030	0.0015	0.0046	0.0047
			0.4	0.0008	0.0080	0.0001	0.0003	0.0003	0.0021	0.0036	0.0019	0.0040	0.0055
			0.5	0.0008	0.0102	0.0002	0.0003	0.0004	0.0031	0.0041	0.0022	0.0033	0.0060
			0.6	0.0008	0.0120	0.0003	0.0003	0.0005	0.0043	0.0042	0.0023	0.0025	0.0060
			0.7	0.0008	0.0138	0.0004	0.0003	0.0006	0.0056	0.0039	0.0022	0.0017	0.0053
			0.8	0.0008	0.0159	0.0005	0.0002	0.0006	0.0074	0.0032	0.0020	0.0009	0.0040
			0.9	0.0008	0.0181	0.0007	0.0001	0.0007	0.0094	0.0019	0.0015	0.0003	0.0022
	40	0.2	0	0.0021	-0.0001	0.0001	0.0000	0.0000	0.0005	0.0000	0.0000	0.0132	-0.0001
			0.1	0.0021	0.0022	0.0001	0.0002	0.0002	0.0007	0.0012	0.0014	0.0129	0.0026
			0.2	0.0021	0.0041	0.0002	0.0004	0.0004	0.0010	0.0022	0.0025	0.0123	0.0045
			0.3	0.0021	0.0061	0.0003	0.0005	0.0006	0.0016	0.0032	0.0036	0.0113	0.0063
			0.4	0.0021	0.0083	0.0004	0.0006	0.0008	0.0025	0.0041	0.0046	0.0098	0.0077
			0.5	0.0021	0.0105	0.0006	0.0007	0.0011	0.0037	0.0046	0.0053	0.0081	0.0084
			0.6	0.0021	0.0126	0.0008	0.0008	0.0013	0.0051	0.0048	0.0056	0.0062	0.0082
			0.7	0.0021	0.0144	0.0010	0.0007	0.0014	0.0066	0.0046	0.0055	0.0044	0.0073
			0.8	0.0021	0.0167	0.0014	0.0006	0.0017	0.0087	0.0038	0.0050	0.0024	0.0053
			0.9	0.0021	0.0187	0.0017	0.0004	0.0019	0.0109	0.0023	0.0038	0.0009	0.0028
	35	0.3	0	0.0040	0.0000	0.0002	0.0000	0.0000	0.0009	0.0000	0.0000	0.0244	0.0000
			0.1	0.0040	0.0018	0.0002	0.0003	0.0003	0.0009	0.0011	0.0020	0.0242	0.0026
			0.2	0.0040	0.0046	0.0004	0.0006	0.0008	0.0015	0.0026	0.0049	0.0229	0.0061
			0.3	0.0040	0.0065	0.0005	0.0009	0.0012	0.0021	0.0036	0.0066	0.0213	0.0082
			0.4	0.0039	0.0086	0.0008	0.0012	0.0016	0.0030	0.0046	0.0084	0.0188	0.0103
			0.5	0.0040	0.0110	0.0011	0.0013	0.0020	0.0044	0.0052	0.0098	0.0157	0.0112
			0.6	0.0040	0.0134	0.0015	0.0014	0.0024	0.0061	0.0056	0.0105	0.0119	0.0110
			0.7	0.0040	0.0154	0.0020	0.0014	0.0028	0.0079	0.0054	0.0104	0.0083	0.0097
			0.8	0.0040	0.0180	0.0026	0.0011	0.0033	0.0105	0.0044	0.0095	0.0046	0.0070
			0.9	0.0040	0.0201	0.0033	0.0007	0.0036	0.0130	0.0027	0.0074	0.0017	0.0036
	30	0.4	0	0.0069	-0.0001	0.0005	0.0000	0.0000	0.0014	-0.0001	-0.0001	0.0418	-0.0002
			0.1	0.0069	0.0023	0.0006	0.0005	0.0006	0.0016	0.0013	0.0038	0.0410	0.0037
			0.2	0.0069	0.0052	0.0008	0.0011	0.0015	0.0022	0.0030	0.0086	0.0390	0.0083
			0.3	0.0068	0.0072	0.0010	0.0015	0.0020	0.0028	0.0041	0.0113	0.0363	0.0109
			0.4	0.0068	0.0099	0.0014	0.0019	0.0028	0.0040	0.0053	0.0149	0.0322	0.0138
			0.5	0.0069	0.0127	0.0021	0.0022	0.0035	0.0057	0.0061	0.0174	0.0267	0.0150
			0.6	0.0068	0.0146	0.0027	0.0023	0.0041	0.0073	0.0064	0.0179	0.0210	0.0145
			0.7	0.0069	0.0173	0.0035	0.0023	0.0048	0.0097	0.0063	0.0178	0.0146	0.0127
			0.8	0.0070	0.0196	0.0044	0.0019	0.0055	0.0123	0.0053	0.0161	0.0083	0.0093
			0.9	0.0069	0.0223	0.0057	0.0012	0.0063	0.0156	0.0032	0.0128	0.0029	0.0046
	25	0.5	0	0.0115	0.0000	0.0012	0.0000	0.0000	0.0025	0.0000	-0.0003	0.0677	-0.0002
			0.1	0.0114	0.0030	0.0012	0.0008	0.0012	0.0025	0.0017	0.0070	0.0677	0.0054
			0.2	0.0114	0.0056	0.0015	0.0016	0.0022	0.0031	0.0032	0.0125	0.0645	0.0096
			0.3	0.0114	0.0091	0.0021	0.0024	0.0036	0.0042	0.0049	0.0200	0.0601	0.0146
			0.4	0.0115	0.0119	0.0027	0.0030	0.0047	0.0056	0.0061	0.0250	0.0530	0.0177
			0.5	0.0114	0.0144	0.0035	0.0034	0.0056	0.0072	0.0069	0.0275	0.0454	0.0187
			0.6	0.0114	0.0174	0.0045	0.0037	0.0068	0.0093	0.0076	0.0301	0.0359	0.0194
			0.7	0.0115	0.0204	0.0059	0.0037	0.0080	0.0121	0.0075	0.0302	0.0253	0.0171
			0.8	0.0114	0.0233	0.0075	0.0031	0.0092	0.0153	0.0064	0.0273	0.0141	0.0124
			0.9	0.0115	0.0261	0.0093	0.0019	0.0103	0.0191	0.0040	0.0210	0.0052	0.0060

Table 13.6: Covariances of the MLEs from partially observed information matrix for $n = 50$

n	k	Proportion of Censoring	ρ	Cov $[\hat{\mu}_X, \hat{\sigma}_X]$	Cov $[\hat{\mu}_X, \hat{\mu}_Y]$	Cov $[\hat{\mu}_X, \hat{\sigma}_Y]$	Cov $[\hat{\mu}_X, \hat{\rho}]$	Cov $[\hat{\sigma}_X, \hat{\mu}_Y]$	Cov $[\hat{\sigma}_X, \hat{\sigma}_Y]$	Cov $[\hat{\sigma}_X, \hat{\rho}]$	Cov $[\hat{\mu}_Y, \hat{\sigma}_Y]$	Cov $[\hat{\mu}_Y, \hat{\rho}]$	Cov $[\hat{\sigma}_Y, \hat{\rho}]$
50	50	0	0	0.0000	-0.0002	0.0000	0.0000	0.0000	0.0002	-0.0001	0.0000	0.0000	-0.0001
			0.1	0.0000	0.0018	0.0000	0.0000	0.0000	0.0003	0.0008	0.0000	0.0000	0.0008
			0.2	0.0000	0.0040	0.0000	0.0000	0.0000	0.0006	0.0018	0.0000	0.0000	0.0018
			0.3	0.0000	0.0059	0.0000	0.0000	0.0000	0.0011	0.0025	0.0000	0.0000	0.0026
			0.4	0.0000	0.0078	0.0000	0.0000	0.0000	0.0017	0.0031	0.0000	0.0000	0.0031
			0.5	0.0000	0.0099	0.0000	0.0000	0.0000	0.0026	0.0035	0.0000	0.0000	0.0035
			0.6	0.0000	0.0118	0.0000	0.0000	0.0000	0.0037	0.0036	0.0000	0.0000	0.0036
			0.7	0.0000	0.0138	0.0000	0.0000	0.0000	0.0049	0.0034	0.0000	0.0000	0.0034
			0.8	0.0000	0.0158	0.0000	0.0000	0.0000	0.0064	0.0028	0.0000	0.0000	0.0028
			0.9	0.0000	0.0177	0.0000	0.0000	0.0000	0.0080	0.0017	0.0000	0.0000	0.0017
	45	0.1	0	0.0008	0.0000	0.0000	0.0000	0.0000	0.0003	0.0000	0.0000	0.0055	0.0000
			0.1	0.0008	0.0018	0.0000	0.0001	0.0001	0.0004	0.0010	0.0005	0.0054	0.0016
			0.2	0.0008	0.0040	0.0001	0.0001	0.0002	0.0008	0.0020	0.0011	0.0051	0.0033
			0.3	0.0008	0.0061	0.0001	0.0002	0.0002	0.0014	0.0030	0.0016	0.0047	0.0047
			0.4	0.0008	0.0080	0.0001	0.0003	0.0003	0.0021	0.0036	0.0019	0.0041	0.0056
			0.5	0.0008	0.0102	0.0002	0.0003	0.0004	0.0031	0.0041	0.0022	0.0034	0.0061
			0.6	0.0008	0.0120	0.0003	0.0003	0.0005	0.0043	0.0042	0.0023	0.0026	0.0060
			0.7	0.0008	0.0138	0.0004	0.0003	0.0006	0.0056	0.0039	0.0023	0.0018	0.0053
			0.8	0.0008	0.0159	0.0005	0.0002	0.0006	0.0074	0.0032	0.0020	0.0010	0.0040
			0.9	0.0008	0.0181	0.0007	0.0001	0.0007	0.0094	0.0019	0.0016	0.0003	0.0022
	40	0.2	0	0.0021	-0.0001	0.0001	0.0000	0.0000	0.0005	0.0000	-0.0001	0.0136	-0.0001
			0.1	0.0021	0.0022	0.0001	0.0002	0.0002	0.0007	0.0013	0.0014	0.0133	0.0026
			0.2	0.0021	0.0041	0.0002	0.0004	0.0004	0.0010	0.0023	0.0026	0.0128	0.0046
			0.3	0.0021	0.0061	0.0003	0.0005	0.0006	0.0016	0.0032	0.0037	0.0116	0.0064
			0.4	0.0021	0.0083	0.0004	0.0006	0.0008	0.0025	0.0041	0.0047	0.0101	0.0079
			0.5	0.0021	0.0105	0.0006	0.0007	0.0011	0.0038	0.0046	0.0054	0.0083	0.0085
			0.6	0.0021	0.0126	0.0008	0.0008	0.0013	0.0052	0.0048	0.0057	0.0064	0.0084
			0.7	0.0021	0.0144	0.0010	0.0007	0.0014	0.0066	0.0046	0.0056	0.0046	0.0074
			0.8	0.0021	0.0167	0.0014	0.0006	0.0017	0.0088	0.0038	0.0051	0.0024	0.0054
			0.9	0.0021	0.0187	0.0017	0.0004	0.0019	0.0109	0.0023	0.0039	0.0009	0.0028
	35	0.3	0	0.0040	0.0000	0.0002	0.0000	0.0000	0.0009	0.0000	0.0000	0.0254	0.0000
			0.1	0.0040	0.0018	0.0002	0.0003	0.0003	0.0009	0.0011	0.0020	0.0250	0.0027
			0.2	0.0040	0.0046	0.0004	0.0006	0.0008	0.0016	0.0026	0.0051	0.0239	0.0064
			0.3	0.0040	0.0065	0.0005	0.0009	0.0012	0.0021	0.0036	0.0069	0.0221	0.0085
			0.4	0.0040	0.0086	0.0008	0.0012	0.0016	0.0030	0.0046	0.0087	0.0195	0.0106
			0.5	0.0040	0.0109	0.0011	0.0013	0.0020	0.0044	0.0053	0.0102	0.0164	0.0116
			0.6	0.0040	0.0133	0.0015	0.0014	0.0024	0.0061	0.0056	0.0108	0.0123	0.0112
			0.7	0.0040	0.0154	0.0020	0.0014	0.0028	0.0079	0.0054	0.0107	0.0087	0.0100
			0.8	0.0040	0.0180	0.0027	0.0011	0.0033	0.0105	0.0044	0.0099	0.0047	0.0072
			0.9	0.0040	0.0201	0.0033	0.0007	0.0036	0.0130	0.0027	0.0076	0.0017	0.0036
	30	0.4	0	0.0069	-0.0001	0.0005	0.0000	0.0000	0.0014	-0.0001	-0.0001	0.0440	-0.0002
			0.1	0.0070	0.0023	0.0006	0.0005	0.0006	0.0016	0.0013	0.0039	0.0432	0.0038
			0.2	0.0070	0.0052	0.0008	0.0011	0.0015	0.0022	0.0030	0.0089	0.0410	0.0086
			0.3	0.0069	0.0072	0.0010	0.0015	0.0020	0.0028	0.0041	0.0118	0.0380	0.0114
			0.4	0.0068	0.0099	0.0015	0.0019	0.0028	0.0040	0.0054	0.0157	0.0340	0.0144
			0.5	0.0070	0.0126	0.0021	0.0022	0.0036	0.0058	0.0062	0.0181	0.0278	0.0154
			0.6	0.0068	0.0146	0.0027	0.0023	0.0041	0.0074	0.0065	0.0188	0.0221	0.0152
			0.7	0.0070	0.0173	0.0036	0.0023	0.0049	0.0098	0.0063	0.0187	0.0153	0.0132
			0.8	0.0070	0.0195	0.0045	0.0019	0.0055	0.0123	0.0054	0.0167	0.0087	0.0096
			0.9	0.0070	0.0223	0.0057	0.0012	0.0063	0.0157	0.0032	0.0132	0.0031	0.0047
	25	0.5	0	0.0116	0.0000	0.0013	0.0000	0.0000	0.0026	0.0000	-0.0004	0.0717	-0.0003
			0.1	0.0116	0.0030	0.0012	0.0008	0.0012	0.0025	0.0017	0.0072	0.0716	0.0054
			0.2	0.0115	0.0056	0.0015	0.0016	0.0022	0.0031	0.0032	0.0133	0.0689	0.0101
			0.3	0.0116	0.0091	0.0021	0.0024	0.0036	0.0043	0.0049	0.0210	0.0637	0.0152
			0.4	0.0116	0.0119	0.0027	0.0030	0.0047	0.0056	0.0062	0.0262	0.0566	0.0184
			0.5	0.0115	0.0143	0.0035	0.0034	0.0057	0.0072	0.0070	0.0290	0.0484	0.0196
			0.6	0.0115	0.0174	0.0046	0.0038	0.0069	0.0094	0.0077	0.0317	0.0378	0.0202
			0.7	0.0116	0.0204	0.0060	0.0037	0.0081	0.0122	0.0076	0.0322	0.0269	0.0181
			0.8	0.0115	0.0233	0.0076	0.0032	0.0093	0.0155	0.0065	0.0288	0.0150	0.0130
			0.9	0.0116	0.0261	0.0094	0.0020	0.0104	0.0193	0.0040	0.0219	0.0055	0.0063

Table 13.7: Means and variances of the MLEs for $n = 20$

n	k	Proportion of Censoring	ρ	$\hat{\mu}_X$	$\hat{\sigma}_X$	$\hat{\mu}_Y$	$\hat{\sigma}_Y$	$\hat{\rho}$	Var $[\hat{\mu}_X]$	Var $[\hat{\sigma}_X]$	Var $[\hat{\mu}_Y]$	Var $[\hat{\sigma}_Y]$	Var $[\hat{\rho}]$
20	20	0	0	-0.0026	0.9604	-0.0017	0.9644	0.0008	0.0524	0.0252	0.0521	0.0231	0.0530
			0.1	0.0096	0.9598	-0.0062	0.9640	0.0987	0.0505	0.0238	0.0482	0.0277	0.0498
			0.2	-0.0079	0.9597	-0.0085	0.9649	0.1986	0.0494	0.0244	0.0494	0.0258	0.0447
			0.3	0.0026	0.9585	0.0056	0.9657	0.2829	0.0516	0.0263	0.0479	0.0262	0.0457
			0.4	0.0082	0.9583	0.0111	0.9694	0.3959	0.0493	0.0248	0.0503	0.0253	0.0378
			0.5	-0.0117	0.9633	-0.0035	0.9498	0.4850	0.0511	0.0240	0.0492	0.0242	0.0307
			0.6	-0.0054	0.9546	-0.0035	0.9598	0.5929	0.0542	0.0246	0.0503	0.0247	0.0214
			0.7	-0.0016	0.9578	0.0070	0.9576	0.6849	0.0499	0.0254	0.0497	0.0263	0.0167
			0.8	0.0074	0.9579	0.0061	0.9614	0.7910	0.0531	0.0276	0.0508	0.0251	0.0089
			0.9	-0.0109	0.9620	-0.0046	0.9561	0.8948	0.0497	0.0242	0.0487	0.0255	0.0022
	18	0.1	0	-0.0167	0.9561	-0.0016	0.9720	0.0049	0.0473	0.0284	0.0592	0.0301	0.0768
			0.1	-0.0071	0.9479	-0.0054	0.9640	0.0936	0.0494	0.0285	0.0575	0.0300	0.0833
			0.2	-0.0169	0.9507	0.0054	0.9589	0.1738	0.0519	0.0280	0.0586	0.0288	0.0708
			0.3	-0.0067	0.9593	0.0066	0.9542	0.2901	0.0509	0.0280	0.0626	0.0294	0.0630
			0.4	-0.0176	0.9596	-0.0039	0.9626	0.3807	0.0500	0.0268	0.0582	0.0299	0.0582
			0.5	-0.0005	0.9637	-0.0022	0.9645	0.4772	0.0472	0.0295	0.0518	0.0324	0.0462
			0.6	-0.0101	0.9612	-0.0102	0.9608	0.5797	0.0499	0.0263	0.0560	0.0324	0.0321
			0.7	0.0053	0.9634	0.0058	0.9578	0.6879	0.0507	0.0279	0.0606	0.0321	0.0209
			0.8	-0.0032	0.9621	0.0003	0.9602	0.7851	0.0486	0.0287	0.0520	0.0332	0.0130
			0.9	-0.0073	0.9534	-0.0069	0.9505	0.8915	0.0510	0.0293	0.0538	0.0324	0.0037
	16	0.2	0	-0.0121	0.9461	-0.0075	0.9725	0.0052	0.0554	0.0359	0.0788	0.0323	0.1015
			0.1	-0.0062	0.9513	0.0101	0.9609	0.0893	0.0533	0.0329	0.0747	0.0330	0.0990
			0.2	-0.0052	0.9511	-0.0010	0.9693	0.1862	0.0521	0.0306	0.0758	0.0350	0.0908
			0.3	-0.0117	0.9596	-0.0028	0.9798	0.2867	0.0542	0.0359	0.0742	0.0374	0.0834
			0.4	-0.0073	0.9459	-0.0106	0.9658	0.3721	0.0539	0.0345	0.0689	0.0376	0.0766
			0.5	-0.0209	0.9396	-0.0154	0.9649	0.4696	0.0534	0.0324	0.0734	0.0356	0.0630
			0.6	0.0084	0.9416	-0.0046	0.9607	0.5735	0.0553	0.0328	0.0663	0.0373	0.0445
			0.7	-0.0186	0.9422	-0.0014	0.9591	0.6681	0.0547	0.0346	0.0636	0.0411	0.0354
			0.8	-0.0119	0.9580	-0.0085	0.9615	0.7850	0.0507	0.0357	0.0593	0.0406	0.0145
			0.9	-0.0092	0.9519	-0.0078	0.9539	0.8923	0.0556	0.0345	0.0591	0.0396	0.0043
	14	0.3	0	-0.0184	0.9412	0.0100	0.9754	0.0259	0.0587	0.0388	0.1067	0.0399	0.1269
			0.1	-0.0141	0.9408	0.0004	0.9694	0.0901	0.0580	0.0382	0.1091	0.0395	0.1171
			0.2	0.0046	0.9371	0.0029	0.9861	0.1745	0.0570	0.0414	0.1058	0.0388	0.1295
			0.3	-0.0008	0.9468	-0.0039	0.9775	0.2852	0.0581	0.0402	0.1006	0.0419	0.1079
			0.4	-0.0154	0.9446	-0.0146	0.9720	0.3540	0.0579	0.0408	0.0960	0.0414	0.1015
			0.5	-0.0261	0.9358	-0.0385	0.9496	0.4385	0.0578	0.0365	0.0956	0.0471	0.0868
			0.6	-0.0146	0.9331	-0.0132	0.9540	0.5473	0.0572	0.0377	0.0942	0.0508	0.0650
			0.7	-0.0253	0.9367	-0.0136	0.9543	0.6607	0.0562	0.0375	0.0847	0.0487	0.0482
			0.8	-0.0164	0.9330	-0.0118	0.9498	0.7673	0.0541	0.0398	0.0658	0.0474	0.0275
			0.9	-0.0297	0.9369	-0.0243	0.9412	0.8856	0.0583	0.0390	0.0657	0.0485	0.0065
	12	0.4	0	-0.0326	0.9291	0.0018	0.9878	-0.0072	0.0623	0.0472	0.1480	0.0491	0.1560
			0.1	-0.0236	0.9294	-0.0193	0.9843	0.0718	0.0581	0.0449	0.1630	0.0557	0.1693
			0.2	-0.0168	0.9393	0.0112	0.9869	0.1671	0.0693	0.0500	0.1599	0.0520	0.1642
			0.3	-0.0306	0.9258	0.0086	0.9833	0.2539	0.0633	0.0470	0.1630	0.0607	0.1511
			0.4	-0.0151	0.9179	-0.0251	0.9648	0.3342	0.0677	0.0489	0.1493	0.0573	0.1396
			0.5	-0.0400	0.9239	-0.0219	0.9686	0.4477	0.0627	0.0479	0.1401	0.0583	0.1136
			0.6	-0.0379	0.9205	-0.0193	0.9580	0.5344	0.0664	0.0493	0.1164	0.0627	0.0887
			0.7	-0.0396	0.9222	-0.0268	0.9469	0.6421	0.0652	0.0450	0.1106	0.0595	0.0626
			0.8	-0.0482	0.9189	-0.0480	0.9303	0.7557	0.0612	0.0467	0.0953	0.0603	0.0377
			0.9	-0.0473	0.9258	-0.0404	0.9351	0.8799	0.0615	0.0433	0.0790	0.0570	0.0105
	10	0.5	0	-0.0422	0.9069	0.0141	0.9986	0.0102	0.0774	0.0613	0.2749	0.0666	0.2054
			0.1	-0.0415	0.9052	-0.0096	1.0027	0.0845	0.0818	0.0608	0.2688	0.0714	0.2171
			0.2	-0.0578	0.9095	-0.0515	0.9899	0.1271	0.0774	0.0618	0.2521	0.0723	0.2036
			0.3	-0.0661	0.8928	0.0057	0.9973	0.2683	0.0768	0.0563	0.2477	0.0713	0.1755
			0.4	-0.0394	0.9206	-0.0138	0.9804	0.3302	0.0749	0.0554	0.2369	0.0760	0.1716
			0.5	-0.0301	0.9102	-0.0172	0.9702	0.4098	0.0774	0.0650	0.2395	0.0824	0.1619
			0.6	-0.0264	0.9238	-0.0160	0.9618	0.5186	0.0795	0.0598	0.2044	0.0831	0.1204
			0.7	-0.0410	0.9152	-0.0364	0.9460	0.6214	0.0783	0.0576	0.1733	0.0869	0.0973
			0.8	-0.0521	0.9057	-0.0456	0.9202	0.7383	0.0753	0.0580	0.1460	0.0869	0.0506
			0.9	-0.0456	0.9158	-0.0292	0.9343	0.8677	0.0817	0.0635	0.1189	0.0799	0.0227

Table 13.8: Variances of the MLEs from the Fisher information matrix for $n = 20$

n	k	Proportion of Censoring	ρ	From I^* Var $[\hat{\mu}_X]$	Var $[\hat{\sigma}_X]$	Var $[\hat{\mu}_Y]$	Var $[\hat{\sigma}_Y]$	Var $[\hat{\rho}]$	From I Var $[\hat{\mu}_X]$	Var $[\hat{\sigma}_X]$	Var $[\hat{\mu}_Y]$	Var $[\hat{\sigma}_Y]$	Var $[\hat{\rho}]$
20	20	0	0	0.0474	0.0237	0.0477	0.0238	0.0451	0.0474	0.0237	0.0477	0.0238	0.0451
			0.1	0.0472	0.0236	0.0478	0.0239	0.0445	0.0472	0.0236	0.0478	0.0239	0.0445
			0.2	0.0473	0.0236	0.0478	0.0239	0.0424	0.0473	0.0236	0.0478	0.0239	0.0424
			0.3	0.0472	0.0236	0.0479	0.0240	0.0390	0.0472	0.0236	0.0479	0.0240	0.0390
			0.4	0.0472	0.0236	0.0482	0.0241	0.0335	0.0472	0.0236	0.0482	0.0241	0.0335
			0.5	0.0476	0.0238	0.0463	0.0232	0.0281	0.0476	0.0238	0.0463	0.0232	0.0281
			0.6	0.0468	0.0234	0.0473	0.0236	0.0209	0.0468	0.0234	0.0473	0.0236	0.0209
			0.7	0.0471	0.0236	0.0472	0.0236	0.0145	0.0471	0.0236	0.0472	0.0236	0.0145
			0.8	0.0473	0.0236	0.0475	0.0237	0.0077	0.0473	0.0236	0.0475	0.0237	0.0077
			0.9	0.0475	0.0237	0.0470	0.0235	0.0023	0.0475	0.0237	0.0470	0.0235	0.0023
	18	0.1	0	0.0481	0.0275	0.0562	0.0284	0.0641	0.0481	0.0275	0.0563	0.0284	0.0647
			0.1	0.0473	0.0271	0.0552	0.0281	0.0620	0.0473	0.0271	0.0553	0.0281	0.0627
			0.2	0.0476	0.0272	0.0545	0.0279	0.0611	0.0476	0.0272	0.0545	0.0280	0.0616
			0.3	0.0484	0.0277	0.0537	0.0283	0.0552	0.0484	0.0277	0.0537	0.0284	0.0556
			0.4	0.0484	0.0277	0.0542	0.0294	0.0484	0.0484	0.0277	0.0542	0.0296	0.0486
			0.5	0.0489	0.0280	0.0540	0.0304	0.0404	0.0489	0.0280	0.0540	0.0304	0.0408
			0.6	0.0485	0.0278	0.0529	0.0308	0.0306	0.0485	0.0278	0.0529	0.0309	0.0308
			0.7	0.0488	0.0279	0.0517	0.0310	0.0198	0.0488	0.0279	0.0517	0.0311	0.0199
			0.8	0.0487	0.0279	0.0511	0.0310	0.0109	0.0487	0.0279	0.0511	0.0312	0.0110
			0.9	0.0479	0.0274	0.0490	0.0295	0.0032	0.0479	0.0274	0.0490	0.0296	0.0032
	16	0.2	0	0.0495	0.0319	0.0709	0.0340	0.0829	0.0495	0.0319	0.0708	0.0339	0.0835
			0.1	0.0499	0.0322	0.0692	0.0336	0.0819	0.0499	0.0322	0.0691	0.0334	0.0822
			0.2	0.0497	0.0321	0.0699	0.0346	0.0784	0.0497	0.0321	0.0699	0.0346	0.0791
			0.3	0.0500	0.0328	0.0706	0.0366	0.0724	0.0509	0.0328	0.0705	0.0365	0.0723
			0.4	0.0494	0.0319	0.0674	0.0365	0.0637	0.0494	0.0319	0.0675	0.0366	0.0641
			0.5	0.0487	0.0314	0.0657	0.0377	0.0539	0.0487	0.0314	0.0656	0.0377	0.0536
			0.6	0.0489	0.0315	0.0633	0.0387	0.0416	0.0489	0.0315	0.0632	0.0387	0.0412
			0.7	0.0491	0.0316	0.0607	0.0391	0.0284	0.0491	0.0316	0.0608	0.0394	0.0286
			0.8	0.0507	0.0327	0.0581	0.0394	0.0141	0.0507	0.0327	0.0581	0.0395	0.0142
			0.9	0.0500	0.0323	0.0540	0.0369	0.0040	0.0500	0.0323	0.0540	0.0370	0.0040
	14	0.3	0	0.0527	0.0377	0.0978	0.0426	0.1049	0.0527	0.0377	0.0973	0.0422	0.1049
			0.1	0.0526	0.0377	0.0970	0.0419	0.1063	0.0527	0.0376	0.0962	0.0415	0.1062
			0.2	0.0524	0.0375	0.0981	0.0448	0.0990	0.0524	0.0375	0.0972	0.0443	0.0984
			0.3	0.0534	0.0382	0.0947	0.0453	0.0918	0.0534	0.0382	0.0942	0.0451	0.0915
			0.4	0.0532	0.0380	0.0915	0.0460	0.0838	0.0532	0.0380	0.0910	0.0459	0.0833
			0.5	0.0520	0.0372	0.0849	0.0457	0.0736	0.0520	0.0372	0.0845	0.0456	0.0734
			0.6	0.0518	0.0371	0.0818	0.0482	0.0581	0.0518	0.0370	0.0812	0.0480	0.0575
			0.7	0.0522	0.0373	0.0757	0.0493	0.0382	0.0522	0.0373	0.0755	0.0493	0.0382
			0.8	0.0519	0.0371	0.0692	0.0487	0.0214	0.0519	0.0371	0.0691	0.0486	0.0212
			0.9	0.0523	0.0374	0.0612	0.0454	0.0062	0.0523	0.0373	0.0610	0.0451	0.0061
	12	0.4	0	0.0580	0.0451	0.1490	0.0559	0.1359	0.0581	0.0449	0.1457	0.0552	0.1332
			0.1	0.0579	0.0450	0.1468	0.0579	0.1304	0.0579	0.0448	0.1432	0.0567	0.1273
			0.2	0.0594	0.0461	0.1447	0.0588	0.1241	0.0594	0.0459	0.1418	0.0577	0.1220
			0.3	0.0576	0.0447	0.1412	0.0610	0.1172	0.0577	0.0446	0.1379	0.0594	0.1158
			0.4	0.0568	0.0441	0.1307	0.0599	0.1076	0.0568	0.0439	0.1283	0.0587	0.1063
			0.5	0.0575	0.0446	0.1250	0.0626	0.0912	0.0575	0.0444	0.1229	0.0616	0.0901
			0.6	0.0572	0.0443	0.1153	0.0628	0.0773	0.0572	0.0442	0.1141	0.0623	0.0763
			0.7	0.0571	0.0443	0.1032	0.0629	0.0559	0.0571	0.0441	0.1018	0.0622	0.0545
			0.8	0.0568	0.0441	0.0885	0.0608	0.0310	0.0568	0.0439	0.0874	0.0600	0.0302
			0.9	0.0574	0.0445	0.0750	0.0569	0.0093	0.0574	0.0444	0.0746	0.0565	0.0091
	10	0.5	0	0.0671	0.0545	0.2394	0.0812	0.1675	0.0671	0.0541	0.2293	0.0777	0.1637
			0.1	0.0668	0.0543	0.2370	0.0824	0.1612	0.0668	0.0539	0.2279	0.0793	0.1575
			0.2	0.0675	0.0549	0.2294	0.0802	0.1633	0.0675	0.0545	0.2217	0.0780	0.1584
			0.3	0.0648	0.0526	0.2245	0.0822	0.1523	0.0648	0.0523	0.2184	0.0806	0.1492
			0.4	0.0685	0.0557	0.2114	0.0843	0.1418	0.0685	0.0553	0.2029	0.0812	0.1370
			0.5	0.0678	0.0551	0.1926	0.0839	0.1249	0.0678	0.0547	0.1869	0.0820	0.1214
			0.6	0.0693	0.0564	0.1788	0.0853	0.1071	0.0693	0.0559	0.1721	0.0827	0.1030
			0.7	0.0680	0.0552	0.1547	0.0847	0.0801	0.0680	0.0548	0.1497	0.0828	0.0752
			0.8	0.0667	0.0542	0.1263	0.0792	0.0503	0.0667	0.0538	0.1234	0.0775	0.0485
			0.9	0.0685	0.0556	0.1018	0.0747	0.0165	0.0685	0.0553	0.1003	0.0734	0.0160

Table 13.9: Means and variances of the MLEs for $n = 100$

n	k	Proportion of Censoring	ρ	$\hat{\mu}_X$	$\hat{\sigma}_X$	$\hat{\mu}_Y$	$\hat{\sigma}_Y$	$\hat{\rho}$	Var $[\hat{\mu}_X]$	Var $[\hat{\sigma}_X]$	Var $[\hat{\mu}_Y]$	Var $[\hat{\sigma}_Y]$	Var $[\hat{\rho}]$
100	100	0	0	-0.0013	0.9896	0.0018	0.9933	-0.0012	0.0100	0.0051	0.0100	0.0050	0.0090
			0.1	0.0047	0.9893	0.0006	0.9899	0.0988	0.0096	0.0049	0.0104	0.0050	0.0101
			0.2	0.0043	0.9896	0.0002	0.9937	0.1964	0.0095	0.0051	0.0107	0.0052	0.0096
			0.3	0.0042	0.9920	0.0037	0.9931	0.2968	0.0096	0.0047	0.0098	0.0049	0.0089
			0.4	0.0003	0.9924	0.0012	0.9974	0.3984	0.0104	0.0051	0.0095	0.0047	0.0074
			0.5	-0.0014	0.9886	-0.0024	0.9919	0.4982	0.0108	0.0049	0.0105	0.0049	0.0057
			0.6	-0.0011	0.9892	-0.0049	0.9914	0.5947	0.0099	0.0051	0.0099	0.0055	0.0044
			0.7	-0.0030	0.9907	-0.0035	0.9950	0.6993	0.0096	0.0053	0.0098	0.0048	0.0027
			0.8	0.0010	0.9894	-0.0038	0.9909	0.7975	0.0092	0.0054	0.0093	0.0053	0.0014
			0.9	0.0018	0.9944	0.0018	0.9917	0.8993	0.0105	0.0052	0.0107	0.0050	0.0004
	90	0.1	0	-0.0076	0.9938	0.0008	0.9902	0.0026	0.0102	0.0065	0.0113	0.0056	0.0147
			0.1	-0.0032	0.9916	0.0027	0.9928	0.0962	0.0105	0.0055	0.0117	0.0058	0.0141
			0.2	-0.0052	0.9921	0.0006	0.9962	0.1976	0.0100	0.0059	0.0130	0.0059	0.0137
			0.3	-0.0045	0.9954	-0.0053	0.9926	0.2949	0.0103	0.0056	0.0119	0.0062	0.0137
			0.4	-0.0006	0.9914	0.0004	0.9898	0.3914	0.0102	0.0060	0.0123	0.0058	0.0099
			0.5	-0.0023	0.9930	-0.0025	0.9924	0.4945	0.0110	0.0060	0.0124	0.0064	0.0085
			0.6	0.0022	0.9889	0.0046	0.9902	0.5950	0.0101	0.0059	0.0115	0.0069	0.0063
			0.7	0.0002	0.9883	-0.0018	0.9878	0.6915	0.0101	0.0059	0.0115	0.0069	0.0040
			0.8	0.0043	0.9942	0.0033	0.9937	0.7973	0.0101	0.0060	0.0114	0.0066	0.0018
			0.9	0.0016	0.9846	0.0012	0.9863	0.8973	0.0101	0.0056	0.0101	0.0068	0.0005
	80	0.2	0	-0.0088	0.9941	0.0035	0.9968	0.0007	0.0103	0.0069	0.0154	0.0063	0.0216
			0.1	-0.0032	0.9905	-0.0009	0.9937	0.0983	0.0102	0.0063	0.0153	0.0067	0.0207
			0.2	-0.0080	0.9878	0.0012	0.9920	0.1934	0.0106	0.0069	0.0140	0.0063	0.0183
			0.3	-0.0049	0.9894	-0.0014	0.9938	0.2985	0.0110	0.0071	0.0145	0.0068	0.0160
			0.4	-0.0018	0.9921	0.0021	0.9961	0.3983	0.0100	0.0072	0.0137	0.0076	0.0147
			0.5	-0.0019	0.9897	0.0015	0.9916	0.4948	0.0106	0.0072	0.0140	0.0079	0.0118
			0.6	-0.0005	0.9916	-0.0062	0.9893	0.5901	0.0108	0.0069	0.0139	0.0080	0.0082
			0.7	-0.0026	0.9903	-0.0068	0.9930	0.6953	0.0097	0.0070	0.0120	0.0082	0.0047
			0.8	-0.0076	0.9899	-0.0058	0.9914	0.7945	0.0110	0.0063	0.0122	0.0089	0.0024
			0.9	-0.0080	0.9855	-0.0067	0.9883	0.8979	0.0098	0.0069	0.0106	0.0078	0.0006
	70	0.3	0	-0.0003	0.9940	-0.0071	0.9959	-0.0039	0.0108	0.0080	0.0201	0.0076	0.0281
			0.1	-0.0004	0.9910	0.0070	0.9994	0.1081	0.0119	0.0080	0.0213	0.0078	0.0281
			0.2	-0.0068	0.9836	-0.0006	0.9936	0.1917	0.0119	0.0083	0.0202	0.0079	0.0260
			0.3	-0.0036	0.9897	-0.0046	0.9917	0.2825	0.0115	0.0081	0.0208	0.0082	0.0232
			0.4	-0.0044	0.9888	0.0027	0.9956	0.3949	0.0114	0.0083	0.0187	0.0087	0.0182
			0.5	-0.0037	0.9854	-0.0022	0.9901	0.4824	0.0115	0.0080	0.0187	0.0100	0.0152
			0.6	0.0005	0.9866	0.0013	0.9934	0.5939	0.0112	0.0087	0.0169	0.0101	0.0100
			0.7	0.0014	0.9869	0.0017	0.9900	0.6927	0.0114	0.0087	0.0156	0.0113	0.0069
			0.8	0.0003	0.9878	-0.0046	0.9853	0.7933	0.0114	0.0084	0.0153	0.0110	0.0030
			0.9	-0.0042	0.9872	-0.0013	0.9902	0.8978	0.0118	0.0076	0.0140	0.0092	0.0007
	60	0.4	0	-0.0075	0.9808	-0.0038	0.9961	-0.0036	0.0115	0.0100	0.0334	0.0091	0.0376
			0.1	-0.0042	0.9844	-0.0005	0.9919	0.0900	0.0131	0.0106	0.0316	0.0090	0.0361
			0.2	-0.0015	0.9904	0.0021	0.9936	0.1957	0.0137	0.0102	0.0301	0.0092	0.0336
			0.3	-0.0105	0.9877	0.0046	0.9977	0.2992	0.0133	0.0091	0.0337	0.0106	0.0303
			0.4	-0.0091	0.9804	-0.0064	0.9915	0.3864	0.0127	0.0092	0.0273	0.0116	0.0241
			0.5	-0.0081	0.9899	-0.0156	0.9880	0.4826	0.0142	0.0100	0.0264	0.0122	0.0195
			0.6	-0.0043	0.9858	-0.0104	0.9889	0.5854	0.0136	0.0105	0.0279	0.0133	0.0151
			0.7	-0.0065	0.9869	-0.0085	0.9886	0.6923	0.0138	0.0097	0.0228	0.0141	0.0075
			0.8	-0.0011	0.9852	0.0032	0.9896	0.7932	0.0126	0.0098	0.0196	0.0149	0.0041
			0.9	-0.0033	0.9901	-0.0013	0.9947	0.8983	0.0132	0.0101	0.0170	0.0127	0.0009
	50	0.5	0	-0.0071	0.9875	0.0047	1.0087	-0.0038	0.0145	0.0118	0.0592	0.0116	0.0540
			0.1	-0.0119	0.9815	-0.0276	0.9992	0.0725	0.0149	0.0120	0.0567	0.0113	0.0525
			0.2	-0.0160	0.9775	0.0010	1.0028	0.1945	0.0157	0.0127	0.0529	0.0122	0.0480
			0.3	-0.0064	0.9856	0.0143	1.0050	0.2970	0.0156	0.0130	0.0483	0.0134	0.0393
			0.4	-0.0096	0.9784	0.0028	1.0031	0.3743	0.0134	0.0119	0.0526	0.0159	0.0392
			0.5	-0.0078	0.9878	-0.0035	0.9980	0.4762	0.0141	0.0122	0.0510	0.0169	0.0314
			0.6	-0.0188	0.9771	-0.0152	0.9885	0.5715	0.0145	0.0124	0.0437	0.0170	0.0221
			0.7	-0.0128	0.9827	-0.0114	0.9865	0.6828	0.0153	0.0121	0.0349	0.0191	0.0133
			0.8	-0.0083	0.9812	-0.0004	0.9890	0.7919	0.0156	0.0124	0.0290	0.0184	0.0056
			0.9	-0.0155	0.9791	-0.0156	0.9803	0.8938	0.0154	0.0120	0.0224	0.0167	0.0013

Table 13.10: Variances of the MLEs from the Fisher information matrix for $n = 100$

n	k	Proportion of Censoring	ρ	From I^* Var $[\hat{\mu}_X]$	Var $[\hat{\sigma}_X]$	Var $[\hat{\mu}_Y]$	Var $[\hat{\sigma}_Y]$	Var $[\hat{\rho}]$	From I Var $[\hat{\mu}_X]$	Var $[\hat{\sigma}_X]$	Var $[\hat{\mu}_Y]$	Var $[\hat{\sigma}_Y]$	Var $[\hat{\rho}]$
100	100	0	0	0.0098	0.0049	0.0099	0.0050	0.0098	0.0098	0.0049	0.0099	0.0050	0.0098
			0.1	0.0098	0.0049	0.0098	0.0049	0.0096	0.0098	0.0049	0.0098	0.0049	0.0096
			0.2	0.0098	0.0049	0.0099	0.0050	0.0091	0.0098	0.0049	0.0099	0.0050	0.0091
			0.3	0.0099	0.0049	0.0099	0.0050	0.0082	0.0099	0.0049	0.0099	0.0050	0.0082
			0.4	0.0099	0.0049	0.0100	0.0050	0.0070	0.0099	0.0049	0.0100	0.0050	0.0070
			0.5	0.0098	0.0049	0.0099	0.0049	0.0056	0.0098	0.0049	0.0099	0.0049	0.0056
			0.6	0.0098	0.0049	0.0099	0.0049	0.0042	0.0098	0.0049	0.0099	0.0049	0.0042
			0.7	0.0099	0.0049	0.0099	0.0050	0.0026	0.0099	0.0049	0.0099	0.0050	0.0026
			0.8	0.0098	0.0049	0.0099	0.0049	0.0013	0.0098	0.0049	0.0099	0.0049	0.0013
			0.9	0.0099	0.0050	0.0099	0.0049	0.0004	0.0099	0.0050	0.0099	0.0049	0.0004
	90	0.1	0	0.0101	0.0058	0.0115	0.0055	0.0150	0.0101	0.0058	0.0115	0.0055	0.0149
			0.1	0.0101	0.0058	0.0116	0.0056	0.0147	0.0101	0.0058	0.0115	0.0056	0.0146
			0.2	0.0101	0.0058	0.0116	0.0058	0.0138	0.0101	0.0058	0.0116	0.0058	0.0137
			0.3	0.0102	0.0058	0.0114	0.0059	0.0123	0.0102	0.0058	0.0114	0.0059	0.0123
			0.4	0.0101	0.0058	0.0113	0.0061	0.0106	0.0101	0.0058	0.0113	0.0061	0.0105
			0.5	0.0101	0.0058	0.0112	0.0063	0.0082	0.0101	0.0058	0.0112	0.0063	0.0082
			0.6	0.0100	0.0058	0.0110	0.0065	0.0059	0.0100	0.0058	0.0110	0.0065	0.0059
			0.7	0.0100	0.0058	0.0108	0.0066	0.0038	0.0100	0.0058	0.0108	0.0066	0.0037
			0.8	0.0101	0.0058	0.0107	0.0066	0.0018	0.0101	0.0058	0.0107	0.0066	0.0018
			0.9	0.0099	0.0057	0.0103	0.0063	0.0005	0.0100	0.0057	0.0103	0.0063	0.0005
	80	0.2	0	0.0106	0.0068	0.0150	0.0064	0.0203	0.0106	0.0068	0.0149	0.0064	0.0201
			0.1	0.0105	0.0068	0.0149	0.0065	0.0200	0.0105	0.0068	0.0148	0.0065	0.0197
			0.2	0.0104	0.0068	0.0147	0.0066	0.0187	0.0104	0.0068	0.0146	0.0066	0.0186
			0.3	0.0105	0.0068	0.0145	0.0070	0.0166	0.0105	0.0068	0.0145	0.0070	0.0164
			0.4	0.0105	0.0068	0.0143	0.0075	0.0139	0.0105	0.0068	0.0142	0.0075	0.0138
			0.5	0.0105	0.0068	0.0138	0.0079	0.0110	0.0105	0.0068	0.0137	0.0078	0.0109
			0.6	0.0105	0.0068	0.0133	0.0082	0.0079	0.0105	0.0068	0.0132	0.0081	0.0079
			0.7	0.0105	0.0068	0.0128	0.0085	0.0048	0.0105	0.0068	0.0128	0.0084	0.0047
			0.8	0.0105	0.0068	0.0121	0.0084	0.0023	0.0105	0.0068	0.0121	0.0084	0.0023
			0.9	0.0104	0.0067	0.0113	0.0079	0.0006	0.0104	0.0067	0.0113	0.0078	0.0006
	70	0.3	0	0.0113	0.0082	0.0211	0.0075	0.0270	0.0114	0.0081	0.0209	0.0075	0.0266
			0.1	0.0113	0.0081	0.0211	0.0077	0.0262	0.0113	0.0081	0.0209	0.0077	0.0259
			0.2	0.0111	0.0080	0.0206	0.0079	0.0248	0.0111	0.0080	0.0204	0.0079	0.0245
			0.3	0.0112	0.0081	0.0201	0.0084	0.0225	0.0113	0.0081	0.0200	0.0084	0.0222
			0.4	0.0112	0.0081	0.0195	0.0092	0.0185	0.0112	0.0081	0.0194	0.0092	0.0184
			0.5	0.0111	0.0080	0.0186	0.0098	0.0151	0.0112	0.0080	0.0185	0.0097	0.0148
			0.6	0.0112	0.0080	0.0176	0.0105	0.0103	0.0112	0.0080	0.0174	0.0104	0.0102
			0.7	0.0112	0.0080	0.0162	0.0108	0.0063	0.0112	0.0080	0.0161	0.0107	0.0063
			0.8	0.0112	0.0081	0.0147	0.0106	0.0030	0.0112	0.0080	0.0146	0.0105	0.0029
			0.9	0.0112	0.0080	0.0131	0.0099	0.0007	0.0112	0.0080	0.0131	0.0099	0.0007
	60	0.4	0	0.0124	0.0097	0.0323	0.0092	0.0359	0.0124	0.0096	0.0317	0.0091	0.0352
			0.1	0.0125	0.0097	0.0318	0.0092	0.0352	0.0125	0.0097	0.0313	0.0092	0.0347
			0.2	0.0126	0.0098	0.0313	0.0098	0.0331	0.0126	0.0098	0.0308	0.0097	0.0324
			0.3	0.0125	0.0098	0.0305	0.0107	0.0292	0.0125	0.0097	0.0301	0.0107	0.0287
			0.4	0.0124	0.0096	0.0290	0.0115	0.0252	0.0124	0.0096	0.0286	0.0114	0.0249
			0.5	0.0126	0.0098	0.0272	0.0124	0.0201	0.0126	0.0098	0.0268	0.0123	0.0197
			0.6	0.0125	0.0098	0.0251	0.0134	0.0142	0.0125	0.0097	0.0248	0.0132	0.0140
			0.7	0.0125	0.0098	0.0225	0.0140	0.0083	0.0125	0.0097	0.0223	0.0138	0.0082
			0.8	0.0125	0.0097	0.0197	0.0139	0.0039	0.0125	0.0097	0.0195	0.0137	0.0039
			0.9	0.0126	0.0098	0.0165	0.0128	0.0009	0.0126	0.0098	0.0164	0.0127	0.0009
	50	0.5	0	0.0150	0.0122	0.0545	0.0121	0.0485	0.0150	0.0122	0.0529	0.0120	0.0470
			0.1	0.0148	0.0121	0.0532	0.0120	0.0478	0.0148	0.0120	0.0517	0.0119	0.0465
			0.2	0.0147	0.0120	0.0520	0.0129	0.0445	0.0147	0.0119	0.0508	0.0128	0.0435
			0.3	0.0149	0.0122	0.0506	0.0141	0.0401	0.0149	0.0121	0.0493	0.0140	0.0391
			0.4	0.0147	0.0120	0.0481	0.0154	0.0350	0.0147	0.0120	0.0469	0.0151	0.0341
			0.5	0.0150	0.0122	0.0442	0.0167	0.0278	0.0150	0.0122	0.0432	0.0165	0.0271
			0.6	0.0147	0.0120	0.0398	0.0177	0.0205	0.0147	0.0119	0.0389	0.0174	0.0200
			0.7	0.0148	0.0121	0.0345	0.0187	0.0123	0.0148	0.0121	0.0338	0.0183	0.0120
			0.8	0.0148	0.0121	0.0288	0.0185	0.0054	0.0148	0.0120	0.0283	0.0182	0.0053
			0.9	0.0147	0.0120	0.0222	0.0165	0.0013	0.0147	0.0120	0.0219	0.0162	0.0013

Table 13.11: Means and variances of the MLEs for $n = 200$

n	k	Proportion of Censoring	ρ	$\hat{\mu}_X$	$\hat{\sigma}_X$	$\hat{\mu}_Y$	$\hat{\sigma}_Y$	$\hat{\rho}$	Var $[\hat{\mu}_X]$	Var $[\hat{\sigma}_X]$	Var $[\hat{\mu}_Y]$	Var $[\hat{\sigma}_Y]$	Var $[\hat{\rho}]$
200	200	0	0	-0.0032	0.9964	-0.0016	0.9946	-0.0039	0.0048	0.0027	0.0052	0.0024	0.0050
			0.1	0.0000	0.9962	0.0017	0.9986	0.0985	0.0051	0.0026	0.0050	0.0025	0.0050
			0.2	0.0005	0.9952	-0.0003	0.9955	0.2032	0.0050	0.0025	0.0048	0.0026	0.0041
			0.3	0.0020	0.9971	0.0012	0.9950	0.2968	0.0051	0.0023	0.0053	0.0025	0.0042
			0.4	-0.0019	0.9951	-0.0007	0.9948	0.3994	0.0052	0.0026	0.0050	0.0026	0.0036
			0.5	0.0023	0.9944	0.0016	0.9957	0.4969	0.0049	0.0024	0.0047	0.0025	0.0030
			0.6	-0.0026	0.9968	0.0020	0.9953	0.5979	0.0049	0.0025	0.0046	0.0024	0.0020
			0.7	-0.0030	0.9960	-0.0013	0.9954	0.6988	0.0051	0.0026	0.0049	0.0026	0.0013
			0.8	0.0004	0.9967	-0.0008	0.9972	0.7998	0.0056	0.0025	0.0050	0.0025	0.0007
			0.9	0.0017	0.9965	0.0019	0.9982	0.8992	0.0052	0.0024	0.0052	0.0024	0.0002
	180	0.1	0	0.0014	1.0014	0.0004	0.9965	0.0006	0.0051	0.0029	0.0053	0.0027	0.0084
			0.1	-0.0040	0.9980	0.0015	0.9956	0.1005	0.0050	0.0030	0.0056	0.0027	0.0072
			0.2	-0.0010	0.9981	0.0030	0.9951	0.1967	0.0054	0.0030	0.0057	0.0030	0.0070
			0.3	-0.0022	0.9977	0.0000	0.9986	0.2970	0.0050	0.0031	0.0060	0.0030	0.0070
			0.4	-0.0013	0.9959	-0.0016	0.9990	0.3990	0.0053	0.0028	0.0054	0.0033	0.0054
			0.5	0.0020	0.9950	0.0027	0.9963	0.4978	0.0048	0.0030	0.0054	0.0034	0.0044
			0.6	0.0000	0.9953	0.0016	0.9959	0.5986	0.0049	0.0028	0.0058	0.0035	0.0031
			0.7	0.0039	0.9977	0.0040	0.9977	0.6989	0.0047	0.0028	0.0054	0.0032	0.0017
			0.8	-0.0026	0.9964	-0.0017	0.9971	0.7984	0.0049	0.0026	0.0051	0.0032	0.0008
			0.9	0.0005	0.9955	0.0001	0.9957	0.8993	0.0049	0.0028	0.0051	0.0032	0.0002
	160	0.2	0	0.0010	0.9972	0.0000	0.9969	-0.0024	0.0051	0.0036	0.0076	0.0033	0.0105
			0.1	-0.0036	0.9954	0.0024	0.9979	0.0995	0.0053	0.0033	0.0073	0.0033	0.0102
			0.2	0.0020	0.9934	0.0016	0.9964	0.1966	0.0056	0.0034	0.0074	0.0032	0.0099
			0.3	-0.0015	0.9954	0.0017	0.9970	0.2973	0.0053	0.0034	0.0071	0.0037	0.0083
			0.4	-0.0048	0.9937	0.0020	0.9963	0.3956	0.0051	0.0034	0.0076	0.0040	0.0072
			0.5	0.0003	0.9926	-0.0019	0.9953	0.4973	0.0055	0.0034	0.0066	0.0037	0.0055
			0.6	-0.0032	0.9930	-0.0020	0.9930	0.5963	0.0053	0.0035	0.0071	0.0037	0.0036
			0.7	-0.0025	0.9923	0.0016	0.9960	0.6980	0.0054	0.0032	0.0066	0.0045	0.0023
			0.8	-0.0025	0.9958	-0.0017	0.9967	0.7995	0.0051	0.0033	0.0061	0.0045	0.0011
			0.9	-0.0007	0.9943	0.0000	0.9951	0.8987	0.0052	0.0036	0.0057	0.0041	0.0003
	140	0.3	0	-0.0020	0.9949	0.0039	0.9948	-0.0002	0.0060	0.0040	0.0106	0.0036	0.0142
			0.1	-0.0040	0.9941	-0.0040	0.9988	0.0964	0.0056	0.0041	0.0113	0.0037	0.0149
			0.2	-0.0032	0.9937	0.0048	1.0007	0.2020	0.0055	0.0039	0.0103	0.0039	0.0138
			0.3	0.0024	0.9928	0.0021	0.9981	0.2935	0.0054	0.0041	0.0101	0.0043	0.0124
			0.4	0.0005	0.9931	0.0097	0.9989	0.3993	0.0055	0.0044	0.0094	0.0045	0.0095
			0.5	-0.0033	0.9956	-0.0049	0.9936	0.4942	0.0057	0.0043	0.0091	0.0048	0.0072
			0.6	-0.0063	0.9915	-0.0013	0.9952	0.5937	0.0056	0.0041	0.0091	0.0053	0.0048
			0.7	-0.0008	0.9941	-0.0022	0.9932	0.6956	0.0059	0.0042	0.0087	0.0059	0.0034
			0.8	-0.0036	0.9942	-0.0044	0.9942	0.7970	0.0054	0.0040	0.0075	0.0054	0.0014
			0.9	0.0009	0.9954	-0.0002	0.9947	0.8990	0.0053	0.0039	0.0059	0.0049	0.0003
	120	0.4	0	-0.0024	0.9921	-0.0044	0.9978	-0.0030	0.0065	0.0050	0.0162	0.0042	0.0187
			0.1	-0.0051	0.9941	0.0010	0.9978	0.0976	0.0065	0.0053	0.0178	0.0045	0.0200
			0.2	-0.0035	0.9948	0.0042	0.9996	0.1994	0.0070	0.0053	0.0166	0.0046	0.0179
			0.3	-0.0051	0.9916	-0.0012	0.9979	0.2980	0.0066	0.0047	0.0166	0.0051	0.0154
			0.4	0.0009	0.9958	-0.0060	0.9978	0.3897	0.0073	0.0049	0.0158	0.0061	0.0143
			0.5	0.0004	0.9978	0.0053	1.0044	0.4960	0.0065	0.0050	0.0128	0.0058	0.0096
			0.6	-0.0045	0.9942	-0.0068	0.9963	0.5929	0.0066	0.0046	0.0116	0.0064	0.0065
			0.7	0.0004	0.9941	0.0022	0.9987	0.6956	0.0060	0.0055	0.0110	0.0075	0.0041
			0.8	-0.0015	0.9960	0.0003	0.9956	0.7980	0.0069	0.0054	0.0105	0.0069	0.0018
			0.9	-0.0060	0.9921	-0.0044	0.9928	0.8982	0.0062	0.0053	0.0084	0.0067	0.0005
	100	0.5	0	-0.0061	0.9883	-0.0014	0.9968	-0.0012	0.0076	0.0056	0.0275	0.0052	0.0258
			0.1	-0.0030	0.9909	0.0019	0.9994	0.0894	0.0075	0.0060	0.0285	0.0049	0.0254
			0.2	-0.0036	0.9926	-0.0054	1.0008	0.1890	0.0075	0.0063	0.0262	0.0058	0.0250
			0.3	-0.0049	0.9923	-0.0010	0.9988	0.2930	0.0071	0.0062	0.0277	0.0063	0.0223
			0.4	-0.0066	0.9940	-0.0039	0.9985	0.3896	0.0078	0.0062	0.0235	0.0074	0.0183
			0.5	-0.0065	0.9912	-0.0040	0.9949	0.4867	0.0082	0.0061	0.0236	0.0078	0.0142
			0.6	-0.0102	0.9920	-0.0055	0.9953	0.5909	0.0071	0.0061	0.0196	0.0087	0.0095
			0.7	-0.0053	0.9925	0.0006	0.9986	0.6944	0.0076	0.0061	0.0184	0.0098	0.0055
			0.8	-0.0059	0.9914	-0.0038	0.9933	0.7961	0.0083	0.0060	0.0154	0.0094	0.0024
			0.9	-0.0064	0.9923	-0.0029	0.9956	0.8996	0.0077	0.0062	0.0105	0.0083	0.0006

Table 13.12: Variances of the MLEs from the Fisher information matrix for $n = 200$

n	k	Proportion of Censoring	ρ	From I^* Var $[\hat{\mu}_X]$	Var $[\hat{\sigma}_X]$	Var $[\hat{\mu}_Y]$	Var $[\hat{\sigma}_Y]$	Var $[\hat{\rho}]$	From I Var $[\hat{\mu}_X]$	Var $[\hat{\sigma}_X]$	Var $[\hat{\mu}_Y]$	Var $[\hat{\sigma}_Y]$	Var $[\hat{\rho}]$
200	200	0	0	0.0050	0.0025	0.0050	0.0025	0.0049	0.0050	0.0025	0.0050	0.0025	0.0049
			0.1	0.0050	0.0025	0.0050	0.0025	0.0049	0.0050	0.0025	0.0050	0.0025	0.0049
			0.2	0.0050	0.0025	0.0050	0.0025	0.0046	0.0050	0.0025	0.0050	0.0025	0.0046
			0.3	0.0050	0.0025	0.0050	0.0025	0.0041	0.0050	0.0025	0.0050	0.0025	0.0041
			0.4	0.0050	0.0025	0.0050	0.0025	0.0035	0.0050	0.0025	0.0050	0.0025	0.0035
			0.5	0.0050	0.0025	0.0050	0.0025	0.0028	0.0050	0.0025	0.0050	0.0025	0.0028
			0.6	0.0050	0.0025	0.0050	0.0025	0.0021	0.0050	0.0025	0.0050	0.0025	0.0021
			0.7	0.0050	0.0025	0.0050	0.0025	0.0013	0.0050	0.0025	0.0050	0.0025	0.0013
			0.8	0.0050	0.0025	0.0050	0.0025	0.0007	0.0050	0.0025	0.0050	0.0025	0.0007
			0.9	0.0050	0.0025	0.0050	0.0025	0.0002	0.0050	0.0025	0.0050	0.0025	0.0002
	180	0.1	0	0.0051	0.0029	0.0058	0.0028	0.0076	0.0051	0.0029	0.0058	0.0028	0.0076
			0.1	0.0051	0.0029	0.0058	0.0028	0.0075	0.0051	0.0029	0.0058	0.0028	0.0074
			0.2	0.0051	0.0029	0.0058	0.0029	0.0070	0.0051	0.0029	0.0058	0.0029	0.0070
			0.3	0.0051	0.0029	0.0058	0.0030	0.0062	0.0051	0.0029	0.0058	0.0030	0.0062
			0.4	0.0051	0.0029	0.0057	0.0031	0.0052	0.0051	0.0029	0.0057	0.0031	0.0052
			0.5	0.0051	0.0029	0.0056	0.0032	0.0041	0.0051	0.0029	0.0056	0.0032	0.0041
			0.6	0.0051	0.0029	0.0055	0.0033	0.0029	0.0051	0.0029	0.0055	0.0033	0.0029
			0.7	0.0051	0.0029	0.0055	0.0034	0.0018	0.0051	0.0029	0.0055	0.0033	0.0018
			0.8	0.0051	0.0029	0.0054	0.0033	0.0009	0.0051	0.0029	0.0054	0.0033	0.0009
			0.9	0.0051	0.0029	0.0052	0.0032	0.0002	0.0051	0.0029	0.0052	0.0032	0.0002
	160	0.2	0	0.0053	0.0034	0.0075	0.0032	0.0104	0.0053	0.0034	0.0075	0.0032	0.0104
			0.1	0.0053	0.0034	0.0075	0.0032	0.0102	0.0053	0.0034	0.0075	0.0032	0.0102
			0.2	0.0053	0.0034	0.0074	0.0033	0.0095	0.0053	0.0034	0.0074	0.0033	0.0095
			0.3	0.0053	0.0034	0.0073	0.0035	0.0084	0.0053	0.0034	0.0073	0.0035	0.0084
			0.4	0.0053	0.0034	0.0072	0.0037	0.0071	0.0053	0.0034	0.0071	0.0037	0.0070
			0.5	0.0053	0.0034	0.0069	0.0039	0.0055	0.0053	0.0034	0.0069	0.0039	0.0054
			0.6	0.0053	0.0034	0.0067	0.0041	0.0038	0.0053	0.0034	0.0067	0.0041	0.0038
			0.7	0.0052	0.0034	0.0064	0.0043	0.0023	0.0052	0.0034	0.0064	0.0043	0.0023
			0.8	0.0053	0.0034	0.0061	0.0043	0.0011	0.0053	0.0034	0.0061	0.0042	0.0011
			0.9	0.0053	0.0034	0.0057	0.0040	0.0003	0.0053	0.0034	0.0057	0.0040	0.0003
	140	0.3	0	0.0057	0.0041	0.0106	0.0036	0.0140	0.0057	0.0041	0.0105	0.0036	0.0139
			0.1	0.0056	0.0041	0.0106	0.0037	0.0136	0.0057	0.0041	0.0105	0.0037	0.0135
			0.2	0.0056	0.0041	0.0105	0.0040	0.0127	0.0056	0.0041	0.0104	0.0040	0.0126
			0.3	0.0056	0.0041	0.0102	0.0042	0.0113	0.0056	0.0041	0.0102	0.0042	0.0112
			0.4	0.0056	0.0041	0.0099	0.0046	0.0094	0.0056	0.0041	0.0098	0.0046	0.0093
			0.5	0.0057	0.0041	0.0093	0.0049	0.0073	0.0057	0.0041	0.0093	0.0049	0.0073
			0.6	0.0056	0.0040	0.0088	0.0053	0.0051	0.0056	0.0040	0.0088	0.0052	0.0051
			0.7	0.0056	0.0041	0.0082	0.0055	0.0031	0.0057	0.0041	0.0081	0.0054	0.0030
			0.8	0.0057	0.0041	0.0074	0.0054	0.0014	0.0057	0.0041	0.0074	0.0054	0.0014
			0.9	0.0057	0.0041	0.0066	0.0050	0.0003	0.0057	0.0041	0.0066	0.0050	0.0003
	120	0.4	0	0.0063	0.0049	0.0163	0.0044	0.0188	0.0063	0.0049	0.0162	0.0044	0.0186
			0.1	0.0063	0.0049	0.0162	0.0045	0.0183	0.0063	0.0049	0.0161	0.0045	0.0182
			0.2	0.0063	0.0049	0.0159	0.0048	0.0170	0.0063	0.0049	0.0158	0.0048	0.0169
			0.3	0.0063	0.0049	0.0154	0.0052	0.0151	0.0063	0.0049	0.0153	0.0052	0.0150
			0.4	0.0063	0.0050	0.0147	0.0057	0.0127	0.0063	0.0049	0.0146	0.0057	0.0127
			0.5	0.0064	0.0050	0.0140	0.0064	0.0097	0.0064	0.0050	0.0139	0.0064	0.0096
			0.6	0.0063	0.0049	0.0128	0.0068	0.0068	0.0063	0.0049	0.0127	0.0068	0.0068
			0.7	0.0063	0.0049	0.0115	0.0072	0.0040	0.0063	0.0049	0.0114	0.0071	0.0040
			0.8	0.0063	0.0050	0.0099	0.0070	0.0018	0.0063	0.0049	0.0099	0.0070	0.0018
			0.9	0.0063	0.0049	0.0082	0.0064	0.0004	0.0063	0.0049	0.0082	0.0063	0.0004
	100	0.5	0	0.0075	0.0061	0.0270	0.0054	0.0258	0.0075	0.0061	0.0266	0.0054	0.0254
			0.1	0.0075	0.0061	0.0269	0.0056	0.0253	0.0075	0.0061	0.0266	0.0056	0.0250
			0.2	0.0075	0.0062	0.0264	0.0060	0.0236	0.0075	0.0061	0.0261	0.0060	0.0233
			0.3	0.0075	0.0061	0.0253	0.0067	0.0208	0.0075	0.0061	0.0250	0.0067	0.0206
			0.4	0.0075	0.0062	0.0240	0.0075	0.0176	0.0075	0.0062	0.0237	0.0074	0.0173
			0.5	0.0075	0.0061	0.0222	0.0083	0.0137	0.0075	0.0061	0.0219	0.0082	0.0135
			0.6	0.0075	0.0061	0.0201	0.0091	0.0095	0.0075	0.0061	0.0198	0.0090	0.0093
			0.7	0.0075	0.0061	0.0176	0.0096	0.0055	0.0075	0.0061	0.0174	0.0095	0.0054
			0.8	0.0075	0.0061	0.0146	0.0094	0.0025	0.0075	0.0061	0.0144	0.0093	0.0024
			0.9	0.0075	0.0061	0.0113	0.0084	0.0006	0.0075	0.0061	0.0112	0.0083	0.0006

Table 13.13: Probability coverages of 95% and 90% CIs for $n = 20$ using the Fisher information matrix

n	k	Proportion of Censoring	ρ	95% CI $\hat{\mu}_X$	95% CI $\hat{\sigma}_X$	95% CI $\hat{\mu}_Y$	95% CI $\hat{\sigma}_Y$	95% CI $\hat{\rho}$	90% CI $\hat{\mu}_X$	90% CI $\hat{\sigma}_X$	90% CI $\hat{\mu}_Y$	90% CI $\hat{\sigma}_Y$	90% CI $\hat{\rho}$
20	20	0	0	0.9260	0.9000	0.9330	0.9160	0.8960	0.8610	0.8510	0.8740	0.8580	0.8400
			0.1	0.9360	0.8980	0.9280	0.8880	0.9080	0.8790	0.8530	0.8780	0.8300	0.8470
			0.2	0.9360	0.9020	0.9280	0.8940	0.9220	0.8810	0.8580	0.8790	0.8560	0.8620
			0.3	0.9210	0.8880	0.9430	0.8930	0.8980	0.8570	0.8350	0.8880	0.8440	0.8470
			0.4	0.9260	0.8940	0.9370	0.8940	0.9030	0.8700	0.8500	0.8740	0.8430	0.8430
			0.5	0.9360	0.9030	0.9240	0.8800	0.9040	0.8750	0.8660	0.8690	0.8270	0.8610
			0.6	0.9130	0.8980	0.9350	0.8960	0.9090	0.8560	0.8360	0.8790	0.8510	0.8720
			0.7	0.9450	0.8860	0.9250	0.8880	0.8960	0.8790	0.8420	0.8830	0.8450	0.8460
			0.8	0.9100	0.8920	0.9140	0.8920	0.8850	0.8550	0.8320	0.8630	0.8340	0.8420
			0.9	0.9210	0.9020	0.9310	0.8870	0.8920	0.8740	0.8470	0.8830	0.8360	0.8490
	18	0.1	0	0.9250	0.8990	0.9320	0.9020	0.8800	0.8730	0.8380	0.8710	0.8560	0.8210
			0.1	0.9270	0.8820	0.9290	0.8920	0.8570	0.8660	0.8250	0.8730	0.8450	0.8050
			0.2	0.9220	0.8890	0.9160	0.9010	0.8920	0.8740	0.8460	0.8670	0.8450	0.8310
			0.3	0.9300	0.8990	0.9190	0.8870	0.8790	0.8780	0.8510	0.8580	0.8480	0.8270
			0.4	0.9160	0.9060	0.9280	0.8930	0.8820	0.8580	0.8590	0.8790	0.8420	0.8250
			0.5	0.9360	0.8920	0.9390	0.8930	0.8770	0.8900	0.8540	0.8940	0.8460	0.8240
			0.6	0.9180	0.9110	0.9270	0.8900	0.9030	0.8780	0.8630	0.8710	0.8490	0.8520
			0.7	0.9260	0.8980	0.9130	0.8930	0.8880	0.8810	0.8410	0.8520	0.8310	0.8440
			0.8	0.9400	0.9040	0.9360	0.8860	0.8780	0.8720	0.8570	0.8750	0.8410	0.8240
			0.9	0.9270	0.8920	0.9290	0.8780	0.8830	0.8740	0.8400	0.8710	0.8340	0.8420
	16	0.2	0	0.9150	0.8660	0.9110	0.9130	0.8550	0.8610	0.8110	0.8650	0.8690	0.7940
			0.1	0.9180	0.8880	0.9190	0.9070	0.8580	0.8730	0.8380	0.8740	0.8510	0.7990
			0.2	0.9170	0.8950	0.9270	0.9010	0.8730	0.8640	0.8490	0.8720	0.8620	0.8270
			0.3	0.9070	0.8820	0.9320	0.8910	0.8560	0.8610	0.8350	0.8730	0.8480	0.8100
			0.4	0.9080	0.8580	0.9250	0.8790	0.8630	0.8450	0.8220	0.8660	0.8330	0.8110
			0.5	0.9040	0.8780	0.9210	0.9040	0.8640	0.8580	0.8300	0.8620	0.8670	0.8160
			0.6	0.9170	0.8640	0.9250	0.8920	0.8890	0.8570	0.8240	0.8760	0.8500	0.8430
			0.7	0.9070	0.8600	0.9150	0.8770	0.8750	0.8540	0.8130	0.8610	0.8330	0.8310
			0.8	0.9260	0.8800	0.9230	0.8960	0.8760	0.8750	0.8340	0.8790	0.8460	0.8380
			0.9	0.9140	0.8840	0.9230	0.9000	0.8650	0.8690	0.8240	0.8730	0.8430	0.8270
	14	0.3	0	0.9060	0.8680	0.9230	0.9040	0.8350	0.8550	0.8210	0.8590	0.8670	0.7750
			0.1	0.9120	0.8750	0.9150	0.9010	0.8550	0.8550	0.8290	0.8580	0.8600	0.8030
			0.2	0.9170	0.8710	0.9270	0.9290	0.8220	0.8600	0.8190	0.8650	0.8900	0.7710
			0.3	0.9100	0.8690	0.9310	0.9160	0.8370	0.8510	0.8230	0.8740	0.8680	0.7860
			0.4	0.9190	0.8790	0.9190	0.9000	0.8330	0.8620	0.8260	0.8780	0.8640	0.7890
			0.5	0.9090	0.8710	0.9110	0.8730	0.8540	0.8510	0.8310	0.8570	0.8250	0.8080
			0.6	0.9180	0.8650	0.9040	0.8800	0.8520	0.8750	0.8200	0.8380	0.8260	0.8050
			0.7	0.9070	0.8670	0.9100	0.8700	0.8540	0.8520	0.8140	0.8590	0.8270	0.8110
			0.8	0.9080	0.8500	0.9370	0.8770	0.8620	0.8620	0.8040	0.8870	0.8340	0.8260
			0.9	0.9090	0.8710	0.9160	0.8690	0.8530	0.8490	0.8140	0.8610	0.8180	0.8230
	12	0.4	0	0.9030	0.8610	0.9320	0.9180	0.8210	0.8500	0.8010	0.8910	0.8750	0.7680
			0.1	0.9200	0.8540	0.9350	0.9000	0.8080	0.8720	0.8150	0.8530	0.8550	0.7490
			0.2	0.9050	0.8500	0.9140	0.9090	0.7930	0.8440	0.7970	0.8570	0.8790	0.7400
			0.3	0.9070	0.8480	0.9150	0.8880	0.8030	0.8470	0.8040	0.8510	0.8420	0.7580
			0.4	0.8850	0.8500	0.9250	0.8790	0.7980	0.8280	0.8020	0.8720	0.8460	0.7570
			0.5	0.8980	0.8510	0.9110	0.8830	0.8200	0.8350	0.8060	0.8580	0.8410	0.7680
			0.6	0.8910	0.8530	0.9220	0.8700	0.8430	0.8350	0.8140	0.8710	0.8250	0.8060
			0.7	0.8910	0.8510	0.9180	0.8540	0.8470	0.8330	0.8090	0.8620	0.8120	0.8100
			0.8	0.9030	0.8490	0.9030	0.8560	0.8470	0.8400	0.8070	0.8470	0.8130	0.8200
			0.9	0.9020	0.8590	0.9070	0.8710	0.8490	0.8430	0.8080	0.8460	0.8280	0.8100
	10	0.5	0	0.8780	0.8110	0.9170	0.9110	0.7680	0.8340	0.7770	0.8430	0.8780	0.7100
			0.1	0.8630	0.8160	0.9050	0.9220	0.7470	0.8180	0.7830	0.8420	0.8880	0.6980
			0.2	0.8750	0.8250	0.9320	0.9090	0.7610	0.8270	0.7830	0.8580	0.8780	0.7180
			0.3	0.8690	0.8230	0.9110	0.9050	0.7900	0.8100	0.7570	0.8540	0.8800	0.7420
			0.4	0.8990	0.8590	0.9110	0.8890	0.7900	0.8430	0.8130	0.8510	0.8520	0.7300
			0.5	0.8810	0.8150	0.8960	0.8690	0.7780	0.8290	0.7530	0.8190	0.8190	0.7290
			0.6	0.8970	0.8390	0.9170	0.8780	0.7920	0.8410	0.8060	0.8480	0.8300	0.7470
			0.7	0.8820	0.8460	0.8950	0.8350	0.8150	0.8270	0.7960	0.8560	0.7950	0.7720
			0.8	0.8800	0.8220	0.8980	0.8250	0.8240	0.8380	0.7750	0.8400	0.7760	0.7830
			0.9	0.8690	0.8200	0.9000	0.8540	0.8100	0.8170	0.7780	0.8520	0.8100	0.7750

Table 13.14: Probability coverages of 95% and 90% CIs for $n = 20$ using the partially observed information matrix

n	k	Proportion of Censoring	ρ	95% CI $\hat{\mu}_X$	95% CI $\hat{\sigma}_X$	95% CI $\hat{\mu}_Y$	95% CI $\hat{\sigma}_Y$	95% CI $\hat{\rho}$	90% CI $\hat{\mu}_X$	90% CI $\hat{\sigma}_X$	90% CI $\hat{\mu}_Y$	90% CI $\hat{\sigma}_Y$	90% CI $\hat{\rho}$
20	20	0	0	0.9260	0.9000	0.9330	0.9160	0.8960	0.8610	0.8510	0.8740	0.8580	0.8400
			0.1	0.9360	0.8980	0.9280	0.8880	0.9080	0.8790	0.8530	0.8780	0.8300	0.8470
			0.2	0.9360	0.9020	0.9280	0.8940	0.9220	0.8810	0.8580	0.8790	0.8560	0.8620
			0.3	0.9210	0.8880	0.9430	0.8930	0.8980	0.8570	0.8350	0.8880	0.8440	0.8470
			0.4	0.9260	0.8940	0.9370	0.8940	0.9030	0.8700	0.8500	0.8740	0.8430	0.8430
			0.5	0.9360	0.9030	0.9240	0.8800	0.9040	0.8750	0.8660	0.8690	0.8270	0.8610
			0.6	0.9130	0.8980	0.9350	0.8960	0.9090	0.8560	0.8360	0.8790	0.8510	0.8720
			0.7	0.9450	0.8860	0.9250	0.8880	0.8960	0.8790	0.8420	0.8830	0.8450	0.8460
			0.8	0.9100	0.8920	0.9140	0.8920	0.8850	0.8550	0.8320	0.8630	0.8340	0.8420
			0.9	0.9210	0.9020	0.9310	0.8870	0.8920	0.8740	0.8470	0.8830	0.8360	0.8490
	18	0.1	0	0.9250	0.8990	0.9310	0.9020	0.8820	0.8730	0.8380	0.8710	0.8560	0.8180
			0.1	0.9270	0.8830	0.9290	0.8910	0.8530	0.8660	0.8240	0.8720	0.8430	0.8020
			0.2	0.9220	0.8890	0.9160	0.9010	0.8920	0.8740	0.8460	0.8660	0.8420	0.8300
			0.3	0.9300	0.8990	0.9210	0.8850	0.8780	0.8780	0.8510	0.8580	0.8500	0.8260
			0.4	0.9160	0.9060	0.9280	0.8920	0.8780	0.8580	0.8580	0.8800	0.8410	0.8260
			0.5	0.9360	0.8920	0.9370	0.8950	0.8730	0.8900	0.8540	0.8930	0.8450	0.8220
			0.6	0.9180	0.9120	0.9280	0.8910	0.9000	0.8780	0.8630	0.8710	0.8470	0.8540
			0.7	0.9260	0.8970	0.9120	0.8910	0.8880	0.8810	0.8400	0.8520	0.8310	0.8380
			0.8	0.9400	0.9040	0.9360	0.8850	0.8800	0.8720	0.8560	0.8750	0.8370	0.8220
			0.9	0.9270	0.8920	0.9290	0.8760	0.8800	0.8740	0.8400	0.8720	0.8330	0.8390
	16	0.2	0	0.9150	0.8660	0.9110	0.9130	0.8490	0.8590	0.8110	0.8660	0.8710	0.7950
			0.1	0.9180	0.8880	0.9170	0.9060	0.8510	0.8730	0.8390	0.8710	0.8500	0.7940
			0.2	0.9180	0.8920	0.9250	0.9010	0.8720	0.8620	0.8470	0.8710	0.8600	0.8220
			0.3	0.9080	0.8820	0.9330	0.8920	0.8540	0.8610	0.8350	0.8710	0.8500	0.8110
			0.4	0.9080	0.8560	0.9210	0.8800	0.8640	0.8450	0.8210	0.8680	0.8290	0.8120
			0.5	0.9040	0.8780	0.9200	0.9040	0.8630	0.8580	0.8310	0.8620	0.8680	0.8170
			0.6	0.9170	0.8650	0.9260	0.8910	0.8910	0.8570	0.8230	0.8770	0.8500	0.8380
			0.7	0.9070	0.8610	0.9140	0.8790	0.8700	0.8550	0.8120	0.8620	0.8290	0.8280
			0.8	0.9260	0.8790	0.9250	0.8980	0.8770	0.8750	0.8350	0.8770	0.8500	0.8380
			0.9	0.9130	0.8840	0.9240	0.8980	0.8630	0.8700	0.8250	0.8730	0.8390	0.8260
	14	0.3	0	0.9060	0.8680	0.9240	0.9070	0.8380	0.8550	0.8210	0.8630	0.8670	0.7770
			0.1	0.9120	0.8760	0.9130	0.9000	0.8590	0.8560	0.8290	0.8590	0.8610	0.8000
			0.2	0.9160	0.8710	0.9270	0.9290	0.8260	0.8590	0.8190	0.8660	0.8870	0.7710
			0.3	0.9100	0.8690	0.9320	0.9170	0.8370	0.8500	0.8230	0.8770	0.8680	0.7850
			0.4	0.9190	0.8790	0.9200	0.9000	0.8330	0.8630	0.8250	0.8770	0.8640	0.7910
			0.5	0.9090	0.8700	0.9120	0.8740	0.8570	0.8510	0.8310	0.8580	0.8270	0.8050
			0.6	0.9180	0.8670	0.9010	0.8800	0.8540	0.8750	0.8190	0.8390	0.8190	0.8030
			0.7	0.9080	0.8680	0.9120	0.8740	0.8550	0.8540	0.8170	0.8510	0.8210	0.8080
			0.8	0.9070	0.8520	0.9370	0.8780	0.8620	0.8620	0.8040	0.8870	0.8330	0.8250
			0.9	0.9090	0.8700	0.9160	0.8730	0.8550	0.8510	0.8130	0.8560	0.8190	0.8230
	12	0.4	0	0.9030	0.8610	0.9280	0.9170	0.8230	0.8500	0.8000	0.8950	0.8730	0.7680
			0.1	0.9200	0.8570	0.9350	0.9000	0.8130	0.8730	0.8140	0.8650	0.8550	0.7470
			0.2	0.9040	0.8490	0.9170	0.9100	0.7910	0.8440	0.7980	0.8670	0.8770	0.7470
			0.3	0.9070	0.8490	0.9180	0.8900	0.8120	0.8490	0.8030	0.8530	0.8430	0.7510
			0.4	0.8850	0.8520	0.9270	0.8790	0.7970	0.8280	0.8050	0.8710	0.8450	0.7490
			0.5	0.8970	0.8510	0.9100	0.8810	0.8200	0.8340	0.8090	0.8580	0.8380	0.7700
			0.6	0.8910	0.8530	0.9220	0.8720	0.8420	0.8350	0.8120	0.8720	0.8240	0.8080
			0.7	0.8910	0.8520	0.9150	0.8540	0.8470	0.8320	0.8110	0.8660	0.8140	0.8070
			0.8	0.9030	0.8500	0.9000	0.8550	0.8530	0.8410	0.8080	0.8460	0.8140	0.8230
			0.9	0.9010	0.8590	0.9080	0.8740	0.8510	0.8440	0.8080	0.8480	0.8320	0.8120
	10	0.5	0	0.8780	0.8140	0.9210	0.9130	0.7730	0.8350	0.7780	0.8510	0.8800	0.7160
			0.1	0.8630	0.8190	0.9060	0.9210	0.7590	0.8180	0.7880	0.8510	0.8930	0.7030
			0.2	0.8750	0.8240	0.9260	0.9090	0.7610	0.8270	0.7860	0.8740	0.8770	0.7120
			0.3	0.8690	0.8240	0.9210	0.9050	0.7940	0.8100	0.7590	0.8530	0.8760	0.7370
			0.4	0.8990	0.8580	0.9080	0.8900	0.7950	0.8420	0.8150	0.8530	0.8550	0.7330
			0.5	0.8820	0.8150	0.9110	0.8730	0.7760	0.8280	0.7540	0.8250	0.8180	0.7310
			0.6	0.8960	0.8390	0.9190	0.8800	0.7960	0.8400	0.8040	0.8610	0.8310	0.7470
			0.7	0.8830	0.8490	0.8990	0.8360	0.8200	0.8270	0.7950	0.8560	0.7970	0.7840
			0.8	0.8800	0.8250	0.9010	0.8290	0.8240	0.8370	0.7760	0.8400	0.7810	0.7830
			0.9	0.8690	0.8220	0.9090	0.8530	0.8100	0.8170	0.7790	0.8490	0.8080	0.7780

Table 13.15: Probability coverages of 95% and 90% CIs for $n = 50$ using the Fisher information matrix

n	k	Proportion of Censoring	ρ	95% CI $\hat{\mu}_X$	95% CI $\hat{\sigma}_X$	95% CI $\hat{\mu}_Y$	95% CI $\hat{\sigma}_Y$	95% CI $\hat{\rho}$	90% CI $\hat{\mu}_X$	90% CI $\hat{\sigma}_X$	90% CI $\hat{\mu}_Y$	90% CI $\hat{\sigma}_Y$	90% CI $\hat{\rho}$
50	50	0	0	0.9490	0.9360	0.9320	0.9370	0.9310	0.9070	0.8950	0.8740	0.8830	0.8640
			0.1	0.9430	0.9310	0.9430	0.9350	0.9440	0.9020	0.8810	0.8810	0.9020	0.8900
			0.2	0.9420	0.9330	0.9410	0.9240	0.9310	0.8950	0.8850	0.8850	0.8760	0.8700
			0.3	0.9410	0.9350	0.9360	0.9370	0.9140	0.8860	0.8890	0.8820	0.8900	0.8710
			0.4	0.9330	0.9130	0.9440	0.9360	0.9280	0.8820	0.8500	0.8810	0.8830	0.8710
			0.5	0.9390	0.9380	0.9300	0.9280	0.9310	0.8980	0.8910	0.8830	0.8830	0.8820
			0.6	0.9400	0.9290	0.9470	0.9320	0.9200	0.8890	0.8700	0.8980	0.8830	0.8620
			0.7	0.9560	0.9310	0.9450	0.9330	0.9180	0.8970	0.8750	0.8910	0.8990	0.8750
			0.8	0.9310	0.9240	0.9460	0.9390	0.9330	0.8780	0.8840	0.8900	0.8860	0.8810
			0.9	0.9490	0.9250	0.9420	0.9380	0.9190	0.8890	0.8760	0.8790	0.8980	0.8820
	45	0.1	0	0.9330	0.9130	0.9450	0.9500	0.9250	0.8690	0.8670	0.8890	0.9010	0.8790
			0.1	0.9340	0.9250	0.9400	0.9250	0.9250	0.8930	0.8780	0.9010	0.8820	0.8610
			0.2	0.9250	0.9250	0.9380	0.9350	0.9040	0.8690	0.8620	0.8780	0.8840	0.8490
			0.3	0.9340	0.9180	0.9450	0.9360	0.9230	0.8780	0.8690	0.8880	0.8850	0.8820
			0.4	0.9500	0.9280	0.9300	0.9310	0.9380	0.9130	0.8840	0.8780	0.8830	0.8910
			0.5	0.9370	0.9160	0.9400	0.9350	0.9200	0.8810	0.8770	0.8950	0.8890	0.8680
			0.6	0.9490	0.9220	0.9410	0.9300	0.9260	0.8980	0.8830	0.8920	0.8860	0.8810
			0.7	0.9330	0.9290	0.9320	0.9110	0.9400	0.8880	0.8780	0.8850	0.8690	0.8890
			0.8	0.9430	0.9240	0.9410	0.9390	0.9350	0.8820	0.8790	0.8820	0.8950	0.8910
			0.9	0.9520	0.9300	0.9430	0.9310	0.9180	0.8970	0.8870	0.9000	0.8920	0.8760
	40	0.2	0	0.9430	0.9410	0.9360	0.9290	0.9200	0.8970	0.8810	0.8930	0.8790	0.8710
			0.1	0.9280	0.9390	0.9430	0.9300	0.9140	0.8770	0.8820	0.8780	0.8690	0.8590
			0.2	0.9230	0.9280	0.9460	0.9320	0.9290	0.8720	0.8750	0.8980	0.8810	0.8710
			0.3	0.9330	0.9240	0.9410	0.9380	0.9130	0.8640	0.8720	0.8820	0.8830	0.8680
			0.4	0.9450	0.9190	0.9330	0.9290	0.9180	0.8840	0.8670	0.8800	0.8750	0.8680
			0.5	0.9480	0.9110	0.9390	0.9170	0.9000	0.8900	0.8610	0.8820	0.8760	0.8490
			0.6	0.9360	0.9340	0.9520	0.9190	0.9090	0.8900	0.8900	0.9090	0.8720	0.8560
			0.7	0.9460	0.9300	0.9330	0.9170	0.9140	0.8960	0.8830	0.8810	0.8600	0.8700
			0.8	0.9270	0.9410	0.9310	0.9340	0.9180	0.8710	0.8930	0.8860	0.8820	0.8740
			0.9	0.9370	0.9270	0.9420	0.9270	0.9240	0.8870	0.8780	0.8810	0.8860	0.8890
	35	0.3	0	0.9360	0.9100	0.9470	0.9320	0.8950	0.8940	0.8640	0.8990	0.8870	0.8360
			0.1	0.9310	0.9080	0.9370	0.9220	0.9010	0.8760	0.8620	0.8930	0.8780	0.8550
			0.2	0.9310	0.9110	0.9400	0.9440	0.8900	0.8820	0.8600	0.8830	0.9040	0.8350
			0.3	0.9280	0.9220	0.9310	0.9400	0.9000	0.8780	0.8710	0.8620	0.8920	0.8300
			0.4	0.9190	0.9140	0.9510	0.9330	0.9080	0.8590	0.8670	0.8950	0.8930	0.8580
			0.5	0.9470	0.9120	0.9310	0.9290	0.9080	0.8940	0.8610	0.8770	0.8800	0.8520
			0.6	0.9400	0.9180	0.9370	0.9110	0.8970	0.8880	0.8700	0.8770	0.8640	0.8530
			0.7	0.9350	0.9220	0.9230	0.9010	0.9200	0.8850	0.8780	0.8670	0.8570	0.8800
			0.8	0.9310	0.9120	0.9460	0.9240	0.9000	0.8790	0.8640	0.8950	0.8730	0.8530
			0.9	0.9380	0.9140	0.9370	0.9240	0.9190	0.8880	0.8580	0.8870	0.8770	0.8690
	30	0.4	0	0.9300	0.9120	0.9290	0.9270	0.8890	0.8730	0.8610	0.8770	0.8820	0.8360
			0.1	0.9300	0.9110	0.9480	0.9350	0.8680	0.8800	0.8660	0.8890	0.8780	0.8230
			0.2	0.9300	0.9190	0.9440	0.9610	0.8800	0.8810	0.8740	0.8860	0.9090	0.8320
			0.3	0.9310	0.9100	0.9380	0.9270	0.8770	0.8650	0.8620	0.8740	0.8870	0.8350
			0.4	0.9230	0.9030	0.9440	0.9300	0.9130	0.8710	0.8570	0.8850	0.8820	0.8510
			0.5	0.9150	0.9260	0.9350	0.9050	0.8840	0.8630	0.8690	0.8890	0.8570	0.8340
			0.6	0.9320	0.8850	0.9300	0.9070	0.8960	0.8800	0.8370	0.8750	0.8540	0.8480
			0.7	0.9340	0.9180	0.9150	0.9030	0.8930	0.8720	0.8660	0.8710	0.8470	0.8570
			0.8	0.9340	0.9200	0.9230	0.8970	0.9000	0.8740	0.8540	0.8640	0.8510	0.8720
			0.9	0.9310	0.9070	0.9320	0.9230	0.9020	0.8900	0.8570	0.8830	0.8700	0.8610
	25	0.5	0	0.9350	0.9080	0.9240	0.9190	0.8460	0.8800	0.8650	0.8660	0.8780	0.7860
			0.1	0.9250	0.9000	0.9430	0.9470	0.8700	0.8620	0.8510	0.8830	0.9060	0.8130
			0.2	0.9200	0.8960	0.9310	0.9420	0.8690	0.8770	0.8430	0.8760	0.9070	0.8240
			0.3	0.9230	0.8870	0.9380	0.9460	0.8780	0.8620	0.8340	0.8740	0.9020	0.8200
			0.4	0.9390	0.9100	0.9320	0.9330	0.8520	0.8880	0.8550	0.8860	0.8960	0.8010
			0.5	0.9120	0.8850	0.9250	0.9200	0.8680	0.8580	0.8400	0.8730	0.8740	0.8180
			0.6	0.9300	0.9040	0.9360	0.8990	0.8730	0.8710	0.8540	0.8780	0.8600	0.8270
			0.7	0.9380	0.9090	0.9220	0.8990	0.8920	0.8940	0.8610	0.8680	0.8490	0.8490
			0.8	0.9160	0.9060	0.9260	0.8980	0.9020	0.8650	0.8660	0.8820	0.8580	0.8610
			0.9	0.9260	0.8980	0.9180	0.9030	0.8820	0.8730	0.8410	0.8790	0.8700	0.8460

Table 13.16: Probability coverages of 95% and 90% CIs for $n = 50$ using the partially observed information matrix

n	k	Proportion of Censoring	ρ	95% CI					90% CI				
				$\hat{\mu}_X$	$\hat{\sigma}_X$	$\hat{\mu}_Y$	$\hat{\sigma}_Y$	$\hat{\rho}$	$\hat{\mu}_X$	$\hat{\sigma}_X$	$\hat{\mu}_Y$	$\hat{\sigma}_Y$	$\hat{\rho}$
50	50	0	0	0.9490	0.9360	0.9320	0.9370	0.9310	0.9070	0.8950	0.8740	0.8830	0.8640
			0.1	0.9430	0.9310	0.9430	0.9350	0.9440	0.9020	0.8810	0.8810	0.9020	0.8900
			0.2	0.9420	0.9330	0.9410	0.9240	0.9310	0.8950	0.8850	0.8850	0.8760	0.8700
			0.3	0.9410	0.9350	0.9360	0.9370	0.9140	0.8860	0.8890	0.8820	0.8900	0.8710
			0.4	0.9330	0.9130	0.9440	0.9360	0.9280	0.8820	0.8500	0.8810	0.8830	0.8710
			0.5	0.9390	0.9380	0.9300	0.9280	0.9310	0.8980	0.8910	0.8830	0.8830	0.8820
			0.6	0.9400	0.9290	0.9470	0.9320	0.9200	0.8890	0.8700	0.8980	0.8830	0.8620
			0.7	0.9560	0.9310	0.9450	0.9330	0.9180	0.8970	0.8750	0.8910	0.8990	0.8750
			0.8	0.9310	0.9240	0.9460	0.9390	0.9330	0.8780	0.8840	0.8900	0.8860	0.8810
			0.9	0.9490	0.9250	0.9420	0.9380	0.9190	0.8890	0.8760	0.8790	0.8980	0.8820
	45	0.1	0	0.9330	0.9130	0.9450	0.9500	0.9240	0.8690	0.8680	0.8890	0.9010	0.8780
			0.1	0.9340	0.9250	0.9400	0.9250	0.9230	0.8930	0.8780	0.9010	0.8820	0.8610
			0.2	0.9250	0.9250	0.9370	0.9350	0.9020	0.8690	0.8610	0.8760	0.8840	0.8510
			0.3	0.9340	0.9170	0.9460	0.9360	0.9240	0.8780	0.8690	0.8880	0.8840	0.8770
			0.4	0.9500	0.9280	0.9300	0.9300	0.9360	0.9130	0.8840	0.8780	0.8850	0.8880
			0.5	0.9370	0.9160	0.9400	0.9350	0.9210	0.8810	0.8790	0.8960	0.8880	0.8700
			0.6	0.9490	0.9220	0.9410	0.9310	0.9290	0.8980	0.8820	0.8930	0.8850	0.8810
			0.7	0.9340	0.9300	0.9320	0.9110	0.9400	0.8880	0.8780	0.8850	0.8670	0.8890
			0.8	0.9430	0.9240	0.9410	0.9380	0.9350	0.8820	0.8790	0.8820	0.8940	0.8930
			0.9	0.9520	0.9310	0.9430	0.9320	0.9180	0.8970	0.8860	0.9000	0.8910	0.8740
	40	0.2	0	0.9440	0.9400	0.9400	0.9290	0.9240	0.8970	0.8820	0.8920	0.8790	0.8760
			0.1	0.9280	0.9390	0.9430	0.9300	0.9150	0.8770	0.8840	0.8810	0.8720	0.8620
			0.2	0.9230	0.9280	0.9460	0.9320	0.9320	0.8720	0.8750	0.9000	0.8800	0.8760
			0.3	0.9330	0.9230	0.9420	0.9370	0.9150	0.8640	0.8710	0.8830	0.8860	0.8680
			0.4	0.9450	0.9200	0.9330	0.9270	0.9210	0.8840	0.8690	0.8810	0.8760	0.8630
			0.5	0.9480	0.9110	0.9410	0.9190	0.9010	0.8900	0.8630	0.8820	0.8780	0.8520
			0.6	0.9360	0.9340	0.9530	0.9210	0.9120	0.8900	0.8890	0.9090	0.8750	0.8550
			0.7	0.9460	0.9310	0.9330	0.9210	0.9160	0.8960	0.8820	0.8820	0.8620	0.8700
			0.8	0.9270	0.9410	0.9300	0.9350	0.9200	0.8710	0.8950	0.8800	0.8820	0.8740
			0.9	0.9370	0.9240	0.9420	0.9240	0.9250	0.8870	0.8780	0.8800	0.8880	0.8900
	35	0.3	0	0.9370	0.9100	0.9490	0.9320	0.9010	0.8940	0.8640	0.9010	0.8870	0.8460
			0.1	0.9310	0.9090	0.9370	0.9200	0.9010	0.8760	0.8620	0.8970	0.8810	0.8560
			0.2	0.9310	0.9110	0.9420	0.9450	0.8910	0.8810	0.8580	0.8810	0.9010	0.8370
			0.3	0.9280	0.9230	0.9320	0.9410	0.9020	0.8780	0.8710	0.8610	0.8950	0.8380
			0.4	0.9190	0.9140	0.9520	0.9380	0.9120	0.8580	0.8690	0.8950	0.8920	0.8620
			0.5	0.9470	0.9130	0.9360	0.9310	0.9150	0.8940	0.8610	0.8790	0.8820	0.8610
			0.6	0.9400	0.9170	0.9370	0.9090	0.8970	0.8880	0.8690	0.8820	0.8640	0.8560
			0.7	0.9350	0.9220	0.9260	0.9030	0.9190	0.8840	0.8800	0.8690	0.8570	0.8830
			0.8	0.9310	0.9140	0.9450	0.9270	0.9020	0.8780	0.8660	0.8980	0.8760	0.8550
			0.9	0.9390	0.9160	0.9390	0.9250	0.9190	0.8880	0.8590	0.8870	0.8760	0.8690
	30	0.4	0	0.9300	0.9120	0.9340	0.9250	0.8920	0.8720	0.8630	0.8790	0.8840	0.8360
			0.1	0.9300	0.9100	0.9520	0.9360	0.8800	0.8800	0.8660	0.8920	0.8750	0.8300
			0.2	0.9300	0.9190	0.9490	0.9600	0.8820	0.8800	0.8770	0.8900	0.9120	0.8290
			0.3	0.9310	0.9110	0.9420	0.9260	0.8840	0.8650	0.8620	0.8820	0.8890	0.8320
			0.4	0.9230	0.9030	0.9470	0.9290	0.9100	0.8700	0.8580	0.8920	0.8860	0.8610
			0.5	0.9160	0.9270	0.9400	0.9060	0.8870	0.8630	0.8690	0.8880	0.8600	0.8380
			0.6	0.9320	0.8850	0.9280	0.9130	0.9010	0.8800	0.8380	0.8790	0.8620	0.8530
			0.7	0.9340	0.9170	0.9210	0.9100	0.8950	0.8720	0.8700	0.8770	0.8500	0.8570
			0.8	0.9340	0.9240	0.9230	0.9030	0.9010	0.8740	0.8570	0.8650	0.8600	0.8720
			0.9	0.9310	0.9100	0.9360	0.9240	0.9030	0.8900	0.8580	0.8820	0.8720	0.8630
	25	0.5	0	0.9350	0.9070	0.9320	0.9190	0.8540	0.8800	0.8670	0.8710	0.8820	0.7860
			0.1	0.9250	0.8990	0.9490	0.9490	0.8710	0.8620	0.8520	0.8870	0.9080	0.8220
			0.2	0.9200	0.8970	0.9380	0.9430	0.8780	0.8750	0.8450	0.8830	0.9100	0.8270
			0.3	0.9220	0.8880	0.9450	0.9450	0.8770	0.8610	0.8340	0.8830	0.9050	0.8220
			0.4	0.9380	0.9120	0.9390	0.9330	0.8600	0.8880	0.8550	0.8860	0.8970	0.8130
			0.5	0.9120	0.8880	0.9330	0.9210	0.8800	0.8580	0.8460	0.8800	0.8780	0.8240
			0.6	0.9300	0.9060	0.9360	0.9020	0.8810	0.8710	0.8590	0.8790	0.8600	0.8300
			0.7	0.9380	0.9080	0.9240	0.9040	0.8930	0.8940	0.8610	0.8780	0.8550	0.8570
			0.8	0.9150	0.9050	0.9280	0.9020	0.9070	0.8640	0.8660	0.8860	0.8580	0.8640
			0.9	0.9260	0.9000	0.9200	0.9020	0.8870	0.8720	0.8410	0.8760	0.8690	0.8490

Table 13.17: Probability coverages of 95% and 90% CIs for $n = 100$ using the Fisher information matrix

n	k	Proportion of Censoring	ρ	95% CI $\hat{\mu}_X$	95% CI $\hat{\sigma}_X$	95% CI $\hat{\mu}_Y$	95% CI $\hat{\sigma}_Y$	95% CI $\hat{\rho}$	90% CI $\hat{\mu}_X$	90% CI $\hat{\sigma}_X$	90% CI $\hat{\mu}_Y$	90% CI $\hat{\sigma}_Y$	90% CI $\hat{\rho}$
100	100	0	0	0.9460	0.9320	0.9440	0.9360	0.9490	0.8970	0.8680	0.8940	0.8850	0.9000
			0.1	0.9500	0.9410	0.9420	0.9380	0.9370	0.9040	0.8960	0.8800	0.8840	0.8820
			0.2	0.9450	0.9310	0.9440	0.9330	0.9390	0.8930	0.8830	0.8890	0.8830	0.8740
			0.3	0.9500	0.9490	0.9540	0.9380	0.9410	0.9000	0.8970	0.9080	0.8900	0.8800
			0.4	0.9380	0.9330	0.9470	0.9560	0.9370	0.8890	0.8850	0.9060	0.9130	0.8840
			0.5	0.9380	0.9320	0.9350	0.9430	0.9420	0.8880	0.8800	0.8820	0.8940	0.8940
			0.6	0.9510	0.9280	0.9390	0.9240	0.9430	0.8940	0.8910	0.8890	0.8690	0.8910
			0.7	0.9490	0.9310	0.9530	0.9490	0.9370	0.9010	0.8790	0.9060	0.9020	0.8810
			0.8	0.9590	0.9340	0.9520	0.9280	0.9400	0.9190	0.8720	0.9090	0.8820	0.8920
			0.9	0.9420	0.9310	0.9420	0.9380	0.9320	0.8940	0.8760	0.8860	0.8850	0.8870
	90	0.1	0	0.9590	0.9320	0.9500	0.9340	0.9400	0.8930	0.8770	0.8910	0.8910	0.8930
			0.1	0.9480	0.9420	0.9480	0.9350	0.9350	0.8950	0.8970	0.8960	0.8830	0.8900
			0.2	0.9460	0.9310	0.9320	0.9380	0.9490	0.8910	0.8870	0.8880	0.8960	0.8920
			0.3	0.9400	0.9430	0.9370	0.9300	0.9270	0.9000	0.9070	0.8650	0.8770	0.8860
			0.4	0.9410	0.9310	0.9420	0.9330	0.9520	0.8870	0.8790	0.8750	0.8930	0.9030
			0.5	0.9420	0.9370	0.9290	0.9400	0.9320	0.8880	0.8900	0.8770	0.8810	0.8740
			0.6	0.9480	0.9350	0.9560	0.9330	0.9270	0.8990	0.8910	0.8800	0.8710	0.8780
			0.7	0.9450	0.9260	0.9370	0.9150	0.9310	0.8930	0.8780	0.8840	0.8690	0.8850
			0.8	0.9510	0.9410	0.9420	0.9420	0.9270	0.9050	0.8930	0.8870	0.8970	0.8780
			0.9	0.9380	0.9290	0.9440	0.9300	0.9320	0.8860	0.8900	0.8900	0.8760	0.8770
	80	0.2	0	0.9520	0.9410	0.9430	0.9390	0.9240	0.9110	0.8940	0.8910	0.8880	0.8700
			0.1	0.9550	0.9460	0.9440	0.9340	0.9310	0.9040	0.8990	0.8990	0.8810	0.8720
			0.2	0.9430	0.9310	0.9490	0.9460	0.9340	0.8910	0.8840	0.9110	0.9150	0.8940
			0.3	0.9350	0.9270	0.9450	0.9310	0.9410	0.8990	0.8760	0.8910	0.8970	0.8910
			0.4	0.9570	0.9380	0.9450	0.9380	0.9160	0.9110	0.8750	0.9030	0.8800	0.8750
			0.5	0.9480	0.9260	0.9460	0.9320	0.9190	0.8880	0.8740	0.8970	0.8810	0.8680
			0.6	0.9500	0.9350	0.9370	0.9200	0.9300	0.9030	0.8890	0.8830	0.8800	0.8930
			0.7	0.9560	0.9290	0.9580	0.9310	0.9440	0.9080	0.8780	0.9110	0.8890	0.8990
			0.8	0.9360	0.9440	0.9520	0.9330	0.9290	0.8840	0.8950	0.9090	0.8700	0.8920
			0.9	0.9490	0.9280	0.9500	0.9330	0.9350	0.8990	0.8810	0.8900	0.8940	0.8850
	70	0.3	0	0.9500	0.9430	0.9430	0.9440	0.9200	0.8900	0.8980	0.8940	0.8860	0.8750
			0.1	0.9410	0.9500	0.9430	0.9430	0.9220	0.8820	0.8880	0.9000	0.8930	0.8670
			0.2	0.9240	0.9300	0.9530	0.9390	0.9280	0.8740	0.8760	0.9050	0.8950	0.8710
			0.3	0.9460	0.9400	0.9460	0.9400	0.9200	0.8980	0.8840	0.8870	0.8820	0.8810
			0.4	0.9450	0.9250	0.9480	0.9420	0.9270	0.9010	0.8830	0.9050	0.9080	0.8780
			0.5	0.9420	0.9330	0.9470	0.9290	0.9420	0.8860	0.8830	0.8780	0.8740	0.8910
			0.6	0.9390	0.9260	0.9530	0.9510	0.9300	0.8930	0.8620	0.9000	0.8960	0.8920
			0.7	0.9360	0.9280	0.9560	0.9260	0.9230	0.8870	0.8780	0.9040	0.8780	0.8690
			0.8	0.9460	0.9330	0.9390	0.9350	0.9340	0.8980	0.8750	0.8950	0.8890	0.8820
			0.9	0.9450	0.9420	0.9360	0.9510	0.9290	0.8900	0.8960	0.8780	0.8970	0.8840
	60	0.4	0	0.9630	0.9200	0.9390	0.9510	0.9090	0.9120	0.8800	0.8730	0.8940	0.8630
			0.1	0.9340	0.9130	0.9440	0.9370	0.9160	0.8890	0.8650	0.8880	0.8910	0.8670
			0.2	0.9240	0.9390	0.9490	0.9320	0.9140	0.8840	0.8770	0.9070	0.8980	0.8680
			0.3	0.9380	0.9440	0.9300	0.9290	0.9100	0.8750	0.9040	0.8720	0.9010	0.8610
			0.4	0.9400	0.9300	0.9480	0.9320	0.9190	0.8840	0.8880	0.8960	0.8790	0.8750
			0.5	0.9240	0.9370	0.9470	0.9180	0.9320	0.8580	0.8950	0.8950	0.8750	0.8810
			0.6	0.9390	0.9240	0.9310	0.9260	0.9170	0.8850	0.8750	0.8730	0.8710	0.8620
			0.7	0.9350	0.9330	0.9440	0.9270	0.9330	0.8660	0.8910	0.8910	0.8770	0.8940
			0.8	0.9390	0.9280	0.9480	0.9270	0.9170	0.8940	0.8830	0.9050	0.8660	0.8770
			0.9	0.9320	0.9360	0.9380	0.9420	0.9290	0.8940	0.8850	0.8940	0.8960	0.8960
	50	0.5	0	0.9360	0.9350	0.9290	0.9470	0.8900	0.8990	0.8880	0.8690	0.9110	0.8300
			0.1	0.9370	0.9240	0.9310	0.9530	0.8980	0.8910	0.8760	0.8720	0.9080	0.8360
			0.2	0.9270	0.9230	0.9410	0.9500	0.8940	0.8750	0.8720	0.9020	0.9120	0.8420
			0.3	0.9360	0.9260	0.9460	0.9300	0.9130	0.8860	0.8700	0.8960	0.8890	0.8670
			0.4	0.9540	0.9260	0.9350	0.9250	0.8960	0.9000	0.8790	0.8760	0.8870	0.8470
			0.5	0.9410	0.9290	0.9310	0.9200	0.9070	0.9010	0.8800	0.8710	0.8800	0.8510
			0.6	0.9350	0.9090	0.9300	0.9210	0.9250	0.8790	0.8580	0.8640	0.8850	0.8760
			0.7	0.9280	0.9300	0.9400	0.9190	0.9230	0.8910	0.8870	0.8850	0.8610	0.8630
			0.8	0.9320	0.9310	0.9410	0.9290	0.9180	0.8870	0.8720	0.8970	0.8870	0.8620
			0.9	0.9290	0.9230	0.9340	0.9290	0.9370	0.8870	0.8740	0.8910	0.8740	0.9030

Table 13.18: Probability coverages of 95% and 90% CIs for $n = 100$ using the partially observed information matrix

n	k	Proportion of Censoring	ρ	95% CI $\hat{\mu}_X$	95% CI $\hat{\sigma}_X$	95% CI $\hat{\mu}_Y$	95% CI $\hat{\sigma}_Y$	95% CI $\hat{\rho}$	90% CI $\hat{\mu}_X$	90% CI $\hat{\sigma}_X$	90% CI $\hat{\mu}_Y$	90% CI $\hat{\sigma}_Y$	90% CI $\hat{\rho}$
100	100	0	0	0.9460	0.9320	0.9440	0.9360	0.9490	0.8970	0.8680	0.8940	0.8850	0.9000
			0.1	0.9500	0.9410	0.9420	0.9380	0.9370	0.9040	0.8960	0.8800	0.8840	0.8820
			0.2	0.9450	0.9310	0.9440	0.9330	0.9390	0.8930	0.8830	0.8890	0.8830	0.8740
			0.3	0.9500	0.9490	0.9540	0.9380	0.9410	0.9000	0.8970	0.9080	0.8900	0.8800
			0.4	0.9380	0.9330	0.9470	0.9560	0.9370	0.8890	0.8850	0.9060	0.9130	0.8840
			0.5	0.9380	0.9320	0.9350	0.9430	0.9420	0.8880	0.8800	0.8820	0.8940	0.8940
			0.6	0.9510	0.9280	0.9390	0.9240	0.9430	0.8940	0.8910	0.8890	0.8690	0.8910
			0.7	0.9490	0.9310	0.9530	0.9490	0.9370	0.9010	0.8790	0.9060	0.9020	0.8810
			0.8	0.9590	0.9340	0.9520	0.9280	0.9400	0.9190	0.8720	0.9090	0.8820	0.8920
			0.9	0.9420	0.9310	0.9420	0.9380	0.9320	0.8940	0.8760	0.8860	0.8850	0.8870
	90	0.1	0	0.9590	0.9330	0.9500	0.9340	0.9370	0.8930	0.8780	0.8910	0.8910	0.8900
			0.1	0.9480	0.9420	0.9480	0.9350	0.9350	0.8950	0.8960	0.8950	0.8820	0.8890
			0.2	0.9460	0.9310	0.9320	0.9390	0.9480	0.8910	0.8870	0.8910	0.8960	0.8950
			0.3	0.9400	0.9430	0.9380	0.9290	0.9300	0.9000	0.9070	0.8660	0.8770	0.8840
			0.4	0.9420	0.9310	0.9430	0.9340	0.9490	0.8860	0.8790	0.8750	0.8950	0.9060
			0.5	0.9410	0.9370	0.9290	0.9410	0.9330	0.8880	0.8900	0.8780	0.8850	0.8730
			0.6	0.9480	0.9350	0.9560	0.9340	0.9310	0.8990	0.8910	0.8800	0.8710	0.8810
			0.7	0.9450	0.9260	0.9370	0.9130	0.9320	0.8930	0.8780	0.8840	0.8710	0.8850
			0.8	0.9510	0.9400	0.9420	0.9410	0.9270	0.9050	0.8940	0.8870	0.8960	0.8780
			0.9	0.9380	0.9290	0.9440	0.9300	0.9310	0.8860	0.8900	0.8900	0.8770	0.8770
	80	0.2	0	0.9520	0.9410	0.9450	0.9400	0.9260	0.9110	0.8950	0.8920	0.8880	0.8680
			0.1	0.9540	0.9480	0.9470	0.9340	0.9360	0.9030	0.9000	0.9000	0.8820	0.8740
			0.2	0.9420	0.9310	0.9470	0.9460	0.9350	0.8900	0.8860	0.9110	0.9150	0.8870
			0.3	0.9350	0.9270	0.9440	0.9330	0.9420	0.8990	0.8750	0.8980	0.8970	0.8940
			0.4	0.9570	0.9370	0.9450	0.9380	0.9200	0.9110	0.8750	0.9040	0.8820	0.8760
			0.5	0.9480	0.9260	0.9480	0.9320	0.9200	0.8880	0.8740	0.8970	0.8810	0.8680
			0.6	0.9500	0.9350	0.9390	0.9190	0.9350	0.9030	0.8880	0.8830	0.8840	0.8920
			0.7	0.9560	0.9300	0.9570	0.9320	0.9460	0.9080	0.8800	0.9110	0.8900	0.8950
			0.8	0.9370	0.9440	0.9530	0.9350	0.9290	0.8830	0.8950	0.9100	0.8720	0.8930
			0.9	0.9490	0.9280	0.9500	0.9330	0.9330	0.8990	0.8820	0.8900	0.8930	0.8860
	70	0.3	0	0.9500	0.9430	0.9480	0.9460	0.9230	0.8900	0.8980	0.8960	0.8870	0.8750
			0.1	0.9410	0.9500	0.9420	0.9430	0.9210	0.8810	0.8870	0.9000	0.8930	0.8710
			0.2	0.9240	0.9310	0.9550	0.9390	0.9260	0.8740	0.8750	0.9020	0.8940	0.8710
			0.3	0.9460	0.9400	0.9450	0.9410	0.9240	0.8980	0.8840	0.8850	0.8830	0.8770
			0.4	0.9440	0.9260	0.9500	0.9440	0.9310	0.9010	0.8830	0.9040	0.9080	0.8810
			0.5	0.9400	0.9330	0.9510	0.9290	0.9420	0.8860	0.8840	0.8810	0.8730	0.8910
			0.6	0.9380	0.9260	0.9540	0.9500	0.9320	0.8930	0.8630	0.9040	0.8950	0.8930
			0.7	0.9350	0.9280	0.9550	0.9270	0.9240	0.8870	0.8770	0.9050	0.8780	0.8650
			0.8	0.9460	0.9330	0.9370	0.9340	0.9330	0.8980	0.8760	0.8960	0.8880	0.8820
			0.9	0.9450	0.9430	0.9370	0.9510	0.9310	0.8910	0.8970	0.8780	0.8970	0.8850
	60	0.4	0	0.9630	0.9210	0.9390	0.9520	0.9150	0.9120	0.8790	0.8760	0.8940	0.8660
			0.1	0.9340	0.9150	0.9440	0.9370	0.9180	0.8890	0.8650	0.8900	0.8910	0.8700
			0.2	0.9240	0.9400	0.9520	0.9330	0.9140	0.8840	0.8780	0.9060	0.9000	0.8710
			0.3	0.9380	0.9450	0.9340	0.9290	0.9060	0.8750	0.9050	0.8740	0.9010	0.8700
			0.4	0.9400	0.9300	0.9550	0.9330	0.9220	0.8840	0.8880	0.8990	0.8800	0.8870
			0.5	0.9240	0.9370	0.9460	0.9190	0.9360	0.8570	0.8940	0.9020	0.8750	0.8810
			0.6	0.9390	0.9240	0.9320	0.9250	0.9180	0.8850	0.8760	0.8760	0.8700	0.8630
			0.7	0.9350	0.9340	0.9430	0.9250	0.9350	0.8650	0.8920	0.8940	0.8790	0.8950
			0.8	0.9390	0.9300	0.9500	0.9280	0.9180	0.8950	0.8850	0.9020	0.8690	0.8780
			0.9	0.9320	0.9360	0.9380	0.9380	0.9290	0.8940	0.8850	0.8940	0.8980	0.8960
	50	0.5	0	0.9360	0.9350	0.9330	0.9470	0.8900	0.8990	0.8900	0.8760	0.9120	0.8330
			0.1	0.9370	0.9240	0.9380	0.9530	0.8960	0.8910	0.8760	0.8760	0.9060	0.8440
			0.2	0.9270	0.9230	0.9420	0.9510	0.9020	0.8750	0.8720	0.8990	0.9150	0.8360
			0.3	0.9360	0.9270	0.9500	0.9300	0.9150	0.8860	0.8700	0.9010	0.8910	0.8690
			0.4	0.9530	0.9270	0.9390	0.9250	0.8990	0.9010	0.8810	0.8790	0.8900	0.8500
			0.5	0.9410	0.9290	0.9280	0.9220	0.9090	0.9010	0.8820	0.8690	0.8810	0.8520
			0.6	0.9350	0.9090	0.9330	0.9240	0.9260	0.8790	0.8590	0.8730	0.8870	0.8800
			0.7	0.9280	0.9300	0.9420	0.9200	0.9240	0.8910	0.8890	0.8890	0.8690	0.8690
			0.8	0.9320	0.9310	0.9410	0.9300	0.9190	0.8870	0.8720	0.9040	0.8900	0.8690
			0.9	0.9300	0.9250	0.9370	0.9310	0.9390	0.8870	0.8740	0.8920	0.8800	0.9010

Table 13.19: Probability coverages of 95% and 90% CIs for $n = 200$ using the Fisher information matrix

n	k	Proportion of Censoring	ρ	95% CI $\hat{\mu}_X$	95% CI $\hat{\sigma}_X$	95% CI $\hat{\mu}_Y$	95% CI $\hat{\sigma}_Y$	95% CI $\hat{\rho}$	90% CI $\hat{\mu}_X$	90% CI $\hat{\sigma}_X$	90% CI $\hat{\mu}_Y$	90% CI $\hat{\sigma}_Y$	90% CI $\hat{\rho}$
200	200	0	0	0.9530	0.9390	0.9400	0.9430	0.9420	0.9040	0.8890	0.8850	0.8990	0.8900
			0.1	0.9480	0.9340	0.9530	0.9510	0.9440	0.8940	0.8990	0.9080	0.9070	0.8890
			0.2	0.9540	0.9480	0.9520	0.9440	0.9590	0.8920	0.9010	0.9070	0.8960	0.9160
			0.3	0.9400	0.9470	0.9420	0.9510	0.9480	0.8810	0.8930	0.8910	0.9030	0.9080
			0.4	0.9460	0.9430	0.9470	0.9370	0.9400	0.8840	0.8900	0.8960	0.8940	0.8900
			0.5	0.9540	0.9410	0.9570	0.9490	0.9450	0.9050	0.8900	0.9110	0.8920	0.8930
			0.6	0.9480	0.9400	0.9530	0.9410	0.9470	0.8950	0.9020	0.9090	0.9000	0.8950
			0.7	0.9430	0.9400	0.9530	0.9360	0.9460	0.8880	0.8810	0.8870	0.8870	0.8970
			0.8	0.9370	0.9500	0.9490	0.9410	0.9420	0.8700	0.8920	0.9030	0.8980	0.8960
			0.9	0.9370	0.9510	0.9480	0.9580	0.9490	0.8880	0.9040	0.8940	0.9080	0.9000
	180	0.1	0	0.9490	0.9490	0.9560	0.9440	0.9390	0.8910	0.8960	0.9120	0.8860	0.8790
			0.1	0.9480	0.9360	0.9530	0.9450	0.9440	0.9110	0.8890	0.9050	0.9060	0.8970
			0.2	0.9450	0.9370	0.9440	0.9350	0.9540	0.8930	0.8910	0.9040	0.8760	0.8960
			0.3	0.9520	0.9380	0.9400	0.9430	0.9300	0.8980	0.8800	0.8870	0.8930	0.8770
			0.4	0.9480	0.9570	0.9530	0.9390	0.9370	0.8990	0.9020	0.9110	0.8900	0.8810
			0.5	0.9550	0.9440	0.9570	0.9370	0.9430	0.9010	0.8860	0.9020	0.8860	0.8840
			0.6	0.9510	0.9480	0.9440	0.9380	0.9320	0.8980	0.8940	0.8820	0.8790	0.8790
			0.7	0.9510	0.9570	0.9440	0.9570	0.9450	0.9120	0.9080	0.9090	0.8990	0.9030
			0.8	0.9490	0.9580	0.9490	0.9500	0.9530	0.9050	0.9110	0.9040	0.9040	0.9070
			0.9	0.9440	0.9530	0.9410	0.9480	0.9510	0.8970	0.9110	0.8880	0.8930	0.8980
	160	0.2	0	0.9470	0.9450	0.9470	0.9420	0.9370	0.9040	0.8930	0.8970	0.8810	0.8730
			0.1	0.9520	0.9560	0.9500	0.9460	0.9390	0.8950	0.8890	0.9090	0.8970	0.8880
			0.2	0.9410	0.9450	0.9430	0.9540	0.9420	0.8820	0.8960	0.8900	0.8920	0.8930
			0.3	0.9450	0.9490	0.9530	0.9470	0.9430	0.9060	0.8930	0.9030	0.8860	0.8930
			0.4	0.9540	0.9440	0.9550	0.9300	0.9410	0.9110	0.8970	0.9040	0.8880	0.8950
			0.5	0.9430	0.9330	0.9590	0.9520	0.9330	0.8820	0.8950	0.8990	0.8940	0.8880
			0.6	0.9440	0.9380	0.9510	0.9490	0.9490	0.8970	0.8840	0.8990	0.9000	0.9020
			0.7	0.9450	0.9530	0.9420	0.9380	0.9430	0.9010	0.8980	0.8940	0.8900	0.9040
			0.8	0.9530	0.9500	0.9390	0.9400	0.9300	0.9060	0.8870	0.8980	0.8900	0.8930
			0.9	0.9500	0.9360	0.9490	0.9370	0.9390	0.9070	0.8830	0.8960	0.8860	0.8860
	140	0.3	0	0.9380	0.9440	0.9460	0.9460	0.9400	0.8810	0.9010	0.8940	0.9070	0.8930
			0.1	0.9470	0.9420	0.9390	0.9500	0.9290	0.8980	0.8940	0.8800	0.8920	0.8840
			0.2	0.9370	0.9360	0.9520	0.9480	0.9240	0.8960	0.8910	0.9030	0.9010	0.8670
			0.3	0.9490	0.9310	0.9490	0.9420	0.9280	0.9070	0.8850	0.8950	0.8950	0.8730
			0.4	0.9440	0.9340	0.9510	0.9490	0.9380	0.8930	0.8790	0.9060	0.8980	0.8890
			0.5	0.9450	0.9340	0.9440	0.9410	0.9400	0.9050	0.8920	0.8920	0.8910	0.9000
			0.6	0.9440	0.9350	0.9440	0.9380	0.9470	0.8980	0.8860	0.8940	0.8900	0.8920
			0.7	0.9350	0.9320	0.9410	0.9270	0.9280	0.8830	0.8850	0.8900	0.8700	0.8890
			0.8	0.9540	0.9420	0.9530	0.9380	0.9420	0.9050	0.8950	0.8910	0.8890	0.8950
			0.9	0.9500	0.9530	0.9560	0.9490	0.9450	0.9020	0.9080	0.9110	0.8970	0.8950
	120	0.4	0	0.9500	0.9380	0.9560	0.9440	0.9360	0.8930	0.8830	0.8970	0.8910	0.8690
			0.1	0.9280	0.9400	0.9410	0.9560	0.9230	0.8880	0.8800	0.8960	0.9040	0.8570
			0.2	0.9360	0.9230	0.9420	0.9560	0.9230	0.8760	0.8740	0.8860	0.9110	0.8690
			0.3	0.9340	0.9350	0.9310	0.9430	0.9320	0.8850	0.9030	0.8870	0.8880	0.8890
			0.4	0.9340	0.9500	0.9380	0.9400	0.9240	0.8640	0.8910	0.8810	0.8900	0.8690
			0.5	0.9460	0.9390	0.9630	0.9560	0.9260	0.8930	0.9000	0.9250	0.9040	0.8880
			0.6	0.9460	0.9470	0.9500	0.9400	0.9400	0.8860	0.9100	0.9180	0.9010	0.9080
			0.7	0.9540	0.9260	0.9520	0.9390	0.9370	0.9140	0.8710	0.8980	0.8960	0.8960
			0.8	0.9370	0.9450	0.9480	0.9380	0.9390	0.8790	0.8780	0.8950	0.8830	0.8990
			0.9	0.9500	0.9270	0.9430	0.9370	0.9320	0.9010	0.8780	0.9000	0.8790	0.8800
	100	0.5	0	0.9430	0.9480	0.9460	0.9570	0.9330	0.8910	0.9000	0.9020	0.8980	0.8830
			0.1	0.9510	0.9350	0.9470	0.9620	0.9220	0.8930	0.8840	0.8960	0.9160	0.8780
			0.2	0.9490	0.9380	0.9520	0.9500	0.9220	0.8820	0.8820	0.9000	0.9020	0.8730
			0.3	0.9550	0.9310	0.9390	0.9590	0.9110	0.8930	0.8890	0.8740	0.9150	0.8590
			0.4	0.9390	0.9440	0.9480	0.9370	0.9340	0.8920	0.8960	0.8940	0.8810	0.8770
			0.5	0.9420	0.9470	0.9320	0.9390	0.9180	0.8740	0.8960	0.8880	0.8970	0.8840
			0.6	0.9470	0.9460	0.9470	0.9450	0.9290	0.9070	0.9040	0.8990	0.8890	0.8850
			0.7	0.9460	0.9350	0.9440	0.9370	0.9280	0.8970	0.8900	0.9020	0.8870	0.8860
			0.8	0.9250	0.9430	0.9500	0.9430	0.9280	0.8740	0.8990	0.8980	0.8890	0.8810
			0.9	0.9430	0.9330	0.9480	0.9400	0.9290	0.8910	0.8780	0.9070	0.8880	0.8770

Table 13.20: Probability coverages of 95% and 90% CIs for $n = 200$ using the partially observed information matrix

n	k	Proportion of Censoring	ρ	95% CI $\hat{\mu}_X$	95% CI $\hat{\sigma}_X$	95% CI $\hat{\mu}_Y$	95% CI $\hat{\sigma}_Y$	95% CI $\hat{\rho}$	90% CI $\hat{\mu}_X$	90% CI $\hat{\sigma}_X$	90% CI $\hat{\mu}_Y$	90% CI $\hat{\sigma}_Y$	90% CI $\hat{\rho}$
200	200	0	0	0.9530	0.9390	0.9400	0.9430	0.9420	0.9040	0.8890	0.8850	0.8990	0.8900
			0.1	0.9480	0.9340	0.9530	0.9510	0.9440	0.8940	0.8990	0.9080	0.9070	0.8890
			0.2	0.9540	0.9480	0.9520	0.9440	0.9590	0.8920	0.9010	0.9070	0.8960	0.9160
			0.3	0.9400	0.9470	0.9420	0.9510	0.9480	0.8810	0.8930	0.8910	0.9030	0.9080
			0.4	0.9460	0.9430	0.9470	0.9370	0.9400	0.8840	0.8900	0.8960	0.8940	0.8900
			0.5	0.9540	0.9410	0.9570	0.9490	0.9450	0.9050	0.8900	0.9110	0.8920	0.8930
			0.6	0.9480	0.9400	0.9530	0.9410	0.9470	0.8950	0.9020	0.9090	0.9000	0.8950
			0.7	0.9430	0.9400	0.9530	0.9360	0.9460	0.8880	0.8810	0.8870	0.8870	0.8970
			0.8	0.9370	0.9500	0.9490	0.9410	0.9420	0.8700	0.8920	0.9030	0.8980	0.8960
			0.9	0.9370	0.9510	0.9480	0.9580	0.9490	0.8880	0.9040	0.8940	0.9080	0.9000
	180	0.1	0	0.9490	0.9490	0.9560	0.9450	0.9410	0.8910	0.8960	0.9120	0.8860	0.8830
			0.1	0.9480	0.9360	0.9530	0.9450	0.9480	0.9100	0.8890	0.9050	0.9060	0.8990
			0.2	0.9450	0.9370	0.9450	0.9350	0.9560	0.8930	0.8910	0.9040	0.8770	0.8960
			0.3	0.9520	0.9390	0.9400	0.9430	0.9310	0.8980	0.8800	0.8870	0.8930	0.8790
			0.4	0.9480	0.9570	0.9520	0.9390	0.9380	0.8990	0.9020	0.9120	0.8890	0.8830
			0.5	0.9550	0.9430	0.9570	0.9370	0.9430	0.9010	0.8860	0.9010	0.8850	0.8850
			0.6	0.9510	0.9480	0.9440	0.9390	0.9340	0.8980	0.8940	0.8830	0.8780	0.8800
			0.7	0.9510	0.9580	0.9440	0.9580	0.9450	0.9120	0.9080	0.9090	0.8980	0.9020
			0.8	0.9490	0.9580	0.9480	0.9490	0.9530	0.9050	0.9120	0.9040	0.9040	0.9090
			0.9	0.9440	0.9530	0.9410	0.9480	0.9500	0.8970	0.9110	0.8880	0.8930	0.9000
	160	0.2	0	0.9470	0.9440	0.9470	0.9420	0.9400	0.9040	0.8930	0.8970	0.8810	0.8740
			0.1	0.9520	0.9560	0.9510	0.9460	0.9390	0.8950	0.8890	0.9060	0.8980	0.8870
			0.2	0.9410	0.9450	0.9450	0.9550	0.9400	0.8820	0.8960	0.8900	0.8920	0.8940
			0.3	0.9450	0.9490	0.9520	0.9470	0.9440	0.9060	0.8930	0.9030	0.8850	0.8890
			0.4	0.9540	0.9430	0.9550	0.9300	0.9470	0.9110	0.8960	0.9050	0.8880	0.8960
			0.5	0.9430	0.9340	0.9600	0.9520	0.9340	0.8820	0.8950	0.8980	0.8970	0.8900
			0.6	0.9440	0.9380	0.9510	0.9490	0.9510	0.8970	0.8840	0.8990	0.9030	0.9020
			0.7	0.9450	0.9530	0.9420	0.9380	0.9430	0.9010	0.8980	0.8930	0.8900	0.9050
			0.8	0.9530	0.9510	0.9400	0.9400	0.9300	0.9050	0.8880	0.8980	0.8900	0.8940
			0.9	0.9500	0.9340	0.9490	0.9370	0.9390	0.9070	0.8830	0.8960	0.8870	0.8860
	140	0.3	0	0.9380	0.9440	0.9470	0.9460	0.9450	0.8810	0.9030	0.8950	0.9080	0.8950
			0.1	0.9470	0.9420	0.9420	0.9500	0.9330	0.8980	0.8940	0.8820	0.8920	0.8840
			0.2	0.9370	0.9360	0.9540	0.9490	0.9350	0.8960	0.8910	0.9010	0.9030	0.8710
			0.3	0.9490	0.9300	0.9500	0.9410	0.9310	0.9070	0.8850	0.8970	0.8960	0.8750
			0.4	0.9440	0.9340	0.9500	0.9500	0.9390	0.8930	0.8790	0.9060	0.8970	0.8900
			0.5	0.9450	0.9340	0.9440	0.9390	0.9380	0.9050	0.8920	0.8920	0.8910	0.9010
			0.6	0.9440	0.9360	0.9420	0.9370	0.9470	0.8980	0.8850	0.8940	0.8900	0.8960
			0.7	0.9350	0.9320	0.9400	0.9250	0.9300	0.8820	0.8870	0.8900	0.8680	0.8900
			0.8	0.9550	0.9430	0.9540	0.9390	0.9430	0.9060	0.8950	0.8910	0.8900	0.8940
			0.9	0.9500	0.9530	0.9590	0.9480	0.9440	0.9020	0.9080	0.9120	0.8980	0.8940
	120	0.4	0	0.9500	0.9380	0.9550	0.9440	0.9400	0.8930	0.8840	0.9000	0.8900	0.8700
			0.1	0.9280	0.9410	0.9410	0.9560	0.9220	0.8880	0.8800	0.8960	0.9040	0.8560
			0.2	0.9360	0.9220	0.9420	0.9560	0.9270	0.8760	0.8740	0.8830	0.9140	0.8730
			0.3	0.9340	0.9350	0.9330	0.9470	0.9300	0.8850	0.9030	0.8890	0.8860	0.8900
			0.4	0.9340	0.9500	0.9390	0.9400	0.9260	0.8640	0.8920	0.8820	0.8920	0.8690
			0.5	0.9460	0.9390	0.9610	0.9540	0.9250	0.8930	0.9010	0.9230	0.9060	0.8860
			0.6	0.9460	0.9490	0.9600	0.9430	0.9420	0.8860	0.9100	0.9160	0.9000	0.9090
			0.7	0.9540	0.9260	0.9520	0.9390	0.9400	0.9140	0.8720	0.9040	0.8960	0.8970
			0.8	0.9370	0.9450	0.9490	0.9380	0.9370	0.8790	0.8780	0.8950	0.8830	0.9000
			0.9	0.9490	0.9280	0.9430	0.9370	0.9310	0.9020	0.8770	0.8990	0.8810	0.8790
	100	0.5	0	0.9430	0.9480	0.9490	0.9570	0.9320	0.8900	0.9010	0.9040	0.9000	0.8840
			0.1	0.9520	0.9350	0.9470	0.9620	0.9250	0.8930	0.8840	0.8990	0.9170	0.8760
			0.2	0.9490	0.9390	0.9570	0.9500	0.9240	0.8820	0.8840	0.8990	0.9020	0.8760
			0.3	0.9550	0.9320	0.9390	0.9590	0.9180	0.8930	0.8900	0.8760	0.9120	0.8610
			0.4	0.9390	0.9440	0.9500	0.9380	0.9370	0.8920	0.8960	0.8950	0.8820	0.8800
			0.5	0.9420	0.9480	0.9300	0.9400	0.9220	0.8740	0.8970	0.8890	0.8970	0.8800
			0.6	0.9470	0.9460	0.9520	0.9470	0.9280	0.9070	0.9050	0.9020	0.8900	0.8850
			0.7	0.9460	0.9360	0.9460	0.9430	0.9290	0.8970	0.8900	0.9020	0.8920	0.8880
			0.8	0.9250	0.9440	0.9520	0.9420	0.9260	0.8740	0.8990	0.8960	0.8920	0.8830
			0.9	0.9430	0.9330	0.9470	0.9380	0.9300	0.8910	0.8780	0.9060	0.8890	0.8790

14

Statistical Estimation Based on Interval Censored Data

M. S. Tikhov

Nizhny Novgorod State University, Russia

Abstract: Let $Z_1, Z_2, \ldots, Z_n$ be independent and identically distributed (iid) random variables (rv) with unknown distribution function $F(x)$; $U_1, U_2, \ldots, U_n$ be iid rv, independent of $\{X_i, 1 \le i \le n\}$, with unknown distribution function $G(x)$. We observe a sequence of identically distributed pairs $\mathcal{U}^{(n)} = \{(U_i, W_i), i = 1, 2, \ldots, n\}$, where $W_i = I(Z_i < X_i)$ is the indicator of the event $\{Z_i < X_i\}$ and (U_i, X_i) have joint density $g(u, x)$. The problem of estimating the distribution function $F(x)$ and its quantile of the order λ based on sample $\mathcal{U}^{(n)}$ is considered. The asymptotic behavior of the constructed estimators is discussed.

Keywords and phrases: Interval censored samples, kernel estimation

14.1 Introduction

In Krischtopenko and Tikhov (1997a) and Tikhov (2000), the methods for estimation of effective doses based on the data of binary responses were offered and analyzed, and associated algorithms were developed too. The effect from the action of one substance on an organism was estimated, and the mathematical model consists of the following. The random dose X of the substance is entered in to an organism. Let Z be the lower bound, at which the effect begins: if $Z \ge X$, the effect is absent; otherwise (i.e. when $Z < X$) the effect is present. In other words, we have a binary rv W which is an indicator of the event $(Z < X)$: $W = I(Z < X)$. As the outcome of the experiment we have a sample pairs of values $\mathcal{X}^{(n)} = \{(X_i, W_i), i = 1, 2, \ldots, n\}$. The characteristic of interest is the regression of W on X, i.e. conditional expectation $\mathbf{E}(W|\, X = x)$ of the rv W with the fixed value $X = x$. If Z and X are independent, then $\mathbf{E}(W|\, X = x) = F(x)$, where $F(x)$ is the cumulative distribution function (cdf) of Z. If random variables Z and X are dependent, the conditional expectation $\mathbf{E}(W|\, X = x) = \mathbf{P}(Z < x|\, X = x) = F(x|\, x) = T(x)$ is a conditional distribution function, which we name *toxicity function* in the case of analyzing toxic substances. For independent random variables, $T(x) = F(x)$ is a nondecreasing function. If Z and X are dependent, $T(x)$ has [0,1] as its domain of values. It can be decreasing on any interval in the domain of variable x. We call such toxicity functions $T(x)$ *paradoxical*: the increase of an entered dose results for some intervals in the domain of definition in a decrease of the effect. Practical toxicity functions are often paradoxical: at first, the increase of a dose of entered substance results in an increase of effect, then with a further increase of a dose the effect decreases and still greater increase of an entered dose increases the effect again. We propose a method of consistent pointwise estimation of the regression function $\mathbf{E}(W|\, X = x)$ whose functional form is completely unknown. Generally, nonparametric estimation methods for the regression function $\mathbf{E}(W|\, X = x)$ will be useful in preliminary analyses of sets of interval censored data, as a guide for parametric formulation of the regression relation; for the estimation of unknown regression coefficients in dose-effect dependence for several substances see Tikhov (1998, 1999) and Krischtopenko and Tikhov (1997b). When the main objective is prediction within the range of data and when the sample size is large, nonparametric estimation method will be useful.

The following class of statistics will be considered:

$$T_n(x) = \frac{S_{2n}(x)}{S_{1n}(x)},$$

where

$$S_{1n}(x) = \frac{1}{nh_n}\sum_{i=1}^{n} K\left(\frac{X_i - x}{h_n}\right) \text{ and } S_{2n}(x) = \frac{1}{nh_n}\sum_{i=1}^{n} W_i K\left(\frac{X_i - x}{h_n}\right).$$

Statistic $T_n(x)$ is used as the estimator of the function $T(x)$ based on the sample $\mathcal{X}^{(n)}$. Here, $K(x)$ is the kernel (i.e., symmetric nonnegative bounded finite function defined on the real line, such that $\int_{-\infty}^{\infty} K(x)\, dx = 1$), $h_n = M\, n^{-1/5}$ is the width of the window, and M is some constant. As the kernel we take an Epanechnikov kernel $K_0(x) = (3/4)(1 - x^2)I(|x| < 1)$. If $\sum_{i=1}^{n} K((U_i - x)/h_n) = 0$, we take $T_n(x) = 0$.

The estimator $T_n(x)$ is defined as the weighed mean of the variable response in a fixed vicinity of the point x, the weight being defined by the kernel $K(x)$ and the width of the window being h_n.

The *k-nearest neighbor estimator* represents a weighed mean in a variable vicinity. This vicinity is defined only by those values of variable X, which are *k-nearest* to x in Euclidean distance. It has the form [see Tikhov (1999)]

$$F_n^*(x) = \frac{1}{nh_n} \sum_{i=1}^{n} W_n^{[i]} K\left(\frac{i/n - G_n(x)}{h_n}\right),$$

where $\{W_n^{[i]}, 1 \leq i \leq n\}$ is induced order statistics: $W_n^{[i]} = W_j$ if $X_n^{(i)} = X_j$, and $X_n^{(i)}$ is the i-th ranked variable $\{X_j,\ j = 1, 2, \ldots, n\}$, and $G_n(x)$ is equal to the proportion of the X_j which are less than x.

In this paper, we are interested in a λ-effective dose, i.e., a value x_λ for which $T(x_\lambda) = \lambda$ $(0 < \lambda < 1)$, as well the asymptotic behavior (as $n \to \infty$) of estimators for the unknown values $T(x)$ and x_λ based on the sample $\mathcal{X}^{(n)}$.

14.2 Results

Theorem 14.2.1 *Suppose that $\{(X_i, Z_i), 1 \leq i \leq n\}$ are independent and identically distributed random two-dimensional vectors with joint distribution function $F(x)G(y)$, where $F(x) = \mathbf{P}(Z < x)$, $G(x) = \mathbf{P}(X < x)$, and joint density function $f(x)g(y)$; $F(x)$ and $g(x) > 0$ are assumed to be twice continuously differentiable.*

Then,

$$\frac{\sqrt{nh_n}(T_n(x) - F(x))}{b(x)\nu} \xrightarrow{d} \zeta_1 \in N(M^{5/2}a(x), 1) \qquad (n \to \infty),$$

where

$$b^2(x) = \frac{F(x)(1 - F(x))}{g(x)},$$

$$a(x) = f'(x) + 2f(x)(g'(x)/g(x)), \quad \nu^2 = \int_{-\infty}^{\infty} K^2(x)\, dx.$$

Let

$$\tilde{S}_{1n}(x) = \frac{1}{nh_n^3} \sum_{i=1}^{n} K''\left(\frac{X_i - x}{h_n}\right) \text{ and } \tilde{S}_{2n}(x) = \frac{1}{nh_n^3} \sum_{i=1}^{n} W_i K''\left(\frac{X_i - x}{h_n}\right),$$

where

(U1) Function $K(x)$ with support $[-B, B]$ has derivatives $K'(x)$, $K''(x)$, and $K'(-B) = K'(B) = 0$, $K(-B) = K(B) = 0$.

Let us consider the statistics

$$V_{1n}(x) = S_{1n}(x) - \frac{1}{\sqrt{nh_n}} M^{5/2} d^2 \tilde{S}_{1n}(x),$$

$$V_{2n}(x) = S_{2n}(x) - \frac{1}{\sqrt{nh_n}} M^{5/2} d^2 \tilde{S}_{2n}(x), \quad d^2 = \int_{-\infty}^{\infty} K^2(x)\, dx.$$

Theorem 14.2.2 *Under the conditions of Theorem 14.2.1 and condition* **(U1)**,

$$\sqrt{nh_n}\left(\frac{V_{2n}(x)}{V_{1n}(x)} - F(x)\right) \xrightarrow{d} \zeta_2 \in N\left(0, \frac{F(x)(1-F(x))\nu^2}{g(x)}\right) \qquad (n \to \infty).$$

Theorem 14.2.3 *Let $\{(X_i, Z_i), 1 \le i \le n\}$ be independent and identically distributed random two-dimensional vectors, and $W_i = I(Z_i < X_i)$. Suppose that the function $F(G^{-1}(t))$ has bounded third derivative.*

Then,

$$\frac{\sqrt{nh_n}\,(F_n^*(x) - F(x))}{\sqrt{F(x)(1-F(x))\nu^2}} \xrightarrow{d} \xi \in N(0,1) \qquad (n \to \infty).$$

The following theorem establishes the asymptotic normality of estimators of the quantile of order λ of distribution $F(x)$ based on the sample $\mathcal{X}^{(n)}$.

Theorem 14.2.4 *Let the conditions of Theorem 14.2.3 be satisfied. Suppose the sample $\{(X_i, W_i), 1 \le i \le n\}$ is observed, $\hat{x}_\lambda$ is a solution of the equation $F_n^*(\hat{x}_\lambda) = \lambda$ $(0 < \lambda < 1)$, and x_λ is quantile of the order λ of distribution $F(x)$.*

Then,

$$\sqrt{nh_n}(\hat{x}_\lambda - x_\lambda) \xrightarrow{d} \zeta_3 \in N\left(0, \frac{\lambda(1-\lambda)\nu^2}{f^2(x_\lambda)}\right) \qquad (n \to \infty).$$

Let $Z_1, Z_2, \ldots, Z_n$ be independent and identically distributed (iid) rv with unknown distribution function $F(x)$; $U_1, U_2, \ldots, U_n$ be iid r.v, independent of $\{X_i, 1 \le i \le n\}$, with unknown distribution function $G(x)$. We observe a sequence of identically distributed pairs $\mathcal{U}^{(n)} = \{(U_i, W_i^*), i = 1, 2, \ldots, n\}$, where $W_i^* = I(Z_i < X_i)$ is the indicator of the event $\{Z_i < X_i\}$, with (U_i, X_i) having a joint density function $g(u, x)$. Then,

$$q(u) = \int_{-\infty}^{\infty} g(u,x)\, dx > 0, \quad g(x) = \int_{-\infty}^{\infty} g(u,x)\, du > 0.$$

Let $q(u|x) = g(u,x)/g(x)$ be the conditional density function of U given the condition $X = x$, and $g(x|u) = g(u,x)/q(u)$ be the conditional density function of X given the condition $U = u$.

We set

$$R(u) = \int_{-\infty}^{\infty} F(x) q(x|\, u)\, dx = \frac{m(u)}{q(u)}.$$

Define $\hat{F}_n(x)$ as

$$\hat{F}_n(x) = \frac{S_{2n}^*(x)}{S_{1n}^*(x)},$$

where

$$S_{1n}^*(x) = \frac{1}{nh_n}\sum_{i=1}^{n} K\left(\frac{U_i - x}{h_n}\right) \text{ and } S_{2n}^*(x) = \frac{1}{nh_n}\sum_{i=1}^{n} W_i^* K\left(\frac{U_i - x}{h_n}\right).$$

Theorem 14.2.5 *Let (Z_i, U_i, X_i), $i = 1, 2, \ldots, n$, be independent and identically distributed three-dimensional random vectors with joint absolutely continuous distribution function $F(z)G(u,x)$, $(z,u,x) \in R^3$ and $h_n = Mn^{-1/5}$. Assume further the validity of the following conditions:*

(A1) *$G(u,x)$ has bounded density function $g(u,x) > 0$;*

(A2) *There exists a constant C_3 such that the third derivative of $q(u)$ satisfies the inequality $|q'''(u)| \leq C_3$.*

Then

$$\sqrt{nh_n}(T_n(x) - R(x)) \xrightarrow{d} N(A(x), B^2(x)\nu^2) \qquad (n \to \infty),$$

where

$$A(x) = \frac{m''(x)q(x) - q''(x)m(x)}{q^2(x)} \text{ and } B^2(x) = \frac{R(x)(1 - R(x))}{q(x)}.$$

The following theorem establishes the asymptotic normality of estimators for the solution x_λ of the equation $T(x_\lambda) = \lambda$ based on the sample $\mathcal{U}^{(n)}$.

Theorem 14.2.6 *Let the conditions of Theorem 14.2.5 hold. Suppose the sample $\{(U_i, W_i), 1 \leq i \leq n\}$ is observed, where $W_i = I(Z_i < X_i)$, $\hat{x}_\lambda$ is the solution of the equation $T_n(\hat{x}_\lambda) = \lambda$ $(0 < \lambda < 1)$, and x_λ is the solution of the equation $T(x_\lambda) = \lambda$.*

Then,

$$\sqrt{nh_n}\,(\hat{x}_\lambda - x_\lambda) \xrightarrow{d} N\left(A(x_\lambda), \frac{\lambda(1-\lambda)\nu^2}{g(x_\lambda)f^2(x_\lambda)}\right) \qquad (n \to \infty).$$

14.3 Examples

Example 14.3.1 Let $U = X + \varepsilon$, where X and ε are independent and normally distributed as $N(a, \sigma^2)$ and $N(0, \sigma_0^2)$ with known variance σ_0^2. Here, the conditional density function of X given the condition $U = u$ is

$$q(x|u) = \frac{1}{\rho_1}\varphi\left(\frac{x-\mu_1(u)}{\rho_1}\right) = \frac{1}{\sqrt{2\pi}\,\rho_1}\exp\left(-\frac{(x-\mu_1(u))^2}{2\rho_1^2}\right),$$

where

$$\mu_1(u) = \frac{a\sigma_0^2}{\sigma_0^2+\sigma^2} + \frac{\sigma^2 u}{\sigma_0^2+\sigma^2} \text{ and } \rho_1^2 = \frac{\sigma_0^2\sigma^2}{\sigma_0^2+\sigma^2}$$

Assume that the random variable Z has normal $N(\mu, \rho^2)$ distribution with distribution function $\Phi((x-\mu)/\rho)$. As $n \to \infty$, the sequence $\{\hat{F}_n(x), n \geq 1\}$ converges to

$$R(v) = \int_{-\infty}^{\infty} \Phi\left(\frac{u-\mu}{\rho}\right)\frac{1}{\rho_1}\Phi\left(\frac{\mu_1(v)-u}{\rho_1}\right)du = \Phi\left(\frac{x-\mu_1(v)}{\sqrt{\rho^2+\rho_1^2}}\right).$$

If $\sigma_0^2\rho^{-2} \ll 1$, for a given x we shall take

$$v^* = x + (x+\bar{u})\frac{\sigma_0^2}{s^2-\sigma_0^2}, \quad \bar{u} = \frac{1}{n}\sum_{i=1}^n u_i \text{ and } s^2 = \sum_{i=1}^n (u_i - \bar{u})^2.$$

The value of $\hat{F}_n(v^*)$ at the point v^* based on the sample $\{(U_i, W_i^*), 1 \leq i \leq n\}$ is used as the estimate $\Phi((x-\mu)/\rho)$ at the point x. Then the estimator $\hat{v}_\lambda^*$ of the quantile v_λ^* of the given order $0 < \lambda < 1$ is determined as the solution of the equation $\hat{F}_n(\hat{v}_\lambda^*) = \lambda$.

We set

$$A_\alpha = t_{1-\alpha/2}(n-1))S_{\hat{v}_\lambda}, \quad S_v = \frac{\hat{\sigma}_n(v)}{f_n(v)}\frac{1}{\sqrt{nh_n}}, \quad \hat{\sigma}_n(v) = \frac{3\lambda(1-\lambda)}{5q_n(v)},$$

$$q_n(v) = \frac{1}{nh_n}\sum_{i=1}^n K_o\left(\frac{u_i - v}{h_n}\right) \text{ and } f_n(v) = \frac{|\hat{F}_n(v+h_n/2) - \hat{F}_n(v-h_n/2)|}{h_n},$$

where $t_\alpha(n)$ is the quantile of order α of Student's t-distribution with n degrees of freedom. We shall then determine the confidence interval for v_λ^* as

$$(\hat{v}_\lambda^* - A_\alpha, \hat{v}_\lambda^* + A_\alpha).$$

Then, a confidence interval for the quantile x_λ of order λ of the $N(\mu;\rho^2)$ distribution, with confidence coefficient $\gamma = 1-(\alpha+\beta)$, is

$$(\hat{v}_\lambda^* - A_\alpha)\frac{s^2-\sigma_0^2}{s^2} + \left(\bar{y} - t_{1-\beta/2}(n-1)\frac{s}{\sqrt{n}}\right)\frac{\sigma_0^2}{s^2} \le x_\lambda$$

$$\le (\hat{v}_\lambda^* + A_\alpha)\frac{s^2-\sigma_0^2}{s^2} + \left(\bar{y} + t_{1-\beta/2}(n-1)\frac{s}{\sqrt{n}}\right)\frac{\sigma_0^2}{s^2}\,.$$

Example 14.3.2 Let $U\cdot\eta = X$, where η has an uniform distribution on the interval $(0,1)$ and η is independent of U and X. Then, the conditional distribution function equals $G(x|\,u) = x/u, \quad 0<x<u$, the conditional density function is $g(x|\,u) = 1/u, \quad 0<x<u$, and $R(v) = \int\limits_{-\infty}^{\infty} F(x)\,g(x|\,v)\,dx = v^{-1}\int\limits_0^v F(x)\,dx$. Hence, if $F(x)$ is a continuous function, then $v\,R(v) = \int\limits_0^v F(x)\,dx$. Differentiating both sides with respect to v, we obtain $R(v) + v\,R'(v) = F(v)$. In this case, as an estimator for $F(v)$, we shall use the statistic

$$\hat{F}_n(v) = T_n(v) + v\,r^*(v),$$

where

$$r^*(v) = \frac{T_n(v+h_n/2) - T_n(v-h_n/2)}{h_n}.$$

The estimator $\hat{x}_\lambda$ of x_λ can be obtained from $\hat{F}_n(\hat{x}_\lambda) = \lambda$, where x_λ is the solution of the equation $F(x_\lambda) = \lambda$.

As

$$\sqrt{nh_n}\,(\hat{x}_\lambda - x_\lambda) = \zeta_n + \delta_n, \quad \delta_n \xrightarrow{p} 0 \quad (n\to\infty),$$

where

$$\delta_n = -\frac{\sqrt{nh_n}\,(\tilde{F}_n(x_\lambda) - F(x_\lambda))}{f(x_\lambda)},$$

the asymptotic behavior (as $n\to\infty$) of the sequence $\sqrt{nh_n}\,(\hat{x}_\lambda - x_\lambda)$ is determined by the asymptotic behavior of sequence $\zeta_n \xrightarrow{d} \zeta_3 \quad (n\to\infty)$.

References

1. Krischtopenko, S. V. and Tikhov, M. S. (1997a). *Toxicometry of effective doses*, NNSU, Nizhny Novgorod.

2. Krischtopenko, S. V. and Tikhov, M. S. (1997b). Statistical estimation of parameters of a multiple regression in dose-effect dependence, *Sixth Conference on "Appl. MSA"*, pp. 69–70, CEMI RAS, Moscow.

3. Tikhov, M. S. (1998). Estimation of regression coefficients in dose-effect dependence, *Surveys in Applied and Industrial Mathematics*, **5**, 288–289.

4. Tikhov, M. S. (1999). Linear functions of induced order statistics and nonparametric estimation of distributions in dose-effect dependence, *Surveys in Applied and Industrial Mathematics*, **6**, 244–245.

5. Tikhov, M. S. (2000). Statistical estimation based on interval censored data, In *Statistical Methods of Estimation and Testing Hypotheses*, pp. 49–70, Perm, PSU.

PART VI

Regression Methods and Applications

15

The Covariate Order Method for Nonparametric Exponential Regression and Some Applications in Other Lifetime Models

Jan Terje Kvaløy[1] **and Bo Henry Lindqvist**[2]

[1] *Department of Mathematics and Natural Science, Stavanger University College, Stavanger, Norway*
[2] *Department of Mathematical Sciences, Norwegian University of Science and Technology, Trondheim, Norway*

Abstract: A new method for nonparametric censored exponential regression, called the covariate order method, is presented. It is shown that the method leads to a consistent estimator of the hazard rate as a function of the covariate. Moreover, interesting applications to more general cases of lifetime regression are presented. Possible applications include the construction of tests for covariate effect and estimation and residual plots in Cox regression models. The key is here to perform suitable transformations to exponentiality before applying the covariate order method.

Keywords and phrases: Hazard regression, nonparametric estimation, kernel estimation, model checking, Cox–Snell residuals, residual plots

15.1 Introduction

Suppose the lifetime of a unit has a distribution which depends on a covariate vector $\mathbf{x}$. Hazard regression means to estimate the hazard as a function of both time and of the covariate vector, based on censored survival data. Exponential regression is the special case when the hazard functions $\lambda(\mathbf{x})$ are constant in time.

Apparently, exponential regression models should be easier to fit than more general hazard regression models because of the time-independence in the hazard. On the other hand, it is often possible to transform lifetime data in a sensible way to follow, at least approximately, an exponential regression model. Thus it might be a good idea to use statistical methods for exponential regression to solve problems in more general hazard regression models. This is a major motivation for the present paper.

The literature contains a number of estimation methods for censored exponential regression. Parametric estimation is most conveniently done by fitting a generalized linear model. Various approaches which can be used for nonparametric estimation of $\lambda(\mathbf{x})$ have furthermore been suggested. For example, [Hastie and Tibshirani (1990)] consider estimation in generalized additive models as a natural nonparametric extension of generalized linear models. Other approaches are reviewed in Kvaløy and Lindqvist (2003).

In this paper we study a new nonparametric method for exponential regression, called the covariate order method. As will be clear from the presentation in the next section, the covariate order method in its basic form rests heavily on the assumption of exponentially distributed lifetimes. In fact, the estimate of $\lambda(\mathbf{x})$ would have no meaning if the same procedure was tried on non-exponential lifetimes. However, as indicated above, many problems can be reduced to exponential regression by transforming the data. The covariate order method has turned out to be a useful approach in such applications. For example, Kvaløy (2002) used the covariate order method to suggest tests for covariate effect in general censored regression models (see Section 15.2.4 of the present paper), while Kvaløy and Lindqvist (2003) used the covariate order method in nonparametric estimation of covariate functions in Cox regression (see Section 15.3).

The main purpose of the present paper is to give a formal presentation of the covariate order method and its practical implementation (Sections 15.2.1-15.2.3), and in addition to give a rigorous proof of consistency of the method in the single covariate case (Section 15.4). In order to illustrate the direct method we give an example with exponential data in Section 15.2.5. Sections 15.3.1 and 15.3.2 illustrate the use of the covariate order method to transformed data. More precisely it is shown how to make illustrative residual plots based on Cox–Snell residuals in Cox regression models, and how the method can be used to suggest possible transformations of covariates.

15.2 The Covariate Order Method for Exponential Regression

The basic formulation of the problem is as follows. Assume that we have n independent observations $(T_1, \delta_1, \mathbf{X}_1), \ldots, (T_n, \delta_n, \mathbf{X}_n)$ of the random triple $(T, \delta, \mathbf{X})$, where $T = \min(Z, C)$, $\delta = I(Z \leq C)$ and $\mathbf{X}$ is a vector of covariates. For given $\mathbf{X} = \mathbf{x}$, Z is assumed to be exponentially distributed with an *unknown* hazard rate $\lambda(\mathbf{x})$, that is $f_Z(t|\mathbf{x}) = \lambda(\mathbf{x}) \exp(-\lambda(\mathbf{x})t)$.

Further, C is distributed according to some unknown censoring distribution $f_C(t|\mathbf{x})$ which may depend on $\mathbf{x}$, and C is assumed to be independent of Z given $\mathbf{X}$. Let Z be called the *lifetime*, C the *censoring time* and T the *observation time*. This terminology is introduced only for convenience; Z can be any kind of exponentially distributed variables.

The domain of the covariate vector $\mathbf{X}$ is a subset $\mathcal{X}$ of $\mathbb{R}^m$, and $\mathbf{X}$ is assumed to be distributed according to some density function $f_{\mathbf{X}}(\mathbf{x})$. The corresponding cumulative distribution function is denoted $F_{\mathbf{X}}(\mathbf{x})$. The covariates are assumed to remain constant over time, and $\lambda(\mathbf{x})$ is assumed to be *continuous* on $\mathcal{X}$. The method is first described for the case of a single covariate, in other words for $m = 1$. Extensions to higher dimensions are discussed in Section 15.2.3.

15.2.1 Method description and main theoretical results

The method proceeds conditionally on $X_1, \ldots, X_n$ and starts by first arranging the observations $(T_1, \delta_1, X_1), \ldots, (T_n, \delta_n, X_n)$ such that $X_1 \le X_2 \le \cdots \le X_n$. Next, for convenience, divide the observation times by the number of observations, n. Then let the scaled observation times $T_1/n, \ldots, T_n/n$, irrespectively if they are censored or not, be subsequent inter-arrival times of an artificial point process on a time axis s. For this process, let points which are endpoints of intervals corresponding to *uncensored* observations be considered as events, occurring at times denoted $S_1, \ldots, S_r$ where $r = \sum_{j=1}^{n} \delta_j$. This is visualized in Figure 15.1, for an example where the ordered observations are $(T_1, \delta_1 = 1), (T_2, \delta_2 = 0), (T_3, \delta_3 = 1), \ldots, (T_{n-1}, \delta_{n-1} = 0), (T_n, \delta_n = 1)$.

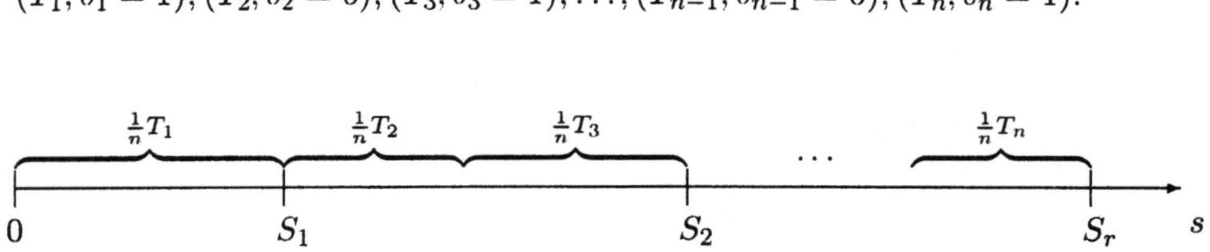

Figure 15.1: Construction of artificial process

More precisely, $S_i = \sum_{j=1}^{k(i)} T_j/n$ where $k(i) = \min\{s | \sum_{j=1}^{s} \delta_j = i\}$. Now the conditional intensity of the process $S_1, \ldots, S_r$ at a point w on the s-axis, given the complete history of the T_j up to s, equals $n\lambda(X_I)$ where I is defined from $\sum_{i=1}^{I-1} T_i/n < w \le \sum_{i=1}^{I} T_i/n$. The basic idea is to estimate this intensity from the process $S_1, \ldots, S_r$, yielding the estimator $\hat{\rho}_n(w)$, and then invert the relation $n\hat{\lambda}(X_I) = \hat{\rho}_n(w)$ to obtain an estimate of $\hat{\lambda}(x)$ at given points x. The key here is the relationship between $X_1, \ldots, X_n$ on the "covariate-axis" and the process $S_1, \ldots, S_r$ on the "s-axis". A possible way of estimating such a relationship is to use the step-function

$$\tilde{s}(x) = \frac{1}{n}\sum_{i=1}^{j} T_i, \quad X_j \leq x < X_{j+1}, \tag{15.1}$$

see Figure 15.2 for an illustration, and then define $\hat{\lambda}(x) = \hat{\rho}_n(\tilde{s}(x))/n$.

The motivating idea of the method is that if $\lambda(x) = \lambda$ is constant, then the process $S_1, \ldots, S_r$ is a homogeneous Poisson process. (The test presented in Section 15.2.4 is in fact based on this observation.) Thus if $\lambda(x)$ is reasonably smooth and not varying too much, then the process $S_1, \ldots, S_r$ could be imagined to be nearly a nonhomogeneous Poisson process for which the intensity can be estimated by for instance kernel density estimation based on the points $S_1, \ldots, S_r$. Combining this kernel estimate and (15.1) leads to an estimate of $\lambda(x)$. The estimator arising from this heuristic reasoning is the one presented below, but more precise arguments are needed to derive the estimator formally and to prove its consistency. All proofs are given in Section 15.4.

Let $\mathcal{F}_s^n$ be the history of the process $S_1, \ldots, S_r$ in the interval $[0, s)$. This history is formally defined as the sub-σ-algebra $\mathcal{F}_s^n = \sigma\{X_1, \ldots, X_n\} \cup \sigma\{S_j : S_j \leq s\}$ for $s \geq 0$. Note that $X_1, \ldots, X_n$ is contained in all the $\mathcal{F}_s^n$. Let $\rho_n(s|\mathcal{F}_s^n)$ be the conditional intensity of the process $S_1, \ldots, S_r$ at the point s [Andersen et al. (1993, p. 75)]. Then the first step in the formal derivation of a consistent estimator for $\lambda(x)$ is Theorem 15.2.1 below. This theorem states that the scaled conditional intensity of the process $S_1, \ldots, S_r$ converges in probability to a deterministic function of $\lambda(\cdot)$, and gives an asymptotic relation between the processes running on the s-axis and the covariate axis respectively.

Theorem 15.2.1 *Let the situation be as described above and in the formulation of the problem at the beginning of the section. Further assume that* $\sup_{x\in\mathcal{X}} \lambda(x) \leq M < \infty$, $\inf_{x\in\mathcal{X}} \lambda(x) \geq a > 0$, *and that* $\sup_{x\in\mathcal{X}} \lambda'(x) \leq D < \infty$. *The conditional distribution of C given x is assumed to have finite first and second order moments and $f_C(t|x)$ is assumed to have bounded first derivative in x for all $x \in \mathcal{X}$. Then*

$$\rho_n(s|\mathcal{F}_s^n)/n \xrightarrow{P} \lambda(\eta(s))$$

as $n \to \infty$ uniformly in s, where $\eta(s)$ is a deterministic function from the s-axis to the covariate axis, the inverse of which is given by

$$s(x) = E(TI(X \leq x)).$$

The function $s(x)$ is called the correspondence function. Note that for the special case of no censoring, $s(x)$ can be written $s(x) = \int_{-\infty}^{x}(f_X(v)/\lambda(v))dv$.

The fact that the scaled conditional intensity of the process $S_1, \ldots, S_r$ converges uniformly to $\lambda(\eta(s))$ can be used to derive an estimator for $\lambda(x)$ by estimating the inverse function $s(x)$ and $\rho_n(s|\mathcal{F}_s^n)/n$. As a first step we state the following lemma.

Lemma 15.2.1 *Let the situation be as in Theorem 15.2.1. Then $\tilde{s}(x)$ in (15.1) is a uniformly consistent estimator of $s(x)$.*

Finally, a uniformly consistent estimator of $\lambda(x)$ is established by the following theorem.

Theorem 15.2.2 *Let the situation be as in Theorem 15.2.1. Further let $K(\cdot)$ be a positive kernel function which vanishes outside [-1,1] and has integral 1, and let h_s be a smoothing parameter which is either constant or varying along the s-axis. Assume that $h_s \to 0$ as $n \to \infty$ for all s. Further assume that there is a sequence h_n such that $h_s \geq h_n$ for all s, n where $nh_n \to \infty$ as $n \to \infty$. Then the estimator*

$$\hat{\lambda}(x) = \frac{1}{nh_s}\sum_{i=1}^{r} K\left(\frac{\tilde{s}(x) - S_i}{h_s}\right) \ ; \quad x \in \mathcal{X} \tag{15.2}$$

is a uniformly consistent estimator of $\lambda(x)$.

15.2.2 Smoothing details

In practical use the estimated correspondence function (15.1) may be replaced by more sophisticated estimators, $\hat{s}(x)$, improving on the smoothness of the estimator (15.2). We have used the super-smoother of Friedman (1984), but in practice this choice of smoother is not important.

To avoid the estimate $\hat{\lambda}(x)$ to be seriously downward biased near endpoints special care must be taken at the boundaries. Viewed only as a problem on the s-axis the estimator (15.2) is simply (scaled) density estimation on the s-axis, and techniques for handling boundary problems in density estimation can be adopted. A common technique is to reflect the data points around both endpoints, see for example Silverman (1986), corresponding to using the estimator

$$\hat{\lambda}(x) = \frac{1}{nh_s}\sum_{i=1}^{r}\left[K\left(\frac{\hat{s}(x) - S_i}{h_s}\right) + K\left(\frac{\hat{s}(x) + S_i}{h_s}\right) + K\left(\frac{\hat{s}(x) + S_i - 2S}{h_s}\right)\right], \tag{15.3}$$

where $S = \sum_{j=1}^{n} T_j/n$.

The smoothing parameter h_s corresponds to smoothing over a certain amount of the data on the s-axis. On the covariate axis, a corresponding smoothing parameter h_x which covers approximately the same amount of the data can be defined via the relation between the points on the s-axis and the covariate axis. See the right plot in Figure 15.2 for a rough description of the idea. If one of

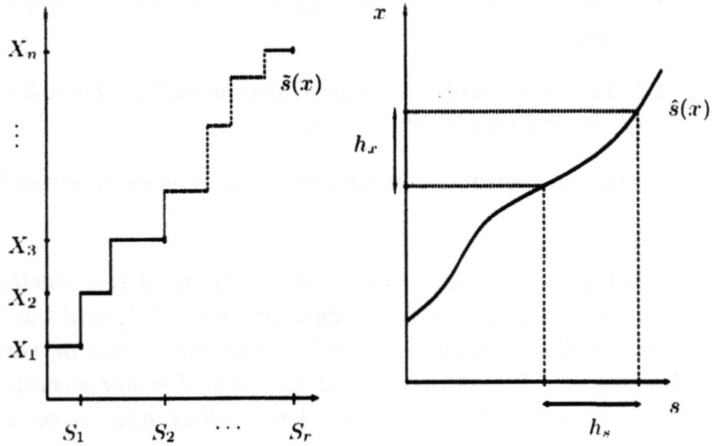

Figure 15.2: The left plot shows an example of what the estimated correspondence function $\tilde{s}(x)$ (15.1) might look like. The right plot illustrates a smoothed correspondence function estimate $\hat{s}(x)$ and the relationship between the smoothing parameter on the covariate axis and the s-axis.

the smoothing parameters, h_s or h_x, is held constant, the other will in general be varying (or both can be varying). Whereas a constant h_s corresponds to ordinary density estimation on the s-axis, a constant h_x corresponds to what is commonly used in nonparametric regression methods. If a constant h_x is used, then (15.3) becomes

$$\hat{\lambda}(x) = \frac{1}{nh_s(\hat{s}(x))} \sum_{i=1}^{r} \left[K(\frac{\hat{s}(x) - S_i}{h_s(\hat{s}(x))}) + K(\frac{\hat{s}(x) + S_i}{h_s(\hat{s}(x))}) + K(\frac{\hat{s}(x) + S_i - 2S}{h_s(\hat{s}(x))}) \right], \tag{15.4}$$

where $h_s(\hat{s}(x)) = \hat{s}(x + h_x/2) - \hat{s}(x - h_x/2)$. For instance likelihood cross-validation can be used as criterion for choosing the "best" value of the smoothing parameter.

15.2.3 Several covariates

The covariate order method is not directly generalizable to higher dimensions, mainly because R^m is not linearly ordered for $m > 1$. Thus instead we suggest to reduce the dimension of the problem by assuming some structure on the covariate space. One way to proceed is to assume that the hazard rate can be written in the form of a generalized additive model

$$\lambda(\mathbf{x}) = \exp(\alpha + g_1(x_1) + \cdots + g_m(x_m)), \tag{15.5}$$

where $\mathbf{x} = (x_1, \ldots, x_m) \in \mathcal{X} \subseteq \mathbb{R}^m$, and where $g_1(\cdot), \ldots, g_m(\cdot)$ are unspecified smooth functions. These functions can be estimated by the covariate order method using an iterative backfitting algorithm. The key point is that if Z is exponentially distributed with parameter $\exp(\alpha + g_1(x_1) + \cdots + g_m(x_m))$, then $Z \exp(\alpha + g_1(x_1) + \cdots + g_{j-1}(x_{j-1}) + g_{j+1}(x_{j+1}) + \cdots + g_m(x_m))$ will be exponentially distributed with parameter $\exp(g_j(x_j))$. Also note that it is possible to let some of the g-functions be parametric, for instance for discrete covariates.

15.2.4 Testing for covariate effect

Recall from Section 15.2.1 that if there is no covariate effect, that is $\lambda(x) \equiv \lambda$, then the process $S_1, \ldots, S_r$ is a homogeneous Poisson process (HPP). This observation suggests that in principle any statistical test for the null hypothesis of an HPP versus various non-HPP alternatives can be applied to test for covariate effect in exponential regression models. Moreover, such an approach can be extended to non-exponentially distributed lifetimes by transforming the observation times to approximately exponentially distributed data.

A detailed account of this approach for testing for covariate effect in lifetime data is given by Kvaløy (2002), who presents a number of different tests based on the covariate order method. The recommendation is to use an Anderson–Darling type test which turns out to have very good power properties against both monotonic and non-monotonic alternatives to constant $\lambda(x)$.

15.2.5 Example: Cardiac arrest versus air temperature

We give an example of direct application of the covariate order method to data for times of out-of-hospital cardiac arrests reported to a Norwegian hospital over a five years period. The relationship between outdoor air temperature and the occurrence of cardiac arrest is investigated. A simple first analysis of this relationship is done by regarding inter-event times to be independent and exponentially distributed with a hazard rate $\lambda(x)$ depending on the temperature x. The average temperature on the day of a cardiac arrest is used as covariate for the next period between cardiac arrests. A total of 449 cardiac arrests were reported during the five years period.

Testing the significance of the covariate effect of temperature by using the Anderson–Darling test for covariate effect mentioned in Section 15.2.4 yielded a p-value of 0.002. Plots of the estimated model are displayed in Figure 15.3. The estimated hazard rate function clearly indicates a decreasing hazard for increasing temperature. The smoothing parameter $h_x = 15$ was chosen by a likelihood cross validation criterion. The bootstrap curves indicate little variability in the estimated hazard rate in the middle temperature range where most of the observations are located, while there is large variability at the boundaries as expected.

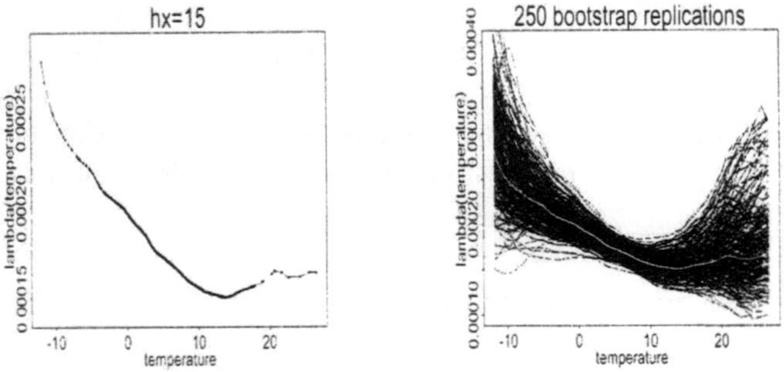

Figure 15.3: Analysis of cardiac arrest occurrence versus air temperature. The left plot shows the estimated hazard rate function obtained using a constant smoothing parameter on the x-axis, with the location of the observations along the curve displayed by the dots. The right plot shows 250 bootstrap curves obtained by resampling observations (original estimate shown as white curve).

15.3 Applications in Cox Regression

The covariate order method has various applications in Cox regression [Cox (1972)]. For instance, consider the generalized Cox model with hazard function $\alpha(t|\mathbf{x}) = \alpha_0(t)\exp(g(\mathbf{x}))$ where $g(\mathbf{x})$ in principle is any smooth function of the covariate vector $\mathbf{x}$. If Z is an uncensored observation from this model, then it is well known that given X the transformed variable $A_0(Z)\exp(g(\mathbf{X}))$ is exponentially distributed with parameter 1, where $A_0(t) = \int_0^t \alpha_0(u)du$. It follows that $A_0(Z)$ is exponentially distributed with parameter $\exp(g(\mathbf{X}))$, which suggests that $g(\mathbf{x})$ can be estimated from data by methods for nonparametric exponential regression. Kvaløy and Lindqvist (2003) show how the covariate order method in this way can be extended to estimation of $g(\mathbf{x})$ in the generalized Cox model (see their paper for details). Here we will concentrate on a similar application to residual plots in the Cox model.

15.3.1 Model checking and model fitting in classical Cox regression

In the Cox proportional hazards model with fixed covariates we have $\alpha(t|\mathbf{x}) = \alpha_0(t)\exp(\boldsymbol{\beta}\mathbf{x})$, where $\boldsymbol{\beta}$ is a vector of regression coefficients. It follows from the above that $r_i = A_0(T_i)\exp(\boldsymbol{\beta}\mathbf{X}_i)$, $i = 1,\ldots,n$, is a censored sample from the exponential distribution with parameter 1. The Cox–Snell residuals [Cox and Snell (1968)] $\hat{r}_i$ are defined by substituting standard estimators $\hat{A}_0(\cdot)$ and $\hat{\boldsymbol{\beta}}$ for $A_0(\cdot)$ and $\boldsymbol{\beta}$ in the expression for r_i. These residuals are mainly used to assess an overall fit by checking whether $(\hat{r}_1, \delta_1), \ldots, (\hat{r}_n, \delta_n)$ is compatible with a (censored) sample from an exponential distribution. However, we shall see that by the covariate order method we can obtain interesting residual plots which are similar to the plots routinely used in ordinary linear regression models. An advantage of our method is that censored observations are treated in a consistent way.

For instance, for each single covariate X_k, say, we may fit an exponential regression model to the data $(\hat{r}_1, \delta_1, X_{1k}), \ldots, (\hat{r}_n, \delta_n, X_{nk})$, where X_{ik} is the kth covariate for the ith observation unit. The covariate order method as described in Section 15.2 gives an estimated hazard rate as a function of X_k which, if the model is correct, is expected to be approximately constant at 1. Deviations from a constant hazard rate indicate a possibly wrong model and can be investigated visually from the plots, or tested more formally by for instance the Anderson–Darling test described in Section 15.2.4.

A related application is to make plots of log hazard rates against covariates not included in the model. Such plots can reveal whether these covariates should be included in the model, and in this case indicate the appropriate functional form of the covariate. This is a simple and intuitive alternative to the plotting of martingale residuals Therneau, Grambsch and Fleming (1990)commonly used for this purpose. A somewhat related approach, but using nonparametric Poisson regression instead of exponential regression, was used by Grambsch, Therneau and Fleming (1995), see also Therneau and Grambsch (2000, chapter 5).

15.3.2 Example: PBC data

We illustrate the use of the covariate order method in the classical Cox model by considering model fitting and model checking for the PBC data from the Mayo Clinic. PBC (primary biliary cirrhosis) is a fatal chronic liver disease, and out of the 418 patients followed in the study, 161 died before study closure. A listing of the data can be found in Fleming and Harrington (1991). The final model proposed by Fleming and Harrington (1991) includes the five covariates age, edema, log(bilirubin), log(protime) and log(albumin).

For a demonstration of residual plotting we will look closer at the covariate bilirubin. First we fitted a Cox model including the five covariates mentioned above, but where the covariate bilirubin was included without making the log transformation. The left plot in Figure 15.4 shows, for this model, the log of the estimated hazard rate of the Cox–Snell residuals against bilirubin. The p-value $2 \cdot 10^{-6}$ reported in the plot was calculated using the Anderson–Darling test described in Section 15.2.4. The low value certainly indicates a significant deviation from constancy, which is also clear from the plot. Thus the covariate is not well modeled. The right plot shows the corresponding plot for a model where the bilirubin covariate is added as log(bilirubin). We see that the bilirubin

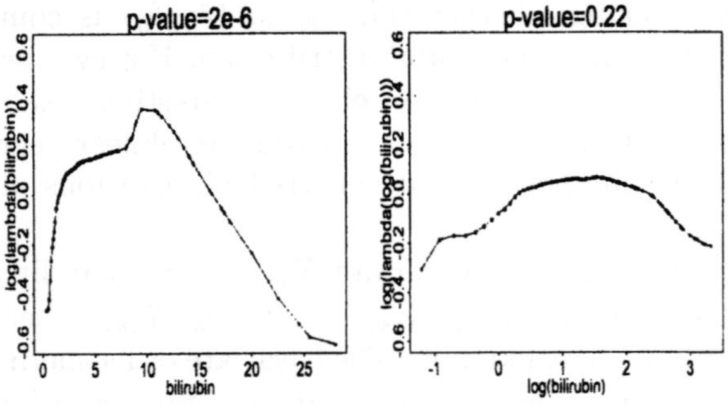

Figure 15.4: Residual analysis of PBC data. Plot of the log of the estimated hazard rate of the Cox-Snell residuals against bilirubin in a model using bilirubin on its original scale (left) and the same plot against log(bilirubin) in a model using log(bilirubin) (right).

covariate now seems to be much better modeled.

As explained in the previous subsection, one may use similar plots to suggest the functional form of covariates before they are entered into the model. Figure 15.5 displays plots of the log of the estimated hazard rate of the Cox–Snell residuals from an empty model versus, respectively, age, bilirubin and log(bilirubin). Note that in this case the Cox–Snell residuals are simply $\hat{A}_0(T_i)$, where $\hat{A}_0(\cdot)$ is the Nelson–Aalen estimator of the cumulative hazard in the empty model. The (approximate) straight line seen for the plot against

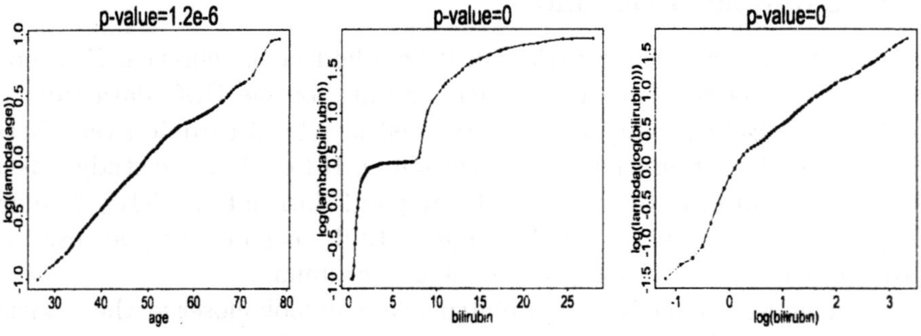

Figure 15.5: Functional form analysis in PBC data. Plots of the log of the estimated hazard rate of the Cox-Snell residuals from an empty model versus respectively age, bilirubin and log(bilirubin). The location of the observations along the curves are displayed by the dots.

age in Figure 15.5 suggests that age can be added directly in the Cox model, while the non-linear behavior of the plot against bilirubin suggests that a transformation should be made for this covariate. The plot against log(bilirubin) indicates that this covariate is much better modeled if it is transformed to log-scale.

15.4 Proofs

15.4.1 Proof of Theorem 15.2.1

In this proof and in the proof of Lemma 15.2.1, the Glivenko–Cantelli theorem, and the Chebychev, Markov and Cauchy–Schwarz inequalities will be used repeatedly.

Define the process $S_1^*, \ldots, S_n^*$ by $S_j^* = \sum_{i=1}^{j} \frac{1}{n} T_i$. Let $N_n^*(s) = \sum_{i=1}^{n} I(S_i^* \leq s)$ be the counting process counting events in this process. Further, let $\mathcal{F}_s^{n^*} = \sigma\{X_1, \ldots, X_n\} \cup \sigma\{(T_j, \delta_j) : \sum_{i=1}^{j} T_i/n \leq s\}$ for $s \geq 0$. The intensity of the process $S_1, \ldots, S_r$ conditional on the history $\mathcal{F}_s^{n^*}$ is $\rho_n(s|\mathcal{F}_s^{n^*}) = n\lambda(X_{N_n^*(s)+1})$. Since $\mathcal{F}_s^n \subseteq \mathcal{F}_s^{n^*}$ it follows from the innovation theorem [Andersen et al. (1993, p. 80)], that

$$\rho_n(s|\mathcal{F}_s^n)/n = \mathrm{E}[\lambda(X_{N_n^*(s)+1})|\mathcal{F}_s^n]. \tag{15.6}$$

Assume that it can be proved that $X_{N_n^*(s)+1} \xrightarrow{p} \eta(s)$ uniformly. Then using Markov's inequality we get

$$\begin{aligned}
&P(\,|\rho_n(s|\mathcal{F}_s^n)/n - \lambda(\eta(s))| > \gamma) \\
&\quad = P(\,|\mathrm{E}[\lambda(X_{N_n^*(s)+1}) - \lambda(\eta(s))|\mathcal{F}_s^n]|\, > \gamma) \\
&\quad \leq \frac{1}{\gamma}\mathrm{E}(\,|\mathrm{E}[\lambda(X_{N_n^*(s)+1}) - \lambda(\eta(s))|\mathcal{F}_s^n]|\,) \\
&\quad \leq \frac{1}{\gamma}\mathrm{E}(\mathrm{E}[\,|\lambda(X_{N_n^*(s)+1}) - \lambda(\eta(s))|\,|\mathcal{F}_s^n]) \\
&\quad \leq \frac{1}{\gamma}\mathrm{E}[\,|\lambda(X_{N_n^*(s)+1}) - \lambda(\eta(s))|\,].
\end{aligned}$$

It now easily follows by the boundedness of $\lambda(x)$ and the assumed uniform convergence of $X_{N_n^*(s)+1}$ that $|\rho_n(s|\mathcal{F}_s^n)/n - \lambda(\eta(s))| \xrightarrow{p} 0$ uniformly in s.

It remains to prove that $X_{N_n^*(s)+1}$ really converges uniformly in probability to $\eta(s)$. Since $T = \min(Z, C)$, given the covariate $X = x$, we have that

$$f_T(t|x) = f_C(t|x)\exp(-\lambda(x)t) + \lambda(x)\exp(-\lambda(x)t)(1 - F_C(t|x)). \tag{15.7}$$

With the assumption $0 < a \leq \lambda(x) \leq M < \infty$ for all x, and the assumption that the censoring distribution for all x has finite first and second order moments, it follows from (15.7) that there exist numbers E_{min}, E_{max} and V_{max} such that

$$\begin{aligned} 0 < E_{min} \leq \quad & \mathrm{E}(T|x) \leq E_{max} < \infty, \quad \text{for all } x, \\ 0 < \quad & \mathrm{Var}(T|x) \leq V_{max} < \infty, \quad \text{for all } x. \end{aligned} \tag{15.8}$$

We proceed by first assuming that X is uniformly distributed on $[0,1]$. Let a point w on the s-axis be fixed in the following, and define I, I_0, I_1 and $\eta(w)$ by the relations

$$\begin{aligned} I: &\quad S^*_{I-1} \leq w < S_{I^*}, \\ I_0: &\quad \textstyle\sum_{i=1}^{I_0-1} \frac{1}{n}\mathrm{E}(T|X_i) \leq w < \sum_{i=1}^{I_0} \frac{1}{n}\mathrm{E}(T|X_i), \\ I_1: &\quad \textstyle\sum_{i=1}^{I_1-1} \frac{1}{n}\mathrm{E}(T|\frac{i}{n+1}) \leq w < \sum_{i=1}^{I_1} \frac{1}{n}\mathrm{E}(T|\frac{i}{n+1}), \\ \eta(w): &\quad \textstyle\int_0^{\eta(w)} \mathrm{E}(T|v)dv = w. \end{aligned} \tag{15.9}$$

In particular $I = N^*_n(w) + 1$. By the triangle inequality

$$\begin{aligned} &|X_I - \eta(w)| \\ \leq &\ |X_I - \frac{I}{n+1}| + |\frac{I}{n+1} - \frac{I_0}{n+1}| + |\frac{I_0}{n+1} - \frac{I_1}{n+1}| + |\frac{I_1}{n+1} - \eta(w)| \\ = &\ A_1 + A_2 + A_3 + A_4. \end{aligned}$$

What remains is to prove that each of A_1, A_2 and $A_3 \xrightarrow{p} 0$ and $A_4 \to 0$ uniformly.

($A_1 \xrightarrow{p} 0$): This follows by the Glivenko–Cantelli theorem which states that if F_n is the empirical distribution function based on n i.i.d. observations from $F \equiv F_X$, then $\sup_x |F_n(x) - F(x)| \xrightarrow{a.s.} 0$. Since $F(x) = x$; $0 \leq x \leq 1$, we have $F(X_i) = X_i$, while $F_n(X_i) = \frac{i}{n}$. Thus, $|X_I - \frac{I}{n}| = |F(X_I) - F_n(X_I)| \leq \sup_x |F(x) - F_n(x)| \xrightarrow{a.s.} 0$, which implies that $A_1 \xrightarrow{p} 0$ uniformly.

($A_2 \xrightarrow{p} 0$): Let $d \geq 2$ be an integer. Then

$$\begin{aligned} &P(I \geq I_0 + d | X_1 = x_1, \ldots, X_n = x_n) \\ = &\ P\left(S^*_{I_0+d-1} \leq w | x_1, \ldots, x_n\right) \leq P\left(S^*_{I_0+d-1} \leq \sum_{i=1}^{I_0} \frac{1}{n}\mathrm{E}(T|x_i) | x_1, \ldots, x_n\right) \\ \leq &\ P\left(|S^*_{I_0+d-1} - \sum_{i=1}^{I_0+d-1} \frac{1}{n}\mathrm{E}(T|x_i)| \geq \sum_{i=I_0+1}^{I_0+d-1} \frac{1}{n}\mathrm{E}(T|x_i) | x_1, \ldots, x_n\right) \\ \overset{Cheb.}{\leq} &\ \frac{\sum_{i=1}^{I_0+d-1} \frac{1}{n^2}\mathrm{Var}(T|x_i)}{\left(\sum_{i=I_0+1}^{I_0+d-1} \frac{1}{n}\mathrm{E}(T|x_i)\right)^2} \leq \frac{V_{max}(I_0 + d - 1)/n^2}{(\frac{d-1}{n}E_{min})^2} \leq \frac{n}{(d-1)^2} \frac{V_{max}}{E^2_{min}}. \end{aligned}$$

Since the upper bound on the conditional probability is not a function of $x_1, \ldots, x_n$ this implies that the inequality also holds for the unconditional probability $P(I \geq I_0 + d)$. By choosing $d = [n^{3/4}]$ we get $P(I \geq I_0 + [n^{3/4}]) \leq cn^{-1/2}$

for a suitable constant c. A similar calculation gives $P(I \leq I_0 - [n^{3/4}]) \leq cn^{-1/2}$. Hence

$$P(|\frac{I}{n+1} - \frac{I_0}{n+1}| \leq \frac{[n^{3/4}]}{n+1}) \geq 1 - \frac{2c}{\sqrt{n}},$$

so $|\frac{I}{n+1} - \frac{I_0}{n+1}| \xrightarrow{p} 0$ uniformly in w.

($A_3 \xrightarrow{p} 0$): A key step in the following is the observation that since $\lambda'(x) \leq D$ and $f_C(t|x)$ by assumption also has finite first derivative, this implies that there exist a B such that $|\mathrm{E}(T|x_1) - \mathrm{E}(T|x_2)| \leq B|x_1 - x_2|$. Also recall that if X_i is the ith order statistic of n independent identically uniformly distributed variables on [0,1], then $\mathrm{Var}(X_i) = \frac{i(n-i+1)}{(n+1)^2(n+2)} \leq \frac{1}{4(n+2)}$. Thus for an integer d,

$$\begin{aligned}
&P(I_0 > I_1 + d) \\
&= P\left(\sum_{i=1}^{I_1+d} \frac{1}{n}\mathrm{E}(T|X_i) < w) \leq P(\sum_{1}^{I_1+d} \frac{1}{n}\mathrm{E}(T|X_i) < \sum_{1}^{I_1} \frac{1}{n}\mathrm{E}(T|\frac{i}{n+1})\right) \\
&\leq P\left(|\sum_{1}^{I_1+d}(\frac{1}{n}\mathrm{E}(T|X_i) - \frac{1}{n}\mathrm{E}(T|\frac{i}{n+1}))| > \sum_{I_1+1}^{I_1+d} \frac{1}{n}\mathrm{E}(T|\frac{i}{n+1})\right) \\
&\overset{Markov}{\leq} \frac{\mathrm{E}|\sum_1^{I_1+d}(\frac{1}{n}\mathrm{E}(T|X_i) - \frac{1}{n}\mathrm{E}(T|\frac{i}{n+1}))|}{\sum_{I_1+1}^{I_1+d}\frac{1}{n}\mathrm{E}(T|\frac{i}{n+1})} \leq \frac{\frac{B}{n}\sum_1^{I_1+d}\mathrm{E}|X_i - \frac{i}{n+1}|}{\frac{d}{n}E_{min}} \\
&\overset{C.-S.}{\leq} \frac{B\sum_1^{I_1+d}\sqrt{\mathrm{E}(X_i - \frac{i}{n+1})^2}}{dE_{min}} = \frac{Bn}{2dE_{min}\sqrt{n+2}}.
\end{aligned}$$

Proving the parallel inequality for $P(I_0 < I_1 - d)$ and letting $d = [n^{3/4}]$ this implies that

$$P\left(|\frac{I_0}{n+1} - \frac{I_1}{n+1}| \leq \frac{[n^{3/4}]}{n+1}\right) \geq 1 - cn^{-1/4}$$

for a suitable constant c. Hence $|\frac{I_0}{n+1} - \frac{I_1}{n+1}| \xrightarrow{p} 0$ uniformly.

($A_4 \to 0$): Observe that $|\sum_{i=1}^{I_1} \frac{1}{n}\mathrm{E}(T|\frac{i}{n+1}) - w| \leq \frac{1}{n}\mathrm{E}(T|\frac{I_1}{n+1}) \leq \frac{1}{n}E_{max}$ which implies that $\sum_{i=1}^{I_1} \frac{1}{n}\mathrm{E}(T|\frac{i}{n+1}) \to w = \int_0^{\eta(w)} \mathrm{E}(T|v)dv$ uniformly. Note that $\eta(w)$ is uniquely defined since $\mathrm{E}(T|v) > 0$ for all v, and it follows that $\frac{I_1}{n+1} \to \eta(w)$ uniformly.

This completes the proof that $\rho_n(w|\mathcal{F})/n \xrightarrow{p} \lambda(\eta(w))$ uniformly in w in the case of uniformly distributed covariates on [0,1].

For covariates $X_1, \ldots, X_n$ drawn from a general continuous distribution $F_X(\cdot)$, let $U_i = F_X(X_i)$ be transformed covariates which are now independent and identically uniformly distributed on [0,1]. Further let $\mathrm{E}^\star(T|u) = \mathrm{E}(T|F_X^{-1}(u))$. Then (15.9) gives $\int_0^{\eta^\star(w)} \mathrm{E}^\star(T|u)du = w$ which by substituting

$u = F_X(x)$ and letting $\eta(w) = F_X^{-1}(\eta^*(w))$ can be written

$$\int_{F_X^{-1}(0)}^{\eta(w)} \mathrm{E}(T|x) f_X(x) dx = w. \tag{15.10}$$

Replacing $\eta(w)$ with x and w with $s(x)$ we get

$$\begin{aligned} s(x) &= \int_{F_X^{-1}(0)}^{x} \mathrm{E}(T|v) f_X(v) dv = \int_{-\infty}^{\infty} I(v \le x) \mathrm{E}(T|v) f_X(v) dv \\ &= \mathrm{E}(I(X \le x)\mathrm{E}(T|X)) = \mathrm{E}(\mathrm{E}(TI(X \le x)|X)) = \mathrm{E}(TI(X \le x)). \end{aligned}$$

15.4.2 Proof of Lemma 15.2.1

We can write

$$\tilde{s}(x) = \frac{1}{n}\sum_{i=1}^{n} T_i I(X_i \le x).$$

Noting that $s(x) = \mathrm{E}(\tilde{s}(x))$ we have by Chebyshev's inequality, for each fixed x and any $\epsilon > 0$,

$$P(|\tilde{s}(x) - s(x)| > \epsilon) \le \frac{\mathrm{Var}(TI(X \le x))}{n\epsilon^2} \le \frac{\mathrm{E}(T^2)}{n\epsilon^2} \le \frac{\mathrm{E}(Z^2)}{n\epsilon^2},$$

which tends to 0 as $n \to \infty$ since $\mathrm{E}(Z^2) < \infty$. In fact, we have $\mathrm{E}(Z^2) = \mathrm{E}[\mathrm{E}(Z^2|X)] = \mathrm{E}[2/\lambda(X)^2] \le 2/a^2$. This proves the result.

15.4.3 Proof of Theorem 15.2.2

Let $N_n(s)$ and m be defined as before. It follows from counting process theory [see, for example, Andersen et al. (1993)], that $M_n(s) = N_n(s) - R_n(s)$, where $R_n(s) = \int_0^s \rho_n(u|\mathcal{F}_u^n)du$, is a local square integrable martingale. The general expression for $\rho_n(s|\mathcal{F}_s^n)$ is given in (15.6). Introduce the notation $\tau_n(s) = \rho_n(s|\mathcal{F}_s^n)/n$ and $\mathcal{T}_n(s) = R_n(s)/n$. The first part of the proof is to find an estimator of $\tau_n(s)$ and to prove that this estimator is a uniformly consistent estimator of $\tau(s) = \lim_{n\to\infty} \tau_n(s) = \lambda(\eta(s))$.

The fact that $M_n(s)$ is a martingale also implies that

$$M^n(s) = M_n(s)/n = N_n(s)/n - \mathcal{T}_n(s) \tag{15.11}$$

is a martingale. Following the same reasoning as in the derivation of the Nelson–Aalen estimator in Andersen et al. (1993, Chap. 4) it follows from (15.11) that a natural estimator for $\mathcal{T}_n(s)$ is $\hat{\mathcal{T}}_n(s) = \int_0^s dN_n(u)/n$ and then a kernel estimator for $\tau_n(s)$ is

$$\hat{\tau}_n(s) = \frac{1}{h_s}\int_0^{\infty} K\left(\frac{s-u}{h_s}\right)\frac{dN_n(u)}{n} = \frac{1}{nh_s}\sum_{i=1}^{r} K\left(\frac{s-S_i}{h_s}\right). \tag{15.12}$$

By this an estimator of $\tau_n(s)$ is motivated, it only remains to prove its consistency as an estimator of $\tau(s)$. It follows from (15.11) that

$$\hat{\tau}_n(s) = \frac{1}{h_s}\int_0^\infty K(\frac{s-u}{h_s})dM^n(u) + \frac{1}{h_s}\int_0^\infty K(\frac{s-u}{h_s})\tau_n(u)du \equiv d_n(s) + \tilde{\tau}_n(s).$$

By showing

$$|\hat{\tau}_n(s) - \tilde{\tau}_n(s)| \xrightarrow{P} 0 \tag{15.13}$$

uniformly and

$$|\tilde{\tau}_n(s) - \tau_n(s)| \xrightarrow{P} 0 \tag{15.14}$$

uniformly, uniform consistency of $\hat{\tau}_n(s)$ follows from the triangle inequality since uniform convergence of $|\tau_n(s)-\tau(s)|$ was proved in Theorem 15.2.1. For (15.13), first notice that by results on stochastic integration and the fact that $< M_n >$ is defined as the compensator of M^{n^2} it follows [Andersen et al. (1993, chap. 4)] that

$$\begin{aligned}\mathrm{E}d_n^2(s) &= \frac{1}{h_s^2}\int_0^\infty K^2(\frac{s-u}{h_s})\mathrm{E}d < M^n > (u) = \frac{1}{h_s^2}\int_{s-h_s}^{s+h_s} K^2(\frac{s-u}{h_s})\frac{1}{n}\mathrm{E}\tau_n(u)du \\ &= \frac{1}{nh_s}\int_{-1}^1 K^2(v)\mathrm{E}\tau_n(s-h_sv)dv \;\leq\; \frac{M}{nh_n}\int_{-1}^1 K^2(v)dv.\end{aligned}$$

Then Markov's inequality gives

$$P(|\hat{\tau}_n(s) - \tilde{\tau}_n(s)| > \epsilon) = P(|d_n(s)| > \epsilon) \leq \frac{\mathrm{E}d_n(s)^2}{\epsilon^2} \leq \frac{M}{\epsilon^2 nh_n}\int_{-1}^1 K^2(v)dv \to 0.$$

For (15.14) the convergence follows from

$$\begin{aligned}|\tilde{\tau}_n(s) - \tau_n(s)| &= |\int_{-1}^1 K(v)(\tau_n(s-h_sv) - \tau_n(s))dv| \\ &\leq \int_{-1}^1 |K(v)||\tau_n(s-h_sv) - \tau_n(s)|dv \xrightarrow{P} 0\end{aligned}$$

uniformly because

$$|\tau_n(s-h_sv)-\tau_n(s)| \leq |\tau_n(s-h_sv)-\tau(s-h_sv)|+|\tau(s)-\tau_n(s)|+|\tau(s-h_sv)-\tau(s)|,$$

where the two first terms converge uniformly to zero in probability by Theorem 15.2.1 and where the last term converges numerically uniformly to 0 by uniform continuity of $\lambda(x)$.

This completes the proof that $\hat{\tau}_n(s)$ given in (15.12) is a uniformly consistent estimator of $\tau(s)$. It now only remains to prove that replacing s by $\tilde{s}(x)$ in (15.12) yields a consistent estimator of $\lambda(x)$. By the triangle inequality

$$|\hat{\tau}_n(\tilde{s}(x)) - \tau(s(x))| \leq |\hat{\tau}_n(\tilde{s}(x)) - \tau(\tilde{s}(x))| + |\tau(\tilde{s}(x)) - \tau(s(x))|,$$

where the second part converges uniformly to 0 in probability by Lemma 15.2.1 and the uniform continuity of $\tau(s)$. This completes the proof that $\hat{\lambda}(x) = \hat{\tau}_n(\tilde{s}(x))$ is a uniformly consistent estimator of $\lambda(x)$.

15.5 Conclusions

We have presented a new method for nonparametric censored exponential regression, and shown some of its applications. While we have given emphasis to applications in Cox regression, one may think of similar applications in any model with (approximately) exponentially distributed residuals, or in other cases where data can be transformed to (approximate) exponentiality.

Notice the flexibility of the covariate order method. Any density estimation method should possibly be usable in the estimation of the scaled intensity, and boundary problems can be handled by adapting various edge correction techniques invented for density estimation. Moreover, different smoothers can be used to estimate the correspondence function. The covariate order method turns out to be numerically very robust, and simulations (not reported here) have shown that the performance in finite samples is comparable to, and often better than, standard local linear likelihood methods.

Acknowledgements. We would like to thank Ørnulf Borgan for helpful comments to the proofs and Eirik Skogvoll for providing the cardiac arrest data. Jan Terje Kvaløy was funded by a Ph.D. grant from the Research Council of Norway during parts of the work on this paper.

References

1. Andersen, P. K., Borgan, Ø., Gill, R. D. and Keiding, N. (1993). *Statistical Models Based on Counting Processes*, Springer-Verlag, New York.

2. Cox, D. R. (1972). Regression models and life-tables, *Journal of the Royal Statistical Society, Series B*, **34**, 187–220.

3. Cox, D. R. and Snell, E. J. (1968). A general definition of residuals, *Journal of the Royal Statistical Society, Series B*, **30**, 248–275.

4. Fleming, T. R. and Harrington, D. P. (1991). *Counting Processes and Survival Analysis*, John Wiley & Sons, New York.

5. Friedman, J. (1984). A variable span smoother, *Technical Report 5*, Department of Statistics, Stanford University.

6. Grambsch, P. M., Therneau, T. M. and Fleming, T. R. (1995). Diagnostic plots to reveal functional form for covariates in multiplicative intensity models, *Biometrics*, **51**, 1469–1482.

7. Hastie, T. J. and Tibshirani, R. J. (1990). *Generalized Additive Models*, Chapman & Hall, London, England.

8. Kvaløy, J. T. (2002). Covariate order tests for covariate effect, *Lifetime Data Analysis*, **8**, 35–52.

9. Kvaløy, J. T. and Lindqvist, B. H. (2003). Estimation and inference in nonparametric Cox-models: Time transformation methods, *Computational Statistics*, **18**, 205–221.

10. Silverman, B. W. (1986). *Density Estimation*, Chapman & Hall, London, England.

11. Therneau, T. M. and Grambsch, P. M. (2000). *Modeling Survival Data: Extending the Cox Model*, Springer-Verlag, New York.

12. Therneau, T. M., Grambsch, P. M. and Fleming, T. R. (1990). Martingale-based residuals for survival models, *Biometrika*, **77**, 147–160.

16

Effect of Ignoring Heterogeneity in Hazards Regression

Hong-Dar Isaac Wu

School of Public Health, China Medical College, Taichung, Taiwan

Abstract: We studied a heteroscedastic hazards regression (HHR) model in order to evaluate bias when intrinsic heterogeneity is ignored. We used the proportional hazards model and the associated estimate to show that bias resulting from neglecting the heterogeneity over different 'treatment' groups or other covariates is an analytically calculated value. We then compared the value with those obtained from simulation results. Further, considering that a 'stratified' proportional hazards (SPH) model could possibly be used as an alternative, we made similar calculations to clarify when the SPH model will succeed or fail in order to account for heterogeneity when the underlying model is the HHR model.

Keywords and phrases: Heteroscedastic hazards regression, heterogeneity, proportional hazards, nonproportional hazards, bias

16.1 Introduction

Cox's proportional hazards (PH) model [Cox (1972)] has been widely used to analyze survival data collected from epidemiologic studies and clinical trials. In order to make reliable inference, validation of the PH model is important. Statistical issues regarding this problem discussed in the literature include the goodness-of-fit test [Schoenfeld (1980), Wei (1984), Gill and Schumacher (1987) and Lin (1991], regression diagnostics [Therneau, Grambsch, and Fleming (1990) and Lin, Wei and Ying (1993)], and the impact of model misspecification [Struthers and Kalbfleisch (1986), Bretagnolle and Huber-Carol (1988) and Lagakos (1988)]. However, a scenario arises when there are nonproportional hazards. In order to deal with this issue, Hsieh (2001) proposed a heteroscedastic hazards regression (HHR) model which accounts for the heterogeneity effect by modeling heteroscedasticity in the framework of a linear transformation model. The cumulative hazard function of the HHR model is written as

$$\Lambda(t; Z(t), X(t)) = \{\Lambda_0(t)\}^{\sigma(t)} \exp\{\beta' Z(t)\}, \qquad (16.1)$$

where $\sigma(t) = \exp\{\gamma' X(t)\}$. Here, $Z(\cdot)$ and $X(\cdot)$ are two covariate vectors, both of which can be time dependent; $\Lambda_0(t)$ is the baseline cumulative hazard. The nonproportional hazards characteristic of the HHR model is as follows. First, if 'relative risk' (RR) is the 'effect-measure' to be estimated, the HHR model allows for a monotonic *time-varying* effect. Second, the RR-measure is *heterogeneous* over the space spanned by the covariates. When the HHR model is used as the underlying model and intrinsic heterogeneity is negelcted, HHR is reduced to the PH model. This is not a typical problem of model misspecification involving the omission of covariates since the omitted part, $X(\cdot)$, can NOT be a subset of $Z(\cdot)$. Under the circumstances, analytic calculation of bias and a qualitative description of its direction are investigated in this paper.

With the assumption of the HHR model (16.1), we see from some calculations shown later in the paper that there is bias even when the omitted covariate $X(\cdot)$ is independent of the covariate $Z(\cdot)$. This is very similar to the results of Gail et al. (1984) and Bretagnolle and Huber-Carol (1988). Two possibilities may occur in the problem of model misspecification of omitting covariates. One is the declination of the effect being estimated. That is, an existing effect is estimated by an attenuated value. In this situation, the bias is said to be *toward the null.* The other is the reverse effect estimate. That is, a treatment with positive effect may be recognized as a negative one, and vice versa. Under model (16.1), disregarding the $X(\cdot)$-component produces both kinds of bias if there is a high correlation between covariables X and Z. In the following context, we give theoretical calculations mimicking those of Gail et al. (1984). Numerical examples will be given for the cases when X and Z are identical or independent. Finally, one may be concerned with whether or not the effect of heterogeneity can be accounted for by 'stratification' through the implementation of a **stratified PH model**. To answer this question, we give some discussion and similar calculation according to two hypothetical cases, one in which the omitted $X(\cdot)$-component is independent of the $Z(\cdot)$-component; the other in which $X(\cdot)$ is a subset of $Z(\cdot)$.

16.2 Calculation of Bias

Let $N_i(t)$ and $Y_i(t)$ be the counting process and at-risk indicator of subject i at time t; τ is a maximal time point which can be selected as the largest failure time. With model (16.1), Hsieh (2001) derived the following estimating functions for the parameters of interests β and γ:

$$\mathcal{E}_\beta = \sum \int_0^\tau \{Z_i - \frac{\sum Y_j(u) Z_j(u) e^{\beta' Z_j + \gamma' X_j} \{\Lambda_0(u)\}^{\exp(\gamma' X_j) - 1}}{\sum Y_j(u) e^{\beta' Z_j + \gamma' X_j} \{\Lambda_0(u)\}^{\exp(\gamma' X_j) - 1}}\} dN_i(u) \quad (16.2)$$

and

$$\mathcal{E}_\gamma = \sum \int_0^\tau \{V_i - \frac{\sum Y_j(u) V_j(u) e^{\beta' Z_j + \gamma' X_j} \{\Lambda_0(u)\}^{\exp(\gamma' X_j) - 1}}{\sum Y_j(u) e^{\beta' Z_j + \gamma' X_j} \{\Lambda_0(u)\}^{\exp(\gamma' X_j) - 1}}\} dN_i(u), \quad (16.3)$$

where $V_i(t) = X_i(t)[1 + \exp(\gamma' X_i)\log\{\Lambda_0(t)\}]$. When the HHR model is misspecified as the PH model, there is no $\mathcal{E}_\gamma$. In this case, the estimating function $\mathcal{E}_\beta$ reduces to

$$\mathcal{E}_b = \sum \int_0^\tau \{Z_i - \frac{\sum Y_j(u) Z_j(u) \exp\{b' Z_j(u)\}}{\sum Y_j(u) \exp\{b Z_j(u)\}}\} dN_i(u), \quad (16.4)$$

where b is the solution to the equation $\mathcal{E}_b = 0$ corresponding to the misspecified model. For simplicity, let dim(Z)=dim(X)=1, and then solve $\mathcal{E}_b = 0$ to obtain the value of b. That is,

$$\sum \int_0^\tau Z_i(u) dN_i(u) = \sum \int_0^\tau \frac{\sum Y_j(u) Z_j(u) \exp\{b Z_j(u)\}}{\sum Y_j(u) \exp\{b Z_j(u)\}} dN_i(u).$$

However, since the true model is the HHR model, and the left-hand side $\sum \int Z_i(u) dN_i(u)$ can only be expressed through $\mathcal{E}_\beta = 0$, we have the following equation for solving b **implicitly** in terms of β and γ:

$$\begin{aligned} &\sum_i \int_0^\tau \frac{\sum Y_j(u) Z_j e^{b Z_j}/n}{\sum Y_j(u) e^{b Z_j}/n} dN_i(u) \\ &= \sum_i \int_0^\tau \frac{\sum Y_j(u) Z_j e^{\beta Z_j + \gamma X_j} \{\Lambda_0(u)\}^{\exp(\gamma X_j) - 1}/n}{\sum Y_j(u) e^{\beta Z_j + \gamma X_j} \{\Lambda_0(u)\}^{\exp(\gamma X_j) - 1}/n} dN_i(u). \quad (16.5) \end{aligned}$$

In the following context, notation and some arguments similar to those of Tsiatis (1981) and Gail et al. (1984) will be employed.

If $\mathbf{v} = (Z, X)'$ is the bivariate random vector with probability density function $f_\mathbf{v}$, and both sides of (16.5) are multiplied by $1/n$, the left- and right-hand sides converge to

$$\int_0^\tau \frac{E_\mathbf{v}[Z e^{bZ}, t]}{E_\mathbf{v}[e^{bZ}, t]} \{-dQ(t)\} \quad (16.6)$$

and

$$\int_0^\tau \frac{E_\mathbf{v}[z e^{\beta Z + \gamma X} \{\Lambda_0(t)\}^{\exp(\gamma X_j) - 1}, t]}{E_\mathbf{v}[e^{\beta Z + \gamma X} \{\Lambda_0(t)\}^{\exp(\gamma X_j) - 1}, t]} \{-dQ(t)\}, \quad (16.7)$$

respectively, where

$$Q(t) = Pr(T \geq t) \quad \text{and} \quad E_{\mathbf{v}}[g(\mathbf{v}), t] = \int g(\mathbf{v}) Pr(T \geq t \mid \mathbf{v}) f_{\mathbf{v}} d\mathbf{v}.$$

By applying this property (which can be proved by Lemmas $A1$ and $A2$ in Tsiatis' paper) and denoting the conditional survivor function as $H(t \mid \mathbf{v}) = Pr(T \geq t \mid \mathbf{v})$, it is obvious that (16.6) and (16.7) are equivalent to

$$\int_0^\tau \frac{P(b,\beta,\gamma;t)}{Q(b,\beta,\gamma;t)} \lambda_0(t) R(\beta,\gamma;t) dt \tag{16.8}$$

and

$$\int_0^\tau \lambda_0(t) S(\beta,\gamma;t) dt \tag{16.9}$$

respectively, where $\lambda_0(t)$ is the baseline hazard function and $P(\cdot), Q(\cdot), R(\cdot)$, and $S(\cdot)$ are defined as

$$\begin{aligned}
P &= \int Z\exp(bZ) H(t|\mathbf{v}) f_{\mathbf{v}} d\mathbf{v}, \\
Q &= \int \exp(bZ) H(t|\mathbf{v}) f_{\mathbf{v}} d\mathbf{v}, \\
R &= \int \exp(\beta Z + \gamma X) \{\Lambda_0(t)\}^{\exp(\gamma X)-1} H(t|\mathbf{v}) f_{\mathbf{v}} d\mathbf{v}, \\
S &= \int Z\exp(\beta Z + \gamma X) \{\Lambda_0(t)\}^{\exp(\gamma X)-1} H(t|\mathbf{v}) f_{\mathbf{v}} d\mathbf{v}.
\end{aligned}$$

16.3 Numerical Study

In this section, we give two worked examples with dichotomous Z and X to illustrate the calculation of (16.8) and (16.9). The theoretical value of parameter b is evaluated analytically and compared to the simulation results. More complex cases when the omitted covariate X is a continuous random variable will also be considered later in the paper.

16.3.1 Examples

Throughout the following examples, we choose $\lambda_0(t) = 1$, i.e., the baseline is the Weibull(1, 1) distribution. In this theoretical calculation, we assume there to be *no censoring*. We let the random failure time variable T, given the set of covariates Z and X, be distributed as Weibull(a, b) with $a = \exp(\beta Z)$ and $b = \exp(\gamma X)$.

Table 16.1: (*Example 16.3.1*) Asymptotic bias $(b - \beta)$, the bias based on 1000 replicates $(\hat{b} - \beta)$ and the standard error $(std.(\hat{b}))$, all with sample sizes of 100

β	γ	b	$(b-\beta)\times 10^2$	$\hat{b}$	$(\hat{b}-\beta)\times 10^2$	$std.(\hat{b})$
0	0	0	0	0.0049	0.5	0.1966
	1	0.2488	24.9	0.2326	23.3	0.2210
	2	0.2891	28.9	0.2609	26.1	0.2403
1	0	1	0	1.0126	1.3	0.2357
	1	0.7038	−29.6	0.6940	−30.6	0.2391
	2	0.4729	−52.7	0.4605	−53.9	0.2513
2	0	2	0	2.0394	3.9	0.3083
	1	1.1576	−84.2	1.1476	−85.2	0.2494
	2	0.6570	−134.3	0.6429	−135.7	0.2623

Example 16.3.1 (Fully identical X and Z.) The covariates X and Z are identical and $Pr(Z = X = 1) = Pr(Z = X = 0) = 1/2$. This example demonstrates the situation when covariate Z is itself the source of heterogeneity effect being neglected. The integrands in (16.8) and (16.9) are

$$\begin{aligned} P &= (1/2)\{\exp(b - \theta t^{\xi})\}, \\ Q &= (1/2)\{\exp(-t) + \exp(b - \theta t^{\xi})\}, \\ R &= (1/2)\{\exp(-t) + \theta\xi t^{\xi-1}\exp(-\theta t^{\xi})\}, \\ S &= (1/2)\{\theta\xi t^{\xi-1}\exp(-\theta t^{\xi})\}, \end{aligned}$$

with $\theta = \exp(\beta)$ and $\xi = \exp(\gamma)$. The theoretical value of parameter b can then be evaluated by equating (16.8) and (16.9). The result is reported in Table 16.1 and compared to $\hat{b}$, the sample mean based on 1000 simulations in which $\mathcal{E}_b = 0$ is solved. The current example offers a chance to clarify that the terminology **bias** is used in this paper only to indicate the difference between the coefficients, according to two different models, of a covariate in the conventional *risk function.* It is worth noting that this usage of bias is in some cases misleading because the coefficient of the same covariate may have different interpretations in different models. In this example, if the HHR model is misspecified as the PH model, parameter b has the interpretation of 'log relative risk'. In the HHR model, however, β is not interpreted in this way [Wu, Hsieh and Chen (2001)]. Therefore, referring to $b - \beta$ as bias is only adequate when b and β are the same 'effect-measure'.

Table 16.2: (*Example 16.3.2*) Asymptotic bias ($b-\beta$), the bias based on 1000 replicates ($\hat{b}-\beta$) and the standard error ($std.(\hat{b})$), all with sample sizes of 100

β	γ	b	$(b-\beta)\times 10^2$	$\hat{b}$	$(\hat{b}-\beta)\times 10^2$	$std.(\hat{b})$
0	0	0	0	0.0074	0.7	0.2084
	1	0	0	−0.0011	−0.1	0.2157
	2	0	0	0.0007	0.1	0.1999
1	0	1	0	1.0082	0.8	0.2290
	1	0.9511	−4.9	0.9742	−2.6	0.2294
	2	0.8926	−10.7	0.9074	−9.3	0.2442
2	0	2	0	2.0497	5.0	0.2955
	1	1.8658	−13.4	1.8531	−14.7	0.2930
	2	1.6170	−38.3	1.6353	−36.4	0.2611

Example 16.3.2 (Independent X and Z.) In this example, the covariates X and Z are assumed to be independent, where $Pr(Z=1)=Pr(Z=0)=1/2$ and $Pr(X=1)=Pr(X=0)=1/2$. Equations (16.8) and (16.9) have the following components:

$$
\begin{aligned}
P &= (1/4)\{\exp(b-\theta t)+\exp(b-\theta t^{\xi})\},\\
Q &= (1/4)\{\exp(-t)+\exp(-t^{\xi})+\exp(b-\theta t)+\exp(b-\theta t^{\xi})\},\\
R &= (1/4)\{\exp(-t)+\xi t^{\xi-1}\exp(-t^{\xi})+\theta\exp(-\theta t)+\theta\xi t^{\xi-1}\exp(-\theta t^{\xi})\},\\
S &= (1/4)\{\theta\exp(-\theta t)+\theta\xi t^{\xi-1}\exp(-\theta t^{\xi})\}.
\end{aligned}
$$

We can also equate the associated integrals and solve b implicitly according to different βs and γs. The result is shown in Table 16.2.

16.3.2 Direction of bias

The hazard function under the HHR model is

$$\lambda(t;Z,X)=\lambda_0(t)e^{\beta Z+\{\gamma X+(e^{\gamma X}-1)\log\Lambda_0(t)\}}. \tag{16.10}$$

When it is misspecified as the PH model with only the covariate Z, it corresponds to a question discussed by Bretagnolle and Huber-Carol (1988). If Z and X are independent, then omitting $\gamma X+(e^{\gamma X}-1)\log\Lambda_0(t)$ results in a biased estimate b of parameter β, and

$$\text{sign }(b-\beta)=-\text{ sign }(\beta).$$

Table 16.3: Direction of bias in 1000 replicates when X is independent of Z; Z is dichotomous and X is *continuously* distributed. There are four configurations of X and Z:
Case I: $Pr(Z=0) = Pr(Z=1) = 1/2, X \sim$ Uniform(0, 1),
Case II: $Pr(Z=0) = Pr(Z=-1) = 1/2, X \sim$ Uniform(0. 1),
Case III: $Pr(Z=0) = Pr(Z=1) = 1/2, X \sim$ Uniform(−1, 0),
Case IV: $Pr(Z=0) = Pr(Z=-1) = 1/2, X \sim$ Uniform(−1, 0)

Case	β	γ	$\hat{b}$	$std.(\hat{b})$	Case	β	γ	$\hat{b}$	$std.(\hat{b})$
I	−0.2	−2	−0.1831	0.2105	II	−0.2	−2	−0.1762	0.2148
		+2	−0.1731	0.2144			+2	−0.1838	0.2151
	0	−2	−0.0074	0.2094		0	−2	−0.0027	0.2084
		+2	−0.0041	0.2083			+2	0.0109	0.2082
	0.2	−2	0.1851	0.2078		0.2	−2	0.1846	0.2069
		+2	0.1817	0.2132			+2	0.1752	0.2119
III	−0.2	−2	−0.1822	0.2082	IV	−0.2	−2	−0.1768	0.2135
		+2	−0.1914	0.2106			+2	−0.1962	0.2155
	0	−2	0.0032	0.2018		0	−2	−0.0188	0.1997
		+2	−0.0001	0.2081			+2	−0.0077	0.2118
	0.2	−2	0.1801	0.2087		0.2	−2	0.1726	0.2087
		+2	0.1973	0.2182			+2	0.1854	0.2091

In other words, the *direction* of bias is *toward the null.* A simulation study is reported in Table 16.3. When Z and X are correlated, the situation is different. The direction of bias depends on the *correlation structure* between Z and X. Simulation studies (not reported in this paper) show that it is possible that the bias is 'towards the null plus a zero crossing' or 'away from the null', if there is some moderate or stronger correlation between Z and X.

16.4 The Stratified PH Analysis Under the HHR Model

When effect heterogeneity is present, sometimes we look for a possibility to use the stratified PH (SPH) model to account for it. If the underlying model is the HHR model, it is appealing to know when the SPH analysis can and cannot replace the HHR analysis.

Heuristically, when the covariate X which appears in the power (hereafter termed as the 'σ-part') of the baseline cumulative hazard, $\Lambda_0(t)$, in model (16.1) is *categorical* and independent of the covariate Z which appears in the risk function (termed as the 'μ-part'), then the parameter β can be consistently estimated using the SPH analysis. Otherwise, these two models (HHR and SPH) cannot replace each other.

To justify the former statement, consider the simple case when X is univariate and dichotomous, $Pr(X=0)=p_0$ and $Pr(X=1)=p_1$. Now apply the argument in Section 16.2,

$$\begin{aligned} S &= \int Z\exp(\beta Z+\gamma X)\{\Lambda_0(t)\}^{\exp(\gamma X)-1}P(T\geq t\mid \mathbf{v})f_{\mathbf{v}}d_{\mathbf{v}} \\ &= p_0\int \lambda_0(t)\left\{\int Ze^{\beta Z}P(T\geq t\mid Z,0)f_{Z,0}dZ\right\}dt \\ &\quad +p_1\int \lambda_0(t)\sigma\{\Lambda_0(t)\}^{\sigma-1}\left\{\int Ze^{\beta Z}P(T\geq t\mid Z,1)f_{Z,1}dZ\right\}dt, \\ &\qquad \text{with } \sigma=e^{\gamma}. \end{aligned} \tag{16.11}$$

Besides, P, Q, and R all split into two components for $X=j$ $(j=0,1)$,

$$\begin{aligned} P_{X=j} &= \int Ze^{bZ}P(T\geq t\mid Z,j)f_{Z,j}dZ, \\ Q_{X=j} &= \int e^{bZ}P(T\geq t\mid Z,j)f_{Z,j}dZ, \\ R_{X=j} &= \mathbf{1}_{\{j=1\}}\sigma\{\Lambda_0(t)\}^{\sigma-1}\int e^{\beta Z}P(T\geq t\mid Z,j)f_{Z,j}dZ. \end{aligned}$$

Then the formula for solving b by equating the expressions associated with (16.8) and (16.9) is

$$\begin{aligned} S &= p_0\int \frac{\int Ze^{bZ}P(T\geq t\mid Z,0)f_{Z,0}dZ}{\int e^{bZ}P(T\geq t\mid Z,0)f_{Z,0}dZ}\lambda_0(t)\int e^{\beta Z}P(T\geq t\mid Z,0)f_{Z,0}dZdt \\ &\quad +p_1\int \frac{\int Ze^{bZ}P(T\geq t\mid Z,1)f_{Z,1}dZ}{\int e^{bZ}P(T\geq t\mid Z,1)f_{Z,1}dZ}\lambda_0(t)\sigma\{\Lambda_0(t)\}^{\sigma-1} \\ &\qquad \times\int e^{\beta Z}P(T\geq t\mid Z,1)f_{Z,1}dZdt. \end{aligned}$$

Therefore, $b=\beta$ is sufficient.

In order to investigate the relationship between the HHR and SPH models, several questions must be asked. First, if the σ-part is categorical and independent of the μ-part, what is the difference in **efficiency** between the SPH analysis and the HHR analysis? Second, if we suppose that the σ-part is continuous and independent of the μ-part, and that a data analyzer will use the SPH analysis via some stratification on the σ-part, what is the *bias*? Finally, if

Table 16.4: Comparison of estimates of the PH, SPH, and HHR analyses for **Model A**. The true parameter values are $\beta_1 = 1, \beta_2 = 1$, and $\gamma = 2$. Values shown in parentheses are the mean square errors.

	$\hat{\beta}_1$	$\hat{\beta}_2$	$\hat{\gamma}$	theoretical b_1	theoretical b_2
PH	0.8164	0.8278		0.8253	0.8253
	(0.1571)	(0.1699)			
SPH	1.0245	0.9996		1.0000	1.0000
	(0.1573)	(0.1568)			
HHR	1.0291	1.0214	2.1280		
	(0.1386)	(0.2149)	(0.3745)		

the σ- and μ-parts are continuous and fully identical ($X = Z$), what is the *bias* using SPH analysis under the HHR model?

To answer these questions, let Z be 'bivariate', $Z = (Z_1, Z_2)'$, for convenience, and consider the following two cases as the underlying HHR model:

$$\text{Model A} : \{\Lambda_0(t)\}^{\exp(\gamma X)} \exp(\beta_1 Z_1 + \beta_2 Z_2),$$
$$\text{Model B} : \{\Lambda_0(t)\}^{\exp(\gamma X)} \exp(\beta_1 Z_1 + \beta_2 X).$$

In Model A, let X, Z_1, and Z_2 be independent; $X = 0$ or 1 with probability 0.5, $Z_1 = 0$ or 1 with probability 0.5, and $Z_2 \sim$U$(0,1)$. Note that Model A is used to study the efficiency of the SPH analysis compared to the HHR analysis. In Model B, the X variable appears in both of the μ- and σ-parts as the source of heterogeneity [Hsieh (2001)]. Let Z_1 and X be independent; $Z_1 = 0$ or 1 with probability 0.5 and $X \sim$U$(0,1)$. Since X is continuous, assume that a stratification variable $X^* = \mathbf{1}_{(X \geq 0.5)}$ is taken in the SPH analysis to deal with the heterogeneity.

This case is designed to study the bias concerned in both of the latter two questions. The results are reported in Tables 16.4 and 16.5 based on 1000 simulations for sample sizes of 100 with *no censoring*. Estimates in the HHR analyses (that is, $\hat{\beta}_1$, $\hat{\beta}_2$, and $\hat{\gamma}$) are obtained from the over-identified estimating equation (OEE) approach of Hsieh's paper. In terms of 'efficiency', since there may be a bias in the misspecified model, we present the *mean square errors* (MSE) of each estimate according to different models and analyses.

Table 16.5: Comparison of estimates of the PH, SPH, and HHR analyses for **Model B**. The true parameter values are $\beta_1 = 1, \beta_2 = 1$, and $\gamma = 2$. Values shown in parentheses are the mean square errors

	$\hat{\beta}_1$	$\hat{\beta}_2$	$\hat{\gamma}$	theoretical b_1	theoretical b_2
PH	0.9455	−0.2497		0.9181	−0.1725
	(0.1514)	(1.7361)			
SPH	0.9906	−0.2664		0.9686	−0.1806
	(0.1519)	(2.1896)			
HHR	1.0295	0.9708	2.0872		
	(0.1523)	(0.3163)	(0.2019)		

16.5 Discussion

This paper discusses the importance of explicit modeling of heterogeneity in hazards regression analysis. With this concern, Hsieh's HHR model is a good choice, in spite of the complex estimating procedures involved. An alternative to Hsieh's OEE approach, the modified partial likelihood estimate of Bagdonavičius, Hafdi, and Nikulin (2002), also can be considered. The calculation of bias reported in Tables 16.1 and 16.2 reveals that omitting the σ-part, even if it is independent of the μ-part, produces bias. Moreover, the direction of bias relies on the correlation and joint distribution of covariates X and Z. When they are independent, the 'magnitude' of the effect estimate tends to be lower than the true value. Qualitative property of the bias is very similar to the situation in the PH model when an important covariate is omitted. Table 16.3 lists a result when X is continuously distributed and X and Z are independent. In all four cases, $\hat{b}$ lies between 0 and the true value of β. The direction of bias is irrespective of the true value of γ and the distribution of X. This is due to the fact that, by inspection of the hazard function (16.10), the HHR model has the same **multiplicative** structure as the PH model. Moreover, considering that the omitted component is a **mixture** of the X variable and the baseline cumulative hazard $\Lambda_0(t)$, future studies of the behavior of the estimate of this baseline function under a misspecified model is appealing. In actual applications, however, X and Z are more likely to be correlated, and the bias can have any direction and the effect estimate may even have a reverse sign.

Under the HHR model, using SPH analysis to 'stratify out' the heterogeneity gives valid estimates only when the variables in the σ-part are totally categorical and independent of the μ-part (Table 16.4). The MSEs of the HHR and SPH models are comparable to each other. This is also a property implied by the aforementioned 'multiplicative' structure of the model. We see from Table 16.5 that if the σ-part contains a continuous variable which also appears in the μ-part, stratification makes the MSE of $\hat{\beta}_2$ even larger. This illustrates a case when the SPH model cannot deal with the heterogeneity effect.

References

1. Bagdonavičius, V., Hafdi, M. A. and Nikulin, M. (2002). The generalized proportional hazards model and its application for statistical analysis of the Hsieh model, Presented at the *Second Euro-Japanese Workshop on Stochastic Modelling for Finance, Insurance, Production and Reliability*, Chamonix, France.

2. Bretagnolle, J. and Huber-Carol, C. (1988). Effects of omitting covariates in Cox's model for survival data, *Scandinavian Journal of Statistics*, **15**, 125–138.

3. Cox, D. R. (1972). Regression models and life-tables (with discussion), *Journal of the Royal Statistical Society, Series B*, **34**, 187-220.

4. Gail, M. H., Wieand, S. and Piantadosi, S. (1984). Biased estimates of treatment effect in randomized experiments with nonlinear regression and omitted covariates, *Biometrika*, **71**, 431–444.

5. Gill, R. D. and Schumacher, M. (1987). A simple test of the proportional hazards assumption, *Biometrika* **74**, 289-300.

6. Hsieh, F. (2001). On heteroscedastic hazards regression models: Theory and application, *Journal of the Royal Statistical Society, Series B*, **63**, 63–79.

7. Lagakos, S. W. (1988). The loss in efficiency from misspecifying covariates in proportional hazards regression models, *Biometrika*, **75**, 156–160.

8. Lin, D. Y. (1991). Goodness of fit for the Cox regression model based on a class of parameter estimators, *Journal of the American Statistical Association*, **86**, 725–728.

9. Lin, D. Y., Wei, L. J. and Ying, Z. (1993). Checking the Cox model with cumulative sums of martingale-based residuals, *Biometrika*, **80**, 557–572.

10. Schoenfeld, D. (1980). Chi-squared goodness-of-fit tests for the proportional hazards regression model, *Biometrika*, **67**, 145–153.

11. Struthers, C. A. and Kalbfleisch, J. D. (1986). Misspecified proportional hazard models, *Biometrika*, **73**, 363–369.

12. Therneau, T. M., Grambsch, P. M. and Fleming, T. R. (1990). Martingale-based residuals for survival models, *Biometrika*, **77**, 147–160.

13. Tsiatis, A. A. (1981). The asymptotic joint distribution of the efficient scores test for the proportional hazards model calculated over time, *Biometrika*, **68**, 311–315.

14. Wei, L. J. (1984). Testing goodness-of-fit for the proportional hazards model with censored observations, *Journal of the American Statistical Association*, **79**, 649–652.

15. Wu, H.-D I., Hsieh, F. and Chen, C.-H. (2001). Validation of a heteroscedastic hazards regression model, *Lifetime Data Analysis*, **8**, 21–34.

PART VII
Time Series Analysis

17

An Introduction to Efficient Estimation for Semiparametric Time Series

Priscilla E. Greenwood[1], **Ursula U. Müller**[2], **and Wolfgang Wefelmeyer**[3]

[1] *Department of Mathematics and Statistics, Arizona State University, Tempe, Arizona, USA*
[2] *Fachbereich Mathematik und Informatik, Universität Bremen, Bremen, Germany*
[3] *Fachbereich Mathematik, Universität Siegen, Siegen, Germany*

Abstract: We illustrate several recent results on efficient estimation for semiparametric time series models with two types of AR(1) models: having independent and centered innovations, and having general and conditionally centered innovations. We consider in particular estimation of the autoregression parameter, the stationary distribution, the innovation distribution, and the stationary density.

Keywords and phrases: Asymptotic efficiency, plug-in principle, pseudo-observation, empirical estimator, weighted least squares estimator, martingale estimating equation, nonparametric model, semiparametric model, constrained model, Markov chain model, linear autoregressive model

17.1 Introduction

The purpose of this paper is to illustrate a number of recent results on efficient estimation for semiparametric time series models in the context of a linear autoregressive process of order one, $X_i = \vartheta X_{i-1} + \varepsilon_i$. In addition, we sketch the construction of efficient estimators in this context. Historically, it was first assumed that the innovations ε_i are i.i.d. with zero mean. We call this model I. For many applications, especially in the recent econometrics literature, independence of the innovations is considered too strong an assumption and is replaced by the weaker condition that the ε_i may depend on the previous observation, with $E(\varepsilon_i \mid X_{i-1}) = 0$. We call this model II. It can be described as a nonparametric Markov chain model fulfilling the constraint $E(X_i \mid X_{i-1}) = \vartheta X_{i-1}$. Structurally, the two models are quite different. Model I is very close to an i.i.d. model, and efficient estimation of ϑ in particular has a long history. Efficient estimation of the stationary law and of the law of the innovations is more recent. Model II is closer to the full nonparametric Markov chain model, for which efficient estimation was considered in the last decade only. Nevertheless, in certain respects model II is simpler than model I because it has less structure. The full nonparametric Markov chain model has just one (infinite-dimensional) parameter, the transition kernel. Model II is described by the transition kernel and the parameter ϑ in the model for the conditional mean. This is not a parametrization, because the transition kernel and the parameter in the constraint on the transition kernel do not vary independently. Model I has two parameters: the innovation law and the autoregression parameter.

This paper does not follow the order of the historical development. Instead, we follow the order of nesting of these models: Sections 17.2 to 17.4 treat first the full nonparametric Markov chain model, then model II, and finally model I. We will only sketch the arguments and refer the reader to the literature for details.

17.2 Nonparametric Efficiency of the Least Squares Estimator

We begin by recalling some results on efficient estimation for general Markov chain models: local asymptotic normality, characterizations of efficient and regular estimators, and efficiency of empirical estimators. Let $X_0, \dots, X_n$ be observations from a real-valued stationary Markov chain with transition kernel $Q(x, dy)$. Let π, $P = \pi \otimes Q$ and $P^{(n)}$ denote the laws of X_0, (X_0, X_1) and $(X_0, \dots, X_n)$, respectively. Write $Ph = \int h(x, y)\, P(dx, dy)$ for the expectation of $h(X_0, X_1)$, and $Q_x h = \int h(x, y) Q(x, dy)$ for the conditional expectation of $h(X_0, X_1)$ given $X_0 = x$.

Local asymptotic normality. To describe asymptotic variance bounds and characterize efficient estimators, we introduce a *local model* through (Hellinger differentiable) perturbations $Q_{nv}(x, dy) \doteq Q(x, dy)(1 + n^{-1/2} v(x, y))$. For Q_{nv} to be a transition kernel, the *local parameter* v must be in the space

$$V = \{v \in L_2(P) : Qv = 0\}.$$

This is the *tangent space* of the full nonparametric model. Write π_{nv}, P_{nv} and $P_{nv}^{(n)}$ for the corresponding laws if Q_{nv} is true. We have *local asymptotic normality*

$$\log \frac{dP_{nv}^{(n)}}{dP^{(n)}}(X_0, \dots, X_n) \quad = \quad n^{-1/2} \sum_{i=1}^{n} v(X_{i-1}, X_i) - \frac{1}{2} P v^2 + o_{P^{(n)}}(1),$$

$$n^{-1/2}\sum_{i=1}^{n} v(X_{i-1}, X_i) \Rightarrow (Pv^2)^{1/2}N \quad \text{under } P^{(n)},$$

with N a standard normal random variable. Proofs under increasingly weaker conditions are given by Roussas (1965), Höpfner, Jacod and Ladelli (1990), Penev (1991) and Höpfner (1993a,b).

Characterization of efficient estimators. Consider now a submodel, described by a subset of the transition kernels. Its tangent space is obtained by perturbing Q within the submodel. The tangent space is a subset of V, say V_*, which we take to be (closed and) linear. A real-valued functional $s(Q)$ is *differentiable* at Q with *gradient* g if $g \in V$ and

$$n^{1/2}(s(Q_{nv}) - s(Q)) \to P[gv] \quad \text{for all } v \in V_*.$$

The *canonical gradient* g_* is the projection of g onto V_*. An estimator $\hat{s}$ of s is *regular* at Q with *limit* L if L is a random variable such that

$$n^{1/2}(\hat{s} - s(Q_{nv})) \Rightarrow L \quad \text{under } P_{nv}^{(n)} \text{ for all } v \in V_*.$$

The convolution theorem says that $L = (Pg_*^2)^{1/2}N + M$ in distribution, with random variable M independent of N. This justifies calling a regular estimator $\hat{s}$ *efficient* for s at Q if $L = (Pg_*^2)^{1/2}N$ in distribution.

An estimator $\hat{s}$ is *asymptotically linear* for s at Q with *influence function* w if $w \in W$ and

$$n^{1/2}(\hat{s} - s(Q)) = n^{-1/2}\sum_{i=1}^{n} w(X_{i-1}, X_i) + o_{P^{(n)}}(1).$$

An asymptotically linear estimator is regular if and only if its influence function is a gradient, and a (regular) estimator is efficient if and only if it is asymptotically linear with influence function equal to the canonical gradient. The convolution theorem is due to Hájek (1970). For the characterizations of regular and efficient estimators we refer to Bickel, Klaassen, Ritov and Wellner (1998, Section 3.3).

Efficiency of empirical estimators. We return to the full nonparametric Markov chain model. Suppose the chain is geometrically ergodic. Let $h(x,y)$ be P-square-integrable. We want to estimate the linear functional $s(Q) = Ph = Eh(X_0, X_1)$. By a perturbation expansion, see Kartashov (1985a,b, 1996),

$$n^{1/2}(P_{nv}h - Ph) \to P[Ah \cdot v] \quad \text{for all } v \in V,$$

where A is a projection from $L_2(P)$ onto V, defined by

$$Ah(x,y) = h(x,y) - Q_x h + \sum_{j=1}^{\infty}(Q_y^j h - Q_x^{j+1}h). \tag{17.1}$$

Hence Ph is differentiable at Q with canonical gradient Ah. On the other hand, the empirical estimator $\frac{1}{n}\sum_{i=1}^n h(X_{i-1}, X_i)$ has the martingale approximation

$$n^{-1/2}\sum_{i=1}^n (h(X_{i-1}, X_i) - Ph) = n^{-1/2}\sum_{i=1}^n Ah(X_{i-1}, X_i) + o_{P(n)}(1). \quad (17.2)$$

This approximation has been found independently by several authors, in particular Gordin (1969), Maigret (1978), Dürr and Goldstein (1986) and Greenwood and Wefelmeyer (1995). The approximation says that the empirical estimator is asymptotically linear with influence function Ah. Hence the estimator is regular and efficient. This was first proved by Penev (1991) without using the martingale approximation. The above proof is due to Greenwood and Wefelmeyer (1995). A shorter proof, not using the perturbation expansion, is obtained if one parametrizes the chain by the joint law P rather than the transition kernel Q; see Bickel (1993) and Bickel and Kwon (2001). The approach is convenient for submodels given by constraints on P rather than Q. For a discussion we refer to Greenwood, Schick and Wefelmeyer (2001). We use the above approach because our submodels will be described in terms of the transition kernel of the chain.

Efficiency of the least squares estimator in the full nonparametric model. Write $\tau^2 = \int x^2\pi(dx) = EX_0^2$ for the stationary second moment. The *least squares functional*

$$\vartheta(Q) = \tau^{-2}\int xy\, P(dx, dy) = E[X_0X_1]/EX_0^2 \quad (17.3)$$

is the minimizer in ϑ of

$$\int (y - \vartheta x)^2\, P(dx, dy) = E[(X_1 - \vartheta X_0)^2].$$

A natural estimator is the empirical version of the functional, the *least squares estimator*

$$\hat{\vartheta}_{LS} = \frac{\sum_{i=1}^n X_{i-1}X_i}{\sum_{i=1}^n X_{i-1}^2}. \quad (17.4)$$

This is the minimizer in ϑ of $\sum_{i=1}^n (X_i - \vartheta X_{i-1})^2$, i.e., the solution of the estimating equation

$$\sum_{i=1}^n X_{i-1}(X_i - \vartheta X_{i-1}) = 0. \quad (17.5)$$

The least squares estimator is the ratio of two empirical estimators. Since continuously differentiable functions of efficient estimators are efficient, $\hat{\vartheta}_{LS}$ is efficient in the full nonparametric model. We have

$$\begin{aligned} n^{1/2}(\hat{\vartheta}_{LS} - \vartheta) &= \frac{n^{-1/2}\sum_{i=1}^n X_{i-1}(X_i - \vartheta X_{i-1})}{\frac{1}{n}\sum_{i=1}^n X_{i-1}^2} \\ &= \tau^{-2}n^{-1/2}\sum_{i=1}^n X_{i-1}(X_i - \vartheta X_{i-1}) + o_{P^{(n)}}(1). \quad (17.6) \end{aligned}$$

Hence, by the martingale approximation (17.2), $\hat{\vartheta}_{LS}$ is asymptotically linear with influence function

$$w = \tau^{-2}Ah_{LS}, \quad \text{where } h_{LS}(x,y) = x(y - \vartheta x). \quad (17.7)$$

The martingale central limit theorem implies that its asymptotic variance is

$$\tau^{-4}E[X_0^2\varepsilon_1^2] + 2\tau^{-4}\sum_{j=2}^{\infty} E[X_0\varepsilon_1 X_{j-1}\varepsilon_j] \quad (17.8)$$

with $\varepsilon_j = X_j - \vartheta X_{j-1}$. We note that since $\hat{\vartheta}_{LS}$ is efficient and regular, its influence function $\tau^{-2}Ah_{LS}$ must be the canonical gradient of ϑ in the full nonparametric model.

17.3 Linear Autoregression

Let $X_0, \ldots, X_n$ be observations from a Markov chain fulfilling the following constraint: There is a number ϑ such that

$$E(X_1 \mid X_0 = x) = \int y\, Q(x, dy) = \vartheta x. \quad (17.9)$$

This is model II. It is a submodel of the full nonparametric Markov chain model of Section 17.2 and can be written as

$$X_i = \vartheta X_{i-1} + \varepsilon_i,$$

where the ε_i are martingale increments, i.e.,

$$E(\varepsilon_1 \mid X_0) = E(X_1 - \vartheta X_0 \mid X_0) = 0.$$

Asymptotic variance of the least squares estimator. The estimating equation (17.5) for $\hat{\vartheta}_{LS}$ is now a *martingale estimating equation* since, in model II, $X_i - \vartheta X_{i-1}$ are martingale increments. This simplifies the influence function and the asymptotic variance of $\hat\vartheta_{LS}$. The expansion (17.6) now says that $\hat\vartheta_{LS}$ has influence function $h_{LS}(x,y) = x(y-\vartheta x)$ and asymptotic variance

$$\tau^{-4} P h_{LS} = \tau^{-4} E[X_0^2 \varepsilon_1^2] = \tau^{-4} E[X_0^2 \rho^2(X_0)],$$

where

$$\rho^2(x) = E(\varepsilon_1^2 \mid X_0 = x) = \int (y-\vartheta x)^2\, Q(x,dy)$$

is the conditional variance of the innovation given that the previous observation is x. By the characterization of regular estimators, h_{LS} is a gradient of ϑ in model II.

Weighted least squares estimators. We will now see that the least squares estimator is not efficient in model II. A large class of alternative estimators in model II is obtained as solutions $\hat\vartheta_W$ of martingale estimating equations

$$\sum_{i=1}^{n} W_\vartheta(X_{i-1})(X_i - \vartheta X_{i-1}) = 0 \tag{17.10}$$

with predictable weights $W_\vartheta(X_{i-1})$. A Taylor expansion argument gives

$$n^{1/2}(\hat\vartheta_W - \vartheta) = -E[W_\vartheta(X_0)X_0]^{-1} n^{-1/2} \sum_{i=1}^{n} W_\vartheta(X_{i-1})(X_i - \vartheta X_{i-1}) + o_{P^{(n)}}(1).$$

Hence $\hat\vartheta_W$ has asymptotic variance

$$E[W_\vartheta(X_0)X_0]^{-2} E[W_\vartheta(X_0)^2 \rho^2(X_0)].$$

By the Cauchy–Schwarz inequality, the asymptotic variance is minimized for

$$W_*(x) = \rho^{-2}(x)x.$$

Since ρ depends on the unknown transition kernel, the weight function W_* cannot be used in the estimating equation (17.10). Suppose we replace ρ by a consistent (kernel) estimator $\hat\rho$. The resulting estimating equation

$$\sum_{i=1}^{n} \hat\rho^{-2}(X_{i-1}) X_{i-1}(X_i - \vartheta X_{i-1}) = 0 \tag{17.11}$$

leads to a weighted least squares estimator

$$\hat\vartheta_{II} = \frac{\sum_{i=1}^n \hat\rho^{-2}(X_{i-1}) X_{i-1} X_i}{\sum_{i=1}^n \hat\rho^{-2}(X_{i-1}) X_{i-1}^2} \tag{17.12}$$

with influence function $M^{-1}\mu$ and asymptotic variance M^{-1}, where

$$\mu(x,y) = \rho^{-2}(x)x(y-\vartheta x) \quad \text{and} \quad M = P\mu^2 = E[\rho^{-2}(X_0)X_0^2]$$

play the roles of *score function* and *Fisher information* for ϑ. The variance M^{-1} is smaller than the asymptotic variance of the least squares estimator $\hat{\vartheta}_{LS}$ unless ρ happens to be constant. Hence $\hat{\vartheta}_{LS}$ is not efficient in model II, in general.

Efficiency of the best weighted least squares estimator. We show that the weighted least squares estimator $\hat{\vartheta}_{II}$ defined in (17.12) is efficient in model II. To calculate the canonical gradient for ϑ in model II, we determine the tangent space of the model. A perturbed transition kernel Q_{nv} must also fulfill the constraint on $E(X_1 \mid X_0)$, possibly with perturbed parameter $\vartheta_{nt} = \vartheta + n^{-1/2}t$:

$$\int y\, Q_{nv}(x, dy) = \vartheta_{nt} x \quad \text{for some } x \in \mathbf{R}.$$

This means that the tangent space V_{II} of model II is the union of the affine spaces

$$V_t = \{v \in V : \int yv(x, y)\, Q(x, dy) = tv\}.$$

For $t = 0$ this is the tangent space when ϑ is known. Since $v(X_0, X_1)$ is conditionally centered, we can write V_0 as

$$V_0 = \{v \in V : \int (y - \vartheta x) v(x, y)\, Q(x, dy) = 0\}.$$

The orthogonal complement of V_0 in V_{II} is spanned by the function $\ell(x, y) = \rho^{-2}(x) x(y - \vartheta x)$.

The canonical gradient of ϑ is obtained as the projection of the gradient $h_{LS}(x, y) = x(y - \vartheta x)$ onto V_{II}. It must lie in the orthogonal complement of the tangent space V_0 for known ϑ. Hence it must be of the form $c\mu(x, y)$, with constant c determined by $P[(h_{LS} - c\mu)\mu] = 0$, i.e., $c = M^{-1}$. Therefore, the canonical gradient of ϑ in model II is $M^{-1}\mu$. This equals the influence function of the best weighted least squares estimator $\hat{\vartheta}_{II}$ defined in (17.12), which is therefore efficient. For generalizations of this result we refer to Wefelmeyer (1996, 1997) and Müller and Wefelmeyer (2002a,b).

Improved empirical estimators. We have seen in Section 17.2 that the empirical estimator $\frac{1}{n}\sum_{i=1}^{n} h(X_{i-1}, X_i)$ is efficient for $Eh(X_0, X_1)$ in the full nonparametric model. We will now construct better, efficient, estimators in the smaller model II, using the constraint (17.9). We follow a heuristic *plug-in principle*, consisting of two steps. First we consider model II with known ϑ and construct an efficient estimator of $Eh(X_0, X_1)$ for each ϑ. Then we replace ϑ by an efficient estimator $\hat{\vartheta}$. Under appropriate conditions, the resulting estimator for $Eh(X_0, X_1)$ will be efficient in model II with unknown ϑ. This construction principle will resurface later in the paper. For general results on the plug-in principle we refer to Müller, Schick and Wefelmeyer (2001a) and Klaassen and Putter (2002).

Suppose first that ϑ is known. We have seen in (17.11) that the constraint $E(\varepsilon_1 \mid X_0) = 0$ leads to new estimators for ϑ. It also leads to new estimators of $Eh(X_0, X_1)$, constructed by adding a correction term to the empirical estimator,

$$\frac{1}{n}\sum_{i=1}^{n}(h(X_{i-1}, X_i) - c(X_i - \vartheta X_{i-1})). \tag{17.13}$$

By the martingale approximation (17.2), they have influence function $Ah(x,y) - c(y - \vartheta x)$ and asymptotic variance $\int (Ah(x,y) - c(y - \vartheta x))^2 \, P(dx, dy)$. By the Cauchy–Schwarz inequality, this variance is minimized for

$$c_*(\vartheta) = \tau^{-2} \int (y - \vartheta x) Ah(x,y) \, P(dx, dy).$$

Since c_* depends on the unknown joint distribution of the Markov chain, it cannot be used in (17.13). We replace c_* by a consistent estimator $\hat{c}_*$, which does not change the asymptotic variance, and obtain the best improved empirical estimator

$$\frac{1}{n}\sum_{i=1}^{n}(h(X_{i-1}, X_i) - \hat{c}_*(\vartheta)(X_i - \vartheta X_{i-1})). \tag{17.14}$$

An explicit construction of $\hat{c}_*$ is in Müller, Schick and Wefelmeyer (2001b).

We have seen in Section 17.2 that $Eh(X_0, X_1)$ has gradient Ah. The canonical gradient in model II with known ϑ is obtained by projecting Ah onto V_0. This amounts to finding c such that $\int (Ah(x,y) - c(y - \vartheta x))^2 \, P(dx, dy)$ is minimized, a problem we already solved when we determined the improved empirical estimator (17.14). It follows that the canonical gradient equals the influence function of the estimator (17.14). Hence this estimator is efficient in model II when ϑ is known.

Now suppose that ϑ is unknown. The plug-in principle says that

$$\hat{E}_{II}h = \frac{1}{n}\sum_{i=1}^{n}(h(X_{i-1}, X_i) - \hat{c}_*(\hat{\vartheta}_{II})(X_i - \hat{\vartheta}_{II} X_{i-1})) \tag{17.15}$$

is efficient for $Eh(X_0, X_1)$ in model II. A version of this result for nonlinear regression is in Müller and Wefelmeyer (2002a).

17.4 Independent Innovations

Let $X_0, \ldots, X_n$ be observations from the AR(1) model $X_i = \vartheta X_{i-1} + \varepsilon_i$, where the innovations ε_i are now i.i.d., with mean zero and finite variance σ^2, and $|\vartheta| < 1$ to ensure geometric ergodicity. This is model I.

Adaptivity. The transition kernel of model I is parametrized by ϑ and by the density, say f, of the innovations:

$$Q(x, dy) = f(y - \vartheta x)\, dy.$$

The tangent space of model I will therefore be expressed in terms of perturbations $\vartheta_{nt} = \vartheta + n^{-1/2}t$ and (Hellinger differentiable) perturbations $f_{nu}(z) \doteq f(z)(1 + n^{-1/2}u(z))$. Since f has mean zero, the functions u vary in

$$U = \{u \in L_2(f) : Eu(\varepsilon) = E[\varepsilon u(\varepsilon)] = 0\}.$$

The transition density is then perturbed as

$$f_{nu}(y - \vartheta_{nt}x) \doteq f(y - \vartheta x)(1 + n^{-1/2}(u(y - \vartheta x) + tx\ell(y - \vartheta x)))$$

with $\ell = -f'/f$. The function ℓ is the score function for location of the innovation distribution and very different from the score function μ in Section 17.3. We obtain that the tangent space of model I is

$$V_I = \{u(y - \vartheta x) + tx\ell(y - \vartheta x) : u \in U, t \in \mathbf{R}\}.$$

Since X_i and $\varepsilon_i = X_i - \vartheta X_{i-1}$ are independent in this model, the decomposition of V_I into functions $u(y - \vartheta x)$ and the linear span of the function $x\ell(y - \vartheta x)$ is orthogonal. Note that the functions $u(y - \vartheta x)$ form the tangent space for fixed ϑ, while the multiples of $x\ell(y - \vartheta x)$ are the tangent space for fixed f. Because they are orthogonal, ϑ can be estimated *adaptively* with respect to f in the sense that the variance bound does not increase if we do not know f.

Estimating the autoregression parameter. Since ϑ is adaptive with respect to f, its canonical gradient can be calculated in the model with fixed f. Hence it is of the form $t_* x\ell(y - \vartheta x)$ with t_* determined by

$$n^{1/2}(\vartheta_{nt} - \vartheta) = t \stackrel{!}{=} t_* t \int x^2 \ell(y - \vartheta x)^2\, P(dx, dy) = t_* t \tau^2 J,$$

where $J = E[\ell(\varepsilon)^2]$ is the Fisher information for location of the innovation distribution. The solution of the above equation is $t_* = \tau^{-2}J^{-1}$. Hence a regular and efficient estimator for ϑ in model I has influence function $\tau^{-2}J^{-1}x\ell(y - \vartheta x)$.

It can be constructed as *one-step improvement* of an $n^{-1/2}$-consistent *initial estimator*, for example the least squares estimator $\hat{\vartheta}_{LS}$:

$$\hat{\vartheta}_I = \hat{\vartheta}_{LS} + \hat{\tau}^{-2}\hat{J}^{-1}\frac{1}{n}\sum_{i=1}^{n} X_{i-1}\hat{\ell}(X_i - \hat{\vartheta}_{LS}X_{i-1}), \tag{17.16}$$

where $\hat{\tau}$, $\hat{J}$ and $\hat{\ell}$ are appropriate estimators of τ, J and ℓ, respectively. For constructions in more general autoregressive models see Kreiss (1987a,b), Jeganathan (1995), Drost, Klaassen and Werker (1997), Koul and Schick (1997) and Schick and Wefelmeyer (2002b).

Of course, in model I the asymptotic variance of the best weighted least squares estimator $\hat{\vartheta}_{II}$ cannot be smaller than the asymptotic variance of $\hat{\vartheta}_I$. Indeed, since $\rho(x) = \sigma$ in model I, the asymptotic variance of $\hat{\vartheta}_{II}$ and of the ordinary least squares estimator $\hat{\vartheta}_{LS}$ both become $\tau^{-2}\sigma^2$, while the asymptotic variance of $\hat{\vartheta}_I$ is $\tau^{-2}J^{-1}$. Since $E[\varepsilon\ell(\varepsilon)] = 1$, the Cauchy–Schwarz inequality gives $\sigma^2 \geq J^{-1}$, with equality only if ε is proportional to $\ell(\varepsilon)$, i.e., for normal errors.

Estimating the innovation distribution. Estimators of functionals of the innovation distribution can be based on residuals, i.e., on estimated innovations $\hat{\varepsilon}_i = X_i - \hat{\vartheta}X_{i-1}$. Consider for example a linear functional $Eh(\varepsilon) = \int h(z)f(z)\,dz$, where h is f-square-integrable. A simple estimator is the empirical estimator $\frac{1}{n}\sum_{i=1}^{n} h(\hat{\varepsilon}_i)$ based on the residuals. It will not be efficient in model I since it uses neither independence nor centeredness of the innovations. To obtain an efficient estimator for $Eh(\varepsilon)$, we use again the plug-in principle employed in Section 17.3 for efficient estimation of $Eh(X_0, X_1)$. If ϑ is known, we can observe the innovations. They are independent with mean zero density f. Similarly as in Section 17.3, we can use the constraint $E\varepsilon = 0$ to introduce modified empirical estimators

$$\frac{1}{n}\sum_{i=1}^{n}(h(\varepsilon_i) - c\varepsilon_i).$$

Their asymptotic variance is $E[(h(\varepsilon) - Eh(\varepsilon) - c\varepsilon)^2]$. The variance is minimized for $c_* = \sigma^{-2}E[\varepsilon h(\varepsilon)]$. The minimal variance is $E[h(\varepsilon)^2] - E[h(\varepsilon)]^2 - \sigma^{-2}E[\varepsilon h(\varepsilon)]^2$. It is not changed if we replace c_* by a consistent estimator $\hat{c}_*$, for example a ratio of empirical estimators. With this choice, the best modified empirical estimator for $Eh(\varepsilon)$ is

$$\frac{1}{n}\sum_{i=1}^{n}\Big(h(\varepsilon_i) - \frac{\sum_{i=1}^{n}\varepsilon_i h(\varepsilon_i)}{\sum_{i=1}^{n}\varepsilon_i^2}\varepsilon_i\Big). \tag{17.17}$$

This estimator is efficient in model I with known ϑ. The result goes back to

Levit (1975). The plug-in principle now says that

$$\frac{1}{n}\sum_{i=1}^{n}\left(h(\hat\varepsilon_i)-\frac{\sum_{i=1}^{n}\hat\varepsilon_i h(\hat\varepsilon_i)}{\sum_{i=1}^{n}\hat\varepsilon_i^2}\hat\varepsilon_i\right) \tag{17.18}$$

is efficient for $Eh(\varepsilon)$ in model I if an efficient estimator $\hat\vartheta_I$ for ϑ is used in the residuals $\hat\varepsilon_i = X_i - \hat\vartheta_I X_{i-1}$. This result is due to Wefelmeyer (1994). For generalizations to nonlinear autoregression and to invertible linear time series see Schick and Wefelmeyer (2002a,b). Generalizations to *nonparametric* autoregression models $X_i = r(X_{i-1})+\varepsilon_i$, with unknown regression function r, are also possible, even though r cannot be estimated at the parametric rate $n^{-1/2}$. See Akritas and Van Keilegom (2001) and Müller, Schick and Wefelmeyer (2004) for corresponding results in nonparametric regression.

Estimating the stationary distribution. The simplest estimator for the expectation $Eh(X_0)$ of a π-square-integrable function h is the empirical estimator discussed in Section 17.2. It will not be efficient in model I since it uses neither independence nor centeredness of the innovations. The condition $EX_0 = 0$ could be used just as for $Eh(\varepsilon)$ in (17.17). To make use of the independence of the innovations, we observe that for $|\vartheta| < 1$ the AR(1) process is invertible and has a moving average representation

$$X_0 = \sum_{j=0}^{\infty}\vartheta^j\varepsilon_{-j} = \sum_{j=1}^{\infty}\vartheta^{j-1}\varepsilon_j$$

in distribution. Hence

$$Eh(X_0) = Eh(S) \quad \text{with } S = \sum_{j=1}^{\infty}\vartheta^{j-1}\varepsilon_j.$$

This is approximated by

$$Eh(S^{(m)}) \quad \text{with } S^{(m)} \doteq \sum_{j=1}^{m}\vartheta^{j-1}\varepsilon_j$$

if m increases with n. To estimate $Eh(X_0)$, we will again use the plug-in principle. Assume first that ϑ is known. Then we can observe the innovations $\varepsilon_i = X_i - \vartheta X_{i-1}$, and we can estimate $Eh(S^{(m)})$ by a U-statistic defined as follows. Let Φ denote the set of one-one functions φ from $\{1,\dots,m\}$ into $\{1,\dots,n\}$. For $\varphi \in \Phi$ set

$$S_\varphi(\vartheta) = \sum_{j=1}^{m}\vartheta^{j-1}\varepsilon_{\varphi(j)} = \sum_{j=1}^{m}\vartheta^{j-1}(X_{\varphi(j)} - \vartheta X_{\varphi(j)-1}).$$

We estimate $Eh(S^{(m)})$ by the average over these sums, the U-statistic

$$U(\vartheta) = \frac{(n-m)!}{n!} \sum_{\varphi \in \Phi} h(S_\varphi(\vartheta)).$$

If m is fixed, and if the stationary distribution is unrestricted, $U(\vartheta)$ is efficient for $Eh(S^{(m)})$. Schick and Wefelmeyer (2002c) show that $U(\vartheta)$ is also efficient for $Eh(X_0)$ if m increases with n at an appropriate rate.

The only constraint on the stationary distribution is $EX_0 = 0$. This constraint can be used to improve the empirical estimator, similarly as in (17.13) and (17.17): Consider

$$U(\vartheta, c) = U(\vartheta) - c\frac{1}{n}\sum_{i=1}^{n}(X_i - \vartheta X_{i-1}).$$

Now note that by the Hoeffding decomposition, $U(\vartheta)$ is asymptotically linear with influence function $w(y - \vartheta x)$, where

$$w = \sum_{j=1}^{\infty} w_j \quad \text{with} \quad w_j(z) = E(h(S) \mid \varepsilon_j = z) - Eh(S).$$

The asymptotic variance of $U(\vartheta, c)$ is therefore minimized for

$$c_*(\vartheta) = \sigma^{-2} E[\varepsilon_1 w(\varepsilon_1)].$$

The asymptotic variance is not changed if we replace c_* by a consistent estimator, e.g., by

$$\hat{c}_*(\vartheta) = \frac{\sum_{i=1}^{n}(X_i - \vartheta X_{i-1}) \sum_{j=1}^{m} H_{ji}(\vartheta)}{\sum_{i=1}^{n}(X_i - \vartheta X_{i-1})^2}$$

with

$$H_{ji}(\vartheta) = \frac{(n-m)!}{(n-1)!} \sum_{\varphi \in \Phi, \varphi(j) = i} h(S_\varphi(\vartheta)).$$

If m increases at an appropriate rate, $U(\vartheta, \hat{c}_*(\vartheta))$ is therefore efficient for $Eh(X_0)$ in model I with ϑ known. The plug-in principle now says that the estimator $U(\hat{\vartheta}_I, \hat{c}_*(\hat{\vartheta}_I))$ is efficient in model I. This result is proved in Schick and Wefelmeyer (2002c) for general causal and invertible linear processes.

Estimating the stationary density. There is a rich literature on kernel estimators for the stationary density of time series: see e.g., Chanda (1983), Yakowitz (1989) and Tran (1992). Such estimators converge more slowly than the parametric rate $n^{-1/2}$, depending on the smoothness of the density. For linear processes one can however use the independence of the innovations to obtain density estimators with convergence rate $n^{-1/2}$. This was first observed by

Saavedra and Cao (1999, 2000). They write the stationary density d of a moving average process $X_i = \varepsilon_i + \vartheta\varepsilon_{i-1}$ with innovation density f in the convolution representation $d(x) = \int f(x-\vartheta y)f(y)\,dy$ and estimate $d(x)$ by plugging in a kernel estimator $\hat{f}$ for f. The resulting plug-in estimator $\hat{d}(x) = \int \hat{f}(x-\hat{\vartheta}y)\hat{f}(y)\,dy$ converges at the parametric rate $n^{-1/2}$. Schick and Wefelmeyer (2002d) show that $\hat{d}(x)$ is efficient if an efficient estimator for ϑ is used, and Schick and Wefelmeyer (2002e) prove functional convergence of such estimators for general MA(q) processes. In the i.i.d. case, related plug-in estimators for other smooth functionals of densities and regression functions have been studied before: see e.g. Hall and Marron (1987), Bickel and Ritov (1988) and Birgé and Massart (1995) for nonlinear integral functionals of densities and their derivatives, and Goldstein and Messer (1992) and Samarov (1993) for analogous results in regression models.

By definition, the stationary density, say k, of our model I also has a convolution representation

$$k(x) = \int f(y - \vartheta x)k(y)\,dy.$$

We therefore obtain a $n^{1/2}$-consistent density estimator

$$\overline{k}(x) = \int \hat{f}(y - \hat{\vartheta}x)\hat{k}(y)\,dy$$

for appropriate choices of $\hat{k}$, $\hat{f}$ and $\hat{\vartheta}$. Such an estimator will however not be efficient, since it does not use the assumption of independent innovations to its full extent. To get an efficient estimator, one would again need to exploit the moving average representation $X_0 = \sum_{j=1}^{\infty} \vartheta^{j-1}\varepsilon_j$ (in distribution) and approximate the density k by the density of $S^{(m)} = \sum_{j=1}^{m} \vartheta^{j-1}\varepsilon_j$ for m increasing with n at an appropriate rate.

References

1. Akritas, M. G. and Van Keilegom, I. (2001). Non-parametric estimation of the residual distribution, *Scandinavian Journal of Statistics*, **28**, 549–567.

2. Bickel, P. J. (1993). Estimation in semiparametric models, In *Multivariate Analysis: Future Directions* (Ed., C. R. Rao), pp. 55–73, North-Holland, Amsterdam.

3. Bickel, P. J., Klaassen, C. A. J., Ritov, Y. and Wellner, J. A. (1998). *Efficient and Adaptive Estimation for Semiparametric Models*, Springer-Verlag, New York.

4. Bickel, P. J. and Kwon, J. (2001). Inference for semiparametric models: Some questions and an answer (with discussion), *Statistica Sinica*, **11**, 863–960.

5. Bickel, P. J. and Ritov, Y. (1988). Estimating integrated squared density derivatives: Sharp best order of convergence estimates, *Sankhyā, Series A*, **50**, 381–393.

6. Birgé, L. and Massart, P. (1995). Estimation of integral functionals of a density, *Annals of Statistics*, **23**, 11–29.

7. Chanda, K. C. (1983). Density estimation for linear processes, *Annals of the Institute of Statistical Mathematics*, **35**, 439–446.

8. Drost, F. C., Klaassen, C. A. J. and Werker, B. J. M. (1997). Adaptive estimation in time-series models, *Annals of Statistics*, **25**, 786–817.

9. Dürr, D. and Goldstein, S. (1986). Remarks on the central limit theorem for weakly dependent random variables, In *Stochastic Processes — Mathematics and Physics* (Eds., S. Albeverio, P. Blanchard and L. Streit), pp. 104–118, Lecture Notes in Mathematics 1158, Springer-Verlag, Berlin.

10. Goldstein, L. and Messer, K. (1992). Optimal plug-in estimators for nonparametric functional estimation, *Annals of Statistics*, **20**, 1306–1328.

11. Gordin, M. I. (1969). The central limit theorem for stationary processes, *Soviet Math. Doklady*, **10**, 1174–1176.

12. Greenwood, P. E., Schick, A. and Wefelmeyer, W. (2001). Comment on Bickel and Kwon, *Statistica Sinica*, **11**, 892–906.

13. Greenwood, P. E. and Wefelmeyer, W. (1995). Efficiency of empirical estimators for Markov chains, *Annals of Statistics*, **23**, 132–143.

14. Hájek, J. (1970). A characterization of limiting distributions of regular estimates, *Z. Wahrsch. Verw. Gebiete*, **14**, 323–330.

15. Hall, P. and Marron, J. S. (1990). On variance estimation in nonparametric regression, *Biometrika*, **77**, 415–419.

16. Höpfner, R. (1993a). On statistics of Markov step processes: Representation of log-likelihood ratio processes in filtered local models, *Probability Theory and Related Fields*, **94**, 375–398.

17. Höpfner, R. (1993b). Asymptotic inference for Markov step processes: Observation up to a random time, *Stochastic Processes and their Applications*, **48**, 295–310.

18. Höpfner, R., Jacod, J. and Ladelli, L. (1990). Local asymptotic normality and mixed normality for Markov statistical models, *Probability Theory and Related Fields*, **86**, 105–129.

19. Jeganathan, P. (1995). Some aspects of asymptotic theory with applications to time series models, *Econometric Theory*, **11**, 818–887.

20. Kartashov, N. V. (1985a). Criteria for uniform ergodicity and strong stability of Markov chains with a common phase space, *Theory of Probability and Mathematical Statistics*, **30**, 71–89.

21. Kartashov, N. V. (1985b). Inequalities in theorems of ergodicity and stability for Markov chains with common phase space, I, *Theory of Probability and Mathematics Statistics*, **30**, 247–259.

22. Kartashov, N. V. (1996). *Strong Stable Markov Chains*, VSP, Utrecht, The Netherlands.

23. Klaassen, C. A. J. and Putter, H. (2002). Efficient estimation of Banach parameters in semiparametric models, *Technical Report*, Korteweg-de Vries Institute for Mathematics, University of Amsterdam, The Netherlands, http://preprint.beta.uva.nl/

24. Koul, H. L. and Schick, A. (1997). Efficient estimation in nonlinear autoregressive time-series models, *Bernoulli*, **3**, 247–277.

25. Kreiss, J.-P. (1987a). On adaptive estimation in stationary ARMA processes, *Annals of Statistics*, **15**, 112–133.

26. Kreiss, J.-P. (1987b). On adaptive estimation in autoregressive models when there are nuisance functions, *Statistics and Decisions*, **5**, 59–76.

27. Levit, B. Y. (1975). Conditional estimation of linear functionals, *Problems Information Transmission*, **11**, 39–54.

28. Maigret, N. (1978). Théorème de limite centrale fonctionnel pour une chaîne de Markov récurrente au sens de Harris et positive, *Annals of the Institute of H. Poincaré for Probability and Statistics*, **14**, 425–440.

29. Müller, U. U., Schick, A. and Wefelmeyer, W. (2001a). Plug-in estimators in semiparametric stochastic process models, In *Selected Proceedings of the Symposium on Inference for Stochastic Processes* (Eds., I. V. Basawa, C. C. Heyde and R. L. Taylor), pp. 213-234, IMS Lecture Notes-Monograph Series 37, Institute of Mathematical Statistics, Beachwood, Ohio.

30. Müller, U. U., Schick, A. and Wefelmeyer, W. (2001b). Improved estimators for constrained Markov chain models, *Statistics & Probability Letters*, **54**, 427-435.

31. Müller, U. U., Schick, A. and Wefelmeyer, W. (2004). Estimating linear functionals of the error distribution in nonparametric regression, *Journal of Statistical Planning and Inference*, **119**, 75–93.

32. Müller, U. U. and Wefelmeyer, W. (2002a). Estimators for models with constraints involving unknown parameters, *Mathematical Methods in Statistics*, **11**, 221–235.

33. Müller, U. U. and Wefelmeyer, W. (2002b). Autoregression, estimating functions, and optimality criteria, In *Advances in Statistics, Combinatorics and Related Areas* (Eds., C. Gulati, Y.-X. Lin, J. Rayner and S. Mishra), World Scientific Publishers, Singapore.

34. Penev, S. (1991). Efficient estimation of the stationary distribution for exponentially ergodic Markov chains, *Journal of Statistical Planning and Inference*, **27**, 105–123.

35. Roussas, G. G. (1965). Asymptotic inference in Markov processes, *Annals of Mathematical Statistics*, **36**, 987–992.

36. Saavedra, A. and Cao, R. (1999). Rate of convergence of a convolution-type estimator of the marginal density of an MA(1) process, *Stochastic Processes and their Applications*, **80**, 129–155.

37. Saavedra, A. and Cao, R. (2000). On the estimation of the marginal density of a moving average process, *Canadian Journal of Statistics*, **28**, 799–815.

38. Samarov, A. (1993). Exploring regression structure using nonparametric functional estimation, *Journal of the American Statistical Association*, **88**, 836–847.

39. Schick, A. and Wefelmeyer, W. (2002a). Estimating the innovation distribution in nonlinear autoregressive models, *Annals of the Institute of Statistical Mathematics*, **54**, 245–260.

40. Schick, A. and Wefelmeyer, W. (2002b). Efficient estimation in invertible linear processes, *Mathematical Methods in Statistics*, **11**, 358–379.

41. Schick, A. and Wefelmeyer, W. (2002c). Estimating invariant laws of linear processes by U-statistics, *Annals of Statistics* (to appear).

42. Schick, A. and Wefelmeyer, W. (2002d). Root n consistent and optimal density estimators for moving average processes, *Scandinavian Journal of Statistics*, **31**, 63–78.

43. Schick, A. and Wefelmeyer, W. (2002e). Functional convergence and optimality of plug-in estimators for stationary densities of moving average processes, *Technical Report*, Department of Mathematical Sciences, Binghamton University, http://math.binghamton.edu/anton/preprint.html.

44. Tran, L. T. (1992). Kernel density estimation for linear processes, *Stochastic Processes and their Applications*, **41**, 281–296.

45. Wefelmeyer, W. (1994). An efficient estimator for the expectation of a bounded function under the residual distribution of an autoregressive process, *Annals of the Institute of Statistical Mathematics*, **46**, 309–315.

46. Wefelmeyer, W. (1996). Quasi-likelihood models and optimal inference, *Annals of Statistics*, **24**, 405–422.

47. Wefelmeyer, W. (1997). Adaptive estimators for parameters of the autoregression function of a Markov chain, *Journal of Statistical Planning and Inference*, **58**, 389–398.

48. Yakowitz, S. (1989). Nonparametric density and regression estimation for Markov sequences without mixing assumptions, *Journal of Multivariate Analysis*, **30**, 124–136.

PART VIII
INFERENTIAL ANALYSIS

18

Distance-based Multivariate Two Sample Tests

Carles M. Cuadras and Josep Fortiana

Department of Statistics, University of Barcelona, Barcelona, Spain

Abstract: Multivariate statistical inference is synonymous with multivariate normal distribution. Most multivariate models and tests are based on the Gaussian hypothesis. But often the variables are not quantitative or fail to follow this hypothesis. We prove that the use of distances between observations can provide alternative tests for comparing two multivariate samples when the data is of general type. This approach is illustrated with two real data examples.

Keywords and phrases: Mixed multivariate data, Mahalanobis distance, joining two parametric models, randomization tests, bootstrap distribution

18.1 Introduction

The classic inference in Multivariate Analysis is mainly based on the hypothesis of normality. Thus the early books by Anderson (1958), Kshirsagar (1972) and Rao (1973) rely on this hypothesis. Normality is also omnipresent in books such as Rencher (1998). This is so because of the key role of the mathematical tractability of the multivariate normal distribution. Efron (1993) showed that, by using bootstrap methods, it is possible to perform useful analyses without invoking normality. Also Flury (1997) approached certain multivariate tests by introducing resampling and bootstrap methods.

It is shown in this paper that testing the hypothesis that two samples have been drawn from populations with the same multivariate distribution may be approached in different ways by using distances between pairs of observations of the variables. This approach is especially useful for handling mixed variables and statistical models where only the marginal parametrizations are available.

18.2 Some Geometrical Concepts

Let Π_1, Π_2 be two populations represented as two independent S-valued random vectors X, Y, with probability density functions f, g with respect to a suitable common measure λ. Suppose that d is a dissimilarity or distance function on S, i.e., $d(x,x') = d(x',x) \geq d(x,x) = 0$ for $x, x' \in S$. To study the differences between Π_1, Π_2 via d, we introduce the *geometric variability* of $\mathbf{X}$ with respect to d,

$$V_d(X) = \frac{1}{2}\int_{S^2} d^2(x,x')\, f(x)f(x')\lambda(dx)\lambda(dx'), \tag{18.1}$$

the *proximity function* of x_0 to the population represented by X with respect to d,

$$\phi_d^2(x_0)_1 = \int_{S^2} d^2(x_0,x)\, f(x)\lambda(dx) - V_d(X), \tag{18.2}$$

and the squared *distance between* Π_1, Π_2,

$$\Delta^2(\Pi_1,\Pi_2) = \int_{S^2} d^2(x,y)\, f(x)g(y)\lambda(dx)\lambda(dy) - V_d(X) - V_d(Y). \tag{18.3}$$

We can likewise consider $V_d(Y)$ and $\phi_d^2(y_0)_2$.

By suitable choices of constants $a > 0$ and b, we may transform the squared distance $d^2 \rightarrow ad^2 + b$, which transforms the proximity function $\phi_d^2(x) \rightarrow a\phi_d^2(x) + b/2$, in order to generate the probability density

$$f_d(x) = \exp[-\phi_d^2(x)], \quad x \in S. \tag{18.4}$$

Then a measure of divergence between f_d and the true density f of X is the Kullback–Leibler divergence, which satisfies

$$I(f\|f_d) = V_d(X) - H(f) \geq 0,$$

with equality if f_d is f, where $H(f)$ is the Shannon entropy.

These geometrical concepts were introduced by Cuadras and Fortiana (1995) and Cuadras et al. (1997a,b). $V_d(X)$ is a version of Rao's quadratic entropy, whereas $\Delta^2(\Pi_1,\Pi_2)$ is a Jensen difference [Rao (1982)].

These definitions depend on distances between observations and, for well-known distances, reduce to some familiar concepts in multivariate analysis:

1. When $S = R^p$ and d is the ordinary Euclidean distance, then $V_d(X) = \mathrm{tr}(\Sigma)$, where Σ is the covariance matrix of X.

2. When $X \sim N_p(\mu_1, \Sigma)$, $Y \sim N_p(\mu_2, \Sigma)$ and d is the Mahalanobis distance between observations, then $V_d(X) = V_d(Y) = p$ and

$$\begin{aligned} \phi_d^2(x)_k &= (x-\mu_k)'\Sigma^{-1}(x-\mu_k),\ k=1,2, \\ \Delta^2(\Pi_1,\Pi_2) &= (\mu_1-\mu_2)'\Sigma^{-1}(\mu_1-\mu_2). \end{aligned}$$

3. When X is a continuous r.v. with range an interval S, cdf F and the distance is $\delta(x, y) = \sqrt{|x-y|}$, then

$$V_\delta(X) = \int_S F(x)(1-F(x))dx = \sum_{n\geq 1} \lambda_n, \tag{18.5}$$

where $\{\lambda_n, n \geq 1\}$ are the eigenvalues of the integral operator $(\mathcal{K}\varphi)(t) = \int_a^b K(s,t)\,\varphi(s)ds$ with symmetric kernel $K(x,y) = \min\{F(x), F(y)\} - F(x)F(y)$, i.e., such that $\mathcal{K}\psi_n = \lambda_n\psi_n$, where $\{\psi_n, n \geq 1\}$ are the eigenfunctions.

4. Liu and Rao (1995) obtained the asymptotic distribution of some statistics based on Rao's quadratic entropy, which is equivalent to $2V_d(X)$. This measure is related to the variance for $d(x,y) = |x-y|$ and to $V_\delta(X)$, see (18.5), for $\delta(x,y) = \sqrt{|x-y|}$, and by introducing the diversity between and within populations, it can provide a generalization of ANOVA called ANOQE. For example, to test $F = G$, where F, G are two univariate cdf's, the asymptotic distribution of

$$\frac{n_1 n_2}{n_1+n_2}\int_S [F_{n_1}(t) - G_{n_2}(t)]^2 dt,$$

is a series of independent chi-squares weighted by the above eigenvalues, where F_{n_1}, G_{n_2} are the empirical cdf's on samples of sizes n_1, n_2. These eigenvalues are well-known for the uniform distribution. The eigenvalues for the exponential, logistic and Pareto distributions were obtained by Cuadras and Fortiana (1995), Cuadras and Lahlou (2000, 2002), respectively. See Cuadras and Cuadras (2002) for an application in goodness of fit assessment.

The tests in ANOQE and other tests depend on the parametric distribution of the variables. The advantage of using distances is more apparent for mixed models, i.e., when the variables are continuous, categorical, binary and/or nominal, and the definition of a probabilistic model is not accessible. However, as pointed out in the Introduction, most multivariate models work well with quantitative variables. This quantification can be done by embedding S in an appropriate space.

We say that d is a Euclidean distance function if there exists a representation $\psi : S \to L$, where L is a Euclidean or a separable Hilbert space, such that for all $x, y \in S$,

$$d^2(x,y) = \| \psi(x) - \psi(y) \|^2 .$$

A necessary and sufficient condition for d to be a Euclidean distance is that d^2 is conditionally negative definite, see Lau (1985).

Theorem 18.2.1 *Assume that $E(\| \psi(X) \|^2)$ and $E(\| \psi(Y) \|^2)$ are finite. Then*

$$\begin{aligned} V_d(X) &= E(\| \psi(X) \|^2) - E(\| \psi(X) \|)^2, \\ \phi_d^2(x)_1 &= \| \psi(x) - E(\psi(X)) \|^2, \\ \Delta^2(\Pi_1, \Pi_2) &= \| E(\psi(X)) - E(\psi(Y)) \|^2 . \end{aligned} \tag{18.6}$$

PROOF. See Theorem 2 for $M = I$ of Cuadras et al. (1997a). ■

Thus the geometric variability generalizes the variance, the proximity function is a measure of the distance between one observation and the population from which it is drawn, and $\Delta^2(\Pi_1, \Pi_2)$ is also a distance between two mean vectors. We may understand $E(\psi(X))$ as a d-mean of X. See Cuadras (1989), Cuadras and Fortiana (1995, 2000), Cuadras et al. (1997a,b).

In practical applications, only finite samples are available, so here we shall review the finite case. Let $D = (d_{ij})$ be an $n \times n$ Euclidean distance matrix. Let J_n be an $n \times n$ matrix whose entries are all 1, $H = I_n - J_n/n$ the centering matrix, $A = (a_{ij})$ a matrix with entries $a_{ij} = -d_{ij}^2/2$ and consider the inner product matrix $G = HAH$. Then if we find the spectral decomposition $G = U\Lambda U'$ and compute $X = U\Lambda^{1/2}$, the rows $x_1', \ldots, x_n'$ of X, called principal coordinates, define n points in a Euclidean space R^p, where p =rank(G), and the Euclidean distance between rows i, j is $d_{ij} = \sqrt{(x_i - x_j)'(x_i - x_j)}$, the original distance.

The columns of X, understood as variables, are centered and can be interpreted as principal components. Metric Multidimensional Scaling or Principal Coordinate Analysis is a well-known multivariate analysis method to represent the individuals in a low-dimensional space. See Rao (1964), Gower (1966), and Cuadras and Fortiana (1998) for a deduction in terms of ANOVA and log-linear models.

Finally note that in the definition of geometric variability and proximity function, we may have a distance and representation for X and another distance and representation for Y, e.g., Mahalanobis distances with different means and variances. However, if we work with $\Pi_1 \cup \Pi_2$, a global distance and representation are necessary.

18.3 DB Approach in Comparing Two Populations

Now assume that two samples $x_1, \ldots, x_{n_1}, y_1, \ldots, y_{n_2}$ of sizes n_1, n_2 coming from Π_1, Π_2, respectively, are available. Our aim is to test

$$H_0 : \Delta^2(\Pi_1, \Pi_2) = 0 \tag{18.7}$$

against $H_1 : \Delta^2(\Pi_1, \Pi_2) > 0$. Acceptance of H_0 means that, in some sense, the two populations are equal, e.g., the d-means are equal:

$$E(\psi(X)) = E(\psi(Y)).$$

The sampling version of (18.1),(18.2),(18.3) as well as (18.6), can easily be obtained:

$$\widehat{V}_d(X) = \frac{1}{2n_1^2}\sum_{i=1}^{n_1}\sum_{j=1}^{n_1} d^2(x_i, x_j), \quad \widehat{V}_d(Y) = \frac{1}{2n_2^2}\sum_{i=1}^{n_2}\sum_{j=1}^{n_2} d^2(y_i, y_j), \tag{18.8}$$

and

$$\widehat{\Delta}^2(\Pi_1, \Pi_2) = \frac{1}{n_1 n_2}\sum_{i=1}^{n_1}\sum_{j=1}^{n_2} d^2(x_i, y_j) - \widehat{V}_d(X) - \widehat{V}_d(Y). \tag{18.9}$$

Let D_{11} and D_{22} be the $n_1 \times n_1$ and $n_2 \times n_2$ distance matrices obtained taking distances between the observations in the first and second sample, respectively, and let D_{12} be the $n_1 \times n_2$ interdistance matrix. The overall distance matrix is

$$D = \begin{pmatrix} D_{11} & D_{12} \\ D_{21} & D_{22} \end{pmatrix} \tag{18.10}$$

and, by using principal coordinate analysis, we can obtain the spectral decomposition $G = U\Lambda U'$, where G is the centered inner product matrix. We then find the coordinate matrix $Z = U\Lambda^{1/2}$ partitioned as

$$Z = \begin{bmatrix} Z_1 \\ Z_2 \end{bmatrix}.$$

Thus $G = ZZ'$. The distances between rows of $Z_1(n_1 \times p)$ give D_{11}, and similarly the rows of $Z_2(n_2 \times p)$ give D_{22}. The entries in D_{12} are the distances between rows of Z_1 and rows of Z_2. We may suppose a common dimension p. The so-called standard coordinates are the rows of U, which can be partitioned as $U' = [U_1', U_2']$.

18.4 The Generalized T^2 Test

A first test may be based in the comparison of the two "data matrices" Z_1 and Z_2, with mean vectors $\overline{z}_1, \overline{z}_2$ and covariance matrices S_1,S_2. Note that $n_1\overline{z}_1+n_2\overline{z}_2 = 0$.

To test H_0 we propose the statistic

$$T^2 = (\overline{z}_1 - \overline{z}_2)'V^{-1}(\overline{z}_1 - \overline{z}_2),$$

where $V = n_1S_1 + n_2S_2 = Z_1'Z_1 + Z_2'Z_2 = \Lambda$. Thus the value of the statistic T^2 is

$$T_0^2 = (\overline{z}_1 - \overline{z}_2)'\Lambda^{-1}(\overline{z}_1 - \overline{z}_2) = \|\overline{u}_1 - \overline{u}_2\|^2,$$

where $\overline{u}_1, \overline{u}_2$ are the mean vectors of the standard coordinates U_1, U_2. Note that T_0^2 could not be the squared distance $\widehat{\Delta}^2(\Pi_1, \Pi_2)$.

For multivariate normal data and ordinary Euclidean distance, the distribution of the above statistic is proportional to Hotelling's T^2. In general, the distribution of T^2 is unknown. A randomization procedure [see Flury (1997)] can be performed by considering the $N = (n_1 + n_2)!/n_1!n_2!$ possible ways of partitioning the $n_1 + n_2$ rows in Z, and obtaining the randomization distribution of the T^2 values obtained. If T_0^2 is in the extreme tail of this distribution, there will be evidence against H_0. If N is too large, we may take only $N' < N$ permutations and follow the same procedure with the N' values.

An alternative test, which gives similar results, can be based on

$$E^2 = \frac{n_1n_2}{n_1+n_2}(\overline{z}_1 - \overline{z}_2)'(\overline{z}_1 - \overline{z}_2).$$

In fact, for a distance like Mahalanobis, the test on E^2 is a test on $\widehat{\Delta}^2(\Pi_1, \Pi_2)$.

Theorem 18.4.1 *Let $x_1', \ldots, x_{n_1}', y_1', \ldots, y_{n_2}'$ be multivariate samples, rows of two data matrices X, Y, with vector means $\overline{x}, \overline{y}$. Suppose that the squared distances between observations are given by*

$$d^2(a,b) = (a-b)'M^{-1}(a-b), \tag{18.11}$$

where $a, b \in \{x_1, \ldots, x_{n_1}, y_1, \ldots, y_{n_2}\}$ and M is a p.d. matrix. Then

$$\widehat{\Delta}^2(\Pi_1, \Pi_2) = (\overline{z}_1 - \overline{z}_2)'(\overline{z}_1 - \overline{z}_2) = (\overline{x}-\overline{y})'M^{-1}(\overline{x}-\overline{y}).$$

PROOF. The inner product matrices for intra and inter distances (18.11) are

$$G_{11} = XM^{-1}X',\ G_{22} = YM^{-1}Y',\ G_{12} = XM^{-1}Y',$$

and the matrices of squared distance can be written as

$$D_{ij}^{(2)} = g_i 1'_{n_j} + 1_{n_i} g'_j - 2G_{ij}, \quad i,j = 1,2,$$

where g_1, g_2 are column vectors with the diagonal entries in G_{11}, G_{22}. Then

$$\widehat{V}_d(X) = \frac{1}{2n_1^2} 1'_{n_1} D_{11}^{(2)} 1_{n_1} = \frac{1}{n_1} g'_1 1_{n_1} - \frac{1}{n_1^2} 1'_{n_1} G_{11} 1_{n_1},$$

similarly $\widehat{V}_d(Y)$, and

$$\begin{aligned}\widehat{\Delta}^2(\Pi_1, \Pi_2) &= (n_1 n_2)^{-1} 1'_{n_1} D_{12}^{(2)} 1_{n_2} - \widehat{V}_d(X) - \widehat{V}_d(Y) \\ &= n_1^{-2} 1'_{n_1} G_{11} 1_{n_1} + n_2^{-2} 1'_{n_2} G_{22} 1_{n_2} - (n_1 n_2)^{-1} 1'_{n_1} G_{12} 1_{n_2}.\end{aligned}$$

The result follows by expanding the right-hand part of

$$(\overline{x} - \overline{y})' M^{-1} (\overline{x} - \overline{y}) = (n_1^{-1} X' 1_{n_1} - n_2^{-1} Y' 1_{n_2})' M^{-1} (n_1^{-1} X' 1_{n_1} - n_2^{-1} Y' 1_{n_2}),$$

and taking into account that $G_{ij} = Z_i Z'_j, i,j = 1,2.$ ■

18.5 The Proximity Function Test

For general mixed data and given a distance d, let ϕ_1^2, ϕ_2^2 be the proximity functions for X, Y, respectively, see (18.2). In view of (18.6), a discriminant rule to classify one individual $\omega \in \Pi_1 \cup \Pi_2$ given the observations x, is:

$$\text{Alllocate} \quad \omega \quad \text{to} \quad \Pi_i \quad \text{if} \quad \phi_i^2(x) = \min\{\phi_1^2(x), \phi_2^2(x)\}. \tag{18.12}$$

This rule, proposed by Cuadras (1989, 1992a), Cuadras et al. (1997a), can deal with general data and is based on the discriminant function

$$D(x) = \frac{1}{2}[\phi_2^2(x) - \phi_1^2(x)],$$

which reduces to well-known discriminators for multinomial and normal data and suitable distances. Note that (18.12) is consistent with the density (18.4), see Cuadras et al. (1997b). Two examples are given.

18.5.1 Linear and quadratic discrimination

If $X \sim N_p(\mu_1, \Sigma_1)$, $Y \sim N_p(\mu_2, \Sigma_2)$ and we use the Mahalanobis distance plus an additive constant

$$d_k^2(x,y) = \begin{cases} (x-y)'\Sigma_k^{-1}(x-y) + \beta_k, & \text{if} \quad x \neq y, \\ 0 & \text{if} \quad x = y, \end{cases}$$

then $D(x)$ reduces to the linear discriminator if $\Sigma_1 = \Sigma_2$, $\beta_k = 0$, and to the quadratic discriminator if $\Sigma_1 \neq \Sigma_2$, $\beta_k = \log|\Sigma_k|/2$.

18.5.2 Marginal parametrization

Suppose that the observation vector w can be partitioned into (u, v), where u is (possibly) discrete and v continuous, (e.g., multinomial and normal), with density functions $p_1(u, \theta_1), p_2(v, \theta_2)$. Assuming regularity conditions we may consider the (column) scores

$$z_1 = \frac{\partial}{\partial\theta_1} \log p_1(u, \theta_1), \quad z_2 = \frac{\partial}{\partial\theta_2} \log p_2(v, \theta_2),$$

and $z = (z_1', z_2')'$. The squared distance between two observations w_a, w_b is now defined as

$$d^2(w_a, w_b) = (z_a - z_b)' F^{-1} (z_a - z_b),$$

where

$$F = \begin{pmatrix} E(z_1 z_1') & E(z_1 z_2') \\ E(z_2 z_1') & E(z_2 z_2') \end{pmatrix},$$

the expectations taken with respect to a density $p(w, \theta_1, \theta_2)$ with marginals $p_1(u, \theta_1), p_2(v, \theta_2)$. The existence of this density is proved by Cuadras (1992b). The proximity function is then

$$\phi^2(w) = z' F^{-1} z.$$

Of course, we would obtain two different proximity functions if the two populations are represented by different values of the parameters.

We need the observed proximity functions to compare two populations. This can be obtained by estimating the parameters. For instance, for the above mixed model, draw a sample of size n of w, obtain the marginal ML estimators of θ_1, θ_2, and replace parameters by estimators. This yields the observed scores $\widehat{z}_i = (\widehat{z}_{i1}', \widehat{z}_{i2}')', i = 1, \ldots, n$. If Z_1 and Z_2 contain the n rows $\widehat{z}_{i1}'$ and $\widehat{z}_{i2}'$, respectively, then Z_1, Z_2 have zero means and F, $\phi^2(w)$ may be estimated by

$$\widehat{F} = \frac{1}{n} \begin{pmatrix} Z_1' Z_1 & Z_1' Z_2 \\ Z_2' Z_1 & Z_2' Z_2 \end{pmatrix}, \quad \widehat{\phi}^2(w) = \widehat{z}' \widehat{F}^{-1} \widehat{z}.$$

This provides the observed $\widehat{\phi}^2(w_i) = \widehat{z}_i' \widehat{F}^{-1} \widehat{z}_i$, $i = 1, \ldots, n$. See Cuadras (1989, 1992a), Cuadras et al. (1997a), Núñez et al. (2003).

As an illustration, suppose that u is Poisson(λ) and v is $N(\mu, \sigma^2)$. The scores are $(u - \lambda)/\lambda$ and $(v - \mu)/\sigma^2$. Assuming stochastic independence, the squared distance and the proximity function are:

$$d^2(w, w') = \frac{(u - u')^2}{\lambda} + \frac{(v - v')^2}{\sigma^2}, \quad \phi^2(w) = \frac{(u - \lambda)^2}{\lambda} + \frac{(v - \mu)^2}{\sigma^2}.$$

The above statistic E^2 for samples $(u_{ik}, v_{ik}), i = 1, \ldots, n_k$, drawn from two populations with parameters $(\lambda_k, \mu_k, \sigma^2), k = 1, 2$, is:

$$E^2 = \frac{n_1 n_2}{n_1 + n_2} [\frac{(\overline{u}_1 - \overline{u}_2)^2}{\overline{u}} + \frac{(\overline{v}_1 - \overline{v}_2)^2}{s^2}],$$

where $\overline{u}$ is the general mean (Poisson data) and s^2 is the pooled variance (normal data). The asymptotic distribution of E^2 is chi-square with 2 d.f. This test is asymptotically equivalent to the likelihood ratio test.

18.5.3 Non-parametric tests with proximity functions

Let us use the proximity functions to test H_0. Under the hypothesis that X and Y follow the same distribution, we have $\phi_1^2 \equiv \phi_2^2$, so each individual has the same chance of being allocated to Π_1 or Π_2. However, given samples of sizes n_1, n_2, the observed proximity functions are different but satisfy, asymptotically,

$$P(\widehat{\phi_1^2}(x) < \widehat{\phi_2^2}(x) \mid \Pi_k) = P(\widehat{\phi_1^2}(x) > \widehat{\phi_2^2}(x) \mid \Pi_k), \quad k = 1, 2, \tag{18.13}$$

and the probability of misclassification or error rate [Lachenbruch (1975)] is 1/2.

We obtain different values for the proximity functions depending on whether the sample is considered as coming from Π_1 or from Π_2. This gives the table:

Sample	Π_1 $\widehat{\phi_1^2}$	Π_2 $\widehat{\phi_2^2}$	Sample	Π_1 $\widehat{\phi_1^2}$	Π_2 $\widehat{\phi_2^2}$
x_1	a_{11}	a_{12}	y_1	b_{11}	b_{12}
$\vdots$	$\vdots$	$\vdots$	$\vdots$	$\vdots$	$\vdots$
x_{n_1}	$a_{n_1 1}$	$a_{n_1 2}$	y_{n_2}	$b_{n_2 1}$	$b_{n_2 2}$

Two non-parametric tests for H_0 are suggested.

The chi-square test

Rule (18.12) gives the allocations:

$$\begin{array}{llll} x_i \to \Pi_1 & \text{if} \quad a_{i1} - a_{i2} < 0, & x_i \to \Pi_2 & \text{if} \quad a_{i1} - a_{i2} > 0, \\ y_i \to \Pi_1 & \text{if} \quad b_{i1} - b_{i2} < 0, & y_i \to \Pi_2 & \text{if} \quad b_{i1} - b_{i2} > 0, \end{array}$$

and under H_0, the error rate is 1/2. Thus, if n_{ij} is the frequency of individuals of Π_i allocated to Π_j, we obtain the following table of observed and expected frequencies

	Observed Π_1	Observed Π_2	Expected Π_1	Expected Π_2
Π_1	n_{11}	n_{12}	$n_1/2$	$n_1/2$
Π_2	n_{21}	n_{22}	$n_2/2$	$n_2/2$

where $n_1 = n_{11} + n_{12}$, $n_2 = n_{21} + n_{22}$. Therefore under H_0 the asymptotic distribution of

$$\chi^2 = \frac{(n_{11} - n_1/2)^2}{n_1/2} + \frac{(n_{12} - n_1/2)^2}{n_1/2} + \frac{(n_{21} - n_2/2)^2}{n_2/2} + \frac{(n_{22} - n_2/2)^2}{n_2/2} \tag{18.14}$$

is chi-square with 2 d.f.

The Mann–Whitney–Wilcoxon test

Under H_0, it turns out that $\phi_1^2(x)$ is iid as $\phi_1^2(y)$, and $\phi_2^2(x)$ is iid as $\phi_2^2(y)$, where x comes from Π_1 and y comes from Π_2. Thus we may use the Mann–Whitney–Wilcoxon test twice, to compare the samples

$$a_{i1},\ i = 1, \ldots, n_1, \quad \text{to} \quad b_{j1},\ j = 1, \ldots, n_2$$

as well as

$$a_{i2},\ i = 1, \ldots, n_1, \quad \text{to} \quad b_{j2},\ j = 1, \ldots, n_2.$$

We should reject H_0 if one or both tests are significant. Note that $\phi_1^2(x)$ may be equally distributed as $\phi_2^2(y)$ even if H_0 is not true, so a_{i1} should not be compared to b_{j2}.

18.6 Bootstrapping Distances

Expression (18.9) is particularly appropriate for obtaining, by nonparametric bootstrap, an estimation of the density of $\Delta^2(\Pi_1, \Pi_2)$ and, from it, either confidence intervals for the distances or critical values to test the null hypothesis $H_0 : \Delta^2(\Pi_1, \Pi_2) = 0$. This section explains the most salient aspects of the algorithm.

We obtain a sequence of B resamplings of the total set of $n = n_1 + n_2$ cases. Each resampling is a concatenation of two separate resamplings (one for each population), keeping constant the population sizes n_1, n_2. Within each population Π_i, a resampling is given by a permutation with repetition of the set of n_i cases in Π_i. For each resampling, we evaluate $\widehat{\Delta}^2(\Pi_1, \Pi_2)$ and the corresponding $n \times n$ matrix $D^{(2)}$ of squared distances, see (18.10).

This theoretical scheme of calculation can be simplified drastically by observing that since each resampling consists of cases from the original sample, all entries in the new $D^{(2)}$ have already appeared in some location of the initial matrix. The main feature of the algorithm is that it can be used instead of having to redo the computation. This is possible because $\Delta^2(\Pi_1, \Pi_2)$, which is built from averages, is invariant to the ordering of cases within each population. Thus we can represent each resampling b within each population π_α as a vector of multiplicities $\mu_{\alpha b}$, with n_α entries, corresponding to the cases in $\Pi_\alpha, \alpha = 1, 2$. The vector $(1, \ldots, 1)$ represents the original sample. Successive resamplings are represented by vectors $\mu_{\alpha b}$ whose entries follow a multinomial distribution of dimension n_α, sum of frequencies n_α, and vector of probabilities $(1/n_\alpha, \ldots, 1/n_\alpha)$. From these vectors of multiplicities it is possible to compute the $\widehat{\Delta}^2(\Pi_1, \Pi_2)$ corresponding to the resampling b with a single sequential reading of the $n\,(n-1)/2$ entries in $D^{(2)}$. The algorithm is as follows:

1) $\widehat{\Delta}^2(\Pi_1, \Pi_2)$ is initialized to 0.

2) Entries in $D^{(2)}$ are read sequentially. The $(i,\, j)$-th entry must be added to $\widehat{\Delta}^2(\Pi_1, \Pi_2)$ with a multiplicity $m\,(i,\, j)$ given by

$$m(i,\, j) = \begin{cases} \mu_{\alpha i_\alpha} \mu_{\beta j_\beta} & \text{if} \quad \alpha \neq \beta, \\ \mu^2_{\alpha i_\alpha} & \text{if} \quad \alpha = \beta \quad \text{and} \quad i = j, \\ 2\,\mu_{\alpha i_v} \mu_{\alpha j_\alpha} & \text{if} \quad \alpha = \beta \quad \text{and} \quad i \neq j, \end{cases}$$

where i_α is an index in the interval $[1, n_\alpha]$ which indicates the number (within population Π_α) of the i-th case of the joint set ($i \in [1, n]$). These two indexes are related by

$$i = n_1 + i_2.$$

This algorithm can be optimized further, since it can be easily vectorized to obtain simultaneously several resamplings, i.e., computing several vectors of multiplicities, for each sequential reading of $D^{(2)}$.

18.7 Comparing Variabilities

The above tests compare means but do not inform about the significant differences between covariance matrices. As our approach uses distances between observations, in accordance with Theorem 18.2.1, the distance-based version of the test for the equality of covariances is

$$H_0' : V_d\,(X) = V_d\,(Y)\,.$$

This test can be performed on the empirical quantities, see (18.8), by computing the statistic $F = \widehat{V}_d\,(X)\,/\widehat{V}_d\,(Y)$. The distribution of F under H_0' is unknown in general, but a resampling test, following the above procedures (randomization or bootstrapping) can be used. If the observed value F_0 is in the extreme (left or right) tail of this distribution, there will be evidence against H_0'.

18.8 Real Data Examples

Example 1. Krzanowski (1980) describes a cancer data set comprising eleven measurements (seven continuous, two binary and two three-state categorical variables) from 137 women with breast tumors, 78 benign (Π_1), 59 malignant (Π_2). As we have mixed variables, we use the distance obtained from Gower's all-purpose similarity coefficient [Gower (1971)].

1. *Discriminant analysis*

The distance based discriminant rule (18.12) yields:

		allocated to	
		Π_1	Π_2
observation	Π_1	60	18
from	Π_2	21	38

Thus the number of misallocations is 39. Notice that the linear discrimination and the location model yield 58 and 45 misallocations, respectively, so this approach works quite well for this data. This allocation is still improved in Nuñez et al. (2003).

2. *Chi-square test*

Application of (18.14) to the above table yields $\chi^2 = 27.50$ with 2 d.f, which is highly significant.

3. *Mann–Whitney–Wilcoxon test*

Computing the proximity functions $\phi_1^2(x), \phi_1^2(y)$ for the 78+59 individuals, and comparing these values using this non-parametric test, we obtain $U = 2583$, yielding a non-significant $z = 1.225 \sim N(0,1)$. Computing similarly $\phi_2^2(x), \phi_2^2(y)$, we obtain $U = 1374$ yielding a significant $z = -4.029 \sim N(0,1)$. Thus there is evidence against equality of the two populations with respect to the observed measurements.

4. *Randomization tests*

Figure 18.1 gives the randomization distribution of T^2 for $N' = 1000$ partitions into subsets of sizes 78 and 59. We obtain $T_0^2 = 0.0109$ which is in the extreme right tail of this distribution and again there is evidence against the null hypothesis of equality of means. On the other hand, for $N' = 1000$ partitions, we obtain $F_0 = 1.2834$, which is also extreme with respect to the distribution shown in Figure 18.2. Thus there is a significant difference between geometric variabilities.

5. *Bootstrap distribution of distances*

We find $\widehat{\Delta}^2(\Pi_1, \Pi_2) = 0.0498$. Then we obtain a sequence of $B = 1000$ resamplings of the total set of $n = 137$ cases, and calculate the bootstrap distribution of $\widehat{\Delta}^2(\Pi_1, \Pi_2)$. This distribution, summarized as a histogram (Figure 18.3), is concentrated around 0.06 and ranges between 0 and 0.12, clearly indicating that $\Delta^2(\Pi_1, \Pi_2) > 0$. Although the conclusion is the same, note the difference in interpreting Figures 18.1 and 18.3.

Example 2. Flury and Riedwyl (1988) consider $p = 6$ measurements taken on two kinds of bank notes: legal notes (Π_1) and false notes (Π_2), with $n_1 = n_2 = 100$. The comparison between legal and false notes is made with respect to the squared distance

$$d_{ij}^2 = |x_{i1} - x_{j1}| + \cdots + |x_{ip} - x_{jp}|.$$

The distance based discriminant rule (18.12) only yields one misallocation, so the discrimination is almost perfect. The chi-square test and the Mann–Whitney–Wilcoxon test ($z = 12.182$ and $z = 11.905$) are consistent with this result: there are significant differences between the two populations.

We reach the same conclusion after computing $T_0^2 = 0.0196$ and $F_0 = 0.8850$ and performing a randomization test with $N' = 1000$ partitions into subsets of sizes 100 and 100, as the two values are right and left extreme, respectively (histograms not displayed here).

The estimated squared distance between both groups is $\widehat{\Delta}^2(\Pi_1, \Pi_2) = 3.5361$ and the bootstrap distribution with $B = 1000$ resamplings (Figure 18.4) also indicates that $\Delta^2(\Pi_1, \Pi_2)$ is positive.

18.9 Conclusions

Some distance-based tests are proposed for comparing two populations on the basis of mixed variables, with respect to a suitable distance d between observations. These tests are compatible with some well-known multivariate tests when there is normality and the distance is Euclidean. In general, the performance of these distance-based tests is satisfactory provided that the probability density generated by the chosen distance d, see (18.4), is close to the true density.

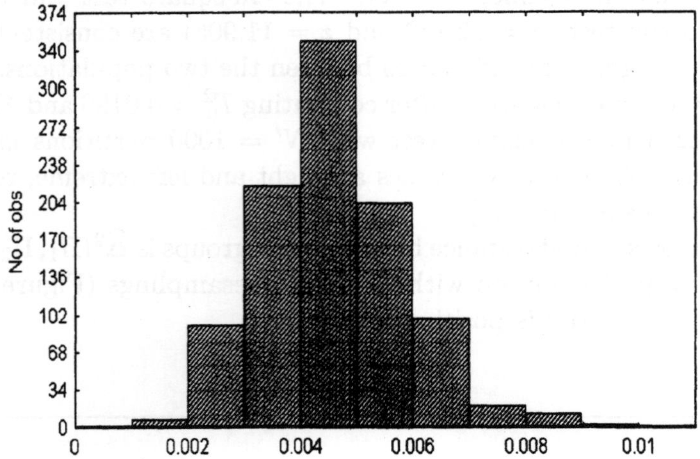

Figure 18.1: Randomization distribution of the statistic T^2 for comparing the $d-$means of two cancer groups, summarized as a histogram.

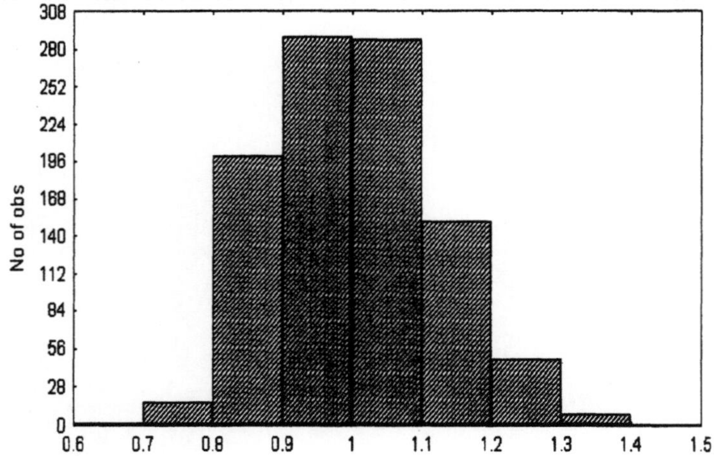

Figure 18.2: Randomization distribution of the statistic F for comparing the geometric variabilities of two cancer groups, summarized as a histogram.

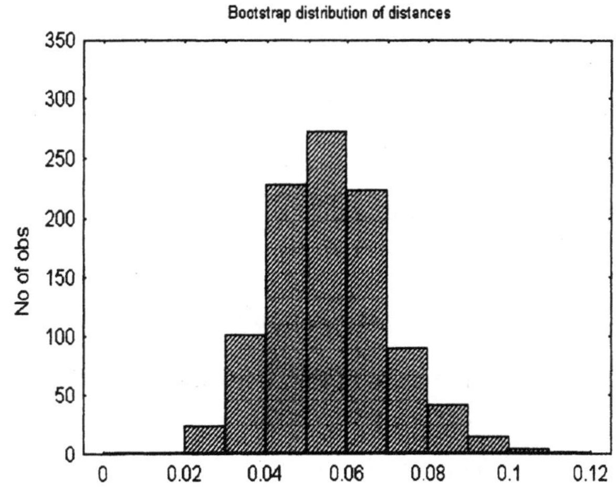

Figure 18.3: Bootstrap estimation of the probability density function of the distance between the two cancer groups, summarized as a histogram.

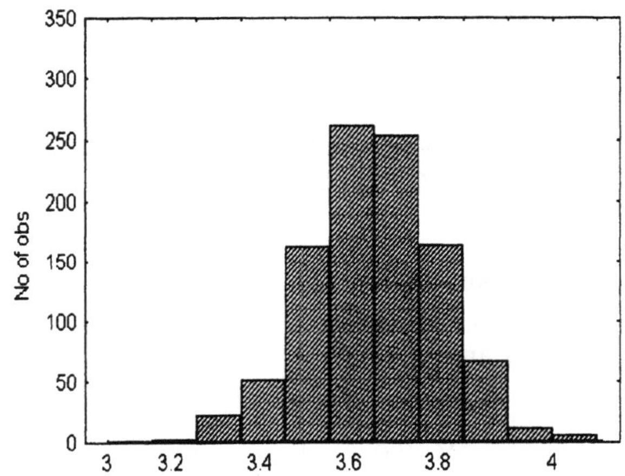

Figure 18.4: Bootstrap estimation of the probability density function of the distance between the two groups of notes, summarized as a histogram.

References

1. Anderson, T. W. (1958). *An Introduction to Multivariate Analysis,* John Wiley & Sons, New York.

2. Cuadras, C. M. (1989), Distance analysis in discrimination and classification using both continuous and categorical variables, In *Statistical Data Analysis and Inference* (Ed., Y. Dodge), pp. 459–473, Elsevier Science Publishers B. V. (North–Holland), Amsterdam.

3. Cuadras, C. M, (1992a), Some examples of distance based discrimination. *Biometrical Letters*, **29**, 3–20.

4. Cuadras, C. M. (1992b). Probability distributions with given multivariate marginals and given dependence structure, *Journal of Multivariate Analysis,* **42**, 51–66.

5. Cuadras, C. M. and Cuadras, D. (2002). Orthogonal expansions and distinction between logistic and normal, In *Goodness-of-fit Tests and Model Validity* (Eds., C. Huber-Carol, N. Balakrishnan, M. S. Nikulin and M. Mesbah), pp. 327–339, Birkhaüser, Boston.

6. Cuadras, C. M. and Fortiana, J. (1995). A continuous metric scaling solution for a random variable, *Journal of Multivariate Analysis,* **52**, 1–14.

7. Cuadras, C. M. and Fortiana, J. (1998). Visualizing categorical data with related metric scaling, In *Visualization of Categorical Data* (Eds., J. Blasius and M. J. Greenacre), Chap. 25, pp. 365-476, Academic Press, London.

8. Cuadras, C. M. and Fortiana, J. (2000). The importance of geometry in multivariate analysis and some applications, In *Statistics for the 21st Century* (Eds., C.R. Rao and G. Szekely), pp. 93–108, Marcel Dekker, New York.

9. Cuadras, C. M. and Lahlou, Y. (2000). Some orthogonal expansions for the logistic distribution, *Communications in Statistics-Theory and Methods,* **29**, 2643–2663.

10. Cuadras, C. M. and Lahlou, Y. (2002). Principal components of the Pareto distribution, In *Distributions with Given Marginals and Statistical Modelling* (Eds. C. M. Cuadras, J. Fortiana and J. A. Rodriguez-Lallena), pp. 43–50, Kluwer Academic Publishers, Dordrecht.

11. Cuadras, C. M., Atkinson, R. A. and Fortiana, J. (1997a). Probability densities from distances and discriminant analysis, *Statistics and Probability Letters,* **33**, 405–411.

12. Cuadras, C. M., Fortiana, J. and Oliva, F. (1997b). The proximity of an individual to a population with applications in discriminant analysis, *Journal of Classification*, **14**, 117–136.

13. Efron, B. (1993). Computer analysis in the computer age, In *Multivariate Analysis, Future Directions 2* (Eds. C. M. Cuadras and C. R. Rao), pp. 451–471, Elsevier Science Publishers B. V. (North–Holland), Amsterdam.

14. Flury, B. (1997). *A First Course in Multivariate Analysis*, Springer-Verlag, New York.

15. Flury, B. and Riedwyl, H. (1988). *Multivariate Statistics: A Practical Approach,* Chapman and Hall, London.

16. Gower, J. C. (1966). Some distance properties of latent root and vector methods in multivariate analysis, *Biometrika,* **53**, 315–328.

17. Gower, J. C. (1971). A general coefficient of similarity and some of its properties, *Biometrics,* **27**, 857–874.

18. Krzanowski, W. J. (1980). Mixtures of continuous and categorical variables in discriminant analysis, *Biometrics,* **36,** 493–499.

19. Kshirsagar, A. M. (1972). *Multivariate Analysis,* Marcel Dekker, New York.

20. Lachenbruch, P. A. (1975), *Discriminant Analysis,* Hafner Press, New York.

21. Lau, K. S. (1985). Characterization of Rao's quadratic entropies, *Sankhya, Series A,* **47**, 295–309.

22. Liu, Z. and Rao, C. R. (1995). Asymptotic distribution of statistics based on quadratic entropy and bootstrapping, *Journal of Statisitcal Planning and Inference,* **43**, 1–18.

23. Nuñez, M., Villarroya, A. and Oller, J. M. (2003). Minimum distance probability discriminant analysis for mixed variables, *Biometrics,* **59**, 248–253.

24. Rao, C. R. (1964). The use and interpretation of principal component analysis in applied research, *Sankhya, Series A,* **26**, 329–358.

25. Rao, C. R. (1973). *Linear Statistical Inference and Its Applications,* John Wiley & Sons, New York.

26. Rao, C. R. (1982). Diversity: Its measurement, decomposition, apportionment and analysis, *Sankhya, Series A,* **44**, 1–21.

27. Rencher, A. C. (1998). *Multivariate Statistical Inference and Applications,* John Wiley & Sons, New York.

19

Empirical Likelihood in Nonparametric and Semiparametric Models

Patrice Bertail

CREST, Laboratoire de Statistiques, Malakoff, France

Abstract: We review some recent results on empirical likelihood in nonparametric and semiparametric models. In particular we show the validity of empirical likelihood for Hadamard differentiable functionals tangential to a well-chosen set. We give the basic ingredients for proving its asymptotic validity and Bartlett correctability: convex duality, LAN properties of the empirical likelihood ratio in its dual form and uniform convergence of the empirical process (over classes of discrete probability). Extensions to semiparametric problems with estimated infinite dimensional parameters are also considered. Taking this point of view we revisit and extend some results in biased sampling models.

Keywords and phrases: Empirical likelihood, semiparametric models, convex duality, LAN, Hadamard differentiability

19.1 Introduction

The likelihood principle has been one of the major tools in parametric statistics. Owen (1988, 1990, 2001) introduced an "empirical likelihood" ratio and obtained in a non-parametric setting a generalization of Wilk's theorem. Nice accounts of empirical likelihood may be found in Hall and La Scala (1990) and Owen (2001)'s recent book, where one may find a huge bibliography. One purpose of this paper is to review in a unifying way the properties of empirical likelihood and summarize some results given in Bertail (2002), who shows that Hadamard differentiability (according to a well-chosen set of functions) is sufficient to validate the use of empirical likelihood. We also give here the basic ingredients to prove the validity and Bartlett correctability [see DiCiccio, Hall and Romano(1988)] of empirical likelihood in quite general semiparametric models. The main tool is the existence of a convex dual representation of the empirical likelihood, which may itself be seen as the log-likelihood ratio associated to a least favorable parametric family passing through the true model. This representation which follows from well-known properties of integral of convex functionals [see Borwein and Lewis (1991)] is closely connected to the important notion of dual likelihood introduced by Mykland (1995). We also discuss extensions to more general semiparametric frameworks with nuisance parameters of infinite dimension. These generalizations are essentially possible when one can exhibit an adequate estimated version of the efficient influence function which may be used as a basis for computing M-generalized estimators. We give an application of these ideas to the bias sampling semiparametric models considered by Qin (1993), extend his results under weaker conditions and obtain the third order validity of the method. The outline of the paper is the following : in the first part we recall the basic ideas of empirical likelihood and the main results of Bertail (2002) concerning the validity of empirical likelihood for Hadamard differentiable functionals; in the second part, we give some extensions to some semiparametric models, when an adequate symmetric estimator of the nuisance parameter is available; in the last part, we apply these ideas to the framework of bias sampling problems.

19.2 Empirical Likelihood for Hadamard Differentiable Functionals

19.2.1 Definitions

Let $X_1, ..., X_n, ...$ be i.i.d. random variables defined on a probability space (Ω, A, P_Ω) with common probability measure P belonging to a convex set $\wp$ of signed measures (containing the Dirac measure). In the following, we are interested in constructing a confidence region for the functional parameter $\theta = T(P)$ defined on $\wp$, taking its value in $\mathbb{R}^q$. For this let $P_n = n^{-1}\sum_{i=1}^n \delta_{X_i}$ be the empirical probability measure. The empirical likelihood ratio evaluated at θ is actually defined by

$$R_{E,n}(\theta) = \sup_{Q \in \mathcal{P}_n} \left\{ \Pi_{i=1}^n \frac{dQ}{dP_n}(X_i),\ T(Q) = \theta \right\}$$

where $\mathcal{P}_n$ is the set of discrete probability measures dominated by P_n. By convention, we put $R_{E,n}(\theta) = 0$ if there is no solution. A better way to see this problem from a probabilistic point of view is to consider $-\log(R_{E,n}(\theta))/n = r_n(\theta)$ as the minimization of the "Kullback distance" $K(Q,P) = -\int \log(\frac{dQ}{dP})dP$, if $Q << P$ and ∞ else, between Q and P_n, over all the probability Q dominated by the empirical distribution P_n (say $Q << P_n$), satisfying the constraint

$T(Q) = \theta$, that is

$$r_n(\theta) = \inf_{Q \in \mathcal{P}_n} (K(Q, P_n),\ T(Q) = \theta,\ \int dQ = 1). \tag{19.1}$$

Other convex distances or divergences [see Liese and Vajda (1987)] may be also used instead of K, leading to similar asymptotic results (except Bartlett correctability). The main tool for this is convex duality.

19.2.2 Some convex duality arguments

We recall the following simple convex duality principle which follows from Borwein and Lewis (1991). For this, let γ be a real convex function and γ^* its convex conjugate (which is thus semicontinuous). γ is normalized in such a way that $\gamma^* \in [0, \infty]$ and $\gamma^*(0) = 0$ (that is γ^* is a Young function). The domain of γ^* is defined by $d(\gamma^*) = \{x, \gamma(x) < \infty\}$ and is assumed to contain an open interval. The γ^*-divergence between Q and P, with respect to the nonnegative reference measure P, on the space $\mathcal{M}$ of signed measure on Ω is defined by

$$\begin{aligned} I_{\gamma^*}(Q, P) &= \int_\Omega \gamma^* \left(\frac{dQ - dP}{dP} \right) dP \; if \; Q << P \\ &= +\infty \quad \text{else.} \end{aligned}$$

Let $f = (f_k)_{1 \le k \le p}$ be measurable functions and consider moments constraints of the form

$$(Q - P)f = (\int_\Omega f_k d(Q - P))_{1 \le k \le q} = (h_k)_{1 \le k \le q} = h_0 \in R^q.$$

Then we have the following theorem which follows from Borwein and Lewis (1991).

Theorem 19.2.1 *If γ is everywhere finite and differentiable, then*

$$\inf_{Q \in \mathcal{M}} (I_{\gamma^*}(Q, P),\ (Q - P)f = h_0\} = \sup_{\lambda \in \mathbb{R}^p} (\lambda' h_0 - \int \gamma(\lambda' f) dP). \tag{19.2}$$

If in addition the following qualification constraint holds:

there exists $R \in \mathcal{M}$ such that $Rf = h_0$ and $\inf d(\gamma^*) < \inf \frac{dR}{dP} \le \sup \frac{dR}{dP} <$ $\sup d(\gamma^*)$,

then the supremum is attained in the right-hand side of (19.2) at some point λ^* and the solution of the initial problem is given by

$$Q^* = (1 + \gamma^{(1)}(\lambda^{*\prime} f)).P$$

where $\gamma^{(1)}$ is the first derivative of γ.

It should be noticed that the dual equality holds even without the qualification constraint which essentially ensures the attainment of the dual program. Much more general dual equalities when there is an infinite number of constraints [see Leonard (2001) and references therein] suggests some extension of empirical likelihood to parameters defined by an infinite number of constraints, that we are currently studying. However the theoretical and computational aspect may be much more complicated in that case and we will not discuss these results here. We essentially focus here on the "Kullback distance" corresponding to $\gamma^*(x) = (-\log(x+1)1_{\{x>-1\}} + x)$ and $\gamma(x) = (-x - \log(1-x)1_{\{x<1\}})$, with a finite number of constraints. In the particular case, when the dominating measure is P_n, the preceding theorem tells us that the original problem (19.1) becomes

$$
\begin{aligned}
&\inf_{Q\in\mathcal{P}_n}\left\{K(Q,P_n),\ Qf=\theta,\ \int dQ=1\right\}\\
&=\inf_{Q\in\mathcal{M}}(K(Q,P_n),\ (Q-P_n)f=-(P_n-P)f,\ (Q-P_n)1=0)\\
&=\sup_{c\in\mathbb{R},\ \lambda\in\mathbb{R}^q}(c+\int\log\{(1-(c+\lambda'(f-\theta)))1_{\{(1-(c+\lambda'(f-\theta))>0\}}dP_n))\\
&=\sup_{\lambda\in\mathbb{R}^q}(\int\log\{(1+\lambda'(f-\theta)))1_{\{(1+\lambda'(f-\theta)>0\}}dP_n)).
\end{aligned}
$$

Thus, the dual representation essentially transforms the original problem into studying a log-likelihood indexed by the Kuhn and Tucker parameter. Notice also that if a solution $\lambda^*(\theta)$ of the last dual equation exists, then the optimum of the empirical likelihood ratio is attained by the preceding theorem at

$$P_n^* = n^{-1}\sum\frac{1}{1+\lambda^*(\theta)'(f(X_i)-\theta)}\delta_{X_i}.$$

The qualification constraint when $\mathcal{M}=\mathcal{P}_n$ essentially means that the only interesting θ for which the dual program has a finite solution are in the convex hull of the $f(X_i)'s$. It should be noted that for some choice of γ, choosing $\mathcal{P}_n$ to be the set of all signed measures instead of probability measure (that is dropping the constraint $(Q-P_n)1=0$) may ease the existence of a solution. It is a very simple way to recover and generalize some results obtained in the empirical likelihood literature.

For linear functionals $T(P)=Pf$ with nondegenerate variance ($0<V_Pf<\infty$), it is easy either by direct arguments (see Owen(2001) or by the preceding convex duality arguments to show that

$$
\begin{aligned}
\Re_{n,1-\alpha} &= \{\theta,\ \Lambda_n(\theta)=-2\log(R_{E,n}(\theta))\le\chi^2_{1-\alpha}(q)\} \qquad (19.3)\\
&= \{T(\widetilde{P}_n),\ \widetilde{P}_n\in\overline{\mathcal{P}_{n.1-\alpha}}\},
\end{aligned}
$$

where

$$\overline{\mathcal{P}_{n,1-\alpha}} = \{Q \in \mathcal{P},\ K(Q, P_n) \leq \frac{\chi^2_{1-\alpha}(q)}{2n},\ \int dQ = 1,\ Q \geq 0\} \subset \mathcal{P}_n$$

(the last equality is only true in the linear case) is a confidence region asymptotically of level $1 - \alpha$,

This essentially means that the image by T of the ball (constructed with the Kullback distance) centered at P_n with diameter $\frac{\chi^2_{1-\alpha}(q)}{n}$ is an asymptotically correct confidence interval for $T(P)$. This explains some of the striking features of empirical likelihood which allows us to construct a confidence region in $\mathbb{R}^q$, which takes into account the asymmetry of the underlying distribution. This result may be generalized to large classes of Hadamard differentiable functionals with respect to well-chosen sets.

19.2.3 Extension to Hadamard differentiable functionals

The representation $\Re_{n,1-\alpha} = \{T(\widetilde{P}_n),\ \widetilde{P}_n \in \overline{\mathcal{P}_{n,1-\alpha}}\}$ in the linear case suggests in the general case to approximate a nonlinear functional T by a linearized version . However, to control this approximation, we need a notion of differentiability for T and some notion of uniformity over $\overline{\mathcal{P}_{n,1-\alpha}}$, which must be topologically compatible. The main tools for doing this are Hadamard differentiability and empirical process arguments over a well-chosen class of functions, which link the structure of T and the uniform asymptotic properties of the process $\widetilde{P}_n - P$ over $\overline{\mathcal{P}_{n,1-\alpha}}$

For sake of generality, we consider the following abstract empirical process framework. We use Hoffmann–Jorgensen's notion of convergence and will make (implicitly) all the necessary assumptions to avoid any measurability problems [see Van der Vaart and Wellner (1996)]. Let $\mathcal{F}$ be a subset of functions of a normed space of a function, here $L^2(P) = \{h,\ E_P h^2 < \infty\}$, endowed with $||f||_{2,P} = (E_P(f)^2)^{1/2}$. Assume that the functional T is defined on $\mathcal{P}$ considered as a subset of $\mathcal{L}_\infty(\mathcal{F})$, equipped with the uniform convergence norm

$$||P - Q||_{\mathcal{F}} = \sup_{h \in \mathcal{F}} |\int h dP - \int h dQ|.$$

It is assumed that $\mathcal{F}$ is Donsker with square integrable envelope $H \geq 1$ and satisfies a uniform entropy condition. More precisely, denote the covering number (the minimal number of balls of radius ε for the seminorm $||.||$ needed to cover $\mathcal{F}$) by $\mathcal{N}(\varepsilon, \mathcal{F}, ||.||)$. We assume the following uniform entropy condition

$$\int_0^\infty \sup_{Q \in \mathcal{D}} \sqrt{\log(\mathcal{N}(\varepsilon ||H||_{2,Q}, \mathcal{F}, ||.||_{2,Q})} d\varepsilon < \infty \tag{19.4}$$

where $\mathcal{D}$ is the set of all discrete probability measures Q with $\infty > \int H^2 dQ > 0$.

We use here a definition of Hadamard differentiability given in Pons and Turckheim (1991) that we recall below, which is well suited to many statistical applications. Let $B(\mathcal{F},P)$ be the subset of $\mathcal{L}_\infty(\mathcal{F})$, seen as paths $f \to \mu f = \int f d\mu$ from $\mathcal{F} \to \mathbb{R}$, which are $\|\ \|_{2,P}-$uniformly continuous and bounded. The Gaussian limit of the empirical process indexed by $\mathcal{F}$ will typically have its trajectory in $B(\mathcal{F},P)$.

Definition 19.2.1 The functional T from $\mathcal{P} \subset \mathcal{L}_\infty(\mathcal{F})$ to $\mathbb{R}^q$ (or any Banach space $(\mathcal{B}_1, \|.\|_{\mathcal{B}_1})$) is said to be Hadamard (or compact) differentiable at $P \in \mathcal{P}$ tangentially to $B(\mathcal{F},P)$, say T is $HDT_{\mathcal{F}} - P$, iff there exists a continuous linear mapping dT_P (defined on $\mathcal{P}$), such that for every sequences $h_n \to h \in B(\mathcal{F},P)$, for every sequences $t_n \to 0$ such that $P + t_n h \in \mathcal{P}$,

$$\frac{T((P+t_nh_n)) - T(P)}{t_n} - dT_P.h \to 0 \text{ as } t_n \to 0.$$

For a Hadamard differentiable functional, we call canonical (or first) gradient $T^{(1)}(.,P)$ (that is an influence function of T) any function from $\mathcal{X}$ to $\mathcal{B}_1$ such that

$$dT_P(Q-P) = \int T^{(1)}(x,P)(Q-P)(dx)$$

with the normalization

$$E_P T^{(1)}(X,P) = 0.$$

In the following we denote by

$$\Re^T_{n,1-\alpha} = \{T(\widetilde{P}_n),\ \widetilde{P}_n \in \overline{\mathcal{P}_{n,1-\alpha}}\}$$

the analog of (19.3) in the general case of a nonlinear functional T. For a general differentiable parameter it is then natural to consider a linearized form of the original problem

$$\overline{R}^L_{E,n}(P) = \sup_{Q_n \in \mathcal{P}_n} \left\{ \Pi_{i=1}^n \frac{dQ_n}{dP_n}(X_i),\ E_{Q_n} T^{(1)}(.,P) = 0 \right\}.$$

The following theorem ensures that such linearization is asymptotically valid for a Hadamard differentiable functional.

Theorem 19.2.2 *Assume that P is dominated by a measure μ. Assume that there exists a Donsker class of function $\mathcal{F}$ with envelope H, satisfying (19.4) such that $T(P)$ defined on $\mathcal{P}$ is Hadamard differentiable tangentially to $B(\mathcal{F},P)$ at P with first gradient $T^{(1)}(.,P)$. If $Var(T^{(1)}(X,P)) < \infty$ is of rank q, then we have*

$$-2\log\left(\overline{R}^L_{E,n}(P)\right) \underset{n\to\infty}{\to} \chi^2(q)$$

and

$$P(T(P) \in \Re^T_{n,1-\alpha}) \underset{n\to\infty}{\to} 1-\alpha.$$

If in addition $\overline{R}^L_{E,n}(P) = \overline{R}^L_{E,n}(\theta)$ *only depends on* $\theta = T(P)$, *under the Cramer condition*

$$\overline{\lim_{t\to\infty}} |E\exp(itT^{(1)}(X,P))| < \infty \tag{19.5}$$

and if $E||T^{(1)}(X_i,P)||^{8+\varepsilon} < \infty,\ \varepsilon > 0$, *the Bartlett corrected confidence region*

$$\Re^B_{1-\alpha} = \left\{\theta,\ q\log\left(\overline{R}^L_{E,n}(\theta)\right)/E(\log\left(\overline{R}^L_{E,n}(\theta)\right)) \le \chi^2_{1-\alpha}(q)\right\}$$

is a third-order correct confidence region for θ, *that is of level* $1-\alpha+O(n^{-2})$.

PROOF. See Bertail (2002) for details. The proof essentially relies on the fact that under the preceding hypotheses, uniformly over $\overline{\mathcal{P}_{n,1-\alpha}} = \{\widetilde{P}_n = \sum_{i=1}^n p_{i,n}\delta_{X_i},\ K(\widetilde{P}_n, P_n) \le \frac{\chi^2_{1-\alpha}(q)}{2n},\ p_{i,n} > 0,\ \sum_{i=1}^n p_{i,n} = 1\}$, we have convergence of the correctly standardized empirical process $\widetilde{P}_n - P$, that is, uniformly over $\overline{\mathcal{P}_{n,1-\alpha}}$,

$$\left(\sum_{i=1}^n p^2_{i,n}\right)^{-1/2} (\widetilde{P}_n - P) \underset{n\to\infty}{\overset{w}{\to}} G_P \text{ in } \mathcal{L}_\infty(\mathcal{F}),$$

where G_P is a Gaussian process in $\mathcal{L}_\infty(\mathcal{F})$. This result follows by showing that for $\widetilde{P}_n \in \mathcal{P}_{n,1-\alpha}$, the $p_{i,n}$ are exactly of order $1/n$ and from empirical process theory [see for instance Van der Vaart and Wellner (1996)]. The uniformity essentially follows from hypothesis (19.4) and the fact that we can lower and upper bound $\sum_{i=1}^n p^2_{i,n}$. The validity and Bartlett correctability of empirical likelihood in its linearized form may be proved by noticing that the dual form of the empirical likelihood ratio $\overline{R}^L_{E,n}(P)$ is a true parametric likelihood ratio [or a dual likelihood in the sense of Mykland (1995)] corresponding to testing $\lambda = 0$ in the least favorable family

$$p_\lambda(.) = \frac{dP}{d\mu}(1+\lambda' T^{(1)}(.,P))\mathbb{I}_{\{1+\lambda' T^{(1)}(.,P)>0\}}.$$

The family $\{p_\lambda(.), \lambda \in \mathbb{R}\}$ is locally asymptotically normal (LAN) because $p_\lambda(x)$ is continuously quadratically differentiable in λ everywhere except on the set $\{x,\ 1+\lambda' T^{(1)}(x,P) = 0\}$ [see LeCam (1986)]. But this set has probability 0 if $Var(T^{(1)}(X,P)) > 0$. Hadamard differentiability insures that $\Re^T_{n,1-\alpha}$ and $\Re^T_{n,1-\alpha}$ do not differ up to $o_P(n^{-1/2})$ ∎

This kind of result may be extended to much more general models, when there is a finite number of Hadamard differentiable constraints on the model. These constraints may be constraints induced by the model (for instance in the case of estimating a mean with a fixed variance) or informational constraints as encountered for instance in survey sampling, for which there is some known margin information [see Deville and Särndal (1992) for connected results]. Margin information and model constraint may be treated simultaneously as shown in a particular case by Chen and Qin (1993). Some examples are also given in Bertail (2002).

19.3 Extensions to Some Semi-parametric Problems

Consider a semi-parametric model $\wp_{\Theta,H} = \{P_{\theta,G} \in \wp\ ,\ \theta \in \Theta \subset \mathbb{R}^q,\ G \in H\}$, where H is an infinite dimensional space. We are interested in constructing a confidence region for the functional $\theta = T(P_{\theta,G})$. The main practical problem is actually that in this framework $T(P_n)$ is not defined or does not make sense. In many problems, it is possible to extend the functional defined on $\wp_{\Theta,H}$ to the whole space $\wp$, that is, to consider, for any $P \in \wp$, a functional $\widetilde{T}(P) = T \circ \Pi(P)$ with $\Pi(P) = P$ on $\wp_{\Theta,H}$. Π may be a projector for a convex model, a pseudo-projector (not uniquely defined) or a convolution operator etc... Assume that such extension exists, then if $\widetilde{T}(P)$ is Hadamard differentiable, we may use the result of the preceding part to construct a confidence region for θ. However, in practice the right "efficient" choice of Π [in the sense of Bickel, Klaasen, Ritov and Wellner(1993)] depends on the geometry of the problem. Ideally it should be chosen in such a way that the gradient of $\widetilde{T}$ at P coincides with the efficient influence function of T in the original semiparametric problem. In many problems, it may be however easier to work directly with the efficient influence function or a nonefficient but tractable one. Let $\widetilde{T}^{(1)}(x_i, P_{\theta,G})$ be such a quantity. The linearized version of the original empirical likelihood problem may not be used directly because the efficient influence function may depend on $P_{\theta,G}$ but may be replaced by an "estimated" version, say

$$\widetilde{R}_{E,n}(\theta) = \sup_{Q_n \in \mathcal{P}_n} \left\{ \Pi_{i=1}^n \frac{dQ_n}{dP_n}(X_i),\ E_{Q_n}\widetilde{T}^{(1)}(X, P_{\theta,\widehat{G}_{\theta,n}}) = 0 \right\}. \tag{19.6}$$

Such construction is implicit in many recent applications [see Owen (2001) and the references]. To prove the validity of empirical likelihood in this framework, we will assume the following hypotheses :

H_1 : Assume that there exists some sequence of estimators $\widehat{G}_{\theta,n}$, which is a symmetric statistic of the observations $X_1, ..., X_n$, converging to G with probability 1.

H_2 : The estimator , $\widehat{G}_{\theta,n}$ is such that

$$E_{P_{\theta,G}}\widetilde{T}^{(1)}(X_i, P_{\theta,\widehat{G}_{\theta,n}}) = o(n^{-1/2}).$$

H_3 : $\widetilde{T}^{(1)}(X_i, P_{\theta,G})$ is a continuous function of G (with respect to a metric metrizing convergence of $\widehat{G}_{\theta,n}$ to G).

H_4 : For every θ and n the functions $\widetilde{T}^{(1)}(., P_{\theta,\widehat{G}_{\theta,n}})$ belong to a Donsker-class of functions with probability 1. The class has an envelope $H(.) > 0$ perhaps be depending on θ with $E_{P_{\theta,G}} H(X)^2 < \infty$ then we show that

Theorem 19.3.1 *Assume that $H_1 - H_4$ holds, then if $Var(\widetilde{T}^{(1)}(X, P_{\theta,G}))$ is of rank q,*

$$-2\log(\widetilde{R}_{E,n}(\theta)) \to \chi^2(q) \text{ as } n \to \infty,$$

yielding asymptotically correct confidence intervals of level $1-\alpha$.

PROOF. See Bertail (2002). The proof relies on the extended backward martingale arguments of Blackwell and Dubins (1965) using the symmetry of the estimator $\widehat{G}_{\theta,n}$ and on some semiparametric arguments very similar to the one used by Van der Vaart (1998, Chap. 25). ∎

Another way to prove this result is to consider the sequence of approximate least favorable models (notice the recentering factor which ensures that we have a density)

$$p_{\lambda,n}(.) = \frac{dP}{d\mu}(.)\left[1 + \lambda'\left(\widetilde{T}^{(1)}(., P_{\theta,\widehat{G}_{\theta,n}}) - E_{P_{\theta,G}}\widetilde{T}^{(1)}(X_1, P_{\theta,\widehat{G}_{\theta,n}})\right)\right]$$
$$\mathbb{I}_{\left\{1+\lambda'\left(\widetilde{T}^{(1)}(.,P_{\theta,\widehat{G}_{\theta,n}})-E_{P_{\theta,G}}\widetilde{T}^{(1)}(X_1,P_{\theta,\widehat{G}_{\theta,n}})\right)>0\right\}}.$$

Conditions which ensure that the maximum likelihood estimator of λ in this family has a good behavior in the presence of a general estimated parameter $\widehat{G}_{\theta,n}$ may be less easy to check. If $\widehat{G}_{\theta,n}$ is not correctly chosen the family may not even be LAN. However it is interesting to see that for the Bartlett correctability of the approximate empirical likelihood (19.6) to hold, the behavior of $E_{P_{\theta,G}}\widetilde{T}^{(1)}(X_1, P_{\theta,\widehat{G}_{\theta,n}})$ is of great importance and somehow explains hypothesis H_2. This will also play an important role for showing Bartlett correctability. In many situations, for instance in some convex models [see Bickel et al. (1993)] we will have

$$E_{P_{\theta,G}}\widetilde{T}^{(1)}(X_i, P_{\theta,\widehat{G}_{\theta,n}}) = 0\,,$$

so that it may be easier to show the Bartlett correctability [at least up to order $O(n^{-3/2})$] in that case [see for instance an important application in Chen and Hall (1993)].

Remark: The existence of an estimator almost surely consistent, symmetric to the observation allows us to avoid the splitting trick or to use U-statistics arguments to prove the asymptotic validity of the procedure. The use of a symmetric statistics allows us to use the extended backward martingale convergence theorem of Blackwell and Dubins (1965). Indeed this allows us to write sums with estimated parameters as

$$\frac{1}{n}\sum_{i=1}^{n} f(\widetilde{T}^{(1)}(X_i, P_{\theta,\widehat{G}_{\theta,n}})) = E\left(f(\widetilde{T}^{(1)}(X_1, P_{\theta,\widehat{G}_{\theta,n}}))|\mathcal{F}_n\right)$$

where $\mathcal{F}_n$ is a symmetric σ-field. This trick may be used in many semiparametric models, for instance when a preliminary estimator of the underlying density is needed. However assumption H_4 may be difficult to check in some difficult cases. In these circumstances, it may be easier to built a split empirical likelihood. For this consider $G^{(1)}_{\theta,n/2}$ and $G^{(2)}_{\theta,n/2}$ two estimators of G respectively based on the first half ($[n/2]$ first values) and second half of the sample. Then we may define the split semiparametric empirical likelihood by

$$\widetilde{\widetilde{R}}_{E,n}(\theta) = \sup\left\{\begin{array}{l} \Pi_{i=1}^{n} np_{i,n}, p_{i,n} \geq 0,\ \sum_{i=1}^{n} p_{i,n}=1 \\ \sum_{i=1}^{[n/2]} p_{i,n}\widetilde{T}^{(1)}(X_i, P_{\theta,G^{(2)}_{\theta,n/2}}) \\ \quad +\sum_{i=[n/2]+1}^{n} p_{i,n}\widetilde{T}^{(1)}(X_i, P_{\theta,G^{(1)}_{\theta,n/2}})=0 \end{array}\right\}.$$

If, instead of H_4, we assume that

H_5 $\widetilde{T}^{(1)}(x_i, P_{\theta,\widehat{G}_{\theta,n}})$ belong to a class of functions with probability one having square integrable envelope $H(.)$

with $E_{P_{\theta,G}}||\widetilde{T}^{(1)}(X_1, P_{\theta,\widehat{G}^{(i)}_{\theta,n}}) - E_{P_{\theta,G}}\widetilde{T}^{(1)}(X, P_{\theta,G})||^2 \to 0$ as $n \to \infty$, for $i = 1,2$

then the result of the preceding theorem still holds. Indeed, the Donsker property is only needed to show the uniformity in the proof and hypothesis H_5 is sufficient for this.

19.4 Some Extensions to Bias Sampling Problems

We refer to Chapter 6 of Owen (2001) and Bickel et al. (1993) for complete references and a complete description of the semiparametric properties of this kind of model.

In biased sampling problems, we have s-independent samples generated by s-biased distributions defined by nonnegative weight functions w_i

$$Q_i(dy) = \frac{w_i(y)}{W_i(P)}P(dy),$$

$$W_i(P) = \int w_i(y)P(dy).$$

We do not assume here that there is a preliminary selection of a "stratum" with known probabilities: this case may be handled quite similarly. We assume for simplicity that P is dominated by a measure μ.

Let

$$X_{1,i}, ..., X_{n_i,i} \text{ i.i.d. } Q_i \ , \ i = 1, ..., s$$

and denote $n = \sum_{i=1}^{s} n_i$ the total sample size.

Let us give some examples :

Example 19.4.1 (Stratified sampling) Let X be a random variable taking its value in R^k. And let $S_1, S_2, ..., S_s$ be a partition of the space: $\overset{s}{\underset{i=1}{U}} S_i = R^k$, $S_j \cap S_i = \emptyset$. Then the weight functions are $w_i(x) = I_{S_i}\{x\}$ where $I_A\{.\}$ is the indicator of the set A. It is known that, unless auxiliary (transverse) information is available, the probability P is not identifiable.

Example 19.4.2 (Enriched sample) It is frequent that a sample obtained by sampling in the population is completed by $s-1$ biased samples (this is for instance the case when a survey is first based on a random sampling scheme and then completed by some additional biased sample), in that case $S_1 = R^k$ and $S_2, ..., S_s$ do not form a partition and we have simply in that case $w_1(x) = 1$. It is generally assumed that the biasing scheme i.e., the w_i, are known. The likelihood of the data in Examples 19.4.1 and 19.4.2 is given by

$$L_n(P, \mu) = \Pi_{i=1}^{s} \Pi_{j=1}^{n_i} \frac{dQ_i}{d\mu}(X_{j,i}) = \Pi_{i=1}^{s} \Pi_{j=1}^{n_i} \frac{w_i(X_{j,i})}{W_i(P)} \frac{dP}{d\mu}(X_{j,i}). \tag{19.7}$$

Example 19.4.3 (Length biased sampling) It happens sometimes that the bias of the sampling scheme is related to the length of the variable [see Vardi (1985)]. In survey sampling this often happens when the inclusion probability is proportional to a positive measure of size. In that case the weight is typically of the form $w(x) = x$.

Vardi (1985), Gill, Vardi and Wellner (1988) have given conditions for the identifiability of P and for the existence and unicity of the non-parametric maximum likelihood estimator (NPMLE) of P say $P_{w,n}$. Qin (1993) has generalized the approach of Owen (1988) in the case of Example 19.4.2 (enriched sample with $s = 2$). We think that it is easier to understand his work in our framework: most of his results may be obtained and generalized in a more straightforward way by using convex duality arguments provided that an adequate (LAN) least favorable family is constructed.

We use in the following the dominating measure

$$P_n = n^{-1} \sum_{i=1}^{s} \sum_{j=1}^{n_i} \delta_{X_{j,i}}.$$

Notice that this is not the non-parametric maximum likelihood estimator for P. The empirical likelihood in a biased sampling model evaluated at θ is defined here similarly to (19.7) by considering only probability dominated by P_n,

$$\begin{aligned} L_{w,n}(\theta) &= \sup_{Q}\left\{ L_n(Q,P_n),\ Q << P_n,\ T(Q)=\theta,\ \int dQ = 1\right\} \\ &= \sup_{\substack{p_{j,i,n}\\ i=1,\dots,s\\ j=1,\dots,n_i}} \left\{ \begin{array}{c} \Pi_{i=1}^{s}\Pi_{j=1}^{n_i}\frac{w_i(X_{j,i})}{\sum_{k=1}^{s}\sum_{l=1}^{n_i} w_i(X_{l,k})p_{l,k,n}} np_{j,i,n}, \\ T(\sum_{i=1}^{s}\sum_{j=1}^{n_i} p_{j,i,n}\delta_{X_{j,i}})=\theta,\ p_{j,i,n}>0, \\ \sum_{k=1}^{s}\sum_{l=1}^{n_i} p_{j,i,n}=1 \end{array}\right\} \\ &= \sup_{\substack{p_{j,i,n},W_i>0\\ i=1,\dots,s\\ j=1,\dots,n_i}} \left\{ \begin{array}{c} \Pi_{i=1}^{s}\Pi_{j=1}^{n_i}\frac{w_i(X_{j,i})}{W_i} np_{j,i,n}, \\ T(\sum_{i=1}^{s}\sum_{j=1}^{n_i} p_{j,i,n}\delta_{X_{j,i}})=\theta,\ p_{j,i,n}>0\ , \\ \sum_{k=1}^{s}\sum_{l=1}^{n_i} p_{j,i,n}=1, \\ \sum_{k=1}^{s}\sum_{l=1}^{n_i} w_i(X_{l,k})p_{l,k,n}=W_i \end{array}\right\} \end{aligned}$$

which we can approximate by the linearized version (we can use the same arguments as in part 1 for an Hadamard differentiable functional). We assume for simplicity that the gradient (or the estimating function) $T^{(1)}(X_{j,i},\theta)$ only depends on θ. The linearized empirical likelihood is given by

$$\widetilde{L}_{w,n}(\theta) = \sup_{\substack{p_{j,i,n},W_i>0\\ i=1,\dots,s\\ j=1,\dots,n_i}} \left\{ \begin{array}{c} \Pi_{i=1}^{s}\Pi_{j=1}^{n_i}\frac{w_i(X_{j,i})}{W_i} np_{j,i,n}, \\ \sum_{i=1}^{s}\sum_{j=1}^{n_i} p_{j,i,n}T^{(1)}(X_{j,i},\theta)=0, \\ p_{j,i,n}\geq 0\ ,\sum_{k=1}^{s}\sum_{l=1}^{n_i} p_{j,i,n}=1, \\ \sum_{k=1}^{s}\sum_{l=1}^{n_i} w_i(X_{l,k})p_{l,k,n}=W_i \end{array}\right\}.$$

We will also assume the following conditions [see Vardi (1985), Owen (2001)] which ensures the existence of a non-parametric maximum likelihood estimator.

(H_1) For every proper subset B of $\{1,...,s\}$,

$$\left(\underset{i\in B}{\cup}\{X_{1,i},...,X_{n_i,i}\}\right)\cap\left(\underset{i\notin B}{\cup}\{X,\ w_i(X)>0\}\right)\neq\emptyset.$$

This condition essentially means that we are not in the situation of Example 19.4.1, that is that we have transverse information or "linking" observations. Actually this condition plays the same role as the qualification constraint of Theorem 19.2.1 ensuring that the original and dual problem have a finite solution.

In the following we define $w(x)=(w_1(x),...,w_s(x))$ and $W=(W_1,...,W_s)$. The next condition also appears in Qin (p. 1183, 1993). In the unidimensional case, it ensures that the sampling bias is not proportional to $T^{(1)}(X,\theta)$. This hypothesis thus excludes Example 19.4.3 in the case of the mean with $T^{(1)}(x,P) = x-\theta$ and $w(x)=x$, a case which can be considered by itself with minor modifications.

(H_2)

$$Var\begin{pmatrix} T^{(1)}(X,P) \\ w(X) \end{pmatrix} \text{ is of rank } q+s.$$

Under the condition H_1, the value of the empirical likelihood calculated at Vardi's non-parametric maximum likelihood is

$$\begin{aligned} L_{w,n} &= \sup_{\substack{p_{j,i,n},W_i>0\\ i=1,\dots,s\\ j=1,\dots,n_i}} \left\{ \begin{array}{c} \Pi_{i=1}^{s}\Pi_{j=1}^{n_i}\frac{w_i(X_{j,i})}{W_i}p_{j,i,n},\ p_{j,i,n}\geq 0\ , \\ \sum_{i=1}^{n} p_{j,i,n}=1, \sum_{k=1}^{s}\sum_{l=1}^{n_i} w_i(X_{l,k})p_{l,k,n} = W_i \end{array}\right\} \\ &= \sup_{Q}\left\{ L_n(Q,P_n),\ Q << P_n,\ \int dQ = 1\right\}. \end{aligned}$$

The non-parametric maximum likelihood estimator $P_{w,n}$ of P is thus the solution of the unconstrained empirical likelihood. The empirical log-likelihood ratio for θ is then

$$\widetilde{R}_{E,n}(\theta) = L_{w,n}(\theta)/L_{w,n}$$

Theorem 19.4.1 *Under H_1 and H_2,*

$$-2\log(\widetilde{R}_{E,n}(\theta)) \rightarrow \chi^2(q) \text{ as } n\rightarrow\infty$$

yielding asymptotically correct confidence intervals of level $1-\alpha$.

If $(T^{(1)}(X,\theta), w(X))$ satisfies the Cramer condition and has moment of order $8+\varepsilon$, then $-2\log(\widetilde{R}_{E,n}(\theta))$ is Bartlett correctable.

PROOF. Following the ideas of the preceding parts, we may now define the least favorable model

$$\begin{aligned} p_{\lambda,\gamma}(x) &= \frac{dP}{d\mu}(x)(1+\lambda' T^{(1)}(x,\theta)+\gamma'(w(x)-W)) \\ &\quad \times \mathbb{I}\left\{1+\lambda' T^{(1)}(x,\theta)+\gamma'(w(x)-W) > 0\right\} \\ &= \frac{dQ_i}{d\mu}(x)(1+\lambda' T^{(1)}(x,\theta)+\gamma'(w(x)-W))\frac{W_i}{w_i(x)}\mathbb{I}\{x\in S_i\} \\ &\quad \times \mathbb{I}\left\{1+\lambda' T^{(1)}(x,\theta)+\gamma'(w(x)-W) > 0\right\} \qquad (19.8) \end{aligned}$$

where the family is indexed by the parameter $(\lambda,\gamma,W)\in\mathbb{R}^q\times\mathbb{R}^s\times\mathbb{R}^{+s}$. The convex duality arguments (used twice) or direct calculus using Kuhn and Tucker coefficients imply that the empirical likelihood ratio is

$$\begin{aligned} -2\log(R_{E,w,n}(\theta)) \\ &= 2\left(\sup_{W,\gamma}\left(\sum_{k=1}^{s}\sum_{l=1}^{n_k}\log(1+\gamma'(w(X_{l,k})-W))+\sum_{k=1}^{s}n_k\log(W_k)\right)\right. \\ &\quad - \sup_{W,\lambda,\gamma}\left(\sum_{k=1}^{s}\sum_{l=1}^{n_k}\log(1+\lambda'T^{(1)}(X_{l,k},\theta)+\gamma'(w(X_{l,k})-W)\right. \\ &\quad \left.\left. +\sum_{k=1}^{s}n_k\log(W_k)\right)\right) \end{aligned} \tag{19.9}$$

which is exactly the log likelihood ratio for testing $\lambda = 0$ in model (19.8). Compare with Qin (1993). Now under H_1 and H_2, (19.8) is quadratically differentiable (if H_2 does not hold, then $P(1+\lambda'T^{(1)}(X,\theta)+\mu'(w(X)-W) = 0) \neq 0$ and the quadratic differentiability may fail). It follows immediately that (19.9) is asymptotically $\chi^2(q)$, yielding a confidence region of the form

$$\Re_{1-\alpha} = \left\{\theta,\ \text{-2log}(R_{E,w,n}(\theta)) \leq \chi^2_{1-\alpha}(q)\right\},$$

asymptotically of level $1-\alpha$. Under the additional moment conditions on $(T^{(1)}(X,P), w(X))$ the log-likelihood ratio admits an Edgeworth expansion up to $O(n^{-2})$. Bartlett correctability also follows from representation (19.9) as a log-likelihood ratio for testing $\lambda = 0$ in the family (19.8). ■

Acknowledgements. I would like to thank Emmanuelle Gautherat for all her corrections, comments and encouragements and Christian Leonard for very stimulating discussions. All errors are mine.

References

1. Bertail, P. (2002). Empirical likelihood in some semiparametric model, *Submitted for publication.*

2. Bickel, P. J., Klaasen, C. A. J., Ritov, Y. and Wellner, J. A. (1993). *Efficient Estimation for Semiparametric Models*, Johns Hopkins University Press, Baltimore, Maryland.

3. Blackwell, D. and Dubins, L. (1965). Merging of opinion with increasing information, *Annals of Mathematical Statistics*, **33**, 882–886.

4. Borwein, J. M. and Lewis, A. S. (1991). Duality relationships for entropy like minimization problem, *SIAM Jouranl of Optimization*, **29**, 325–338.

5. Chen, S. X. and Hall, P. (1993). Smoothed empirical likelihood confidence intervals for quantiles, *Annals of Statistics*, **21**, 1166–1181

6. Chen, S. X. and Qin, J. (1993). Empirical likelihood estimation for finite populations and the effective usage of auxiliary information, *Biometrika*, **80**, 107–116.

7. Deville, J. C. and Särndal, C. E. (1992). Calibration estimators in survey sampling, *Journal of the American Statistical Association*, **87**, 376–382.

8. DiCiccio,T., Hall, P. and Romano, J.(1988). Empirical likelihood is Bartlett correctable, *Annals of Statistics*, **19**, 1053–1061.

9. Gill, R. D., Vardi, Y. and Wellner, J.A. (1988). Large sample theory of empirical distributions in biased sampling models, *Annals of Statistics*, **16**, 1069–1112.

10. Hall, P. and La Scala, B. (1990). Methodology and algorithms of empirical likelihood, *International Statistical Review*, **58**, 109–127.

11. LeCam, L. (1986). *Asymptotic Methods in Statistical Decision Theory*, Springer-Verlag, New York.

12. Liese, F. and Vajda, I. (1987). *Convex Statistical Distances*, Teubner, Leipzig, Germany.

13. Leonard, C. (2001). Minimizers of Energy functionals, *Acta Math. Hungar.*, **93**, 281–325.

14. Mykland, P. (1995). Dual likelihood, *Annals of Statistics*, **23**, 396–421.

15. Owen, A. B. (1988). Empirical likelihood ratio confidence intervals for a single functional, *Biometrika*, **75**, 237–249.

16. Owen, A. B. (1990). Empirical likelihood ratio confidence regions, *Annals of Statistics*, **18**, 90–120.

17. Owen, A. B. (2001). *Empirical Likelihood*, Chapman & Hall/CRC, Boca Raton, Florida.

18. Pons, O. and Turckheim E. (1991). Von Mises method, Bootstrap and Hadamard differentiability, *Statistics*, **22**, 205–214.

19. Quin, J. (1993). Empirical likelihood in biased sample problems, *Annals of Statistics*, **21**, 1182–1196.

20. Van der Vaart, A. W. (1998). *Asymptotic Statistics*, Cambridge University Press, London, England.

21. Van der Vaart, A. W. and Wellner, J. A. (1996). *Weak Convergence and Empirical Process*, Springer-Verlag, New York.

22. Vardi, Y. (1985). Empirical distributions in selection bias models, *Annals of Statistics*, **13**, 178–203.

20

Goodness of Fit Tests of L_2-Type

Hannelore Liero

Institute of Mathematics, University of Potsdam, Germany

Abstract: We give a survey on procedures for testing functions which are based on quadratic deviation measures. The following problems are considered: Testing whether a density function lies in a parametric class of functions, whether continuous random variables are independent; testing cell probabilities and independence in sparse data sets; testing the parametric fit of a regression homoscedasticity in a regression model and testing the hazard rate in survival models with censoring and with and without covariates.

Keywords and phrases: Density tests, sparse data, testing independence, regression fit, homoscedasticity, hazard rate under censoring with covariates

20.1 Introduction

We give a survey on procedures for testing functions in several nonparametric setups. The common idea of all considered tests is to express the deviation of the alternative from the hypothesis by a quadratic distance measure between a nonparametric kernel type estimator for the function of interest and a smoothed function characterizing the hypothetical function. Based on limit theorems stating that these quadratic functionals are asymptotically normally distributed we formulate asymptotic α-tests. Several aspects of the application of these test procedures are investigated.

So, in Section 20.2.1 after introducing a test statistic for checking whether a density function belongs to a parametric class we discuss the behavior of the power of the resulting test in detail. In Section 20.2.2 we apply similar ideas to test independence of two continuous random variables. Here the main point is to find good estimators for the standardizing terms in the limit theorem to avoid bias problems in the application of this limit statement for the formulation of the test.

In Section 20.3 the discrete analogues of the density test problems are considered. For estimating cell probabilities in sparse multinomial data sets Simonoff (1996) introduced local polynomial estimators. We use a special case of these kernel estimators to test hypothetical cell probabilities and compare our approach with the "classical" test procedure based on frequencies. Furthermore, the connection between testing in sparse data sets and testing a density is investigated. The case of testing independence in a sparse contingency table completes these considerations for sparse data.

In the following two sections we consider the nonparametric regression model. In Section 20.4 about testing whether a regression function has a parametric form we review the results of Härdle and Mammen (1993) to show, that bootstrap methods can be useful to apply tests of L_2-type in practice.

Section 20.5 deals with testing homoscedasticity in a regression model. Here we show, how the conditional variance can be estimated nonparametrically. Further, we mention the problem of estimating the variance in a nonparametric homoscedastic regression model with random design.

In the last section tests for testing the hazard function in survival models for censored observations are given. Firstly the case without covariates is investigated; here the main point is to handle the maximum likelihood estimator for the unknown parameter in the hypothetical hazard function. In Section 20.6.2 the model with fixed covariates is studied. Here, following the approach of Van Keilegom and Veraverbeke (2001), we construct our test statistic on the basis of a weighted estimator for the hazard function, where the weights depend on the covariates.

20.2 Tests for Densities

20.2.1 Testing whether a density has a parametric form

Let $Z_1, \ldots, Z_n$ be independent and identically distributed (i.i.d.) random variables with Lebesgue density f. We wish to test whether f lies in the parametric class

$$\mathcal{F} = \{ f_\vartheta = f(\cdot, \vartheta) \; : \; \vartheta \in \Theta \subseteq \mathbb{R}^d \}$$

against the alternative that f does not belong to $\mathcal{F}$, i.e.,

$$\mathcal{H} : \quad f \in \mathcal{F} \quad \text{against} \quad \mathcal{K} : \quad f \notin \mathcal{F}.$$

The idea of the test procedure is to compare an estimator $\hat{f}_n$, which is "good for all possible densities f", with the hypothetical one. It is well-known that the Rosenblatt–Parzen kernel estimator

$$\hat{f}_n(t) = \frac{1}{nb_n}\sum_{i=1}^{n} K\left(\frac{t-Z_i}{b_n}\right)$$

is such a good nonparametric estimator. Here K is the kernel function satisfying some regularity conditions and $\{b_n\}$ is a sequence of bandwidths tending to zero as n tends to infinity. As deviation measure we choose the weighted L_2-distance. This approach was studied among others by Bickel and Rosenblatt (1973), Ghosh and Huang (1991), Liero, Läuter and Konakov (1998). Note, that the kernel estimator $\hat{f}_n$ is not an unbiased estimator. Thus, deriving the limiting distribution of this distance one has to handle the bias. To avoid this problem it seems to be useful to take instead of the difference between $\hat{f}_n$ and a hypothetical f_ϑ the difference of $\hat{f}_n$ from its expectation under $\mathcal{H}$, that is from

$$\mathsf{E}_{\mathcal{H}}\hat{f}_n(t) = \frac{1}{b_n}\int K\left(\frac{t-x}{b_n}\right) f(x,\vartheta)\,\mathrm{d}x = \int K(x)f(t-xb_n,\vartheta)\,\mathrm{d}x.$$

In other words, we compare the smoothed data with a smoothed version of the hypothetical density. Since this expectation depends on the unknown parameter ϑ one has to replace it by a suitable estimator. Liero et al. (1998) propose to use the maximum likelihood estimator, say $\hat{\vartheta}_n$, which is $\sqrt{n}$-consistent under $\mathcal{H}$. Thus, finally we define the following test statistic:

$$\mathbb{Q}_n = \int \left(\hat{f}_n(t) - e_n(t, f_{\hat{\vartheta}_n})\right)^2 a(t)\,\mathrm{d}t$$

with $e_n(t, f_{\hat{\vartheta}_n}) = \int K(x)f(t-xb_n,\hat{\vartheta}_n)\,\mathrm{d}x$ and a weight function a, which is introduced to control the region of integration and has to be chosen by the statistician. Before we formulate the basic limit statement let us introduce some notation, which is used also in the next sections: For $b > 0$ we write $K_b(t) = \frac{1}{b}K(t/b)$. Further, we define $\kappa^2 = \int K^2(x)\,\mathrm{d}x$ and the convolution $\kappa^*(z) = \int K(u)K(z+u)\,\mathrm{d}u$. Throughout the paper we assume

(K) The kernel K is a Lipschitz continuous density function with finite support.

(W) The weight function a is nonnegative, piecewise continuous and bounded on $\mathbb{R}$; (resp. on $\mathbb{R}^2$).

Theorem 20.2.1 *Suppose that* (K), (W) *and the following assumptions are satisfied: Any density $f \in \mathcal{F}$ is bounded on $\mathbb{R}$, Lipschitz continuous and partially differentiable w.r.t. ϑ; $\nabla_\vartheta f(\cdot,\cdot)$, the vector of the partial derivatives, is bounded and uniformly continuous in both arguments. The estimator $\hat{\vartheta}_n$ is $\sqrt{n}$-consistent under $\mathcal{H}$. Further, $\int |\nabla_\vartheta f(t,\vartheta)| a(t)\,\mathrm{d}t < \infty$ for each $\vartheta \in \Theta$ and*

$nb_n \to \infty$, $b_n \to 0$ and $b_n(\log n)^\zeta \to 0$ for some $\zeta > d/2$. Then under $\mathcal{H}$,

$$\frac{nb_n^{1/2}}{\sigma_{fn}}\left(\mathbb{Q}_n - \mu_{fn}\right) \xrightarrow{\mathcal{D}} \mathsf{N}(0,1)$$

where

$$\mu_{fn} = (nb_n)^{-1}\kappa^2 \int f(t,\hat{\vartheta}_n)\, a(t)\, dt \quad \text{and}$$

$$\sigma_{fn}^2 = 2\int f(t,\hat{\vartheta}_n)^2\, a^2(t)\, dt \int (\kappa^*(z))^2\, dz.$$

Applying this limit statement we obtain an asymptotic α-test of $\mathcal{H}$ against $\mathcal{K}$ by the rule: Reject $\mathcal{H}$ if $\mathbb{Q}_n \geq \mu_{fn} + z_\alpha\, \sigma_{fn}/(nb_n^{1/2})$, where z_α is the $(1-\alpha)$-quantile of the standard normal distribution.

Some Remarks. 1. This test may be regarded as an analogue of a modified Cramér–von Mises test for testing whether an unknown distribution function lies in a parametric family of distribution functions. In contrast to the test for densities the limit distribution under the null hypothesis of the Cramér–von Mises test statistic with estimated parameter depends on the error of the parameter estimation. This is due to the fact that the normalizing factor n in the Cramér–von Mises test statistic is of the same order as the square of the rate of consistency of the parameter estimation, while in the density case this factor is $nb_n^{1/2}$ which tends to infinity moer slowly. Therefore the error of the parameter estimation can be neglected in the problem presented here.

2. One can show (see Liero (1999)) that the limit statement formulated in Theorem 20.2.1 holds true if the bandwidth b_n is replaced by an adaptively chosen bandwidth $\hat{b}_n$ as long as $\hat{b}_n/b_n \xrightarrow{\mathrm{P}} c$ for an arbitrary but fixed deterministic bandwidth b_n, satisfying the conditions of Theorem 20.2.1 and some positive constant c.

3. Theorem 20.2.1 says nothing about the order of convergence of the distribution of the standardized test statistic to its limit. Simulations show that the approximation of the critical values by those of the standard normal distribution may fail for moderate sample size n. Therefore this limit theorem should be considered more as a theoretical result which gives an insight into the behavior of the test statistic, but it is not recommended for the approximate calculation of the critical values. (See also Section 20.4.)

Power considerations. It is easy to show, that the proposed L_2-test is consistent, that is, if the alternative holds, then the probability for rejecting $\mathcal{H}$ tends to 1. Therefore, for a characterization of the test and the comparison with other tests it is useful to investigate the asymptotic behavior of the power under local alternatives. In the literature there are different approaches to that problem. Here we will follow the "classical" approach and consider local alternatives of the form

$$\mathcal{K}_n : \; f_n(\cdot) = f(\cdot, \vartheta) + \Delta_n(\cdot)$$

where $\{\Delta_n\}$ is a sequence of functions tending to zero and ϑ is arbitrarily fixed. The aim is to study how the power depends on the convergence behavior of the disturbing function Δ_n. Such investigations were done under different aspects by Bickel and Rosenblatt (1973), Rosenblatt (1975), Ghosh and Huang (1991) and Liero et al. (1998). They considered the following types of alternatives: The so-called *Pitman alternatives*, *sharp peak alternatives* and alternatives with *rapidly oscillating disturbing functions.* To derive the behavior of the power $\Pi(\Delta_n) = \mathsf{P}_{\mathcal{K}_n}\left(\mathbb{Q}_n \geq \mu_{fn} + z_\alpha \sigma_{fn}(nb_n^{1/2})^{-1}\right)$, where $\mathsf{P}_{\mathcal{K}_n}$ is the probability measure with respect to the local alternative, one has to study the asymptotic properties of the parameter estimator $\hat{\vartheta}_n$ under local alternatives. This is done in the paper of Liero et al.(1998), where an asymptotic expansion of the maximum likelihood estimator is given.

Generally speaking, the results in the paper mentioned above say that the L_2-test is sensitive against local alternatives $\mathcal{K}_n$, where the weighted L_2-norm of the disturbing function Δ_n behaves asymptotically as $n^{-1/2}b_n^{-1/4}$, in other words $\Pi(\Delta_n)$ tends to a number between α and 1, if $n^{1/2}b_n^{1/4}\|\Delta_n a^{1/2}\|_2 \to c \neq 0$.

In more detail one can prove the following results: 1. The error of the parameter estimation has an influence on the value of the limit of the power under Pitman alternatives and rapidly oscillating disturbing terms. Under sharp peak alternatives the value of the power does not depend on that estimation error.

2. Measured in the L_2-norm all three types of alternatives tend to the hypothesis at the same rate of convergence.

3. The highly oscillating disturbing function can be interpreted as a function with a growing number of peaks. But, here more sharpness of the peaks is compensated by a larger number of peaks. Thus, the L_2-norm of the disturbing function does not depend on the sharpness of the peaks, and the asymptotic behavior of the power under highly oscillating alternatives and under Pitman alternatives does not differ qualitatively.

4. If we translate our problem of testing a density function into a problem of testing distribution functions we get the following results: Pitman alternatives remain Pitman type alternatives also in the context of distribution functions. Therefore, our L_2-density test is worse than the Cramér–von Mises test, if we compare both with respect to this type of alternatives. The sharp peak disturbing function yields for distribution functions a disturbing function of sharp peak type, but with other "less sharp peaks". That means that there exist alternatives of sharp peak type which are detected by the test based on density estimators, but not by the classical Cramér–von Mises test. The reason is that

integration of the alternative density smoothes the sharp peak away. Integration of the rapidly oscillating disturbing function leads to the following result: Although the behavior of the power of the L_2-test under Pitman and highly oscillating alternatives is qualitatively the same, we can find highly oscillating disturbing functions where the Cramér–von Mises test fails, but the L_2- density test does not. The explanation is, that also these 'infinitely many peaks' are smoothed away by the translation from density to distribution function, despite their growing number.

5. The investigations show that a larger bandwidth improves the power. Heuristically speaking, this means, that the rate of convergence of the alternative, measured in the L_2-norm, may increase if the variance of the kernel estimator tends to zero faster. This feature, incidentally, conflicts with the fact discussed before stating that the approximation of the distribution of the test statistic by the standard normal distribution improves if the bandwidth tends to zero faster.

6. A test based on the integrated difference of $\hat{f}_n$ from the hypothetical $f_{\hat{\vartheta}_n}$ is discussed by Liero (1999). Here the additional bias term can lead to an increase or an decrease of the power.

7. In the paper of Liero et al. (1998) a so-called L_∞-test, which is based on the (normalized) maximal deviation of $\hat{f}_n$ from $\mathsf{E}_{\mathcal{H}}\hat{f}_n(t)$ is studied. The power considerations carried out there show that with respect to Pitman alternatives the L_2-test behaves better than the L_∞-test. Further, it is proved that there exist local alternatives of sharp peak type for which the L_∞-test distinguishes between hypothesis and alternative, but the L_2-test does not.

20.2.2 Testing independence

Let $(U_1, V_1), \ldots, (U_n, V_n)$ be a sample of i.i.d. $(\mathbb{R} \times \mathbb{R})$-valued random variables with density f. We wish to test whether U_i and V_i are independent, that is the test problem has the form

$$\mathcal{H}: \quad f = g \cdot h \quad \text{against} \quad \mathcal{K}: \quad f \neq g \cdot h,$$

where g and h are the marginal densities of U_i and V_i, respectively. Again we will use a kernel estimator for the construction of our test statistic. It is defined by

$$\hat{f}_n(s,t) = \frac{1}{nb_n^2} \sum_{i=1}^{n} K\left(\frac{U_i - s}{b_n}, \frac{V_i - t}{b_n}\right).$$

Here $K : \mathbb{R}^2 \longrightarrow \mathbb{R}$ is the kernel function and b_n is the bandwidth sequence. We take a kernel of product form, i.e.,

$$K(x,y) = K_1(x)\, K_2(y) \quad \text{with} \quad \int K_j(x)\,\mathrm{d}x = 1 \quad \text{for} \quad j = 1,2. \tag{20.1}$$

The estimators of the marginal densities are given by

$$\hat{g}_n(s) = \frac{1}{nb_n}\sum_{i=1}^{n} K_1\left(\frac{s-U_i}{b_n}\right) \quad \text{and} \quad \hat{h}_n(t) = \frac{1}{nb_n}\sum_{i=1}^{n} K_2\left(\frac{t-V_i}{b_n}\right).$$

The formulation of the test procedure goes back to Rosenblatt (1975). His idea was to compare a kernel estimator of f with the estimator of f under the hypothesis, that is with the product of kernel estimators of the marginal densities. This leads to the test statistic

$$\mathbb{I}_n = \int (\hat{f}_n(s,t) - \hat{g}_n(s)\cdot \hat{h}_n(t))^2 a(s,t)\,\mathrm{d}s\,\mathrm{d}t,$$

where a is again a suitable weight function and the integration is taken over $\mathbb{R}^2$. Let us denote the expectations of $\hat{f}_n, \hat{g}_n$, and $\hat{h}_n$, by $\overline{f}_n, \overline{g}_n$ and $\overline{h}_n$, respectively. Further, define

$$\xi_{fn} = (nb_n^2)^{-1}D_{1n} - (nb_n)^{-1}D_{2n}$$

with

$$\begin{aligned} D_{1n} &= \int\int K_1^2(u)g(s-ub_n)\,\mathrm{d}u \int K_2^2(v)h(t-vb_n)\,\mathrm{d}v\, a(s,t)\,\mathrm{d}s\,\mathrm{d}t, \\ D_{2n} &= \int \Big(\overline{g}_n^2(s)\int K_2^2(v)h(t-vb_n)\,\mathrm{d}v \\ &\qquad + \overline{h}_n^2(t)\int K_1^2(u)g(s-ub_n)\,\mathrm{d}u\Big)\, a(s,t)\,\mathrm{d}s\,\mathrm{d}t, \end{aligned}$$

and

$$\tau_f^2 = 2\int g^2(s)h^2(t)a^2(s,t)\,\mathrm{d}s\,\mathrm{d}t \int (K_1 * K_1)^2(r)\,\mathrm{d}r \int (K_2 * K_2)^2(v)\,\mathrm{d}v.$$

With this notation we can formulate:

Theorem 20.2.2 *Suppose that the marginal densities g and h are Lipschitz continuous and bounded. Let the kernel K be of product type* (20.1). *The K_j's ($j = 1, 2$) satisfy condition* (K) *and the weight function a condition* (W). *If $f = g\cdot h$, then*

$$nb_n(\mathbb{I}_n - \xi_{fn}) \xrightarrow{\mathcal{D}} \mathsf{N}(0, \tau_f^2) \tag{20.2}$$

as $b_n \to 0$ and $nb_n^2 \to \infty$.

To apply limit statement (20.2) for the construction of the test procedure we have to replace the unknown terms D_{1n} and D_{2n} by estimators which are consistent with a certain rate of convergence. To avoid bias problems we do not follow the proposal of Rosenblatt (1975), who replaced the unknown functions g and h by the kernel estimators $\hat{g}_n$ and $\hat{h}_n$. Observe, that

$$D_{1n} = b_n^{-2} \int \Omega_{1n}(s)\, \Omega_{2n}(t)\, a(s,t)\, \mathrm{d}s\, \mathrm{d}t \quad \text{and}$$

$$D_{2n} = b_n^{-1} \int \left([\overline{g}_n(s)]^2\, \Omega_{2n}(t) + \left[\overline{h}_n(t)\right]^2 \Omega_{1n}(s) \right) a(s,t)\, \mathrm{d}s\, \mathrm{d}t$$

where $\Omega_{1n}(s) = \mathsf{E}K_1^2\left(\frac{s-U_1}{b_n}\right)$ and $\Omega_{2n}(t) = \mathsf{E}K_2^2\left(\frac{t-V_1}{b_n}\right)$.

We estimate these quantities by

$$\tilde{\Omega}_{1n}(s) = \frac{1}{n}\sum_{i=1}^{n} K_1^2\left(\frac{s-U_i}{b_n}\right) \quad \text{and} \quad \tilde{\Omega}_{2n}(t) = \frac{1}{n}\sum_{i=1}^{n} K_2^2\left(\frac{t-V_i}{b_n}\right)$$

and obtain as estimator for D_{1n},

$$\tilde{D}_{1n} = b_n^{-2} \int_A \tilde{\Omega}_{1n}(s)\, \tilde{\Omega}_{2n}(t)\, a(s,t)\, \mathrm{d}s\, \mathrm{d}t,$$

which is consistent and unbiased under $\mathcal{H}$. Estimators of D_{2n} and τ_f^2 are given by

$$\tilde{D}_{2n} = b_n^{-1} \int_A \left([\hat{g}_n(s)]^2\, \tilde{\Omega}_{2n}(t) + [\hat{h}_n(t)]^2\, \tilde{\Omega}_{1n}(s) \right) a(s,t)\, \mathrm{d}s\, \mathrm{d}t \quad \text{and}$$

$$\tilde{\tau}_{fn}^2 = 2 \int \hat{g}_n^2(s)\, \hat{h}_n^2(t)\, a^2(s,t)\, \mathrm{d}s\, \mathrm{d}t \int (K_1 * K_1)^2(u)\, \mathrm{d}u \int (K_2 * K_2)^2(v)\, \mathrm{d}v.$$

Set $\tilde{\xi}_{fn} = (nb_n^2)^{-1}\tilde{D}_{1n} - (nb_n)^{-1}\tilde{D}_{2n}$. It is easy to verify that the limit statement formulated in Theorem 20.2.2 remains valid if the unknown terms ξ_{fn} and τ_{fn} are replaced by these estimators. Thus, an asymptotic α-test is provided by: Reject $\mathcal{H}$, if $\mathbb{I}_n \geq \tilde{\xi}_{fn} + z_\alpha\, \tilde{\tau}_{fn}/(nb_n)$.

Some remarks. (1) In difference to the approach of Rosenblatt (1975) we propose another estimator of the standardizing terms in the limit theorem for $\mathbb{I}_n$. The advantage of our method is, that this estimator is unbiased. So we do not need additional assumptions on the smoothness of the underlying densities to ensure that the limit theorem remains valid with the estimated standardizing terms.

(2) The behavior of the power of this test is qualitatively very similar to that of the L_2-test considered in Section 20.2.1. As there one can show that the power tends to a nontrivial limit, i.e., a number between α and 1, if $\sqrt{nb_n}||\Delta_n a^{1/2}||_2 \to c \neq 0$, where the disturbing function Δ_n describes the deviation from independence.

20.3 Tests for Sparse Data Sets

In the classical case the number of cells, say k, in a multinomial distribution or a contingency table is assumed to be fixed. But there are data sets where the total number of observations is moderate in comparison to the total number of cells. Consequently the number of observations falling in each cell is rather small. We describe such sparseness mathematically by assuming $k = k_n \to \infty$ as n tends to infinity. It is known that for sparse data nonparametric smoothing techniques provide estimators of the cell probabilities, which have a better asymptotic performance than the frequency estimators, see for example Aerts et al. (1997) and Simonoff (1996). Here we use such smoothed estimators to define a test statistic of L_2-type.

20.3.1 Testing cell probabilities in sparse multinomial data

Let $p_n = (p_{n1}, \ldots, p_{nk_n})^t$ be the vector of cell probabilities of a k_n-cell multinomial distribution, where n is the total sample size. The simplest test problem is to test

$$\mathcal{H}: \quad p_{ni} = \pi_{ni} \text{ for all } i = 1, \ldots, k_n \quad \text{against} \quad \mathcal{K}: \quad p_{ni'} \neq \pi_{ni'} \text{ for some } i', \tag{20.3}$$

where $\pi_n = (\pi_{n1}, \ldots, \pi_{nk_n})^t$ is a vector of given cell probabilities. To formulate the test procedure we start with the definition of the estimators of the cell probabilities p_{ni}, $i = 1, \ldots, k_n$. As smoothed estimators we propose local constant estimators, which are the simplest local polynomial estimators introduced by Simonoff (1996). For the definition of these estimators let $x_{nj} = (j - \frac{1}{2})/k_n$ be equidistant design points on the interval $[0, 1]$ and denote the relative frequency of cell i by p^*_{ni}. The data (x_{nj}, p^*_{nj}) can be considered as regression type data. Following the idea of smoothing in the regression set-up we estimate the cell probability p_{ni} by

$$\hat{p}_{ni} = \frac{\sum\limits_{\mu=1}^{k_n} K\left(\frac{x_{n\mu} - x_{ni}}{b_n}\right) p^*_{n\mu}}{\sum\limits_{\mu=1}^{k_n} K\left(\frac{x_{n\mu} - x_{ni}}{b_n}\right)},$$

where K is a kernel function and b_n is a sequence of bandwidths introduced already in the previous section. For simplicity of writing we skip the subscript n in the notation of the cell probabilities and the design points; furthermore we write

$$\hat{p}_i = \frac{1}{k_n b_n} \sum_{\mu=1}^{k_n} L_i\left(\tfrac{x_\mu - x_i}{b_n}\right) p_\mu^* \quad \text{with} \quad L_i(u) = \frac{K(u)}{\frac{1}{k_n b_n}\sum_{\mu=1}^{k_n} K\left(\frac{x_\mu - x_i}{b_n}\right)}.$$

As test statistic we propose the sum of squared differences between the estimators and their expectations under the hypothesis:

$$\mathbb{T}_n = \sum_{i=1}^{k_n} (\hat{p}_i - \mathsf{E}_{\mathcal{H}}\hat{p}_i)^2 \quad \text{with} \quad \mathsf{E}_{\mathcal{H}}\hat{p}_i = \frac{1}{b_n k_n} \sum_{j=1}^{k_n} L_i\left(\frac{x_j - x_i}{b_n}\right) \pi_j.$$

Set

$$\mu_{\pi n} = \frac{1}{n k_n^2} \sum_{i=1}^{k_n} \left[\sum_{j=1}^{k_n} L_{i,b_n}^2 (x_j - x_i)\, \pi_j - \left(\sum_{j=1}^{k_n} L_{i,b_n}(x_j - x_i)\, \pi_j \right)^2 \right]$$

where $L_{i,b}(u) = \frac{1}{b} L_i(\frac{u}{b})$, and

$$\sigma_{\pi n}^2 = 2k_n^2 b_n \sum_{l=1}^{k_n} \sum_{m=1}^{k_n} C_{lm}^2 \pi_l \pi_m \quad \text{with} \quad C_{lm} = k_n^{-2} \sum_{i=1}^{k_n} L_{i,b_n}(x_l - x_i)\, L_{i,b_n}(x_m - x_i).$$

The following limit theorem shows that $\mathbb{T}_n$ is asymptotically normal under $\mathcal{H}$.

Theorem 20.3.1 *Assume* (K) *and* $|\pi_{ni} - \pi_{nj}| \le L\, k_n^{-1}\, |x_{ni} - x_{nj}|$ *for all* i, j *and some constant* L. *If* $b_n \to 0$, $b_n k_n \to \infty$ *and* $n b_n \to \infty$ *as* $n \to \infty$, *then under* $\mathcal{H}$,

$$\frac{n k_n \sqrt{b_n}}{\sigma_{\pi n}} (\mathbb{T}_n - \mu_{\pi n}) \xrightarrow{\mathcal{D}} \mathsf{N}(0, 1).$$

Under the assumptions of Theorem 20.3.1 we get the test: Reject the hypothesis $\mathcal{H}$ if $\mathbb{T}_n \ge \mu_{\pi n} + z_\alpha\, \sigma_{\pi n}/(n k_n \sqrt{b_n})$.

Comparison to the "classical" approach. Another possibility to test (20.3) is the quadratic deviation of the unsmoothed estimators of the cell probabilities, that is to use the test statistic

$$\mathbb{S}_n = \sum_{i=1}^{k_n} (p_i^* - \pi_i)^2 .$$

Applying results proved by Holst (1972) and Burman (1987) one can show that this statistic, properly standardized, is also asymptotically normally distributed. The asymptotic α-test based on this limit statement has the following form: Reject the hypothesis $\mathcal{H}$, if $\mathbb{S}_n \ge \mathsf{E}_{\mathcal{H}}\mathbb{S}_n + \rho_{\pi n}\, z_\alpha/(n\sqrt{k_n})$, where

$\mathsf{E}_{\mathcal{H}}\mathbb{S}_n = \frac{1}{n}\left(1 - \sum_{i=1}^{k_n} \pi_i^2\right)$, and $\rho_{\pi n}^2 = 2k_n \sum_{j=1}^{k_n} \pi_j^2$ is a sequence of positive numbers tending to a positive constant.

To compare both test procedures we consider the behavior of the power under local alternatives of the form: $\mathcal{K}_n :\ p_i := \pi_i + \delta_i$ with $\sum_{i=1}^{k} \delta_i = 0$.

Let us denote the power of the test based on the (unsmoothed) frequency estimators by β_{1n}, and that of the test based on the $\hat{p}_{ni}$'s by β_{2n}. Then under mild conditions on the disturbing terms δ_i in Liero (2001) it is proved that

$$\lim_{n\to\infty} \frac{\beta_{1n}}{\beta_{2n}} \leq 1,$$

where "=" holds if $nk_n\sqrt{b_n}\sum_{i=1}^{k_n} \delta_i^2 \to 0$ or $n\sqrt{k_n}\sum_{i=1}^{k_n} \delta_i^2 \to \infty$.

That means, roughly speaking, the test based on the quadratic deviation of the local polynomial estimator is better than the test based on the frequency estimators. Only in the case that the square of the L_2-norm of the disturbing terms, i.e., $\sum_i \delta_i^2$, is very large, the power of both tests tends to 1; and in the case that the bandwidth b_n is very small, i.e. we smooth only "a little bit", both tests behave poorly.

Connection to the goodness-of-fit test using densities. Suppose that the cell probabilities p_i and π_i are generated by latent densities f and f_0, respectively, which are defined on $[0, 1]$:

$$p_i = \int_{I_i} f(x)\,\mathrm{d}x, \qquad \pi_i = \int_{I_i} f_0(x)\,\mathrm{d}x \quad \text{where} \quad I_i = [(i-1)/k_n, i/k_n].$$

Then the test problem (20.3) corresponds to the simple problem

$$\mathcal{H} :\ \ f = f_0 \ \text{ against } \ \mathcal{K} :\ \ f \neq f_0. \tag{20.4}$$

Furthermore it follows from results proved by Augustyns (1997) that

$$\mu_{\pi n} = (nk_n b_n)^{-1}(\kappa^2 + o(1)) \quad \text{and} \quad \lim_{n\to\infty} \sigma_{\pi n}^2 = 2\int f_0^2(x)\,\mathrm{d}x \int (\kappa^*(z))^2\,\mathrm{d}z.$$

Thus we have the following correspondence between the test for the cell probabilities and the modified test for testing (20.4) (where the weight function a is the indicator of $[0,1]$): $k_n\,\mathbb{T}_n = k_n \sum_{i=1}^{k_n} (\hat{p}_i - \mathsf{E}_{\mathcal{H}}\hat{p}_i)^2$ corresponds to $\mathbb{Q}_n = \int (\hat{f}_n(t) - \mathsf{E}_{\mathcal{H}}\hat{f}_n(t))^2\,\mathrm{d}t$, $k_n\mu_{\pi n}$ to $\mu_{fn} = (nb_n)^{-1}\kappa^2 \int_0^1 f_0(t)\,dt$ and the variance term $\sigma_{\pi n}^2$ to $\sigma_{fn}^2 = 2\int_0^1 f_0^2(t)\,\mathrm{d}t \int (\kappa^*(z))^2\,\mathrm{d}z$.

Moreover, let us consider the behavior of the power from the viewpoint of the existence of a latent density. For that purpose we write the local alternative in the form $\mathcal{K}'_n :\ f_n := f_0 + \Delta_n$. Suppose that $\sum_{i=1}^{k_n} \delta_i^2 \sim \frac{1}{k_n}\int_0^1 \Delta_n^2(u)\,\mathrm{d}u$. Then, expressed in terms of densities, the power of the second test tends to a nontrivial limit, if the square of the L_2-norm of the disturbing function Δ_n is asymptotically equivalent to $(n\sqrt{b_n})^{-1}$. Note, that this is the same rate of convergence as in the problem of testing a density obtained before.

20.3.2 Testing independence in sparse contingency tables

We consider a two-dimensional contingency table with $k_n = l_n \cdot m_n$ cells, where $l_n \to \infty$ and $m_n \to \infty$. The (joint) cell probabilities are denoted by p_{nij}, the marginal cell probabilities by q_{ni} and r_{nj}. To test independence we have to check the hypothesis

$$\mathcal{H}: \; p_{nij} \; = \; q_{ni}\, r_{nj} \;\text{ for all }\; (i,j) \;\text{ against }\; \mathcal{K}: \; p_{ni'j'} \neq q_{ni'}\, r_{nj'} \;\text{ for some }\; (i',j'). \tag{20.5}$$

For testing (20.5) we will use

$$\mathbb{M}_n = \sum_{i=1}^{m_n} \sum_{j=1}^{l_n} \left(\hat{p}_{nij} - \hat{q}_{ni} \cdot \hat{r}_{nj} \right)^2, \tag{20.6}$$

where $\hat{p}_{nij}$, $\hat{q}_{ni}$ and $\hat{r}_{nj}$ are the local constant estimators of p_{nij}, q_{ni} and r_{nj}. Following the ideas presented in the previous sections these estimators have the following form (we skip the n in the subscript if appropriate):

$$\hat{p}_{ij} = \frac{\sum_{\mu=1}^{m_n} \sum_{\nu=1}^{l_n} K\left(\frac{x_\mu - x_i}{b_n}, \frac{y_\nu - y_j}{b_n}\right) p^*_{\mu\nu}}{\sum_{\mu=1}^{m_n} \sum_{\nu=1}^{l_n} K\left(\frac{x_\mu - x_i}{b_n}, \frac{y_\nu - y_j}{b_n}\right)} = \frac{1}{m_n l_n b_n^2} \sum_{\mu=1}^{m_n} \sum_{\nu=1}^{l_n} L_{ij}\left(\frac{x_\mu - x_i}{b_n}, \frac{y_\nu - y_j}{b_n}\right) p^*_{\mu\nu},$$

where $x_\mu = (\mu - \frac{1}{2})/m_n$ and $y_\nu = (\nu - \frac{1}{2})/l_n$ are equidistant design points on the interval $[0,1] \times [0,1]$, p^*_{ij} is the relative frequency of cell (i,j) and

$$L_{ij}(u,v) = \frac{K(u,v)}{\frac{1}{m_n l_n b_n^2} \sum_{\mu=1}^{m_n} \sum_{\nu=1}^{l_n} K\left(\frac{x_\mu - x_i}{b_n}, \frac{y_\nu - y_j}{b_n}\right)}.$$

Again we take a kernel of product type. Then as estimators of the marginal cell probabilities $q_i = \sum_j p_{ij}$ and $r_j = \sum_i p_{ij}$ we obtain straightforward

$$\hat{q}_i = \frac{1}{k_n b_n} \sum_{\mu=1}^{m_n} L_{1i}\left(\frac{x_\mu - x_i}{b_n}\right) q^*_\mu \quad \text{and} \quad \hat{r}_j = \frac{1}{l_n b_n} \sum_{\nu=1}^{l_n} L_{2j}\left(\frac{y_\nu - y_j}{b_n}\right) r^*_\nu$$

with weight functions

$$L_{1i}(u) = \frac{K_1(u)}{\frac{1}{m_n b_n} \sum_{\mu=1}^{m_n} K_1\left(\frac{x_\mu - x_i}{b_n}\right)} \quad \text{and} \quad L_{2j}(v) = \frac{K_2(v)}{\frac{1}{l_n b_n} \sum_{\nu=1}^{l_n} K_2\left(\frac{y_\nu - y_j}{b_n}\right)},$$

and marginal frequencies q^*_μ and r^*_ν.

To formulate the asymptotic normality of $\mathbb{M}_n$ under $\mathcal{H}$ we make use of the following notation: $\overline{p}_{ij}, \overline{q}_i$, and $\overline{r}_j$, are the expectations of $\hat{p}_{ij}, \hat{q}_i$, and $\hat{r}_j$, respectively. Define

$$\xi_{pn} = (nm_nl_nb_n^2)^{-1}\, d_{1n} - (nm_nl_nb_n)^{-1}\, d_{2n}$$

with

$$d_{1n} = (m_nl_n)^{-1}b_n^2 \sum_{i=1}^{m_n}\sum_{\mu=1}^{m_n} K_{1i,b_n}^2(x_\mu - x_i)q_\mu \sum_{j=1}^{l_n}\sum_{\nu=1}^{l_n} K_{2j,b_n}^2(y_\nu - y_j)\, r_\nu,$$

$$d_{2n} = b_n \left(m_n^{-1}l_n \sum_{j=1}^{l_n} \overline{r}_j^2 \sum_{i=1}^{m_n}\sum_{\mu=1}^{m_n} K_{1i,b_n}^2(x_\mu - x_i)q_\mu \right.$$
$$\left. + l_n^{-1}m_n \sum_{i=1}^{m_n} \overline{q}_i^2 \sum_{j=1}^{l_n}\sum_{\nu=1}^{l_n} K_{2j,b_n}^2(y_\nu - y_j)r_\nu \right)$$

and

$$\tau_{pn}^2 = 2\, m_nl_n \sum_{i=1}^{m_n}\sum_{j=1}^{l_n} q_i^2\, r_j^2 \int\!\!\int (K * K)^2(x,y)\, \mathrm{d}x\, \mathrm{d}y.$$

The basis of our test is the following theorem:

Theorem 20.3.2 *Suppose that the marginal probabilities satisfy for all i, μ, j and ν, and some constants L_1 and L_2,*

$$|q_i - q_\mu| \le L_1\, m_n^{-1}\, |x_i - x_\mu| \quad \text{and} \quad |r_j - r_\nu| \le L_2\, l_n^{-1}\, |y_j - y_\nu|.$$

Further, the kernel K is of product type (20.1), *and the K_j's $(j = 1,2)$ satisfy* (K). *If the hypothesis $\mathcal{H}$ holds, then*

$$\frac{nm_nl_nb_n}{\tau_{pn}} \left(\mathbb{M}_n - \xi_{pn} \right) \xrightarrow{\mathcal{D}} \mathsf{N}(0,1)$$

as $b_n \to 0$, $m_nl_nb_n^2 \to \infty$ and $nb_n^2 \to \infty$.

To apply this limit statement for the construction of an asymptotic α-test we have to replace the unknown terms d_{1n} and d_{2n} in ξ_{pn} by suitable estimators. To avoid bias problems we choose

$$\tilde{d}_{1n} = (m_nl_n)^{-1}b_n^2 \sum_{i=1}^{m_n}\sum_{\mu=1}^{m_n} K_{1i,b_n}^2(x_\mu - x_i)\, q_\mu^* \sum_{j=1}^{l_n}\sum_{\nu=1}^{l_n} K_{2j,b_n}^2(y_\nu - y_j)\, r_\nu^*,$$

which is an unbiased $\sqrt{n}$-consistent estimator of d_{1n}. The second term is estimated by the consistent estimator

$$\tilde{d}_{2n} = b_n \left(m_n^{-1} l_n \sum_{j=1}^{l_n} r_j^{*2} \sum_{i=1}^{m_n} \sum_{\mu=1}^{m_n} K_{1i,b_n}^2 (x_\mu - x_i) q_\mu^* \right.$$

$$\left. + l_n^{-1} m_n \sum_{i=1}^{m_n} q_i^{*2} \sum_{j=1}^{l_n} \sum_{\nu=1}^{l_n} K_{2j,b_n}^2 (y_\nu - y_j) r_\nu^* \right),$$

and the variance term can be replaced by

$$\tilde{\tau}_{pn}^2 = 2 m_n l_n \sum_{i=1}^{m_n} \sum_{j=1}^{l_n} q_i^{*2} r_j^{*2} \int\int (K * K)^2(x,y) \, \mathrm{d}x \, \mathrm{d}y.$$

It is easy to verify that under $\mathcal{H}$

$$b_n^{-1} \left(\tilde{d}_{1n} - d_{1n} \right) \xrightarrow{\mathrm{P}} 0, \qquad \tilde{d}_{2n} - d_{2n} \xrightarrow{\mathrm{P}} 0 \quad \text{and} \quad \tilde{\tau}_{pn}^2 - \tau_{pn}^2 \xrightarrow{\mathrm{P}} 0.$$

Thus, with $\tilde{\xi}_{pn} = (n m_n l_n b_n^2)^{-1} \tilde{d}_{1n} - (n m_n l_n b_n)^{-1} \tilde{d}_{2n}$ an asymptotic α-test is provided by: Reject $\mathcal{H}$, if $\mathbb{M}_n \geq \tilde{\xi}_{pn} + z_\alpha \, \tilde{\tau}_{pn} / (n m_n l_n b_n)$.

20.4 Parametric Versus Nonparametric Regression Fit

Härdle and Mammen (1993) investigated the problem of testing whether a regression function has a parametric form. Let us briefly review their results. We have the following model: The pairs (X_i, Y_i), $i = 1, \ldots, n$, are i.i.d. $(\mathbb{R} \times \mathbb{R})$-valued random variables satisfying

$$Y_i = r(X_i) + \sqrt{v(X_i)} \, \varepsilon_i, \qquad i = 1, \ldots, n, \tag{20.7}$$

with the unknown regression function $r(\cdot) = \mathsf{E}(Y_1 | X_1 = \cdot)$ and the conditional variance $v(\cdot) = \mathsf{E}\left((Y_1 - r(X_1))^2 | X_1 = \cdot \right)$. Conditionally on $X_1, \ldots, X_n$ the errors ε_i are independent and identically distributed with expectation zero and variance 1. The test problem is:

$$\mathcal{H} : r \in \{ r(\cdot, \vartheta) : \vartheta \in \Theta \subseteq \mathbb{R}^d \} \quad \text{against} \quad \mathcal{K} : r \notin \{ r(\cdot, \vartheta) : \vartheta \in \Theta \subseteq \mathbb{R}^d \},$$

and as test statistic Härdle and Mammen propose the L_2-distance

$$\mathbb{R}_n(r_{\hat{\vartheta}_n}) = \int (\hat{r}_n(t) - \mathcal{K}_n \, r_{\hat{\vartheta}_n}(t))^2 a(t) \, \mathrm{d}t.$$

Here $\hat{r}_n$ is the Nadaraya–Watson kernel estimator with bandwidth b_n, and $\mathcal{K}_n r_{\hat{\vartheta}_n}$ denotes its smoothed version under $\mathcal{H}$, $\hat{\vartheta}_n$ is a suitable parameter estimator. It is proved that the properly standardized test statistic is asymptotically normal. The main point of the paper is to investigate different *bootstrap procedures* for the approximation of the critical values of the test. As already pointed out, the convergence of the distribution of an ISE-type statistic to the normal distribution is very slow (see Remark 3 in Section 20.2), therefore quantiles of the normal distribution are not appropriate for testing in practice. Since this problem arises not only in the context of regression testing it seems to be useful to think about, whether it is possible to apply similar bootstrap approaches also in other setups.

Suppose that (X_i^*, Y_i^*), $i = 1,\ldots,n$, is a bootstrap sample, then create $\mathbb{R}_n^*(r_{\hat{\vartheta}_n^*})$ like $\mathbb{R}_n(r_{\hat{\vartheta}_n})$ by the squared deviation between the parametric fit $r_{\hat{\vartheta}_n^*}$ and the nonparametric fit $\hat{r}_n^*$ (both computed from the bootstrap sample). The conditional distribution of $\mathbb{R}_n^*(r_{\hat{\vartheta}_n^*})$ under the (X_i, Y_i) can be approximated by Monte Carlo simulations. From this Monte Carlo approximation $(1-\alpha)$ quantile $\hat{q}_\alpha$ is defined, and one rejects $\mathcal{H}$ if $nb_n^{1/2}\mathbb{R}_n(r_{\hat{\vartheta}_n}) > \hat{q}_\alpha$. Härdle and Mammen show that the *naive resampling* does not work. The same is true for the so-called *adjusted residual bootstrap*. As an alternative they propose the *wild bootstrap*. The idea is to construct a bootstrap sample (X_i^*, Y_i^*), $i = 1,\ldots,n$, such that $\mathsf{E}^*(Y_i^*|X_i^*) = r_{\hat{\vartheta}_n}(X_i^*)$, where E^* denotes the conditional expectation $\mathsf{E}(\cdot|(X_i, Y_i), i = 1,\ldots,n)$. In simulation studies Härdle and Mammen consider parametric models of polynomials of different degree. It turns out that in all cases wild bootstrap estimates the distribution of $nb_n^{1/2}\,\mathbb{R}_n(r_{\vartheta_n})$ quite well. The normal approximation with estimated standardizing terms is totally misleading. The inaccuracy of the normal approximation increases with the dimension of the parametric model. Moreover, the authors give Monte Carlo estimates for the power of the test with bootstrapped quantiles and consider the influence of the bandwidth on the level of the test.

20.5 Testing Homoscedasticity in Nonparametric Regression

Again, assume model (20.7). Now, we wish to check whether the model is heteroscedastic, that is, we wish to test the hypothesis

$$\begin{aligned}\mathcal{H} &: v(t) = v \quad \text{for some } v > 0 \text{ and all } t \in [0,1] \quad \text{against}\\ \mathcal{K} &: v(t) \not\equiv v \quad \text{for all } v > 0.\end{aligned}$$

In the paper of Liero (2003a) the following approach is proposed: As test statistic we take the L_2- distance between a nonparametric kernel estimator of v in the underlying heteroscedastic model (20.7) and an estimator of the conditional variance in the hypothetical homoscedastic model

$$Y_i = r(X_i) + \sqrt{v}\,\varepsilon_i, \qquad i = 1, \ldots, n, \tag{20.8}$$

with $v \in \mathbb{R}_+$. To avoid bias problems in the limit theorem we modify this difference and use the following statistic:

$$\mathbb{V}_n = \int (\hat{v}_n(t) - \eta_{nv}(t))^2\, a(t)\, dt.$$

The estimator $\hat{v}_n$ has the form

$$\hat{v}_n(t) = \sum_{i=1}^{n} \sum_{j=1, i \neq j}^{n} \frac{1}{2}(Y_i - Y_j)^2\, W_{nij}(t, X_1, \ldots, X_n)$$

with weights

$$W_{nij}(t, X_1, \ldots, X_n) = \frac{K\left(\frac{t-X_i}{b_n}\right) K\left(\frac{t-X_j}{b_n}\right)}{\sum_{\mu=1}^{n} \sum_{\nu=1, \nu\neq\mu}^{n} K\left(\frac{t-X_\mu}{b_n}\right) K\left(\frac{t-X_\nu}{b_n}\right)},$$

and the term η_{nv} is defined by

$$\eta_{nv}(t) = v + \sum_{i=1}^{n} \sum_{j=1, i \neq j}^{n} \frac{1}{2}(r(X_i) - r(X_j))^2\, W_{nij}(t, X_1, \ldots, X_n).$$

Liero (1999) showed that $\mathbb{V}_n$ (properly standardized) is asymptotically normally distributed. But to apply this limit result for the construction of a test it is necessary to replace the unknown terms r and $v > 0$ by suitable estimators. This leads to the problem of estimating the variance in a homoscedastic regression model (20.8). For the fixed design model this problem is investigated by several authors. For the present random design model three different estimators for v are given by Liero (1999). It turns out that these estimators are $\sqrt{n}$-consistent under the hypothesis, which is sufficient to use them for the construction of the desired asymptotic α-test. Furthermore, it seems to be useful to replace the unknown regression function r in the term η_{nv} by a Nadaraya–Watson kernel estimator with a suitable bandwidth. The question of an appropriate choice of this bandwidth is discussed in Liero (2003a). Moreover, power considerations with respect to different types of local alternatives complete the approach presented there.

20.6 Testing the Hazard Function Under Censoring

20.6.1 Survival model without covariates

Firstly we consider a survival model without covariates, that is: Let $Y_1, \ldots, Y_n$ be a sequence of i.i.d. survival times with absolutely continuous distribution function F. As often occurs in applications the Y_i's are subject to random right censoring, i.e., the observations are

$$T_i = \min(Y_i, C_i) \quad \text{and} \quad \delta_i = 1(Y_i \leq C_i)$$

where $C_1, \ldots, C_n$ are i.i.d. continuous random censoring times which are independent of the Y- sequence. The δ_i indicates whether Y_i has been censored or not. The function of interest is the hazard rate λ which is defined by

$$\lambda(t) = \lim_{s \downarrow 0} \frac{1}{s} \mathsf{P}(t \leq Y_i \leq t + s | Y_i \geq t).$$

We wish to test whether λ lies in a parametric class of functions, i.e.,

$$\mathcal{H}: \quad \lambda \in \mathcal{L} = \{\lambda(\cdot, \vartheta) \,|\, \vartheta \in \Theta \subseteq \mathbb{R}^d\} \quad \text{against} \quad \mathcal{K}: \quad \lambda \notin \mathcal{L}.$$

Since no parametric form of the alternative is assumed we will use a nonparametric estimator of λ for testing $\mathcal{H}$ against $\mathcal{K}$. The idea for the construction of such a nonparametric estimator goes back to the paper of Watson and Leadbetter (1964), who considered the case without censoring. The censored case was investigated, for example by Lo and Singh (1986) and by Diehl and Stute (1988). To describe the estimation procedure we introduce the distribution function of the observations T_i and the subdistribution function of the uncensored observations:

$$H(t) := \mathsf{P}(T_i \leq t) \quad \text{and} \quad H^U(t) := \mathsf{P}(T_i \leq t, \delta_i = 1).$$

Since

$$1 - H(t) = (1 - G(t))\,(1 - F(t)) \quad \text{and} \quad H^U(t) = \int_0^t (1 - G(s))\, \mathrm{d}F(s),$$

where G is the distribution function of the censoring times C_i, the cumulative hazard function $\Lambda(t) := \int_0^t \lambda(s)\, \mathrm{d}s$ can be written as

$$\Lambda(t) = \int_0^t \frac{\mathrm{d}F(s)}{1 - F(s_-)} = \int_0^t \frac{\mathrm{d}H^U(s)}{1 - H(s)}.$$

Now, for estimating Λ we replace H^U and H by their empirical versions, that is by

$$\hat{H}_n^U(t) = \frac{1}{n}\sum_{i=1}^n 1(T_i \le t, \delta_i = 1) \quad \text{and} \quad \hat{H}_n(t) = \frac{1}{n}\sum_{i=1}^n 1(T_i \le t). \tag{20.9}$$

The resulting estimator

$$\hat{\Lambda}_n(t) := \int_0^t \frac{d\hat{H}_n^U(s)}{1 - \hat{H}_n(s_-)} = \sum_{i=1}^n \frac{1(T_{(i)} \le t)\, \delta_{[i]}}{n - i + 1}$$

is the Nelson–Aalen estimator of Λ. Here $T_{(1)} \le \cdots \le T_{(n)}$ are the ordered observations and $\delta_{[i]} = \delta_j$ if $T_j = T_{(i)}$. As estimator of the derivative of Λ we define the kernel smoothed Nelson–Aalen estimator

$$\hat{\lambda}_n(t) := \frac{1}{b_n}\int K\left(\frac{t-s}{b_n}\right) d\hat{\Lambda}_n(s) = \frac{1}{b_n}\sum_{i=1}^n \frac{K\left(\frac{t-T_{(i)}}{b_n}\right) \delta_{[i]}}{n-i+1}, \tag{20.10}$$

where K is a kernel function and $\{b_n\}$ is a sequence of bandwidths tending to zero at an appropriate rate. As before we choose as test statistic the L_2-distance of $\hat{\lambda}_n$ from the "smoothed version of the hypothesis"

$$e_n(t, \lambda_\vartheta) := \int K_{b_n}(t-s)\, \lambda(s, \vartheta)\, ds = \int K_{b_n}(t-s)\, d\Lambda(s, \vartheta),$$

where $\Lambda(t, \vartheta) = \int_0^t \lambda(s, \vartheta)\, ds$. Since the parameter ϑ is unknown we have to replace it by a suitable estimator. We propose to take the maximum likelihood estimator. The likelihood function is given by

$$\begin{aligned}
&\prod_{i=1}^n (1 - G(T_i))^{\delta_i}\, (1 - F(T_i, \vartheta))^{1-\delta_i}\, f(T_i, \vartheta)^{\delta_i}\, g(T_i)^{1-\delta_i} \\
&\qquad = \prod_{i=1}^n \lambda(T_i, \vartheta)^{\delta_i} \exp(-\Lambda(T_i, \vartheta))\, (1 - G(T_i))^{\delta_i}\, g(T_i)^{1-\delta_i},
\end{aligned}$$

where g is the density of the censoring times. Thus, the maximum likelihood estimator $\hat{\vartheta}_n$ is a (measurable) maximizer of

$$l_n(\vartheta) = \sum_{i=1}^n (\delta_i \log \lambda(T_i, \vartheta) - \Lambda(T_i, \vartheta)).$$

The test statistic is given by

$$\mathbb{L}_n := \int \left(\hat{\lambda}_n(t) - e_n(t, \lambda_{\hat{\vartheta}_n})\right)^2 a(t)\, dt.$$

To formulate the test procedure we state the following limit theorem, proved in Liero (2003b). Let T_H be the right end point of the distribution H and fix an arbitrary point $T' < T_H$. Further set

$$\begin{aligned}\mu_n(\lambda_\vartheta) &= (nb_n)^{-1}\kappa^2 \int \frac{\lambda(t,\vartheta)}{1-H(t)}\, a(t)\, \mathrm{d}t,\\ \sigma^2(\lambda_\vartheta) &= 2\int \left(\frac{\lambda(t,\vartheta)}{1-H(t)}\right)^2 a^2(t)\, \mathrm{d}t \int (\kappa^*(z))^2\, \mathrm{d}z.\end{aligned}$$

Theorem 20.6.1 *Suppose that the kernel satisfies* (K), *that the weight function* a *fulfills* (W) *and vanishes outside* $[0, T']$ *and that the distribution function* H *is Lipschitz continuous. Further, let any hazard rate* $\lambda \in \mathcal{L}$ *be bounded, Lipschitz continuous and partially differentiable w.r.t.* ϑ; $\nabla_\vartheta \lambda(\cdot,\cdot)$ *is bounded and uniformly continuous in both arguments. If* $b_n \to 0$ *and* $nb_n^2 \to \infty$, *then under* $\mathcal{H}$ *we have for all* $\lambda \in \mathcal{L}$,

$$nb_n^{1/2}\left(\int \left(\hat\lambda_n(t) - e_n(t,\lambda_\vartheta)\right)^2 a(t)\, \mathrm{d}t \;-\; \mu_n(\lambda_\vartheta)\right) \xrightarrow{\mathcal{D}} \mathsf{N}(0, \sigma^2(\lambda_\vartheta)). \quad (20.11)$$

To conclude from Theorem 20.6.1 to the asymptotic normality of our test statistic we use the consistency of the maximum likelihood estimator. For that purpose we formulate the following regularity conditions:

(i) For all $t \in [0,\infty)$ the second derivatives $\frac{\partial^2}{\partial\vartheta_i\partial\vartheta_j} f(t,\vartheta)$ and $\frac{\partial^2}{\partial\vartheta_i\partial\vartheta_j} F(t,\vartheta)$ exist and are continuous for all $\vartheta \in \Theta^o$, $i,j = 1,\ldots,d$, where Θ^o is the open kernel of Θ.

(ii) $\int \nabla_\vartheta f(t,\vartheta)(1-G(t))\, \mathrm{d}t - \int \nabla_\vartheta F(t,\vartheta)\, \mathrm{d}G(t) = 0$,
$\int \frac{\partial^2}{\partial\vartheta_i\partial\vartheta_j} f(t,\vartheta)(1-G(t))\, \mathrm{d}t - \int \frac{\partial^2}{\partial\vartheta_i\partial\vartheta_j} F(t,\vartheta)\, \mathrm{d}G(t) = 0.$

(iii) For any $\vartheta \in \Theta^o$ there exist a $\nu = \nu(\vartheta)$ and a sphere $U(\nu) \subset \Theta^o$ with the radius ν such that for all $i,j = 1,\ldots,d$,

$$\left|\frac{\partial^2}{\partial\vartheta_i\partial\vartheta_j}\left(\log f(t,\vartheta') - \log(1-F(t,\vartheta'))\right)\right| \le m(x,\vartheta) \quad \text{for all } \vartheta' \in U(\nu)$$

for a measurable function m with $\mathsf{E}m(T_i,\vartheta)\, \mathrm{d}x < \infty$.

(iv) The determinant of the Fisher information

$$\begin{aligned} I(\vartheta) &= \mathsf{E}\big([\delta_1 \nabla_\vartheta \log\lambda(T_1,\vartheta) + \nabla_\vartheta \log(1-F(T_1,\vartheta))]\\ &\qquad \times [\delta_1 \nabla_\vartheta \log\lambda(T_1,\vartheta) + \nabla_\vartheta \log(1-F(T_1,\vartheta))]^t\big)\end{aligned}$$

is nonzero for any $\vartheta \in \Theta^o$.

Under these conditions we have: Under $\mathcal{H}$

$$\sqrt{n}\left(\hat{\vartheta}_n - \vartheta\right) \xrightarrow{\mathcal{D}} \mathsf{N}(0, I(\vartheta)^{-1})$$

for any $\vartheta \in \Theta^o$. That is, the maximum likelihood estimator $\hat{\vartheta}_n$ is $\sqrt{n}$-consistent. Therefore the limit statement (20.11) remains true for $\mathbb{L}_n$. Furthermore, in the standardizing terms the unknown distribution function H can be replaced by $\hat{H}_n$. Thus, finally we obtain an asymptotic α-test by the rule: Reject $\mathcal{H}$, if $\mathbb{L}_n \geq \mu_{\lambda n} + z_\alpha \sigma_{\lambda n}/(nb_n^{1/2})$, where $\mu_{\lambda n} = (nb_n)^{-1} \kappa^2 \int \frac{\lambda(t,\hat{\vartheta}_n)}{1-\hat{H}_n(t)} a(t)\,\mathrm{d}t$ and $\sigma^2_{\lambda n} = 2 \int \left(\frac{\lambda(t,\hat{\vartheta}_n)}{1-\hat{H}_n(t)}\right)^2 a^2(t)\,\mathrm{d}t \int (\kappa^*(z))^2\,\mathrm{d}z$.

20.6.2 Survival model with fixed covariates

Now, let us extend the previous approach to survival models with covariates. That is, at fixed design points $x_1 \leq x_2 \leq \cdots \leq x_n$ we have nonnegative survival times $Y_1, \ldots, Y_n$. For simplicity we assume that the support of the covariates x_i is the interval $[0,1]$. Consequently, from the mathematical point of view, the Y_i's are no longer identically distributed. We define, analogously to section 20.6.1,

$$F_{x_i}(t) = \mathsf{P}(Y_i \leq t), \quad H_{x_i}(t) := \mathsf{P}(T_i \leq t), \quad H^U_{x_i}(t) := \mathsf{P}(T_i \leq t, \delta_i = 1) \quad \text{and}$$

$$\Lambda_{x_i}(t) := \int_0^t \lambda_{x_i}(s)\,\mathrm{d}s = \int_0^t \frac{\mathrm{d}H^U_{x_i}(s)}{1 - H_{x_i}(s)}.$$

The problem of nonparametric estimation of λ, the survival function $1-F$ and the hazard λ has been studied by several authors. We mention here: Gonzalez-Manteiga and Cadarso-Suarez (1994) and Van Keilegom and Veraverbeke (1997, 2001, 2002). Roughly speaking, the main aim of these papers is to approximate the distance between the function of interest and its nonparametric estimator by a sum of independent random variables. Based on such an approximation, consistency properties are established and asymptotic normality at fixed points t and x is derived. A modification of a result proved by Van Keilegom and Veraverbeke (2001) leads to a limit statement for the quadratic deviation. First, let us define the estimators. The idea is the same as before — Λ is estimated by a Nelson–Aalen type estimator. But to take into account the covariates we take instead of the empirical distribution functions (20.9) *weighted* empirical distribution functions:

$$\hat{H}_{xn}(t) = \sum_{j=1}^n w_{nj}(x) 1(T_j \leq t), \quad \hat{H}^U_{xn}(t) = \sum_{j=1}^n w_{nj}(x) 1(T_j \leq t, \delta_j = 1).$$

Following Van Keilegom and Veraverbeke (2001) we will use Gasser–Müller type kernel weights $w_{nj}(x)$. They are defined as

$$
\begin{aligned}
{}_{nj}(x) &= \frac{1}{c_n(x)} \int_{x_{j-1}}^{x_j} \frac{1}{a_n} W\left(\frac{x-z}{a_n}\right) \mathrm{d}z \quad \text{with} \\
c_n(x) &= \int_0^{x_n} \frac{1}{a_n} W\left(\frac{x-z}{a_n}\right) \mathrm{d}z.
\end{aligned}
$$

Here $x_0 = 0$, W is a symmetric kernel function and a_n is a sequence of bandwidths. Then a nonparametric estimator of $\Lambda_x(t)$ is given by

$$
\hat{\Lambda}_{xn}(t) := \int_0^t \frac{\mathrm{d}\hat{H}_{xn}^U(s)}{1 - \hat{H}_{xn}(s_-)} = \sum_{i=1}^n \frac{1(T_{(i)} \leq t)\, \delta_{[i]}\, w_{n[i]}(x)}{1 - \sum_{k=1}^{i-1} w_{n[k]}(x)}.
$$

Now, further smoothing with a kernel K and bandwidth b_n leads to the estimator of the hazard function:

$$
\begin{aligned}
\hat{\lambda}_{xn}(t) &= \frac{1}{b_n} \int K\left(\frac{t-s}{b_n}\right) \mathrm{d}\hat{\Lambda}_{xn}(s) \\
&= \frac{1}{b_n} \sum_{i=1}^n K\left(\frac{t-T_{(i)}}{b_n}\right) \delta_{[i]} \frac{w_{n[i]}(x)}{1 - \sum_{k=1}^{i-1} w_{n[k]}(x)}.
\end{aligned}
$$

Note that if we take the weights all equal to n^{-1}, then the estimator becomes the estimator defined in (20.10) for the case without covariates. Now, consider the problem of testing the simple hypothesis

$$
\mathcal{H} : \lambda_x(t) = \lambda_x^o(t) \text{ for all } t, x \quad \text{against} \quad \mathcal{K} : \lambda_{x'}(t') \neq \lambda_{x'}^o(t') \text{ for some } t', x'.
$$

As test statistic we propose the following quadratic deviation:

$$
\mathrm{W}_n = \frac{1}{n} \sum_{i=1}^n \int \left(\hat{\lambda}_{x_i n}(t) - e_n(t, \lambda_{x_i}^o)\right)^2 a(t)\, \mathrm{d}t,
$$

where

$$
e_n(t, \lambda_{x_i}^o) = \int K_{b_n}(t-s)\, \lambda_{x_i}^o(s)\, \mathrm{d}s.
$$

To formulate the limit theorem for this functional we introduce the following quantities.

$$
\begin{aligned}
\zeta_n(\lambda) &= (n^2 a_n b_n)^{-1} \kappa^2 \omega^2 \sum_{i=1}^n \int \frac{\lambda_{x_i}(t)}{1 - H_{x_i}(t)} a(t)\, \mathrm{d}t, \qquad (20.12) \\
\rho_n^2(\lambda) &= 2n^{-1} \sum_{i=1}^n \int \left(\frac{\lambda_{x_i}(t)}{1 - H_{x_i}(t)}\right)^2 a^2(t)\, \mathrm{d}t \int (\kappa^*(z))^2\, \mathrm{d}z \int (\omega^*(z))^2\, \mathrm{d}z,
\end{aligned}
$$

where ω^* denotes the convolution of the kernel W. For the design points set $\underline{s}_n = \min_{1\le i\le n}(x_i - x_{i-1})$ and $\overline{s}_n = \max_{1\le i\le n}(x_i - x_{i-1})$. Further define $r_n = (na_nb_n)^{-1}\log n + (b_n^{1/2} + a_n^2 b_n^{-1})(na_n)^{-1/2}(\log n)^{1/2} + a_n^4 b_n^{-1}$.

The following assumptions are used:

(i) $x_n \to 1$, $\overline{s}_n = O(n^{-1})$, $\overline{s}_n - \underline{s}_n = o(n^{-1})$

(ii) The derivatives $\frac{\partial^2 H}{\partial x^2}$, $\frac{\partial^2 H}{\partial t^2}$, $\frac{\partial^2 H}{\partial x \partial t}$ and $\frac{\partial^3 H}{\partial x^2 \partial t}$ exist and are continuous in the interval $[0,1] \times [0,T']$.

(iii) The derivatives $\frac{\partial^2 H^U}{\partial x^2}$, $\frac{\partial^2 H^U}{\partial t^2}$ and $\frac{\partial^2 H^U}{\partial x \partial t}$ exist and are continuous in the interval $[0,1] \times [0,T']$.

With these assumptions we can state the following theorem:

Theorem 20.6.2 *Suppose that the kernels K and W satisfy* (K). *If assumptions* (i)–(iii) *and* (W) *are fulfilled and $r_n^2 n(a_nb_n)^{1/2} \to 0$, $r_n(a_nb_n)^{-1/2} \to 0$, then under $\mathcal{H}$*

$$\frac{n(a_nb_n)^{1/2}}{\rho_n(\lambda^o)}\left(\mathbb{W}_n - \zeta_n(\lambda^o)\right) \xrightarrow{\mathcal{D}} \mathsf{N}(0,1).$$

The only unknown function in the standardizing terms is H_x. We replace H_x by its consistent estimator $\hat{H}_{xn}$. If $((na_n)^{-1/2}(\log n)^{1/2} + a_n^2)(a_nb_n)^{-1/2} \to 0$, then the estimation error tends to zero fast enough such that the limit statement remains valid with the estimated distribution function. So, finally we get the rule:

Reject $\mathcal{H}$, if $\mathbb{W}_n \ge \tilde{\zeta}_n(\lambda^o) + z_\alpha\,\tilde{\rho}_n(\lambda^o)/(n(a_nb_n)^{1/2})$. Here $\tilde{\zeta}_n(\lambda^o)$ and $\tilde{\rho}_n(\lambda^o)$ are defined as in (20.12), where H_x is replaced by $\hat{H}_{xn}$.

Some remarks. ·(1) The investigation of the power of this test requires more technical effort than that carried out in Section 2, but roughly speaking it leads to the same conclusion. Namely that the power tends to a nontrivial limit, if the squared L_2-norm of the disturbing function tends to a nonnegative constant with a rate $n(a_nb_n)^{1/2}$.

(2) It seems to be not very difficult to extend the presented approach to the problem of testing, whether the unknown hazard rate λ_x lies in a parametric class. A more complicated problem is to test a semiparametric hypothesis. For example, suppose that the hypothetical class of hazard functions is the class of proportional hazard functions with unknown baseline hazard function $\alpha(\cdot)$ and a parametric function describing the influence of the covariates. For the construction of the test statistic one has to estimate both functions. Using the partial likelihood method one obtains a suitable estimator for the parametric part. To estimate the baseline function it seems to be useful to apply an approach via the Breslow estimator for the cumulative baseline hazard. But for this estimator rates of convergence are not derived. That means, it is not

clear whether the estimation error tends to zero fast enough such that the limit theorem remains valid with the estimated hypothetical hazard function.

References

1. Aerts, M. and Augustyns, I. and Janssen, P. (1997). Smoothing sparse multinomial data using local polynomial fitting, *Journal of Nonparametric Statistics*, **8**, 127–147.

2. Augustyns, I. (1997). Local polynomial smoothing of sparse multinomial data, Thesis, Limburgs Universitar Centrum.

3. Bickel, P. and Rosenblatt, M. (1973). On some global measures of deviation of density function estimators, *Annals of Statistics*, **1**, 1071–1095.

4. Burman, P. (1987). Central limit theorem for quadratic forms for sparse tables, *Journal of Multivariate Analysis*, **22**, 258–277.

5. Diehl, S. and Stute, W. (1988). Kernel density and hazard function estimation in the presence of censoring, *Journal of Multivariate Analysis*, **25**, 299–310.

6. Ghosh, B. and Huang, W-M. (1991). The power and optimal kernel of the Bickel-Rosenblatt test for goodness of fit, *Annals of Statistics*, **19**, 999–1009.

7. González-Manteiga, W. and Cadarso-Suarez, C. (1996). Asymptotic properties of a generalized Kaplan-Meier estimator with some applications, *Journal of Nonparametic Statistics*, **4**, 65–78.

8. Härdle, W. and Mammen, E. (1993). Comparing nonparametric versus parametric regression fits, *Annals of Statistics*, **21**, 1926–1947.

9. Holst, L. (1972). Asymptotic normality and efficiency for certain goodness-of-fit tests, *Biometrika*, **59**, 137–145.

10. Liero, H. (1999). Global deviation measures in nonparametric curve estimation, *Habilitationsschrift an der Mathematisch-Naturwissenschaftlichen Fakultät der Universität Potsdam*, Germany.

11. Liero, H. (2001). L_2-tests for sparse multinomials, *Statistics & Probability Letters*, **55**, 147-158.

12. Liero, H. (2003a). Testing homoscedasticity in nonparametric regression, *Journal of Nonparametric Statistics*, **15**, 31-51.

13. Liero, H. (2003b). Testing the hazard rate, *Preprint*, Institut für Mathematik, Universität Potsdam, Germany.

14. Liero, H., Läuter, H. and Konakov, V. D. (1998). Nonparametric versus parametric goodness of fit, *Statistics*, **31**, 115–149.

15. Lo, S.-H. and Singh, K. (1986). The product-limit estimator and the bootstrap: Some asymptotic representations,*Probability Theory and Related Fields*, **71**, 455–465.

16. Rosenblatt, M. (1975). A quadratic measure of derivation of two-dimensional density estimates and a test of independence, *Annals of Statistics*, **3**, 1–14.

17. Simonoff, J. S. (1996). *Smoothing Methods in Statistics*, Springer-Verlag, New York.

18. Van Keilegom, I. and Veraverbeke, N. (1997). Estimation and bootstrap with censored data in fixed design nonparametric regression, *Annals of the Institute of Statistical Mathematics*, **49**, 467–401.

19. Van Keilegom, I. and Veraverbeke, N. (2001). Hazard rate estimation in nonparametric regression with censored data, *Annals of the Institute of Statistical Mathematics*, **53**, 730–745.

20. Van Keilegom, I. and Veraverbeke, N. (2002). Density and hazard estimation in censored regression models, *Bernoulli*, **8**, 607–625.

21. Watson, G. S. and Leadbetter, M. R. (1964). Hazard analysis I, *Biometrika*, **51**, 175–184.

PART IX
Multicentre Studies

21

Asymptotic Properties of the CRT Estimators for Multicentre Studies

Vladimir V. Anisimov[1] and Valerii V. Fedorov[2]

Research Statistics Unit, GlaxoSmithKline
[1]*Harlow, Essex, UK*
[2]*Upper Providence, Philadelphia, USA*

Abstract: We discuss asymptotic properties of the maximum likelihood estimators for a combined response to treatment (CRT) of two-arm multicentre clinical trials under random recruitment. Responses to different treatments are described by a random effects model. The conditions of consistency and asymptotic normality of the estimator of a CRT are given under rather mild assumptions on the recruitment process. Approximation of the variance of the CRT estimator in the case of well-populated centres is considered.

Keywords and phrases: Multicentre clinical trials, combined response to treatment, maximum likelihood estimator, asymptotic properties, approximation

21.1 Introduction

An important part of the drug development process in the pharmaceutical industry is the testing of a new therapeutic drug in a large clinical trial. It is almost always the case that the number of patients needed for the trial is too large for the trial to be run at a single medical centre. Consequently, the patients are recruited by multiple centres and the trial is called a multicentre clinical trial.

We consider the case where two treatments are to be compared in every centre. Each centre provides a value for the treatment difference and these need to be combined into a single overall value. For the case where treatments and centres are fixed effects, Dragalin et al. (2001) proposed that the Combined Response to Treatment (CRT) be used as this single value and described the properties of three well-known estimators of the CRT. Fedorov et al. (2002)

extended the definition of the CRT to the case where centre and treatment effects are random and derived an estimator of the CRT for this case.

We note that a similar estimator for the one-dimensional case has appeared previously in the meta-analysis and multicentre trial literature [e.g., DerSimonian and Laird (1986), Hardy and Thompson (1996) and Senn (1997, Ch. 14)]. See Senn (2000) for a comparison of the linear models used in meta-analysis and multicentre trials.

We discuss asymptotic properties of the estimator of a CRT under random recruitment. While only the case of two-arm multicentre clinical trials with a scalar response is considered, the extension to more general settings is relatively straightforward. Our derivation includes trials where many centres may have observations for only one treatment.

The rest of the paper is organized as follows. In Section 21.2 the estimator of a CRT is introduced. In Section 21.3 the asymptotic properties of estimators are studied. The approximation of the variance of the estimator is provided in Section 21.4.

21.2 Estimator of CRT

Assume that to test properties of a drug we need to treat n patients which should be recruited to N centres. Consider a two-arm treatment and suppose that the response for the k-th patient on the j-th treatment in the i-th centre can be presented in the form:

$$y_{ijk} = \mu_{ij} + \varepsilon_{ijk}, \quad i = 1, \ldots, N, \; j = 1, 2, \; k = 1, \ldots, n_{ij}, \tag{21.1}$$

where n_{ij} is the number of patients on the j-th treatment.

Here ε_{ijk} are independent random variables with zero means and common variance σ^2. For design purposes it is often assumed that σ^2 is given. Usually ε_{ijk} is labelled as "within centre" variability. It is also assumed that the vector $\boldsymbol{\mu}_i^T = (\mu_{i1}, \mu_{i2})$ is random with mean $\boldsymbol{\mu}^T = (\boldsymbol{\mu}_1, \boldsymbol{\mu}_2)$ and the covariance matrix $\boldsymbol{V}$. Introduction of random μ_{ij} allows one to model "between centre" variability.

Note that the model (21.1) is a linear regression model with random coefficients, which is often referred to as a random effects model. It has been addressed many times in the literature [see for example, Brown and Prescott (1999), Pinheiro and Bates (2000), Rao (1965), Searle (1987), Spjotvoll (1977) and Verbeke and Molenberghs (2000)].

At the design stage it is assumed that $\boldsymbol{V}$ is given. For convenience denote $\boldsymbol{V} = \sigma^2 \boldsymbol{G}$, where $\boldsymbol{G} = \begin{pmatrix} \rho_1 & \rho \\ \rho & \rho_2 \end{pmatrix}$. Everywhere in this paper we assume that $\boldsymbol{G}$ is positive definite. Let $\bar{y}_{ij} = n_{ij}^{-1} \sum_{k=1}^{n_{ij}} y_{ijk}$, if $n_{ij} > 0$, and $\bar{y}_{ij} = 0$ if $n_{ij} = 0$, $j = 1, 2$. Put $\bar{\boldsymbol{y}}_i = (\bar{y}_{i1}, \bar{y}_{i2})^\top$.

Given n_{i1}, n_{i2} the best linear estimator of $\boldsymbol{\mu}$ is (Fedorov et al. 2002):

$$\hat{\boldsymbol{\mu}} = \begin{bmatrix} \hat{\mu}_1 \\ \hat{\mu}_2 \end{bmatrix} = \left\{ \sum_{i=1}^{N} \boldsymbol{W}_i \right\}^{-1} \sum_{i=1}^{N} \boldsymbol{W}_i \bar{\boldsymbol{y}}_i, \tag{21.2}$$

where in case of nonzero n_{i1}, n_{i2},

$$\boldsymbol{W}_i = (\boldsymbol{G} + \boldsymbol{M}_i^{-1})^{-1},\ \boldsymbol{M}_i = \begin{pmatrix} n_{i1} & 0 \\ 0 & n_{i2} \end{pmatrix}. \tag{21.3}$$

If in the i-th centre there are no patients on treatment j, then jj-th entry for the matrix $\boldsymbol{W}_i$ is $(\rho_{jj} + 1/n_{ij})^{-1}$ and all other entries are zeros.

Given $\{n_{ij}\}$ a variance-covariance matrix of $\hat{\boldsymbol{\mu}}$ has the form

$$\mathbf{Var}[\hat{\boldsymbol{\mu}} \mid \{n_{ij}\}] \;=\; \mathbf{S}^2[\{n_{ij}\}] = \sigma^2 \left\{ \sum_{i=1}^{N} \boldsymbol{W}_i \right\}^{-1}, \tag{21.4}$$

where we denote by symbols $\mathbf{E}[\cdot \mid \{n_{ij}\}]$, and $\mathbf{Var}[\cdot \mid \{n_{ij}\}]$ the conditional expectation and the variance taken at fixed values $n_{i1}, n_{i2}, i = 1, \ldots, N$ with respect to all other random variables ε_{ijk} and μ_{ij}. In the analysis of multicentre trials the variables n_{ij} are often assumed to be random (random recruitment).

The best linear estimator of a CRT $\boldsymbol{\delta} = \boldsymbol{\mu}_1 - \boldsymbol{\mu}_2$ is given by the expression $\widehat{\boldsymbol{\delta}} = \hat{\mu}_1 - \hat{\mu}_2 = \ell^\top \widehat{\boldsymbol{\mu}}$ with variance $\mathbf{Var}[\widehat{\boldsymbol{\delta}}] = \ell^\top \mathbf{Var}[\widehat{\boldsymbol{\mu}}]\ell$, where $\ell^\top = (1, -1)$.

It is easy to verify that for any fixed $n_{ij} \geq 0, i = 1, \ldots, N$, $j = 1, 2$, the estimator $\widehat{\boldsymbol{\mu}}$ defined by (21.2) is unbiased, i.e., $\mathbf{E}[\widehat{\boldsymbol{\mu}} \mid \{n_{ij}\}] = \boldsymbol{\mu}$. Thus, the estimator (21.2) is also unbiased with respect to $\{n_{ij}\}$, i.e.,

$$\mathbf{E}[\widehat{\boldsymbol{\mu}}] = \mathbf{E}\Big[\mathbf{E}[\widehat{\boldsymbol{\mu}} \mid \{n_{ij}\}]\Big] = \boldsymbol{\mu}.$$

Correspondingly, the unconditional variance of $\widehat{\boldsymbol{\mu}}$ has the form:

$$\mathbf{Var}[\widehat{\boldsymbol{\mu}}] = \sigma^2 \mathbf{E}\left[\sum_{i=1}^{N} \boldsymbol{W}_i \right]^{-1}, \tag{21.5}$$

where the expectation is taken with respect to all random variables included in $(\{\varepsilon_{ijk}, \mu_{ij}, n_{ij}\})$.

21.3 Asymptotic Properties of CRT Estimator

Consider the case when the number of centres and the number of patients are large. Denote by $\chi_{ij} = \chi(n_{ij} > 0)$ the indicator of the event that the i-th centre has a positive number of patients for the j-th arm, $j = 1, 2$. Let

$$N_1 = \sum_{i=1}^{N} \chi_{i1},\ N_2 = \sum_{i=1}^{N} \chi_{i2},\ N_{12} = \sum_{i=1}^{N} \chi_{i1}\chi_{i2},$$

where N_j is the number of centres with a positive number of patients for the j-th arm (N_{12} - for both arms).

21.3.1 Deterministic setting

Assume first that n_{ij} are deterministic. Denote by $\xrightarrow{\text{P}}$ ($\stackrel{\text{w}}{\Rightarrow}$) the convergence in probability (weak convergence), respectively.

Theorem 21.3.1 *If $N_1 \to \infty$ and $N_2 \to \infty$ as $N \to \infty$, $n \to \infty$, then the estimator $\widehat{\boldsymbol{\mu}}$ is consistent, and asymptotically normal, that is*

$$\zeta_N = \sigma^{-1}\boldsymbol{R}^{-1/2}(\widehat{\boldsymbol{\mu}} - \boldsymbol{\mu}) \stackrel{\text{w}}{\Rightarrow} \mathcal{N}(0,1), \tag{21.6}$$

where $\mathcal{N}(0,1)$ is a standard multivariate normal random variable.

PROOF. Denote by $\boldsymbol{R} = \{\sum_{i=1}^N \boldsymbol{W}_i\}^{-1}$ the covariance matrix of the estimator. As the estimator is unbiased, for consistency it is enough to prove that the elements of $\boldsymbol{R}$ tend to zero. Put $D_i = (\rho_1 + 1/n_{i1})(\rho_2 + 1/n_{i2}) - \rho^2\chi_{i1}\chi_{i2}$. Then for any values $n_{i1} \geq 0$, $n_{i2} \geq 0$,

$$\boldsymbol{W}_i = \begin{pmatrix} (\rho_2 + 1/n_{i2})\chi_{i1}/D_i & -\rho\chi_{i1}\chi_{i2}/D_i \\ -\rho\chi_{i1}\chi_{i2}/D_i & (\rho_1 + 1/n_{i1})\chi_{i2}/D_i \end{pmatrix}, \tag{21.7}$$

where $(\rho_2 + 1/n_{i2})\chi_{i1}/D_i = (\rho_1 + 1/n_{i1})^{-1}$ if $n_{i2} = 0$, and correspondingly $(\rho_1 + 1/n_{i1})\chi_{i2}/D_i = (\rho_2 + 1/n_{i2})^{-1}$ if $n_{i1} = 0$. Put

$$S_{1N} = \sum_{i=1}^N \frac{(\rho_1 + \frac{1}{n_{i1}})}{D_i}\chi_{i2}, \quad S_{2N} = \sum_{i=1}^N \frac{(\rho_2 + \frac{1}{n_{i2}})}{D_i}\chi_{i1}, \quad S_N = \sum_{i=1}^N \frac{1}{D_i}\chi_{i1}\chi_{i2}.$$

Then

$$\sum_{i=1}^N \boldsymbol{W}_i = \begin{pmatrix} S_{2N} & -\rho S_N \\ -\rho S_N & S_{1N} \end{pmatrix}, \quad \boldsymbol{R} = \begin{pmatrix} \Delta^{-1}S_{1N} & \rho\Delta^{-1}S_N \\ \rho\Delta^{-1}S_N & \Delta^{-1}S_{2N} \end{pmatrix},$$

where $\Delta = S_{N1}S_{N2} - \rho^2 S_N^2$. Consider the element $r_N(11) = \Delta^{-1}S_{N1}$ of the matrix $\boldsymbol{R}$. After some algebra we can get the following inequalities:

$$\begin{aligned} S_{1N} &\geq \rho_1 S_N + (\rho_2 + 1)^{-1}(N_2 - N_{12}), \\ S_{2N} &\geq \rho_2 S_N + (\rho_1 + 1)^{-1}(N_1 - N_{12}), \\ S_{1N} &\leq (\rho_1 + 1)S_N + \rho_2^{-1}(N_2 - N_{12}). \end{aligned}$$

As $N_1 \to \infty$, $N_2 \to \infty$, we have for N_{12} two possibilities: either $N_{12} \to \infty$ or N_{12} is bounded. Consider the case when $N_{12} \to \infty$. Then

$$S_N \geq \left((\rho_1 + 1)(\rho_2 + 1) - \rho^2\right)^{-1} N_{12} \to \infty,$$

and finally as $S_N \to \infty$,

$$r_N(11) \;\le\; \frac{(\rho_1+1)S_N + \rho_2^{-1}(N_2 - N_{12})}{(\rho_1\rho_2 - \rho^2)S_N^2 + \rho_1(\rho_1+1)^{-1}S_N(N_1 - N_{12})} \to 0. \quad (21.8)$$

Let now N_{12} be bounded. Then S_N is bounded and

$$r_N(11) \le (\rho_1+1)(\rho_2+1)\frac{(\rho_1+1)S_N + \rho_2^{-1}(N_2 - N_{12})}{(N_1 - N_{12})(N_2 - N_{12})} \to 0, \quad (21.9)$$

as $N_1 \to \infty$, $N_2 \to \infty$. By analogy other elements of matrix $\boldsymbol{R}$ also tend to zero, which implies the consistency.

Let us prove asymptotic normality. Note that in case of normally distributed μ_{ij} and ε_{ijk} (21.6), is trivial as the estimator $\widehat{\boldsymbol{\mu}}$ is normal itself. Further, we represent $\bar{\zeta}_N$ as a sum of independent random variables:

$$\bar{\zeta}_N = \sigma^{-1}\boldsymbol{R}^{1/2}\sum_{i=1}^{N}\boldsymbol{W}_i(\bar{\boldsymbol{y}}_i - \mathbf{E}\bar{\boldsymbol{y}}_i). \quad (21.10)$$

As the covariance matrix of $\bar{\zeta}_N$ is equal to $\boldsymbol{I}$, where $\boldsymbol{I}$ is a unit matrix, it is enough to prove that the variables $\bar{\xi}_{Ni} = \boldsymbol{R}^{1/2}\boldsymbol{W}_i(\bar{\boldsymbol{y}}_i - \mathbf{E}\bar{\boldsymbol{y}}_i)$ satisfy the Lindeberg condition (for instance, see Anderson, 1958, p. 74): for any $\varepsilon > 0$,

$$\boldsymbol{L}_N(\varepsilon) = \sum_{i=1}^{N}\mathbf{E}\,||\bar{\xi}_{Ni}||^2\,\chi(||\bar{\xi}_{Ni}|| \ge \varepsilon) \to 0, \quad (21.11)$$

where for any vector $\bar{a}$, we define $||\bar{a}||^2 = \bar{a}^\top\bar{a}$.

Let us prove an auxiliary result: if n_{ij} are deterministic, then uniformly in $N > 1$, n_{ij},

$$\alpha_N(L) = \lim_{L\to\infty}\max_{i=1,\ldots,N}\mathbf{E}\,||(\bar{y}_i - \mathbf{E}\bar{y}_i)||^2\,\chi(||\bar{y}_i - \mathbf{E}\bar{y}_i|| \ge L) \to 0. \quad (21.12)$$

Indeed, as $\bar{\boldsymbol{y}}_i = (\bar{y}_{i1}, \bar{y}_{i2})^\top$, it is enough to prove a one-dimensional variant of (21.12), that is, for any $j = 1, 2$,

$$\lim_{L\to\infty}\max_{i=1,\ldots,N}\mathbf{E}\,(\bar{y}_{ij} - \mathbf{E}\bar{y}_{ij})^2\,\chi(|\bar{y}_{ij} - \mathbf{E}\bar{y}_{ij})| \ge L) \to 0. \quad (21.13)$$

Put $j = 1$. If $n_{i1} > 0$, then $\bar{y}_{i1} = \mu_{i1} + n_{i1}^{-1}\sum_{k=1}^{n_{i1}}\varepsilon_{i1k}$, otherwise $\bar{y}_{i1} = 0$. For $n_{i1} > 0$, denote $\delta_{i1} = n_{i1}^{-1}\sum_{k=1}^{n_{i1}}\varepsilon_{i1k}$. As $(x_1 + \cdots + x_n)^2 \le n(x_1^2 + \cdots + x_n^2)$, then for any independent random variables $\xi_1, \ldots, \xi_n$ with distribution functions $F_k(x)$, $k = 1, \ldots, n$, we get:

$$\mathbf{E}\Big(\sum_{k=1}^{n}\xi_k\Big)^2\chi\Big(|\sum_{k=1}^{n}\xi_k| \ge L\Big)$$

$$= \int\limits_{|x_1+\ldots+x_n|\geq L} (x_1+\cdots+x_n)^2 \mathrm{d}F_1(x_1)\ldots\mathrm{d}F_n(x_n)$$
$$\leq \int\limits_{|x_1|\geq L/n} \cdots \int\limits_{|x_n|\geq L/n} n(x_1^2+\cdots+x_n^2)\mathrm{d}F_1(x_1)\ldots\mathrm{d}F_n(x_n)$$
$$\leq n^2 \max_{k=1,\ldots,N} \mathbf{E}\xi_k^2\chi\Big(|\xi_k| \geq L/n\Big). \tag{21.14}$$

Now we apply this inequality to the variable $\bar{y}_{i1}$. As $\mathbf{E}\delta_{i1} = 0$, we get that

$$\mathbf{E}(\bar{y}_{i1} - \mathbf{E}\bar{y}_i)^2\chi(|\bar{y}_{i1} - \mathbf{E}\bar{y}_i)| \geq L)$$
$$\leq 4\max\Big\{\mathbf{E}(\mu_{i1} - \mathbf{E}\mu_{i1})^2\chi(|\mu_{i1} - \mathbf{E}\mu_{i1})| \geq L/2),\ \mathbf{E}\delta_{i1}^2\chi(|\delta_{i1}| \geq L/2)\Big\}.$$

As $\mathbf{E}\mu_{i1}^2 < \infty$, then $\mathbf{E}(\mu_{i1} - \mathbf{E}\mu_{i1})^2\chi(|\mu_{i1} - \mathbf{E}\mu_{i1})| \geq L) \to 0$ as $L \to \infty$. Therefore, we need to check the relation

$$\max_{n_{i1}>0} \mathbf{E}\delta_{i1}^2\chi(|\delta_{i1}| \geq L) \to 0, \tag{21.15}$$

as $L \to \infty$. Applying (21.14) to variable δ_{i1} we get uniformly in $n_{i1} > 0$,

$$\mathbf{E}\delta_{i1}^2\chi(|\delta_{i1}| \geq L) = \mathbf{E}\left(\Big(n_{i1}^{-1}\sum_{k=1}^{n_{i1}}\varepsilon_{i1k}\Big)^2\chi\Big(|n_{i1}^{-1}\sum_{k=1}^{n_{i1}}\varepsilon_{i1k}| \geq L\Big)\right)$$
$$\leq \mathbf{E}\varepsilon_{i11}^2\chi(|\varepsilon_{i11}| \geq L) \to 0, \tag{21.16}$$

as $L \to \infty$. This implies (21.12).

For any square matrix A with elements a_{ij} we define the norm of A by the relation $||A|| = [tr(A^\top A)]^{1/2} = \left(\sum_{i,j} a_{ij}^2\right)^{1/2}$, where $tr(A) = \sum_i a_{ii}$ (Harville, 1997, p.60).

Let us now estimate the Lindeberg function (21.11) in the following way:

$$L(\varepsilon) \leq ||\boldsymbol{R}^{1/2}||^2 \sum_{i=1}^{N} ||\boldsymbol{W}_i||^2 \mathbf{E}||\bar{\boldsymbol{y}}_i - \mathbf{E}\bar{\boldsymbol{y}}_i||^2$$
$$\times\ \chi\Big(||\boldsymbol{W}_i(\bar{\boldsymbol{y}}_i - \mathbf{E}\bar{\boldsymbol{y}}_i)|| \geq \varepsilon||\boldsymbol{R}^{1/2}||^{-1}\Big). \tag{21.17}$$

A matrix $\boldsymbol{R}^{-1}$ as a sum of covariance matrices is a positive symmetric definite matrix. That means, the matrices $\boldsymbol{R}$ and $\boldsymbol{R}^{1/2}$ are also symmetric matrices. Then

$$||\boldsymbol{R}^{1/2}|| = [tr(\boldsymbol{R}^{1/2}\boldsymbol{R}^{1/2})]^{1/2} = [tr(\boldsymbol{R})]^{1/2} \to 0.$$

Now using (21.7) we get for any n_{ij},

$$||\boldsymbol{W}_i||^2 \leq (\rho_1\rho_2 - \rho^2)((\rho_1+1)^2 + (\rho_2+1)^2 + 2\rho^2). \tag{21.18}$$

Denote the right-hand side in (21.18) by K^2. As $||\boldsymbol{R}^{1/2}||^{-1} \to \infty$, then for any fixed $\varepsilon > 0$, $L > 0$, at large enough N the following inequality holds: $\varepsilon||\boldsymbol{R}^{1/2}||^{-1}K^{-1} \geq L$. Thus, the event $\left\{||\boldsymbol{W}_i(\bar{\boldsymbol{y}}_i - \mathbf{E}\bar{\boldsymbol{y}}_i)|| \geq \varepsilon||\boldsymbol{R}^{1/2}||^{-1}\right\}$ at large enough N is included in the event $\left\{||(\bar{\boldsymbol{y}}_i - \mathbf{E}\bar{\boldsymbol{y}}_i)|| \geq L\right\}$ and the right-hand side in (21.17) we can estimate in such a way: $\boldsymbol{L}(\varepsilon) \leq \alpha_N(L)||\boldsymbol{R}^{1/2}||^2 \sum_{i=1}^N ||\boldsymbol{W}_i||^2$. Using (21.18), we get that

$$\boldsymbol{L}(\varepsilon) \leq \alpha_N(L)K^2(S_{1N} + S_{2N})(S_{1N}S_{2N} - \rho^2 S_N^2)^{-1}(N_1 + N_2 - N_{12}).$$

Similar to proof of (21.8), we obtain: $\boldsymbol{L}(\varepsilon) \leq \alpha_N(L)K(N_1, N_2, N_{12})$, where

$$\begin{aligned} K(N_1, N_2, N_{12}) &\\ = \Big(\alpha_1 N_{12} + \alpha_2(N_1 - N_{12}) + \alpha_3(N_2 - N_{12})\Big)(N_1 + N_2 - N_{12}) &\\ \times\Big(\beta_1 N_{12}^2 + \beta_2 N_{12}(N_1 - N_{12}) + \beta_3 N_{12}(N_2 - N_{12}) &\\ + \beta_4(N_1 - N_{12})(N_2 - N_{12})\Big)^{-1}, & \end{aligned}$$

and α_i, β_j, are some positive constants.

Assume first that the variables N_1 and N_2 have the same order ($N_2 = O(N_1)$). Then it is easy to verify that $K(N_1, N_2, N_{12})$ is bounded and according to (21.12) the Lindeberg condition (21.11) is satisfied.

Consider the case when $N_2 = o(N_1)$. Introduce the following sets of centres: $A_1 = \{i : n_{i1} > 0, n_{i2} = 0\}$, $A_2 = \{i : n_{i2} > 0, n_{i1} = 0\}$, $A_{12} = \{i : n_{i1} > 0, n_{i2} > 0\}$. Denote $\bar{Z}_N = \sum_{i=1}^N \boldsymbol{W}_i(\bar{\boldsymbol{y}}_i - \boldsymbol{\mu}) = (Z_{N1}, Z_{N2})$. Then

$$\begin{aligned} Z_{N1} &= \sum_{i \in A_1} (\rho_1 + 1/n_{i1})^{-1}(\bar{y}_{i1} - \mu_1) \\ &\quad + \sum_{i \in A_{12}} \Big(w_{11}^{(i)}(\bar{y}_{i1} - \mu_1) + w_{12}^{(i)}(\bar{y}_{i2} - \mu_2)\Big), \end{aligned}$$

(similar for Z_{N2}), where $w_{kl}^{(i)}$ are the elements of a matrix $\boldsymbol{W}_i$. Note that $w_{kl}^{(i)}$ are bounded uniformly in i in the set A_{12} given that $\rho^2 < \rho_1\rho_2$.

Denote by $\widetilde{r}_N(kl)$, $k, l = 1, 2$, the elements of a matrix $\boldsymbol{R}^{1/2}$. We need to check the Lindeberg condition for the variables $\widetilde{r}_N(11)Z_{N1}$, $\widetilde{r}_N(12)Z_{N2}$ (similarly for the second component of a vector ζ_N).

In this case one can verify that $S_{1N} = O(N_2)$, $S_{2N} = O(N_1)$, $S_N = O(N_{12})$. When $N_2 = o(N_1)$, then also $N_{12} = o(N_1)$, and $S_{1N}S_{2N} - S_N^2 = O(N_1 N_2)$, and we have the following relations for the elements $r_N(kl)$: $r_N(11) = O(N_1^{-1})$, $r_N(22) = O(N_2^{-1}), r_N(12) = r_N(21) = O(N_{12}N_1^{-1}N_2^{-1})$.

Now as $\boldsymbol{R}^{1/2}\boldsymbol{R}^{1/2} = \boldsymbol{R}$, then $r_N(11) = \widetilde{r}_N(11)^2 + \widetilde{r}_N(12)^2$, $r_N(12) = \widetilde{r}_N(11)\widetilde{r}_N(12) + \widetilde{r}_N(22)\widetilde{r}_N(12)$, $r_N(22) = \widetilde{r}_N(22)^2 + \widetilde{r}_N(12)^2$, and

$$\widetilde{r}_N(11) = O(N_1^{-1/2}), \widetilde{r}_N(22) = O(N_2^{-1/2}), \widetilde{r}_N(12) = \widetilde{r}_N(21) = O(N_1^{-1/2}).$$

As the values $\widetilde{r}_N(11)^2N_1, \widetilde{r}_N(12)^2N_2, \widetilde{r}_N(12)^2N_1, \widetilde{r}_N(22)^2N_2$, are bounded, one can verify that the variables $\widetilde{r}_N(11)Z_{N1}, \widetilde{r}_N(12)Z_{N2}, \widetilde{r}_N(12)Z_{N1}, \widetilde{r}_N(22)Z_{N2}$, satisfy the Lindeberg condition. ■

The consistency of the estimator is in agreement with Kiefer and Wolfowitz, (1956), where the consistency of the maximum likelihood estimator for a random effects model in a scalar case was proved under the additional assumption that in our case would mean there is an identical number of patients in each centre. Note we don't assume the existence of a probability density function for the variable $\boldsymbol{\mu}$, as was done in the above mentioned paper. This variable in our case can be also non-regular when some of the n_{ij} can be equal to zero, so that we assume only the existence of the first two moments of $\boldsymbol{\mu}$.

If $\rho^2 = \rho_1\rho_2$, then $\mu_{i2} = c_1\mu_{i2} + c_2$ and the correlation coefficient between μ_{i2} and μ_{i2} is equal to 1. Then we can estimate μ_1 and μ_2 for each treatment separately and these estimators are consistent and asymptotically normal as $N_1 \to \infty$, $N_2 \to \infty$.

Note that the elements of the matrix $\boldsymbol{R}$ are bounded by some deterministic constants and the variance (21.5) always exists.

21.3.2 Random setting

Consider now the case when the values n_{ij} are random.

Theorem 21.3.2 *If $N_1 \xrightarrow{\text{P}} \infty$ and $N_2 \xrightarrow{\text{P}} \infty$ as $N \to \infty$, $n \to \infty$, then the estimator $\widehat{\boldsymbol{\mu}}$ is consistent and asymptotically normal, that means,*

$$\bar{\zeta}_N = \sigma^{-1}\left\{\sum_{i=1}^{N} \boldsymbol{W}_i\right\}^{1/2} (\widehat{\boldsymbol{\mu}} - \boldsymbol{\mu}) \overset{\text{w}}{\Rightarrow} \mathcal{N}(0,1). \tag{21.19}$$

PROOF. To prove consistency it is again enough to show that the variance of estimator in (21.5) tends to zero. Using (21.8) we get that $\mathbf{E}\, r_N(11) \to 0$ as $N_1 \xrightarrow{\text{P}} \infty$, $N_2 \xrightarrow{\text{P}} \infty$. The same is true for the other elements of matrix $\boldsymbol{R}$.

To prove asymptotic normality we show that the characteristic function of $\bar{\zeta}_N$ converges to $\exp\{-(\bar{s},\bar{s})/2\}$. Using (21.10) we get

$$\mathbf{E}\exp\{-\mathbf{i}(\bar{s},\bar{\zeta}_N)\} = \mathbf{E}\Big(\mathbf{E}\Big[\exp\{-\mathbf{i}\sum_{i=1}^{N}(\bar{s},\bar{\zeta}_{Ni})\} \mid \{n_{ij}\}\Big]\Big),$$

where $\mathbf{i} = +\sqrt{-1}$. Note that $\mathbf{E}\,\bar{\zeta}_{Ni}\bar{\zeta}_{Ni}^{\mathsf{T}} = \sigma^2\boldsymbol{R}^{-1/2}\boldsymbol{W}_i\boldsymbol{R}^{-1/2}$.

As the Lindeberg condition (21.12) is satisfied uniformly in n_{ij} and $||\boldsymbol{R}|| \xrightarrow{\text{P}} 0$, then for any i uniformly in n_{i1}, n_{i2} we have the representation

$$\mathbf{E}\Big[\exp\{-\mathbf{i}(\bar{s},\bar{\zeta}_{Ni})\} \mid \{n_{i1},n_{i2}\}\Big] = 1 - \frac{\sigma^2}{2}\Big(\boldsymbol{R}^{-1/2}\boldsymbol{W}_i\boldsymbol{R}^{-1/2}\bar{s},\bar{s}\Big)(1+o_i),$$

where o_i converges uniformly in i to zero. Using the relation $1-x \asymp e^{-x}$, valid at small x, and the relation $\boldsymbol{R}^{-1/2}\Big(\sum_{i=1}^N \boldsymbol{W}_i\Big)\boldsymbol{R}^{-1/2} = \boldsymbol{I}$, we get

$$\begin{aligned}
\mathbf{E}\exp\{-\mathrm{i}(\bar{s},\bar{\zeta}_N)\} &= \mathbf{E}\Big(\prod_{i=1}^N \mathbf{E}\Big[\exp\{-\mathrm{i}(\bar{s},\bar{\zeta}_{Ni})\} \mid \{n_{i1},n_{i2}\}\Big]\Big) \\
&\asymp \mathbf{E}\exp\Big\{-\frac{1}{2}\Big(\boldsymbol{R}^{-1/2}\Big(\sum_{i=1}^N \boldsymbol{W}_i\Big)\boldsymbol{R}^{-1/2}\bar{s},\bar{s}\Big)\Big\} \\
&= \exp\Big\{-\frac{1}{2}(\bar{s},\bar{s})\Big\}.
\end{aligned}$$

■

This means that $\widehat{\boldsymbol{\mu}} \asymp \boldsymbol{\mu} + \sigma\Big(\sum_{i=1}^N \boldsymbol{W}_i\Big)^{-1/2}\mathcal{N}(0,\boldsymbol{I})$, where $\boldsymbol{W}_i$ are in general random. These results can be easy extended to the case when the variance of ε_{ijk} may depend on j, that is $\mathbf{E}\varepsilon_{ijk}^2 = \sigma_j^2$, $j=1,2$.

Consider now a sufficient condition when $N_1 \xrightarrow{\mathrm{P}} \infty$ and $N_2 \xrightarrow{\mathrm{P}} \infty$. Assume that the patients arrive at centre i according to a Poisson process with rate λ_i, $i=1,\ldots,N$. Consider the trial on the time interval $[0,T]$ and suppose that N and T are changing in such a way that there exists a set $A_N = \{j_1, j_2, \ldots, j_{L_N}\}$, $L_N \to \infty$, and for some $c > 0$,

$$T \min_{i\in A_N} \lambda_i \geq c. \tag{21.20}$$

Lemma 21.3.1 *If (21.20) is true, then as $N\to\infty$, $N_1 \xrightarrow{\mathrm{P}} \infty$, $N_2 \xrightarrow{\mathrm{P}} \infty$.*

PROOF. Assume for simplicity that each patient arriving at centre i is assigned with probability $1/2$ to one of two treatments. Then the probability that in centre i on the interval $[0,T]$ we have at least one patient in the first treatment is $p_i = 1 - e^{-\lambda_i T/2}$. According to (21.20), for any $i \in A_N$, $p_i \geq 1 - e^{-c} = \alpha < 1$. Denote by $N_1(A_N)$ the number of centres in the region A_N with a positive number of patients in the first arm. Then at $k \leq L_N$,

$$\mathbf{P}(N_1(A_N) = k) = \sum_{(i_1\ldots i_k)\in A_N} \prod_{l=1}^k p_{i_l} \prod_{s\notin(i_1\ldots i_k)} (1-p_s) \leq \binom{L_N}{k}(1-\alpha)^{L_N-k}.$$

Using Stirling's formula we can show that the right-hand side behaves as $L_N^k(1-\alpha)^{L_N} \to 0$ for any $k \geq 0$ as $L_N \to \infty$. Then for any fixed $K > 0$,

$$\mathbf{P}(N_1(A_N) > K) = 1 - \sum_{k=0}^K \mathbf{P}(N_1 = k) \to 1.$$

As $N_1 \geq N_1(A_N)$, this relation means that $N_1 \xrightarrow{\mathrm{P}} \infty$ as $N \to \infty$. In the same way we get that $N_2 \xrightarrow{\mathrm{P}} \infty$. ■

Let us study now how the condition (21.20) is transformed if we consider the trail on the time interval needed to recruit in total n patients to all centres. Assume that for all i, $\lambda_i < C < \infty$. Denote $\Lambda_N = \sum_{i=1}^N \lambda_i$. Then the recruitment time $T(n, N)$ is Gamma distributed with parameters (n, Λ_N). Using the normal approximation for Gamma distribution, we get at large n that with probability close to 1 for any $\varepsilon > 0$, $T(n, N) > n(1-\varepsilon)/(NC)$. Thus, if (21.20) is satisfied with $T = n/N$, then $N_1 \xrightarrow{\mathrm{P}} \infty$, $N_2 \xrightarrow{\mathrm{P}} \infty$ as $n \to \infty$ and $N \to \infty$.

21.4 Approximation of Variance of the Estimator

Consider the approximation of $\mathbf{Var}(\widehat{\boldsymbol{\mu}})$ at large n and N. Let patients arrive at centre i according to a Poisson process $\Pi_i(t)$ with rate λ_i. We consider the trial on the time interval needed to recruit in total n patients to all centres. Denote $p_i = \lambda_i/\Lambda_N$, $i = 1, \ldots, N$, where $\Lambda_N = \sum_{i=1}^N \lambda_i$. Then the numbers of patients n_i in centres have a joint multinomial distribution. Assume that in each centre the patients with equal probabilities 1/2 are assigned to treatment j, $j = 1, 2$. Let n_{ij} be the number of patients in i-th centre at treatment j.

We cannot use directly the normal approximation of the vector $\{n_{ij},\ i = 1, \ldots, N,\ j = 1, 2\}$, because its dimension is increasing. An additional problem is that $p_i \to 0$. Therefore, if we would like to apply some kind of Law of Large Numbers for the function of vector $\{n_{ij}\}$, then N cannot be of the order n.

Let $B(n, a_n)$ be a binomial random variable with parameters n, a_n, and $n \to \infty, a_n \to 0$ and $na_n \to \infty$. Then for any integer $r \geq 1$,

$$\mu_{2r} = \mathbf{E}\Big(B(n, a_n) - na_n\Big)^{2r} = C_r(na_n)^r(1 + o(1)), \tag{21.21}$$

where C_r are constants.

Indeed, using formula (3.8) from Johnson et al, (1993), we get: $\mu_2 = na_n(1+o(1))$, $\mu_3 = na_n(1+o(1))$, $\mu_4 = 3(na_n)^2(1+o(1))$. Using the recurrent relation (3.13) from this book we get by induction (21.21).

Lemma 21.4.1 *Assume that* $N = O(n^\beta)$, *where* $\beta < 1$, *and* $0 < b_0 \leq \lambda_i \leq B_0$, $i = 1, \ldots, N$. *Then for any* $0 < \varepsilon < (1-\beta)/2$ *there exists a sequence* $g_n = O(1/n^{(3-\beta)/2-\varepsilon})$ *such that*

$$\mathbf{P}\Big(||\mathbf{Var}(\widehat{\boldsymbol{\mu}}) - \frac{1}{N}\boldsymbol{V} - \frac{2\Lambda_N}{nN}\frac{\sigma^2}{N}\sum_{i=1}^{N}\frac{1}{\lambda_i}\boldsymbol{I}|| < g_n\Big) \to 1. \tag{21.22}$$

PROOF. Let us choose $N \approx an^\beta$, $\beta < 1$. For any $L_n > 0$, using the Bonferroni inequality, we get:

$$\mathbf{P}(\max_{i,j}|n_{ij} - np_i/2| \leq L_n) \geq 1 - 2N\max_i \mathbf{P}(|n_{i1} - np_i/2| \geq L_n). \tag{21.23}$$

Using (21.21) and the Chebyshev inequality for the function $(x-a)^{2r}$, we get for any $i = 1, \ldots, N$ and any integer $r \geq 1$:

$$\mathbf{P}(|n_{i1} - np_i/2| \geq L_n) \leq \mu_{2r}/L_n^{2r} \leq \max_i \left(np_i/L_n^2\right)^r \leq \left(B_0 n^{1-\beta}/L_n^2\right)^r.$$

Now at $L_n = n^{(1-\beta)/2+\varepsilon}$ and large enough r such that $\beta/r < 2\varepsilon$, one can verify that the right-hand side of (21.23) tends to 1, i.e., with probability close to one for all i, j, $|n_{ij} - np_i/2| < L_n$ and $||\boldsymbol{M}_i^{-1} - 2\mathbf{I}/(np_i)|| < 4\sqrt{2}L_n/(np_i)^2$. Using the relation $(\boldsymbol{I} + \alpha\mathbf{A})^{-1} \approx \boldsymbol{I} - \alpha\mathbf{A}$ as $\alpha \to 0$, we get after some algebra (21.22). Note that $g_n = o(1/n)$. ■

The asymptotic expression for the variance reaches its minimum in a stochastically balanced case (all λ_i are equal). In this case

$$\mathbf{Var}(\widehat{\boldsymbol{\mu}}) \approx \boldsymbol{V}/N + 2\sigma^2\boldsymbol{I}/n,$$

which coincides with the completely balanced case ($n_{ij} = n/(2N)$ for all i, j), Fedorov et al. (2002).

Another types of the approximation of the variance related to different models of the recruitment and optimization models are studied in Anisimov et al. (2003).

21.5 Conclusions

The weighted estimator of $\widehat{\boldsymbol{\mu}}$ has good asymptotic properties as $n \to \infty$ and $N \to \infty$, even in such practically important cases when most centres may not have patients in both arms, but naturally we need to have asymptotically many centres with non-zero number of patients for the first and for the second arms correspondingly. Therefore we may effectively apply all inferential tools developed for the multivariate normal distribution at the analysis of a CRT.

The variance of $\widehat{\boldsymbol{\mu}}$ for a large number of patients in each centre can be approximated by simpler deterministic expression. In general it contains two terms of the order $O(1/N)$ and $O(1/n)$. This result can be used on the stage of a design of a trial, calculation of the necessary number of patients needed to estimate the CRT with given accuracy and solving optimization problems involving cost constraints related to recruitment and restrictions on the variance.

References

1. Anderson, T. W. (1958). *Introduction to Multivariate Statistical Analysis*, John Wiley & Sons, New York.

2. Anisimov, V., Fedorov, V. and Jones, B. (2003). Optimization of clinical trials with random enrollment, *GSK BDS Technical Report*, Chichester.

3. Brown, H. and Prescott, R. (1999). *Applied Mixed Models in Medicine*, John Wiley & Sons, Chichester, England.

4. DerSimonian, R. and Laird, N. M. (1986). Meta-analysis in clinical trials, *Controlled Clinical Trials*, **7**, 177–188.

5. Dragalin, V., Fedorov, V., Jones, B. and Rockhold, F. (2001). Estimation of the combined response to treatment in multicentre trials, *Journal of Biopharmaceutical Statistics*, **11(4)**, 275–295.

6. Fedorov, V., Jones, B. and Rockhold, F. (2002). The design and analysis of multicentre trials in the random effects setting, *GSK BDS Technical Report*, 2002-03.

7. Hardy, R. and Thompson, S. G. (1996). A likelihood approach to meta-analysis with random effects, *Statistics in Medicine*, **15**, 619–629.

8. Harville, D. A. (1997). *Matrix Algebra From a Statistician's Perspective*, Springer-Verlag, New York.

9. Johnson, N. L., Kotz, S. and Kemp, A. W. (1993). *Univariate Discrete Distributions*, Second edition, John Wiley & Sons, New York.

10. Kiefer, J. and Wolfowitz, J. (1956). Consistency of the maximum likelihood estimator in the presence of infinitely many incidental parameters, *Annals of Mathematical Statistics*, **27**, 887–906.

11. Pinheiro, J. C. and Bates, D. M. (2000). *Mixed-Effects Models in S and S-PLUS*, Springer-Verlag, New York.

12. Rao, C. R. (1965). The theory of least squares when the parameters are stochastic and its applications to the analysis of growth curves, *Biometrika*, **52**, 447–458.

13. Searle, S. (1987). *Linear Models for Unbalanced Data*, John Wiley & Sons, New York.

14. Senn, S. (1997). *Statistical Issues in Drug Development*, John Wiley & Sons, Chichester, England.

15. Senn, S. (2000). The many modes of meta, *Drug Information Journal*, **34**, 535–549.

16. Spjotvoll, E. (1977). Random coefficients regression models: A review. *Mathematische Operationsforschung und Statistik, Serie Statistik*, **8**, 69–93.

17. Verbeke, G. and Molenberghs, G. (2000) *Linear Mixed Models for Longitudinal Data*, Springer-Verlag, New York.

PART X
Quality of Life Studies

22

HRQoL and Concomitant Adjusted Mean Residual Life Analysis

Pranab Kumar Sen

Departments of Biostatistics and Statistics, University of North Carolina, Chapel Hill, NC U.S.A.

Abstract: For aging properties of life-time distributions in reliability and survival analysis, the DMRL property plays a vital role. Generally, there are multiple auxiliary or explanatory variables, and in HRQoL studies, QAL perspectives are nonignorable. These factors introduce complications in parametric modeling of MRL; for QLAMRL functions it is anticipated, though not yet established, that semiparametrics may suit well. As an alternative approach, matrix-valued counting processes for repeated measurement data models, involving clusters of random sizes, nonexchangeable within-cluster dependence, and QAL adjustments, are appraised.

Keywords and phrases: Aging, DMRL, IFR, QLAMRL, recurrent events, renewal theorem, weak convergence

22.1 Introduction

We are concerned here with *repeated measures survival analysis* (RMSA) where, typically, the clusters are formed by individual subjects, and the measurements relate to recurrent events which take place in a *time-sequential* setup. In any fixed study-period, the number of events occurring within each cluster is stochastic, and so also are the respective episode times. Further, in the context of *health related quality of life* (HRQoL) studies, there may be possibly a *degradation* of the health condition, affecting the survival status; even otherwise, the inter-episode time intervals may not be independent or even marginally identically distributed. Also, in many cases, the primary focus is on the *mean remaining life* (MRL), and keeping in mind the *aging* aspects, it is natural to bring the *nonparametric family of aging distributions* incorporating the characterization *decreasing mean remaining life* (DMRL) to depict the *quality adjusted life* (QAL) pattern. In the presence of a (possibly large) number of auxiliary or explanatory variables, usually referred to as *concomitants*, simple parametric models may not suit real applications, and often, nonparametric and *semi-parametric* approaches are advocated where also QAL adjustments pose some genuine and more challenging statistical problems; some of these problems are to be appraised here.

In Section 22.2, the nonparametric family of survival functions based on the DMRL property are appraised with a view to examine how far such characterizations percolate under semiparametrics. Section 22.3 brings the concept of *quality of life adjusted mean remaining life* (QLAMRL), and an appraisal of the DMRL property is made in this framework as well. Section 22.4 is devoted to a class of *matrix-valued counting processes* for some repeated measurement data models, and the relevance of the classical *renewal process* is appraised in this perspective. Section 22.5 deals with a *nondecreasing counting process* typically arising in such studies, and its *weak convergence* properties are studied. The need for the semiparametric modeling in this context is highlighted.

22.2 DMRL Characterizations

The MRL $\mu(x)$, at an age $x(\geq 0)$, occupies a focal point in *life table analysis*, and is formally defined as

$$\mu(x) = E\{X - x|X \geq x\}, \quad x \geq 0, \tag{22.1}$$

where X stands for the survival time. The MRL function $\mu = \{\mu(x),\ x \in R^+\}$ has its genesis in reliability theory; it is also extensively used in survival analysis. In the characterization of *aging properties* of life distributions the DMRL property, namely, $\mu(x)$ being nonincreasing in x, is particularly advocated for its its interpretability as well as mathematical convenience. In survival analysis, however, there may be generally a (large) number of auxiliary variables, some nonstochastic while the others stochastic. We denote such concomitant variables by $\mathbf{Z}$, and define MRL with a concomitant $\mathbf{Z} = \mathbf{z}$ as

$$\mu(x|\mathbf{z}) = E\{X - x|X \geq x; \mathbf{Z} = \mathbf{z}\},\ x \geq 0,\ \mathbf{z}\ \in\ R^q, \tag{22.2}$$

where q stands for the dimension of the concomitants. Since not all the coordinates of $\mathbf{Z}$ may be continuous variables (some may even be binary or categorical), expressing $\mu(x|\mathbf{z})$ in a simple parametric form may require quite stringent regularity assumptions on the conditional survival function of X, given $\mathbf{Z}$, thus limiting the scope of parametric modeling in live applications. For this reason, both nonparametric and semiparametric modelings are more commonly advocated.

The nonparametric modeling, albeit quite flexible from a practical usage point of view, is based on nonparametric estimation of the conditional survival function of X, given $\mathbf{Z}$. Generally, *kernel, spline* and *nearest neighbor* methods are incorporated in this respect. At the cost of a generally slower rate of convergence, the estimators are consistent under very general regularity conditions (though requiring a relatively much larger sample size to achieve a reasonable margin of statistical error). In many practical applications, the sample size may not be large enough to have an effective statistical precision in a nonparametric formulation of the MRL or DMRL studies. Semiparametric modeling may suit a comparatively smaller sample size situation if the assumed model is close enough to the true one. This *local robustness* aspect of semiparametrics has to be contrasted with the *global robustness* aspect of nonparametrics, especially when the sample size is large enough.

For a nonnegative random variable Y, treated as *failure time*, the survival function $S_o(t)$ is defined as

$$S_o(t) = P\{Y > t\} = e^{-\Lambda_o(t)} \ , \ t \in R^+, \tag{22.3}$$

where

$$\Lambda_o(t) = \int_0^t \lambda_o(u)du, \ t \geq 0, \tag{22.4}$$

is the *integrated hazard function* and $\lambda_o(t)$, the *hazard function*, is nonnegative. Thus, $\Lambda_o(0) = 0$ and $\Lambda_o(t)$ is nondecreasing in $t \geq 0$ with $\Lambda_o(\infty) = +\infty$. In this setup, we have

$$P\{Y > x | Y \geq t\} = S_o(x)/S_o(t), \ x \geq t. \tag{22.5}$$

Therefore, we have the simple representation

$$\begin{aligned} \mu_o(x) &= \text{mean remaining life at age } x \\ &= \int_0^\infty \{S_o(x+u)/S_o(x)\}\, du \\ &= \int_0^\infty e^{-[\Lambda_o(x+u)-\Lambda_o(x)]}\, du \ , \ x \geq 0. \end{aligned} \tag{22.6}$$

The DMRL property, namely that $\mu_o(x)$ nonincreasing in x, is isomorphic to $\frac{\partial}{\partial x}\mu_o(x) \leq 0, \ \forall x \geq 0$, so that we have the following characterization of DMRL: $\forall x \in R^+$,

$$\int_0^\infty \{\lambda_o(x+u) - \lambda_o(x)\} e^{-[\Lambda_o(x+u)-\Lambda_o(x)]}\, du \ \geq 0. \tag{22.7}$$

In passing we may note that the *increasing failure rate* (IFR) property is characterized by $\lambda_o(x+u) \geq \lambda_o(x), \ \forall x, u \geq 0$, and hence, IFR $\Rightarrow$ DMRL.

Consider now a simple semiparametric model due to Cox (1972), wherein we express the conditional hazard function, given the concomitant $\mathbf{Z}$ as

$$\lambda(t|\mathbf{Z}) = \lambda_o(t) e^{\boldsymbol{\beta}'\mathbf{Z}}, \tag{22.8}$$

so that for the conditional survival function, given $\mathbf{Z}$, we have

$$\begin{aligned} S(t|\mathbf{Z}) &= [S_o(t)]^{e^{\boldsymbol{\beta}'\mathbf{z}}} \\ &= e^{-e^{\boldsymbol{\beta}'\mathbf{z}}\Lambda_o(t)}. \end{aligned} \tag{22.9}$$

As a result, we obtain that

$$\begin{aligned} \mu(x|\mathbf{Z}) &= \int_0^\infty \{S(x+u|\mathbf{Z}) - S(x|\mathbf{Z})\}\, du \\ &= \int_0^\infty e^{-e^{\boldsymbol{\beta}'\mathbf{z}}[\Lambda_o(x+u)-\Lambda_o(x)]}\, du. \end{aligned} \tag{22.10}$$

As such, by differentiation, we have

$$-\frac{d}{dx}\mu(x|\mathbf{Z}) = \int_0^\infty e^{\boldsymbol{\beta}'\mathbf{z}}[\lambda_o(x+u) - \lambda_o(x)]e^{-e^{\boldsymbol{\beta}'\mathbf{z}}[\Lambda_o(x+u)-\Lambda_o(x)]}\, du. \tag{22.11}$$

Note that when $S_o(x)$ belongs to the IFR family, $\lambda_o(x+u) \geq \lambda_o(x)$, $\forall x, u \geq 0$, and hence, from the above, we arrive at the following.

Theorem 22.2.1 *Whenever $S_o(x)$ belongs to the IFR family and the PHM pertains, for all $\boldsymbol{\beta}$ and all $\mathbf{Z}$, the conditional MRL $\mu(x|\mathbf{Z})$ is nonincreasing in x, and hence, the DMRL property holds for the PHM.*

In the same way, it follows that if $S_o(.)$ belongs to the IFR family and PHM pertains, then $S(t|\mathbf{Z})$ belongs to the IFR family for all $\mathbf{Z}$ and $\boldsymbol{\beta}$. This characterization provides much convenience to the PHM formulations preserving the desired aging property.

Let me pose an open question in this setup: *Is it true or not that when $S_o(.)$ belongs to the DMRL family and the PHM pertains, then $S(x|\mathbf{z})$ also belongs to the DMRL family for all $\mathbf{Z}$ and $\boldsymbol{\beta}$?* To resolve this issue, it suffices to verify whether or not

$$\int_0^\infty \{\lambda_o(x+u) - \lambda_o(x)\}e^{-[\Lambda_o(x+u)-\Lambda_o(x)]}\, du \;\geq 0,\ \forall x \in R^+ \tag{22.12}$$

implies that for every $k \in (0, \infty)$

$$\int_0^\infty \{\lambda_o(x+u) - \lambda_o(x)\}e^{-k[\Lambda_o(x+u)-\Lambda_o(x)]}\, du \;\geq 0,\ \forall x \in R^+. \tag{22.13}$$

If we define for $k > 0$,

$$\mu_{(k)}(x) = \int_0^\infty e^{-k[\Lambda_o(x+u)-\Lambda_o(x)]} du, \tag{22.14}$$

then

$$-\frac{\partial}{\partial k}\mu_{(k)}(x) = \int_0^\infty [\Lambda_o(x+u) - \Lambda_o(x)]e^{-k[\Lambda_o(x+u)-\Lambda_o(x)]}du$$
$$\geq 0, \ \forall x \geq 0, \ k \geq 0. \qquad (22.15)$$

Therefore, for any given x, $\mu_{(k)}(x)$ is nonincreasing in $k \geq 0$. On the other hand, if it is assumed that $S_o(.)$ belongs to the DMRL family, then $\mu_{(1)}(x)$ is nonincreasing in $x \geq 0$. Hence, we may claim that for every $k \geq 1$, $\mu_{(k)}(x)$ is dominated by $\mu_{(1)}(x)$, which is nonincreasing in x. However, by the same argument, we may claim that for every $k : 0 \leq k \leq 1$, $\mu_{(k)}(x)$ dominates $\mu_{(1)}(x)$. Hence, it is not clear how we can establish the DMRL property of the conditional survival function $S(x|\mathbf{z})$, allowing $\boldsymbol{\beta}$ to be arbitrary. If $S_o(x)$ belongs to the DMRL family but not necessarily in the IFR family, noting that as $k \to 0$, $e^{-k[\Lambda_o(x+u)-\Lambda_o(x)]}$ becomes very slowly convergent to zero, we may utilize this flatness condition to show that (22.11) may not be everywhere nonnegative if the IFR condition does not hold. The same pathological examples which exhibit that IFR is a subclass of the DMR family can be used and then allowing $k \to 0$, we can show that sans IFR, the DMRL property may not percolate under the PHM with arbitrary $\boldsymbol{\beta}$. Thus, in the presence of concomitant variables, even under the PHM, it seems that the IFR formulation is statistically much more convenient to handle, albeit it is a bit more stringent than the DMRL formulation.

Let us now consider the case of *time-dependent* concomitant variables where we let

$$\lambda(x| \ \mathbf{Z}(t) = \mathbf{z}(t) \ ) = \lambda_o(x)e^{\boldsymbol{\beta}'\mathbf{Z}(t)}, \qquad (22.16)$$

so that the concomitant is allowed to be possibly time-dependent. In this setup, we have

$$\Lambda(x|\mathbf{Z}(s), s \leq x) = \int_0^x \lambda_o(s)e^{\boldsymbol{\beta}'\mathbf{Z}(s)} \, ds. \qquad (22.17)$$

Unfortunately, this may not be expressible as $\Lambda_o(x).g(\mathbf{z}(t), \boldsymbol{\beta})$ with a suitable nonnegative $g(.)$. As a result, much of the simplicity of the PHM will be lost in this formulation, and the characterization results for the time-independent concomitants may not percolate. Let us consider a bit less general time-dependent covariate model where we set for all $x \geq 0$,

$$\Lambda(x|\mathbf{Z}(s), s \leq x) = \Lambda_o(x)e^{\boldsymbol{\beta}'\mathbf{Z}(x)}, \qquad (22.18)$$

so that the conditional survival function behaves as

$$S(x|\mathbf{Z}(s), s \leq x) = e^{-e^{\boldsymbol{\beta}'\mathbf{Z}(x)}\Lambda_o(x)}. \qquad (22.19)$$

Since the conditional survival function has to be nonincreasing in x, in order that the above is a valid survival function, we must have then

$$\Lambda_o(t)e^{\boldsymbol{\beta}'\mathbf{Z}(x)} \ \nearrow \ in \ x \geq 0. \qquad (22.20)$$

Though, such a condition holds for some degradation models, it may not be generally true for survival analysis models where the concomitants $\mathbf{Z}(t)$ may not have monotone (in t) nature. However, this semiparametric model has a Markovian flavor in the sense that the integrated hazard function depends on the time-dependent covariates through their current states.

For the above model, we note that

$$\frac{S(x+u|\mathbf{Z}(x+u))}{S(x|\mathbf{Z}(x))} = \frac{e^{e^{\boldsymbol{\beta}'\mathbf{z}(x)}\Lambda_o(x)}}{e^{-e^{\boldsymbol{\beta}'\mathbf{z}(x+u)}\Lambda_o(x+u)}}, \tag{22.21}$$

for all $x, u \geq 0$. As a result,

$$\mu(x|\mathbf{Z}(t), t \geq x) = e^{e^{\boldsymbol{\beta}'\mathbf{z}(x)}\Lambda_o(x)} \int_0^\infty e^{-e^{\boldsymbol{\beta}'\mathbf{z}(x+u)}\Lambda_o(x+u)} du. \tag{22.22}$$

Recall that $\mathbf{Z}(x+u)$, $u > 0$, may not be observable at time x, and hence, what really we are to study is $\mu(x|\mathbf{Z}(x))$. Toward this, note that

$$\mu(x|\mathbf{Z}(x) = \mathbf{z}) = E_{\{\mathbf{Z}(u), u > x\}}[\mu(x|\mathbf{Z}(t),\ t \geq x)], \tag{22.23}$$

where the expectation ranges over the sample space of the covariate set $\mathbf{Z}(t), t > x$. How easy is it to evaluate this even if we have a single (time-dependent) covariate? This involves the probability law for the entire process $\{\mathbf{Z}(t),\ t > x\}$, and hence, much of the charm of the PHM is lost. QAL adjustments are one-step harder for the study of MRL and we study this in the next section.

22.3 QLAMRL: Statistical Perspectives

Typically, in HRQoL studies, based on composite scores (item analysis) a QAL score at time t, denoted by $Q(t),\ t \geq 0$, is conceived. Such a QAL score also takes into account the covariates, and hence, to start with, we consider the most simple setup where a QAL score is well defined and no concomitant variates are externally associated. For notational simplicity, we also define

$$f_o(t) = -(d/dt)S_o(t),\ t\ \in\ R^+. \tag{22.24}$$

The QLAMRL at age x, denoted by $\mu_Q(x)$, is given by

$$\begin{aligned} \mu_Q(x) &= \{\int_0^\infty Q(x+u)uf_o(x+u)du\}/S_o(x) \\ &= -\{\int_0^\infty Q(x+u)udS_o(x+u)\}/S_o(x) \end{aligned}$$

$$
\begin{aligned}
&= \{\int_0^\infty S_o(x+u)Q(x+u)du\}/S_o(x) \\
&\quad +\{\int_0^\infty S_o(x+u)udQ(x+u)\}/S_o(x) \\
&= \int_0^\infty Q(x+u)e^{-[\Lambda_o(x+u)-\Lambda_o(x)]}du \\
&\quad +\int_0^\infty ue^{-[\Lambda_o(x+u)-\Lambda_o(x)]}dQ(x+u). \qquad (22.25)
\end{aligned}
$$

At this stage, we note that for a *degradation process* (relating to the QAL score), we would have

$$
Q(t) \searrow \text{ in } t \geq 0 \Rightarrow dQ(t) \leq 0, \ \forall t > 0. \qquad (22.26)
$$

Consequently, for a degradation QAL process,

$$
\begin{aligned}
\mu_Q(x) &\leq \int_0^\infty Q(x+u)e^{-[\Lambda_o(x+u)-\Lambda_o(x)]}du \\
&\leq Q(x)\int_0^\infty e^{-[\Lambda_o(x+u)-\Lambda_0(x)]}du \\
&= Q(x).\mu_o(x), \qquad (22.27)
\end{aligned}
$$

where $\mu_o(x)$ stands for the MRL at age x for the baseline $S_o(.)$. The last equation demonstrates that for a QAL degradation process, making adjustment to the MRL $\mu_o(x)$ by the current QAL score $Q(x)$ may result in positive bias for the QLAMRL function. In the same way, if we have a regradation QAL process, $Q(t), t \geq 0$ is nondecreasing, and hence, we would have the opposite inequality in (22.27), depicting a negative bias. Even for other than degradation/ regradation processes, bias corrections can be made with (22.25), provided suitable modeling of $\{Q(t),\ t \in R^+\}$ can be made in an appropriate manner that conforms to the particular application.

Note that the QAL score at a time point t may generally depend on the set of covariates $\mathbf{Z}(t)$ at that time point, and is possibly nonstationary over time. For simplicity, we consider a discrete *state space* $\mathcal{A}$ and denote a typical state by A, so that the QAL score at time point t and state A is denoted by $Q(t,A),\ A \in \mathcal{A}$. Further, let $\pi_t(A),\ A \in \mathcal{A}$ be the probability elements associated with the state space at time point t. Then

$$
\{Q(t,A),\ A \in \mathcal{A}\} \text{ and } \{\pi_t(A),\ A \in \mathcal{A}\}
$$

are the determining factors for $Q(t)$. We write

$$
Q(t) = \sum_{\{A \in \mathcal{A}\}} \pi_t(A)Q(t,A),\ \forall t \geq 0. \qquad (22.28)
$$

Let us go back to (22.25) and appraise the situation in the presence of such time-dependent concomitants. We need to formulate the conditional survival function and QAL score at time $x+u$, given the survival up to the time point x and that the state at that time point is A_x. Thus, we need to model suitably the following:

$$P\{X \geq x+u, A_{x+u} = A^*|X \geq x, A_x = A\}, \forall A^*, A \in \mathcal{A}, \tag{22.29}$$

and all $x, u \geq 0$. We express the above as

$$P\{X \geq x+u|X \geq x, A_x = A\}P\{A_{x+u} = A^*|A_x = A, X \geq x+u\}. \tag{22.30}$$

Thus, if we denote the QLAMRL at age x and state A, by $\mu_Q(x; A)$, we can use (22.29) and write

$$\begin{aligned} &\mu_Q(x; A) \\ &= -\sum_{\{A^* \in \mathcal{A}\}} \int_0^\infty u\, Q(x+u, A^*)\, dP\{X \geq x+u, A_{x+u} = A^*|X \geq x, A_x = A\}, \end{aligned} \tag{22.31}$$

for $x \geq 0,\ A \in \mathcal{A}$. Of course, we have then

$$\mu_Q(x) = \sum_{\{A \in \mathcal{A}\}} \pi_x(A)\mu_Q(x, A),\ x \geq 0. \tag{22.32}$$

The utility of a Markov chain modeling (albeit in a possibly nonstationary setup) for the state space and a semiparametric modeling for the survival function is quite apparent from (22.30) and (22.31). Further, when $Q(t), t \geq 0$ can be regarded as a degradation process, then the transition probabilities (or the stochastic matrix) has to be formulated carefully so as to satisfy the degradation condition. The QAL scores are so basic in such a formulation that a critical appraisal is necessary to justify their adaption as well as to satisfy the postulated regularity assumptions. We intend to communicate some of these findings in a subsequent study. However, we like to conclude this section with a remark that there are some HRQoL problems where a degradation model for the $Q(t)$ may not be very appropriate. There could be some periodicity with stochastic wavelets, and in such a case, a formulation along the above lines could be much more complex. This brings us to some alternative approaches which do not rest on QLAMRL measures to that extent.

22.4 Matrix-Valued Counting Process in HRQoL

Our focus is on matrix-valued counting processes for repeated measurement data models or RMSA. This type of dataset is seen in measuring the time between repeated hospitalizations for some chronic disease, or time between pediatric infections, or successive study-defined events, such as a laboratory defined toxicity. The correlated *clusters* are thus, typically, represented by individual subjects wherein the number of *events* seen in individuals is random and can vary widely. Thus, the within-cluster correlations or dependence are generated by all the measurements coming from one individual. Successive event times for an individual are neither independent nor marginally identically distributed. In many cases, QAL concept is quite pertinent, and hence, HRQoL adjustments are desirable.

Since a multivariate survival data model typically consists of clusters with a predetermined size limit, parameterizing a model in the presence of cluster sizes that may range from no event to a possibly large number of events poses a challenge to most (parametric) multivariate survival analysis; the primary difficulty stems from the fact that the time to successive events are neither independent nor identically distributed, and as such, these models can not adjust for the additional event times and thereby eliminate valuable statistical information by only looking at the time to the prespecified *first few events*. Moreover, some of these models control for the dependence without describing it, yielding no inference or statistical description of the dependence parameters. *Within-cluster resampling plans* attempt to handle some clusters with varying sizes, albeit under some sort of *exchangeability* or interchangeability assumptions [Hoffman, Sen and Weinberg (2001)] that might not be true for such HRQoL adjusted models.

Let us consider matrix-valued counting processes for multivariate sequential survival data. Let $\mathbf{T}_k = (T_{k1}, \ldots, T_{kj_k})'$ be the random vector of times between the *sequential events* for the kth cluster (individual) with j_k events taking place during the span of the study, for $k = 1, \ldots, K$. If no event takes place, we let $j_k = 0$, $\mathbf{T}_k = \mathbf{0}$. Thus, the random variable j_k varies from cluster to cluster, with distributions depending on the auxiliary or explanatory variable(s) as well as any underlying HRQoL factor. The vector $\mathbf{T}_k$ may not be totally observable, and may be subject to an independent censoring (random) variable C. Due to the sequential nature, we assume that the censoring affects the event times only through the last (random) j_kth member.

Let us now consider the above setup in a temporal mode. Corresponding to a time interval $(0, t]$, we denote the vector of episode times for the kth individual by $\mathbf{T}_k(t) = (T_{k0}, T_{k1}, \ldots, T_{kj_k(t)})'$, for $k = 1, \ldots, K$. Then

$$j_k(t) = \max\{j(\geq 0) : T_{kj} \leq t\}, \forall t > 0, \tag{22.33}$$

for $k = 1, \ldots, K$. Note that

$$[j_k \geq m] \Leftrightarrow [T_{km} \leq t], \forall t > 0, m \geq 0, \tag{22.34}$$

for $k = 1, \ldots, K$. Let's introduce the notation:

$$\begin{aligned} Z_{ki} &= T_{ki},\ i = 0, \ldots, j_k(t) - 1,\ Z_{kj_k(t)} = \min\{T_{kj_k(t)},\ C_k\}; \\ \delta_{ki} &= 1,\ i \le j_k(t) - 1,\ \delta_{kj_k(t)} = I(Z_{kj_k(t)} = T_{kj_k(t)}). \end{aligned} \tag{22.35}$$

Due to the time-sequential nature of the dataset, censoring affects only the $j_k(t)$th episode time. As such, if we define the counting process for the ith member of the kth cluster as

$$N_{ki}(t) = I\{T_{ki} \le t\},\quad i \le j_k(t),\ t \ge 0, \tag{22.36}$$

we get a vector-valued counting process for the kth cluster as

$$\mathbf{N}_k(t) = (N_{k1}(t), \ldots, N_{kj_k(t)}(t))', \tag{22.37}$$

where $j_k(t)$, nondecreasing in t, may vary (stochastically) over k. If $j_k(t) = 0$ for some k, then the corresponding $N_k(t) = 0$. Furthermore, we introduce a predictable vector of the *at risk process* for the kth cluster as

$$\mathbf{Y}_k(t) = (Y_{k1}(t), \ldots, Y_{kj_k(t)}(t))',\ t \ge 0, \tag{22.38}$$

where the individual elements are defined as

$$Y_{ki}(t) = I(T_{ki-1} < t \le T_{ki}),\ \text{for } i \ge 0,\ k = 1, \ldots, K \tag{22.39}$$

with $T_{k0} = 0$. In conventional multivariate survival data the clusters are of the same size with the time to events running concurrently. However, for repeated measurement models, the times run subsequent to one another. Also, the number of events an individual can have is not prespecified, so that the cluster size is random. Further, according to our formulation, the second event is not at risk until after the first event occurs (at time T_{k1}), and it is then at risk only until the second event occurs (at time T_{k2}). In the same way, for the kth individual, the ith even is not at risk until the $(i-1)$th event occurs (at time T_{ki-1}) and it continues to be at risk until the ith event occurs (at the random time point T_{ki}), for $i \ge 1,\ k = 1, \ldots, K$. For this reason, a *rescaling* of these (stochastic) time points may be made in order to conform to conventional multivariate survival models. Note that by definition, for every $k(= 1, \ldots, K)$, and every $t \ge 0$,

$$0 = T_{k0} \le T_{k1} \le T_{k2} \le \cdots \le T_{kj_k(t)},$$

so that we may define the *rescaled times* as

$$T^*_{ki} = T_{ki} - T_{ki-1},\ i = 1, \ldots, j_k(t),\ t \ge 0. \tag{22.40}$$

Thus, the T_{ki}^* are the *gap times* or inter-event times. However, as censoring is affecting only the $j_k(t)$th (random) episode, these rescaled times are not of much help in our analysis. Consider now the matrix-valued counting process:

$$\mathbf{N}(t) = \begin{matrix} N_{11}(t) & \cdots & N_{1j_1(t)}(t) \\ \cdot & \cdots & \cdot \\ N_{K1}(t) & \cdots & N_{Kj_K(t)}(t) \end{matrix}, \quad t \geq 0. \tag{22.41}$$

Note that if the $j_k(t)$ were all equal to a prefixed $j \geq 1$, this would have been the regular matrix-valued counting process considered by Pedroso de Lima and Sen (1997), and others. However, in our case, not only these $j_k(t)$ are stochastic and vary from one individual to another but also they vary over the time t. The random variable $N_k(t)$ tells us the number of recurrent events for the kth individual occurring in the time interval $(0, t]$. Thus, $N_k(t)$ is a nondecreasing stochastic process (in t), and for different k they are stochastically independent too. This is a type of process that we intend to handle in HRQoL studies. We proceed via the related counting processes $j_k(t), t \geq 0; k \geq 1$.

22.5 Stopped Counting Processes

We are primarily interested in the following *stopped counting process*:

$$\mathbf{J}^*(t) = (j_k(t),\ k = 1, \ldots, K).\ t \geq 0, \tag{22.42}$$

which contains a lot of statistical information. In view of the fact that the $j_k(t)$ are themselves stochastic, first we set to find the centering of this process. For simplicity, we ignore the censoring random variables $C_k, k \geq 1$, and denote the cumulative distribution (and survival) function of T_{ki} by $F_{(ki)}$ (and $\bar{F}_{(ki)}$), respectively, for $i \geq 0,\ k = 1, \ldots, K$. Then note that

$$\begin{aligned} \nu_k(t) &= E\{j_k(t)\} = \sum_{r \geq 0} r P\{j_k(t) = r\} \\ &= \sum_{r \geq 1} P\{j_k(t) \geq r\} \\ &= \sum_{r \geq 1} P\{T_{kr} \leq t\} \\ &= \sum_{r \geq 1} F_{(kr)}(t), \forall t \geq 0,\ k \geq 1. \end{aligned} \tag{22.43}$$

The analogy of this expression with the classical *Renewal Theorem* is evident, though in the present context, the inter-episode times are neither independent nor identically distributed (more so in the presence of auxiliary or explanatory variables). Let us denote then

$$\bar{J}_K(t) = K^{-1} \sum_{k=1}^{K} j_k(t), \; t \geq 0. \tag{22.44}$$

We are primarily interested in a suitable stochastic process relating to this average stopped counting process, in the presence of a plausible censoring variable as well as other explanatory variables.

For simplicity, first, we consider the case of a *progressive censoring scheme* [Chatterjee and Sen (1973)] wherein we have a time-sequential setup with a right censoring mechanism caused by shifting the time parameter t continuously over $\mathbf{R}^+$ or a finite compact subset of that. Keeping that in mind, we define

$$\bar{\nu}_K(t) = K^{-1} \sum_{k=1}^{K} \sum_{r \geq 1} F_{(kr)}(t), \; t \geq 0. \tag{22.45}$$

Consider then the stochastic process

$$W_K(t) = K^{1/2}\{\bar{J}_K(t) - \bar{\nu}_K(t)\}, \; t \geq 0. \tag{22.46}$$

We intend to study the asymptotic (as K becomes large) properties of $W_K = \{W_K(t), t \in \mathbf{R}^+\}$. The very basic fact is that for different k, the counting processes $\{j_k(t) - \nu_k(t), \; t \in \mathbf{R}^+\}$ are independent, albeit not necessarily identically distributed (in the presence of explanatory variables). So, it should be comparatively simpler to handle this problem.

We keep in mind the HRQoL problem for which the QAL adjustments can be made effectively as in Sen (2002). Given that the residual life time following the occurrence of a disease or disorder is finite, one assumption we can make confidently is the following:

$$\lim_{t \to \infty} \nu_k(t) = \nu_k^* \text{ exists,} \tag{22.47}$$

for every $k = 1, \ldots, K$. Of course, these limits may generally depend on the auxiliary variables, and we shall make provision for that as well. As a result, we also make the following assumption:

$$\lim_{t \to \infty} \bar{\nu}_K(t) = \bar{\nu}_K^* \text{ exists.} \tag{22.48}$$

Note that the $j_k(t)$ are nonnegative integer-valued random variables, nondecreasing in $t \in \mathbf{R}^+$, whereas the $\nu_k(t)$, though nondecreasing in t, are generally continuous functions of t. Therefore our situation is quite similar to the usual jump-process encountered with empirical processes (but involving possibly nonidentically distributed components). As such, the convergence of *finite-dimensional distributions* (f.d.d.) of $W_K(.)$ to an appropriate Gaussian process follows by routine analysis. The crux of the problem is therefore to establish the *compactness* or tightness part. Towards this, we assume that

$\bar{\nu}_K(t)$ is uniformly continuous on any compact interval.

Further, for convergence of the f.d.d. to that of a Gaussian function, we need that the second order moment of the $j_k(t)$ exists, for every $k \geq 1$. In the same spirit as in above, we define

$$\bar{\gamma}_K^2(t) = K \ \mathrm{Var}\{\bar{J}_K(t)\}, \ t \geq 0, \ K \geq 1$$

and assume that

$\bar{\gamma}_K^2(t)$ is uniformly continuous on any compact interval.

Since $\bar{J}_K(t)$ and $\bar{\nu}_K(t)$ are both nondecreasing in t, using the above ucc condition, it follows that $W_K(.)$ is compact. The proof is quite analogous to the classical Bahadur (1966) representation of sample quantiles, and hence, is omitted. As a result, we conclude that as K increases, $W_K(.)$ converges weakly to a Gaussian function. In spite of the analogy with the renewal process, here $W_K(t)$ may not have generally independent and stationary increments. However, a martingale property is not difficult to verify.

In order to relate the above results to HRQoL studies, we need to relate the distributions $F_{(kr)}(t)$, albeit in a conditional form, to the concomitant variables, including the QAL scores, and then estimate the regression effects of these concomitants in a semiparametric way. It is also possible to do it in a nonparametric way treating the QAL factor as a random variable. We intend to communicate these findings in an integrated and unified manner in future communications.

References

1. Bahadur, R. R. (1966). A note on quantiles in large samples, *Annals of Mathematical Statistics*, **37**, 577–580.

2. Chatterjee, S. K. and Sen, P. K. (1973). Nonparametric testing for the bivariate two-sample problem, *Calcutta Statistical Association Bulletin*, **22**, 13–58.

3. Cox, D. R. (1972). Regression models and life tables (with discussion), *Journal of the Royal Statistical Society, Series B*, **34**, 187–220.

4. Hoffman, E.B., Sen, P. K. and Weinberg, C. R. (2001). Within-cluster resampling, *Biometrika*, **88**, 1121–1134.

5. Pedroso de Lima, A. C. and Sen, P. K. (1997). A matrix valued counting process with first-order interactive intensities, *Annals of Applied Probability*, **7**, 494–507.

6. Sen, P. K. (2002). Measures of QAL and QOLD: Statistical perspectives, In *Statistical Methods for Quality of Life Studies* (Eds., M. Mesbah, B. Coles and M. L. T. Lee), Kluwer Academic Publishers, Norwell, Massachusetts.

23

Semiparametric Approach to the Multivariate Mixed Rasch Model

Mohand-Larbi Feddag and Mounir Mesbah

University of South Brittany, IUP de Vannes Tohannic, Vannes, France

Abstract: The multivariate mixed Rasch model is used to analyse binary reponses of questionnaires with several subscales. The responses of our model are two-way correlated. First, at a given subscale of questionnaire, the binary responses of a single individual are correlated; second, they are repeated over the subscales and also become correlated. It is, however, well known that a full likelihood analysis for such mixed models is hampered by the need for numerical integrations. To overcome such integration problems, we propose the generalized estimating equations approach. Approximations of the joint moments of the variables are proposed. The estimators of the fixed effects parameters and variance components are consistent and asymptotically normal. We illustrate the usefulness method with simulations and with an analysis of real data from quality of life.

Keywords and phrases: Generalized linear mixed model, fixed effects and variance components, correlated data, generalized estimating equations, IRT models, Rasch model, quality of life

23.1 Introduction

Generalized linear mixed models (GLMMs) are an extension of generalized linear models (GLMs) that accommodate correlated and overdispersed data by adding random effects to the linear predictor. Their broad applications are useful in various disciplines, such as the analysis of clustered data including longitudinal data or repeated measures. These models are useful for accommodating the overdispersion often observed among outcomes that nominally have binomial or Poisson distribution; for modelling the dependence among outcome variables in longitudinal or repeated measures designs. Such generalized linear mixed models are also increasingly used in various fields where subjective variables need to be measured using questionnaires with polychotomous items. This is usual in health sciences and clinical trials, where these subjective variables could be pain, depression or quality of life. Other examples come from marketing where satisfaction or attitudes need to be well measured and educational testing services where well calibrated exams need to be produced. In such fields, the Rasch model, the most popular IRT model, is very often used. Statistical properties of this model are well known. When the primary interest is the population or to compare treatment groups, we consider a Rasch model with random latent trait. It belongs to the family of GLMMs.

Generalized linear models are based on the exponential family of distributions. Members of this family include the Gaussian or normal, gamma, inverse Gaussian, Poisson and geometric. As likelihood-based models, GLMs are based on the assumption that individual subjects or observations are independent. However, there are many common data situations for which responses are correlated. Such correlated data embraces multivariate observations, clustered data, repeated measurements and longitudinal data. In these examples, the individual data records are not independent, and, hence, violate the independence within subjects assumption. As an alternative, Liang and Zeger (1986) proposed the generalized estimating equations (GEE) approach. It is defined as an extension of quasi-likelihood method [see Wedderburn (1974)] to the correlated data using GLMs. This semi-parametric approach does not require the complete specification of the joint distribution of the repeated responses but rather only the first two moments. The correlation parameter is considered as a nuisance in their method. They introduced a class of estimating equations that give consistent estimates of regression parameters and of their variances regardless of whether or not the assumed correlation matrix is correct

Several modifications and extensions of the GEE methods have been studied. Prentice (1988) extended Liang and Zeger's work for correlated data by specifying second generalized estimating equations, based on the empirical pairwise covariances that permit the estimation of the correlation parameters. This approach, referred to as generalized estimating equations of order 1 (GEE1), consider the two estimating equations orthogonal to one another even when they are not. Zhao and Prentice (1990) and Prentice and Zhao (1991) proposed a unified approach to regression and correlation parameter estimation that typically leads to greater efficiency. In their generalized estimating equations of order 2 (GEE2), the orthogonality of the GEE1 approach is dropped, and the functional and statistical dependence between the first and second moment elementary estimating functions is incorporated into their weight matrices. These two methods require the third and fourth-order moments which are, generally, estimated with a working structure.

Estimation in GLMMs is complicated by the fact that these models are typically nonlinear with respect to the random effects and have no closed form expressions for the marginal likelihood or indeed for the marginal moments. Several approximate methods have been proposed. These include exact integration based methods using either numerical integration methods or Markov chain Monte Carlo techniques. Such integration based techniques work extremely well when the number of random effects is small. However, these techniques become increasingly more difficult to use when the number of random effects increases and the computations are intensive. As an alternative to numerical integration, two different first order Taylor series expansion methods have been used to approximate the marginal likelihood function, marginal moments or both. The first expand the conditional mean or likelihood about the mean of the random effects and the estimation is obtained by the GEE approach. The second method entails expanding about the maximum of current estimates of the random effects, meaning the Laplace approximation to the integrals. The estimation of different parameters is obtained by conditional GEE for the fixed parameters and by the restricted estimation maximum likelihood (REML) or profiled maximum likelihood for the variance components. Feddag et al. (2003) have used GEE2 to estimate simultaneously the fixed effects parameters and the variance components for the logistic mixed models with univariate random effects. This work is extended to the longitudinal mixed Rasch model [see, Feddag and Mesbah (2002)].

This article considers the GEE2 approach to estimate the fixed effects parameters and variance components of multivariate mixed Rasch models. This method involves using the approximations of the higher order moments which are derived from the approximate marginal likelihood. The estimators obtained are consistent and asymptotically normal. Thus our results may be regarded as an extension of Feddag and Mesbah's (2002) methods.

The specific outline of the paper is as follow. In Section 23.2, we present the model considered. In Section 23.3, we give a review on estimation methods in GLMMs. Thereafter in Section 23.4, we give approximations for the marginal likelihood to the considered model and we derive the higher order moments. Next, the estimating equations for the different parameters are constructed and we establish the asymptotic properties of the estimators. In Section 23.5, we present some simulation results in order to illustrate our methods. A real data set from quality of life experiment is also used. A summary and discussions are presented in Section 23.6.

23.2 Model Description

IRT (Item Response Theory) models first appeared in the field of psychometry and educational sciences to quantify human behavior. They are now increasingly used in medicine to study psychological traits in psychiatry and more recently, to assess quality of life in clinical trials or epidemiology. Generally, the quality of life of the patients is evaluated using questionnaires with dichotomous items. One of the most popular IRT models is the Rasch model [see Fischer and Molennar (1995)]. In our study, we consider responses of questionnaires with several subscales. In such case, the binary responses are two-way correlated. First, at a given subscale, the responses of a single individual are correlated and second, they are repeated over the subscales, they, also become correlated. The multivariate mixed Rasch model which permits the analysis of such data is defined below:

From now on, we will consider a sample of N independent $(\sum_{l=1}^{Q} J_l \times 1)$ random multivariate binary observations $Y_i = \left(Y_i^{1'}, \ldots, Y_i^{Q'}\right)'$, $i = 1, \ldots, N$, where $Y_i^l = (Y_{i1}^l, \ldots, Y_{iJ_l}^l)'$ is the response variable of individual i to the subscale l of the questionnaire and Y_{ij}^l is the binary variable response of individual i to item j at subscale l $(l = 1, \ldots, Q)$.

Let $Y = (Y_1, \ldots, Y_N)$ the vector of the variables, b_{il} the latent trait associated to subject i at subscale l and $b_i = (b_{i1}, \ldots, b_{iQ})'$ the multivariate latent trait for subject i to the questionnaire. We denote by y a response variable to the variable Y. The multivariate mixed Rasch model satisfies the following assumptions:

- Given the latent trait $b_i, i = 1, \ldots, N$, we have

$$P\left(Y_i^1 = y_i^1, \ldots, Y_i^Q = y_i^Q \mid b_i, \beta\right) = \prod_{l=1}^{Q} \prod_{j=1}^{J_l} P(Y_{ij}^l = y_{ij}^l \mid b_{il}, \beta_j^l), \quad (23.1)$$

  where β_j^l is the fixed effects parameter associated to item j of the subscale l and b_{il} the random effects associated to the subscale l. In quality of life [see Fischer and Molenaar (1995)], these parameters are respectively called difficulty parameters and latent trait.

- For all i, j, l: $i = 1, \ldots, N$, $j = 1, \ldots, J_l$, $l = 1, \ldots, Q$, The probability of the variable Y_{ij}^l is given by

$$P\left(Y_{ij}^l = y_{ij}^l \mid b_{il}, \beta_j^l\right) = \frac{\exp\left\{(b_{il} - \beta_j^l) y_{ij}^l\right\}}{1 + \exp(b_{il} - \beta_j^l)}. \quad (23.2)$$

- The latent variables $b_1, \ldots, b_N$, are independent and identically distributed with a multivariate normal with mean vector 0 and covariance matrix $\Sigma = (\sigma_{jl})_{j,l=1,\ldots,Q}$.

This mixed Rasch model is a particular case of the GLMM [see McCullagh and Nelder (1989)] with the link and variance functions respectively defined by

$$h(t) = \ln(t/(1-t)) \ , \ v(t) = t(1-t),$$

and the covariates associated with the fixed effects and with the random effects respectively given for all i, j, by

$$x_{ij}^l = (0, \ldots, 0, -1, 0, \ldots, 0) \ , \ z_{ij}^l = 1.$$

We are interested in estimating the difficulty parameters $\beta^l = (\beta_1^l, \ldots, \beta_{J_l}^l)'$, l=1,...,Q, the vector of the variances of the latent traits α=$(\sigma_{11}, \sigma_{22}, \ldots, \sigma_{QQ})'$ and the vector of their covariances $\gamma = (\sigma_{12}, \ldots, \sigma_{1Q}, \sigma_{23}, \ldots, \sigma_{Q-1,Q})'$. Let $\beta = (\beta^1, \ldots, \beta^Q)$ the difficulty parameter associated to all subscales.

The marginal likelihood of y is given by

$$L(\beta, \alpha, \gamma \mid y) = \prod_{i=1}^{N} \int_{\mathbb{R}^Q} \prod_{l=1}^{Q} \prod_{j=1}^{J_l} \frac{\exp\left\{(b_{il} - \beta_j^l) y_{ij}^l\right\}}{1 + \exp(b_{il} - \beta_j^l)} \phi(b_i, \alpha, \gamma) db_i, \qquad (23.3)$$

where

$$\phi(b_i, \alpha, \gamma) = \frac{1}{(2\pi)^{Q/2} |\Sigma|^{1/2}} \exp\left\{-\frac{1}{2} b_i' \Sigma^{-1} b_i\right\}$$

is the distribution of the multivariate normal with mean vector 0 and covariance matrix Σ.

23.3 An Overview of Estimating Methods in GLMMs

As we previously noted, the multivariate mixed Rasch model belongs to the family of GLMMs with link and variance functions respectively defined by

$$h(t) = Logit(t) = \ln(t/(1-t)) \ , \ v(t) = t(1-t).$$

The maximization of the function given by equation (23.3) is computationally difficult and requires evaluation of integrals where the integral's dimension is equal to the number of random effects. In our case, this number is equal to number of subscales in the questionnaire. Various methods have been proposed to circumvent this problem. These include exact integration based methods using either numerical integration methods or Markov chain Monte Carlo techniques and approximations to the marginal likelihood or marginal moments.

The numerical methods are centered on the adaptive quadrature [see Pinheiro and Bates (2000)]. An alternative to numerical integration is based on Monte Carlo Markov Chain methods (MCMC) [see Gilks and Wild (1992)]. However, these techniques become increasingly more difficult to use when the number of random effects increases and the computations are intensive.

As an alternative to these methods, different approaches using approximations of the marginal likelihood or marginal moments have been proposed. The methods proposed by Schall (1991) and Engel and Keen (1994) are based on the transformation of GLMM to classical linear mixed model (LMM). The link function $h(.)$ applied to the data y_i, $i = 1, \ldots, N$ is linearized to first order as given in McCullagh and Nelder (1989),

$$h(y_i) = h(\mu_i) + (y_i - \mu_i)h'(\mu_i) = y_i^*.$$

Then, we obtain a linear random effects model for the variables y_i^*, $i = 1, \ldots, N$. The estimation of the parameters are obtained by classical maximum likelihood or restricted maximum likelihood (REML). Lavergne and Trottier (2000) have compared these methods with the approach proposed by Gilmour et al. (1985). Zeger et al. (1988) have used an approximate mean vector and a working covariance matrix, meaning an expansion to the link function $h(.)$ in a Taylor series about $b_i = 0$, to estimate the regression parameters by GEE. Breslow and Clayton (1993) have used the same approximations in the marginal quasi-likelihood (MQL) approach. The regression parameters are estimated as in Zeger et al. (1988), while the variance components are estimated by the REML or the profiled maximum likelihood. The second approximate method of Breslow and Clayton (1993) namely, the penalized quasi-likelihood (PQL), is based on first order Taylor expansions about the maximum of current estimates of the random effects via the first order Laplace approximations to the integrals. These approaches produce biased estimates for the regression parameters and variance components parameters. Breslow and Lin (1995) provided a correction factor for the estimates of the univariate variance components derived from the second order Laplace approximations. Lin and Breslow (1996) extends this bias correction to the GLMM with multivariate random effects. Many of these methods are implemented in the SAS macro GLIMMIX macro and in Splus macro GEEX. Feddag et al. (2003) have used GEE2 to estimate simultaneously the fixed effects parameter and the variance components for the logistic mixed models. The estimators obtained are consistent and asymptotically normal. This method produces unbiased estimates for the regression effects and small bias for the variance components larger than 0.5. This approach is extended to the longitudinal mixed Rasch model [see Feddag and Mesbah (2002)].

23.4 GEE to the Multivariate Mixed Rasch Model

In this section we give the approximation of the marginal likelihood, then we derive the higher order joint moments of the variables. Next, we propose estimating equations to estimate the different parameters.

23.4.1 Approximations of marginal likelihood and joint moments

The aim of this section is to give the approximations of the joint moments up to order 4 of the observed variable y_i, which we shall use later on in the GEE. Their computation requires an approximation of the marginal likelihood of $y = (y_1, \dots, y_N)$, given in Equation (23.3).

Up to now, we assume that the joint moments from order 6 of the random effects b_i satisfies the following condition:

$$\mathbb{E}(\|b_i\|^r) = o\left(f_r(\alpha, \gamma)\right) \quad \text{for all } r \geq 6, \tag{23.4}$$

where f_r is function of r and the parameters (α, γ) of the matrix Σ.

Theorem 23.4.1 (Feddag and Mesbah (2003)) *Under the assumption made in (23.4), we have:*

1. *The marginal likelihood $L(\beta, \sigma^2 \mid y)$ defined in (23.3), is approximated by*

$$L^*(\beta, \alpha, \gamma \mid y) \;=\; \prod_{i=1}^{N} (1 + P_Q(y_i, \beta, \alpha, \gamma)) \prod_{l=1}^{T} \prod_{j=1}^{J_l} g^*_{ij}(y^l_{ij}; \beta^l_j), \tag{23.5}$$

where

$$P_Q(y_i, \beta, \alpha, \gamma)$$
$$= \int_{\mathbb{R}^Q} \sum_{t=1}^{Q} \sum_{\substack{1 \leq t_1 < t_2 < \dots < t_T \leq Q \\ \sum_j t_j = t}} \left[\prod_{j=1}^{Q} h_i^{t_j}(y_i, b_i, \beta) \right] \phi(b_i, \alpha, \gamma) db_i. \tag{23.6}$$

2. *The joint density of* $Y^h_{i,m} = (Y^h_{ij_1}, \ldots, Y^h_{ij_m})$, $i = 1, \ldots, N$, $m = 1, \ldots, J_h$, $h = 1, \ldots, Q$ *is approximated by*

$$\begin{aligned} & L^*_{i,(m)}\left(y^h_{i,m}; \beta, \alpha, \gamma\right) \\ &= \prod_{l=1}^{m} g^*_{ij_l}(y^h_{ij_l}; \beta^l_j)\left[1 + \frac{1}{2}\sigma_{hh} R^h_{i,j_1\ldots j_m} + \frac{1}{8}\sigma^2_{hh} Q^h_{i,j_1\ldots j_m}\right]. \end{aligned} \quad (23.7)$$

3. *The density of* $(Y^k_{i,m}, Y^h_{i,s}) = (Y^k_{ij_1}, \ldots, Y^k_{ij_m}, Y^h_{ij_1}, \ldots, Y^h_{ij_s})$, $i = 1, \ldots, N$, $m = 1, \ldots, J_k$, $s = 1, \ldots, J_h$, $1 < k \neq h < Q$, *is given by*

$$\begin{aligned} & L^*_{i,(m,s)}\left(y^k_{i,m}, y^h_{i,s}; \beta, \alpha, \gamma\right) \\ &= \prod_{l=1}^{m} g^*_{ij_l}(y^k_{ij_l}; \beta^l_j) \prod_{l=1}^{s} g^*_{ij_l}(y^h_{ij_l}; \beta^l_j) \\ &\times \Big[1 + \frac{1}{2}\left(\sigma_{kk} R^k_{i,j_1\ldots j_m} + \sigma_{hh} R^h_{i,j_1\ldots j_s}\right) + \sigma_{kh} A^k_{i,j_1\ldots j_m} A^h_{i,j_1\ldots j_s} \\ &\quad + \frac{1}{2}\left(A^k_{i,j_1\ldots j_m} P^h_{i,j_1\ldots j_s}\sigma_{kh}\sigma_{hh} + P^t_{i,j_1\ldots j_m} A^h_{i.j_1\ldots j_s}\sigma_{kk}\sigma_{kh}\right) \\ &\quad + \frac{1}{4} R^k_{i,j_1\ldots j_m} R^h_{i,j_1\ldots j_s}\left(\sigma_{kk}\sigma_{hh} + 2\sigma^2_{kh}\right) \\ &\quad + \frac{1}{8}\left(\sigma^2_{kk} Q^k_{i,j_1\ldots j_m} + \sigma^2_{hh} Q^h_{i,j_1\ldots j_s}\right)\Big]. \end{aligned} \quad (23.8)$$

The quantities involved in the points 1, 2 and 3 are defined below:

$$\begin{aligned} h^{t_j}_i(y_i, b_i, \beta) &= A^{t_j}_i b_{it_j} + \frac{1}{2} R^{t_j}_i b^2_{it_j} + \frac{1}{6} P^{t_j}_i b^3_{it_j} + \frac{1}{24} Q^{t_j}_i b^4_{it_j}, \\ g^*_{ij}(y^t_{ij}; \beta^t_j) &= \exp\left\{-y^t_{ij}\beta^t_j - \ln(1 + e^{-\beta^t_j})\right\}, \\ a_{i.lj} &= \ln\left\{1 + \exp\left(b_{il} - \beta^l_j\right)\right\}, \\ A^t_{i,j} &= y^t_{ij} - a^{(1)}_{tj},\ B^t_j = a^{(2)}_{tj},\ C^t_j = a^{(3)}_{tj},\ F^t_j = a^{(4)}_{tj}, \\ R^t_{i,j} &= (A^t_{i,j})^2 - B^t_j,\ P^t_{i,j} = (A^t_{i,j})^3 - 3A^t_{i,j} B^t_j - C^t_j, \\ Q^t_{i,j} &= (A^t_{i,j})^4 - 6(A^t_{i,j})^2 B^t_j - 4A^t_{i,j} C^t_j + 3(B^t_j)^2 - F^t_j, \end{aligned}$$

$$
\begin{aligned}
A^t_{i,j_1\cdots j_m} &= \sum_{l=1}^{j_m} A^t_{i,j_l}, \quad B^t_{j_1\cdots j_m} = \sum_{l=1}^{j_m} B^t_{j_l}, \\
C^t_{j_1\cdots j_m} &= \sum_{l=1}^{j_m} C^t_{j_l}, \; F^t_{j_1\cdots j_m} = \sum_{l=1}^{j_m} F^t_{j_l}, \\
R^t_{i,j_1\cdots j_m} &= (A^t_{i,j_1\cdots j_m})^2 - B^t_{j_1\cdots j_m}, \; P^t_{j_1\cdots j_m} \\
&= (A^t_{j_1\cdots j_m})^3 - 3A^t_{j_1\cdots j_m} B^t_{j_1\cdots j_m} - C^t_{j_1\cdots j_m}, \\
Q^t_{i,j_1\cdots j_m} &= (A^t_{i,j_1\cdots j_m})^4 - 6(A^t_{i,j_1\cdots j_m})^2 B^t_{j_1\cdots j_m} - 4A^t_{i,j_1\cdots j_m} C^t_{j_1\cdots j_m} \\
&\quad +3(B^t_{j_1\cdots j_m})^2 - F^t_{j_1\cdots j_m},
\end{aligned}
$$

and $a^{(t)}_{lj}$ *is the t-order derivative of* $a_{i,lj}$ *at* $b_{il} = 0$, $t = 1, \ldots, 4$.

PROOF. The proof of this theorem is similar to the one given in Feddag and Mesbah (2002). ■

We use this theorem to derive the joint moments up to order 2 within and between subscales, given in the following proposition.

Proposition 23.4.1 *Under the assumption given in (23.4), we have the following approximation of the joint moments:*

1. *The marginal mean of* Y^l_{ij} *is given by*

$$
\mathbb{E}\left(Y^l_{ij}\right) = \mu^l_j = \frac{1}{1+e^{\beta^l_j}} + \frac{\sigma_{ll}}{2}\frac{e^{\beta^l_j}(e^{\beta^l_j}-1)}{(1+e^{\beta^l_j})^3} + \frac{\sigma^2_{ll}}{8}\frac{e^{\beta^l_j}(e^{3\beta^l_j} - 11e^{2\beta^l_j} + 11e^{\beta^l_j} - 1)}{(1+e^{\beta^l_j})^5}. \tag{23.9}
$$

2. *The covariance (within subscale) of* (Y^l_{ij}, Y^l_{ik}) *is given by*

$$
\begin{aligned}
&Cov\left(Y^l_{ij}, Y^l_{ik}\right) \\
&= \frac{\sigma_{ll}}{2}\frac{1}{(1+e^{\beta^l_j})^2(1+e^{\beta^l_k})^2} \times \left[2e^{\beta^l_j}e^{\beta^l_k}\right. \\
&\quad +\sigma_{ll}\frac{(e^{3\beta^l_j} - 4e^{2\beta^l_j} + e^{\beta^l_j})e^{\beta^l_k}}{(1+e^{\beta^l_j})^2} + \sigma_{ll}\frac{e^{\beta^l_j}(e^{\beta^l_j}-1)e^{\beta^l_k}(e^{\beta^l_k}-1)}{(1+e^{\beta^l_j})(1+e^{\beta^l_k})} \\
&\quad \left. +\sigma_{ll}\frac{e^{\beta^l_j}(e^{3\beta^l_k} - 4e^{2\beta^l_k} + e^{\beta^l_k})}{(1+e^{\beta^l_k})^2}\right].
\end{aligned} \tag{23.10}
$$

3. The covariance (between subscales) of (Y^l_{ij}, Y^h_{ik}) is given by

$$
\begin{aligned}
&Cov\left(Y^l_{ij}, Y^h_{ik}\right) \\
&= \frac{\sigma_{lh}}{2}\frac{1}{(1+e^{\beta^l_j})^2(1+e^{\beta^h_k})^2} \times \left[2e^{\beta^l_j}e^{\beta^h_k}\right. \\
&+\sigma_{ll}\frac{(e^{3\beta^l_j}-4e^{2\beta^l_j}+e^{\beta^l_j})e^{\beta^h_k}}{(1+e^{\beta^l_j})^2} + \sigma_{lh}\frac{e^{\beta^l_j}(e^{\beta^l_j}-1)e^{\beta^h_k}(e^{\beta^h_k}-1)}{(1+e^{\beta^l_j})(1+e^{\beta^h_k})} \\
&\left.+\sigma_{hh}\frac{e^{\beta^l_j}(e^{3\beta^h_k}-4e^{2\beta^h_k}+e^{\beta^h_k})}{(1+e^{\beta^h_k})^2}\right]. \qquad (23.11)
\end{aligned}
$$

Graphical illustrations

The mean μ^l_j and the variance σ^l_{jj} of Y^l_{ij} given by equations (23.9) and (23.10) respectively, are now compared graphically with the approximation obtained by standard procedures given in Splus based on Gauss–Kronrod quadrature [see Pissens et al. (1983)] and the logit approximation by the probit function [see Johnson et al. (1995)]. The last approximation given by:

$$\hat{\mu}_{ij} = \frac{\exp\left(-|c^2\Sigma + \mathbb{I}_q|^{-q/2}\beta_j\right)}{1+\exp\left(-|c^2\Sigma + \mathbb{I}_q|^{-q/2}\beta_j\right)}, \quad \text{where } c = \frac{16\sqrt{3}}{15\pi},$$

had been used by Zeger et al. (1988) to estimate the regression parameters of GLMM by GEE approach. We display the approximate means in Figures 23.1 and 23.2 respectively for $\beta = 1$ and $\beta = -2$. In Figures 23.3 and 23.4, we have the approximate variances for $\beta = 1$ and $\beta = -2$. These figures show that the means and the variances are very similar for small σ ($\sigma < 1$). For large value of σ, the approximations of our mean and variance are different from the two others. These graphs confirm the good quality of these approximations for small σ.

23.4.2 Estimation of the parameters

Consider $y_i = (y^1_i, \ldots, y^Q_i)'$, $i = 1, \ldots, N$, to be the outcomes of the multivariate mixed Rasch model defined in Section 23.2. Our approach for estimating parameters β, α and γ is as follows. Along with the basic estimating equations for the mean of y_i, $i = 1, \ldots, K$, which gives estimators for the difficulty parameter β, we shall use supplementary equations to estimate the parameters α and γ. These equations are based on the empirical covariances within and between subscales which are unbiased estimators of the true covariance of the vector y_i.

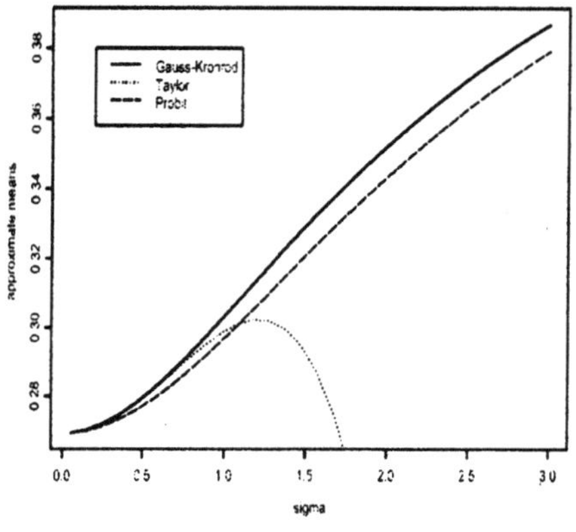

Figure 23.1: Approximate means for $\beta = 1$

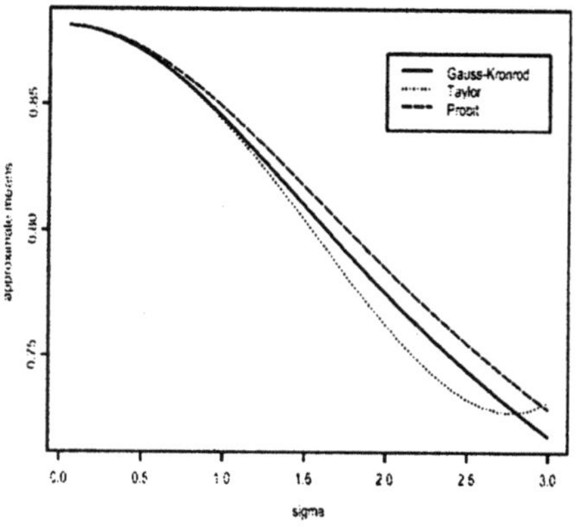

Figure 23.2: Approximate means for $\beta = -2$

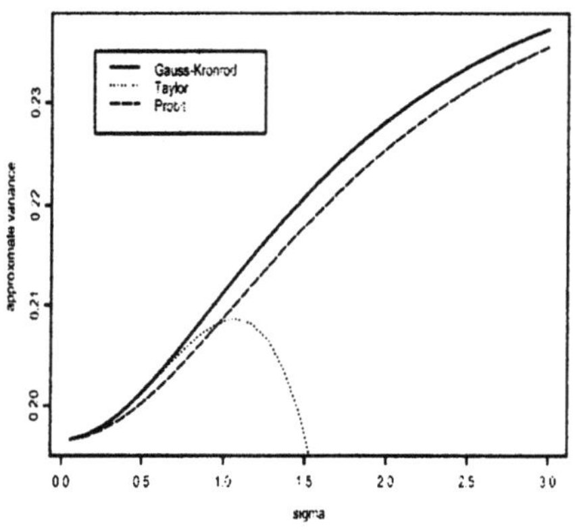

Figure 23.3: Approximate variances for $\beta = 1$

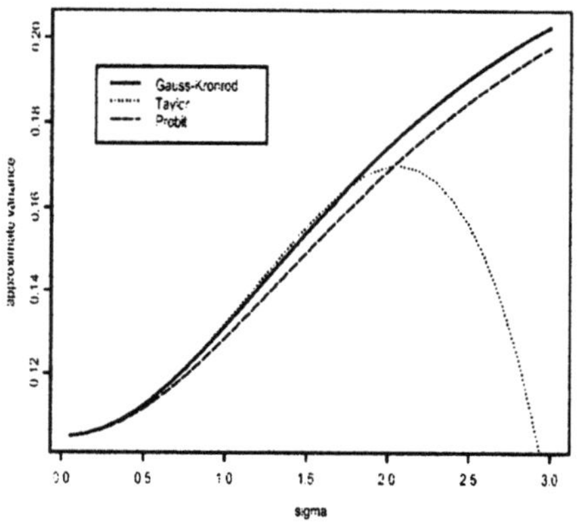

Figure 23.4: Approximate variances for $\beta = -2$

Let $S_i' = (S_i^l)_{l=1,\dots,Q}$ be a vector where S_i^l is a $\frac{J_l(J_l-1)}{2} \times 1$ vector of empirical pairwise covariances (within subscale) associated to y_i^l defined by

$$S_i^l = \left(S_{i,jk}^l\right)_{1\le j<k\le J_l} \quad \text{where} \quad S_{i,jk}^l = (y_{ij}^l - \mu_j^l)(y_{ik}^l - \mu_k^l), \tag{23.12}$$

and $W_i' = (W_i^{lh})_{1\le l<h\le Q}$, a vector with elements W_i^{lh}, be a $J_l J_h \times 1$ vector of empirical pairwise covariance associated to (y_i^l, y_i^h), defined by

$$W_i^{lh} = \left(W_{i,km}^{lh}\right)_{\substack{1\le k\le J_l\\ 1\le m\le J_m}} \quad \text{where} \quad W_{i,km}^{lh} = (y_{ik}^l - \mu_k^l)(y_{im}^h - \mu_m^h). \tag{23.13}$$

To apply the GEE approach we consider as observations the vectors (y_i, S_i, W_i), $i = 1, \dots, N$ with means

$$\begin{aligned} \mu &= \mathbb{E}(Y_i) = \left(\mu^l\right)_{l=1,\dots,Q}, \\ \eta &= \mathbb{E}(S_i) = \left(\eta^l\right)_{l=1,\dots,Q}, \\ \nu &= \mathbb{E}(W_i) = \left(\nu^{lh}\right)_{1\le l<h\le Q}, \end{aligned} \tag{23.14}$$

where

$$\mu^l = \mathbb{E}(Y_i^l) = \left(\mu_j^l\right)_{j=1,\dots,J_l}, \quad \eta^l = \left(\eta_{jk}^l\right)_{1\le j<k\le J_l}, \quad \nu^{lh} = \left(\nu_{jk}^{lh}\right)_{\substack{1\le j\le J_l\\ 1\le k\le J_h}},$$

and with covariance matrix

$$V = \begin{pmatrix} V_{11} & V_{12} & V_{13} \\ V_{21} & V_{22} & V_{23} \\ V_{31} & V_{32} & V_{33} \end{pmatrix}, \tag{23.15}$$

where

$$V_{11} = \mathbb{V}(Y_i), \; V_{22} = \mathbb{V}(S_i), \; V_{33} = \mathbb{V}(W_i),$$

$$V_{12} = \mathbb{C}(Y_i, S_i), \; V_{13} = \mathbb{C}(Y_i, W_i), V_{23} = \mathbb{C}(S_i, W_i).$$

All these quantities are computed using the marginal joint moments up to order 4 of the variables. The quantities μ_j^l, η_{jk}^l, ν_{jk}^{lh} are respectively given by the equations (23.9), (23.10) and (23.11). The elements of the matrix V are computed using the joint moments of Y_i up to order 4. The elements of V_{11} are obtained by formulas (23.10)–(23.11). The computation of V_{12}, V_{13}, V_{21} and V_{31} requires the joint moments up to order three and the elements of the other matrices requires the joint moments up to order 4.

The parameters (β, α, γ) will be estimated by the solution of the generalized estimating equations

$$U(\beta, \alpha, \gamma) = D'V^{-1}\sum_{i=1}^{N} \xi_i = 0, \tag{23.16}$$

where

$$\xi_i = \begin{pmatrix} y_i - \mu \\ S_i - \eta \\ W_i - \nu \end{pmatrix}, \quad D = \begin{pmatrix} D_{11} & D_{12} & D_{13} \\ D_{21} & D_{22} & D_{23} \\ D_{31} & D_{32} & D_{33} \end{pmatrix},$$

with

$$D_{11} = \frac{\partial \mu}{\partial \beta}, \; D_{12} = \frac{\partial \mu}{\partial \alpha}, \; D_{13} = \frac{\partial \mu}{\partial \gamma}, \; D_{21} = \frac{\partial \eta}{\partial \beta},$$

$$D_{22} = \frac{\partial \eta}{\partial \alpha}, \; D_{23} = \frac{\partial \eta}{\partial \gamma}, \; D_{31} = \frac{\partial \nu}{\partial \beta}, \; D_{32} = \frac{\partial \nu}{\partial \alpha}, \; D_{33} = \frac{\partial \nu}{\partial \gamma}.$$

All these quantities are computed using the expressions (23.9)–(23.11) of Proposition 23.4.1.

Let $\left(\hat{\beta}, \hat{\alpha}, \hat{\gamma}\right)$ be the root of the equation (23.16).

23.4.3 Asymptotic properties

As consequence of the theorem of Zhao and Prentice (1990), we derive the following asymptotic properties of the estimator $\left(\hat{\beta}, \hat{\alpha}, \hat{\gamma}\right)$.

- $\left(\hat{\beta}, \hat{\alpha}, \hat{\gamma}\right)$ is consistent for (β, α, γ),
- $$N^{1/2}\left\{\left(\hat{\beta} - \beta\right)', (\hat{\alpha} - \alpha)', (\hat{\gamma} - \gamma)'\right\}' \xrightarrow[N \longmapsto \infty]{} N(0, W); \tag{23.17}$$

where

$$W = \lim_{N \longmapsto \infty} N\left(A_1^{-1} A_2 A_1^{-1}\right), \tag{23.18}$$

with

$$A_1 = N\left(D'V^{-1}D\right), \; A_2 = D'V^{-1}\left(\sum_{i=1}^{N} \xi_i \xi_i'\right) V^{-1}D.$$

The covariance matrix W is consistently estimated by

$$\hat{W} = \lim_{N \longmapsto \infty} N\left(\hat{A}_1^{-1} \hat{A}_2 \hat{A}_1^{-1}\right), \tag{23.19}$$

where $\hat{A}_1, \hat{A}_2$ are respectively the values of A_1 and A_2 at $(\hat{\beta}, \hat{\alpha}, \hat{\gamma})$.

The computations of $(\hat{\beta}, \hat{\alpha}, \hat{\gamma})$ are easily obtained by the Fisher-scoring algorithm.

23.5 Illustrations

23.5.1 Simulations

In this section we study the sample performance in estimating the difficulty parameters β and the parameters α and γ of the latent trait by the GEE2 approach. We performed 500 replications of the multivariate mixed Rasch model with parameters fixed as

- $Q = 3,\ J_1 = 4,\ J_2 = 5,\ J_3 = 6,$
- $\beta^1 = (-2, -1, 1, 2)$, $\beta^2 = (-2, -1, -0.5, 1, 2)$, $\beta^3 = (-2, -1, -0.5, .5, 1, 2)$,
- two sample size $N = 100, 300$ and two covariance matrices defined as follows:

$$\Sigma_1 = \begin{pmatrix} 0.4 & 0.1 & 0.2 \\ 0.1 & 0.5 & 0.3 \\ 0.2 & 0.3 & 0.6 \end{pmatrix}, \qquad \Sigma_2 = \begin{pmatrix} 0.8 & 0.5 & 0.6 \\ 0.5 & 0.9 & 0.7 \\ 0.6 & 0.7 & 1 \end{pmatrix}.$$

We denote by $\alpha_1 = (0.4, 0.5, 0.6)$, $\gamma_1 = (0.1, 0.2, 0.3)$ the parameters of Σ_1 and by $\alpha_2 = (0.8, 0.9, 1)$, $\gamma_2 = (0.5, 0.6, 0.7)$ the parameters of Σ_2. The software package Splus is used for the simulations.

The simulation results for the classical mixed Rasch model [see Feddag et al. (2002, 2003)], shows that the specification of the third and fourth joint moments of the variables do not improve the estimates of different parameters. Then we have considered in this case a V matrix where all elements containing third and fourth order joint moments are supposed equal to zero. The different matrices $V_{jl},\ j, l = 1, 2, 3$, are defined as follows:

- V_{11} matrix with elements obtained by formulas (23.9)–(23.11).
- V_{12} where the elements $\text{Cov}(Y_{ij}^h, S_{i,kl}^h) = \text{Cov}(Y_{ij}^h, S_{i,kl}^t) = 0$ for $j \neq k, l$ and $h \neq t$.
- V_{13} where $\text{Cov}(Y_{ij}^t, W_{i,kl}^{sh}) = 0$ $(t \neq s \neq h)$, $\text{Cov}(Y_{ij}^t, W_{i,kl}^{th}) = \text{Cov}(Y_{ij}^t, W_{i,kl}^{ht})$ $= 0$ $(j \neq k, l)$.
- V_{22}, V_{33} considered as diagonal matrices and $V_{23} = 0$.

We give in the Tables 23.1 and 23.2 the mean and the standard error (s.e) of the 500 estimates values obtained for each parameter.

From Tables 23.1 and 23.2, it is seen that all estimate are unbiased. Table 23.1 shows that the bias for GEE estimation of both difficulty parameters and component of variance relative to their true values is less than 2 per cent. From Table 23.2, we note that the largest bias for the difficulty parameter estimates

Table 23.1: Simulation results for multivariate mixed Rasch model with parameters $\beta^1 = (-2, -1, 1, 2)$, $\beta^2 = (-3, -1, 0, 1, 3)$, $\beta^3 = (-2, -1, -0.5, 0.5, 1, 2)$, $\alpha = (0.4, 0.5, 0.6)$, $\gamma = (0.1, 0.2, 0.3)$

N								
100	$\hat{\beta}^1$	mean	-2.048	-1.003	1.021	2.033		
		s.e	0.248	0.169	0.186	0.231		
	$\hat{\beta}^2$	mean	-3.053	-0.989	-0.010	0.999	3.074	
		s.e	0.365	0.166	0.162	0.174	0.401	
	$\hat{\beta}^3$	mean	-2.013	-0.986	-0.503	0.479	0.986	2.030
		s.e	0.220	0.186	0.162	0.155	0.170	0.243
	$(\hat{\alpha}, \hat{\gamma})$	mean	0.419	0.546	0.630	0.109	0.211	0.313
		s.e	0.244	0.268	0.198	0.142	0.130	0.144
300	$\hat{\beta}^1$	mean	-2.014	-1.000	0.997	2.006		
		s.e	0.186	0.140	0.145	0.194		
	$\hat{\beta}^2$	mean	-3.037	-0.997	-0.009	1.007	3.021	
		s.e	0.294	0.141	0.125	0.147	0.290	
	$\hat{\beta}^3$	mean	-2.030	-0.992	-0.499	0.505	0.992	2.010
		s.e	0.184	0.140	0.131	0.131	0.147	0.182
	$(\hat{\alpha}, \hat{\gamma})$	mean	0.389	0.524	0.618	0.097	0.202	0.303
		s.e	0.213	0.222	0.160	0.114	0.109	0.102

is equal to 3 per cent and less than 3 per cent for the variance components estimates. We note non-surprisingly that the standard errors (s.e), of all estimates increases slightly as the elements of the covariance matrix Σ increase or N decrease. These results confirm the theoretical asymptotic properties of these estimators when the elements of the covariance matrix are small. We note also that this approach is less intensive with large value of N.

23.5.2 Example

In this section, we illustrate the application of this methodology to the analysis from a quality of life data. The sample is composed of 470 depressive patients who answer at three occasions to a French version of the Sickness Impact Profile (SIP) questionnaire [see Bergner et al. (1973), for the international version]. This questionnaire is composed of 12 subscales of items, each related to a particular aspect from a quality of life. We present here the result for the Emotional Behavior dimension (the item contents are given in Table 23.3). Each person has to choose the items that describe surely his actual situation and that are related to his health. If an item is checked, then we code the response 1. So this

Table 23.2: Simulation results for multivariate mixed Rasch model with parameters $\beta^1 = (-2, -1, 1, 2)$, $\beta^2 = (-3, -1, 0, 1, 3)$, $\beta^3 = (-2, -1, -.5, .5, 1, 2)$, $\alpha = (.8, .9, 1)$, $\gamma = (.5, .6, .7)$

N								
100	$\hat{\beta^1}$	mean	-2.020	-0.979	0.970	2.009		
		s.e	0.190	0.158	0.160	0.206		
	$\hat{\beta^2}$	mean	-3.026	-0.987	-0.0317	0.976	3.005	
		s.e	0.262	0.145	0.134	0.148	0.290	
	$\hat{\beta^3}$	mean	-2.028	-0.991	-0.489	0.495	0.985	2.005
		s.e	0.182	0.171	0.137	0.137	0.160	.200
	$(\hat{\alpha}, \hat{\gamma})$	mean	0.760	0.881	0.969	0.478	0.585	0.706
		s.e	0.242	0.244	0.182	0.138	0.132	0.137
300	$\hat{\beta^1}$	mean	-2.005	-0.987	0.971	2.012		
		s.e	0.194	0.151	0.149	0.198		
	$\hat{\beta^2}$	mean	-3.031	-.984	0.000	.977	3.045	
		s.e	0.264	0.158	0.136	0.159	0.275	
	$\hat{\beta^3}$	mean	-2.007	-0.974	-0.481	0.490	0.995	2.003
		s.e	0.197	0.161	0.134	0.136	0.147	0.205
	$(\hat{\alpha}, \hat{\gamma})$	mean	0.779	0.894	0.965	0.498	0.587	0.710
		s.e	0.239	0.269	0.177	0.142	0.124	0.132

subscale measures the decrease of the quality of life. Only 131 individuals have responded at all three occasions. In our case, we assimilate the longitudinal repetition of this subscale to several subscales.

We report in Table 23.4, the estimations of different parameters and their standard errors (s.e). The standard errors are obtained by the estimate of the covariance matrix, given by formula (23.19).

From Table 23.4, we note that the difficulty items are different from one subscale to another. The most difficult items (with largest estimation) are numbers 5, 1 and 6 respectively to subscale 1,2 and 3. And the easiest items (with smaller estimation)is number 6 for the first subscale and number 4 for the other subscales. All measures of quality of life are correlated. The first and the second latent traits are positively correlated and the correlation between the third and the two first latent traits are negative.

Table 23.3: Description of the Emotional Behavior subscale of SIP ($N = 131,\ J = 6,\ T = 3$)

1. I say how bad or useless I am, for example, that I am a burden on others
2. I laugh or cry suddenly
3. I often moan and groan in pain or discomfort
4. I act nervous or restless
5. I keep rubbing or holding areas of my body that hurt or are uncomfortable
6. I talk about the future in a hopeless way

Table 23.4: Parameter estimates for (β, α, γ) and their standard errors (s.e)

$\hat{\beta}^1$	-1.102	-0.571	-0.331	-1.021	0.663	-2.579
s.e	0.229	0.204	0.203	0.218	0.213	0.328
$\hat{\beta}^2$	1.772	1.341	1.028	0.451	1.388	1.307
s.e	0.300	0.258	0.224	0.188	0.260	0.249
$\hat{\beta}^3$	2.473	2.545	1.885	1.183	2.045	2.591
s.e	0.349	0.338	0.361	0.298	0.381	0.349
$(\hat{\alpha}, \hat{\gamma})$	0.788	1.483	1.816	0.256	-0.374	-0.529
s.e	0.277	0.182	0.258	0.196	0.171	0.241
$\hat{\rho}$				0.237	-0.317	-0.322

23.6 Conclusions

In this paper, we used the GEE2 approach for the multivariate mixed Rasch models. This method has been used by several authors for the generalized linear models with fixed effects (GLMs). Zeger et al. (1988) have used this approach to estimate only the regression parameter in GLMMs through the use of approximations to the mean and the covariance matrix. Breslow and Clayton (1993); Breslow and Lin (1995) have used the penalized quasi-likelihood approach and the marginal quasi-likelihood method to estimate the regression parameters and the univariate component of the random effects in GLMMs. Recently Sutradhar and Rao (2001) have proposed approximations of the joint moments up to order 4 to estimate separately the regression parameters and the small univariate component of the random effects. Feddag et al. (2003) have proposed estimating equations to estimate simultaneously the fixed effects parameters and the variance of the random effects to the logistic mixed models. This later work is extended by Feddag and Mesbah (2002) to the longitudinal mixed Rasch model. This paper illustrates how the estimating equations proposed by Feddag and Mesbah (2002), can be extended to the multivariate mixed Rasch model. The approximations of the higher order moments are based on the generalizations of those used by Sutradhar and Rao (2001) to GLMMs with univariate random effects. The obtained estimators are consistent and asymptotically normal. The simulation results confirm the theoretical asymptotic properties of the estimators and the computations are less intensive. In summary, various models for repeated binary data have been considered. It would be interesting to further compare our current results with those of Sutradhar and Sinha (2002) models.

References

1. Bergner, M. et al. (1973). The Sickness Impact Profile: Validation of a health status measure, *Medical Care*, **14**, 56–67.

2. Breslow, N. E. and Clayton, N. E. (1993). Approximate inference in generalized linear mixed models, *Journal of the American Statistical Association*, **88**, 9–25.

3. Breslow, N. E. and Lin, N. (1995). Bias correction in generalized linear mixed models with a single components of dispersion, *Biometrika*, **82**, 81–91.

4. Engel, B. and Keen, A. (1994). A simple approach for the analysis of generalized linear mixed models, *Statistica Neelandica*, **48**, 1–22.

5. Feddag, M. L., Grama, I. and Mesbah, M. (2002). Estimation des paramtres d'un modle de Rasch mixte par la mthode GEE. Application en Qualit de Vie, *Journal de la SFDS*, **143**, 169–175.

6. Feddag, M. L., Grama, I. and Mesbah, M. (2003). Generalized Estimating Equations for Mixed Logistic Models, *Communications in Statistics—Theory and Methods*, **32**, 851-874.

7. Feddag, M. L. and Mesbah, M. (2002). Estimating Equations for Parameters of Longitudinal Mixed Rasch Model, In *Abstract book of the 2nd Euro-Japanese Workshop on Stochastic Risk Modelling for Finance, Insurance, Production and Reliability*, pp. 91–100, Chamonix, France.

8. Feddag, M. L. and Mesbah, M. (2002). Generalized estimating equations for longitudinal mixed Rasch model, *Submitted for publication.*

9. Fischer, G. H. and Molenaar, I. W. (1995). *Rasch Models: Foundations, Recent Developments and Applications*, Springer-Verlag, New York.

10. Gilks, W. R. and Wild, P. (1992). Adaptative rejection sampling for Gibbs sampling, *Applied Statistics*, **4**, 337–348.

11. Gilmour, A. R., Anderson, R. D. and Rae, A. R. (1985). The analysis of binomial data by a generalized linear mixed models, *Biometrika*, **72**, 593–599.

12. Johnson, N. L., Kotz, S. and Balakrishnan, N. (1995). *Continuous Univariate Distributions – Vol. 2*, Second edition, John Wiley & Sons, New York.

13. Lavergne, C. and Trottier, C. (2000). Sur l'estimation dans les modles linaires gnraliss effets alatoires, *Revue de Statistique Applique*, **48**, 49–67.

14. Liang, K. Y. and Zeger, S. L. (1986). Longitudinal data analysis using generalized linear models, *Biometrika*, **73**, 13–22.

15. Lin, N. and Breslow, N. E. (1996). Bias correction in generalized linear mixed models with a multiple components of dispersion, *Journal of the American Statistical Association*, **91**, 1007–1016.

16. McCullagh, P. and Nelder, J. A. (1989). *Generalized Linear Models*, Second edition, Chapman & Hall, London, England.

17. Piessens, R., deDoncker-Kapenga, E., Uberhuber, C. and Kahaner, D. (1983). *Quadpack : A Subroutine Package for Automatic Integration*, Springer-Verlag, Berlin, Germany.

18. Pinheiro, J. C. and Bates, D. M. (2000). *Mixed Effects Models in S and S-Plus*, Springer-Verlag, New-York.

19. Prentice, R. L. (1988). Correlated binary regression with covariates specific to each binary observations, *Biometrics*, **44**, 1033–1048.

20. Prentice, R. L. and Zhao, L. P. (1991). Estimating equation for parameters in means and covariances of multivariate discrete and continuous responses, *Biometrics*, **47**, 825–839.

21. Schall, R. (1991). Estimation in generalized linear mixed models with random effects, *Biometrika*, **78**, 719–727.

22. Sutradhar, B. C. and Rao, R. P. (2001). On marginal quasi-likelihood inference in generalized linear mixed models, *Journal of Multivariate Analysis*, **76**, 1–34.

23. Sutradhar, B. C. and Sinha, S. K. (2002). On pseudo-likelihood inference in the binary longitudinal mixed model, *Communications in Statistics—Theory and Methods*, **31**, 397–417.

24. Wedderburn, R. W. M. (1974). Quasi-likelihood function, generalized linear models and the Gauss-Newton method, *Biometrika*, **48**, 439–447.

25. Zeger, S. L., Liang, K. Y. and Albert, P. S. (1988). Models for longitudinal data: A generalized estimating equation approach, *Biometrics*, **44**, 1049–1060.

26. Zhao, L. P. and Prentice, R. L. (1990). Correlated binary regression using a quadratic exponential model, *Biometrika*, **77**, 642–648.

PART XI

Breast Cancer Studies

24

Breast Cancer Prognosis Using Survival Forests

Thu M. Hoàng[1] **and Van L. Parsons**[2]

[1] *Université René Descartes, Paris, France*
[2] *National Center for Health Statistics, Hyattsville, Maryland, U.S.A.*

Abstract: Combinations of survival regression trees called survival forests (SF) have been proposed by Breiman to estimate survival functions and assess variable importance. We examine some operating characteristics of the method on simulated data, and we apply it to breast cancer data for prognosis.

Keywords and phrases: Survival forests, prognosis, Bayesian neural networks, breast cancer

24.1 Introduction

Following Cox's seminal paper (1972) the semiparametric proportional hazards model has become extremely popular for the analysis of right censored survival data. Violations of the proportional hazards assumption in certain applications has led to alternative approaches such as tree-structured survival analysis [Leblanc and Crowley (1992), Intrator and Kooperberg (1995), Keles and Segal (2002) and Xu and Adak (2002)], neural networks models [Boracchi et al. (2001), Ripley and Ripley (2001) and Trinh et al. (2002, 2003)], or both [Hoàng et al. (2002)]. Recently, Breiman (2002 Wald lecture) has suggested combinations of survival regression trees called survival forests (SF) for estimating survival functions and assessing variable importance. Here, a survival tree is constructed using a bootstrap sample, and an ensemble of trees is created over many such samples. In view of prognosis, a new case can be run through the ensemble of trees to obtain an averaged survival function, which in turn can be used to produce survival prediction. In this paper we examine some of the operating characteristics of the survival forests method on both simulated and real data. In particular for prognostic prediction of advanced breast cancer we

[1]Research results and conclusions expressed in this paper are those of the authors and do not necessarily indicate concurrence by the National Center for Health Statistics.

apply SF to survival data on 1208 patients with metastases followed up at the Institut Curie in France.

24.2 Survival Forests

Given a random survival time, $T(\mathbf{x})$, depending on a covariate vector $\mathbf{x}$ define the hazard function, $h(t,\mathbf{x})$, and the survival function, $S(t,\mathbf{x})$, by

$$\begin{aligned} h(t,\mathbf{x}) &= P(T(\mathbf{x}) \in (t, t+dt)|T(\mathbf{x}) > t)/dt, \\ S(t,\mathbf{x}) &= P(T(\mathbf{x}) \geq t) = \exp\left(-\int_0^t h(\tau,\mathbf{x})d\tau\right), \\ &\equiv \exp(-H(t,\mathbf{x})). \end{aligned}$$

An experiment consists of N independent observations of survival times subject to random independent right censoring. Such a sample will be represented as $(\mathbf{x}_i, t(\mathbf{x}_i), c_i), i = 1, 2, ..., N$ where $c_i = 1$ (0) if the survival outcome is observed (censored). To grow a survival tree consider maximizing the loglikelihood

$$\ell = \sum c_i \log h(t_i, \mathbf{x}_i) - \sum \int_0^{t_i} h(\tau, \mathbf{x}_i) d\tau$$

for fixed covariate effects and partitioning the product time-covariate space into L rectangles $r_l = I_l \otimes R_l$ where I_l is a time interval and R_l a rectangle in the covariate space, for $l = 1, 2, \ldots L$. These rectangles are the nodes of the tree. The hazard function is assumed to be stepwise constant

$$h(t,\mathbf{x}) = \exp\left(\sum_l 1_{r_l}(t,\mathbf{x})\alpha_l\right)$$

where $1_{r_l}(t,\mathbf{x}) = 1$ if $(t,\mathbf{x}) \in r_l$, and 0 otherwise. Maximizing over the α_l yields

$$\alpha_l = \log \frac{D_l}{T_l} \text{ and } \ell = \sum_l D_l \log \frac{D_l}{T_l} - \sum_l D_l$$

where $D_l = \sum_i c_i 1_{r_l}(t, \mathbf{x}_i)$ is the number of events, e.g., deaths, in the node l, and $T_l = \sum_i 1_{R_l}(\mathbf{x}_i) \cdot \text{length}[(0, t_i) \cap I_l]$.

The tree is grown so that each node l is split which most increases $\sum_l D_l \log \frac{D_l}{T_l}$. However, estimates based upon one tree may be unstable [e.g., see Hoàng et al. (2002)]. Recent experience in combining or integrating an ensemble of models computed by the same or different algorithms over multiple data subsets are shown to improve efficiency by executing the estimation processes in parallel and on (possibly disjoint) subsets of the data [Breiman (1996a) and Leblanc and Tibshirani (1993)]. Breiman (1996b) introduced the ensemble technique for classification, based on bootstrap, called bagging (**b**ootstrap **agg**regat**ing**), which has become a successful statistical tool. Here similarly, Breiman proposes to use a bootstrap sample to create each survival tree and then average over many such trees, hence the name forest, in the hope of reducing variability and improving accuracy.

A bootstrap sample of size N from the original N data points is selected and a survival tree is grown until each terminal node has exactly one uncensored event. For each observation not used in the bootstrap sample, a so-called out-of-bag case (oob), its covariate $\mathbf{x}_{\text{oob}}$ is put through this single tree to get an estimate of $S(t, \mathbf{x})$. A bootstrap sample serves as a training set, and training over many bootstrap samples obtains the corresponding trees. The averaged survival function estimated for all the oob cases is de facto a test set estimate. If an independent test set is also available, for each new observation, its covariate can be put through all the trees to get an estimate of the individual survival curve.

24.3 Choosing and Tuning SF Parameters

Here, we focus upon the practical implementation of the survival forests algorithm. While the concept of survival forests seems promising, the available software is still somewhat coarse (Fortran code generated from Breiman's original), and the experience for using it limited. At this time we view survival forests as an exploratory tool, and we discuss some issues for which the practitioner should be aware in their applications. Preliminary results indicate that there are several SF parameters whose selection influences the outcome.

Selecting the number of time points, K, on the survival curves. The same time points are used for each curve, so a sufficient number must be selected to adequately represent the dispersion of survival times for each covariate $\mathbf{x}$ in the data. Breiman suggested selecting uniformly over the order statistics of noncensored survival times. This seems to work well. As the number of points increases, so do the required computer resources. In simulations where some survival time distributions had little overlap with others, to achieve good results the number of time points had to be close to the number of noncensored observations.

Selecting the estimator of $S(t, \mathbf{x})$. Now, the tree formed by a single bootstrap sample, say, b, will produce an estimator of the cumulative hazard $\hat{H}_b(t, \mathbf{x})$, for the K selected time points and any $\mathbf{x}$ put through the tree. The resulting single sample bootstrap survival estimator is $\hat{S}_b(t, \mathbf{x}) = \exp(-\hat{H}_b(t, \mathbf{x}))$. For B bootstrap trees let $\hat{S}_B(t, \mathbf{x}) = \sum_b \hat{S}_b(t, \mathbf{x})/B$, and $\hat{H}(t, \mathbf{x}) = \sum_b \hat{H}_b(t, \mathbf{x})/B$. A convex combination, $(1 - \lambda(t))\hat{S}_B(t, \mathbf{x}) + \lambda(t) \exp(-\hat{H}(t, \mathbf{x}))$, could be used to produce a tree ensemble estimator of the survival curve. Breiman suggested $\lambda(t) = k(t)/K$, where $k(t)$ is the position of t in relation to the K ordered time points, considering that this will provide the most accurate estimator of the cross-section of the survival curves at a given time point. We refer to this estimator as $\hat{S}_A(t, \mathbf{x})$ and discuss it later.

Probability p_s to choose between splitting the covariate or the time variable. In the tree building process, we need to specify the probability p_s to choose between splitting along the covariate or along the time. Breiman suggested 0.50 and 0.75 for the target estimators $\hat{S}_B(t, \mathbf{x})$ and $\hat{S}_A(t, \mathbf{x})$ respectively. Note that if $p_s = 1.0$, the covariates are ignored and the data are treated as a sample from a common distribution. The tree building then optimizes on time alone and the estimators $\hat{S}_B(t, \mathbf{x})$ tend to a commonly defined curve. At the other extreme, if $p_s = 0.0$ the tree building optimizes on the covariates $\mathbf{x}$ alone. The resulting $\hat{S}_B(t, \mathbf{x})$ curves tend to be diverse, and poor predictive properties observed. Experience with real data suggests that a robust interval for p_s can be found by experimentation.

Numerical issues in optimizing the likelihood. Recall that a node l is split to increase $D_l \log(D_l/T_l)$. A small value of T_l may lead to a large single bootstrap estimate of the cumulative hazard, and the average $\hat{H}(t, \mathbf{x})$ may be strongly influenced by relatively few observations. If the estimator $\hat{H}(t, \mathbf{x})$ is too large at some t, then the curve estimator $\hat{S}_A(t, \mathbf{x})$ may appear to have too short a tail. The estimator $\hat{S}_B(t, \mathbf{x})$ seems more robust. This issue is highly data dependent and experimentation with bounds for such extreme conditions and estimation consequences should be considered. Perhaps because of the discretized nature of real data, this topic would be more of an issue for simulated data than real data.

24.4 Assessing the Fit by SF

From the discussion of SF parameters above, it seems clear that many survival forests can be fit to a given data set. Unlike the classical modelling approach, assessment of the quality of fit does not rely on significance tests but takes the following steps: (i) compute $\hat{S}_g(t, \mathbf{x})$ on tree ensembles for different parameters, say $g \in G$, (ii) for each g compute predictive functions of interest, e.g., means, median, probability of $t(\mathbf{x})$ exceeding some specified threshold, on the estimated

g-defined curve, (iii) choose the SF parameters on the basis of goodness-of-fit statistics, and (iv) test on an independent set. Some quality of fit related criteria are discussed below.

Cox–Snell-type residuals. General curve assessment uses the estimated Cox–Snell-type residuals, $\hat{H}_g(T(\mathbf{x}), \mathbf{x})$ where $T(\mathbf{x})$ is the observed survival time. Under ideal conditions, these residuals should exhibit similar behavior to that of a sample from a censored exponential(1) distribution. Now, if $\hat{H}_{\mathrm{NA}}(t)$ is the Nelson–Aalen estimator of cumulative hazard, then it should be an estimate of t when exponential(1) survival times are the input. This suggests the statistics

$$\sum_{i(\mathbf{x})=1}^{N} |\hat{H}_{\mathrm{NA}}(\hat{H}_g(T(\mathbf{x}), \mathbf{x})) - \hat{H}_g(T(\mathbf{x}), \mathbf{x})|/N$$

can be compared for different SF parameters g to determine consistency with exponential(1) residuals. Similarly, we can do assessment on $\hat{S}_g(T(\mathbf{x}), \mathbf{x})$, treating the statistics as censored uniform variables and use the Kaplan–Meier statistic, $\hat{S}_{\mathrm{KM}}(t)$, to compare.

Martingale residuals. Averages of the censoring variable, $c(x)$, minus the Cox–Snell residual, $c(\mathbf{x}) - \hat{H}_g(T(\mathbf{x}), \mathbf{x})$, are compared for different g parameters. Negative values often result when the estimated cumulative hazard, $\hat{H}_g(t, \mathbf{x})$ is too large, and positive values result when it is too small.

Harrell c-index. The c-index [Harrell, Lee and Mark (1996)] is the proportion of predictions that are concordant out of all pairs of observations for which ordering of the survival times can be determined. This measure uses both non-censored and censored survival times. It assumes values between 0.0 and 1.0, and the constant predictor has c-index = 0.5. A value of the c-index near 0.5 means that the model is not predictive while a value of c-index near 1 indicates that the model is highly predictive. Each curve may produce several predictors, e.g., means, medians, for which a c-index can be computed. Predictors with a large c-index may be used to order the underlying populations.

Accuracy, sensitivity and specificity Here, we select a target time t_0 of interest and for each i estimate the probability $P(T(\mathbf{x}_i) \geq t_0)$ by $\hat{S}_g(t_0, \mathbf{x}_i)$. Assessment of prediction may be based on the following measures (Ripley and Ripley, 2001). Let p represent some threshold, $A^+ = \{i : \hat{S}_g(t_0, \mathbf{x}_i) > p\}$, and $A^- = \{i : \hat{S}_g(t_0, \mathbf{x}_i) < p\}$. Define

$$\text{Accuracy} = \frac{\sum_{i \in A^+} q_i + \sum_{i \in A^-} (1 - q_i)}{N},$$
$$\text{Specificity} = \frac{\sum_{i \in A^-} (1 - q_i)}{\sum_i (1 - q_i)},$$
$$\text{Sensitivity} = \frac{\sum_{i \in A^+} q_i}{\sum_i q_i}$$

where $q_i = 1(0)$ if $t(\mathbf{x}_i) \geq t_0 (< t_0)$ if $t(\mathbf{x}_i)$ is a noncensored response. For censored $t(\mathbf{x}_i) < t_0$, the variable q_i is predicted by $\hat{q}_i = \frac{\hat{S}_g(t_0, \mathbf{x}_i)}{\hat{S}_g(t(\mathbf{x}_i), \mathbf{x}_i)}$, but for censored $t(\mathbf{x}_i) \geq t_0$ we let $\hat{q}_i = 1$.

To keep some degree of balance in the comparisons among different SF parameters we defined p as the general population (not a function of the covariates) Kaplan–Meier estimate of $P(T \geq t_0)$. These measures are only recommended if the estimates are well-calibrated, otherwise the results can be very inaccurate.

Standard measures of loss Here we consider typical absolute and squared error loss functions, and only for the noncensored cases. In the presence of light censoring, this provides a useful comparison.

24.5 Sample Runs Through SF

To consider the feasibility of SF as an exploratory tool, we test on two generated data.

- Siml: 1200 survival times were generated from a proportional hazards model
$$h(t, \mathbf{x}) = \exp(x_1 - 2x_2 + 2x_3)$$
where x_1, x_2, and x_3 were uniform variates. Two additional uniform variables, x_4 and x_5 which had no influence were included as covariates. Censoring was about 16%.
- Sim2: 1200 survival times were generated from a variable $T(\mathbf{x}) = \ln(\text{Uniform}[a, b])/h(\mathbf{x})$ where $a = 50, b = 10^6$,
$$h(\mathbf{x}) = 2x_1 + 2x_2 + .5x_3 + x_1x_2 + x_1x_3 + x_2x_3$$
x_1, x_2, were Bernoulli variables with probabilities .25 and .50 respectively and x_3 a standard normal. About 30% censoring was applied. This is a non-proportional hazards model, with some populations exhibiting little overlap with others.

In both examples training and test sets of size 600 were used. For SF parameters $p_s = 0.50$ and 0.75, the basic bootstrap survival curve $\hat{S}_B(t, \mathbf{x})$ and the convex combination estimator $\hat{S}_A(t, \mathbf{x})$ were computed on the training set and fitted to the test set. The results are displayed in Table 24.1 and Table 24.2 using as predictor the mean, $\hat{m}(\mathbf{x})$, of the estimated survival $\hat{S}_g(t, \mathbf{x})$ for different SF parameters g. Our objective was not to optimize the process, but to examine behavior over a reasonable range of SF parameters. We generated 1000 trees for Sim1 and 300 trees for Sim2.

For Sim1 we found that using more than 100 time points did not improve the goodness of fit, but for Sim2 the severe non-overlap of populations required about 425 time points.

Observations and comments. Looking at the residuals and the true average $S_0(t_0)$ it appears that the estimated average survival curve $\hat{S}_A$ has too light a right tail, possibly for reasons discussed earlier for numerical issues. Note that the c-index for $\hat{S}_A$ is comparable to that of $\hat{S}_B$. It seems that $\hat{S}_A$ needs additional calibration, and consequently the current measures of accuracy, specificity and sensitivity should be interpreted with caution. Also note that in estimating the mean of the curve an extrapolation beyond the last endpoint was used, so this measure may be considered coarse.

The choice of the time-covariate SF parameter p_s is data dependent. If the populations survival times overlap to a high degree, e.g., as in Sim1, $p_s = 0.75$ seems to give somewhat better closeness measures, i.e., absolute and squared error distances, but the accuracy measures appear comparable to that for $p_s = 0.50$. For Sim2, which fit some rather diverse distributions, the results for $p_s = 0.50$ appear superior to those for $p_s = 0.75$. At this time, given the rather limited experience in calibrating the estimates, we feel that practitioners should focus on the estimator $\hat{S}_B$.

As mentioned earlier, the sequential bootstrap in SF embeds into the full training set both training and test sets. This seems to work well in avoiding overfitting. Results not displayed but similar to Tables 24.1 and 24.2 show consistency when using an independent test set. For Sim1 the c-indices were ≈ 0.69 vs 0.68 and the measure $\text{Avg}|T - \hat{m}|/\text{Avg}|T - \bar{\hat{m}}|$ was 0.80 vs 0.81.

24.6 Using SF for Cancer Prognosis

The metastatic breast cancer data. A database was constructed including 1477 patients with metastatic breast cancer from eight randomized trials of anthracycline-based first-line chemotherapy performed at the Institut Curie from 1977 to 1998 [Pierga et al. (2001)] . Protocol entry criteria, pretreat-

Table 24.1: Fitting SF to simulated data Sim1 on the test set

Probability p_s for splitting between t and $\mathbf{x}$	0.75	0.50	0.75	0.50
Survival Curve type	$\hat{S}_B$	$\hat{S}_B$	$\hat{S}_A$	$\hat{S}_A$
$\overline{\hat{m}}$	0.80	0.86	0.61	0.49
Avg $\lvert T - \hat{m}\rvert$	0.47	0.49	0.41	0.39
Avg $\lvert T - \overline{\hat{m}}\rvert$	0.58	0.61	0.48	0.45
$\frac{\text{Avg } \lvert T-\hat{m}\rvert}{\text{Avg } \lvert T-\overline{\hat{m}}\rvert}$	0.81	0.81	0.84	0.88
$\text{Avg} \frac{\lvert T-\hat{m}\rvert}{\hat{m}}$	0.63	0.64	0.73	0.91
Avg $\lvert T - \hat{m}\rvert^2$	0.41	0.49	0.36	0.38
Avg $\lvert T - \overline{\hat{m}}\rvert^2$	0.50	0.53	0.44	0.44
$\frac{\text{Avg } \lvert T-\hat{m}\rvert^2}{\text{Avg } \lvert T-\overline{\hat{m}}\rvert^2}$	0.82	0.91	0.83	0.86
$\text{Avg} \frac{\lvert T-\hat{m}\rvert^2}{\hat{m}^2}$	0.60	0.65	0.98	1.77
Harrell c-index	0.68	0.68	0.67	0.68
Target $t_0 = 1$,				
True Avg $S_0(t_0, \mathbf{x})$	0.26	0.26	0.26	0.26
Avg $\hat{S}(t_0, \mathbf{x})$	0.24	0.25	0.19	0.14
Avg $\lvert\hat{S}(t) - S_0(t)\rvert$	0.07	0.06	0.07	0.09
Accuracy	0.67	0.69	0.73	0.78
Sensitivity	0.76	0.74	0.67	0.53
Specificity	0.64	0.67	0.75	0.86
Residuals				
Avg $\lvert\hat{H}_{\mathrm{NA}}(\hat{H}(T)) - \hat{H}(T)\rvert$	0.08	0.06	0.23	0.60
Avg $\lvert\hat{S}_{\mathrm{KM}}(\hat{S}(T)) - \hat{S}(T)\rvert$	0.03	0.02	0.05	0.10
Avg martingale residual	-0.02	-0.02	-0.23	-0.61

Table 24.2: Fitting SF to simulated data Sim2 on the test set

Probability p_s for splitting between t and $\mathbf{x}$	0.75	0.50	0.75	0.50
Survival Curve type	$\hat{S}_B$	$\hat{S}_B$	$\hat{S}_A$	$\hat{S}_A$
$\overline{\hat{m}}$	6.79	7.71	5.00	5.62
Avg $\|T-\hat{m}\|$	1.76	0.88	1.80	1.40
Avg $\|T-\overline{\hat{m}}\|$	5.54	5.99	4.80	5.03
$\frac{\text{Avg } \|T-\hat{m}}{\text{Avg } \|T-\overline{\hat{m}}\|}$	0.32	0.15	0.37	0.28
Avg$\frac{\|T-\hat{m}\|}{\hat{m}}$	0.49	0.31	0.61	0.53
Avg $\|T-\hat{m}\|^2$	5.90	2.20	9.66	6.11
Avg $\|T-\overline{\hat{m}}\|^2$	41.00	45.81	36.49	37.32
$\frac{\text{Avg } \|T-\hat{m}\|^2}{\text{Avg } \|T-\overline{\hat{m}}\|^2}$	0.14	0.05	0.26	0.16
Avg$\frac{\|T-\hat{m}\|^2}{\hat{m}^2}$	0.39	0.20	1.08	0.66
Harrell c-index	0.91	0.95	0.92	0.95
Target $t_0 = 15$,				
True Avg $S_0(t_0, \mathbf{x})$	0.17	0.17	0.17	0.17
Avg $\hat{S}(t_0, \mathbf{x})$	0.15	0.17	0.11	0.11
Avg $\|\hat{S}(t) - S_0(t)\|$	0.15	0.09	0.15	0.11
Accuracy	0.91	0.93	0.94	0.95
Sensitivity	0.95	0.96	0.83	0.82
Specificity	0.90	0.93	0.96	0.97
Residuals				
Avg $\|\hat{H}_{\text{NA}}(\hat{H}(T)) - \hat{H}(T)\|$	0.22	0.22	0.37	0.48
Avg $\|\hat{S}_{\text{KM}}(\hat{S}(T)) - \hat{S}(T)\|$	0.08	0.09	0.14	0.18
Avg martingale residual	0.07	0.04	-0.37	-0.48

ment evaluation, assessment of response and statistical methods were the same for the trials. We consider the influence of 14 variables on the survival time distribution, including

- four patients characteristics as age, menopausal status and the Karnovsky performance score;
- nine disease characteristics as previous X-ray treatment and adjuvant chemotherapy, the disease free interval, the number and sites of metastases (liver, lung, pleura, bone and skin);
- one chemistry test, the serum lactic dehydrogenase (LDH) rate.

We omit 269 patients with missing values on any of these variables, thus retaining 1208 patients. To keep in line with published results [Pierga et al. (2001)] all variables were dichotomized except the LDH rate (continuous), age (continuous), the grade of nodal involvement, *nuicc* (categorical, three classes), and the number of metastatic sites, *nbs* (count ≤ 5). The two categories of the biphasic disease free interval (*dfi*) were 1 if *dfi* < 24 months or 0 otherwise. Survival was computed as the time in months from randomization until death or the last day of follow-up. In the latter case or in the case of a death by other risk, the survival time is censored.

Fitting SF to the metastatic breast cancer data. For the cancer data the SF splitting parameter value $p_s = .70$ worked reasonably well for the basic bootstrap curve estimator $\hat{S}_B(t, \mathbf{x})$, and $p_s = .75$ was used for the convex combination curve $\hat{S}_A(t, \mathbf{x})$. The results are exhibited in Table 24.3 and Table 24.4. As in the simulation cases, the residuals for $\hat{S}_B$ seem to indicate better fit and the right tail of $\hat{S}_A(t, \mathbf{x})$ may be too light. Predictors were the means and medians of $\hat{S}_A(t, \mathbf{x})$ and $\hat{S}_B$, along with the averaged median of $\hat{S}_b$. It should be noted that the Harrell c-index was about 0.68 in all cases. Depending upon the selected loss function for assessing goodness of fit, we will have different minimizing curves and predictors. The average median for $\hat{S}_b$ would seem a good compromise. As a comparison we obtained the training and test sets for this data that were used in developing a Bayesian neural network (BNN) predictor of the survival time T [Trinh et al. (2002, 2003)] evaluated with the loss function

$$\frac{\text{Avg}|T - \hat{m}|}{\text{Avg}|T - \bar{\hat{m}}|}.$$

Assessment of the provided BNN predictor showed a c-index of 0.70 and a loss value of 0.84. While the BNN predictor may appear to be slightly superior, it should be kept in mind that the survival forests predictors were basic estimators with no attempt to optimize any loss function. Reasonable SF estimators seem to be within $2\% - 3\%$ of a near optimal loss. With calibration toward a specific loss function, one may see closer agreement to the neural network optimum.

The SF method produces a survival curve for every individual whose $\mathbf{x}$ covariate is run through the SF. The set of covariates and curves $(\mathbf{x}, \hat{S}(t, \mathbf{x}))$ can be further analyzed to determine covariate importance and used to form prognostic groups. For the former, Breiman suggests graphing the estimators $\widehat{\text{Corr}}(x_p, -\log(\hat{S}(t, \mathbf{x})))$ as a function of t for each component x_j of the covariate $\mathbf{x}$, and performing an ordinary least squares regression of $-\log(\hat{S}(t, \mathbf{x}))$ on the set of covariates $\mathbf{x}$ for each time t. These plots displayed in Figure 24.1 and Figure 24.2 show time-varying effects from 0 to approximately 5 years (60 months) for at least the five most important variables (*LDH*, liver metastasis, disease free interval *dfi*, Karnovsky score of general status *kar*, and nodal involvement *nuicc*). It is of interest to note that the initial influence of the LDH rate and the Karnovsky score appears to decrease and level after four years while the other three variables appear to have increasing influence up to four or five years.

Table 24.3: Comparing Bayesian neural network to SF for cancer data

	Bayesian NN predictor	Mean $\hat{S}_B$	Median $\hat{S}_B$	Bootstrap Avg median $\hat{S}_b$	mean $\hat{S}_A$	Median $\hat{S}_A$
$\overline{\hat{m}}$	28.01	36.63	25.22	28.15	29.48	23.90
Avg $\lvert T - \hat{m}\rvert$	15.90	18.74	16.10	16.37	16.65	16.11
Avg $\lvert T - \overline{\hat{m}}\rvert$	18.92	21.70	18.52	18.94	19.24	18.44
$\frac{\text{Avg } \lvert T-\hat{m}\rvert}{\text{Avg } \lvert T-\overline{\hat{m}}\rvert}$	0.84	0.86	0.87	0.86	0.87	0.87
$\text{Avg} \frac{\lvert T-\hat{m}\rvert}{\hat{m}}$	0.65	0.56	0.69	0.62	0.60	0.71
Avg $\lvert T - \hat{m}\rvert^2$	613	683	647	625	624	666
Avg $\lvert T - \overline{\hat{m}}\rvert^2$	740	782	758	739	737	772
$\frac{\text{Avg } \lvert T-\hat{m}\rvert^2}{\text{Avg } \lvert T-\overline{\hat{m}}\rvert^2}$	0.83	0.87	0.85	0.85	0.85	0.86
$\text{Avg} \frac{\lvert T-\hat{m}\rvert^2}{\hat{m}^2}$	1.04	0.58	1.14	0.87	0.79	1.22
Harrell c-index	.70	0.68	0.68	0.68	0.68	0.68

Table 24.4: SF prediction for cancer data

	$\hat{S}_B$	$\hat{S}_A$
Avg$\hat{S}(t_0, x)$	0.15	0.11
Accuracy	0.68	0.76
Sensitivity	0.70	0.56
Specificity	0.68	0.80
Avg $\lvert\hat{H}_{NA}(\hat{H}(T)) - \hat{H}(T)\rvert$	0.08	0.30
Avg $\lvert\hat{S}_{KM}(\hat{S}(T)) - \hat{S}(T)\rvert$	0.02	0.05
Avg martingale residuals	-0.07	-0.30

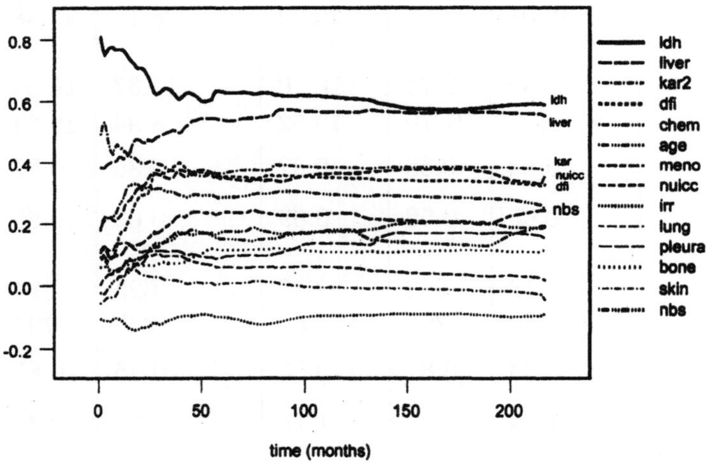

Figure 24.1: Correlations. Metastatic cancer data

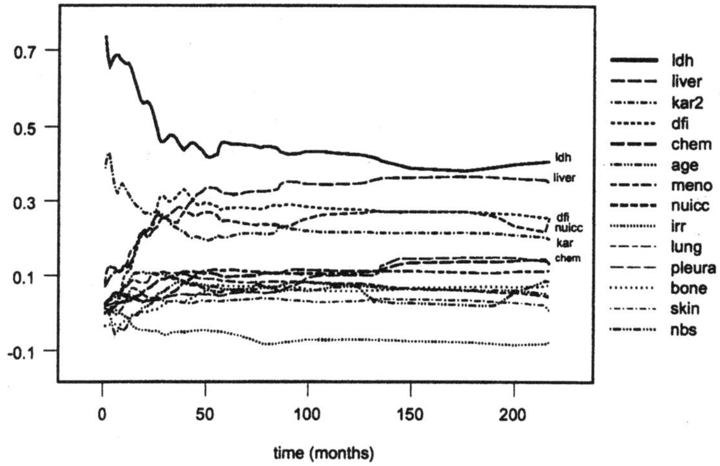

Figure 24.2: Coefficients traces. Metastatic cancer data

References

1. Boracchi P., Biganzoli E. and Marubini E. (2001). Modelling cause-specific hazards with radial basis function artificial neural networks: Application to 2233 breast cancer patients, *Statistics in Medicine*, **20**, 3677–3694.

2. Breiman, L. (1996a). Stacked regressions, *Machine Learning*, **24**, 41–48.

3. Breiman, L. (1996b). Bagging predictors, *Machine Learning*, **24**, 123–140.

4. Breiman, L. (2002). Wald III Lecture: Software for the masses, *Lecture notes available at* `http://stat-www.berkeley.edu/users/breiman/wald2002-3.pdf`.

5. Cox D. R. (1972). Regression model and life tables (with discussion), *J. Royal Statistical Society, Series B*, **34**, 187–220.

6. Harrell F. E., Lee K. L. and Mark D. B. (1996). Multivariable prognostic models: Issues in developing models, evaluating assumptions and adequacy, and measuring and reducing errors, *Statistics in Medicine*, **15**, 361–387.

7. Hoàng, M. T., Trinh, Q. A. and Asselain, B. (2002). Construction and validation of a prognostic model for metastatic breast cancer using Bayesian neural network and regression tree, *Intelligent Data Analysis in Medicine and Pharmacology, Workshop Notes*, 37–43, `http://www.cs.uu.nl/ lucas/idamap2002/idamap2002-proc.pdf`.

8. Intrator, O. and Kooperberg, C. (1995). Trees and splines in survival analysis, *Statistical Methods in Medical Research*, **4**, 237–261.

9. Keles, S. and Segal, M. R. (2002). Residual-based tree-structured survival analysis, *Statistics in Medicine*, **21**, 313–326.

10. LeBlanc, M. and Crowley, R. (1992). Survival trees by goodness of fit, *Journal of the American Statistical Association*, **88**, 457–467.

11. LeBlanc, M. and Tibshirani, R. (1993). Combining estimates in regression and classification, *Technical Report 9318*, Department of Statistics, University of Toronto, Toronto, Ontario, Canada.

12. Pierga, J. Y. et al. (2001). Effect of adjuvant chemotherapy on outcome in patients with metastatic breast carcinoma treated with first-line Doxorubicin-containing chemotherapy, *Cancer*, **91**, 1079–1089.

13. Ripley, B. D. and Ripley, R. M. (2001). Neural networks as statistical methods in survival analysis, In *Artificial Neural Networks: Prospects for Medicine* (Eds., R. Dybowski and V. Gant), pp. 237–255, Cambridge University Press, Cambridge, England.

14. Trinh, Q. A., Hoàng, M. T. and Asselain, B. (2002). Neural network models for survival analysis using backpropagation or Bayesian learning, In *STAIRS 2002* (Eds., T. Vidal and P. Liberatore), pp. 167–176, IOS Press, Amsterdam, The Netherlands.

15. Trinh, Q. A., Hoàng, M. T., Dorizzi, B. and Asselain, B. (2003). The effect of sample size and architecture on Bayesian learning of cancer prognosis, In *Medical Informatics Europe* (Eds., P. Lebeux et al.), IOS Press, Amsterdam, The Netherlands (to appear).

16. Xu, R. and Adak, S. (2002). Survival analysis wih time-varying regression effects using a tree-based approach, *Biometrcs*, **58**, 305–315.

25

Semiparametric Versus Parametric Regression Analysis Based on the Bounded Cumulative Hazard Model: An Application to Breast Cancer Recurrence

Kenneth M. Boucher[1], **Bernard Asselain**[2], **Alexander D. Tsodikov**[1], **and Andrei Y. Yakovlev**[3]

[1] *Huntsman Cancer Institute, University of Utah, Salt Lake City, Utah, U.S.A.*
[2] *Biostatistical Unit, Institut Curie, Paris, France*
[3] *Department of Biostatistics and Computational Biology, University of Rochester, Rochester, New York, U.S.A.*

Abstract: The bounded cumulative hazard model is a generalization of the proportional hazard model with several distinct advantages: (1) It has superior flexibility, allowing a fit to data where the proportional hazards assumption does not apply; (2) It has a form that is suitable for semiparametric inference; and (3) It may offer an interpretation in terms of biologically meaningful parameters. In this paper the bounded cumulative hazard model is discussed. Several versions of the model are applied to breast cancer recurrence data from the Curie Institute.

Keywords and phrases: Cure models, survival models, bounded cumulative hazard, semiparametric inference, parametric models, breast cancer

25.1 Introduction

In many clinical and epidemiological settings, investigators encounter cause-specific survival curves that tend to level off at a value strictly greater than zero as time increases. This plateau may be taken as an indication of the presence of a proportion of patients for whom the disease under study will never recur. One can consider such patients to be effectively cured. The probability of (biological) cure, variously referred to as the cure rate or the surviving fraction, is defined as an asymptotic value of the survival function $\overline{G}(t)$ as t tends to infinity. Let X be the survival time with cumulative distribution function (c.d.f.) $G(t) = 1-\bar{G}(t)$.

The existence of a non-zero surviving fraction, p, is determined by the behavior of the hazard function, $\lambda(t)$, by virtue of the equality

$$p = \lim_{t \to \infty} \overline{G}(t) = \exp\left\{ - \int_0^\infty \lambda(u) du \right\}. \tag{25.1}$$

Whenever $p > 0$ (the integral in (25.1) converges), the underlying survival time distribution is said to be improper. Clearly, $\lambda(u) \to 0$ as $u \to \infty$ if $p > 0$ and the limit of $\lambda(u)$ (as $u \to \infty$) exists.

Boag (1949) and later Berkson and Gage (1952) proposed a two-component (binary) mixture model for the analysis of survival data when a proportion of patients is cured. Since then, the binary mixture-based approach has become the dominant one in the literature on cure models [Miller (1981) and Maller and Zhou (1996)]. The main idea behind this approach is that any improper survival function can be represented as

$$\bar{G}(t) = \mathrm{E}\left\{ [\bar{G}_0(t)]^M \right\} = p + (1-p)\bar{G}_0(t), \tag{25.2}$$

where M is a binary random variable taking on the values of 0 and 1 with probability p and $1 - p$, respectively, with

$$p = \Pr\{X = \infty\},$$

and $\bar{G}_0(t)$ is defined as the survival function for the time to failure conditional upon ultimate failure, i.e.,

$$\bar{G}_0(t) = \Pr\{X \geq t | X < \infty\}. \tag{25.3}$$

An alternative, but equally general, representation of an improper survival time distribution can be obtained by assuming that the cumulative hazard $\Lambda(t) = \int_0^t \lambda(t) dt$ has a finite positive limit, say θ, as t tends to infinity. In this case, one can write

$$\bar{G}(t) = e^{-\theta F(t)}, \quad \theta > 0, \quad t \geq 0, \tag{25.4}$$

where $F(t) = \Lambda(t)/\theta$ is the c.d.f. of some non-negative random variable such that $F(0) = 0$. In what follows, we will call the model given by (25.4) the bounded cumulative hazard (BCH) model.

Clearly, estimating the proportion of cured patients may have important medical implications. In addition, clinical covariates may exert dissimilar effects on the probability of cure and the timing of tumor relapse or other events of interest. There are at least two advantages of the BCH model: (1) they enrich our ability to interpret survival analysis in terms of characteristics that have a clear biomedical meaning; (2) they lead to more general regression models, thereby extending our ability to describe actual data. It took a long time to realize these virtues of the BCH model. The first move in this direction was due to Haybittle (1959, 1965). The author proceeded from the observation that in some clinical data on cancer survival, an actuarial estimate of the hazard function tends to decrease exponentially with time. If the same property holds for the true hazard, expression (25.4) assumes the form

$$\bar{G}(t) = e^{-\theta(1-e^{-\alpha t})}, \quad \alpha > 0. \tag{25.5}$$

A comprehensive treatment of this model was given by Cantor and Shuster (1992). Clearly, the Gompertz-like model given by formula (25.5) is a special case of formula (25.4) with the function $F(t)$ specified by an exponential c.d.f. with parameter α.

The representation (25.4) for the BCH model was first introduced in the paper of Yakovlev et al. (1993) and discussed later as an alternative to the binary mixture model by Yakovlev (1994). Interestingly enough, Yakovlev et al. (1993) proceeded from purely biological considerations; the idea of imposing a constraint on the behavior of the hazard function was introduced much later in Tsodikov et al. (1998). In fact, the authors proposed a simple mechanistically motivated model of tumor recurrence yielding an improper survival time distribution. Under this model, the probability of tumor cure is defined as the probability of no clonogenic tumor cells surviving by the end of treatment. A comprehensive account of the BCH model and associated statistical methods is given by Tsodikov, Ibrahim, and Yakovlev (2003).

In this paper, we discuss methods of semiparametric inference based on the BCH model that has been recently developed. We then illustrate the method by applying several (semiparametric and parametric) versions of the BCH model to breast cancer recurrence data from the Curie Institute.

25.2 BCH Regression Models

25.2.1 Mixture models and generalizations

In this subsection we discuss how certain BCM models, and in particular, the double proportional hazards model, arise naturally within a more general class of regression models, called Nonlinear Transformation Models.

Perhaps the simplest bounded cumulative hazard model is obtained by making the assumption that the cumulative hazard is bounded in a proportional hazards (PH) model. In this way we obtain the so-called improper PH model

$$\bar{G}(t|z) = \exp\{-\theta(\beta, z)F(t)\}, \tag{25.6}$$

where β is a vector of regression coefficients, and z is a vector of covariates, and $-\theta(\beta, z)$ is a fixed (known) function relating β to z. A general class of regression models, Nonlinear Transformation Models (NTM), was proposed in Tsodikov (2002, 2003), in the context of semiparametric models:

$$\bar{G}(t|z) = \gamma(\bar{F}(t)|\beta, z), \tag{25.7}$$

where $\gamma(x|\beta, z)$ is some parametrically specified cumulative distribution function in x with support on $[0, 1]$. In the examples we assume γ is parametrized through a set of parameters/predictors $\theta, \eta, ...$, where each predictor is further parametrized using (generally different) sets of regression coefficients $\beta_1, \beta_2, ...$, so that $\theta = \exp\{\beta_1' z\}, \eta = \exp\{\beta_2' z\}$. Linear transformation models considered by Cheng et al. (1995) represent a subclass of NTM. In order to make (25.7) a BCH model, the following assumptions are made to enforce the limit $\bar{G}(t|\beta, z) \to p > 0$:

$$\gamma(0|\beta, z) > 0, \ \lim_{t\to\infty} \bar{F}(t) = 0. \tag{25.8}$$

The restriction $\bar{F}$(last failure) $= 0$ proposed by Taylor (1995) in the context of the two-component mixture model removes the over-parameterization of the description of the baseline cure rate through F and h. Following the restriction, the estimate $\hat{\bar{F}}$ is assumed to satisfy $\hat{\bar{F}}$(last failure) $= 0$. Moreover, it is necessary for separation of the long- and short-term covariate effects on survival.

Alternatively, a BCH model can be formulated as a two-stage model. First, an unobservable random variable M is postulated with distribution $p(m|z)$ that depends on covariates. Second, the observed survival function is formed as

$$\bar{G}(t|\beta, z) = \mathrm{E}\left[\bar{F}(t)^M \middle| z\right], \tag{25.9}$$

where the expectation is taken with respect to M given z, and the distribution of M depends on β and z . The model (25.9) represents a generalized PH frailty model, which we simply call the PH mixture model [Tsodikov (2002, 2003)], where M is assumed to be an arbitrary non-negative random variable. In this context, M can be interpreted as a missing covariate in a PH model. Since all models considered previously in this paper are particular cases of (25.9), an estimation procedure designed for the PH mixture model represents a universal tool for statistical inference with such models.

The discrete mixture model (i.e., (25.9) with a discrete random variable M) can be linked to the NTM class. Observe that the probability generating function of a discrete random variable M,

$$\varphi_M(x) = \sum_{m=0}^{\infty} p_m x^m,$$

is a distribution function in x with the support [0,1]. Indeed, since p_m is non-negative, $\varphi_M(x)$ is increasing in x. Also, $\varphi_M(1) = \sum_{m=0}^{\infty} p_m = 1$. If M is discrete and depends on covariates, (25.9) can be rewritten as an NTM with

$$\gamma(x|z) = \varphi_M(x|z).$$

With the discrete mixture model, the probability of cure is given by $\gamma(0|z) = \varphi_M(0|\beta, z) = p_0(\beta, z)$.

We can now represent both mixture and BCH models discussed earlier in this paper as members of the Mixture - NTM model family. Consider an extension of the improper PH model allowing for dissimilar covariate effects on long- and short-term survival. To construct an extended hazard model [the term was introduced by Etezadi-Amoli and Ciampi (1987)], we employ the fact that $\bar{F}$ is a survival function. Incorporating covariates into $\bar{F}$, we can add a short-term effect to the improper PH model. The class of extended hazard models

$$\bar{G}(t|\beta, z) = \exp\left\{-\theta(\beta_1, z)[1 - \tilde{\gamma}(\bar{F}(t)|\beta_2, z)]\right\}, \tag{25.10}$$

where $\tilde{\gamma}$ is an NTM, $\beta = (\beta_1, \beta_2)$, $\beta_i = (\beta_{0i}, \beta_{1i}, ...), i = 1, 2$ was introduced in Tsodikov (2002). If $\tilde{\gamma}$ is a mixture model itself, then (25.10) is also a mixture model with $\gamma = \exp\{-\theta[1 - \tilde{\gamma}]\}$. The mixing variable M that generates the family (25.10) has a well defined structure of the compound Poisson variable

$$M = \sum_{k=1}^{\nu} \xi_k, \tag{25.11}$$

where ν is a Poisson random variable, and ξ_k are i.i.d. copies of a random variable ξ with Laplace transform $\tilde{\gamma}(e^{-\lambda})$. The binary distribution for ν gives rise to the two-component mixture class of models

$$\bar{G}(t|\beta, z) = p(\beta_1, z) + \bar{p}(\beta_1, z)\tilde{\gamma}[\bar{F}(t)|\beta_2, z], \bar{p} = 1 - p. \tag{25.12}$$

Kuk and Chen (1992) proposed a regression model in the form of (25.12) with p being a logistic regression, and $\tilde{\gamma}$ a proportional hazards regression. This model was further studied by Sy and Taylor (2000) and Peng and Dear (2000). The Poisson distribution for ν leads to the bounded hazard family of mixture models. For example, let ξ be degenerate (nonrandom): $\Pr(\xi = \eta(z)) = 1$. Then $M = \eta(\beta_2, z)\nu$, where ν is Poisson with expectation $\theta(z)$. With the parametrization $\theta(\beta_1, z) = \exp(\beta_1' z)$ and $\eta(\beta_2, z) = \exp(\beta_2' z)$, we have the PHPH model

$$G(t|\beta, z) = \exp\left[-\exp(\beta_1' z)\left\{1 - \bar{F}^{\exp(\beta_2' z)}\right\}\right]. \tag{25.13}$$

The model (25.13) was proposed by Broët et al. (2001) in the context of two sample score tests for long- and short-term covariate effects. An intercept term (log(cure rate)) is included in β_1 but not β_2. With β_{20} set to zero, β_{10} may be interpreted as the baseline log cure rate. With the exponential parametrization of predictors (25.13), the baseline survival function takes the form $G_b(t|\beta_0, z) = \exp\left[-\exp(\beta_{10} z)\left\{1 - \tilde{\gamma}(\bar{F}|\mathbf{0})\right\}\right]$, where $\mathbf{0}$=(0, 0, ...) and $\beta_0 = ((\beta_{10}, 0, 0, ...,), \mathbf{0})$. As in the Cox model [Cox (1972)], the regression coefficient β_{ij}, $i = 1, 2$, $j = 1, 2, ...$ correspond to the relative effects of the covariate z_{ij} on the long- or short-term predictor, $i = 1, 2$, respectively.

At times an extension of (25.7) may be convenient. For brevity, we describe the extension only for the case we need later. Let $\bar{F}_1$ and $\bar{F}_2$ be two (usually parametrically specified) survival functions. Consider the model

$$G(t|\beta, z) = \exp\left[-\exp(\beta_1' z)\left\{1 - \bar{F}_1\right\} - \exp(\beta_2' z)\left\{1 - \bar{F}_2\right\}\right]. \tag{25.14}$$

Models of this type are called competing risks models, since F_1 and F_2 may clearly be interpreted as the c.d.f's for competing causes of failure. If we stipulate that the means μ_1 and μ_2 of the corresponding c.d.f's F_1 and F_2 satisfy $\mu_1 < \mu_2$, then θ_1 and θ_2 may be interpreted in terms short-term and long-term effects, respectively. In this way we can form a very different model from the PHPH model with a similar interpretation. Note that if we impose the pair of restrictions $\lim_{t\to\infty} \bar{F}_1(t) = 0$, and $\lim_{t\to\infty} \bar{F}_2(t) = 0$, the cumulative hazard is bounded, so that (25.14) is in the class of models (25.4), with $\theta = \theta_1 + \theta_2$.

25.2.2 An EM-based estimation procedure for the PH mixture model

The issue of potentially unlimited dimension has been the most critical deterrent to the use of maximum likelihood estimation (MLE) in semiparametric regression models. Methods based on partial likelihood are specific to the proportional hazards model, and do not extend to other models. The Newton–Raphson procedure requires taking the inverse of the information matrix, which gets computationally prohibitive and unstable with increasing dimension. Characterizing the future directions in survival analysis in their recent editorial in *Biometrics*, Fleming and Lin (2000) pointed out that "it would be highly useful to develop efficient and reliable numerical algorithms for the semiparametric estimation...".

Recently, Tsodikov (2003) generalized the EM algorithm for the frailty model into a universal 'distribution-free' procedure applicable to the NTM class (25.7). This family of algorithms (Quasi-EM) is a subclass of the so-called MM algorithms based on surrogate objective functions [Lange, Hunter, and Yang (2000)]. Broadly defined, an MM algorithm substitutes a computationally simpler surrogate objective function for the target function on each step of the procedure (similar to the E-step of the EM). Maximizing the surrogate objective function drives the target function in the correct direction. Thus, a difficult maximization procedure is replaced by a simpler one. The idea of the QEM approach is to obtain a surrogate objective function that does not depend on the entire distribution of the missing data, that is, the derivation of a 'distribution-free' E-step for the PH mixture model. The following statement is the key result.

Proposition 25.2.1 (Tsodikov (2003)) *Let τ be the observed event time, and c the observed censoring indicator ($c = 1$ for failures, and $c = 0$ otherwise).*

Under the PH mixture model

$$G(t|\cdot) = \gamma[\bar{F}(t)|\cdot] = E\{[\bar{F}(t)]^{M(\cdot)}\},$$

and if $F'(t) > 0$, *the conditional expectation of* M, *given the observed event, is given by*

$$E\{M(\cdot)|\cdot,\tau,c\} = \Theta[\bar{F}(\tau)|\cdot,c],$$

where

$$\Theta[x|\cdot,c] = c + x\frac{\gamma^{(c+1)}(x|\cdot)}{\gamma^{(c)}(x|\cdot)},$$

and $\gamma^{(i)}(x|\cdot) = \partial^i\gamma(x|\cdot)/\partial x^i$, $i = 1,2$ *and* $\gamma^{(0)}(x|\cdot) = \gamma(x|\cdot)$.

The above statement indicates that the E-step can be constructed using the first two derivatives of the NTM-generating function γ without any knowledge or even existence of the mixing random variable M. Specification of the algorithm for a particular model requires evaluation of Θ for that model.

Cure models require a correction of Θ at the last observation (see 25.16) below). Introduce a set of times t_i, $i = 1,\ldots,n$, arranged in increasing order, where $t_{n+1} = \infty$. Associated with each t_i is a set of individuals $\mathcal{D}_i$ with covariates z_{ij}, $j \in \mathcal{D}_i$ who fail at t_i, and a similar set of individuals $\mathcal{C}_i$ with covariates z_{ij}, $j \in \mathcal{C}_i$ who are censored at t_i.

According to the nonparametric maximum likelihood method, the model can be fitted by maximizing the generalized loglikelihood ℓ with respect to the regression parameters β and unspecified survival function $\bar{F}$ (or the corresponding distribution function F). An argument similar to that adopted by Kalbfleisch and Prentice (1980) can be used to verify that ℓ is maximized by a step-function F with steps at the times of failures. Let τ_k, $k = 1,\ldots,K$ be the time points in ascending order where failures occur ($\mathcal{D}$ is not empty). Denote by r_k the rank of τ_k in the set $\{t_i\}$. The ranks r_k locate the points τ_k on the net $\{t_i\}$: $\tau_k = t_{r_k}$. By definition we set $\tau_{K+1} = n+1$ and $\tau_0 = 0$. For any $A(t)$ let $A_i = A(t_i)$.

Differentiating the generalized loglikelihood with respect to $\Delta H_k = H_k - H_{k-1}$, $k = 1,\ldots,r_{K-1}$, we obtain the score equations for F in the form

$$\Delta H_k = \frac{D_k}{\sum_{ij\in\mathcal{R}_k}\Theta(\bar{F}_i|\beta,z_{ij})}, \tag{25.15}$$

where $\mathcal{R}_k = \bigcup_{i=k}^{n}\mathcal{D}_i \cup \mathcal{C}_i$ is the risk set at t_k, D_k is the number of failures at t_k, and

$$\Theta(\bar{F}_i|\beta,z_{ij}) = \begin{cases} \frac{\gamma'(\bar{F}_i|\beta,z_{ij})}{\gamma(\bar{F}_i|\beta,z_{ij})}\bar{F}_i, & j \in \mathcal{C}_i,\ i \le r_K - 1, \\ 1 + \frac{\gamma''(\bar{F}_i|\beta,z_{ij})}{\gamma'(\bar{F}_i|\beta,z_{ij})}\bar{F}_i, & j \in \mathcal{D}_i,\ i \le r_{K-1}, \\ \frac{\gamma'(\bar{F}_{r_{K-1}}|\beta,z_{ij})}{\gamma(\bar{F}_{r_{K-1}}|\beta,z_{ij}) - \gamma(0|\beta,z_{ij})}\bar{F}_{r_{K-1}}, & j \in \mathcal{D}_{r_K},\ i = r_K, \\ 0, & j \in \mathcal{C}_i,\ i \ge r_K. \end{cases} \tag{25.16}$$

The fact that the contributions of the observations ij to the likelihood as well as Θ at, or after the last failure ($i \geq r_K$), are different from their counterparts before the last failure ($i \leq r_K - 1$) is due to the restriction $\bar{F}_i = 0$, $i = r_K, \ldots, n$ representing the fact that $\bar{F}$ is a proper survival function. The above distinction is not made if the model is a general NTM with unrestricted F.

It is interesting to note that the score equations have the form of the Nelson–Aalen–Breslow estimator for the PH model with the usual predictor replaced by Θ. Given β, iterations with respect to F can be carried out as follows:

qE: With kth-iteration $\bar{F}^{(k)}$, compute $\Theta^{(k)}$ for each subject.

MA: Update F using the Nelson–Aalen–Breslow estimator (25.15),

It can be shown [Tsodikov (2003)] that if $\Theta(x, \cdot)$ is a nondecreasing function of x, or if γ is a PH mixture model (a stronger assumption), then each iteration described above improves the likelihood. Also, this assumption makes the above procedure a member of the MM family (Lange, Hunter, and Yang, 2000), and the convergence properties follow from the general MM theory.

There are many ways to build a particular model fitting algorithm based on the principles above, and this depends on how the maximization with respect to β is incorporated into the procedure. One possibility is to maximize the profile likelihood

$$\ell_{pr}(\beta) = \ell(\beta, \hat{\bar{F}}(\beta)), \tag{25.17}$$

with respect to β, where $\hat{\bar{F}}(\beta)$ is determined at each step of maximization by iterating the steps (qE) and (MA) until convergence. This method was used for the semiparametric model used in the analysis presented below.

25.3 Analysis of Data from the Curie Institute

Two bounded cumulative hazards models, the double proportional hazards (25.13) and the competing risks model (25.14) were applied to data from women diagnosed with primary breast carcinoma between 1981 and 1991 at the Curie Institute. The endpoint of interest was recurrence in the ipsilateral breast. Both models admit interpretation in terms of long and short term effects, so sensitivity to model specification could be explored. In addition we used three versions of the PHPH model, differing in the way $\bar{F}$ was specified, to explore sensitivity to specification of $\bar{F}$.

The general treatment policy for the Curie Institute data was aimed towards breast conservation through a combined application of radiotherapy and limited surgery. A detailed description of subcohorts of these patients is given in Fourquet et al. (1989). The strategy of the analysis was to choose important predictors using backward elimination. Women with missing tumor size, grade, nodal status, estrogen receptor status, progesterone receptor status, or treatment information were excluded from the analysis. Women for whom grade was coded as 'not done' were included however. After these exclusions, a sample of 6232 women out of an initial total of 9899 was available for analysis. The vast majority (3278, or 89%) of the excluded patients had either missing estrogen or progesterone receptor status, or both.

The initial model from which backward elimination proceeded included both long- and short-term effects for all variables. The variables included were: primary tumor size at detection, estrogen and progesterone receptor status, clinical axillary lymph node status (nodal involvement), age, and tumor size, histological grade, and treatment. Estrogen receptor status (ER), progesterone receptor status (PgR) and nodal involvement were taken to be indicator variables. Treatment was taken to be a categorical variable with four levels: radiotherapy alone (RT), mastectomy alone, tumorectomy plus radiotherapy, and mastectomy plus radiotherapy. Age (in years) at diagnosis and primary tumor size (in millimeters) were considered as continuous covariates in this analysis. Histological grading (Scarff Bloom) was taken as a variable with three categories (I, IIa and IIb combined, and III). Tumor grade was highly correlated with stage, and, in combination with the other covariates, was chosen preferentially over stage as predictive of local recurrence in a preliminary analysis with all covariates. Each of the predictors included in the backward elimination was highly significant in univariate analysis.

The three PHPH models differed in the way in which $\bar{F}$ was specified. Our most flexible model used a semiparametric specification of $\bar{F}$. We refer to this model as the 'semiparametric PHPH model'. Our intermediate form specified that $\bar{F}$ be a linear spline. For our spline models we use fixed, equally spaced knots between $t = 0$ and $t = t_{max}$, where t_{max} is the last failure time. $\bar{F}$ is parametrized by the values at the (ten) knots, subject to the conditions that $\bar{F}$ is monotone decreasing, $\bar{F}(0) = 1$, and $\bar{F}(t_{max}) = 0$. Linear interpolation is used between the knots to ensure continuity. We refer to this model as the 'spline PHPH model'. The most restrictive parametric form specified that $\bar{F}$ have a Weibull distribution. This model uses only two parameters, (a median and shape parameter), to specify $\bar{F}$. We refer to this model as 'Weibull PHPH model'. Weibull distributions were also used for both $\bar{F}_1$ and $\bar{F}_2$ of the competing risks model. A semiparametric or spline-based form of the competing risks model was not used, as it is difficult to retain identifiability with such a model.

Estimates for semiparametric PHPH model were obtained by maximizing the profile likelihood (25.17) using the Powell algorithm [Himmelblau (1972)], as described in Section 25.2.2. We fit the parametric and spline PHPH models, as well as the competing risks model, by backward elimination as well. For these models, the complete likelihood was maximized via the Powell algorithm.

Variables with several levels were considered in blocks. Table 25.1 gives the significance levels of variables at the step of removal or inclusion in the final model, calculated using the likelihood ratio test. The cycle in which predictors were removed is given for all variables not in the final model. The significance level $\alpha = 0.05$ was chosen for inclusion in the final model. Although predictors are removed in a different order, in the end the final list of predictors selected for the semiparametric PHPH and Weibull-based PHPH models were exactly the same. One more short-term predictor (size) was selected as significant for the spline PHPH model. The competing risks model contained the additional short-term effect of age (but no size parameters). We note that in each case, there are highly significant short-term effects, so that the proportional hazards assumption is easily rejected.

Parameter estimates for each of the models are provided in Table 25.2. The parametrization for the PHPH and competing risks models are given by (25.13) and (25.14) respectively. Note that the cure rate estimates, although comparable to each other, apply to the extrapolated 'baseline' subject with age equal to zero and tumor size equal to zero. For the cure rate model, the median for one component, interpreted as long term, was 123 years, while the median for the other component, labeled short term, was 23 years.

The models were generally in very good agreement. For all three PHPH models and the competing risks model, positive progesterone receptor status and older age all significantly increased the probability of 'local cure', while lack of nodal involvement, positive estrogen receptor status, and low histological grade significantly increased the mean time to local recurrence for uncured patients. Interestingly, tumor size was chosen as (marginally) significant only for the spline PHPH model. For the spline PHPH model, there was a significant decrease in the mean time to recurrence with increasing primary tumor size. For the competing risk model, older age had a significant short term benefit as well as a long term benefit. The cure rate parameters were in good agreement for all the PHPH models, but the cure rate parameter for the competing risks model varied somewhat, perhaps because of the quite different model specification.

Treatment had both a highly significant long term and short term effect. Compared to the 'baseline' group of patients receiving radiotherapy alone, every other group of patients had a greater chance of local cure for all the models, with those receiving mastectomy (with or without RT) having the best chance of local cure. Curiously, according to the all the PHPH models, compared to patients receiving only radiotherapy, patients receiving mastectomy plus radiotherapy or mastectomy alone, and who have not been locally cured, had smaller mean time to recurrence. On the other hand, patients who received lumpectomy plus radiotherapy were less likely to have been cured locally than those receiving mastectomy or mastectomy plus radiotherapy, but the mean time to recurrence was increased. For the competing risk model mastectomy, mastectomy plus

Table 25.1: Step at which predictors were removed, and significance level (in parentheses) for the three PHPH models and the competing risks model fitted to breast cancer data using backward elimination. Predictors that were not removed are indicated by *. The significance level applies either to the removal step (for predictors that were eliminated by backward selection) or the final model (for all the other predictors).

Effect	Predictor	Model Semi. PHPH	 Spline PHPH	 Weibull PHPH	 Comp. Risks
Long Term	Nodes	1 (0.92)	1 (0.78)	2 (0.88)	4 (0.18)
	PrR	* (0.02)	* (0.02)	* (0.02)	* (0.01)
	ER	4 (0.23)	6 (0.09)	5 (0.20)	2 (0.94)
	Histology	5 (0.16)	5 (0.14)	6 (0.15)	1 (0.90)
	Treatment	* (E-13)	* (E-12)	* (E-15)	* (E-13)
	Age	* (E-13)	* (E-13)	* (E-13)	* (1E-9)
	Size	2 (0.86)	3 (0.51)	1 (0.99)	3 (0.69)
Short Term	Nodes	* (3E-4)	* (2E-4)	* (1E-6)	* (7E-7)
	PgR	3 (0.22)	4 (0.19)	4 (0.18)	5 (0.14)
	ER	* (5E-9)	* (4E-9)	* (E-10)	* (1E-9)
	Histology	* (1E-8)	* (8E-9)	* (2E-9)	* (7E-9)
	Treatment	* (9E-5)	* (5E-5)	* (6E-7)	* (1E-6)
	Age	6 (0.12)	2 (0.58)	3 (0.25)	* (8E-4)
	Size	7 (0.10)	* (0.04)	7 (0.07)	6 (0.09)

RT, and lumpectomy + RT all had comparable beneficial effect on time to recurrence compared to RT alone.

25.4 Discussion

Bounded cumulative hazards regression models have advantages over other models: they are readily interpretable in terms of characteristics with biomedical meaning, and they have more flexibility than more common modeling tools, and can be fit using semiparametic methods. These advantages have been illustrated by our application to the Curie Institute data, presented in the preceding section. Even though our results appear to be in agreement with the literature, our approach, which includes both long and short term effects, provides more information than the typical proportional hazards modeling. From our model we can see that estrogen receptor status, nodal involvement, and histological grade appear to largely affect the timing of local recurrence, rather than the probability of local cure. Age, on the other hand, has a strong effect on the probability of local cure, and treatment on both the probability of local cure as well as the timing of recurrence.

The results of our application of the PHPH and competing risks models to the Curie Institute data suggest that the long and short terms effects are largely robust to changes in model specification (from PHPH to competing risks model) or changes in specification of time to failure distributions. In addition, our results appear to be largely consistent with the breast cancer literature: decreased probability of local relapse is associated with older age at onset, negative nodes, positive estrogen receptor status, and low histological grade, particularly after breast conserving therapy. [See, for example, McReady et al. (1996), who analyzed breast cancer treated with lumpectomy alone]. Young age is often regarded as an important risk factor not only for local recurrence, but distant metastasis and shorter overall survival as well [Elkhuizen et al. (1998)]. Irradiation of the breast in patients receiving lumpectomy is regarded as reducing risk of recurrence, except, perhaps, for patients at very low risk [Clark et al. (1996)].

Association of small tumor size with decreased risk of local recurrence has been observed in the literature. [See, for example, Asselain et al. (1996)]. We do not see much effect of tumor size, as size was absent from two out of three of our final PHPH models, was only marginally significant in the short-term component of the other models, and never appeared in the long-term component of the models. A further exploration of this is discussed below.

A tumor recurrence data set containing a subset of 877 patients from the Curie Institute data was analyzed by Asselain et al. (1996) using the acceler-

Table 25.2: Parameter estimates for each of the three versions of the PHPH model, and the competing risks model, fit to breast cancer recurrence data from the Curie Institute. The baseline group is indicated by —. Components that are absent from the model are indicated by ***.

Effect	**Predictor**	**Level**	**Model** Semi. PHPH	Spline PHPH	Weibull PHPH	Comp. Risks
Long Term	PgR	Negative	—	—	—	—
		Positive	-0.16	-0.16	-0.16	-0.25
	Treatment	RT	—	—	—	—
		Mastectomy	-1.29	-1.27	-1.28	-1.26
		RT+Lump.	-0.40	-0.38	-0.38	-0.61
		RT+Mast.	-1.57	-1.57	-1.53	-1.55
	Age (yrs.)	(continuous)	-0.023	-0.023	-0.023	-0.026
	Cure Rate		0.064	0.064	0.068	0.043
Short Term	Nodes	Negative	—	—	—	—
		Positive	0.47	0.42	0.50	0.66
	ER	Negative	—	—	—	—
		Positive	-0.60	-0.60	-0.64	-0.81
	Histology	Not done	—	—	—	—
		I	-0.57	-0.53	-0.64	-1.11
		II	-0.17	-0.13	-0.20	0.15
		III	0.29	0.34	0.25	0.44
	Treatment	RT	—	—	—	—
		Mastectomy	0.36	0.39	0.37	-0.74
		RT+Lump.	-0.47	-0.39	-0.52	-0.87
		RT+Mast.	0.67	0.70	0.63	-0.90
	Age (yrs.)	(continuous)	***	***	***	-0.020
	Size (mm.)	(continuous)	***	0.0049	***	***

ated failure time model to describe short-term covariate effects. Although recurrence in both the ipsilateral and contralateral breast were analyzed together, the authors state that analysis of ipsilateral recurrence alone gave similar results. The model incorporated components that could be interpreted in terms of long-term effects (mean number of surviving clonogens) and short-term effects (time of progression). A more limited set of covariates (age, tumor size, nodal involvement, and treatment) was considered in this analysis. After backward elimination, treatment, the short-term effect of tumor size, and the long-term effects of both age and tumor size. In accordance with the analysis presented in the previous section, age appeared to have a pronounced long-term effect, but little effect on the time of tumor recurrence. The authors indicate that the long-term effect of age appeared to be the predominant one.

The apparent lack of significance of tumor size in the current analysis appears to contradict the results of the Asselain et al. (1996), but it is possible that this is due to different lists of predictors. To check whether the inclusion of receptor status in the current analysis is responsible for the discrepancy, we removed estrogen and progesterone receptor status from the list of predictors and fit the semiparametric PHPH model to the data by backwards elimination. There is a significant short-term effect of tumor size when this is done ($p=0.026$). There is, in addition, a significant short-term effect of age ($p = 0.027$), in addition to the long-term effect that was present in the extended model. The other selected predictors are the same. We see that even after a more similar list of predictors are included, there remains some slight discrepancy between the effects of tumor size. We speculate that this discrepancy may be caused by the inclusion of more detail on radiotherapy in the earlier analysis. Shorter follow-up analysis and smaller sample size may also contribute to the discrepancy.

Finally, we note that if a different endpoint is analyzed, such as cause-specific survival or time to metastatic spread, the effect of age at onset may be different. For example, in an analysis of survival in women diagnosed with localized breast cancer in Utah using a similar model, with age and stage as predictors, Tsodikov (2002) found that the long term effect of age at diagnosis was not significant, while the short term effect remained significant, at least for younger patients.

Acknowledgment. This research was supported by NIA/NIH Grant RO1 AG14650-01A2, NCI/NIH U01 CA88177-01, NIH/NCI 1 UO1 CA97414-01 and a Department of Defense grant DAMD17-03-1-0034.

References

1. Asselain, B., Fourquet, A., Hoang, T., Tsodikov, A. D. and Yakovlev, A. Y. (1996). A parametric regression model of tumor recurrence: An application to the analysis of clinical breast cancer data, *Statistics & Probability Letters*, **29**, 271–278.

2. Berkson, J. and Gage, R. P. (1952). Survival curves for cancer patients following treatment, *Journal of the American Statistical Association*, **47**, 501–515.

3. Boag, J. M. (1949). Maximum likelihood estimates of the proportion of patients cured by cancer therapy, *Journal of the Royal Statistical Society, Series B*, **11**, 15–44.

4. Broët, P., Rycke, Y. D., Tubert-Bitter, P., Lellouch, J., Asselain, B., and Moreau, T. (2001). A semi-parametric approach for the two-sample comparison of survival times with long-term survivors, *Biometrics*, **57**, 844–852.

5. Cantor, A. B. and Shuster, J. J. (1992), Parametric versus nonparametric methods for estimating cure rates based on censored survival data, *Statistics in Medicine*, **11**, 931–937.

6. Cheng, S., Wei, L. and Ying, Z. (1995). Analysis of transformation models with censored data, *Biometrika*, **82**, 835–845.

7. Clark, R. M., Whelan, T., Levin, M., Roberts, R., Willan, A., McCulloch, P., Lipa, M., Wilkinson, R. H. and Mahoney, L.J. (1996). Randomized clinical trial of breast irradiation following lumpectomy and axillary dissection for node-negative breast cancer: an update. Ontario Clinical Oncology Group, *Journal of National Cancer Institute*, **88**(22), 1659–1664.

8. Cox, D. R. (1972), Regression models and life tables (with discussion), *Journal of the Royal Statistical Society, Series B*, **34**, 187–220.

9. Elkhuizen, P. H., van de Vijver, M. J., Hermans, J., Zonderland, H. M., van de Velde, C. J. and Leer, J. W. (1998). Local recurrence after breast-conserving therapy for breast cancer: high incidence in young patients and association with poor survival, *Int. J. Rad. Oncol. Biol. Phys.*, **40**,) 859–867.

10. Etezadi-Amoli, L. and Ciampi, A. (1987). Extended hazard regression for censored survival data with covariates: A spline approximation for the baseline hazard function, *Biometrics*, **43**, 181–192.

11. Fleming, T. and Lin, D. (2000). Survival analysis in clinical trials: Past developments and future directions, *Biometrics*, **56**, 971–983.

12. Fourquet, A. F., Campana, F., Zafrani, V, Mosseri, P., Vielh, P., Durand, J.-C., and Vilcoq, J. R. (1989). Prognostic factors of breast recurrence in the conservative management of early breast cancer: A 25 year follow-up, *Int. J. Radiat. Onc. Biol. Phys.*, **17**, 719–725.

13. Haybittle, J. L. (1959). The estimation of the proportion of patients cured after treatment for cancer of the breast, *British Journal of Radiology*, **32**, 725–733.

14. Haybittle, J. L. (1965). A two-parameter model for the survival curve of treated cancer patients, *Journal of the American Statistical Association*, **60**, 16–26.

15. Himmelblau, D. M. (1972). *Applied Nonlinear Programming*, McGraw-Hill, Austin.

16. Kalbfleisch, J. D. and Prentice, R. L. (1980). *The Statistical Analysis of Failure Time Data*, John Wiley & Sons, New York.

17. Kuk, A. Y. C. and Chen, C.-H. (1992). A mixture model combining logistic regression with proportional hazards regression, *Biometrika*, **79**, 531–541.

18. Lange, K., Hunter, D. R. and Yang, I. (2000). Optimization transfer using surrogate objective functions (with discussion), *Journal of Computational and Graphical Statistics*, **9** 1–59.

19. McCready, D. R., Hanna, W., Kahn, H., Chapman, J. A., Wall, J., Fish, E. B. and Lickley, H. L. (1996). Factors associated with local breast cancer recurrence after lumpectomy alone, *Ann. Surg. Oncol.*, **3**, 358–366.

20. Maller, R. A. and Zhou, S. (1996). *Survival Analysis with Long-Term Survivors*, John Wiley & Sons, Chichester, England.

21. Miller, R. G. (1981). *Survival Analysis*, John Wiley & Sons, New York.

22. Peng, Y., and Dear, K. B. G. (2000). A nonparametric mixture model for cure rate estimation, *Biometrics*, **56**, 237–243.

23. Sy, J. P. and Taylor, J. M. G. (2000). Estimation in a Cox proportional hazards cure model, *Biometrics*, **56**, 227–236.

24. Taylor, J. M. G. (1995). Semi-parametric estimation in failure time mixture models, *Biometrics*, **51**, 899–907.

25. Tsodikov, A. (1998). Asymptotic efficiency of a proportional hazards model with cure, *Statistics & Probability Letters*, **39**, 237–244.

26. Tsodikov, A. (2002). Semiparametric models of long- and short-term survival: An application to the analysis of breast cancer survival in Utah by age and stage, *Statistics in Medicine*, **21**, 895–920.

27. Tsodikov, A. (2003). Semiparametric models: A generalized self-consistency approach, *Journal of the Royal Statistical Society, Series B*, **65**, 759–774.

28. Tsodikov, A. D., Ibrahim, J. G. and Yakovlev, A. Y. (2003). Estimating cure rates from survival data: An alternative to two-component mixture models, *Journal of the American Statistica Association*, **98**, 1063–1078.

29. Yakovlev, A. Yu. (1994). Letter to the editor, *Statistics in Medicine*, **13**, 983–986.

30. Yakovlev, A. Y., Asselain, B., Bardou, V. J., Fourquet, A., Hoang, T., Rochefordiere, A. and Tsodikov, A. D. (1993). A simple stochastic model of tumor recurrence and its application to data on premenopausal breast cancer, In *Biometrie et Analyse de Donnees Spatio–Temporelles* No. 12, (Eds., B. Asselain, M. Boniface, C. Duby, C. Lopez, J. P. Masson and J. Tranchefort), pp. 66–82, Société Française de Biométrie, ENSA Rennes, France.

PART XII
Inference for Processes

Estimation of Analytic Spectral Density of Gaussian Stationary Processes

I. Ibragimov

St. Petersburg Branch of Steklov Mathematical Institute Russian Academy of Sciences, St. Petersburg State University, Russia

Abstract: In this paper we consider the problem of estimation of the spectral density $f(\lambda)$ of a Gaussian stationary sequence (process) on the base of observations $X_t, 0 \le t \le T$. We suppose that f belongs to the class of spectral densities analytic and bounded inside a bounded region $G \ni [a, b]$. We found the rate of the asymptotic minimax risk in $L_p[a, b], 1 \le p \le \infty$ when $T \to \infty$. [1]

Keywords and phrases: Spectral density, Gaussian stationary process, periodogram

26.1 Introduction

Let $\{X_j\}$ be a real-valued stationary Gaussian sequence with mean zero and spectral density $f(\lambda)$. We assume that the spectral density f is unknown and should be estimated on the base of observations

$$X_1, X_2, \ldots, X_T. \tag{26.1}$$

Assume further that $f \in F$ where F is a given (known) class of spectral densities. Let $[a, b] \subseteq [-\pi, \pi]$ be an interval and assume that we would like to estimate the restriction of f on $[a, b], \{f(\lambda), a \le \lambda \le b\}$. If f_T is an estimator for f, we measure its quality by the average $L_p[a, b]$ distance from f, namely

$$\mathbf{E}_f|f - f_T|_p = \mathbf{E}_f\Big(\int_a^b |f(\lambda) - f_T(\lambda)|^p d\lambda\Big)^{1/p}, \quad 1 \le p < \infty,$$

[1]This work began when the author was visiting the CMA of the Australian National University. I am very grateful to CMA and to Prof. P. Hall personally for their hospitality and the financial support of this work. I wish to thank Bret Presnell who read the first version of the manuscript and made many helpful comments. The work is partly supported by Russian Foundation for Basic Research, grants, 96-01-00684, 96-15-96199.

$$\mathbf{E}_f|f - f_T|_\infty = \mathbf{E}_f\{\sup_{a\le\lambda\le b}|f(\lambda) - f_T(\lambda)|\}. \tag{26.2}$$

We denote by $|.|_p$ the norms in $L_p(a,b)$ spaces; the norms in $L_p(-\pi,\pi)$ are denoted $||.||_p$.

Let us define

$$\Delta_p(T;F) = \Delta_p(T) = \inf\sup \mathbf{E}_f|f - f_T|_p \tag{26.3}$$

where sup is taken over all $f \in F$ and inf is taken over all possible estimators f_T of f. We are interested in the asymptotic behavior of the minimax risk $\Delta_p(T;F)$ when $T \to \infty$. The rate of convergence of $\Delta_p(T)$ to zero depends on F.

For example, let F consist of all spectral densities f with uniformly bounded in L_p fractional derivatives of order β. Then [see, for example, Bentkus (1985a,b), Golubev (1991) and references there]

$$\begin{aligned} \Delta_p &\asymp (T)^{-\frac{\beta}{1+2\beta}}, \quad 1 \le p < \infty, \\ \Delta_\infty &\asymp (T)^{-\frac{\beta}{1+2\beta}}(\ln T)^{\frac{\beta}{1+2\beta}}. \end{aligned} \tag{26.4}$$

If F consists of all 2π-periodic functions analytic and uniformly bounded inside a strip $|\Im z| < c, z = \lambda + i\mu$, then

$$\begin{aligned} \Delta_p(T) &\asymp \sqrt{\frac{\ln T}{T}}, \quad 1 \le p < \infty, \\ \Delta_\infty(T) &\asymp \sqrt{\frac{\ln T}{T}}\sqrt{\ln\ln T}. \end{aligned} \tag{26.5}$$

In this paper we study the asymptotic behavior of $\Delta_p(T)$ for the case when spectral densities f are analytic in a *bounded* region G of the complex plane, $[a,b] \subset G$. Some similar problems of curve estimation for curves analytic in a bounded region has been considered in the author's paper Ibragimov (2003). We use the same approach in this paper.

Theorem 26.1.1 *Let the expression $\Delta_p(T;F)$ be defined by (26.2) where the set F consists of all spectral densities $f(\lambda)$ such that*

$$\int_\pi^\pi f(\lambda)d\lambda \le \sigma^2,$$

$f(\lambda)$ are analytic in some bounded region $G, [a,b] \subset G$, and bounded there by a common constant M. Then

$$\Delta_p(T) \asymp \sqrt{\frac{\ln T}{T}}, \quad 1 \le p < 4,$$

$$\Delta_4(T) \asymp \sqrt{\frac{\ln T}{T}}(\ln\ln T)^{1/4},$$
$$\Delta_p(T) \asymp \sqrt{\frac{1}{T}}(\ln T)^{1-2/p}, \quad 4 < p \leq \infty. \tag{26.6}$$

The constants on the right depend on G, M, σ only.

Sections 26.2 and 26.3 are devoted to the proof of the theorem. Section 26.4 treats continuous time processes.

26.2 Proof of Theorem 26.1.1. Upper Bounds

Without loss of generality we may and will suppose that $[a,b] = [-1,1]$. The only exception arises in the special case when one or both endpoints of the interval $[a,b]$ coincides with $-\pi$ or π. A spectral density is naturally defined on the unit circle or as a function 2π-periodic on the real line. If $[a,b] \subset (-\pi,\pi)$, the function f is smooth in a vicinity of $[a,b]$, but if a (or b) is $-\pi$ (or π), the function f may have jumps at this points. Sometimes we have to consider this case separately.

If O is a domain in the complex plane we denote by $A(O,M)$ the class of functions analytic in O and uniformly bounded there by a common constant M. The set G contains inside itself an ellipsis E with the foci at the points ± 1 and the sum of half-axes equal to $R > 1$. Evidently $A(G,M) \subseteq A(E,M)$ and hence any function $f \in F$ belongs to $A(E,M)$. Functions $f \in A(E,M)$ can be represented by the Fourier series with respect to the orthonormal Legendre polynomials $P_j(\lambda)$, namely

$$f(\lambda) = \sum_0^\infty a_j P_j(\lambda), \quad a_j = \int_{-1}^1 P_j(\lambda) f(\lambda) d\lambda. \tag{26.7}$$

The coefficients a_j decrease exponentially fast,

$$\left(\sum_{n+1}^\infty |a_j|^2\right) \leq \left(\frac{\pi}{R^2-1}\right)^{1/2} \frac{M}{R^n},$$

see, for example, Szegö (1959). Hence

$$|a_n| \leq cR^{-n} = ce^{-\gamma n}. \tag{26.8}$$

Here and below let c, C represent constants which depend on G, M, σ only; they may be different even inside the same formula.

Denote $I_T(\lambda)$ the periodogram

$$I_T(\lambda) = \frac{1}{2\pi}\left|\sum_1^T X_j e^{-ij\lambda}\right|^2.$$

Because of (26.7),

$$\hat{a}_k = \int_{-1}^{1} I_T(\lambda) P_k(\lambda)$$

are natural estimators for a_k. Consider estimators $f_N(\lambda)$ for $f(\lambda)$ defined by the formula

$$f_N(\lambda) = \sum_1^N \hat{a}_k P_k(\lambda).$$

The main result of this section is the following:

Theorem 26.2.1 *The estimators f_N satisfy the following inequalities:*

$$\begin{aligned}
\mathbf{E}_f|f - f_N|_p &\le C(N^{1/2}T^{-1/2} + NT^{-3/4} + T^{-1/2} + e^{-\gamma N}),\\
&\qquad 1 \le p < 4,\\
\mathbf{E}_4|f - f_N|_4 &\le C(N^{1/2}(\ln N)^{1/4}T^{-1/2} + NT^{-3/4} + T^{-1/2} + e^{-\gamma N}),\\
\mathbf{E}_p|f - f_N|_p &\le C(N^{1/2}(\ln N)^{1-2/p}T^{-1/2} + NT^{-3/4} + T^{-1/2} + e^{-\gamma N}),\\
&\qquad p > 4.
\end{aligned} \tag{26.9}$$

The upper bounds of Theorem 26.1.1 now follow easily from (26.9).

Corollary 26.2.1 *There exist estimators $\hat{f}_T$ such that*

$$\begin{aligned}
\mathbf{E}_f|f - \hat{f}_T|_p &\le C\sqrt{\frac{\ln T}{T}}, \quad 1 \le p < 4,\\
\mathbf{E}_f|f - \hat{f}_T|_4 &\le C\sqrt{\frac{\ln T}{T}}(\ln\ln T)^{1/4},\\
\mathbf{E}_f|f - \hat{f}_T|_p &\le CT^{-1/2}(\ln T)^{1-2/p}.
\end{aligned} \tag{26.10}$$

The corollary follows immediately from (26.9). Indeed, we can define $\hat{f}_T$ as $f_{N(T)}$, where $N(T) = [\frac{1}{\gamma}\ln T]$.

Proof of Theorem 26.2.1. Split the difference $f_N(\lambda) - f(\lambda)$ into the bias part

$$b_N(\lambda) = \sum_1^N P_k(\lambda)\int_{-1}^{1} P_k(\lambda)(\mathbf{E}I_T(\lambda) - f(\lambda))d\lambda - \sum_{N+1}^{\infty} a_k P_k(\lambda)$$

and the random part

$$Z_N(\lambda) = \sum_1^N (\hat{a}_k - \mathbf{E}\hat{a}_k) P_k(\lambda)$$

and bound them separately. ■

It is well known that the expectation $\mathbf{E}I_T(\lambda)$ of the periodogram is the Fejer integral of the spectral density [see, for example, Hannan (1970)],

$$\mathbf{E}I_T(\lambda) = \frac{1}{2\pi T} \int_{-\pi}^{\pi} \frac{\sin^2 T\frac{\lambda - l}{2}}{\sin^2 \frac{\lambda - l}{2}} f(l) dl. \tag{26.11}$$

Lemma 26.2.1 *Let $\Phi_T(g)(\lambda)$ denote the Fejer integral of a function g.*
(i) If $g \in L_p(-\pi, \pi)$ and satisfies there the Hölder condition of order α, then

$$\begin{aligned} ||g - \Phi_T(g)||_p &\leq CT^{-\alpha}, & 0 < \alpha < 1, \\ ||g - \Phi_T(g)||_p &\leq CT^{-1} \ln T, & \alpha = 1. \end{aligned} \tag{26.12}$$

(ii) If g satisfies a Lipschitz condition on an interval $[a, b]$, then for any interval $[\alpha, \beta] \subset (a, b)$,

$$\sup_{[\alpha,\beta]} |g(\lambda) - \Phi_T(g)(\lambda)| \leq CT^{-1}. \tag{26.13}$$

The proof of Lemma 26.2.1 can be found in books on approximation theory, see, for example, Timan (1963).

Lemma 26.2.2 *For all $1 \leq p \leq \infty$ the following inequality holds:*

$$|b_N|_p \leq C(N^{3/2} T^{-3/4} + T^{-1/2} + e^{-\gamma N}). \tag{26.14}$$

Proof of the Lemma. The derivative $f'(\lambda)$ is bounded in a vicinity of $[-1, 1]$ by a constant C which depends on M and G only. Hence by (26.11) and (26.13) for all $\lambda \in [-1, 1]$,

$$|\mathbf{E}I_T(\lambda) - f(\lambda)| \leq CT^{-1} \ln T. \tag{26.15}$$

It follows that

$$\left| \int_{-1}^{1} P_n(\lambda)(\mathbf{E}I_T(\lambda) - f(\lambda)) d\lambda \right| \leq C|P_n|_2 T^{-1} \ln T = CT^{-1} \ln T. \tag{26.16}$$

On the interval $[-1, 1]$ the Legendre polynomials satisfy the inequalities [see Szegö (1959)]

$$|P_n(\lambda)| \leq P_n(1) = (n + 1/2)^{1/2}. \tag{26.17}$$

Hence

$$|b_N|_p \le CT^{-1}\ln T\sum_{1}^{N} k^{1/2} + \sum_{N+1}^{\infty} |a_k| \le C(N^{3/4}T^{-3/4} + e^{-\gamma N})$$

and the lemma is proved except the special case which has been mentioned above. Indeed if the interval $[a, b]$ of analyticity coincides with the interval $[-\pi, \pi]$,the arguments proving (26.16) do not work because the *periodic* function $f(\lambda)$ may have a jump at the points $\pm\pi$. This jump is the only possible singularity of f so in any case it satisfies Holder condition of order $1/2$ in the L_2-norm and by Lemma 26.2.1,

$$\left|\int_{-\pi}^{\pi} P_n(\lambda)(\mathbf{E}I_T(\lambda) - f(\lambda))d\lambda\right| \le ||P_n||_2||\Phi_T(f) - f||_2 \le CT^{-1/2}.$$

(Of course P_n here are the polynomials orthonormal on $[\pi, \pi]$). The lemma is proved. ■

We pass now to the investigations of the random polynomials Z_N.

Lemma 26.2.3 *The norms of the polynomials Z_N satisfy the following inequalities:*

$$\begin{aligned}
\mathbf{E}|Z_N|_p &\le C(N^{1/2}T^{-1/2} + NT^{-1}), \quad p < 4,\\
\mathbf{E}|Z_N|_4 &\le C(N^{1/2}T^{-1/2}(\ln N)^{1/4} + NT^{-3/4}),\\
\mathbf{E}|Z_N|_p &\le C(N^{1-2/p}T^{-1/2} + NT^{-3/4}), \quad p > 4. \qquad (26.18)
\end{aligned}$$

PROOF. We begin with $p = 2$. Evidently

$$\begin{aligned}
\mathbf{E}|Z_N(\lambda)|^2 &= \sum_{k,l=1}^{N} P_k(\lambda)P_l(\lambda)\\
&\times \int_{-1}^{1}\int_{-1}^{1} P_k(\mu_1)P_l(\mu_2)\mathrm{cov}(I_T(\mu_1), I_T(\mu_2))d\mu_1 d\mu_2. \qquad (26.19)
\end{aligned}$$

The covariation of I_T has the following integral representation [see Hannan (1970) and Ibragimov (1963)]

$$\begin{aligned}
&\mathrm{cov}(I_T(\lambda), I_T(\mu))\\
&= \left[\frac{1}{2\pi T}\int_{-\pi}^{\pi} \frac{\sin(T(l-\lambda)/2)\sin(T(l-\mu)/2)}{\sin((l-\lambda)/2)\sin((l-\mu)/2)} f(l)dl\right]^2\\
&+ \left[\frac{1}{2\pi T}\int_{-\pi}^{\pi} \frac{\sin(T(l-\lambda)/2)\sin(T(l+\mu)/2)}{\sin((l-\lambda)/2)\sin((l+\mu)/2)} f(l)dl\right]^2. \qquad (26.20)
\end{aligned}$$

■

Lemma 26.2.4 *Let $k, l \le N$. Then*

$$\begin{aligned} I_{kl} &= \int_{-1}^{1}\int_{1}^{1} P_k(\lambda)P_l(\mu)cov(I_T(\lambda), I_T(\mu)d\lambda d\mu \\ &= \frac{2\pi}{T}\int_{-1}^{1}(P_k(\lambda)P_l(\lambda) + P_k(\lambda)P_l(-\lambda))f^2(\lambda)d\lambda + r_{TN} \qquad (26.21)\end{aligned}$$

where $|r_{TN}| \le CN^3T^{-3/2}$.

Proof. Using the representation (26.20) we find that I_{kl} is the sum of two summands

$$I_{kl}^{(1)} = \int_{-1}^{1}\int_{-1}^{1} P_k(\lambda)P_l(\mu)\left(\frac{1}{2\pi T}\int_{-\pi}^{\pi}\frac{\sin T(l-\lambda)/2}{\sin(l-\lambda)/2}\frac{\sin T(l-\mu)/2}{\sin(l-\mu)/2}f(l)dl\right)^2 d\lambda d\mu$$

and

$$I_{kl}^{(1)} = \int_{-1}^{1}\int_{-1}^{1} P_k(\lambda)P_l(\mu)\left(\frac{1}{2\pi T}\int_{-\pi}^{\pi}\frac{\sin T(l-\lambda)/2}{\sin(l-\lambda)/2}\frac{\sin T(l+\mu)/2}{\sin(l+\mu)/2}f(l)dl\right)^2 d\lambda d\mu.$$

It is enough to study the first summand, $I_{kl}^{(1)}$. Uniformly in $\lambda \in [-1, 1]$ the integral

$$\begin{aligned} &\frac{1}{2\pi T}\int_{-\pi}^{\pi}\frac{\sin T(l-\lambda)/2}{\sin(l-\lambda)/2}\frac{\sin T(l-\mu)/2}{\sin(l-\mu)/2}f(l)dl \\ &\quad = \frac{f(\lambda)}{2\pi T}\int_{-\pi}^{\pi}\frac{\sin T(l-\lambda)/2}{\sin(l-\lambda)/2}\frac{\sin T(l-\mu)/2}{\sin(l-\mu)/2}dl \\ &\quad\quad + O(1)T^{-1}\int_{-\pi}^{\pi}\left|\frac{\sin T(l-\lambda)/2}{\sin(l-\lambda)/2}\right| dl \\ &\quad = \frac{f(\lambda)}{T}\frac{\sin T(\lambda-\mu)/2}{\sin(\lambda-\mu)/2} + O\left(\frac{\ln T}{T}\right),\end{aligned}$$

and we have that

$$I_{kl}^{(1)} = \frac{2\pi}{T}\int_{-1}^{1} P_k(\lambda)f^2(\lambda)d\lambda\frac{1}{2\pi T}\int_{-1}^{1} P_l(\mu)\frac{\sin^2\frac{T(\lambda-\mu)}{2}}{\sin^2\frac{\lambda-\mu}{2}}d\mu + O(T^{-3/2}).$$

Let us define the 2π-periodic function $\tilde{P}_l(\mu)$ as

$$\begin{aligned} \tilde{P}_l(\mu) &= P_l(\mu), \qquad & |\mu| &\le 1, \\ \tilde{P}_l(\mu) &= 0, \qquad & 1 < |\mu| &\le \pi. \end{aligned}$$

It follows easily from properties of the Fejer integral and the inequalities (26.17) that

$$\int_{-1}^{1}\left|\frac{1}{2\pi T}\int_{-1}^{1} P_l(\mu)\frac{\sin^2\frac{T(\lambda-\mu}{2}}{\sin^2\frac{\lambda-\mu}{2}}d\mu - P_l(\lambda)\right|^2 d\lambda$$

$$\leq \int_{-1}^{1} d\lambda \left| \frac{1}{2\pi T} \int_{-\pi}^{\pi} \left[\tilde{P}_l(\mu) - \tilde{P}_l(\lambda) \right] \frac{\sin^2 \frac{T(\lambda-\mu)}{2}}{\sin^2 \frac{\lambda-\mu}{2}} d\mu \right|^2$$
$$\leq CNT^{-1/2}.$$

Hence

$$I_{kl}^{(1)} = \frac{2\pi}{T} \int_{-1}^{1} P_k(\lambda) P_l(\lambda) f^2(\lambda) d\lambda + O(N^2 T^{-3/2}). \tag{26.22}$$

In the same way

$$I_{kl}^{(2)} = \frac{2\pi}{T} \int_{-1}^{1} P_k(\lambda) P_l(-\lambda) f^2(\lambda) d\lambda + O(N^2 T^{-3/2}). \tag{26.23}$$

The relations (26.22) and (26.23) prove the lemma if the endpoints a, b do not coincide with $\pm\pi$. The exceptional case can be treated as in the proof of Lemma 26.2.2. We omit the arguments. ■

It follows from Lemma 26.2.4 that

$$\begin{aligned}
\mathbf{E}|Z_N|_2^2 &= \sum_1^N \mathbf{E}|\hat{a}_k - \mathbf{E}\hat{a}_k|^2 \\
&\leq \frac{4\pi}{T} \int_{-1}^{1} \left(\sum_1^N P_k^2(\lambda)\right) f^2(\lambda) d\lambda + O(N^3 T^{-3/2}) \\
&\leq CNT^{-1} + O(N^3 T^{-3/2}).
\end{aligned}$$

Hence for all $1 \leq p \leq 2$,

$$\mathbf{E}|Z_N|_p \leq CN^{1/2} T^{-1/2} + O(N^2 T^{-3/4}).$$

To consider the case $p > 2$ we need the following lemma.

Lemma 26.2.5 *Let ξ be a Gaussian random variable with values in a Euclidian or Hilbert space H and $\mathbf{E}\{\xi\} = 0$. Let $A : H \to H$ be a self adjoint bounded linear operator. Then for any $p > 2$,*

$$\mathbf{E}|(A\xi, \xi) - \mathbf{E}(A\xi, \xi)|^p \leq c_p \mathbf{E}^{p/2} |(A\xi, \xi) - \mathbf{E}(A\xi, \xi)|^2.$$

PROOF. The quadratic form $(A\xi, \xi)$ has the same distribution as the sum $\sum \mu_i \xi_i^2$, where ξ_j are iid normal random variables with mean 0 and variance 1, while μ_j are the eigenvalues of the operator RA where R is the correlation operator of ξ [see Ibragimov (1963, Appendix)]. Hence

$$\begin{aligned}
\mathbf{E}|(A\xi, \xi) - \mathbf{E}(A\xi, \xi)|^p &= \mathbf{E}\left|\sum \mu_j (\xi_j^2 - 1)\right|^p \leq c_p \left(\sum \mu_j^2\right)^{p/2} \\
&= c_p (2^{-1} (\mathrm{Var}(A\xi, \xi))^{p/2}.
\end{aligned}$$

The lemma is proved. ■

Notice now that the value $Z_N(\lambda)$ of the polynomial Z_N at a point λ is a quadratic form of the variables $X_1, \dots, X_T$. Applying Lemma 2.5 we find that

$$\mathbf{E}|Z_N|_p \le \left(\int_{-1}^{1} \mathbf{E}|Z_N(\lambda)|^p d\lambda\right)^{1/p} \le \left(\int_{-1}^{1} (\mathbf{E}|Z_N(\lambda)|^2)^{p/2} d\lambda\right)^{1/p}. \quad (26.24)$$

It follows from (26.19) and Lemma 26.2.4 that

$$\begin{aligned}\mathbf{E}|Z_N(\lambda)|^2 &\le \frac{C}{T}\int_{-1}^{1}\left|\sum_1^N P_k(\lambda)P_k(\mu)\right|^2 d\mu + O(N^5T^{-3/2}) \\ &= CT^{-1}\sum_1^N P_k^2(\lambda) + O(N^5T^{-3/2}). \end{aligned} \quad (26.25)$$

Substitute the last estimate into (26.24). We find that

$$\mathbf{E}|Z_N|_p \le CT^{-1/2}\left(\int_{-1}^{1} (\sum_1^N P_k^2(\lambda))^{p/2} d\lambda\right)^{1/p} + O(N^3T^{-3/4}). \quad (26.26)$$

The integrals on the right can be bounded as follows.

Lemma 26.2.6 *For* $2 \le p < 4$,

$$\int_{-1}^{1}\left(\sum_1^N P_k^2(\lambda)\right)^{p/2} d\lambda \le c_p N^{p/2}.$$

For $p = 4$,

$$\int_{-1}^{1}\left(\sum_1^N P_k^2(\lambda)\right)^{2} d\lambda \le cN^2 \ln N.$$

For $p > 4$,

$$\int_{-1}^{1}\left(\sum_1^N P_k^2(\lambda)\right)^{p/2} d\lambda \le c_p N^{p-2}.$$

PROOF. The inequalities of the lemma follow immediately from the inequalities [see Szegö (1959)]

$$|P_k(\lambda)| \le (k+1/2)^{1/2}, |P_k(\lambda)| \le \sqrt{\frac{3}{\pi}}(1-\lambda^2)^{-1/4}.$$

The inequalities of Lemma 26.2.5 together with (26.26) prove the bounds (26.18) for $p < \infty$. To treat the case $p = \infty$ we apply the following result [see Timan (1963)]. ■

Lemma 26.2.7 *Let Q be an algebraic polynomial of degree n. Then*

$$|Q|_\infty \leq (p+1)^{1/p} n^{2/p} |Q|_p.$$

The last lemma gives that for any $p > 4$,

$$\begin{aligned} \mathbf{E}|Z_N|_\infty &\leq CN^{2/p}\mathbf{E}|Z_N|_p \leq CN^{2/p}(T^{-1/2}N^{1-2/p} + O(N^3T^{-3/4})) \\ &\leq C(NT^{-1/2} + N^4T^{-3/4}). \end{aligned}$$

The inequalities (26.18) are thus proved and we have finished the proof of Theorem 26.2.1. ∎

26.3 Lower Bounds

Theorem 26.3.1 *Let the set F consist of all spectral densities of real-valued stationary sequences analytic in some bounded region G, $[a,b] \subset G$ and bounded there by a common constant M. Then there exist positive constants $c > 0$ such that*

$$\begin{aligned} \Delta_p(T) &\geq c\sqrt{\frac{\ln T}{T}}, \quad 1 \leq p < 4 \\ \Delta_4(T) &\geq c\sqrt{\frac{\ln T}{T}}(\ln\ln T)^{1/4}), \\ \Delta_p(T) &\geq cT^{-1/2}(\ln T)^{1-2/p}, \quad 4 < p \leq \infty. \end{aligned}$$

Let E be an ellipsis with the foci at the points ± 1 and such that $G \subset E$. Then $A(G,M) \supset A(E,M)$ and any spectral density $f \in A(E,M)$ will also belong to F. Hence it is enough to establish the lower bounds for the case $G = E$ and below we suppose that $G = E$.

We prove at first the necessary lower bounds for $p \neq 4$ applying methods developed by Ibragimov and Khasminskii (1981, 1989) and based on the use of Fano's lemma [see Ibragimov and Khasminskii (1981)]. Namely, assume that there is a set $S = \{f_1, \ldots, f_M\}$ of spectral densities from F such that $|f_i - f_j|_p \geq 2\delta$ for any $i \neq j$. Then evidently for any estimator $\hat{f}$ for f,

$$\begin{aligned} \sup_{f\in F} \mathbf{P}_f\{|\hat{f} - f|_p \geq \delta\} &\geq \sup_{1\leq i\leq M} \mathbf{P}_{f_i}\{|\hat{f} - f|_p \geq \delta\} \\ &\geq \frac{1}{M}\sum_{1\leq i\leq M} \mathbf{P}_{f_i}\{|\hat{f} - f|_p \geq \delta\}. \qquad (26.27) \end{aligned}$$

Let θ be a random variable taking values $1, \ldots, M$ with equal probabilities $1/M$. It follows from the inequality (26.27) that

$$\begin{aligned}\inf_{\hat f} \sup_{f \in F} \mathbf{P}_f\{|\hat f - f|_p \geq \delta\} \\ \geq \inf_{\hat\theta} \frac{1}{M} \sum_{i=1}^{M} \mathbf{P}\{\hat\theta \neq i | \theta = i\} = p_e, \end{aligned} \tag{26.28}$$

where the last inf is taken over all estimators $\hat\theta = \hat\theta(X_1, \ldots, X_T)$ of θ. By Fano's lemma [see Ibragimov and Khasminskii (1981)]

$$p_e \geq 1 - \frac{I(\theta; X) + \ln 2}{\ln M - 1}, \tag{26.29}$$

where $I(\theta; X)$ denotes Shannon's information in $X = (X_1, \ldots X_T)$ about θ.

The estimates (26.27)–(26.29) show that

$$\begin{aligned}\inf_{\hat f} \sup_{f \in F} \mathbf{E}_f |\hat f - f|_p &\geq \inf_{\hat f} \sup_{f \in S} \mathbf{E}_f |\hat f - f|_p \\ &\geq \delta \inf_{\hat f} \sup_{f \in S} \mathbf{P}_f\{|\hat f - f|_p \geq \delta\} \\ &\geq \delta p_e \geq \delta \left(1 - \frac{\sup_\theta I(\theta, X) + 1}{\ln M - 1}\right). \end{aligned} \tag{26.30}$$

This inequality will serve as an initial point in our arguments for the cases $p \neq 4$. To use it we study at first the asymptotic behaviour of $I(\theta, X)$ when $T \to \infty$.

By the definition of Shannon information [see Gallager (1968)]

$$I(\theta, X) = \mathbf{E}(\ln \frac{dP_{\theta,X}}{dP_\theta \times dP_X}(\theta, X)).$$

Let now P_0 denote the probability distribution generated by the stationary sequence $\{X_j\}$ of iid Gaussian random variables with the spectral density equal to 1. Then

$$I(\theta, X) = \mathbf{E} \ln \frac{dP_{X|\theta}}{dP_X} = \mathbf{E} \ln \frac{dP_{X|\theta}}{dP_0} + \mathbf{E} \ln \frac{dP_0}{dP_X}.$$

By Jensen's inequality

$$\mathbf{E} \ln \frac{dP_0}{dP_X} \leq \ln \mathbf{E} \frac{dP_0}{dP_X} \leq 0.$$

Hence

$$I(\theta, X) \leq \mathbf{E} \ln \frac{dP_{X|\theta}}{dP_0}. \tag{26.31}$$

Lemma 26.3.1 *Let $K(m, M)$ denote the class of all spectral densities $f(\lambda)$ such that*

$$f(\lambda) \geq m > 0, \quad f(\lambda) \leq M, \quad \int_{-\pi}^{\pi} |f'(\lambda)|^2 \leq M.$$

Let $P_f^T = P_f$ be the distribution of the sequence $X_1, \ldots, X_T$ with the spectral density $f(\lambda)$. Then

$$\ln \frac{dP_f^T}{dP_0^T}(X) = -\frac{T}{4\pi} \int_{\pi}^{\pi} \ln f(\lambda d\lambda + \frac{T}{4\pi} \int_{-\pi}^{\pi} I_T(\lambda) \left(\frac{1}{f(\lambda)} - 1\right) d\lambda + R_T,$$

where $\sup_{f \in K} \mathbf{E}|R_T| < \infty$.

For the proof of the lemma see Dzhaparidze (1986, pp. 50–59). If we apply the result of Lemma 26.3.1 to the inequality (26.31), we find that for $f_j \in K$,

$$\begin{aligned} & I(\theta, X) \\ \leq\ & CT \sup_j \left\{ \left| \int_{-\pi}^{\pi} \ln f_j(\lambda) d\lambda \right| + \int_{-\pi}^{\pi} |1 - (f_j(\lambda))^{-1}| \mathbf{E} I_T(\lambda) d\lambda \right\}. \end{aligned} \tag{26.32}$$

We begin with the case $1 \leq p < 4$, construct the set S and prove the lower bounds for this case. Introduce the function $\phi(\lambda) = \frac{\sin^3 \lambda}{\lambda^2}$. Evidently

$$\int_{-c}^{c} \phi(\lambda) d\lambda = 0, \quad \int_{-\infty}^{\infty} |\phi(\lambda)| d\lambda < \infty. \tag{26.33}$$

The function $\phi(z)$ is analytic in the whole complex plane and on any strip $|\Im z| \leq R$ it satisfies the inequality

$$|\phi(z)| \leq Ce^{3R}. \tag{26.34}$$

Define the number κ from the relation

$$\int_{-\kappa}^{\kappa} |\phi(\lambda)| d\lambda = 100 \int_{|\lambda| \geq \kappa} |\phi(\lambda)| d\lambda. \tag{26.35}$$

Take an integer N and introduce the functions

$$\phi_{jN}(\lambda) = \phi_j(\lambda) = \left\{ \frac{\sin^3 N(\lambda - 2j\kappa/N)}{(\lambda - 2j\kappa/N)^2} - \frac{\sin^3 N(\lambda + 2j\kappa/N)}{(\lambda + 2j\kappa/N)^2} \right\} - c_{jN},$$

where the indices j run through the integers of the interval $1 \leq j \leq M, M = [\frac{\pi N}{4\kappa}]$ and the constants $c_{jN} = c_j$ are defined in a such way that the integrals

$$\int_{-\pi}^{\pi} \phi_j(\lambda) d\lambda = 0, \qquad j = 1, \ldots, M.$$

Notice that

$$|c_j| \leq \left| 2N \int_{N\pi-2j\kappa}^{N\pi+2j\kappa} \frac{\sin^3(\lambda)}{\lambda^2} d\lambda \right| \leq \frac{C}{N}.$$

Consider vectors $\mathbf{a} = (a_1, \ldots, a_M)$ where a_j takes the values $-1, 1$. Define spectral densities $f_{\mathbf{a}}(\lambda)$ as

$$f_{\mathbf{a}}(\lambda) = b + e^{-\gamma N} \sum_1^M a_j \phi_{jN}(\lambda).$$

The functions $f_{\mathbf{a}}$ are analytic and in the strip $|\Im z| \leq R$ they are bounded by $b + Ce^{-(N\gamma - 3R)}$. Hence one can always choose $b > 0$ and γ in a such way that all $f_{\mathbf{a}}$ will be spectral densities of a real stationary Gaussian sequence($f_{\mathbf{a}}(\lambda) \geq 0, \lambda \in [-\pi, \pi], f_{\mathbf{a}}(\lambda) = f_{\mathbf{a}}(-\lambda)$) from F ($|f_{\mathbf{a}}(z)| \leq M, |\Im z| \leq R$). Below for the sake of simplicity we take $\theta = 1$.

Denote by $\mathbf{A}$ the subset of the vectors $\mathbf{a}$ such that for any two vectors $\mathbf{a}, \mathbf{b} \in \mathbf{A}$,

$$\sum_1^M |a_j - b_j| \geq M/2.$$

Define the set $S = \{f_{\mathbf{a}}, \mathbf{a} \in \mathbf{A}\}$.

Lemma 26.3.2 [see Ibragimov and Khasminskii (1989) and Ibragimov (2003)] *The cardinality of the set S is*

$$card(S) = card(\mathbf{A}) \geq \mathbf{2^{M/8}}. \tag{26.36}$$

The L_1-distance between any two functions $f_{\mathbf{a}}, f_{\mathbf{b}} \in S$,

$$\begin{aligned} |f_{\mathbf{a}} - f_{\mathbf{b}}|_1 &= e^{-\gamma N} \int_{-1}^{1} \left| \sum_j (a_j - b_j) \phi_{jN}(\lambda) \right| d\lambda \\ &\geq e^{-\gamma N} \sum_j \left(\int_{(2j-1)\kappa/N}^{(2j+1)\kappa/N} |\phi_j(\lambda)| d\lambda - \sum_{r \neq j} \int_{(2j-1)\kappa/N}^{(2j+1)\kappa/N} |\phi_r(\lambda)| d\lambda \right) \\ &\geq ce^{-\gamma N} N \left(\int_{-\kappa}^{\kappa} |\phi(\lambda)| d\lambda - \int_{\lambda \geq \kappa} |\phi(\lambda)| d\lambda \right) \sum_j |a_j - b_j|, \quad c > 0. \end{aligned} \tag{26.37}$$

Taking into account inequalities (26.35) and (26.36) we find from (26.37) that there exists a positive constant $c > 0$ such that

$$|f_{\mathbf{a}} - f_{\mathbf{b}}|_1 \geq cN^2 e^{-\gamma N}. \tag{26.38}$$

For $f_{\mathbf{a}} \in S$ the Shannon information $I(\theta, X)$ is bounded by [see (26.32)]

$$CT(\int_{-\pi}^{\pi}(|f_{\mathbf{a}}-1|^2 d\lambda + O(\sup_{\lambda}|f_{\mathbf{a}}(\lambda)-1|^3)) + C.$$

Now

$$\begin{aligned}\int_{-\pi}^{\pi}|1-f_{\mathbf{a}}(\lambda)|^2 d\lambda &\leq Ce^{-\gamma N}N^4 \\ &\times \sum_{-(N-1)\leq j\leq (N-1)} \int_{-\infty}^{\infty} \frac{|\sin^3(\lambda)|}{\lambda^2}\frac{|\sin^3(\lambda-j)|}{(\lambda-j)^2}d\lambda \\ &\leq CN^4 e^{-2\gamma N}. \qquad (26.39)\end{aligned}$$

Following the program outlined above, we return to (26.30) and use the bounds (26.38) for δ in (26.30) and the estimates of $I(\theta, X)$ given by (26.39). We find that

$$\begin{aligned}\Delta_1(T) &\geq cN^2 e^{-\gamma N}\left(1 - C\frac{\sup_{\mathbf{a}} I(\theta, X)}{N}\right) \\ &\geq cN^2 e^{-\gamma N}(1 - CTe^{-2\gamma N}N^3 + O(N^{-1})).\end{aligned}$$

Now take $N \asymp \ln T$ in a such way that $CTN^3 e^{-2\gamma N} \leq 1/2$. We find then that for sufficiently large N,

$$\Delta_1(T) \geq c\left(\frac{\ln T}{T}\right)^{1/2}, \qquad /c > 0.$$

Hence for $1 \leq p < 4$,

$$\Delta_p(T) \geq 2^{-3/4}\Delta_1(T) \geq c\left(\frac{\ln T}{T}\right)^{1/2}.$$

Consider now the case $p > 4$. Define the set S of spectral densities f_j. Let τ be a small positive number. Define $f_j(\lambda)$ for $|\lambda| \leq 1$ as follows:

$$\begin{aligned}f_0(\lambda) &= 1 + e^{-\gamma N}\sum_{N/2\leq j\leq N} P_{2j}(\lambda), \\ f_1(\lambda) &= 1 + e^{-\gamma N}\sum_{N/2\leq j\leq N-N\tau} P_{2j}(\lambda), \ \ldots \qquad (26.40)\end{aligned}$$

where as above P_n are the Legendre polynomials. Notice that $P_{2k}(\lambda)$ and hence $f_j(\lambda)$ are even functions. To any f_j we associate the numbers

$$\lambda_+(j) = |f_j(1) - 1|(\sqrt{2}+1), \quad \lambda_-(j) = |f_j(-1) - 1|(\sqrt{2}+1) \qquad (26.41)$$

and set

$$f_j(\lambda) = 1, \lambda \in [-\pi, \pi] \setminus [-1 - \lambda_-(j), 1 + \lambda_+(j)].$$

At the small intervals $[-1-\lambda_-(j),-1],[1,1+\lambda_+(j)]$ the function $f_j(\lambda)$ is defined as the broken line with slopes 1 and the knots at the points

$$(-1,f_j(-1)),(-1-|f(-1)-1|(1+\frac{1}{\sqrt{2}}),1-\frac{1}{\sqrt{2}}(f_j(-1)-1)),(-1-\lambda_-(j),1) \tag{26.42}$$

and

$$(1,f_j(1)),(1+|f_j(1)-1|(1+\frac{1}{\sqrt{2}}),1-\frac{1}{\sqrt{2}}(f_j(1)-1)),(1+\lambda_+(j),1) \tag{26.43}$$

respectively.

The collection of functions $\{f_j\}$ will constitute the set S. The number of points in the set is close to $N/2\tau$. For large N all these functions are positive and thus they are spectral densities. The restriction of f_j on the interval $[-1,1]$ can be continued analytically into the whole complex plane as the function $\psi_j(z) = 1+e^{-\gamma N}\sum P_k(z)$. The Legendre polynomials $P_n(z)$ for $z \notin [-1,1]$ satisfy the inequality [see Szegö (1959)]

$$|P_n(z)| \le Cn^{1/2}|z+(z^2-1)^{1/2}|^n.$$

Hence it is possible to choose γ in a such way that all $f_j \in F$, in other words, $|\psi_j(z)| \le M, z\in E$ (if M is too small we change 1's in the definition (26.40) of f_j to small δ's etc).

For any two functions f_i, f_j the distance $|f_i - f_j|_p$ satisfies

$$|f_i - f_j|_p \ge e^{-\gamma}\Big(\int_{-1}^{1} |\sum_{k\in I(i,j)} P_k(\lambda)|^p d\lambda\Big)^{1/p}, \tag{26.44}$$

where $I(i,j)$ is an interval of the type $[xN,yN]$ and $y-x>\tau$.

By V. Markov's inequality

$$|P_n'|_\infty \le cn^2|P_n|_\infty \le cn^{5/2}. \tag{26.45}$$

It follows from (26.17) and (26.45) that

$$\int_{-1}^{1}\Big|\sum_{k\in I(i,j)} P_k(\lambda)\Big|^p d\lambda \ge \int_{1-N^{-2}}^{1}\Big|\sum_{k\in I(i,j)} P_k(\lambda)\Big|^p d\lambda \ge c\tau N^{3p/2-2}, \quad c>0.$$

Hence

$$|f_i - f_j|_p \ge c\tau N^{3/2-2/p}e^{-\gamma N}. \tag{26.46}$$

By the definition of f_j,

$$\int_{-\pi}^{\pi}(f_j(\lambda)-1)d\lambda = \sum_k \int_{-1}^{1} P_k(\lambda)d\lambda = 0.$$

Hence by formula (26.32) Shannon's information satisfies

$$I(\theta, X) \le CTe^{-2\gamma N} \sup_j \left(\int_{-\pi}^{\pi} |f_j(\lambda) - 1|^2 d\lambda + O\left(\sup_\lambda |f_j(\lambda) - 1|^3 \right) + C \right).$$

It follows that

$$I(\theta, X) \le CTe^{-2\gamma N} N + C. \tag{26.47}$$

Returning to (26.30), we find with the help of (26.46) and (26.47) that

$$\Delta_p(T) \ge c\tau N^{3/2-2/p} e^{-\gamma N} \left(1 - \frac{C_1 Te^{-2\gamma N} + C_2}{|\ln \tau|} \right), \qquad c > 0.$$

Take at first τ so small that $C_2(|\ln \tau|)^{-1} \le 1/4$ and then choose $N \asymp \ln T$ in such a way that $C_1 Te^{-2\gamma N} |\ln \tau|^{-1} \le 1/4$. We then obtain

$$\Delta_p(T) \ge cT^{-1/2} (\ln T)^{1-2/p}, \qquad c > 0.$$

The case $p = 4$. This case is more complicated and needs a special treatment. We will only outline the proof omitting technical details concerning the estimation of remainders. Consider the set $\mathbf{A}$ of vectors $\mathbf{a} = (a_1, \dots, a_N)$ such that $\mathbf{A} = \{\mathbf{a} : |\mathbf{a}| \le N^2\}$. Set up in correspondence to a vector $\mathbf{a} \in \mathbf{A}$ the spectral density $f_{\mathbf{a}}$ defined in the following way. In the interval $[-1, 1]$,

$$f_{\mathbf{a}}(\lambda) = 1 + T^{-1/2} \sum_1^N a_j P_{2j}(\lambda).$$

Define now the numbers $\lambda_-(\mathbf{a}), \lambda_+(\mathbf{a})$ as in (26.41) and set $f_{\mathbf{a}}(\lambda) = 1$ outside the interval $[-1 - \lambda_-(\mathbf{a}), 1 = \lambda - +(\mathbf{a})]$. In the small intervals $f_{\mathbf{a}}$ is defined by (26.42) and (26.43). We denote the set $\{f_{\mathbf{a}}, \mathbf{a} \in \mathbf{A}\}$ also $\mathbf{A}$. If we relate N and T by $N \sim 1/\gamma \ln T$, one can choose γ in a such way that $\mathbf{A} \subset F$.

We have then that

$$\begin{aligned} \inf_{\hat{f}} \sup_{f \in F} \mathbf{E}_f |f - \hat{f}|_4 &\ge \inf_{\hat{f}} \sup_{f \in \mathbf{A}} \mathbf{E}_{\mathbf{a}} |f_{\mathbf{a}} - \hat{f}|_4 \\ &\ge T^{-1/2} \inf_t \frac{1}{mes\mathbf{A}} \int_{\mathbf{A}} \mathbf{E}_{\mathbf{a}} |\sum_1^N (a_j - t_j) P_{2j}|_4 \, d\mathbf{a}. \end{aligned} \tag{26.48}$$

Let P_0 denote the distribution corresponding to the Gaussian stationary sequence with the spectral density 1. It follows from (26.48) that

$$\Delta_4(T) \ge T^{-1/2} \inf_t \frac{1}{mes\mathbf{A}} \int_{\mathbf{A}} \mathbf{E}_0 \left\{ \left| \sum_1^N (a_j - t_j) P_{2j} \right|_4 \frac{dP_{\mathbf{a}}}{dP_0} \right\} d\mathbf{a}. \tag{26.49}$$

By Lemma 26.3.1 the likelihood ratio can be rewritten as

$$\ln \frac{dP_{\mathbf{a}}}{dP_0} = -\frac{T}{4\pi}\int_{-\pi}^{\pi} \ln f_{\mathbf{a}}(\lambda)d\lambda - \frac{T}{4\pi}\int_{-\pi}^{\pi} I_T(\lambda)\left(\frac{1}{f_{\mathbf{a}}(\lambda)} - 1\right)d\lambda + R_T. \quad (26.50)$$

The integrals

$$\int_{-\pi}^{\pi}(1 - f_{\mathbf{a}}(\lambda)d\lambda = 0.$$

Hence

$$-\frac{T}{4\pi}\int_{-\pi}^{\pi} \ln f_{\mathbf{a}}(\lambda)d\lambda = \frac{|\mathbf{a}|^2}{8\pi} + O(N^4T^{-1/2}).$$

Further

$$\begin{aligned}\frac{T}{4\pi}\int_{-\pi}^{\pi} I_T(\lambda)\left(\frac{1}{f_{\mathbf{a}}(\lambda)} - 1\right)d\lambda &= \frac{T}{4\pi}\int_{-\pi}^{\pi}(I_T(\lambda) - \mathbf{E}I_T(\lambda)(f_{\mathbf{a}}(\lambda) - 1)d\lambda \\ &\quad -\frac{T}{4\pi}\int_{-\pi}^{\pi} I_T(\lambda)(f_{\mathbf{a}}(\lambda) - 1)^2 d\lambda + \Delta.\end{aligned} \quad (26.51)$$

The first integral on the right is asymptotically normal with mean zero and variance $|\mathbf{a}|^2/4\pi$. The expectation of the second integral is

$$\frac{T}{4\pi}\int_{-\pi}^{\pi} \mathbf{E}I_T(\lambda)(f_{\mathbf{a}}(\lambda - 1)^2 d\lambda = \frac{|\mathbf{a}|^2}{4\pi}(1 + o(1))$$

and the variance of this integral is $O(N^5T^{-1})$. The expectation

$$\mathbf{E}|\Delta| \leq CN^5T^{-1/2}.$$

It follows then from (26.49) and (26.50) that for large T

$$\Delta_4(T) \geq \frac{c}{\sqrt{T}}\int_{\mathbf{A}} \mathbf{E}\left\{\left|\sum_1^N (a_j - t_j)P_{2j}\right| \exp\{-\frac{|\mathbf{a}|^2}{8\pi} + (\mathbf{a}, \xi)\}\right\},$$

where $c > 0$ is a positive constant, $\xi = (\xi_1, \ldots, \xi_N)$ and $\xi_1, \ldots, \xi_N$ are iid Gaussian random variables with mean zero and variance $(4\pi)^{-1}$

It is shown in Ibragimov (2003) that the expression on the right is bigger than

$$\frac{c}{\sqrt{T}}\mathbf{E}\left\{\left|\sum_1^N \xi_j P_{2j}\right|_4\right\} \geq c\sqrt{\frac{N}{T}}(\ln)^{1/4} \geq c\sqrt{\frac{\ln T}{T}}(\ln\ln T)^{1/4}.$$

Theorem 26.3.1 is thus proved. Theorem 26.1.1 follows from Corollary 26.2.1 and Theorem 26.3.1. ∎

26.4 Processes with Continuous Time

In this section we suppose that $X(\phi)$ is a stationary (generalized) Gaussian process with mean zero and the spectral density $f(\lambda)$ observed at the interval $[0,T]$. [See about generalized stationary processes Ito (1953) and Gelfand and Vilenkin (1964)]. If $X(\phi)$ is a generalized stationary process with the spectral density $f(\lambda)$, then there exists an integer $r \geq 0$ that

$$\int_{\infty}^{\infty} \frac{f(\lambda)}{(1+\lambda^2)} < \infty.$$

For stationary (not generalized) processes $r = 0$. Denote $C_r(\sigma)$ the class of generalized stationary processes for which

$$\int_{\infty}^{\infty} \frac{f(\lambda)}{(1+\lambda^2)} \leq \sigma^2.$$

In the case of generalized processes observed at the interval $[0,T]$ the set of observable (=statistics) consists of random variables $X(\phi)$, where ϕ runs over the set of all test functions with the support in $[0,T]$. We estimate the restriction of the spectral density f on an interval $[a,b]$ and measure the deviation of estimates from f in $L_p(a,b)$ norms $|.|_p$. As above we define the minimax risk function $\Delta_p(T;F)$ as

$$\Delta_p(T;F) = \Delta_p(T) = \inf \sup \mathbf{E}_f |\hat{f}_T - f|_p,$$

where sup is taken over all $f \in F$ and inf is taken over all possible estimators $\hat{f}_T$.

Theorem 26.4.1 *Let the set F consist of all spectral densities of generalized processes $X \in C_r(\sigma)$ analytic in some bounded region G, $[a,b] \subset G$, and bounded there by a common constant M. Then there exist estimators $\hat{f}_T$ such that*

$$\begin{aligned}
\sup_{f\in F} \mathbf{E}_f |\hat{f}_T - f|_p &\leq C\sqrt{\frac{\ln T}{T}}, \quad 1 \leq p < 4, \\
\sup_{f\in F} \mathbf{E}_f |\hat{f}_T - f|_4 &\leq C\sqrt{\frac{\ln T}{T}}(\ln\ln T)^{1/4}, \\
\sup_{f\in F} \mathbf{E}_f |\hat{f}_T - f|_p &\leq T^{-1/2}(\ln T)^{1-2/p}, \quad 4 < p \leq \infty. \qquad (26.52)
\end{aligned}$$

The constants C depend on G, M, r, σ, p only.

Theorem 26.4.2 *Let the set F consist of all spectral densities of real-valued stationary processes analytic in some bounded region G, $[a,b] \subset G$ and bounded there by a common constant M. Then there exist positive constants $c > 0$ such that*

$$\begin{aligned}\Delta_p(T) &\geq c\sqrt{\frac{\ln T}{T}}, \quad 1 \leq p < 4, \\ \Delta_4(T) &\geq c\sqrt{\frac{\ln T}{T}}(\ln\ln T)^{1/4}), \\ \Delta_p(T) &\geq cT^{-1/2}(\ln T)^{1-2/p}, \quad 4 < p \leq \infty. \end{aligned} \tag{26.53}$$

The constants c depend on G, M, p only.

Proof of Theorem 26.4.1. As above for the sake of simplicity we suppose that $[a,b] = [-1,1]$. To prove Theorem 26.4.1 we again expand $f(\lambda)$ in the interval $[-1,1]$ into the Fourier series with respect to the Legendre polynomials

$$f(\lambda) = \sum_0^N a_j P_j(\lambda),$$

estimate the coefficients a_j by $\hat{a}_j$ and then consider as estimators of $f(\lambda)$ the sums

$$f_N(\lambda) = \sum_0^N \hat{a}_j P_j(\lambda).$$

Now, to estimate a_j we proceed in the following way. If the process $X(t)$ is not a generalized one, we define

$$\hat{a}_k = \int_{-1}^1 I_T(\lambda) P_k(\lambda) d\lambda,$$

where the periodogram

$$I_T(\lambda) = \frac{1}{2\pi T}\left|\int_0^T e^{-it\lambda} X(t)dt\right|^2.$$

The other arguments coincide absolutely with the proof of Corollary 26.2.1.

If $X(\phi)$ is a generalized process, we define an analogue of the periodogram as

$$I_T(\lambda;\phi_T) = \frac{1}{2\pi T}|X(e_\lambda \phi_T)|^2, \; e_\lambda(t) = e^{-it\lambda},$$

where ϕ_T is a test function with the support in $[0,T]$ and set again

$$\hat{a}_k = \int_{-1}^1 I_T(\lambda;\phi_T) P_k(\lambda) d\lambda, \; f_N(\lambda) = \sum_1^N \hat{a}_k P_k(\lambda).$$

The natural way to construct functions ϕ_T is the following one. Take an infinite times differentiable function $\phi(t)$ with the support in $[0,1]$ and such that

$$\int_0^1 |\phi(t)|^2 dt = 1.$$

Set now $\phi_T(t) = \phi(t/T)$. Denote

$$\psi(\lambda) = \int_{-\infty}^{\infty} e^{it\lambda}\phi(t)dt$$

the Fourier transform of ϕ. The Fourier transform of ϕ_T is $T\psi(\lambda T)$ and

$$\frac{1}{2\pi}\int_{-\infty}^{\infty} |\psi(\lambda)|^2 d\lambda = \frac{1}{2\pi T}\int_{-\infty}^{\infty} |\psi_T(\lambda)|^2 d\lambda = \int_0^1 |\phi(t)|^2 dt = 1.$$

If the function ϕ is chosen and fixed, we write $I_T(\lambda)$ instead of $I_T(\lambda;\phi_T)$.

To prove Theorem 26.4.1 we repeat the steps which have been used to prove Theorem 26.2.1.

As above we denote c, C constants which may now depend on $M, G, D_r(\sigma)$.

Lemma 26.4.1 *For all* $1 \le p \le \infty$,

$$|f - \mathbf{E}I_T|_p \le CT^{-1}. \tag{26.54}$$

PROOF. The difference

$$\begin{aligned} f(\lambda) - \mathbf{E}I_T(\lambda) &= f(\lambda) - \frac{T}{2\pi}\int_{-\infty}^{\infty} |\psi(T(\mu - \lambda)|^2 f(\mu) \\ &= \frac{1}{2\pi}\int_{-\infty}^{\infty} |\psi((\mu-\lambda)|^2 (f(\lambda) - f(\lambda+\mu/T))d\lambda. \end{aligned}$$

For $\mu/T \in [-1-\delta, 1+\delta] \subset G$ the difference

$$|f(\lambda) - f(\lambda+\mu/T)| \le C\mu T^{-1}.$$

The function $\psi(\lambda)$ decreases when $|\lambda| \to \infty$ faster than any power $|\lambda|^{-a}$. Hence

$$\begin{aligned} |f(\lambda) - \mathbf{E}I_T(\lambda)| &\le \frac{C}{2\pi T}\int_{\{|\mu|\le T(1+\delta)\}} |\mu||\psi(\lambda-\mu)|^2 d\mu \\ &+ \int_{\{|\mu|\ge T(1+\delta)\}} |\psi(\lambda-\mu)|^2 (C + f(\lambda+\mu/T))d\mu \le CT^{-1}. \end{aligned}$$

The lemma is proved. ∎

Lemma 26.4.2 *The norm of the bias term*

$$|b_N|_p \le C(N^{3/2}T^{-1} + e^{-\gamma N}),\ \gamma > 0. \tag{26.55}$$

The proof of the lemma coincides with the proof of Lemma 26.2.2. The only difference is that instead of Lemma 26.2.1 we use Lemma 26.4.1.

Lemma 26.4.3 *The norms of the random polynomials*

$$Z_N(\lambda) = \sum_0^N (\hat{a}_k - \mathbf{E}a_k)P_k(\lambda)$$

satisfy the following inequalities:

$$\begin{aligned} \mathbf{E}|Z_N|_p &\leq C(N^{1/2}T^{-1/2} + NT^{-1}), \qquad p < 4, \\ \mathbf{E}|Z_N|_4 &\leq C(N^{1/2}T^{-1/2}(\ln N)^{1/4} + NT^{-3/4}), \\ \mathbf{E}|Z_N|_p &\leq C(N^{1-2/p}T^{-1/2} + NT^{-3/4}), \qquad p > 4. \end{aligned} \tag{26.56}$$

PROOF. We copy the proof of Lemma 26.2.3. As above [see (26.19)]

$$\begin{aligned} \mathbf{E}|Z_N(\lambda)|^2 &= \sum_{k,l=1}^N P_k(\lambda)P_l(\lambda) \\ &\times \int_{-1}^1 \int_{-1}^1 P_k(\mu_1)P_l(\mu_2)\mathrm{cov}(I_T(\mu_1), I_T(\mu_2))d\mu_1 d\mu_2. \end{aligned} \tag{26.57}$$

Instead of (26.20) we have now the representation for the covariances,

$$\begin{aligned} \mathrm{cov}(I_T(\lambda), I_T(\mu)) &= \left[\frac{T}{2\pi}\int_{-\infty}^{\infty} \psi(T(l-\lambda)\psi(T(l-\mu)f(l)dl\right]^2 \\ &+ \left[\frac{T}{2\pi}\int_{-\infty}^{\infty} \psi(T(l-\lambda)\psi(T(l+\mu)f(l)dl\right]^2 \end{aligned} \tag{26.58}$$

The last formula gives us the possibility to prove the following analogue of Lemma 26.2.4.

Lemma 26.4.4 *Let $k, l \leq N$. Then*

$$\begin{aligned} I_{kl} &= \int_{-1}^1 \int_1^1 P_k(\lambda)P_l(\mu)\,cov(I_T(\lambda), I_T(\mu)d\lambda d\mu \\ &= \frac{A}{T}\int_{-1}^1 (P_k(\lambda)P_l(\lambda) + P_k(\lambda)P_l(-\lambda))f^2(\lambda)d\lambda + r_{TN}, \end{aligned} \tag{26.59}$$

where $A = 2\pi\int_0^1 |\phi(t)|^4 dt$ and the remainder term $|r_{TN}| \leq CN^{5/2}T^{-3/2}$

PROOF. Using the representation (26.58) we find that I_{kl} is the sum of two summands

$$I_{kl}^{(1)} = \int_{-1}^{1}\int_{-1}^{1} P_k(\lambda)P_l(\mu)\left[\frac{T}{2\pi}\int_{-\infty}^{\infty} \psi(T(l-\lambda)\psi(T(l-\mu)f(l)dl\right]^2$$

and

$$I_{kl}^{(2)} = \int_{-1}^{1}\int_{-1}^{1} P_k(\lambda)P_l(\mu)\left[\frac{T}{2\pi}\int_{-\infty}^{\infty} \psi(T(l-\lambda)\psi(T(l+\mu)f(l)dl\right]^2.$$

It is enough to study $I_{kl}^{(1)}$. We have

$$\begin{aligned}\frac{T}{2\pi}&\int_{-\infty}^{\infty} \psi(T(l-\lambda)\psi(T(l-\mu)f(l)dl\\ &= \frac{1}{2\pi}\int_{-\infty}^{\infty} \psi(l)\psi(l-T(\mu-\lambda))f(l/T+\lambda)dl\\ &= \frac{f(\lambda)}{2\pi}\int_{-\infty}^{\infty} \psi(l)\psi(l-T(\mu-\lambda)dl + R,\end{aligned}$$

where

$$|R| \le C\int_{-\infty}^{\infty} |\psi(l)||f(\lambda)-f(\lambda+\mu/T)|dl \le CT^{-1}.$$

Thus

$$\frac{T}{2\pi}\int_{-\infty}^{\infty} \psi(T(l-\lambda)\psi(T(l-\mu)f(l)dl = f(\lambda)h(T(\mu-\lambda)+O(T^{-1}), \quad (26.60)$$

where the function

$$h(\mu) = \frac{1}{2\pi}\int_{-\infty}^{\infty} \psi(l)\psi(l-\mu)dl.$$

For any $a > 0$ the function $|h(\mu)| \le c_a(1+\mu^2)^{-a}$ and the integral

$$\frac{1}{2\pi}\int_{-\infty}^{\infty} |h(\mu)|^2 d\mu = \int_{-\infty}^{\infty} |\phi(t)|^4 dt = 1.$$

It follows from (26.60) that

$$I_{kl}^{(1)} = \frac{1}{T}\int_{-1}^{1}\int_{-1}^{1} P_k(\lambda)P_l(\mu)f^2(\lambda)|h(T(\mu-\lambda)d\lambda d\mu + O(T^{-2}). \quad (26.61)$$

Set

$$\begin{aligned}\tilde{P}_k(\mu) &= P_k(\mu), \quad |\mu| \le 1,\\ \tilde{P}_k(\mu) &= 0, \qquad\quad |\mu| > 1.\end{aligned}$$

Write the integral with respect to μ in (26.61) as

$$\int_{-\infty}^{\infty} [\tilde{P}_l(\lambda+\mu/T) - \tilde{P}_l(\lambda)]|h(\mu)|^2 d\mu + P_l(\lambda)\int_{-\infty}^{\infty} |h(\mu)|^2 d\mu.$$

It follows from the properties of the function h and the inequalities (26.17) that

$$\int_{-1}^{1}|P_k(\lambda)|d\lambda\int_{-\infty}^{\infty}|\tilde{P}_l(\lambda+\mu/T)-\tilde{P}_l(\lambda)||h(\mu)|^2d\mu$$
$$\leq\sqrt{N}\int_{-1}^{1}d\lambda\int_{-\infty}^{\infty}|\tilde{P}_l(\lambda+\mu/T)-\tilde{P}_l(\lambda)||h(\mu)|^2d\mu$$
$$\leq C(N^3T^{-1}+N^{1/2}T^{-1/2}).$$

Lemma 26.4.4 is proved ∎

The rest of the proof of Lemma 26.4.3 and the theorem coincides absolutely with the same part of the proof of Lemma 26.2.3 and Theorem 26.2.1 and we omit it.

The proof of Theorem 26.4.1 coincides up to technical details with the proof of Theorem 26.3.1 and we only sketch it. As in Section 26.3 the proof is based on (26.30). Notice at first that if spectral densities f_1, f_2 of Gaussian stationary processes coincide outside an interval $[-a,a]$, the probability measures P_1^T, P_2^T generated by the random variables $\{X_1(t), 0\leq t\leq T\}, \{X_2(t), 0\leq t\leq T\}$ are equivalent [see, for example, Ibragimov and Rozanov (1978, Ch. 3)].

Let $g_0(\lambda)=1$ if $|\lambda|\leq\pi$ and $g_0(\lambda)=2(1+\lambda^2)^{-1}$ if $|\lambda|>\pi$. Let the spectral densities $g_{\mathbf{a}}(\lambda)$ coincide with the spectral densities $f_{\mathbf{a}}(\lambda)$ of Section 26.3 for $|\lambda|\leq\pi$ and $g_{\mathbf{a}}$ coincide with g_0 for $|\lambda|>\pi$. Denote $P_0^T, P_{\mathbf{a}}^T$ the distributions of Gaussian processes $\{X_0(t), 0\leq t\leq T\}, \{X_{\mathbf{a}}(t), 0\leq t\leq T\}$ with spectral densities $g_0, g_{\mathbf{a}}$ respectively.

It follows from the formula of Lemma 26.3.1 that

$$\mathbf{E}\frac{dP_{\mathbf{a}}^T}{dP_0^T}(X)$$
$$=-\frac{T}{4\pi}\int_{-\infty}^{\infty}\ln\frac{g_{\mathbf{a}}(\lambda)}{g_0(\lambda)}d\lambda+\frac{T}{4\pi}\int_{-\infty}^{\infty}\mathbf{E}I_T(\lambda)\left(\frac{1}{g_{\mathbf{a}}(\lambda)}-\frac{1}{g_0(\lambda)}\right)d\lambda+R_{T\mathbf{a}}$$

where now the periodogram

$$I_T(\lambda)=\frac{1}{2\pi T}\left|\int_0^T e^{-it\lambda}X(t)dt\right|^2$$

and $\sup_{\mathbf{a},\mathbf{T}} R_{T\mathbf{a}}\leq C$. To get this result one can consider the Gaussian sequences $X^{(N)}(k)=X(\frac{k}{N})$, apply to $X^{(N)}(k)$ Lemma 26.3.1 and pass to the limit when $N\to\infty$. We omit the details. Thus we get for Shannon's information absolutely the same bounds as in Section 26.3 and hence we found for $1\leq p<4$ that $\Delta_p(T)\geq c(T^{-1}\ln T$. If $p\geq 4$ we apply the same method as above to the constructions of Section 26.3. These arguments prove Theorem 26.4.2. ∎

References

1. Bentkus, R. (1985a). On the mean square minimax risk asymptotic of the statistical estimators of the spectral density, *Litovskii Mat. Sbornik*, **25**, 42–56 (in Russian).

2. Bentkus, R. (1985b). Uniform convergence rate of the spectral density statistical estimators, *Litovskii Mat. Sbornik*, **25**, 31–37 (in Russian).

3. Dzhaparidze, K. (1986). *Parameter Estimation and Hypothesis Testing in Spectral Analysis of Stationary Time Series*, Springer-Verlag, New York.

4. Gallager, R. (1968). *Information Theory and Reliable Communications*, John Wiley & Sons, New York.

5. Gelfand, I. and Vilenkin, N. (1964). *Generalized Functions – Vol. 4: Applications of Harmonic Analysis*, Academic Press, New York.

6. Golubev, G. K. (1991). LAN in problems of nonparametric estimation of functions and lower bounds for quadratic risk, *Theory of Probability and its Applications*, **36**, 152–157.

7. Hannan, E. J. (1970). *Multiple Time Series*, John Wiley & Sons, New York.

8. Ibragimov, I. A. (1963). On the estimation of the spectral function of a stationary Gaussian process, *Theory of Probability and its Applications*, **8**, 366–400.

9. Ibragimov, I. (2003). On estimation of analytic functions, *Studia Sci. Math. Hungarica* (to appear).

10. Ibragimov, I. A. and Khasminskii, R. Z. (1981). *Statistical Estimation: Asymptotic Theory*, Springer-Verlag, New York.

11. Ibragimov, I. A. and Khasminskii, R. Z. (1989). On density estimation in the view of Kolmogorov's ideas in approximation theory, *Annals of Statistics*, **18**, 999–1010.

12. Ibragimov, I. and Rozanov, A. (1978). *Gaussian Random Processes*, Springer-Verlag, New York.

13. Ito, K. (1953). Stationary random distributions, *Mem. Coll. Sci. Univ. Kyoto*, **28**, 209–223.

14. Szegö, S. (1959). *Orthogonal polynomials*, AMS Colloq., Publ vol. XXIII.

15. Timan,A. F. (1963). *Theory of Approximation of Functions of a Real Variable*, McMillan, New York.

27

Sub-optimal Estimation of an Unknown Function from Stationary Noisy Data

V. Solev

Steklov Mathematical Institute, Saint Petersburg, Russia

Abstract: Generally the problem looks as follows. On a large time interval $[-T, T]$ we observe an unknown function $s \in \mathcal{L}_*$ in stationary noise. The spectral density f of the noise process is unknown and belongs to a given class of nonnegative functions $\mathcal{K}$. The main interest of the paper is the search for a simple condition on the parametric set $\mathcal{L}_*$ and the class $\mathcal{K}$ under which there exists an asymptotically sub-optimal estimator $\hat{s}_T$.

Keywords and phrases: Sub-optimal estimator, estimation in stationary noise, pseudo-periodic function, minimax risk

27.1 Statistical Problem

Generally the problem looks as follows. On a large time interval $[-T, T]$ we observe a process $Y(t)$,

$$dY(t) = s(t)dt + dX(t), \tag{27.1}$$

where an unknown function s belongs to a given set $\mathcal{L}_*$,

$$\mathcal{L}_* \subset L^2_{loc},$$

$X(t)$ is a zero-mean Gaussian process with stationary increments and with the spectral density f. The spectral density f of the noise process is unknown and belongs to a given class of nonnegative functions $\mathcal{K}$.

To estimate an unknown function s one makes observations (for some collection $\mathcal{D}_T$ of smooth functions φ supported on interval $[-T, T]$)

$$\mathbf{y}[\varphi] = \mathbf{s}[\varphi] + \mathbf{x}[\varphi],$$

and constructs an estimator $\hat{s}_T$, based on these observations. Here we set

$$\mathbf{y}[\varphi] = \int \varphi(t)\, dY(t),$$

and define in the same way random variables $\mathbf{s}[\varphi]$ and $\mathbf{x}[\varphi]$ for $\varphi \in \mathcal{D}_T$.

We denote by $\mathcal{L}$ the Banach space of locally square integrable functions with the norm $\|s\|^2_{\mathcal{L}}$:

$$\|s\|^2_{\mathcal{L}} = \sup_x \int_x^{x+1} |s(t)|^2\, dt,$$

and assume that $\mathcal{L}_* \subset \mathcal{L}$. We shall use the notation $\mathfrak{F}$ for the σ-algebra

$$\mathfrak{F} = \sigma\left\{\mathbf{y}[\varphi],\ \varphi \in \mathcal{D}\right\}$$

and define the σ-algebra $\mathfrak{F}_T$ by

$$\mathfrak{F}_T = \sigma\left\{\mathbf{y}[\varphi],\ \varphi \in \mathcal{D}_T\right\}.$$

An estimator $\hat{s}_T(\cdot)$ is $\mathfrak{F}_T$-measurable random element of the space $\mathcal{L}$ such that $\hat{s}_T \in \mathcal{L}_*$. The set of all such estimators we denote by $\mathcal{N}(T)$.

Consider as the loss function of an estimator $\hat{s}$ for s the value $\mathbf{E}_f \|\hat{s} - s\|^2_{\mathcal{L}}$. The risk function $\delta(\hat{s})$ is defined by

$$\delta(\hat{s}) = \delta(\hat{s}, \mathcal{L}_*, f) = \sup_{s \in \mathcal{L}_*} \mathbf{E}_f \|\hat{s} - s\|^2_{\mathcal{L}},$$

and the minimax risk is

$$\Delta(T, \mathcal{L}_*, f) = \inf_{\hat{s} \in \mathcal{N}(T)} \delta(\hat{s}, \mathcal{L}_*, f). \tag{27.2}$$

For a given estimator $\hat{s} \in \mathcal{N}_T$, we can take the ratio

$$\rho(\hat{s}, T, \mathcal{L}_*, f) = \frac{\delta(\hat{s}, T, \mathcal{L}_*, f)}{\Delta(T, \mathcal{L}_*, f)}, \tag{27.3}$$

in order to compare an estimator $\hat{s}$ and the minimax estimator. It is convenient to think that an estimator $\hat{s}_T$ is the value of a function $\tilde{s}$ on a point T:

$$\tilde{s}(T) = \hat{s}_T.$$

So it is reasonable to write

$$\delta(\tilde{s}, T) = \delta(\hat{s}_T), \quad \rho(\tilde{s}, T, \mathcal{L}_*, f) = \rho(\hat{s}_T, T, \mathcal{L}_*, f).$$

We set

$$\varrho(\tilde{s}, \mathcal{L}_*, \mathcal{K}) = \limsup_{T \to \infty} \sup_{f \in \mathcal{K}} \rho(\tilde{s}, T, \mathcal{L}_*, f)$$

An estimator $\hat{s}_T$ will be called asymptotically sub-optimal if

$$\varrho(\tilde{s}, \mathcal{L}_*, \mathcal{K}) < \infty.$$

The main interest of the talk is the search for a simple condition on the parametric set $\mathcal{L}_*$ and the class $\mathcal{K}$ under which there exists an asymptotically sub-optimal estimator $\hat{s}_T$. We investigate also the asymptotic behavior of the minimax risk $\Delta(T, \mathcal{L}_*, f)$.

27.2 Process with Stationary Increments

If $X(t)$ is a zero-mean Gaussian process with stationary increments, then there exists such a measure μ that

$$\mathbf{E}\,|\mathbf{x}[\varphi]|^2 = \int_{-\infty}^{\infty} |\hat{\varphi}(u)|^2\, d\mu, \ \varphi \in \mathcal{D}. \tag{27.4}$$

Here $\hat{\varphi}$ is the Fourier transformation of function φ. The measure ν in (27.4) satisfies the condition

$$\int_{-\infty}^{\infty} \frac{d\nu}{1+u^2} < \infty \tag{27.5}$$

and is called the spectral measure of the process $X(t)$ [see for details Ibragimov and Rosanov (1974)]. If spectral measure ν is absolutely continuous with respect to the Lebesgue measure du, then the function

$$f(u) = \frac{d\nu}{du}$$

is called the spectral density of the process $X(t)$. Clearly,

$$\int_{-\infty}^{\infty} \frac{f(u)}{1+u^2}\, du < \infty. \tag{27.6}$$

Therefore, for any $\varphi_1, \varphi_2 \in \mathcal{D}_T$,

$$\mathbf{E}\,\mathbf{y}[\varphi_1]\,\overline{\mathbf{y}[\varphi_2]} = \mathbf{s}[\varphi_1]\,\overline{\mathbf{s}[\varphi_2]} + \int_{-\infty}^{\infty} \hat{\varphi}_1(u)\overline{\hat{\varphi}_1(u)} f(u)\, du.$$

For a nonnegative function f denote by L^2_f the L^2-space which is constructed on a measure with density f. We use notations $(\cdot,\cdot)_f$, $\|\cdot\|_f$ for the inner product and the norm of the Hilbert space L^2_f. In the special case as $f(t)\equiv 1$ we shall use notations $L^2, (\cdot,\cdot), \|\cdot\|$. So we have

$$\mathbf{E}\,\mathbf{y}[\varphi_1]\,\overline{\mathbf{y}[\varphi_2]} = \mathbf{s}[\varphi_1]\,\overline{\mathbf{s}[\varphi_2]} + (\hat{\varphi}_1,\hat{\varphi}_2)_f \quad (\varphi_1,\varphi_2\in\mathcal{D}_T)\,, \tag{27.7}$$

and

$$\mathbf{E}\,|\mathbf{y}[\varphi]|^2 = |\mathbf{s}[\varphi]|^2 + \|\hat{\varphi}\|_f^2 \quad (\varphi\in\mathcal{D}_T)\,. \tag{27.8}$$

27.3 Linear Estimators

For the set M denote by $\operatorname{sp} M$ the linear span of M, and by $\overline{\operatorname{sp}}\, M$ the closure of $\operatorname{sp} M$. We set

$$Y_T = \overline{\operatorname{sp}}\,\{\mathbf{y}[\varphi],\ \varphi\in\mathcal{D}_T\}.$$

Here the closure is taken in the space $L^2(dP)$. It is evident that the linear set $\{\mathbf{y}[\varphi],\ \varphi\in\mathcal{D}_T\}$ is dense in Y_T.

Definition 27.3.1 The estimator $\hat{s}_T\in\mathcal{N}(T)$ is called a linear estimator if

$$\hat{s}_T[\varphi] = \int \hat{s}_T(t)\varphi(t)\,dt \in Y_T, \text{ for any } \varphi\in\mathcal{D}_T. \tag{27.9}$$

The class of all such estimators will be denoted by $\mathcal{L}(T)$.

Further we shall write $\Delta_{\mathcal{N}}(T,\mathcal{L}_*,f)$ instead of $\Delta(T,\mathcal{L}_*,f)$ and set

$$\Delta_{\mathcal{L}}(T,\mathcal{L}_*,f) = \inf_{\hat{s}\in\mathcal{L}(T)} \delta(\hat{s},\mathcal{L}_*,f).$$

The estimator $\hat{s}_T$ is called the linear minimax estimator if

$$\delta(\hat{s}) = \delta(\hat{s},\mathcal{L}_*,f) = \sup_{s\in\mathcal{L}_*} \mathbf{E}_f\,\|\hat{s}-s\|^2_{\mathcal{L}} = \Delta_{\mathcal{L}}(T,\mathcal{L}_*,f).$$

It is clear that $\Delta_{\mathcal{L}}(T,\mathcal{L}_*,f)\geq\Delta_{\mathcal{N}}(T,\mathcal{L}_*,f)$. We shall prove, under some conditions on $\mathcal{L}_*$ and $\mathcal{K}$, that

$$\limsup_{T\to\infty}\ \sup_{f\in\mathcal{K}} \frac{\Delta_{\mathcal{L}}(T,\mathcal{L}_*,f)}{\Delta_{\mathcal{N}}(T,\mathcal{L}_*,f)} < \infty.$$

27.4 Pseudo-periodic Functions

Let s be a function, $s \in \mathcal{L}$. Consider the smallest shift invariant linear set $L[s]$ which contains s. Function s is called a pseudo-periodic function with the period $d > 0$ if there exists the constant $C = C(s)$ such that, for any $g \in L[s]$,

$$\sup_{x,\,y} \frac{\int_x^{x+A} |g(t)|^2\, dt}{\int_y^{y+A} |g(t)|^2\, dt} \leq C, \text{ as } A > d. \tag{27.10}$$

Paley and Wiener (1934) proved that, under the condition (27.10), there exists the set

$$\Lambda = \Lambda(s) = \{u_j, j = 1, 2, \cdots\} \subset \mathbb{R}^1$$

such that

$$\kappa = \kappa\,(\Lambda) = \inf_{i \neq j} |u_i - u_j| > 0, \tag{27.11}$$

and the series

$$\sum_{u \in \Lambda} a(u) e^{itu}$$

converges in the space $\mathcal{L}$ to the function s. We denote by $\mathcal{L}\,(\Lambda)$ the set of all functions s such that

$$s(t) = \sum_{u \in \Lambda} a(u) e^{itu}.$$

The set Λ is called the spectral set of a function s if $a(u) \neq 0$, as $u \in \Lambda$. The following propositions belong to Paley and Wiener.

Proposition 27.4.1 *Suppose that the spectral set Λ satisfies the condition $\kappa(\Lambda) > 0$. Then there exist positive constants C_1, C_2, which depend only on κ, such that*

$$C_1 \sum_{u \in \Lambda} |a(u)|^2 \leq \|s\|^2_{\mathcal{L}} \leq C_2 \sum_{u \in \Lambda} |a(u)|^2,$$

for any $s \in \mathcal{L}(\Lambda)$ with the representation

$$s(t) = \sum_{u \in \Lambda} a(u) e^{iut}.$$

Proposition 27.4.2 *Suppose that the spectral set* Λ *satisfies the condition* $\kappa(\Lambda) > 0$. *Then there exist positive constants* d *and* C, *which depend only on* κ, *such that, for any* $x, y \in \mathbf{R}$, *and any* $s \in \mathcal{L}(\Lambda)$,

$$\int_x^{x+A} |s(u)|^2 \, du \leq C \int_y^{y+A} |s(u)|^2 \, du, \quad \text{as } d < A < \infty.$$

27.5 Parametric Set $\mathcal{L}_*$

For a given the spectral set $\Lambda = \{u_j, j = 1, 2, \cdots\}$ such that

$$\kappa = \kappa(\Lambda) = \inf_{i \neq j} |u_i - u_j| > 0,$$

we assume that parametric set $\mathcal{L}_* \subset \mathcal{L}(\Lambda)$. So, if a function $s \in \mathcal{L}_*$, then there exists the representation

$$s(t) = \sum_{u \in \Lambda} \theta(u) \exp\{iut\}. \tag{27.12}$$

Denote by $l^2(\Lambda)$ the L^2-space of functions $\theta(u)$ (which defined on Λ) with respect to the counting measure. That is $l^2(\Lambda)$ is the Hilbert space with the inner product $(\cdot, \cdot)_{l^2}$ and the norm $\|\cdot\|_{l^2}$ that is defined by

$$(\theta_1, \theta_2)_{l^2} = \sum_{u \in \Lambda} \theta_1(u) \overline{\theta_2(u)}, \quad \|\theta\|_{l^2}^2 = \sum_{u \in \Lambda} |\theta(u)|^2.$$

By (27.12) we can define a linear operator

$$V : \mathcal{L}(\Lambda) \to l^2(\Lambda), \ Vs = \theta.$$

For analytical reasons it is convenient to deal with the Hilbert norm $\|\cdot\|_T$, and the inner product $(\cdot, \cdot)_T$,

$$\|s\|_T^2 = \frac{1}{2T} \int_{-T}^{T} |s(t)|^2 \, dt, \quad (s_1, s_2)_T = \frac{1}{2T} \int_{-T}^{T} s(t) \overline{s_2(t)} \, dt.$$

By Proposition 27.4.1 and Proposition 27.4.2, the Banach norm $\|\cdot\|_{\mathcal{L}}^2$ and the Hilbert norm $\|\cdot\|_T^2$ are topologically equivalent on $\mathcal{L}(\Lambda)$ as $\kappa(\Lambda) > 0$ and T is sufficiently large.

Proposition 27.5.1 *Under the condition* $\kappa > 0$*, there exist positive constants* $C_1 = C_1(\kappa) \le C_2 = C_2(\kappa)$ *and* $T_* = T_*(\kappa)$ *such that, for any* $s \in \mathcal{L}(\Lambda)$,

$$C_1 \, \|s\|_{\mathcal{L}}^2 \le \frac{1}{2T} \int_{-T}^{T} |s(t)|^2 \, dt \le C_2 \, \|s\|_{\mathcal{L}}^2, \quad as \; T \ge T_*.$$

Denote

$$\|s\|_*^2 = \lim_{T \to \infty} \|s\|_T^2, \quad (s_1, s_2)_* = \lim_{T \to \infty} (s_1, s_2)_T \,.$$

It is clear that, for $s \in \mathcal{L}(\Lambda)$,

$$\|s\|_*^2 = \sum_{u \in \Lambda} |\theta(u)|^2, \quad \text{as} \;\; s(t) = \sum_{u \in \Lambda} \theta(u) \exp\{iut\},$$

and, for $s_1, s_2 \in \mathcal{L}(\Lambda)$,

$$(s_1, s_2)_* = \sum_{u \in \Lambda} \theta_1(u) \overline{\theta_2(u)}, \quad \text{as} \;\; s_j(t) = \sum_{u \in \Lambda} \theta_j(u) \exp\{iut\}, \; (j = 1, 2).$$

Therefore the operator V is the isometry : $\mathcal{L}(\Lambda) \to l^2(\Lambda)$, if $\mathcal{L}(\Lambda)$ is considered as the Hilbert space with the norm $\| \cdot \|_*$.

Proposition 27.5.2 *Under the condition* $\kappa > 0$*, there exist positive constants* $C_1 = C_1(\kappa) \le C_2 = C_2(\kappa)$ *such that, for any* $s \in \mathcal{L}(\Lambda)$,

$$C_1 \, \|s\|_{\mathcal{L}}^2 \le \|s\|_*^2 \le C_2 \, \|s\|_{\mathcal{L}}^2, \quad as \;\; T \ge T_*.$$

Suppose that an estimator $\hat{s}_T$ is a random element with values in $\mathcal{L}(\Lambda)$. It is convenient to deal with the loss function $\mathbf{E} \, \|\hat{s}_T - s\|_*^2$. Denote

$$\delta_*(\hat{s}_T) = \sup_{s \in \mathcal{L}_*} \mathbf{E} \, \|\hat{s}_T - s\|_*^2.$$

The following proposition (which can be proved from Proposition 27.5.2) is useful for this purpose.

Proposition 27.5.3 *Suppose an estimator* $\hat{s}_T$ *is a random element with values in* $\mathcal{L}(\Lambda)$*. Under the condition* $\kappa > 0$*, there exist positive constants* $C_1 = C_1(\kappa) \le C_2 = C_2(\kappa)$ *such that*

$$C_1 \, \delta_*(\hat{s}_T) \le \delta(\hat{s}_T) \le C_2 \, \delta_*(\hat{s}_T).$$

Now suppose that the system $\{\psi_1, \psi_2, \ldots\}$ is an orthonormal basis in $\mathcal{L}(\Lambda)$ if $\mathcal{L}(\Lambda)$ is considered as the Hilbert space with the norm $\| \cdot \|_*$. It must be noted that the system $\{\exp\{iut\}, \; u \in \Lambda\}$ is an orthonormal basis in $\mathcal{L}(\Lambda)$. Let $\hat{s}_T$ be an estimator for $s \in \mathcal{L}_* \subset \mathcal{L}(\Lambda)$. We assume that $\hat{s}_T$ is a random element with values in $\mathcal{L}_*$,

$$\hat{s}_T(t) = \sum_{j=1}^{\infty} \hat{\theta}_T^j \psi_j(t). \tag{27.13}$$

Consider the representation

$$s(t) = \sum_{j=1}^{\infty} \theta_j \psi_j(t),$$

and define operator V by the relation: $Vs = \theta$, where $\theta = (\theta_1, \theta_2, \ldots)$. Denote $\Theta_* = V\mathcal{L}_*$. It is clear that $\hat{\theta}_T = \left(\hat{\theta}_T^1, \hat{\theta}_T^2, \ldots\right)$ may be considered as an estimator for $\theta = Vs$. We set

$$\delta_*(\hat{\theta}_T) = \sup_{\theta \in \Theta_*} \mathbf{E}\, \|\hat{\theta}_T - \theta\|_{l^2}^2 = \sup_{\theta \in \Theta_*} \mathbf{E} \sum_{j=1}^{\infty} |\hat{\theta}_T^j - \theta_j|^2.$$

It is clear that $\delta_*(\hat{\theta}_T) = \delta_*(\hat{s}_T)$. Thus, we obtain from Proposition 27.5.3.

Proposition 27.5.4 *Suppose an estimator $\hat{s}_T$ is a random element with values in $\mathcal{L}(\Lambda)$, and the estimator $\hat{\theta}_T$ is defined on $\hat{s}_T$ by (27.13). Under the condition $\kappa > 0$, there exist positive constants $C_1 = C_1(\kappa) \le C_2 = C_2(\kappa)$ such that*

$$C_1\, \delta_*(\hat{\theta}_T) \le \delta(\hat{s}_T) \le C_2\, \delta_*(\hat{\theta}_T).$$

27.6 Example

Consider the case that an unknown function s is observed in white noise on an interval $[-T, T]$:

$$dY(t) = s(t)dt + \upsilon\, dW(t). \tag{27.14}$$

Given observations $\mathbf{y}[\varphi]$, $\varphi \in \mathcal{D}_T$, we wish to estimate s as we know that s, belongs to a given $\mathcal{L}_*$. Here $W(t)$ is the Wiener process, positive υ is known.

Let $\mathcal{P}(s)$ be the measure on $\mathfrak{F}$ induced by the process

$$\mathbf{y}[\varphi] = \mathbf{s}[\varphi] + \mathbf{x}[\varphi], \quad \varphi \in \mathbf{D}$$

and $\mathcal{P}_T(s)$ be the restriction of $\mathcal{P}(s)$ on $\mathfrak{F}_T$.

Suppose for simplicity, that we try to estimate an unknown function s from the observation $\mathbf{y}[\varphi], \varphi \in \mathbf{D}_T$, given that $s = s_1$ or $s = s_2$:

$$\mathcal{L}_* = \{s_1, s_2\}. \tag{27.15}$$

It is clear that a consistent estimator $\hat{s}$ for s exists only in the case when Gaussian measures

$$\mathcal{P}(s_1) \text{ and } \mathcal{P}(s_2) \text{ are orthogonal.}$$

For example, it is impossible to estimate s consistently if $\mathcal{L}_* \subset L^2_{loc}$ and for an interval $[a, b]$,

$$s_1(t) = s_2(t), \text{ as } t \notin [a, b].$$

Our main conclusion is as follows. *Generally speaking, it is impossible to estimate consistently any local characteristic of the unknown function s.*

27.6.1 Hyperrectangle

Now, for a given system $\{\psi_j,\ j = 1, 2, \ldots\}$ of real-valued functions $\psi_j \in \mathcal{L}$, and fixed T, suppose

$$\mathcal{L}_* = \{s :\ s(t) = \sum_{j=1}^{\infty} \theta_j \psi_j(t),\ \theta = (\theta_1, \theta_2, \ldots) \in \Theta \subset l^2\}. \tag{27.16}$$

At the beginning we assume that

$$\int_{-T}^{T} \psi_i(t)\psi_j(t)\, dt = 0, \text{ if } i \neq j. \tag{27.17}$$

We have observations

$$\mathbf{y}_j = \theta_j + \mathbf{x}_j, \tag{27.18}$$

$$\mathbf{y}_j = \frac{\int_{-T}^{T} \psi_i(t)\, dY(t)}{\int_{-T}^{T} |\psi_i(t)|^2\, dt}, \qquad \mathbf{x}_j = \frac{v \int_{-T}^{T} \psi_i(t)\, dW(t)}{\int_{-T}^{T} |\psi_i(t)|^2\, dt}. \tag{27.19}$$

Here $\mathbf{x}_1, \mathbf{x}_2, \ldots$ are i.i.d. normal variables, $\mathbf{x}_j \in N(0, \sigma_j^2)$, where

$$\sigma_j^2 = \mathbf{E}\, \mathbf{x}_j^2 = \frac{v^2}{\int_{-T}^{T} |\psi_j(t)|^2\, dt}. \tag{27.20}$$

It must be noted that, if $\psi_j \in \mathcal{L}(\Lambda)$ with $\kappa(\Lambda) > 0$, then σ_j^2 tends to zero as $T \to \infty$ with the rate $1/T$. The situation is not so complicated [see details in Donoho et al. (1990)] if we take in (27.16) the set

$$\Theta = \Theta(\tau) = \{\theta : \theta = (\theta_1, \theta_2, \ldots) : |\theta_j| \le \tau_j, \ (j = 1, \ldots)\},$$

where $\boldsymbol{\tau} = (\tau_1, \ldots) \in l^2$.

27.6.2 The one-dimensional parametric set

Consider estimating an unknown coefficient θ when we observe on interval $[-T, T]$ the random process $Y(t)$,

$$dY(t) = \theta\psi(t)dt + v\,dW(t),$$

given that $|\theta| \le \tau$. Here ψ is a known function from the space $\mathcal{L}$. So unknown function $s(t) = \theta\psi(t)$ belongs to the parametric set

$$\mathcal{L}_* = \{s : s(t) = \theta\psi(t), |\theta| \le \tau\}.$$

It is clear that statistics

$$\mathbf{y} = \frac{\int_{-T}^{T} \psi(t)dY(t)}{\int_{-T}^{T} |\psi(t)|^2 dt}$$

is the sufficient statistics. In other words we need to know only the observation

$$\mathbf{y} = \mathbf{y} = \theta + \mathbf{x}, \quad \text{where } \mathbf{x} \in N(0, \sigma^2). \tag{27.21}$$

$$\text{Here } \mathbf{x} = \frac{v \int_{-T}^{T} \psi(t)dW(t)}{\int_{-T}^{T} |\psi(t)|^2 dt}, \quad \sigma^2 = \frac{v^2}{\int_{-T}^{T} |\psi(t)|^2 dt}. \tag{27.22}$$

It is clear that any estimator $\hat{s}_T$ for s with value in $\mathcal{L}_*$ may be represented in the form

$$\hat{s}_T(t) = \hat{\theta}_T\psi(t),$$

where $\hat{\theta}_T$ is an estimator for coefficient θ. The risk $\delta(\hat{s}_T)$ as we use the estimator $\hat{s}_T$ is

$$\delta(\hat{s}_T) = \delta(\hat{\theta}_T)\|\psi\|^2_{\mathcal{L}}. \tag{27.23}$$

Here

$$\delta(\hat{\theta}_T) = \sup_{\theta \in \Theta} \mathbf{E} \left(\hat{\theta}_T - \theta\right)^2.$$

In this case the linear minimax estimator $\hat{\theta}_T$ is

$$\hat{\theta}_T = \frac{\tau^2}{\tau^2 + \sigma^2} \mathbf{y}. \tag{27.24}$$

So

$$\Delta_{\mathcal{L}}(T, \mathcal{L}_*) = \frac{\tau^2 \sigma^2}{\tau^2 + \sigma^2} \|\psi\|^2_{\mathcal{L}}.$$

As it was proved by Ibragimov and Has'minskii (1984), the minimax linear risk in this problem is not so different from the nonlinear minimax risk. Namely, consider the ratio

$$\mu(\tau, \sigma) = \frac{\Delta_{\mathcal{L}}(T, \mathcal{L}_*)}{\Delta_{\mathcal{N}}(T, \mathcal{L}_*)}.$$

Ibragimov and Has'minskii proved that

$$\mu^* = \sup_{\tau, \sigma} \frac{\Delta_{\mathcal{L}}(T, \mathcal{L}_*)}{\Delta_{\mathcal{N}}(T, \mathcal{L}_*)} < \infty. \tag{27.25}$$

The linear minimax estimator $\hat{\theta}_T$ in (27.24) depends on ratio τ/σ. But it is possible to construct an estimator θ_T^* for θ without any information about the power of noise. Consider the estimator θ_T^*,

$$\theta_T^* = \begin{cases} \mathbf{y}, & \text{if } |\mathbf{y}| \le \tau, \\ \tau, & \text{if } \mathbf{y} > \tau, \\ -\tau, & \text{if } \mathbf{y} < -\tau. \end{cases} \tag{27.26}$$

We shall prove that

$$\delta(\theta_T^*) = \sup_{|\theta| \le \tau} \mathbf{E} \left(\theta_T^* - \theta\right)^2 \le 4 \left(\tau^2 \wedge \sigma^2\right), \tag{27.27}$$

where $x \wedge y = \min\{x, y\}$. It is clear that

$$\begin{aligned} \mathbf{E} \left(\theta_T^* - \theta\right)^2 &= \frac{\sigma^2}{\sqrt{2\pi}} \int\limits_{|\theta + \sigma x| \le \tau} x^2 \exp\left\{-\frac{x^2}{2}\right\} dx \\ &+ \frac{(\tau - \theta)^2}{\sqrt{2\pi}} \int\limits_{\theta + \sigma x \ge \tau} \exp\left\{-\frac{x^2}{2}\right\} dx \\ &+ \frac{(-\tau - \theta)^2}{\sqrt{2\pi}} \int\limits_{\theta + \sigma x \le -\tau} \exp\left\{-\frac{x^2}{2}\right\} dx. \end{aligned} \tag{27.28}$$

Since

$$\frac{(\tau-\theta)^2}{\sqrt{2\pi}} \int\limits_{\theta+\sigma x \geq \tau} \exp\left\{-\frac{x^2}{2}\right\} dx + \frac{(-\tau-\theta)^2}{\sqrt{2\pi}} \int\limits_{\theta+\sigma x \leq -\tau} \exp\left\{-\frac{x^2}{2}\right\} dx$$
$$\leq \frac{\sigma^2}{\sqrt{2\pi}} \int\limits_{x \geq \frac{(\tau-\theta)}{\sigma}} x^2 \exp\left\{-\frac{x^2}{2}\right\} dx + \frac{\sigma^2}{\sqrt{2\pi}} \int\limits_{x \leq \frac{(-\tau-\theta)}{\sigma}} x^2 \exp\left\{-\frac{x^2}{2}\right\} dx,$$

then we obtain

$$\mathbf{E}\ (\theta_T^* - \theta)^2 \leq \sigma^2. \tag{27.29}$$

On the other hand

$$\frac{\sigma^2}{\sqrt{2\pi}} \int\limits_{|\theta+\sigma x| \leq \tau} x^2 \exp\left\{-\frac{x^2}{2}\right\} dx$$
$$\leq \frac{\sigma^2}{\sqrt{2\pi}} \int\limits_{\{|x| \leq \frac{2\tau}{\sigma}\} \bigcap \{|\theta+\sigma x| \leq \tau\}} x^2 \exp\left\{-\frac{x^2}{2}\right\} dx$$
$$\leq 4\tau^2 \frac{1}{\sqrt{2\pi}} \int\limits_{|\theta+\sigma x| \leq \tau} \exp\left\{-\frac{x^2}{2}\right\} dx,$$

and

$$\frac{(\tau-\theta)^2}{\sqrt{2\pi}} \int\limits_{\theta+\sigma x \geq \tau} \exp\left\{-\frac{x^2}{2}\right\} dx + \frac{(-\tau-\theta)^2}{\sqrt{2\pi}} \int\limits_{\theta+\sigma x \leq -\tau} \exp\left\{-\frac{x^2}{2}\right\} dx$$
$$\leq 4\tau^2 \frac{1}{\sqrt{2\pi}} \int\limits_{|\theta+\sigma x| \geq \tau} \exp\left\{-\frac{x^2}{2}\right\} dx.$$

Therefore,

$$\mathbf{E}\ (\theta_T^* - \theta)^2 \leq 4\tau^2. \tag{27.30}$$

From (27.29), (27.30) we obtain (27.27). Since

$$\delta(\hat{\theta}_T) = \frac{\tau^2\sigma^2}{\tau^2+\sigma^2} \geq \frac{1}{2}\left(\tau^2 \wedge \sigma^2\right),$$

then

$$\sup_{\tau,\sigma} \frac{\sup\limits_{|\theta| \leq \tau} \mathbf{E}\ (\theta_T^* - \theta)^2}{\sup\limits_{|\theta| \leq \tau} \mathbf{E}\ \left(\hat{\theta}_T - \theta\right)^2} < \infty. \tag{27.31}$$

If we take estimator $s^*(t) = \theta_T^* \psi(t)$ for s, then, by (27.25), we can prove that there exists absolute constant C such that

$$\delta(s^*) \leq C\Delta_{\mathcal{N}}(T, \mathcal{L}_*). \tag{27.32}$$

27.6.3 General case

In the general case (as the set Θ in (27.16) is a hyperrectangle: $\Theta = \Theta(\boldsymbol{\tau}), \boldsymbol{\tau} \in l^2$), the minimax linear estimator $\hat{\theta}_T = \left(\hat{\theta}_T^1, \hat{\theta}_T^2, \dots\right)$ for $\theta = (\theta_1, \theta_2, \dots)$ is defined by [see Donoho et al. (1990)]

$$\hat{\theta}_T^j = \frac{\tau_j^2}{\tau_j^2 + \sigma_j^2} \mathbf{y}_j.$$

Thus,

$$\delta(\hat{\theta}_T) = \sup_{\theta \in \Theta(\boldsymbol{\tau})} \mathbf{E} \sum_{j=1}^{\infty} \left(\hat{\theta}_T^j - \theta_j\right)^2 = \sum_{j=1}^{\infty} \frac{\tau_j^2 \sigma_j^2}{\tau_j^2 + \sigma_j^2}.$$

Clearly,

$$\frac{1}{2} \sum_{j=1}^{\infty} \left(\tau_j^2 \wedge \sigma_j^2\right) \leq \delta(\hat{\theta}_T) \leq \sum_{j=1}^{\infty} \left(\tau_j^2 \wedge \sigma_j^2\right).$$

Therefore if we define estimator $\theta_T^* = (\theta_T^*(1), \theta_T^*(2), \dots)$ by the relation

$$\theta_T^*(j) = \begin{cases} \mathbf{y}_j, & \text{if } |\mathbf{y}_j| \leq \tau_j, \\ \tau_j, & \text{if } \mathbf{y}_j > \tau_j, \\ -\tau_j, & \text{if } \mathbf{y}_j < -\tau_j, \end{cases}$$

then, following from (27.31),

$$\sup_{\tau, \sigma} \frac{\sup\limits_{|\theta| \leq \tau} \mathbf{E}\, (\theta_T^* - \theta)^2}{\sup\limits_{|\theta| \leq \tau} \mathbf{E} \left(\hat{\theta}_T - \theta\right)^2} < \infty. \tag{27.33}$$

As an estimator s_T^* for s one can take

$$s_T^* = \sum_{j=1}^{n} \theta_T^*(1) \psi_j(t).$$

It can be proved that there exists absolute constant C such that

$$\delta(s^*) \leq C\Delta_{\mathcal{N}}(T, \mathcal{L}_*). \tag{27.34}$$

27.7 Main Result

For a given function $r: r(u) > 0,\ r(u) \to \infty,$ as $u \to \infty$, *value* $p \geq 1$ and the spectral set $\Lambda = \{u_j, j = 1, 2, \cdots\}$ such that

$$\kappa = \kappa(\Lambda) = \inf_{i \neq j} |u_i - u_j| > 0,$$

we denote by $\mathcal{L}_p(\Lambda, r)$ the set of all pseudo-periodic functions $s(t) = \sum\limits_{u \in \Lambda} a(u) e^{iut}$ such that

$$\sum_{u \in \Lambda} r(u) |a(u)|^p \leq 1.$$

We take as parametric set $\mathcal{L}_* = \mathcal{L}_p(\Lambda, r) \subset \mathcal{L}(\Lambda)$.

For a nonnegative function g we denote

$$M(g) = \sup_I \frac{1}{|I|} \int_I g(u)\, du \times \frac{1}{|I|} \int_I \frac{1}{g(u)}\, du.$$

Here inf is taken over all intervals I. We assume, that the spectral density f may be represented in the form

$$f(u) = \frac{1}{(1+u^2)^n} f_*(u), \quad \text{where} \quad n = 0, 1, \ldots, \quad M(f_*) < \infty.$$

We set $G(C) = \{g:\ g\text{-nonnegative function},\ M(g) \leq C\}$,

$$\begin{aligned} \mathcal{K} &= \mathcal{K}(n, C) \\ &= \left\{ f:\ f(u) = \frac{1}{(1+u^2)^n} f_*(u),\ f_* \in G(C) \right\}. \end{aligned}$$

Theorem 27.7.1 *Suppose that*
1) unknown function $s \in \mathcal{L}_* = \mathcal{L}_p(\Lambda, r),\ p > 1,\ \kappa(\Lambda) > 0$;
2) spectral density $f \in \mathcal{K}(n, C)$.
Then there exists an asymptotically sub-optimal estimator.

Acknowledgements. The research was supported in part by Russian Foundation for Basic Research, grant 02-01-00262, 2258-2003.01, RFBR–DFG grant 99-01-04027.

References

1. Donoho, D. L., Liu, R. C. and MacGibbon, B. (1990). Minimax risk over hyperrectangles, and implications, *Annals of Statistics*, **18**, 1416–1437.

2. Ibragimov, I. A. and Has'minskii, R. Z. (1984). Nonparametric estimation of the value of a linear functional in Gaussian white noise, *Theory of Probability and its Applications*, **29**, 1–32.

3. Ibragimov, I. A. and Rozanov, Yu. A. (1974). *Gaussian Stationary Processes*, MIR, Moscow.

4. Paley, R. and Wiener, N. (1934). *Fourier Transforms in the Complex Domain*, American Mathematical Society, New York.

28

On Parameter Estimation by Contaminated Observations of Ergodic Diffusion Processes

Yu. A. Kutoyants

Laboratoire de Statistique & Processus, Université du Maine, Le Mans, France

Abstract: We consider several problems of parameter estimation by observations of ergodic diffusion processes in the situations when the underlying model depends on an unknown function (under misspecification). We propose some consistent and asymptotically efficient estimators.

Keywords and phrases: Ergodic diffusion, semiparametric estimation, change-point, cusp estimation

28.1 Introduction

We consider several problems of parameter estimation by observations of diffusion process

$$\mathrm{d}X_t = S(\vartheta, X_t)\,\mathrm{d}t + h(X_t)\,\mathrm{d}t + \sigma(X_t)\,\mathrm{d}W_t, \quad X_0, \quad 0 \le t \le T \tag{28.1}$$

where we suppose that functions $S(\cdot,\cdot)$ and $\sigma(\cdot)^2 > 0$ are known and the parameter $\vartheta \in \Theta = (\alpha, \beta)$ and the function $h(\cdot)$ (contamination) are unknown. The initial value X_0 is a random variable which does not depend on the Wiener process $\{W_t, 0 \le t \le T\}$. Therefore we have a semiparametric estimation problem [Bickel et al. (1998)]. To have a consistent estimation we suppose that the function $h(\cdot)$ belongs to some classes $\mathcal{H}$ and the goal of this work is to describe such classes in regular (smooth w.r.t. parameter) and nonregular (cusp and change-point) estimation problems.

We suppose that the functions $S(\cdot)$, $\sigma(\cdot)$ and $\sigma(\cdot)^{-1}$ are locally bounded, there exists a unique weak solution of the underlying stochastic differential equation and the condition

$$\mathcal{A}_0. \qquad \lim_{|x|\to\infty} \sup_{\vartheta\in\Theta} \operatorname{sgn}(x)\, \frac{S(\vartheta, x) + h(x)}{\sigma(x)^2} < 0$$

holds for all possible functions $h(\cdot) \in \mathcal{H}$.

By this condition the process $\{X_t, t \geq 0\}$ has ergodic properties with the density of the invariant law:

$$f_h(\vartheta, x) = G(\vartheta, h)^{-1} \sigma(x)^{-2} \exp\left\{2\int_0^x \frac{S(\vartheta, y)}{\sigma(y)^2}\, dy + 2\int_0^x \frac{h(y)}{\sigma(y)^2}\, dy\right\}$$

where

$$G(\vartheta, h) = \int_{-\infty}^{\infty} \sigma(x)^{-2} \exp\left\{2\int_0^x \frac{S(\vartheta, y)}{\sigma(y)^2}\, dy + 2\int_0^x \frac{h(y)}{\sigma(y)^2}\, dy\right\} dx$$

[see, e.g., Gikhman and Skorohod (1972)].

We are interested in asymptotic ($T \to \infty$) properties of estimators of the parameter ϑ.

Recall that if $h(\cdot) \equiv 0$ and the function $S(\cdot)$ is sufficiently smooth w.r.t. ϑ, then the MLE $\hat{\vartheta}_T$ and a wide class of Bayesian estimators $\tilde{\vartheta}_T$ are consistent, asymptotically normal:

$$\sqrt{T}\left(\hat{\vartheta}_T - \vartheta\right) \Longrightarrow \mathcal{N}\left(0, \mathrm{I}(\vartheta)^{-1}\right)$$

and asymptotically efficient [see, e.g., Kutoyants (2003)].

Here $\mathrm{I}(\vartheta)$ is Fisher information:

$$\mathrm{I}(\vartheta) = \mathbf{E}_\vartheta \left(\frac{\dot{S}(\vartheta, \xi)}{\sigma(\xi)}\right)^2 > 0$$

and ξ is a random variable with invariant density

$$f(\vartheta, x) = \frac{1}{G(\vartheta)\, \sigma(x)^2} \exp\left\{2\int_0^x \frac{S(\vartheta, v)}{\sigma(v)^2}\, dv\right\}.$$

In the case of observations (28.1) with unknown $h(\cdot)$ these estimators (constructed on the basis of the model with $h(\cdot) \equiv 0$) converge to the value

$$\hat{\vartheta} = \arg\inf_{\vartheta \in \Theta} \mathbf{E}_{\vartheta_0, h} \left(\frac{S(\vartheta, \xi) - S(\vartheta_0, \xi) - h(\xi)}{\sigma(\xi)}\right)^2 \tag{28.2}$$

where we denote the true value as ϑ_0 [see McKeague (1984) and Kutoyants (2003)]. Here ξ is a random variable with the density function $f_h(\vartheta_0, \cdot)$ Of course, usually $\hat{\vartheta} \neq \vartheta_0$ and the consistent estimation in general is impossible.

Our goal in this work is to present several models of diffusion processes (28.1) and the classes $\mathcal{H}$ of functions $h(\cdot)$ such that the consistent and sometimes asymptotically efficient estimation of the parameter ϑ is nevertheless possible.

28.2 Regular Case

Suppose that the function $h(\cdot)$ has a compact support $\mathbb{A}$. Let us denote by $\mathcal{H}_{\mathbb{A}}$ the class of such functions.

Regularity conditions $\mathcal{F}^*$. The function $S(\vartheta,\cdot)$ is three times continuously differentiable on ϑ, the derivative $\dot{S}(\vartheta,\cdot)$ is bounded and the functions

$$S(\vartheta,\cdot)\,,\ddot{S}(\vartheta,\cdot)\,,\dddot{S}(\vartheta,\cdot)\,\sigma(\cdot)\,,\sigma(\cdot)^{-1}\in\mathfrak{P}$$

where we denote by gothP the class of functions having polynomial majorants.

For any $\nu>0$ and $\vartheta_0\in\Theta$,

$$\inf_{|\vartheta-\vartheta_0|>\nu}\mathbf{E}_{\vartheta_0}\left(\frac{S(\vartheta,\xi)-S(\vartheta_0,\xi)}{\sigma(\xi)}\right)^2\chi_{\{\xi\in\mathbb{A}^c\}}>0$$

and

$$\bar{\mathrm{I}}(\vartheta_0)=\mathbf{E}_{\vartheta_0}\left(\frac{\dot{S}(\vartheta_0,\xi)}{\sigma(\xi)}\right)^2\chi_{\{\xi\in\mathbb{A}^c\}}>0.$$

Fix ϑ_0 and a function $h_0(\cdot)\in\mathcal{H}_A$ and introduce their vicinity

$$V_\delta=\left\{\vartheta,h(\cdot):\ |\vartheta-\vartheta_0|\le\delta,\sup_x|h(x)-h_0(x)|<K\delta\right\}$$

where the constant $K>0$ is such that $\left|\dot{S}(\vartheta,x)\right|\le K$. Then we have the following lower bound on the risks of all estimators.

Theorem 28.2.1 *For all estimators $\bar{\vartheta}_T$,*

$$\underline{\lim}_{\delta\to0}\,\underline{\lim}_{T\to\infty}\sup_{\vartheta,h(\cdot)\in V_\delta}\mathbf{E}_{\vartheta,h}\left(\sqrt{T}\left(\bar{\vartheta}_T-\vartheta\right)\right)^2\ge\mathrm{I}_*(\vartheta_0,h_0)^{-1}\tag{28.3}$$

where

$$\mathrm{I}_*(\vartheta,h)=\mathbf{E}_{\vartheta,h}\left(\frac{\dot{S}(\vartheta,\xi)}{\sigma(\xi)}\right)^2\chi_{\{\xi\in\mathbb{A}^c\}}.$$

PROOF. The proof is based on the obvious estimate

$$\sup_{\vartheta,h(\cdot)\in V_\delta}\mathbf{E}_{\vartheta,h}\left(\sqrt{T}\left(\bar{\vartheta}_T-\vartheta\right)\right)^2\ge\sup_{|\vartheta-\vartheta_0|\le\delta}\mathbf{E}^*_\vartheta\left(\sqrt{T}\left(\bar{\vartheta}_T-\vartheta\right)\right)^2$$

where $\mathbf{E}_\vartheta^*$ is expectation corresponding to the following family of diffusion processes

$$\mathrm{d}X_t = S(\vartheta, X_t)\,\chi_{\{X_t\in\mathbb{A}^c\}}\,\mathrm{d}t + \left[h(X_t) + S(\vartheta_0, X_t)\,\chi_{\{X_t\in\mathbb{A}\}}\right]\mathrm{d}t + \sigma(X_t)\,\mathrm{d}W_t.$$

Note that this parametric family of trends is a subfamily of V_δ. Then using LAN of this parametric family we can apply the Hajek–Le Cam lower bound. This last bound gives us the right-hand side of the inequality (28.3). This result is based on our work with Höpfner and Kutoyants (2003) and can be found as well in Kutoyants (2003). ■

Recallthat the MLE due to misspecification converges to the value $\hat{\vartheta} \neq \vartheta_0$. Therefore to construct a consistent estimator we *modify the likelihood ratio*. Introduce the *"pseudo likelihood ratio"* $\bar{L}(\vartheta, X^T)$:

$$\ln \bar{L}(\vartheta, X^T) = \int_0^T \frac{S(\vartheta, X_t)\,\chi_{\{X_t\in\mathbb{A}^c\}}}{\sigma(X_t)^2}\,\mathrm{d}X_t - \frac{1}{2}\int_0^T \frac{S(\vartheta, X_t)^2\,\chi_{\{X_t\in\mathbb{A}^c\}}}{\sigma(X_t)^2}\,\mathrm{d}t$$

and the corresponding *"pseudo maximum likelihood estimator"* (pseudo-MLE) $\hat{\vartheta}_T$:

$$\sup_{\vartheta\in\Theta} \bar{L}(\vartheta, X^T) = \bar{L}\left(\hat{\vartheta}_T, X^T\right). \tag{28.4}$$

Of course, $\bar{L}(\vartheta, X^T)$ is not a likelihood ratio but just a contrast function. Note that the estimator $\hat{\vartheta}_T$ does not depend on observations in the *contaminated window* $\mathbb{A}$.

Theorem 28.2.2 *The estimator $\hat{\vartheta}_T$ is consistent, asymptotically normal*

$$\sqrt{T}\left(\hat{\vartheta}_T - \vartheta\right) \Longrightarrow \mathcal{N}\left(0, \mathrm{I}_*(\vartheta, h)^{-1}\right)$$

and asymptotically efficient.

PROOF. The proof follows the usual steps of the proof of the consistency and asymptotic normality of the minimum contrast estimator. We have just to add the uniform convergence on the set V_δ and convergence of moments. The details can be found in Höpfner and Kutoyants (2003) or Kutoyants (2003). ■

Note that these two theorems are true for a wide class of loss functions including polynomial functions.

28.3 Nonregular Cases

We consider two other problems of parameter estimation but in the situations when Fisher information does not exist. We can see that here the statistical problems are *much more robust* in some sense. That means that the MLE and Bayesian estimators constructed with the help of the wrong likelihood ratio (misspecification) or constructed on the basis of observations in the vanishing windows can be consistent and even asymptotically efficient.

28.3.1 Change-point estimation

Suppose that the observed process (28.1) has a discontinuous trend coefficient. In particular, let $S(\vartheta, x) = S(x - \vartheta)$ and there exists a point x_* such that $S(x_*-) - S(x_*) = r \neq 0$, i.e., the limits from the left and from the right at this point are different. One of the simplest models of such processes is the process of switching:

$$\mathrm{d}X_t = -\operatorname{sgn}(X_t - \vartheta)\ \mathrm{d}t + h(X_t)\ \mathrm{d}t + \mathrm{d}W_t, \quad 0 \le t \le T. \tag{28.5}$$

Recall now the properties of the MLE and Bayesian estimators (quadratic loss function and positive continuous prior density) for this model without contamination:

$$\mathrm{d}X_t = -\operatorname{sgn}(X_t - \vartheta)\ \mathrm{d}t + \mathrm{d}W_t, \quad 0 \le t \le T.$$

Let us denote

$$Z(u) = \exp\{2W(u) - 2|u|\}, \qquad u \in \mathbf{R}$$

and put

$$\hat{u} = \arg\sup_{u \in \mathbf{R}} Z(u), \qquad \tilde{u} = \frac{\int uZ(u)\mathrm{d}u}{\int Z(u)\mathrm{d}u}.$$

Here $W(\cdot)$ is a two-sided Wiener process.

We have the following lower bound:

$$\underline{\lim}_{\delta \to 0}\ \underline{\lim}_{T \to \infty}\ \sup_{|\vartheta - \vartheta_0| \le \delta} \mathbf{E}_\vartheta \left(T\left(\bar{\vartheta}_T - \vartheta\right)\right)^2 \ge \mathbf{E}\tilde{u}^2$$

and as usual we call *asymptotically efficient estimator* any estimator attaining this bound.

Theorem 28.3.1 *The MLE $\hat{\vartheta}_T$ and BE $\tilde{\vartheta}_T$ are uniformly consistent,*

$$\mathcal{L}_\vartheta\left\{T\left(\hat{\vartheta}_T - \vartheta\right)\right\} \Longrightarrow \mathcal{L}\{\hat{u}\},$$

$$\mathcal{L}_\vartheta\left\{T\left(\tilde{\vartheta}_T - \vartheta\right)\right\} \Longrightarrow \mathcal{L}\{\tilde{u}\}$$

and the moments of $T\left(\hat{\vartheta}_T - \vartheta\right)$ *and* $T\left(\tilde{\vartheta}_T - \vartheta\right)$ *converge too. Moreover, the Bayesian estimators are asymptotically efficient.*

PROOF. The proof can be found in Kutoyants (2000) and (2003). ■

Now we return to the model (28.5) with unknown function $h(\cdot)$. Denote by $\mathcal{H}_\gamma$ with $\gamma < 1$ the set

$$\mathcal{H}_\gamma = \left\{h(\cdot) : \sup_x |h(x)| < \gamma\right\}.$$

The condition $h(\cdot) \in \mathcal{H}_\gamma$ is quite natural in this problem because if we allow the equality $|h(x)| = 1$, then in the case of the function $h(x) = \operatorname{sgn}(x - \vartheta)$ we observe a Wiener process $\{W_t, 0 \leq t \leq T\}$ only and the consistent estimation of the parameter ϑ is impossible.

We study the *"pseudo-MLE"* $\hat{\vartheta}_T$ constructed by a "naive statistician", i.e., by someone who supposes that there is *no contamination* $(h(\cdot)) \equiv 0$. Therefore, he (or she) introduces the *"pseudo likelihood ratio"*

$$\bar{L}\left(\vartheta, X^T\right) = \exp\left\{\int_0^T \operatorname{sgn}\left(\vartheta - X_t\right) \mathrm{d}X_t - \frac{T}{2}\right\}$$

which is obtained by putting $h(\cdot) \equiv 0$ in the true LR. Hence

$$\hat{\vartheta}_T = \arg\max_\vartheta \int_0^T \operatorname{sgn}\left(\vartheta - X_t\right) \mathrm{d}X_t. \tag{28.6}$$

To study the asymptotic properties of the pseudo-MLE $\hat{\vartheta}_T$ we [following Ibragimov and Khasminskii (1981)] describe the behavior of the normalized process

$$Z_T(u) = L\left(\vartheta + u/T, \vartheta, X^T\right), \qquad u \in \mathbf{R}.$$

The limit of $Z_T(u)$ is the random function

$$Z_*(u) = \exp\left\{\Gamma(h,\vartheta) W(u) + \frac{u}{2}\Gamma_-(h,\vartheta)^2\right\}, \qquad u \leq 0$$

and

$$Z_*(u) = \exp\left\{\Gamma(h,\vartheta) W(u) - \frac{u}{2}\Gamma_+(h,\vartheta)^2\right\}, \qquad u > 0.$$

Here

$$\Gamma(\vartheta,h)^2 = \frac{4}{G(h,\vartheta)}, \quad \Gamma_\pm(h,\vartheta)^2 = \frac{4\left(1 \pm h(\vartheta)\right)}{G(\vartheta,h)}.$$

Define the random variable

$$u^* = \arg\sup_{u \in \mathbf{R}} Z_*(u).$$

Theorem 28.3.2 *The "pseudo-MLE" $\hat{\vartheta}_T$ is uniformly in $h(\cdot) \in \mathcal{H}_\gamma$ consistent,*

$$\mathcal{L}_{h,\vartheta}\left\{T\left(\hat{\vartheta}_T - \vartheta\right)\right\} \Longrightarrow \mathcal{L}_{h,\vartheta}\{u^*\},$$

and the moments of $T\left(\hat{\vartheta}_T - \vartheta\right)$ converge too.

PROOF. The proof follows from the weak convergence of the random function $Z_T(\cdot)$ to the process $Z_*(\cdot)$. The details can be found in Kutoyants (2000) and (2003). ∎

Note that we can go further and exclude all observations outside of Θ, i.e., it can be shown that the estimator

$$\hat{\vartheta}_T = \arg\max_{\vartheta} \int_0^T \operatorname{sgn}(\vartheta - X_t)\, \chi_{\{\alpha < X_t < \beta\}}\, \mathrm{d}X_t$$

is consistent and has the same limit distribution as pseudo-MLE [see Kutoyants (2003)].

It seems surprising that this estimator is consistent even for misspecified model with (*wrong "likelihood ratio"*). The same is true for the Bayesian estimators but its limit distribution is different.

A similar result is true in the general case of diffusion processes

$$\mathrm{d}X_t = S(X_t - \vartheta)\, \mathrm{d}t + h(X_t)\, \mathrm{d}t + \mathrm{d}W_t, \quad 0 \le t \le T$$

with discontinuous trend coefficients $S(\cdot)$ having several jumps along some curves. We have the consistent estimation if the jumps are greater than $h(\cdot)$ [see Kutoyants (2003)].

Let us consider two examples.

The first example is the diffusion process

$$\mathrm{d}X_t = -X_t\left(a + b\,\chi_{\{\vartheta < X_t < c+\vartheta\}}\right)\mathrm{d}t + h(X_t)\,\mathrm{d}t + \mathrm{d}W_t, \quad X_0,$$

where $a > 0,\ b \neq 0,\ c > 0,\ \vartheta \in (\alpha, \beta)$. If the function $|h(x)| < \alpha b$, then the pseudo-MLE is consistent.

The next example is the problem of the estimation parameter $\vartheta \in (\alpha, \beta)$ by observations of an ergodic diffusion process

$$\mathrm{d}X_t = -X_t \operatorname{sgn}\left(X_t^2 - 2\vartheta X_t + 1\right)\mathrm{d}t + h(X_t)\,\mathrm{d}t + \sigma\,\mathrm{d}W_t, \quad X_0.$$

If we suppose that $\alpha > 1$ and $|h(x)| < 2\alpha - 2\sqrt{\alpha^2 - 1}$, then the pseudo-MLE is consistent. Of course, it is sufficient that this condition on $h(x)$ holds for the values of x in the vicinity of the jump only and outside we require just that the condition $\mathcal{A}_0$ holds.

We have a similar result in the case of multi-dimensional parameter ϑ. For example, suppose that the observed diffusion process is

$$\mathrm{d}X_t = -X_t \left(a + b\,\chi_{\{\vartheta^{(1)}<X_t<\vartheta^{(2)}\}}\right) \,dt + h\left(X_t\right)\mathrm{d}t + \mathrm{d}W_t, \qquad X_0,$$

where

$$a>0,\ b\neq 0,\quad \vartheta^{(1)}\in\left(\alpha_1,\beta_1\right),\quad \vartheta^{(2)}\in\left(\alpha_2,\beta_2\right),\quad \alpha_2>\beta_1.$$

Then the pseudo-MLE is consistent for the corresponding class of functions $h(\cdot)$.

Note that the integral (28.6) is similar to that used in the definition of the local time estimator

$$f_T^{\circ}(x) = \frac{1}{T}\int_0^T \operatorname{sgn}\left(x - X_t\right)\mathrm{d}X_t + \frac{\left|X_T - x\right| - \left|X_0 - x\right|}{T}$$

of the invariant density studied in Kutoyants (2003). Hence it can be approximated by the *maximum local time estimator*

$$\vartheta_T^* = \arg\sup_{\vartheta\in\Theta} f_T^{\circ}(\vartheta).$$

Its construction is based on the local time $\Lambda_T(\Theta) = \left\{\Lambda_T(\vartheta), \vartheta\in\Theta\right\}$ only. Therefore, $\Lambda_T(\Theta)$ is sufficient statistics and, as it uses the observations of the process X_t fitting in the window $[\alpha,\beta]$ only, all the other observations can be canceled.

Note that the random function

$$\eta_T(u) = \Lambda_T\left(\vartheta + \frac{u}{T}\right) - \Lambda_T(\vartheta), \qquad u\in\mathbf{R}$$

converges in distribution on compacts $|u|\leq K$ to the double-sided Wiener process with shift

$$\eta(u) = W(u) - \frac{|u|}{2}, u\in\mathbf{R}.$$

Here ϑ is the true value. In the general case we have a similar effect. For example, if $\vartheta\in(\alpha,\beta)$ and the trend coefficient $S(\vartheta,x)$ is discontinuous along the strictly increasing curve $x_*(\vartheta), \vartheta\in(\alpha,\beta)$, then we can introduce the set $\mathbb{B} = \left[x_*(\alpha), x_*(\beta)\right]$ (*window*) and to restrict ourselves by the observations $X_t\in\mathbb{B}$. This means, that we can introduce the *pseudo likelihood ratio*

$$\ln \bar{L}\left(\theta, X^T\right) = \int_0^T \frac{S\left(\theta, X_t\right)}{\sigma\left(X_t\right)^2}\chi_{\{X_t\in\mathbb{B}\}}\,\mathrm{d}X_t - \frac{1}{2}\int_0^T \left(\frac{S\left(\theta, X_t\right)}{\sigma\left(X_t\right)}\right)^2 \chi_{\{X_t\in\mathbb{B}\}}\,\mathrm{d}t \tag{28.7}$$

and then with the help of this function to construct the MLE and BE. These estimators will have the same asymptotic properties as *the true MLE and BE*, i.e., the same limit distributions and convergence of moments.

Obviously, if the function $S(\vartheta, x)$ has two jumps along the increasing curves $x_*^{(i)}(\cdot)$, $i = 1, 2$, then the window is

$$\mathbb{B} = \left[x_*^{(1)}(\alpha), x_*^{(1)}(\beta)\right] \cup \left[x_*^{(2)}(\alpha), x_*^{(2)}(\beta)\right].$$

If the set $\mathbb{B}$ nevertheless seems to be quite large, it is possible to seek a narrower one in two steps. First we estimate consistently the parameter ϑ by observations $\left\{X_t, 0 \leq t \leq \sqrt{T}\right\}$ and then having the estimator $\bar{\vartheta}_{\sqrt{T}}$ we choose the window

$$\mathbb{B}_T = \left[x_*\left(\bar{\vartheta}_{\sqrt{T}}\right) - b_T, x_*\left(\bar{\vartheta}_{\sqrt{T}}\right) + b_T\right]$$

with $b_T \to 0$ slowly, say, $b_T = T^{-1/4}$. Then we use the "likelihood ratio" (28.7) with $\mathbb{B} = \mathbb{B}_T$ starting at time $\sqrt{T}$ and construct the estimators. The *first properties* of the MLE and BE (consistency and limit distributions) can be the same [see Kutoyants (2003)].

For example, let the observed process be

$$\mathrm{d}X_t = S(X_t - \vartheta)\ \mathrm{d}t + \sigma(X_t)\ \mathrm{d}W_t,$$

where the function $S(x)$ has a jump at the point $x = 0$ and $\bar{\vartheta}_{\sqrt{T}}$ be an estimator of the method of moments (say, for simple switching it can be the empirical mean $\bar{X}_T$). This estimator is consistent and asymptotically normal

$$T^{1/4}\left(\bar{\vartheta}_{\sqrt{T}} - \vartheta\right) \Longrightarrow \mathcal{N}\left(0, \sigma_*^2\right).$$

Then

$$\mathbf{P}_\vartheta^{(T)}\left\{\left|\bar{\vartheta}_{\sqrt{T}} - \vartheta\right| > T^{-1/8}\right\} \to 0$$

as $T \to \infty$. Hence we can take the window $\mathbb{B}_T = \left[\bar{\vartheta}_{\sqrt{T}} - T^{-1/4}, \bar{\vartheta}_{\sqrt{T}} + T^{-1/4}\right]$ and the pseudo likelihood ratio

$$\bar{L}\left(\theta, X^T\right) = \exp\left\{\int_{\sqrt{T}}^T \frac{S(\theta, X_t)}{\sigma(X_t)^2} \chi_{\{X_t \in \mathbb{B}_T\}}\ \mathrm{d}X_t - \frac{1}{2}\int_{\sqrt{T}}^T \left(\frac{S(\theta, X_t)}{\sigma(X_t)}\right)^2 \chi_{\{X_t \in \mathbb{B}_T\}}\ \mathrm{d}t\right\}.$$

It can be shown that this function provides the same properties of the normalized likelihood ratio $Z_T(\cdot)$ as the true likelihood ratio. Hence the MLE and BE constructed with the help of this function have the same asymptotic properties [see Kutoyants (2003)].

28.3.2 Cusp estimation

Suppose that the observed diffusion process

$$dX_t = [a\,|X_t - \vartheta|^{\kappa} + h\,(X_t - \vartheta)]\ dt + \sigma\,(X_t)\,dW_t,$$

where $\kappa \in (0, 1/2)$ and the function $h\,(\cdot)$ satisfies the Hölder condition of order $\mu > \kappa + 1/2$. The parameter $\vartheta \in (\alpha, \beta) = \Theta$ and the condition $\mathcal{A}_0$ is fulfilled too. Note that Fisher information does not exist and we have a nonregular problem of parameter estimation.

Introduce the fractional Brownian motion $W^H\,(\cdot)$ with the Hurst constant $H = \kappa + 1/2$,

$$\mathbf{E}W^H\,(u)\,W^H\,(v) = \frac{1}{2}\left[|u|^{2H} + |v|^{2H} - |u-v|^{2H}\right]$$

and the stochastic process

$$Z\,(u) = \exp\left\{ W^H\,(u) - \frac{1}{2}\,|u|^{2H} \right\}.$$

Define two random variables

$$Z\,(\hat{u}) = \sup_{u \in R} Z\,(u), \quad \tilde{u} = \frac{\int u\, Z(u)\, du}{\int Z(v)\, dv}.$$

We introduce as well the function

$$\Gamma_\vartheta^2 = \frac{\mathrm{B}\big(\kappa+1\,,\,\kappa+1\big)}{G\,(\vartheta)\ \sigma\,(\vartheta)^4}\ \left[\frac{a^2+b^2}{\cos(\pi\kappa)} - 2ab\right]$$

where $\mathrm{B}\big(\kappa+1\,,\,\kappa+1\big)$ is a Beta-function. Finally, we put $\gamma_\vartheta = \Gamma_\vartheta^{1/H}$.

Suppose first that the function $h\,(\cdot)$ is known and recall the following two theorems:

Theorem 28.3.3 *We have*

$$\lim_{\delta \to 0}\ \varliminf_{T\to\infty}\ \inf_{\bar{\vartheta}_T}\ \sup_{|\vartheta - \vartheta_0| < \delta} T^{1/H}\mathbf{E}_\vartheta\left(\bar{\vartheta}_T - \vartheta\right)^2 \geq \frac{\mathbf{E}\widetilde{u}^2}{\gamma_{\vartheta_0}^2}.$$

For the proof see Dachian and Kutoyants (2003) and Kutoyants (2003). This inequality allows us to define asymptotically efficient estimators as estimators attaining this bound.

Theorem 28.3.4 *The MLE and BE are, uniformly on compacts* $\mathbf{K} \subset \Theta$, *consistent,*

$$\mathcal{L}_\vartheta\left\{T^{1/2H}\left(\hat{\vartheta}_T - \vartheta\right)\right\} \Longrightarrow \mathcal{L}_\vartheta\left\{\frac{\hat{u}}{\gamma_\vartheta}\right\},$$

$$\mathcal{L}_\vartheta\left\{T^{1/2H}\left(\tilde{\vartheta}_T - \vartheta\right)\right\} \Longrightarrow \mathcal{L}_\vartheta\left\{\frac{\tilde{u}}{\gamma_\vartheta}\right\},$$

and we have the convergence of all moments. Moreover, the BE are asymptotically efficient.

The proofs can be found in Dachian and Kutoyants (2003) and Kutoyants (2003).

Suppose that the function $h(x)$ is unknown for $x < \alpha$ and $x > \beta$. Introduce the pseudo likelihood ratio

$$\ln \bar{L}\left(\vartheta, X^T\right) = \int_0^T S(\vartheta, X_t)\ \chi_{\{\alpha < X_t < \beta\}}\ \mathrm{d}X_t - \\ - \int_0^T S(\vartheta, X_t)^2\ \chi_{\{\alpha < X_t < \beta\}}\ \mathrm{d}t$$

and corresponding pseudo MLE $\hat{\vartheta}_T^*$ and pseudo BE $\tilde{\vartheta}_T^*$

Theorem 28.3.5 *The MLE and BE are, uniformly on compacts* $\mathbf{K} \subset \Theta$, *consistent,*

$$\mathcal{L}_\vartheta\left\{T^{1/2H}\left(\hat{\vartheta}_T^* - \vartheta\right)\right\} \Rightarrow \mathcal{L}_\vartheta\left\{\frac{\hat{u}}{\gamma_\vartheta}\right\},$$

$$\mathcal{L}_\vartheta\left\{T^{1/2H}\left(\tilde{\vartheta}_T^* - \vartheta\right)\right\} \Rightarrow \mathcal{L}_\vartheta\left\{\frac{\tilde{u}}{\gamma_\vartheta}\right\},$$

and we have the convergence of all moments. Moreover, the BE are asymptotically efficient.

Therefore these pseudo-MLE and pseudo-BE have the same asymptotic properties as if all the observations X^T were used. This *pseudo-likelihood ratio* corresponds to the *misspecified model* of observations

$$\mathrm{d}X_t = S(\vartheta, X_t)\ \chi_{\{X_t \in \mathbb{B}\}}\ \mathrm{d}t + \sigma(X_t)\ \mathrm{d}W_t, \quad X_0, 0 \le t \le T. \tag{28.8}$$

Of course, we do not suppose that the observations come from Equation (28.8) and just use the *contrast function* $\overline{L}\left(\theta, X^T\right)$ to construct the estimators. As one of the consequences we need not remember all observations X^T and it is sufficient to save X_t which fits the *window* $\mathbb{B}$ only. Remember that the local time estimator of the density is sufficient statistics as well and in this problem we can save the values of $X_0, X_T,\ f_T^\circ(x), x \in [\alpha, \beta]$ only. Moreover, the detailed analysis shows that, similar to change-point estimation problems, we can consider a two-step procedure with a preliminary consistent estimator $\bar{\vartheta}_{\sqrt{T}}$ constructed by the observations $Y^{\sqrt{T}} = \left\{X_t\ \chi_{\{X_t \in \mathbb{B}\}}, 0 \le t \le \sqrt{T}\right\}$. Then we introduce another window $\mathbb{B}_T = \left[\bar{\vartheta}_{\sqrt{T}} - h_T, \bar{\vartheta}_{\sqrt{T}} + h_T\right]$, where $h_T \to 0$ is a *slowly decreasing function* and construct the MLE and BE on the base of the *pseudo-likelihood ratio* $\overline{L}\left(\theta, X^T_{\sqrt{T}}\right)$ and the window $\mathbb{B}_T$. In this case the length of the window tends to zero (in the scheme of series), but nevertheless the MLE and the BE have the same asymptotic properties as those described in Theorem 28.3.5. Hence we have even an asymptotically efficient Bayesian estimator constructed by the observations in the *vanishing window*. Note that the lower bound of Theorem 28.3.3 is based on the whole set of observations.

References

1. Bickel, P. J., Klaassen, C. A. J., Ritov, Y. and Wellner, J. A. (1998). *Efficient and Adaptive Estimation for Semiparametric Models*, Springer-Verlag, New York.

2. Dachian, S. and Kutoyants, Yu. A. (2003). On cusp estimation of ergodic diffusion process, *Journal of Statistical Planning and Inference*, **117**, 153–166.

3. Gikhman, I. I. and Skorohod, A. V. (1972). *Stochastic Differential Equations*, Springer-Verlag, New York.

4. Höpfner, R. and Kutoyants, Yu. A. (2003). On a problem of statistical inference in null recurrent diffusions, *Statistical Inference for Stochastic Processes*, **6**, 25–42.

5. Ibragimov, I. and Khasminskii, R. (1981). *Statistical Estimation*, Springer-Verlag, New York.

6. Kutoyants, Yu. A. (2003). *Statistical Inference for Ergodic Diffusion Processes*, Springer-Verlag, London, England.

7. Kutoyants, Yu. A. (2000). On parameter estimation for switching ergodic diffusion processes, *CRAS Sér. 1*, **330**, 925–930.

8. McKeague, I. (1984) Estimation for diffusion processes under misspecified models, *Journal of Applied Probability*, **21**, 511–520.

29

On Parameter Estimation for a Position-Dependent Marking of a Doubly Stochastic Poisson Process

Heide Wendt and Waltraud Kahle

Faculty of Mathematics, Otto-von-Guericke-University, Magdeburg, Germany

Abstract: For analyzing reliability of technical systems it is often important to investigate damage processes. In this paper we describe a damage process (Z_t) which is assumed to be generated by a position-dependent marking of a doubly stochastic Poisson process. For some parametric intensity kernels of the corresponding marked point process we determine maximum-likelihood estimations. Censored observations are taken into account. Furthermore, the large sample case is considered.[1]

Keywords and phrases: Marked point process, doubly stochastic Poisson process, position-dependent marking, shock model, parametric models, maximum-likelihood estimation, asymptotic properties

29.1 Introduction

In connection with the investigation of the reliability of systems it is often necessary to consider the development of damage processes at elements of such systems. These include processes of wear, corrosion and crack-growth. This paper deals with a damage process (Z_t) whose paths are monotone increasing step functions. For modeling it, we use marked point processes $\Phi = ((T_n, X_n))_{n\geq 1}$, presented in detail e.g., by Last and Brandt (1995) or Anderson et al. (1993). The cumulative process (Z_t) is assumed to be generated by a position-dependent marking of a doubly stochastic Poisson process (T_n). The doubly stochastic Poisson process was introduced by Cox (1955). Cramer (1969) applied it in risk theory, and Grandell (1991) gave a detailed discussion of these processes

[1]This research is supported by DFG # Ka 1011/3-1.

and their impact on risk theory. Further applications of the doubly stochastic Poisson process (T_n) may be found in reliability theory, medicine and queueing theory [Bremaud (1981), Anderson et al. (1993), Grandell (1997), König and Schmidt (1992)].

Here we will consider a shock model where each damage from a shock at the random time T_n is described by the scalar non-negative random variable X_n $(n = 1, 2, \ldots)$. There is a stream of papers concerning shock models. Applications were given — among others — by Sobczyk (1987), Esary, Marshall, Proshan (1973) and Pieper and Tiedge (1983). Further papers dealing with shock models are Feng, Adachi, Kowada (1994), Shaked (1983), Aven and Jensen (1998) and Wendt (1999). In the present paper we assume cumulated intensities of (T_n) given as a random multiple Y of a deterministic function η. Usually, the nonnegative random variable Y cannot be observed.

Our aim is to describe suitable parametric models for damage accumulation, to calculate the corresponding cumulative damage at time t, and to determine estimators for the parameters of the deterministic part, the distribution of Y, and the distribution of the marks X_n $(n \geq 1)$ using maximum likelihood theory. We compare the estimators in the deterministic term when a doubly stochastic Poisson process is assumed, with those when a nonhomogeneous Poisson process is assumed. Furthermore, the problem of estimating the parameters by means of a censored sample is considered.

29.2 A Cumulative Damage Process

Let $[\Omega, \mathcal{F}, P]$ be a fixed probability space and let $\{\mathcal{F}_t\}$ be a filtration in $\mathcal{F}$. The random variable T_n $(n \geq 1)$ is the time of the n-th shock. We suppose

$$T_n < T_{n+1} \quad \text{if } T_n < \infty \quad \text{and} \quad T_n = T_{n+1} = \infty \quad \text{otherwise} .$$

The size of the n-th increment of the cumulative damage process $(Z(t))_{t\geq 0}$, is given by a nonnegative random variable X_n $(n \geq 1)$. Thus,

$$Z(t) = \sum_{n=1}^{\infty} I(T_n \leq t) \cdot X_n$$

describes the total amount of damage at time t. The sequence $\Phi = ((T_n, X_n))$ is called a marked point process, and $\Phi(t)$ is defined as the random variable representing the number of events occurred up to time t. Typical courses of realizations of the processes $(\Phi(t))$ and $(Z(t))$ are sketched in Figure 29.1.

Let us consider the filtration $\{\mathcal{F}_t\} = \{\mathcal{F}_t^{\Phi}\} \vee \sigma(Y)$, where $\{\mathcal{F}_t^{\Phi}\}$ describes the internal history of Φ, and Y (generating $\mathcal{F}_0$) is a nonnegative random variable

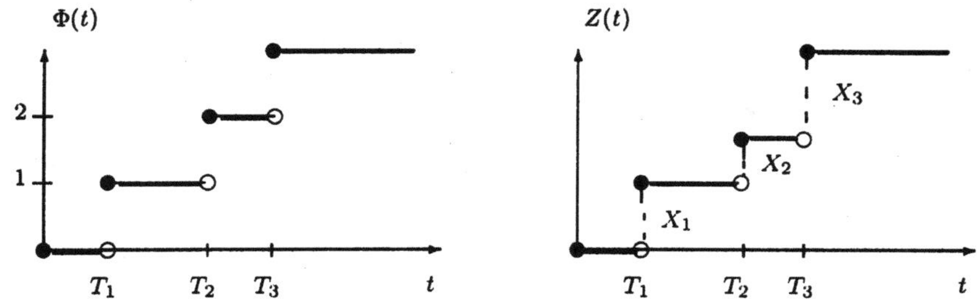

Figure 29.1: A realization of $\Phi(t)$ and $Z(t)$

with finite expectation. The cumulated $(P, \mathcal{F}_t)$-stochastic intensity $\bar{\nu}(t)$ of (T_n) is assumed to be given by $\bar{\nu}(t) = Y \cdot \eta(t)$, where $\eta(t)$ is a deterministic function with derivative $\xi(t) \geq 0$. Hence, given the outcome $Y = y$ the random variable $\Phi(t)$ is Poisson distributed with mean $y \cdot \eta(t)$. But usually, Y cannot be observed. And the unconditional distribution of $\Phi(t)$ is given by

$$P(\Phi(t) = k) = E\left[\frac{[Y\eta(t)]^k}{k!}\exp(-Y\eta(t))\right], \qquad k = 0, 1, \ldots, \tag{29.1}$$

where the expectation is formed with respect to Y. The sequence (T_n) is called a $(P, \mathcal{F}_t)$-**doubly stochastic Poisson process**. Both the deterministic function η and the distribution of Y can be specified. For instance, if $\eta(t) = t$ then (T_n) is called a **mixed Poisson process**, and if $P(Y = y_0) = 1$ the sequence (T_n) represents a **nonhomogeneous Poisson process** with the deterministic intensity function $y_0 \cdot \xi(t)$.

The following types belong to the most common models for η:

- Weibull type: $\eta(t) = t^{\alpha+1}$ $\quad(\alpha > -1)$
- log-linear type: $\eta(t) = t \cdot e^{\alpha t^\gamma}$ $\quad(\alpha \geq 0,\ \gamma \geq 0)$
- logistic type: $\eta(t) = t \cdot [1 + \ln(1 + \alpha t^\gamma)]$ $\quad(\alpha \geq 0,\ \gamma > -1)$.

Some examples for the distribution of Y are given in Wendt (1999). We want to present two cases:

1. Let $Y - y_0$ be a Gamma distributed random variable with the density

$$f_Y(y) = I(y \geq y_0)\,\frac{c^b}{\Gamma(b)}(y - y_0)^{b-1}e^{-c(y-y_0)} .$$

We use that

$$\int_{y_0}^{\infty} \frac{[c + \eta(t)]^{k-u+b}}{\Gamma(k - u + b)}(y - y_0)^{k-u+b-1}e^{-[c+\eta(t)]\cdot(y-y_0)}dy = 1.$$

Then we get (with $0^0 := 1$)

$$
\begin{aligned}
P(\Phi(t)=k) &= \left[\frac{c}{c+\eta(t)}\right]^b \left[\frac{\eta(t)}{c+\eta(t)}\right]^k \cdot e^{-y_0\eta(t)} \\
&\cdot \sum_{u=0}^{k} \frac{\Gamma(k-u+b)}{\Gamma(b)\,\Gamma(k+1)} \binom{k}{u} \left(y_0\,[c+\eta(t)]\right)^u . \qquad (29.2)
\end{aligned}
$$

2. Let Y be inverse Gaussian distributed with the density

$$
f_Y(y) = I(y \geq 0) \sqrt{\frac{\beta}{2\pi\, y^3}} \exp\left(-\frac{1}{2}\frac{\beta(y-\mu)^2}{\mu^2\, y}\right) .
$$

Considering the k-th ordinary moment of an inverse Gaussian distributed random variable some arithmetic yields

$$
\begin{aligned}
P(\Phi(t)=k) &= \exp\left(-\frac{\beta}{\mu}(\sqrt{1+2\eta(t)\mu^2/\beta}-1)\right) \left[\frac{\mu\,\eta(t)}{\sqrt{1+2\eta(t)\mu^2/\beta}}\right]^k \frac{1}{k!} \\
&\cdot \sum_{u=0}^{k-1} \frac{(k-1+u)!}{(k-1-u)!\,u!} \left[\frac{\mu}{2\beta\sqrt{1+2\eta(t)\mu^2/\beta}}\right]^u . \qquad (29.3)
\end{aligned}
$$

Assuming the mixed Poisson process further expressions can be found in Albrecht (1981), Schröter (1995) and Grandell (1997).

Next we consider a marking of the sequence (T_n) proposed by Last and Brandt (1995). At every time point T_n a shock causes a random damage. We describe the damage increment at T_n by the mark X_n. Let $(\mathbf{R}^+, \mathcal{B}^+)$ be the space of marks and let G be a stochastic kernel from $(\mathbf{R}^+, \mathcal{B}^+)$ to $(\mathbf{R}^+, \mathcal{B}^+)$. Then $\Phi = ((T_n, X_n))$ is said to be a **position-dependent G-marking** of (T_n) if $X_1, X_2, \ldots$ are conditionally independent given (T_n) where for all $B \in \mathcal{B}^+$ and $n \in \mathbf{N}$ the following relation holds true:

$$
P(X_n \in B \,|\, (T_n)) = G(T_n, B) \qquad P-a.s. \quad \text{on } \{T_n < \infty\}. \qquad (29.4)
$$

Moreover, we assume that each mark X_n and Y are conditionally independent given (T_n), i.e., $P(X_n \in B \,|\, (T_n), Y) = P(X_n \in B \,|\, (T_n))$. With a position-dependent-marking it is possible to describe a damage model where the increment at the $n-th$ shock depends on the time T_n of the $n-th$ shock. We want to give a simple example.

Example 29.2.1 Let (U_n) be a sequence of non-negative iid random variables with the density f_U and let $\delta \in \mathbf{R}$. We assume that the sequence (U_n) is independent on (T_n). The sequence (X_n) is defined by $X_n = U_n \cdot e^{\delta T_n}$. That means we get damage increments which tend to be increasing ($\delta > 0$) or decreasing ($\delta < 0$). The stochastic kernel G is given by

$$G(t,B) = \int_B f_U(x \cdot e^{-\delta t}) \cdot e^{-\delta t} \mathrm{d}x, \qquad B \in \mathcal{B}^+, \quad t \geq 0.$$

For $\delta = 0$ we have the special case of independent marking. In this case G defines a probability measure which is independent of the time t.

29.3 Maximum Likelihood Estimations

Now we assume the existence of a family of probability measures $P_\Theta = (P_\theta, \theta \in \Theta)$ on $(\Omega, \mathcal{F})$. Let θ be given as $\theta = (\theta^Y, \theta^T, \theta^X) \in R^p$ with $p = u+v+w$. Here, $\theta^Y \in R^u$ is a parameter of the distribution function F_Y of Y, $\theta^T \in R^v$ denotes a parameter of the deterministic terms η and ξ, respectively. And $\theta^X \in R^w$ represents a parameter of the kernel G.

Remark 29.3.1 For the reason of more simple notation we deal with discrete random variables X_n. Here we use the notation

$$P_\theta(X_n = x|(T_n)) = G(T_n, \{x\}; \theta^X).$$

In the continuous case we have to replace the terms $G(T_n, \{x\}; \theta^X)$ by conditional densities of the marks X_n, $n \in \mathbf{N}$.

Under the assumptions of section 29.2 we get the following $(P_\theta, \mathcal{F}_t)$-stochastic intensity kernel of the marked point process $\Phi = ((T_n, X_n))$

$$\lambda(t,B;\theta) = Y \cdot \xi(t;\theta^T) \cdot G(t,B;\theta^X), \qquad B \in \mathcal{B}^+.$$

Let the observed history be given by $\{\mathcal{F}_t^\Phi\}$. In this case the σ-field $\sigma(Y)$ is unknown. Considering the $\{\mathcal{F}_t^\Phi\}$-likelihood process $(L(t;\theta))_{t\geq 0}$ we have to determine the $(P_\theta, \mathcal{F}_t^\Phi)$-stochastic intensity kernel $\tilde{\lambda}$. According to Anderson et al. (1993) and Last and Brandt (1995) it follows that $\tilde{\lambda}$ is given as

$$\begin{aligned} \tilde{\lambda}(t,B;\theta) &= E_\theta[\lambda(t,B;\theta)|\mathcal{F}_{t-}^\Phi] = \xi(t;\theta^T)\, G(t,B;\theta^X)\, E_\theta[Y \,|\, \mathcal{F}_{t-}^\Phi] \\ &= \xi(t;\theta^T)\, G(t,B;\theta^X) \frac{\int_0^\infty y^{\Phi(t-)+1} e^{-y\,\eta(t;\theta^T)} F_Y(\mathrm{d}y;\theta^Y)}{\int_0^\infty y^{\Phi(t-)} e^{-y\,\eta(t;\theta^T)} F_Y(\mathrm{d}y;\theta^Y)} \end{aligned} \tag{29.5}$$

for $B \in \mathcal{B}^+$.

Now we are able to present the likelihood function $L(t;\theta)$ based on the sequence $((T_n, X_n))$ up to time t. In Last and Brandt (1995) it is shown that (the essential part of) $L(t;\theta)$ is given as

$$L(t;\theta) = \left[\prod_{n=1}^{\Phi(t)} \tilde{\lambda}(T_n, X_n;\theta)\right] \exp\Big(\int_0^t \tilde{\lambda}(s, \mathbf{R}^+;\theta)\mathrm{d}s\Big).$$

Here $\tilde{\lambda}(t, \mathbf{R}^+;\theta) = \xi(t;\theta^T)\, E_\theta[\,Y \mid \mathcal{F}^\Phi_{t-}\,]$ denotes the $(P_\theta, \mathcal{F}^\Phi_t)$-intensity function of the sequence (T_n). We use that

$$\Phi(s-) = n-1 \quad \text{on } (T_{n-1}, T_n] \qquad \text{and} \qquad \Phi(s-) = \Phi(t) \quad \text{on } (T_{\Phi(t)}, t].$$

Furthermore, it holds for any $k = 0, 1, \ldots$ that

$$\begin{aligned} &\frac{\mathrm{d}}{\mathrm{d}s} \ln\Big(\int_0^\infty y^k e^{-y\,\eta(s;\theta^T)} F_Y(\mathrm{d}y;\theta^Y)\Big) \\ &\quad = \; -\xi(s;\theta^T)\, \frac{\int_0^\infty y^{k+1} e^{-y\,\eta(s;\theta^T)} F_Y(\mathrm{d}y;\theta^Y)}{\int_0^\infty y^k e^{-y\,\eta(s;\theta^T)} F_Y(\mathrm{d}y;\theta^Y)}\,. \end{aligned}$$

Hence, by elementary calculations we get the log-likelihood function

$$\begin{aligned} \ln L(t;\theta) \;=\; &\sum_{n=1}^{\Phi(t)} \Big(\ln \xi(T_n;\theta^T) + \ln G(T_n, \{X_n\};\theta^X)\Big) \\ &+ \ln\Big(\int_0^\infty y^{\Phi(t)} e^{-y\,\eta(t;\theta^T)} F_Y(\mathrm{d}y;\theta^Y)\Big). \end{aligned} \tag{29.6}$$

But in many situations the sequence $((T_n, X_n))$ cannot be fully observed up to fixed time t, e.g., if the system fails whenever the process $(Z(t))$ exceeds a certain value h resulting in censored data. Consider the $\{\mathcal{F}^\Phi_t\}$-stopping time $\tilde{T} \wedge Z^h = \min\{\tilde{T}, Z^h\}$ where Z^h is given by

$$Z^h := \inf\Big\{\, t:\; Z(t) = \sum_{n=1}^\infty I(T_n \le t)\, X_n \ge h \Big\}$$

and $\tilde{T}$ denotes an arbitrary $\{\mathcal{F}^\Phi_t\}$-stopping time.

Our aim is to determine an estimator of θ based on $m \ge 1$ independent copies of the censored process Φ, where $T_{n,i}$, $X_{n,i}$, $\tilde{T}_i \wedge Z^h_i$, $\tilde{\lambda}_i$ as well as $\Phi_i(t)$ refer to the i-th copy ($i = 1, \ldots, m$). The $(P_\theta, \mathcal{F}^{\Phi_i}_t)$-intensity kernels of the censored processes are given by

$$I(t \le \tilde{T}_i \wedge Z^h_i) \cdot \tilde{\lambda}_i(t, B;\theta), \quad B \in \mathcal{B}^+.$$

Anderson et al. (1993) (or direct calculations) show that we can replace in equation (29.6) t by $t \wedge \tilde{T}_i \wedge Z^h_i$. It is possible that the distributions of the independent stopping times $\tilde{T}_i$ contain unknown parameters. In this case $L(t,\theta)$ is called a *partial likelihood function.* Using the notation

$$\Phi_i^c(t) = \Phi_i(t \wedge \tilde{T}_i \wedge Z_i^h) \quad \text{and} \quad \tilde{\lambda}_i(\cdot;\theta) = \tilde{\lambda}_i(\cdot, \mathbf{R}^+;\theta)$$

we get the maximum (partial) likelihood equations $\frac{\partial}{\partial \theta_r} \ln(L(t;\theta)) = 0$ with

$$\frac{\partial \ln L}{\partial \theta_j^Y} = \sum_{i=1}^m \frac{\partial}{\partial \theta_j^Y} \ln \left(\int_0^\infty y^{\Phi_i^c(t)} \exp(-y\, \eta(t \wedge \tilde{T}_i \wedge Z_i^h; \theta^T)) F_Y(\mathrm{d}y;\theta^Y) \right), \tag{29.7}$$

$$\frac{\partial \ln L}{\partial \theta_k^T} = \sum_{i=1}^m \left\{ \sum_{n=1}^{\Phi_i^c(t)} \frac{\partial \ln(\xi(T_{n,i};\theta^T))}{\partial \theta_k^T} - \frac{\partial \eta(t \wedge \tilde{T}_i \wedge Z_i^h;\theta^T)}{\partial \theta_k^T} \frac{\tilde{\lambda}_i(t+ \wedge \tilde{T}_i \wedge Z_i^h;\theta)}{\xi(t \wedge \tilde{T}_i \wedge Z_i^h;\theta^T))} \right\}, \tag{29.8}$$

$$\frac{\partial \ln L}{\partial \theta_l^X} = \sum_{i=1}^m \sum_{n=1}^{\Phi_i^c(t)} \frac{\partial}{\partial \theta_l^X} \ln \left(G(T_{n,i}, \{X_{n,i}\};\theta^X) \right), \tag{29.9}$$

provided the derivatives exist. In general these equations can be solved only by numerical methods. It can be shown that the estimator $\hat{\theta}^X$ does not depend on θ^T and θ^Y. Hence, the equation system (29.9) can be solved separately.

A simplification arises in the uncensored case where we have $t = t \wedge \tilde{T}_i \wedge Z_i^h$ for all $i = 1,\ldots,m$. We use that the likelihood function is sufficient for θ. Exploiting the resulting sufficiency of the statistic

$$E_\theta \left[T^\star \mid \sigma(\hat{\theta}) \right] \qquad \text{with} \qquad T^\star = \frac{1}{m} \sum_{i=1}^m Y_i \cdot \eta(t;\theta^T)$$

we get

$$E_\theta \left[T^\star | \sigma(\hat{\theta}) \right] = \frac{\eta(t;\hat{\theta}^T)}{\xi(t;\hat{\theta}^T)} \cdot \frac{1}{m} \sum_{i=1}^m \tilde{\lambda}_i(t+;\hat{\theta}^Y,\hat{\theta}^T)$$

and

$$E_\theta \left[T^\star | \sigma(\hat{\theta}) \right] = \frac{1}{m} \sum_{i=1}^m \Phi_i(t).$$

Hence, the following condition is satisfied $P_\theta - a.s.$:

$$\sum_{i=1}^m \frac{\tilde{\lambda}_i(t+;\hat{\theta}^Y,\hat{\theta}^T)}{\xi(t;\hat{\theta}^T)} = \sum_{i=1}^m \frac{\int_0^\infty y^{\Phi_i(t)+1}\, e^{-y\,\eta(t;\hat{\theta}^T)} F_Y(\mathrm{d}y;\hat{\theta}^Y)}{\int_0^\infty y^{\Phi_i(t)}\, e^{-y\,\eta(t;\hat{\theta}^T)} F_Y(\mathrm{d}y;\hat{\theta}^Y)} = \sum_{i=1}^m \frac{\Phi_i(t)}{\eta(t;\hat{\theta}^T)}. \tag{29.10}$$

For more details we refer to Wendt (1999). An immediate consequence is the possibility to determine the estimator $\hat{\theta}^T$ of the system (29.8) without knowing the distribution of the random variable Y. Actually, if we take into consideration the relation (29.10), then the system of equations for $\hat{\theta}^T$ is given by

$$\sum_{i=1}^{m} \left\{ \sum_{n=1}^{\Phi_i(t)} \frac{\partial}{\partial \theta_k^T} \ln(\xi(T_{n,i};\theta^T)) - \Phi_i(t) \cdot \frac{\partial}{\partial \theta_k^T} \ln(\eta(t;\theta^T)) \right\} \Bigg|_{\theta^T=\hat{\theta}^T} = 0,$$

$k = 1, \ldots, v$. Thus, for any distribution function F_Y the same estimator $\hat{\theta}^T$ is obtained as for the assumption of a nonhomogeneous Poisson process. Having determined $\hat{\theta}^T$, we can solve the system of equations (29.7) for θ^Y.

In the censored case, however, the ML-estimator $\hat{\theta}^T$ depends on the distribution function F_Y. Let us deal with an example.

Example 29.3.1 We consider the position-dependent marking $X_n = U_n \cdot e^{\delta T_n}$ from Example 29.2.1 where we make the assumptions:

- the iid random variables U_n are exponentially distributed with the density $f_U(u) = I(u \geq 0) \cdot p e^{-pu}$,
- the $(P_\theta, \mathcal{F}_t)$-intensity of the (uncensored) sequence (T_n) is given by $Y \cdot \xi(t;\theta^T)$ where the random variable Y is Gamma distributed and the deterministic part is from Weibull type. That means, $\theta^Y = (c, b)$, $\theta^T = \alpha > -1$ with

$$f_Y(y) = I(y \geq 0) \frac{c^b}{\Gamma(b)} y^{b-1} e^{-cy} \qquad \text{and} \qquad \xi(t;\alpha) = (\alpha+1) \cdot t^\alpha.$$

With respect to Remark 29.3.1 we have to replace in (29.9) the terms

$$\ln\Big(G(T_{n,i}, \{X_{n,i}\}; \theta^X\Big)$$

by

$$\ln\Big(f_U(X_{n,i} \cdot e^{-\delta T_{n,i}})\Big) - \delta T_{n,i}, \quad n = 1, \ldots, \Phi_i^c(t)\ ,\ i = 1, \ldots, m.$$

After some calculations it can be shown that the ML-estimator $\hat{\theta}^X = (\hat{p}, \hat{\delta})$ is a unique solution of the system (29.9). Here, $\hat{p}$ is given as

$$\hat{p} = \frac{\sum_{j=1}^{m} \Phi_j^c(t)}{\sum_{i=1}^{m} \sum_{n=1}^{\Phi_i^c(t)} X_{n,i} \exp(-\hat{\delta}\, T_{n,i})}$$

where $\hat{\delta}$ is the solution of a non-linear equation.

From (29.7) and (29.8) we get the following equation system for the parameters α, c and b:

$$
\begin{aligned}
0 &= \sum_{i=1}^{m}\left\{\frac{\Phi_i^c(t)}{\hat{\alpha}+1}+\sum_{n=1}^{\Phi_i^c(t)}\ln(T_{n,i})\right. \\
&\qquad \left.-(t\wedge\tilde{T}_i\wedge Z_i^h)^{\hat{\alpha}+1}\ln(t\wedge\tilde{T}_i\wedge Z_i^h)\,\frac{\Phi_i^c(t)+\hat{b}}{\hat{c}+(t\wedge\tilde{T}_i\wedge Z_i^h)^{\hat{\alpha}+1}}\right\}, \\
0 &= \sum_{i=1}^{m}\left\{\frac{\hat{b}}{\hat{c}}-\frac{\Phi_i^c(t)+\hat{b}}{\hat{c}+(t\wedge\tilde{T}_i\wedge Z_i^h)^{\hat{\alpha}+1}}\right\}, \\
0 &= \sum_{i=1}^{m}\left\{\ln(\hat{c})-\ln\left(\hat{c}+(t\wedge\tilde{T}_i\wedge Z_i^h)^{\hat{\alpha}+1}\right)+\sum_{n=1}^{\Phi_i^c(t)}\frac{1}{n-1+\hat{b}}\right\}.
\end{aligned}
$$

In the uncensored case, however, we find after elementary calculations the estimator

$$
\hat{\alpha}=\frac{\Phi_\bullet(t)}{\ln(t)\cdot\Phi_\bullet(t)-\sum_{i=1}^{m}\sum_{n=1}^{\Phi_i(t)}\ln(T_{n,i})}-1 \qquad \text{with } \Phi_\bullet(t)=\sum_{i=1}^{m}\Phi_i(t)
$$

which does not contain $\hat{\theta}^Y$. Let us again mention, the ML-estimator $\hat{\alpha}$ is not changed considering another distribution of Y.

Furthermore, we get in the uncensored case $\hat{c}=m\cdot\hat{b}\,\dfrac{t^{\hat{\alpha}+1}}{\Phi_\bullet(t)}$ where $\hat{b}$ is the solution of

$$
\sum_{i=1}^{m}\left\{\sum_{n=1}^{\Phi_i(t)}\frac{1}{n-1+\hat{b}}+\ln\frac{m\,\hat{b}}{\Phi_\bullet(t)+m\,\hat{b}}\right\}=0.
$$

29.4 The Large Sample Case

In this section we deal with the large sample properties of the maximum (partial) likelihood estimator $\hat{\theta}=(\hat{\theta}^Y,\hat{\theta}^T,\hat{\theta}^X)$ based on observations up to time $\tau<\infty$. It is used that the score vector process, i.e., the vector of the first derivatives of the log-likelihood process $(\ln L(t;\theta))$ with respect to $\theta\in R^p$ is a local square integrable $(P_\theta,\mathcal{F}_t^\Phi)$-martingale. Under mild differentiation and continuity requirements on η, ξ, G as well as on F_Y with respect to θ, Taylor series expansions on the score vector process are valid. And further conditions ensure a proper behaviour of the remaining terms in these Taylor expansions.

In particular, we shall assume that $\frac{1}{m}\sum_{i=1}^{m}\tilde{F}_i(s)\to\tilde{F}(s)$ where $\tilde{F}_i$ denotes the distribution function of $\tilde{T}_i$ and $\tilde{F}$ is a distribution function, for $m\to\infty$ and

$s \in [0, \infty]$. Furthermore, we assume the existence of positive constants $k_j < 1$, $j = 1, 2$, such that the conditions $P_\theta(Z^h \le \tau) \le k_1$ and $\tilde{F}(\tau) \le k_2$ are satisfied.

For additional information we have to distinguish between the regularity conditions for the distributions of the marks and for the distribution of the process (T_n). Here we want to give some notes for the sequence (T_n). As for the other case we refer to Wendt (1999).

An essential but hard to prove part of the regularity conditions is a condition on convergence in probability of the predictable covariation processes of the score vector process. Let $\tilde{\lambda}^m(t;\theta)$ denote the $(P_\theta, \mathcal{F}_t^\Phi)$-stochastic intensity of the counting process $(\sum_{i=1}^m \Phi_i^c(t))$, i.e.,

$$\tilde{\lambda}^m(t;\theta) = \sum_{i=1}^m I(t \le \tilde{T}_i \wedge Z_i^h) \cdot \tilde{\lambda}_i(t;\theta) \quad \text{with} \quad \tilde{\lambda}_i(t;\theta) = \xi(t;\theta^T)\, E_\theta[\, Y_i | \mathcal{F}_{t-}^{\Phi_i}\,] \,.$$

Then we have to show that for $m \to \infty$,

$$\frac{1}{m} \int_0^\tau \left(\frac{\partial}{\partial \theta_j} \ln \tilde{\lambda}^m(s,\theta) \right) \left(\frac{\partial}{\partial \theta_l} \ln \tilde{\lambda}^m(s,\theta) \right) \tilde{\lambda}^m(s,\theta) ds \bigg|_{\theta=\theta_0} \xrightarrow{P_\theta} \sigma_{j,l}(\theta_0) \;,$$

where $\sigma_{j,l}(\theta)$ are finite functions defined on a neighbourhood Θ_0 of the true parameter θ_0 $(j, l = 1, \ldots, u + v)$. We make use of suitable criteria on the pointwise convergence in probability of a stochastic process to a deterministic function and then we apply *Helland*'s condition to show the existence of the functions $\sigma_{j,l}(\theta)$ [Anderson et al.(1993, p. 85)]. But in general we are not able to determine these functions exactly. Additionally, we have to check the *Lindeberg*-type condition ensuring the jumps of normalized versions of the score vector process are asymptotically neglectable. Here we can again apply *Helland*'s condition.

According to Anderson et al. (1993) (for (θ^Y, θ^T)) and Wendt (1999) (for θ^X) it follows that under the regularity conditions with a probability tending to 1 the likelihood equations have a consistent solution $\hat{\theta} = (\hat{\theta}^Y, \hat{\theta}^T, \hat{\theta}^X)$. Further, $\hat{\theta}$ is asymptotically multinormally distributed around the true value θ_0. The asymptotic covariance matrix $[\sigma_{j,l}(\theta_0)]^{-1}$ may be estimated consistently by $\left[-\frac{1}{m} \frac{\partial^2 \ln L(\tau;\theta)}{\partial\theta_j \partial\theta_l} \Big|_{\theta=\hat{\theta}} \right]^{-1}$

These propositions can be used to calculate confidence regions and acceptance areas for tests on the parameter vector θ applying the well-known Wald test statistic and the (partial) likelihood ratio test statistic, respectively.

Let us give some examples for such confidence regions. At first we consider the position-dependent marking of (T_n) from Example 29.3.1. We determine confidence estimations for the parameter $\theta^X = (p, \delta)$ based on $m = 25$ simulated realizations of $\Phi = ((T_n, X_n))$. The parameter α of the function $\eta(t) = t^{\alpha+1}$ is varying ($\alpha = -0.3$, $\alpha = 0$ and $\alpha = 0.3$). Hence, α has an influence on the sequence (T_n) and also on the confidence regions of θ^X.

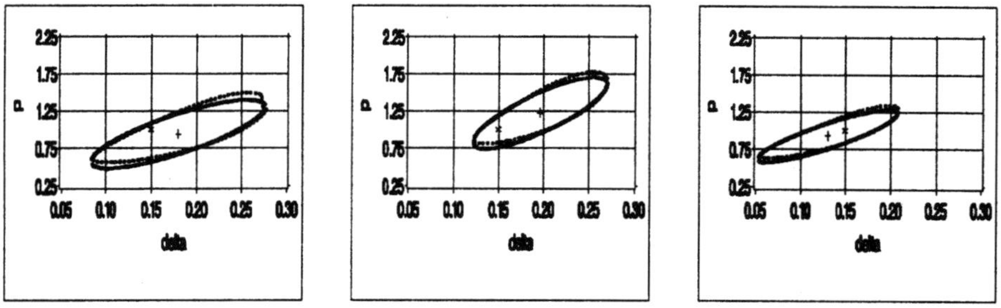

Figure 29.2: Confidence estimates for $\theta^X = (p, \delta)$ and different η

In Figure 29.2 the solid line represents the confidence region based on the Wald test statistic and the broken line those based on the likelihood ratio test statistic ($p_0 = 1$, $\delta_0 = 0.15$). Let us mention that $\frac{1}{m}\Phi^c_\bullet(10)$ was given by 2.1 – 7.1 (depending on α). The both statistics yield only little differences for the considered small sample size $m = 25$.

Next we want to give an example for θ^Y and θ^T. Here we assume an inverse Gaussian distribution for the random variable Y and a function η of logistic type, i.e.,

$$f_Y(y) = I(y \geq 0)\sqrt{\frac{\beta}{2\pi\, y^3}}\exp\left(-\frac{1}{2}\,\frac{\beta(y-\mu)^2}{\mu^2\, y}\right)$$

and

$$\eta(t) = t\,[1 + \ln(1 + \alpha t^\gamma)]\ .$$

In Figure 29.3 confidence estimates are plotted for μ, β and α, γ, respectively ($\mu_0 = 0.45$, $\beta_0 = 1$, $\alpha_0 = 0.5$, $\gamma_0 = 0.75$). Again, the solid line represents the confidence region based on the Wald test statistic. We have chosen the sample size $m = 100$ and $m = 250$.

In the considered censored case we observed $\frac{1}{m}\Phi^c_\bullet(10) = 10.2$. Note that we assumed Y is not observable and the estimates are found from the observation of (T_n). Furthermore, always two nuisance parameters have to be estimated.

Figure 29.3 shows that both confidence regions nearly coincide only in case of a large size m of realizations.

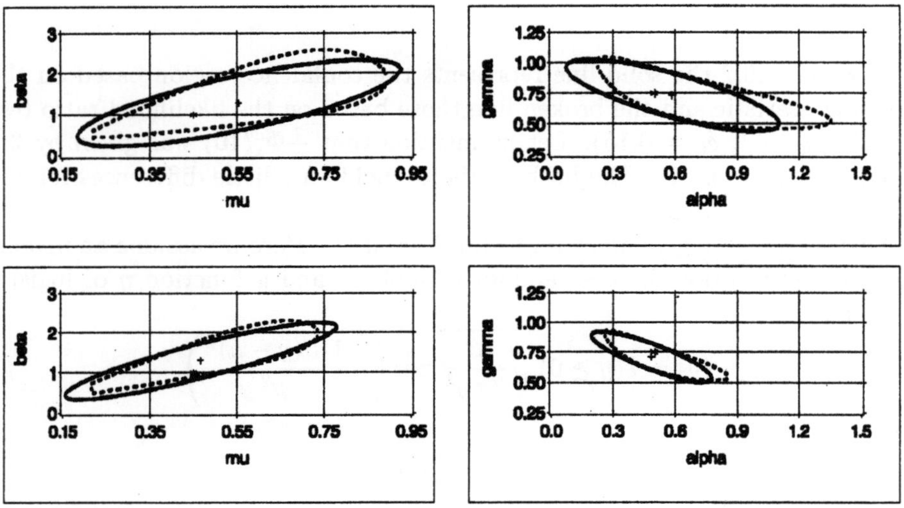

Figure 29.3: Confidence estimates for $\theta^Y = (\mu, \beta)$ and $\theta^T = (\alpha, \gamma)$

References

1. Albrecht, P. (1981). *Dynamische Statistische Entscheidungsverfahren für Schadenzahlprozesse*, VVW Karlsruhe, Germany.

2. Anderson, P., Borgan, Ø., Gill, R. and Keiding, N. (1993). *Statistical Models Based on Counting Processes*, Springer-Verlag, New York.

3. Aven, T. and Jensen, U. (1998). *Stochastic Models in Reliability*, Springer-Verlag, New York.

4. Bremaud, P. (1981). *Point Processes and Queues*, Springer-Verlag, New York.

5. Cox, D. R. (1955). Some statistical methods connected with series of events. *Journal of the Royal Statistical Society, Series B*, **17**, 129–164.

6. Cramer, H. (1969). On streams of random events. *Skandinavisk Aktuarietidskrift Suppl.*, **85**, 13–23.

7. Esary, J. D., Marshall, A. W. and Proshan, F. (1973). Shock models and wear processes, *Annals of Probability*, **1**, 627–649.

8. Feng, W., Adachi, K. and Kowada, M. (1994). Optimal replacement under additive damage in a Poisson random environment, *Communications in Statistics—Stochastic Models*, **10**, 679–700.

9. Grandell, J. (1991). *Aspects of Risk Theory*, Springer-Verlag, New York.

10. Grandell, J. (1997). *Mixed Poisson Processes*, Chapman & Hall, London, England.

11. König, D. and Schmidt, V. (1992). *Zufällige Punktprozesse*, B.G. Teubner, Stuttgart, Germany.

12. Last, G. and Brandt, A. (1995). *Marked Point Processes on the Real Line*, Springer-Verlag, New York.

13. Pieper, V. and Tiedge, J. (1983). Zuverlässigkeitsmodelle auf der Grundlage stochastischer Modelle von Verschleißprozessen, *Mathenatuscg Operationsforschung und Statistik, Series Statistics*, **14**, 485–502.

14. Schröter, K.J. (1995). *Verfahren zur Approximation der Gesamtschadenverteilung*, VVW Karlsruhe, Germany.

15. Shaked, M. (1983). Wear and damage processes from shock models, In *Proceedings of the Symposium on Reliability Theory and Models*, pp. 43–64, Charlotte, North Carolina.

16. Sobczyk, K. (1987). Stochastic models for fatigue damage of materials, *Advances in Applied Probability*, **19**, 652–673.

17. Wendt, H. (1999). Parameterschätzungen für eine Klasse doppelt–stochastischer Poisson Prozesse bei unterschiedlichen Beobachtungsinformationen, Ph.D. Thesis.

30

Discrete Time Semi-Markov Processes for Reliability and Survival Analysis – A Nonparametric Estimation Approach

Vlad Barbu and Nikolaos Limnios

Université de Technologie de Compiègne, Compiègne, France

Abstract: In the first part of this paper, after recalling the definition and properties of a discrete time semi-Markov model with a finite state space, we propose a computation procedure for solving the corresponding Markov renewal equation, appearing in survival analysis. This provides us an easily computable formula for the survival function. In the second part, we consider a censored history and we propose empirical estimators for the semi-Markov kernel, semi-Markov transition matrix and survival or reliability. We study the strong consistency and the asymptotic normality of the estimator of survival function. We apply the results to a three state system.

Keywords and phrases: Discrete time Markov renewal process, discrete time semi-Markov process, discrete time semi-Markov kernel, survival function, reliability, empirical estimator, asymptotic properties

30.1 Introduction

Semi-Markov processes and Markov renewal processes are a class of stochastic processes which generalize Markov processes and renewal processes. The main advantage of the semi-Markov processes is that the sojourn time can be arbitrarily distributed. It renders them more suitable to applications than the Markov processes; see, e.g., Pyke (1961), Çinlar (1969), Limnios and Oprişan (2001).

The inference problems for continuous time semi-Markov processes have been widely presented in the literature [see, e.g., Moore and Pyke (1968), Lagakos et al. (1978), Gill (1980), Ouhbi and Limnios (1999)]. Comparatively, the discrete time semi-Markov processes have been less studied [see, for instance, Mode and Sleeman (2000), Csenki (2002), Barbu et al. (2004)].

Nevertheless, the discrete time framework offers, especially for applications, some important advantages as compared to the continuous time framework. Indeed, a discrete time semi-Markov process cannot explode and the Markov renewal function is expressed as a finite series of semi-Markov kernel convolution products, instead of an infinite series in the continuous case. Consequently, all the numerical computations for a functional of a discrete semi-Markov kernel are much faster and more accurate than for a continuous one. Moreover, the discrete time Markov renewal processes constitute a good support for the numerical calculus of the continuous time Markov renewal processes, after their discretization [see, e.g., Janssen and Manca (2001), Csenki (2002)].

Many authors have studied discrete models for lifetimes in a general i.i.d. context; see, e.g., Barlow et al. (1963), Roy and Gupta (1992), Bracquemond et al. (2001), Bracquemond (2001). The discrete time reliability modeling via homogeneous and non-homogeneous Markov chains can be found in Balakrishnan et al. (2001), Platis et al. (2002). Statistical estimation and asymptotic properties for reliability metrics, using discrete time homogeneous Markov chains, are presented in Sadek and Limnios (2002). The continuous time semi-Markov model for survival analysis and reliability can be found in Limnios and Oprişan (1999, 2001), Ouhbi and Limnios (2003), Limnios and Ouhbi (2003). The discrete time semi-Markov model for survival or reliability and the associated inference problems are presented in Barbu et al. (2004), Barbu and Limnios (2003). The aim of the present paper is to give unified presentation of our previous results concerning the survival of discrete time semi-Markov systems.

The paper is organized as follows. In Section 30.2 we present some basic definitions and results from Barbu et al. (2004) concerning the homogeneous discrete time semi-Markov process. We also give a simple forward algorithm which allows easy computation of measures appearing in Reliability and Survival analysis. In Section 30.3 we propose empirical estimators of the conditional sojourn time distribution, discrete time semi-Markov kernel, transition matrix and survival function. We give the asymptotic properties of the survival function estimator and we construct confidence intervals. The main steps of the proof of the principal result are given in Section 30.4. In Section 30.5 we illustrate our results by a numerical example.

30.2 Discrete Time Semi-Markov Processes: Definitions and Basic Properties

Let us consider:

- E, the state space. We suppose E to be finite, with $|E| = s$.
- The stochastic process $J = (J_n;\ n \in \mathbb{N})$ with state space E for the system state at the n-th jump.
- The stochastic process $S = (S_n;\ n \in \mathbb{N})$ with state space $\mathbb{N}$ for the n-th jump of the process. We suppose $S_0 = 0$ and $0 < S_1 < S_2 < \ldots < S_n < S_{n+1} < \ldots$.
- The stochastic process $X = (X_n;\ n \in \mathbb{N}^*)$ with state space $\mathbb{N}^*$ for the sojourn time X_n in state J_{n-1} before the n-th jump. Thus, we have for all $n \in \mathbb{N}^*$
$$X_n = S_n - S_{n-1}.$$
- $\mathcal{M}_E$, the set of nonnegative matrices on $E \times E$. If $A \in \mathcal{M}_E$, we set $A = (A_{ij};\ i, j \in E)$, where A_{ij} is the entry of the i-th row, j-th column, for all $i, j \in E$; we also denote by $I_E \in \mathcal{M}_E$ the identity matrix and by $0_E \in \mathcal{M}_E$ the null matrix.
- $\mathcal{M}_E(\mathbb{N})$, the set of matrix-valued functions : $\mathbb{N} \to \mathcal{M}_E$. If $A \in \mathcal{M}_E(\mathbb{N})$, we set $A = (A(k);\ k \in \mathbb{N})$, where, for any fixed $k \in \mathbb{N}, A(k) = (A_{ij}(k);\ i, j \in E) \in \mathcal{M}_E$.

Definition 30.2.1 The stochastic process $(J, S) = ((J_n, S_n);\ n \in \mathbb{N})$ is said to be a discrete time Markov renewal process (DTMRP) if for all $n \in \mathbb{N}$, for all $i, j \in E$ and for all $k \in \mathbb{N}$ it almost surely satisfies

$$\begin{aligned} &\mathbb{P}(J_{n+1} = j, S_{n+1} - S_n = k \mid J_0, \ldots, J_n; S_0, \ldots, S_n) \\ &\qquad = \mathbb{P}(J_{n+1} = j, S_{n+1} - S_n = k \mid J_n). \end{aligned} \tag{30.1}$$

Moreover, if (30.1) is independent of n, (J, S) is said to be homogeneous, with discrete semi-Markov kernel $q(k) = (q_{ij}(k);\ i, j \in E) \in \mathcal{M}_E$ defined by

$$q_{ij}(k) := \mathbb{P}(J_{n+1} = j, X_{n+1} = k \mid J_n = i).$$

Definition 30.2.2 The transition function of the embedded Markov chain $(J_n;\ n \in \mathbb{N})$ is the matrix-valued function $V \in \mathcal{M}_E$ defined by

$$V = (p_{ij})_{i,j \in E}, \quad p_{ij} := \mathbb{P}(J_{n+1} = j \mid J_n = i), \quad i, j \in E,\ n \in \mathbb{N}. \tag{30.2}$$

Definition 30.2.3 For all $i, j \in E$ such that $p_{ij} \neq 0$, let us denote by:

1. $f_{ij}(\cdot)$, the conditional distribution of the sojourn time in state i before going to state j :
$$f_{ij}(k) = \mathbb{P}(X_{n+1} = k \mid J_n = i, J_{n+1} = j), \quad k \in \mathbb{N}, \tag{30.3}$$

2. $h_i(\cdot)$, the sojourn time distribution in state i:

$$h_i(k) = \mathbb{P}(X_{n+1} = k \mid J_n = i) = \sum_{l \in E} q_{il}(k), \quad k \in \mathbb{N}^*,$$

3. $H_i(\cdot)$, the sojourn time cumulative distribution function in state i :

$$H_i(k) = \mathbb{P}(X_{n+1} \leq k \mid J_n = i) = \sum_{l \geq 1}^{k} h_i(l), \quad k \in \mathbb{N}^*.$$

Obviously, for all $i, j \in E$ such that $p_{ij} \neq 0$ and for all $k \in \mathbb{N}$, we have

$$q_{ij}(k) = p_{ij} f_{ij}(k). \tag{30.4}$$

Let us give some definitions and results from Barbu et al. (2004).

Definition 30.2.4 (Discrete time convolution product) Let $A, B \in \mathcal{M}_E(\mathbb{N})$ be two matrix-valued functions. The matrix convolution product $A * B$ is the matrix-valued function $C \in \mathcal{M}_E(\mathbb{N})$ defined by

$$C_{ij}(k) := \sum_{k \in E} \sum_{l=0}^{k} A_{ik}(k-l)\, B_{kj}(l), \quad i, j \in E, \quad k \in \mathbb{N}.$$

Lemma 30.2.1 *Let $\delta I = (d_{ij}(k);\ i, j \in E) \in \mathcal{M}_E(\mathbb{N})$ be the matrix-valued function defined by*

$$d_{ij}(k) := \begin{cases} 1 & \textit{if } i = j \textit{ and } k = 0, \\ 0 & \textit{elsewhere.} \end{cases}$$

Then, δI is the neutral element for the discrete time matrix convolution product, i.e., δI satisfies

$$\delta I * A = A * \delta I = A, \quad A \in \mathcal{M}_E(\mathbb{N}).$$

Definition 30.2.5 Let $A \in \mathcal{M}_E(\mathbb{N})$ be a matrix-valued function. If there exists some $B \in \mathcal{M}_E(\mathbb{N})$ such that

$$B * A = \delta I, \tag{30.5}$$

then B is called the left inverse of A in the convolution sense and it is denoted by $A^{(-1)}$.

We stress the fact that the left inverse of A is not always defined. The next proposition gives a sufficient condition for the existence and uniqueness of the left inverse.

Proposition 30.2.1 *Let $A \in \mathcal{M}_E(\mathbb{N})$ be a matrix-valued function. If* $\det A(0) \neq 0$, *then the left inverse of A exists and is unique.*

Definition 30.2.6 (Discrete time n-fold convolution) Let $A \in \mathcal{M}_E(\mathbb{N})$ be a matrix-valued function and $n \in \mathbb{N}$. The n-fold convolution $A^{(n)}$ is the matrix function $C \in \mathcal{M}_E(\mathbb{N})$ defined recursively by:

$$\begin{aligned} A_{ij}^{(0)}(k) &:= \begin{cases} 1 & \text{if} \quad k=0 \quad \text{and} \quad i=j, \\ 0 & \text{else,} \end{cases} \\ A_{ij}^{(1)}(k) &:= A_{ij}(k) \end{aligned}$$

and

$$A_{ij}^{(n)}(k) := \sum_{l \in E} \sum_{s=0}^{k} A_{lj}^{(n-1)}(k-s)\, A_{il}(s), \quad n \geq 2, k \geq 1.$$

For a DTMRP (J, S), the n-fold convolution of the semi-Markov kernel can be expressed as follows.

Proposition 30.2.2 *For all $i, j \in E$, for all n and $k \in \mathbb{N}$, we have*

$$q_{ij}^{(n)}(k) = \mathbb{P}(J_n = j, S_n = k \mid J_0 = i). \tag{30.6}$$

Let us consider the matrix-valued functions $Q = (Q(k);\ k \in \mathbb{N}) \in \mathcal{M}_E(\mathbb{N})$, defined by

$$Q_{ij}(k) := \mathbb{P}(J_{n+1} = j, X_{n+1} \leq k \mid J_n = i) = \sum_{l=1}^{k} q_{ij}(l), \quad i, j \in E,\ k \in \mathbb{N} \tag{30.7}$$

and $\psi = (\psi(k);\ k \in \mathbb{N}) \in \mathcal{M}_E(\mathbb{N})$, defined by

$$\psi_{ij}(k) := \sum_{n=0}^{k} q_{ij}^{(n)}(k), \quad i, j \in E,\ k \in \mathbb{N}. \tag{30.8}$$

Proposition 30.2.3 *The matrix-valued function $\psi = (\psi(k);\ k \in \mathbb{N})$ is given by*

$$\psi(k) = (\delta I - q)^{(-1)}(k), \tag{30.9}$$

where $(\delta I - q)^{(-1)}$ denotes the left convolution inverse of the matrix function $(\delta I - q)$ and is computed using the forward algorithm

$$\begin{cases} (\delta I - q)^{(-1)}(k) = I_E & \text{if } k = 0, \\ (\delta I - q)^{(-1)}(k) = -\sum_{s=0}^{k-1} (\delta I - q)^{(-1)}(s)\, (\delta I - q)(k-s) & \text{if } k \in \mathbb{N}^*. \end{cases} \tag{30.10}$$

Definition 30.2.7 The matrix renewal function $\Psi = (\Psi(k);\ k \in \mathbb{N}) \in \mathcal{M}_E(\mathbb{N})$ of the DTMRP is defined by

$$\Psi_{ij}(k) := \mathbb{E}_i[N_j(k)], \quad i,j \in E,\ k \in \mathbb{N}, \tag{30.11}$$

where $N_j(k)$ is the number of visits to state j before time k.

The matrix renewal function can be expressed in the following form:

$$\Psi_{ij}(k) = \sum_{n=0}^{k} Q_{ij}^{(n)}(k) = \sum_{l=0}^{k} \psi_{ij}(l), \quad i,j \in E,\ k \in \mathbb{N}. \tag{30.12}$$

Let us now define the discrete time semi-Markov transition matrix and we propose a computation procedure.

Definition 30.2.8 A stochastic process $Z = (Z_k;\ k \in \mathbb{N})$ is called the discrete time semi-Markov process associated with the DTMRP (J,S), if

$$Z_k = J_{N(k)}, \quad k \in \mathbb{N},$$

where $N(k) := \max\{n \geq 0; S_n \leq k\}$ is the discrete time counting process of the number of jumps in $[1, k] \subset \mathbb{N}$.

Thus, Z_k gives the system state at time k. We have also $J_n = Z_{S_n}, n \in \mathbb{N}$.

Definition 30.2.9 The transition matrix of the semi-Markov process Z is the matrix-valued function $P \in \mathcal{M}_E(\mathbb{N})$ defined by

$$P_{ij}(k) := \mathbb{P}(Z_k = j \mid Z_0 = i), \quad i,j \in E,\ k \in \mathbb{N}.$$

The Markov renewal equation for the semi-Markov transition function P is [see Barbu et al. (2004)]

$$P = I - H + q * P, \tag{30.13}$$

where $H(k) := diag(H_i(k);\ i \in E)$.

Solving the Markov renewal equation (30.13) [see Barbu et al. (2004)] we obtain that the unique solution is

$$P(k) = \Big[(\delta I - q)^{(-1)} * (I - H)\Big](k) = \Big[\psi * (I - diag(Q \cdot \mathbf{1}))\Big](k), \tag{30.14}$$

where $\mathbf{1}$ denotes the s-column vector all of whose elements are 1.

30.3 Estimation of Survival Function and Asymptotic Properties

Let us consider a history $\boldsymbol{H}(M)$ of the discrete time MRP $((J_n, S_n);\ n \in \mathbb{N})$, censored at time $M \in \mathbb{N}$,

$$\boldsymbol{H}(M) := (J_0, X_1, \ldots, J_{N(M)-1}, X_{N(M)}, J_{N(M)}, u_M),$$

where we set $N(M) := \max\{n \mid S_n \leq M\}$ and $u_M := M - S_{N(M)}$.

Definition 30.3.1 For all $i, j \in E$ and $k \leq M$, we define:

1. $N_i(M) := \sum_{n=0}^{N(M)-1} \mathbf{1}_{\{J_n = i\}}$, the number of visits to state i, up to time M;

2. $N_{ij}(M) := \sum_{n=1}^{N(M)} \mathbf{1}_{\{J_{n-1}=i, J_n=j\}}$, the number of transitions from i to j, up to time M;

3. $N_{ij}(k, M) := \sum_{n=1}^{N(M)} \mathbf{1}_{\{J_{n-1}=i, J_n=j, X_n=k\}}$, the number of transitions from i to j, up to time M, with sojourn time in state i equal to $k, 1 \leq k \leq M$.

Taking a history $\boldsymbol{H}(M)$ of a discrete time MRP, for all $i, j \in E$ and $k \in \mathbb{N}, k \leq M$, we define the empirical estimators of the probability transition function p_{ij}, sojourn conditioned time $f_{ij}(k)$ and discrete semi-Markov kernel $q_{ij}(k)$ by

$$\hat{p}_{ij}(M) := \frac{N_{ij}(M)}{N_i(M)}, \tag{30.15}$$

$$\hat{f}_{ij}(k, M) := \frac{N_{ij}(k, M)}{N_{ij}(M)} = \frac{1}{N_{ij}(M)} \sum_{n=1}^{N(M)} \mathbf{1}_{\{J_{n-1}=i, J_n=j, X_n=k\}}, \tag{30.16}$$

$$\hat{q}_{ij}(k, M) := \frac{N_{ij}(k, M)}{N_i(M)} = \frac{1}{N_i(M)} \sum_{n=1}^{N(M)} \mathbf{1}_{\{J_{n-1}=i, J_n=j, X_n=k\}}. \tag{30.17}$$

Let us also set $\hat{q}(k, M) := (\hat{q}_{ij}(k, M);\ i, j \in E)$.

Remark The above empirical estimators are approached nonparametric maximum likelihood estimators, i.e., they maximize an approached likelihood function, obtained by neglecting the part corresponding to u_M.

Let us denote by $\hat{Q}(k, M)$ the estimator of $Q(k)$ defined by

$$\hat{Q}(k, M) := \sum_{l=1}^{k} \hat{q}(l) \tag{30.18}$$

and by $\hat{\psi}(k, M)$ the estimator of $\psi(k)$ defined by

$$\hat{\psi}(k, M) := \sum_{n=0}^{k} \hat{q}^{(n)}(k, M), \tag{30.19}$$

where $\hat{q}^{(n)}(k, M)$ is the n-fold convolution of $\hat{q}(k, M)$ (see Definition 30.2.6).

Using relation (30.12), we denote by $\hat{\Psi}(k, M)$ the estimator of $\Psi(k)$ defined by

$$\hat{\Psi}(k, M) := \sum_{l=0}^{k} \hat{\psi}(l, M) = \sum_{l=0}^{k} \sum_{n=0}^{l} \hat{q}^{(n)}(l, M).$$

We propose the following estimator for $P(\cdot)$:

$$\begin{aligned} \hat{P}(k, M) &:= \left[\left(\delta I - \hat{q}(\cdot, M) \right)^{(-1)} * \left(I - diag(\hat{Q}(\cdot, M) \cdot \mathbf{1}) \right) \right](k) \\ &= \left[\hat{\psi}(\cdot, M) * \left(I - diag(\hat{Q}(\cdot, M) \cdot \mathbf{1}) \right) \right](k). \end{aligned}$$

In the sequel of this section, we give a method for computing the survival function (or reliability) of a discrete time semi-Markov system, we propose an empirical estimator with asymptotic properties.

Let E be partitioned into two subsets U and D, respectively for the up states and for the down states, where $E = U \cup D$ and $U \cap D = \emptyset$. Without loss of generality, we can suppose that

$$U = \{1, \ldots, s_1\} \text{ and } D = \{s_1 + 1, \ldots, s\}, \text{ with } 0 < s_1 < s.$$

We will partition all matrix-valued functions according to this partition. For instance, the transition matrix P of the semi-Markov process and the initial distribution α can be written as:

$$P(k) = \begin{matrix} & \begin{matrix} U & \quad D \end{matrix} & \\ & \begin{pmatrix} P_{11}(k) & P_{12}(k) \\ P_{21}(k) & P_{22}(k) \end{pmatrix} & \begin{matrix} U \\ D \end{matrix} \end{matrix} \qquad \alpha = \begin{matrix} U \quad D \\ (\alpha_1 \quad \alpha_2) \end{matrix}.$$

For $m, n \in \mathbb{N}^*$ such that $m > n$, let $\mathbf{1}_{m,n}$ denote the m-column vector whose first n elements are 1 and the last $m - n$ elements are 0; for $m \in \mathbb{N}^*$, let $\mathbf{1}_m$ denote the m-column vector all of whose elements are 1.

Let T_D denote the first passage time in the subset D, i.e., $T_D := \inf\{n \in \mathbb{N};\quad Z_n \in D\}$. The survival function at time $k \in \mathbb{N}$ is given by

$$S(k) := \mathbb{P}(T_D > k) = \mathbb{P}(Z_n \in U, n \in \{0, \dots, k\}).$$

We define a new semi-Markov process $Y = (Y_n;\ n \in \mathbb{N})$ with state space $E_Y = U \cup \{\Delta\}$, where Δ is an absorbing state

$$Y_n := \begin{cases} Z_n & \text{if } n < T_D \\ \Delta & \text{if } n \geq T_D \end{cases}, \quad n \in \mathbb{N}.$$

The semi-Markov kernel of the process Y is

$$q_Y(k) = \begin{bmatrix} q_{11}(k) & q_{12}(k)\,\mathbf{1}_{s-m} \\ \mathbf{0}_{1m} & 0 \end{bmatrix}, \quad k \in \mathbb{N},$$

where $\mathbf{0}_{1m}$ is an m-dimensional row vector all of whose elements are 0. Let P_Y denote the transition function of Y. Thus, for all $k \in \mathbb{N}$, the survival function is given by

$$\begin{aligned} S(k) &= \mathbb{P}(Y_k \in U) = \sum_{j\in U}\sum_{i \in U} \mathbb{P}(Y_k = j \mid Y_0 = i)\,\mathbb{P}(Y_0 = i) \\ &= [\alpha_1, 0]\,P_Y(k)\,\mathbf{1}_{m+1,m} = \alpha_1 \cdot P_{11}(k) \cdot \mathbf{1}_{s_1} \\ &= \alpha_1\,\psi_{11} * (I - H_1)(k)\,\mathbf{1}_{s_1} = \alpha_1\,\psi_{11} * (I - diag(Q\cdot\mathbf{1})_{11})\mathbf{1}_{s_1}. \end{aligned}$$

We propose the following estimator for the system survival function:

$$\begin{aligned} \hat{S}(k, M) &:= \alpha_1 \cdot \hat{P}_{11}(k, M) \cdot \mathbf{1}_{s_1} \\ &= \alpha_1 \Big[\hat{\psi}_{11}(\cdot, M) * \Big(I - diag(\hat{Q}(\cdot, M)\cdot \mathbf{1})_{11}\Big)\Big](k)\mathbf{1}_{s_1}, \quad (30.20) \end{aligned}$$

where the estimators $\hat{\psi}$ and $\hat{Q}$ are respectively defined by (30.19) and (30.18).

Let us give now the theorems concerning the consistency and the normality of the survival function estimator.

Theorem 30.3.1 *The estimator of the survival function of a discrete time semi-Markov system is strongly consistent, i.e.,*

$$\max_{0\leq k \leq M} \mid \hat{S}(k, M) - S(k) \mid \xrightarrow[M\to\infty]{a.s.} 0.$$

Theorem 30.3.2 *For any fixed $k \in \mathbb{N}$ we have*

$$\sqrt{M}[\hat{S}(k, M) - S(k)] \xrightarrow[M\to\infty]{\mathcal{D}} \mathcal{N}(0, \sigma_S^2(k)),$$

where

$$\begin{aligned}
\sigma_S^2(k) &= \frac{\mu_{jj}^*}{\mu_{jj}} \sum_{i=1}^{s} \frac{\mu_{ii}^2}{\mu_{ii}^*} \Big\{ \sum_{j=1}^{s} \Big[D_{ij} - \mathbf{1}_{\{i \in U\}} \sum_{t \in U} \alpha(t) \Psi_{ti} \Big]^2 * q_{ij}(k) \\
&\quad - \Big[\sum_{j=1}^{s} \Big(D_{ij} * q_{ij} - \mathbf{1}_{\{i \in U\}} \sum_{t \in U} \alpha(t) \psi_{ti} * Q_{ij} \Big) \Big]^2 (k) \Big\}, \qquad (30.21) \\
D_{ij} &:= \sum_{n \in U} \sum_{r \in U} \alpha(n) \psi_{ni} * \psi_{jr} * \Big(I - diag(Q \cdot \mathbf{1}) \Big)_{rr},
\end{aligned}$$

and μ_{ii}^* (resp. μ_{ii}) denotes the mean recurrence time of state i for the embedded Markov chain $J = (J_n;\ n \in \mathbb{N})$ (resp. for the semi-Markov process $Z = (Z_k;\ k \in \mathbb{N})$).

Let us now give the confidence interval of the survival function, using the above asymptotic results.

For $k \in \mathbb{N}, k \leq M$, replacing $q(k)$ (resp. $Q(k), \psi(k), \Psi(k)$) by $\hat{q}(k, M)$ (resp. $\hat{Q}(k, M)$, $\hat{\psi}(k, M)$, $\hat{\Psi}(k, M)$) in Equation (30.21), we obtain an estimator $\hat{\sigma}_S^2(k)$ of the variance $\sigma_S^2(k)$. From the strong consistency of the estimators $\hat{q}(k, M), \hat{Q}(k, M)$, $\hat{\psi}(k, M)$ and $\hat{\Psi}(k, M)$ [see Barbu and Limnios (2003)], we obtain that $\hat{\sigma}_S^2(k)$ converges almost surely to $\sigma_S^2(k)$, as M tends to infinity.

For $k \in \mathbb{N}, k \leq M$, the confidence interval of $S(k)$ at level $100(1-\gamma)\%$, $\gamma \in (0,1)$, is given by

$$\hat{S}(k, M) - u_{1-\gamma/2} \frac{\hat{\sigma}_S(k)}{\sqrt{M}} \leq S(k) \leq \hat{S}(k, M) + u_{1-\gamma/2} \frac{\hat{\sigma}_S(k)}{\sqrt{M}}, \qquad (30.22)$$

where u_γ is the γ-quantile of an $N(0,1)$-distributed variable.

30.4 Proof

PROOF OF THEOREM 30.3.2. (MAIN STEPS)

$$\begin{aligned}
&\sqrt{M}\ \ [\hat{S}(k, M) - S(k)] \\
&= \sqrt{M} \sum_{i \in U} \sum_{j \in U} \alpha(i) \Big[(\hat{\psi}_{ij} - \psi_{ij}) * \Big(I - diag(Q \cdot \mathbf{1}) \Big)_{jj} \\
&\qquad\qquad - \psi_{ij} * \Big(diag(\hat{Q} - Q) \cdot \mathbf{1} \Big)_{jj} \Big] \\
&\quad - \sqrt{M} \sum_{i \in U} \sum_{j \in U} \alpha(i) \Big[(\hat{\psi}_{ij} - \psi_{ij}) * \Big(diag(\hat{Q} - Q) \cdot \mathbf{1} \Big)_{jj} \Big] (k).
\end{aligned}$$

For every $l \in \mathbb{N}, l \leq M, \sqrt{M}(\hat{\psi}_{ij} - \psi_{ij})(l)$ converges in distribution to a normal random variable, when M tends to infinity, and $\Delta Q_{jr}(k-l)$ converges in probability to zero, when M tends to infinity [see Barbu and Limnios (2003)]. Thus, using Slutsky's theorem, we obtain that the last term of the previous equality converges in distribution to zero when M tends to infinity. So, $\sqrt{M}[\hat{S}(k,M) - S(k)]$ has the same limit in distribution as

$$\sqrt{M}\sum_{i\in U}\sum_{j\in U}\alpha(i)\Big[(\hat{\psi}_{ij}-\psi_{ij})*\Big(I-diag(Q\cdot\mathbf{1})\Big)_{jj}-\psi_{ij}*\Big(diag(\hat{Q}-Q)\cdot\mathbf{1}\Big)_{jj}\Big](k).$$

On the other hand, we can show that $\sqrt{M}[\hat{\psi}_{ij}(k,M) - \psi_{ij}(k)]$ has the same limit in distribution as $\sqrt{M}[\psi*\Delta q*\psi]_{ij}(k)$. We obtain that $\sqrt{M}[\hat{S}(k,M)-S(k)]$ has the same limit in distribution as

$$\begin{aligned}
&\sqrt{M}\sum_{i\in U}\sum_{j\in U}\alpha(i)\sum_{r=1}^{s}\sum_{l=1}^{s}\Big[\Big(I-diag(Q\cdot\mathbf{1})\Big)_{jj}*\psi_{ir}*\psi_{lj}*\Delta q_{rl}\Big](k)\\
&\quad-\sqrt{M}\sum_{i\in U}\sum_{j\in U}\alpha(i)\sum_{l=1}^{s}\psi_{ij}*\Delta Q_{jl}(k)\\
&= \frac{1}{\sqrt{M}}\sum_{n=1}^{N(M)}\sum_{r=1}^{s}\sum_{l=1}^{s}\frac{M}{N_r(M)}\Big[D_{rl}*\Big(\mathbf{1}_{\{J_{n-1}=r,J_n=l,X_n=\cdot\}}\\
&\qquad\qquad -q_{rl}(\cdot)\mathbf{1}_{\{J_{n-1}=r\}}\Big)(k)\\
&\quad-\mathbf{1}_{\{r\in U\}}\Big(\sum_{t\in U}\alpha(t)\psi_{tr}\Big)*\Big(\mathbf{1}_{\{J_{n-1}=r,J_n=l,X_n\leq\cdot\}}-Q_{rl}(\cdot)\mathbf{1}_{\{J_{n-1}=r\}}\Big)(k)\Big]\\
&= \frac{1}{\sqrt{M}}\sum_{l=1}^{N(M)}f(J_{l-1},J_l,X_l),
\end{aligned}$$

where, for fixed $k \in \mathbb{N}$, we have defined the function $f : E \times E \times \mathbb{N} \to \mathbb{R}$ by

$$\begin{aligned}
f(i,j,x) &:= \sum_{r=1}^{s}\sum_{l=1}^{s}\frac{M}{N_r(M)}\Big[D_{rl}*\Big(\mathbf{1}_{\{i=r,j=l,x=\cdot\}}-q_{rl}(\cdot)\mathbf{1}_{\{i=r\}}\Big)(k)\\
&\quad-\mathbf{1}_{\{r\in U\}}\Big(\sum_{t\in U}\alpha(t)\psi_{tr}\Big)*\Big(\mathbf{1}_{\{i=r,j=l,x\leq\cdot\}}-Q_{rl}(\cdot)\mathbf{1}_{\{i=r\}}\Big)(k)\Big]\\
&= \frac{M}{N_i(M)}\Big\{D_{ij}*\mathbf{1}_{\{x=\cdot\}}(k)-\mathbf{1}_{\{i\in U\}}\Big(\sum_{t\in U}\alpha(t)\psi_{ti}\Big)*\mathbf{1}_{\{x\leq\cdot\}}(k)\\
&\quad-\sum_{l=1}^{s}\Big[D_{il}*q_{il}(k)-\mathbf{1}_{\{i\in U\}}\Big(\sum_{t\in U}\alpha(t)\psi_{ti}\Big)*Q_{il}(k)\Big]\Big\}.
\end{aligned}$$

Since $N_i(M)/M \xrightarrow[M\to\infty]{a.s.} 1/\mu_{ii}$ [see Limnios and Oprişan (2001)], we obtain from Pyke and Schaufeles' central limit theorem [see Pyke and Schaufele (1964) and Moore and Pyke (1968)] that $\sqrt{M}[\hat{S}(k,M)-S(k)]$ converges in distribution, as M tends to infinity, to a normal random variable with zero mean and variance $\sigma_S^2(k)$ given by (30.21). And hence, the proof is finished. ■

30.5 Numerical Example

In this section we apply the previous results to a three-state discrete time semi-Markov process.

Let us consider that the state space $E = \{1,2,3\}$ is partitioned into the up-state set $U = \{1,2\}$ and the down-state set $D = \{3\}$. The system is defined by the initial distribution $\mu := (1\ 0\ 0)$, by the transition probability matrix V of the embedded Markov chain $(J_n;\ n \in \mathbb{N})$

$$V := \begin{pmatrix} 0 & 1 & 0 \\ 0.9 & 0 & 0.1 \\ 1 & 0 & 0 \end{pmatrix}$$

and by the conditional distributions of the sojourn time

$$f(k) := \begin{pmatrix} 0 & f_{12}(k) & 0 \\ f_{21}(k) & 0 & f_{23}(k) \\ f_{31}(k) & 0 & 0 \end{pmatrix},\ k \in \mathbb{N}.$$

We consider the following distributions for the conditional sojourn time:

- f_{12} is a geometric distribution defined by

  $$\begin{aligned} f_{12}(0) &:= 0, \\ f_{12}(k) &:= p(1-p)^{k-1}, k \geq 1, \end{aligned}$$

  where we take $p = 0.6$.

- $f_{21} := W_{q_1,b_1}, f_{23} := W_{q_2,b_2}$ and $f_{31} := W_{q_3,b_3}$ are discrete time, first type Weibull distributions, defined by

  $$\begin{aligned} W_{q,b}(0) &:= 0, \\ W_{q,b}(k) &:= q^{(k-1)^b} - q^{k^b}, k \geq 1, \end{aligned}$$

  where we take $q_1 = 0.7, b_1 = 1, q_2 = 0.6, b_2 = 0.9, q_3 = 0.5, b_3 = 1$.

By a Monte Carlo method we get a trajectory of the DTSMP in the time interval $[0,M]$ and from this trajectory we estimate $S(k), k \leq M$.

In Figure 30.1 is plotted the empirical estimator of the survival function for some values of the censoring time M. We notice that when M becomes large,

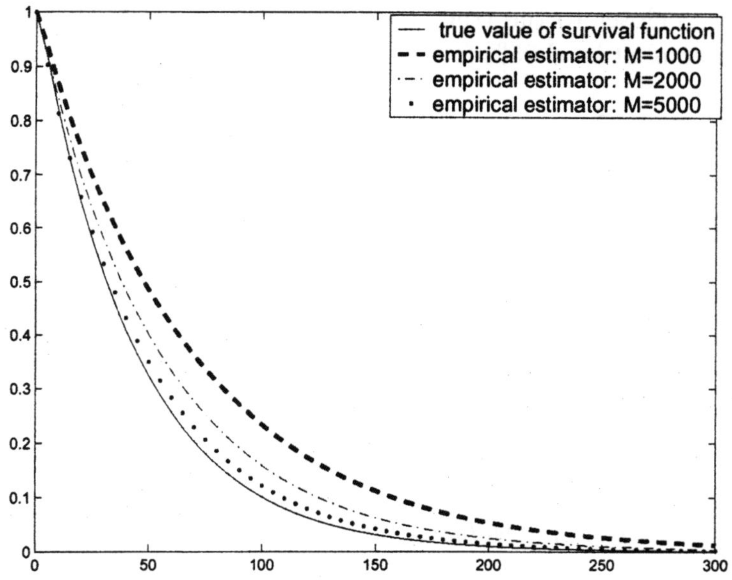

Figure 30.1: Survival function estimator

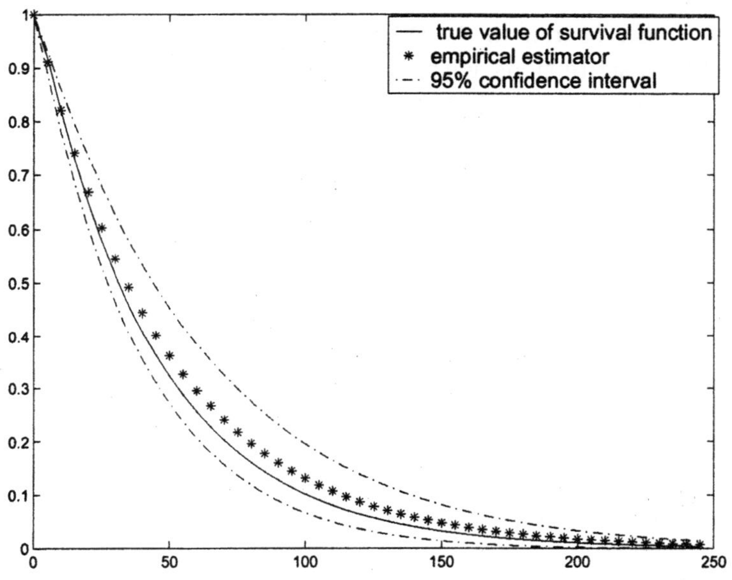

Figure 30.2: Survival function estimator and confidence interval at level 95%

the estimator is closer to the true value, so the consistency is validated. The confidence interval at level 95% for the survival function is given in Figure 30.2.

References

1. Balakrishnan, N., Limnios, N. and Papadopoulos, C. (2001). Basic probabilistic models in reliability, In *Handbook of Statistics, Vol. 20: Advances in Reliability* (Eds., N. Balakrishnan and C.R. Rao), pp. 1–42, Elsevier, North-Holland, Amsterdam, The Netherlands.

2. Barbu, V., Boussemart, M. and Limnios, N. (2004). Discrete time semi-Markov model for reliability and survival analysis, *Communications in Statistics–Theory and Methods*, **33** (to appear).

3. Barbu, V. and Limnios, N. (2003). Nonparametric estimation for discrete time semi-Markov processes with applications in reliability, *Submitted for publication.*

4. Barlow, R. E., Marshall, A. W. and Prochan, F. (1963). Properties of probability distributions with monotone hazard rate, *Annals of Mathematical Statistics*, **34**, 375–389.

5. Bracquemond, C. (2001). *Modélisation stochastique du vieillissement en temps discret*, Thesis, Laboratoire de Modélisation et Calcul, Grenoble, France.

6. Bracquemond, C., Gaudoin, O. and Xie, M. (2001). Towards a new definition of failure rate for discrete distributions, *Proceedings of the 10th International Symposium on Applied Stochastic Models and Data Analysis*, Vol. 2, pp. 266–270.

7. Çinlar, E. (1969). Markov renewal theory, *Advanced in Applied Probability*, **1**, 123–187.

8. Csenki, A. (2002). Transition analysis of semi-Markov reliability models - A tutorial review with emphasis on discrete-parameter approaches, In *Stochastic Models in Reliability and Maintenance* (Ed., S. Osaki), pp. 219–251, Springer-Verlag, Berlin.

9. Gill, R. D. (1980). Nonparametric estimation based on censored observations of Markov renewal processes, *Z. Wahrsch. verw. Gebiete.*, **53**, 97–116.

10. Janssen, J. and Manca, R. (2001). Numerical solution of non homogeneous semi-Markov processes in transient case, *Methodolgy and Computing in Applied Probability*, **3**, 271–293.

11. Lagakos, S. W., Sommer, C. J. and Zelen, M. (1978). Semi-Markov models for partially censored data, *Biometrika*, **65**, 311–317.

12. Limnios, N. and Oprişan, G. (1999). A unified approach for reliability and performability evaluation of semi-Markov systems, *Applied Stochastic Models in Business and Industry*, **15**, 353–368.

13. Limnios, N. and Oprişan, G. (2001). *Semi-Markov Processes and Reliability*, Birkhäuser, Boston, Massachusetts.

14. Limnios, N. and Ouhbi, B. (2003). Empirical estimators of reliability and related functions for semi-Markov systems, In *Mathematical and Statistical Methods in Reliability* (Eds., B. Lindqvist and K. A. Doksum), pp. 469–483, World Scientific Publishing, Singapore.

15. Mode, C. J. and Sleeman, C. K. (2000). *Stochastic Processes in Epidemiology*, World Scientific, New Jersey.

16. Moore, E. H. and Pyke, R. (1968). Estimation of the transition distributions of a Markov renewal process, *Annals of the Institute of Statistical Mathematics*, **20**, 411–424.

17. Ouhbi, B. and Limnios, N. (1999). Nonparametric estimation for semi-Markov processes based on its hazard rate functions, *Statistical Inference for Stochastic Processes*, **2**, 151–173.

18. Ouhbi, B. and Limnios, N. (2003). Nonparametric reliability estimation for semi-Markov processes, *Journal of Statistical Planning and Inference*, **109**, 155–165.

19. Platis, A., Limnios, N. and Le Du, M. (2002). Hitting times in a finite nonhomogeneous Markov chain with applications, *Applied Stochastic Models on Data Analysis*, **14**, 241–253.

20. Pyke, R. (1961). Markov renewal processes: Definitions and preliminary properties, *Annals of Mathematical Statistics*, **32**, 1231–1241.

21. Pyke, R. and Schaufele, R. (1964). Limit theorems for Markov renewal processes, *Annals of Mathematical Statistics*, **35**, 1746–1764.

22. Roy, D. and Gupta, R. (1992). Classification of discrete lives, *Microelectronics and Reliability*, **32**, 1459–1473.

23. Sadek, A. and Limnios, N. (2002). Asymptotic properties for maximum likelihood estimators for reliability and failure rates of Markov chains, *Communications in Statistics—Theory and Methods*, **31**, 1837–1861.

31

Non-parametric Estimation on Lifetimes of Subjects Exposed to Radiation from a Semi-Markov Process

Anne-Laure Afchain

Université Paris-Sud XI, France

Abstract: Distinct problems in the analysis of death age with competing causes of death include the study of interrelations among death cause types. In this report, we establish the lifetime of a male mouse exposed to radiation during a clinical program, by suggesting to model the competing death causes by a semi-Markov process. A non-parametric estimation method for the semi-Markov kernel from lifetime empirical distributions bound to the different causes and for the survival function assessed from the kernel is proposed. Then, we establish the consistency of the semi-Markov kernel estimator.

Keywords and phrases: Semi-Markov process, lifetime and remission empirical distributions, survival function, non-parametric estimation, competing risks

31.1 Experimental Context and Its Modelling

Let us assume a situation where several death causes are identified after having observed X-ray exposed mice. A classical illustration of this situation is provided in Andersen et al. (1993, p. 27): In a laboratory experiment, two groups of Radio-Frequency-Modular (RFM) strain male mice were given a radiation dose of 300 rads at an age of 5–6 weeks. The first group of 95 mice lived in a conventional laboratory environment, while the second group had a germ-free environment. Here, we consider the first group.

After having spooled the ages at death in days for all the mice, the considered observation period begins at the minimum of death times and ends at their maximum. Each individual has a different starting time and was followed until death. The supposed independent 95 mice define the possible different trajectories of the semi-Markov process as in Pérez-Ocón et al. (1998). They were divided into m ($m = 3$) death-cause-type groups as far as the thymic lymphoma, the reticulum cell sarcoma (both specific types of cancer) and the other causes are concerned. The percents of three groups which are 23%, 40% and 37% respectively correspond to components of the initial probability vector estimator. Each cause of death corresponds typically to each state of the semi-Markov graph (see Fig. 31.1). Each state has two types of transitions, the one towards the alive state (state 0) and the other towards the death state (state 4). The alive and death states are absorbent states. The health status of 95 subjects varies from the ill status (due to the different death cause types) either to worsening, or to remission.

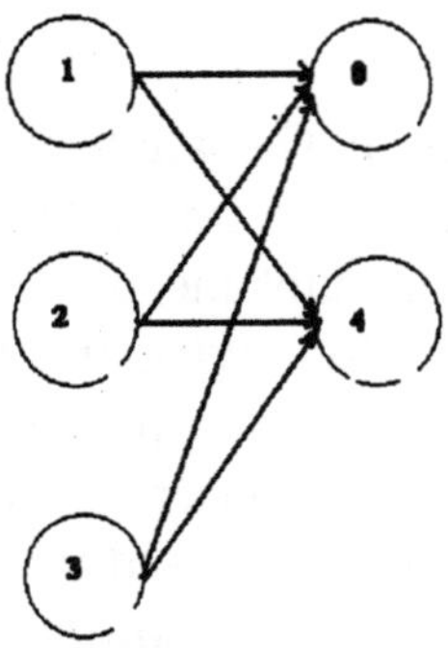

Figure 31.1: Semi-Markov graph with 5 states (3 causes)

When the time spent in a specific state is censored by the death or the remission, one may consider it as a competing risk problem. The analysis of failure times in the presence of competing risks is usually focused on the proportional hazard model by using the likelihood function of Cox in Heutte and Huber-Carol (2002) or on the use of an empirical transition probability matrix for non-homogeneous Markov processes in Fleming and Harrington (1991, p. 287). In this paper, a non-parametric estimation approach for the semi-Markov kernel and for the survival function is presented. Then, the consistency of the semi-Markov kernel estimator is studied.

31.2 Preliminaries

Let us here give some definitions on finite semi-Markov process according to Limnios and Oprisan (2001). Consider a finite set, say $E = \{0, \dots, m+1\}$, $(m < \infty)$ and an E-value stochastic process $Z = (Z_t)_{t \in \mathbb{R}_+}$. Let $0 = S_0 \leq S_1 \leq S_2 \leq \cdots \leq S_n \leq S_{n+1} \leq \cdots$ be the jump times of Z which will be in $\overline{R}_+ = [0, \infty]$ and $J_0, J_1, \dots, J_n, \dots$ be the successive visited states of Z. Let us also define $X_0 = 0$ and $X_n = S_n - S_{n-1}$, $n \geq 1$, the inter-jump times of Z.

Definition 31.2.1 The stochastic process, $(J, S) = (J_n, S_n)_{n \in \mathbb{N}}$, defined on a probability complete space $(\Omega, \mathcal{A}, \mathcal{P})$ is called a continuous time Markov renewal process, with state space E, if for all $n \in \mathbb{N}$, for $j \in E$ and for $t \in \mathbb{R}^+$, it verifies a.s., the following relation:

$$\begin{aligned} \mathbb{P}(J_{n+1} &= j, S_{n+1} - S_n \leq t | J_0, \dots, J_n; S_1, \dots, S_n) \\ &= \mathbb{P}(J_{n+1} = j, S_{n+1} - S_n \leq t | J_n) \end{aligned}$$

for all $j \in E$ and all $t \in \mathbb{R}_+$. In that case, Z is called a semi-Markov process.

Thus, the stochastic process, $(J, S) = (J_n, S_n)_{n \in \mathbb{N}}$ is a Markov chain with state space $E \times \mathbb{R}_+$ and transition kernel $Q = (Q_{ij}(t))_{i,j \in E, t \in \mathbb{R}^+}$ called the semi-Markov kernel such as:

$$Q_{ij}(t) = \mathbb{P}(J_{n+1} = j, S_{n+1} - S_n \leq t | J_n = i) \tag{31.1}$$

for all $t \in \overline{R}_+$, $1 \leq i \leq m$ and $j = 0, m+1$ and $Q_{ij}(t) = 0$, otherwise. In the sequel, we will assume that, $Q_{ii}(t) \equiv 0$, for all $i \in E$. A Markov renewal process can be completely determined if we know its initial law $p(k) = \mathbb{P}(J_0 = k)$ and its semi-Markov kernel $Q = (Q_{ij}(t))_{i,j \in E, t \in \mathbb{R}^+}$ (which we shall estimate).

Definition 31.2.2 The stochastic process $(J_n)_{n \in \mathbb{N}}$ is a Markov chain with state space E and transition probabilities $P(i,j) = Q_{ij}(\infty) = \mathbb{P}(J_{n+1} = j | J_n = i)$, which is called the embedded Markov Chain of Z.

The semi-Markov Z is connected to $(J_n, S_n)_{n \in \mathbb{N}}$ through

$$Z_t = J_n, \text{ if } S_n \leq t \leq S_{n+1}, \quad t \geq 0, \quad \text{and } J_n = Z_{S_n}, \quad n \in \mathbb{N}.$$

31.3 Semi-Markov Kernel Estimator

Definition 31.3.1 A random censure model is defined by copulae (Z_i, δ_i) where

- $Z_i = \min(S_i, D_i)$ is the sojourn time in state i with distribution K_i:
  - S_i is the lifetime (days between exposure & death) of the subject, whose death is due to the i-th illness cause type with distribution F_i,
  - D_i represents the number of days until the moment where the cause disappeared after exposure with distribution G_i,

  where S_i is independent of D_i,
- δ_i defines the dummy variable, taking values in the set $\{0, 1\}$:

$$\delta_i = 1_{\{S_i \leq D_i\}}$$

  equal to 0 for the transition $i \to 0$ and 1 for the transition $i \to m+1 = 4$.

Definition 31.3.2 From random censure model (Z_i, δ_i), the semi-Markov kernel $Q = (Q_{ij}(t))_{i,j \in E, t \in \mathbb{R}^+}$ is defined as follows:

$$Q_{ij}(t) = \begin{cases} 0 & \forall i = 0, \ldots, m, \quad j \neq 0, m+1, \\ \mathbb{P}(Z_i \leq t, \delta_i = 0) & \forall i = 1, \ldots, m, \quad j = 0, \\ \mathbb{P}(Z_i \leq t, \delta_i = 1) & \forall i = 1, \ldots, m, \quad j = m+1. \end{cases} \tag{31.2}$$

From semi-Markov kernel Q given by (31.2), we construct the estimator of $Q = (Q_{ij}(t))_{i,j \in E, t \in \mathbb{R}^+}$, defined by $\hat{Q} = (\hat{Q}_{ij}(t))_{i,j \in E, t \in \mathbb{R}^+}$.

From the definition $Z_i = \min(S_i, D_i)$, we obtain the following result:

$$1 - K_i(t) = [1 - F_i(t)][1 - G_i(t)]. \tag{31.3}$$

From Equation (31.3), an m-***differential equation system*** is provided as in Kalbfleisch and Prentice (1980, p. 170) or in Klein and Moeschberger (1997, p. 68) with respect to known distributions F_i and G_i written as:

$$\begin{cases} dQ_{i0}(t) = [1 - F_i(t)]dG_i(t) = \left[\dfrac{1 - K_i(t)}{1 - G_i(t)}\right] dG_i(t), \\ \\ dQ_{i4}(t) = [1 - G_i(t)]dF_i(t) = \left[\dfrac{1 - K_i(t)}{1 - F_i(t)}\right] dF_i(t) \end{cases} \tag{31.4}$$

with $K_i(t) = Q_{i0}(t) + Q_{i4}(t)$ from Definition 31.3.2.

As we can not analytically integrate the previous system with respect to Q_{i0} and Q_{i4}-distributions, we discretize it according to a random step, defined thanks to the intervals between the consecutive observations of sojourn times. This stage consists of splitting up the interval $[0;T]$ into irregular sub-intervals. In each sub-interval beginning at the failure time and ending before the true one which follows, the function is piecewise constant.

The modified differential equation system (31.4) must undergo the ***estimation procedure*** on time interval $[0;T]$ of $\mathbb{R}^+$ as in Fleming and Harrington (1991, p. 26) or in Kalbfleish and Prentice (1980, p. 40, p. 90). Distributions Q_{i0} and of Q_{i4} can be observable from random sample $(Z_{i(k)}, \delta_{i(k)})_{1\le i\le m, 1\le k\le n}$ with size n defined by $n = \sum_{i=1}^{m} n_i = \sum_{i,j=1}^{m} n_{ij}$ where

- n_i is the number of all the transitions from state i,
- n_{ij} is the number of transitions from state i to state j (here, $n_{ij} \neq 0$ for $j = 0, m+1 = 4$ and $n_{ij} = 0$, otherwise),
- $i(k)$ means that, at the k-th jump, the process stays in state i,
- the realization $T_{i(k)}$ (respectively, $\delta_{i(k)}$) is defined as the k-th observation of random variable T_i (respectively, indicator δ_i) drawn from the sample with size n when the process visits n_i times the state i (respectively, and when it makes a transition towards the following states); here, T_i ($T_{i(k)}$, respectively) can be replaced by S_i, D_i or Z_i ($S_{i(k)}$, $D_{i(k)}$ or $Z_{i(k)}$, respectively).

After the discretization stage and the estimation procedure, we deduce that:

$$\begin{cases} \dfrac{\Delta \hat{Q}_{i0}(k)}{1-\hat{K}_i(k)} = \dfrac{\Delta \hat{G}_i(k)}{1-\hat{G}_i(k)}, \\[2ex] \dfrac{\Delta \hat{Q}_{i4}(k)}{1-\hat{K}_i(k)} = \dfrac{\Delta \hat{F}_i(k)}{1-\hat{F}_i(k)}, \end{cases}$$

with

$$\Delta \hat{F}_i(k) = \hat{F}_i(k+1) - \hat{F}_i(k), \ \Delta \hat{G}_i(k) = \hat{G}_i(k+1) - \hat{G}_i(k)$$

and

$$\Delta \hat{Q}_{ij}(k) = \hat{Q}_{ij}(k+1) - \hat{Q}_{ij}(k)$$

for $i = 1, \ldots, m$, $j = 0, m+1$ and $k = 1, \ldots, n$.

Let us take for example the case of $\hat{Q}_{i4}$ (the case of $\hat{Q}_{i0}$ is treated in the same way) as follows:

$$\begin{aligned}\Delta\hat{Q}_{i4}(k) &= [1-\hat{K}_i(k)]\frac{\Delta\hat{F}_i(k)}{1-\hat{F}_i(k)},\\ \hat{Q}_{i4}(k+1) &= \hat{Q}_{i4}(k)+[1-\hat{K}_i(k)]\frac{\Delta\hat{F}_i(k)}{1-\hat{F}_i(k)}.\end{aligned}$$

From the previous result, we deduce that:

$$\begin{array}{lcll} k=0 & \hat{Q}_{i4}(1) & = & \hat{Q}_{i4}(0)+[1-\hat{K}_i(0)]\dfrac{\Delta\hat{F}_i(0)}{1-\hat{F}_i(0)},\\ k=1 & \hat{Q}_{i4}(2) & = & \hat{Q}_{i4}(1)+[1-\hat{K}_i(1)]\dfrac{\Delta\hat{F}_i(1)}{1-\hat{F}_i(1)},\\ \vdots & \vdots & \vdots & \vdots\\ k=l-1 & \hat{Q}_{i4}(l) & = & \hat{Q}_{i4}(l-1)+[1-\hat{K}_i(l-1)]\dfrac{\Delta\hat{F}_i(l-1)}{1-\hat{F}_i(l-1)}. \end{array}$$

By simplification member to member (with $\hat{Q}_{i4}(t)=0$), we obtain the following result:

$$\hat{Q}_{i4}(k) = \sum_{l=0}^{k-1}\left[\frac{1-\hat{K}_i(l)}{1-\hat{F}_i(l)}\right]\Delta\hat{F}_i(l).$$

Finally, the estimator $\hat{Q}$ of the semi-Markov kernel Q defined by (31.2) is written as follows:

$$\hat{Q}_{ij}(t,n)=\begin{cases} 0 & \forall i=0,\dots,m,\ \ j\neq 0,m+1,\\ \displaystyle\sum_{s\leq t}\left\{\frac{1-\hat{K}_i(s,n)}{1-\hat{G}_i(s,n)}\right\}\Delta\hat{G}_i(s,n) & \forall i=1,\dots,m,\ \ j=0,\\ \displaystyle\sum_{s\leq t}\left\{\frac{1-\hat{K}_i(s,n)}{1-\hat{F}_i(s,n)}\right\}\Delta\hat{F}_i(s,n) & \forall i=1,\dots,m,\ \ j=m+1=4,\end{cases} \tag{31.5}$$

with:

- $\hat{F}_i(t,n)=\frac{1}{n_{i4}}\sum_{k=1}^{n}1_{\{S_{i(k)}\leq t\}}$ and $\hat{G}_i(t,n)=\frac{1}{n_{i0}}\sum_{k=1}^{n}1_{\{D_{i(k)}\leq t\}}$,
- $\Delta\hat{F}_i(t,n)=\frac{1}{n_{i4}}$ and $\Delta\hat{G}_i(t,n)=\frac{1}{n_{i0}}$,
- $\hat{K}_i(t,n)=\frac{1}{n_i}\sum_{k=1}^{n}1_{\{Z_{i(k)}\leq t\}}$.

Nevertheless, we can only simulate the distributions $(\hat{Q}_{i4}(t,n))_{1\leq i\leq m, t\in[0;T]}$, given that we do not know in the clinical example the remission times from a specific illness cause. After having determined the semi-Markov kernel estimator, we study the estimator consistency in the following proposition.

Proposition 31.3.1 *For all $i \in E-\{0, m+1\}$, the semi-Markov kernel element estimators given by (31.5) are consistent on compact $[0;T]$, for $T \in \mathbb{R}^+$, in the sense that, when n tends to infinity,*

$$\begin{cases} \hat{Q}_{i0}(t,n) - Q_{i0}(t) & \rightarrow & 0 \quad (a.s.), \\ \hat{Q}_{i4}(t,n) - Q_{i4}(t) & \rightarrow & 0 \quad (a.s.). \end{cases}$$

PROOF. When n tends to infinity, from the three almost sure convergence relations as follows,

$$\begin{cases} \hat{F}_i(t,n) & \rightarrow & F_i(t), \\ \hat{G}_i(t,n) & \rightarrow & G_i(t), \\ \hat{K}_i(t,n) & \rightarrow & K_i(t), \end{cases}$$

we deduce that:

$$\left\{\frac{1-\hat{K}_i(t,n)}{1-\hat{F}_i(t,n)}\right\} \Delta \hat{F}_i(t,n) \quad \rightarrow \quad \left\{\frac{1-K_i(t)}{1-F_i(t)}\right\} \Delta F_i(t).$$

According to differential equation system (31.4), we have the following relation:

$$\Delta Q_{i4}(t) \quad = \quad \left\{\frac{1-K_i(t)}{1-F_i(t)}\right\} \Delta F_i(t).$$

Thus, we deduce, when n tends towards the infinity, the almost sure convergence of $\hat{Q}_{i4}(t,n)$, such as:

$$\hat{Q}_{i4}(t,n) \quad \rightarrow \quad \sum_{s \leq t} \Delta Q_{i4}(s) = Q_{i4}(t).$$

By analogy, we obtain the same proof for $\hat{Q}_{i0}(t,n)$ expression. ■

31.4 Survival Function Estimator

Proposition 31.4.1 *If semi-Markov kernel $(Q_{ij}(t))_{1 \leq i \leq m, j=0,4, t \geq 0}$ has for estimator $(\hat{Q}_{ij}(t))_{1 \leq i \leq m, j=0,4, t \geq 0}$ given by (31.5), then, whatever the death cause is, the survival function estimator has for expression:*

$$\hat{S}(t,n,m) = 1 - \sum_{i=1}^{m-1} \frac{n_i}{n} [2\hat{Q}_{i0}(t,n) + \hat{Q}_{i4}(t,n)] - \frac{n_m}{n} \hat{Q}_{m4}(t,n). \qquad (31.6)$$

Remark. $\hat{S}$, the estimator of S is consistent, when n tends to infinity, by linear combination of consistent functions (see Proposition 31.3.1).

PROOF. To simplify, let us take $m = 3$ and recall that S is given according to Ouhbi and Limnios (1998, eq. (7)) by the following expression:

$$S(t) \quad = \quad p^{\mathcal{A}}.P^{\mathcal{A}}(t).1$$

with

- $\mathcal{A} = \{0, \dots, m\}$ is the set of alive states,
- $p^{\mathcal{A}}$ is the initial law restricted to $\mathcal{A}$, such as:

$$p^{\mathcal{A}} = \begin{bmatrix} 0 \\ \frac{n_1}{n} \\ \frac{n_2}{n} \\ \frac{n_3}{n} \end{bmatrix},$$

- $1 = (1, \dots, 1)^T$ is the unity vector,
- $Q^{\mathcal{A}}$ is the semi-Markov kernel Q restricted to $\mathcal{A}$, by suppressing the last column and the last row, such as:

$$Q(t) = \left[\begin{array}{cccc|c} 0 & 0 & 0 & 0 & 0 \\ Q_{10}(t) & 0 & 0 & 0 & Q_{14}(t) \\ Q_{20}(t) & 0 & 0 & 0 & Q_{24}(t) \\ Q_{30}(t) & 0 & 0 & 0 & Q_{34}(t) \\ \hline 0 & 0 & 0 & 0 & 0 \end{array}\right],$$

- $P^{\mathcal{A}}(t) = [I - Q^{\mathcal{A}}(t)]^{(-1)} * [I - diag(Q(t)1)^{\mathcal{A}}]$ given according to Ouhbi and Limnios (1998, p. 4).

Also, after computing of the matrix inverse, we obtain the following results:

$$[I - Q^{\mathcal{A}}(t)]^{(-1)} = \begin{bmatrix} 1 & 0 & 0 & 0 \\ -Q_{10}(t) & 1 & 0 & 0 \\ -Q_{20}(t) & 0 & 1 & Q_{14}(t) \\ Q_{30}(t) & 0 & 0 & 1 \end{bmatrix}$$

and

$$[I - diag(Q(t)1)^{\mathcal{A}}] = \begin{bmatrix} 1 & 0 & 0 & 0 \\ 0 & 1-[Q_{10}(t)+Q_{14}(t)] & 0 & 0 \\ 0 & 0 & 1-[Q_{20}(t)+Q_{24}(t)] & 0 \\ 0 & 0 & 0 & 1-[Q_{30}(t)+Q_{34}(t)] \end{bmatrix}.$$

After matrix product and estimation procedure, we deduce the wanted result:

$$P^{\mathcal{A}}(t) = \begin{bmatrix} 1 & 0 & 0 & 0 \\ -Q_{10}(t) & 1-[Q_{10}(t)+Q_{14}(t)] & 0 & 0 \\ -Q_{20}(t) & 0 & [1-(Q_{20}(t)+Q_{24}(t)] & Q_{14}(t) \\ Q_{30}(t) & 0 & 0 & 1-[Q_{30}(t)+Q_{34}(t)] \end{bmatrix}.$$

■

31.5 Numerical Application

31.5.1 Raw data set

Table 31.1: Necropsy data (individual ages at death in days) for RFM male mice exposed to 300 rads X-rays at 5-6 weeks of age in Andersen et al. (1993, p. 27)

k	δ_1	δ_2	δ_3	$Z_{i(k)}$	$Z^*_{i(k)}$
1	1	0	0	$Z_{1(1)}$	159
2	0	0	1	$Z_{3(2)}$	163
3	0	0	1	$Z_{3(3)}$	179
4	1	0	0	$Z_{1(4)}$	189
⋮	⋮	⋮	⋮	⋮	⋮
$n = 95$	$n_{14} = 22$	$n_{24} = 38$	$n_{34} = 35$		

If we assume that the necropsy data sample is indeed generated by the semi-Markov process, then we can consider Table 31.1, whose column k corresponds to 95 individuals, whose columns δ_1, δ_2, δ_3 characterize the indicator variables associated to the specific death causes, whose column $Z_{i(k)}$ defines the k-th sojourn time in state i and whose column $Z^*_{i(k)}$ gives the numerical value of $Z_{i(k)}$. For instance, $Z_{3(2)}$ corresponds to the second value equal to 163 of the sojourn time sample, when the process reaches state 3. In the last row of Table 31.1, for $i = 1, 2, 3$, n_{i4} represents the number of transitions from state i to state 4 and n is defined by $n = \sum_{i=1}^{3} n_{i4}$.

31.5.2 Results

The raw data set presented here allows us to determine the empirical lifetime distributions, but not the empirical distributions bound to the sojourn times from specific ill status towards the remission state. However, these remission times do not appear in the experiment plan. The knowledge of the empirical lifetime distributions permits us to compute the semi-Markov kernel estimator $(\hat{Q}_{i4}(t, n))_{1 \leq i \leq m, t \geq 0}$ (see Fig. 31.2) from Formulae (31.5). Then, from coefficients of the estimated semi-Markov kernel, we build the survival function estimator $\hat{S}(t, n, m)$ versus age for all given subject, whatever the death cause is (see Fig. 31.3) from Formulae (31.6).

31.6 Conclusion

In our study, the semi-Markov process built up is very similar to the concrete evolution of the subject pathology. It reflects well the succession of events, the remission or the death of the subject, and we consider that at each time, the various processes have the same law, as if they would start again at this same time. The non-parametric estimation method proposed for the semi-Markov kernel and for survival function has an interest which widely exceeds the classical frame of the semi-Markov representation of several types of failures. It can test not only the safety analysis of industrial systems with repair rigorously described by Afchain (2001), but also the efficiency of medical treatments, the medical trials carried out on patients whose health status varies from remission to worsening. For instance, we can study the quality of life (scoring level) of cancerous patients or the age influence on the illness progress of infected (positive VIH) patients to the AIDS.

Acknowledgments: The author is indebted to an anonymous referee and Professor Nikolaos Limnios, Université de Technologie de Compiègne, Compiègne, France for their valuable comments.

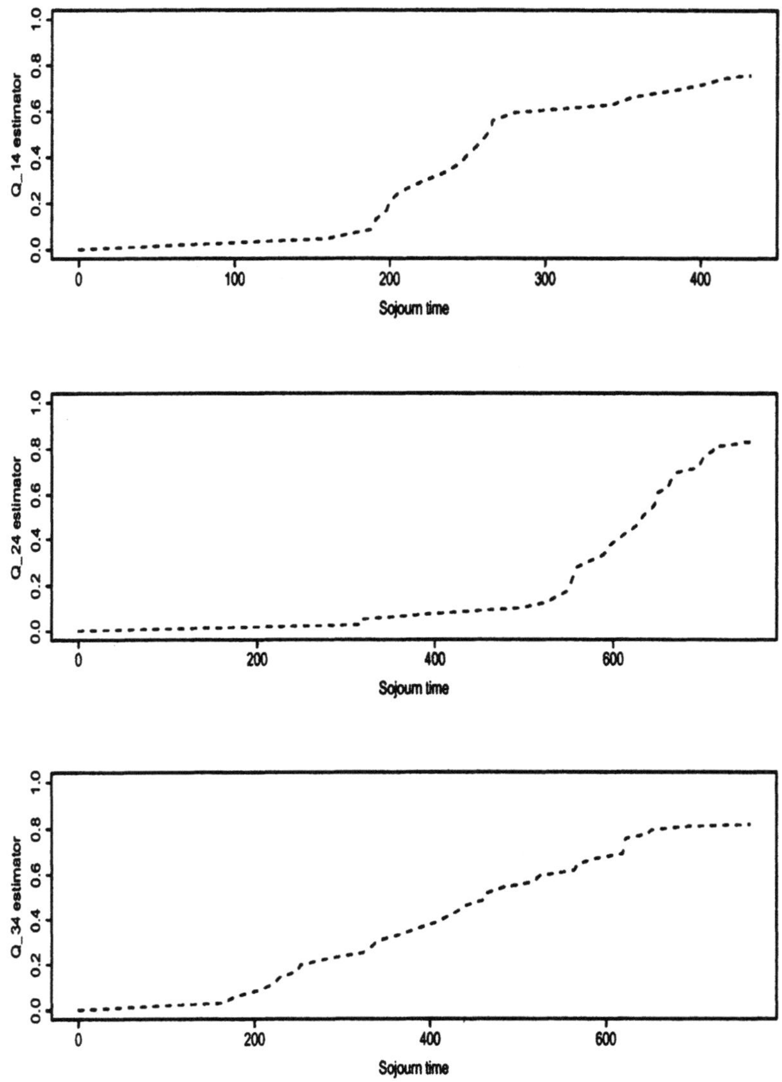

Figure 31.2: Curves of the semi-Markov kernel element estimators $\hat{Q}_{14}$, $\hat{Q}_{24}$, $\hat{Q}_{34}$ at the level of the death state versus sojourn time (or individual age at death in days)

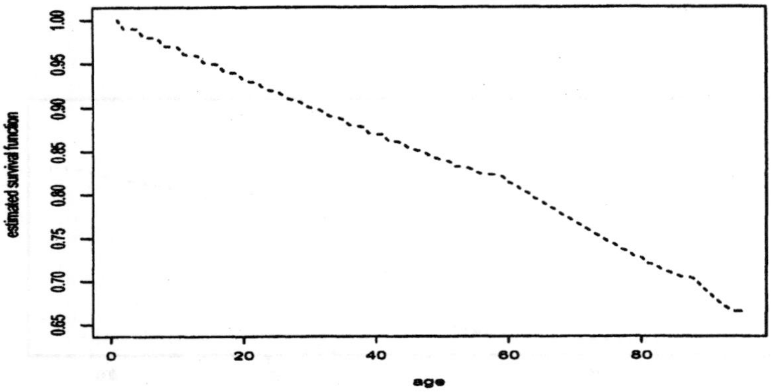

Figure 31.3: Curves of the survival function estimator $\hat{S}(t, n, m)$ versus time t for a given subject, whatever the death cause is, ($n = 95$ and $m = 3$)

References

1. Afchain, A. L. (2001). Robustesse d'évaluation appliquée à des problèmes de fiabilité markovienne en Sûreté de Fonctionnement, *Thèse de l'Université de Nantes*, France.

2. Andersen, P. K., Borgan, O., Gill, R. D. and Keiding, N. (1993). *Statistical Models based on Counting Processes*, Springer-Verlag, New York.

3. Fleming, T. R. and Harrington, D. P. (1991). *Counting Processes and Survival Analysis*, John Wiley & Sons, New York.

4. Heutte, N. and Huber-Carol, C. (2002). Semi-Markov models for quality of life data with censoring, In *Statistical Methods for Quality of Life Studies: Design, Measurements and Analysis* (Eds., M. Mesbah, B. Coles and M. L. T. Lee), Kluwer Academic Publishers, Norwell, Massachusetts.

5. Kalbfleisch, J. D. and Prentice, R. L. (1980). *The Statistical Analysis of Failure Time Data*, John Wiley & Sons, New York.

6. Klein, J. P. and Moeschberger, M. L. (1997). Survival analysis: techniques for censored and truncated data, In *Statistics for Biology and Health*, Springer-Verlag, New York.

7. Limnios, N. and Oprisan G. (2001). *Semi-Markov Processes and Reliability*, Birkhäuser, Boston, Massachusetts.

8. Ouhbi, B. and Limnios, N. (1998).Failure rate non-parametric estimation in Semi-Markov systems, In *Proceedings of the 2nd International Symposium on Semi-Markov Models: Theory and Applications*, pp. 1–10.

9. Pérez-Ocón, R., Ruiz-Castro, J. E. and Gámiz-Pérez, M. L. (1998). Semi-Markov models for lifetime data analysis, In *Semi-Markov Models and Applications* (Eds., J. Janssen and N. Limnios), pp. 229–239, Kluwer Academic Publishers, Dordrecht, The Netherlands.

PART XIII
PROBABILITY THEORY AND APPLICATIONS

32

An Extension of Lévy's Formula to Weighted Wiener Processes

Paul Deheuvels

L.S.T.A., Université Paris VI, Bourg-la-Reine, France

Abstract: A well-known formula of Paul Lévy gives the Laplace transform of the $L^2(0,1)$ norm of the Wiener process $\{W(t) : t \geq 0\}$ conditional on $W(1) = x$. In this paper, we extend Lévy's formula to the $L^2(0,1)$ norm of the weighted Wiener process $\{t^\beta W(t) : 0 < t \leq 1\}$, conditional on $W(1) = x$. Our result has a simple closed-form expression in terms of Bessel functions.

Keywords and phrases: Wiener process, Brownian motion, Brownian bridge, quadratic functionals, Lévy's formula, Bessel functions

32.1 Introduction and Results

A celebrated formula due to Paul Lévy (1951, 1953) shows that, for each $x \in \mathbb{R}$ and $u \in \mathbf{C}$ with $|\text{Re}(u)| < \frac{\pi}{2}$,

$$\begin{aligned}\mathbb{E}\Big(\exp\Big\{\frac{u^2}{2}\int_0^1 W^2(t)dt\Big\}\Big|W(1) = x\Big) \qquad (32.1)\\ = \Big\{\frac{\sin u}{u}\Big\}^{-1/2}\exp\Big\{-\frac{x^2}{2}\Big(u\cot u - 1\Big)\Big\},\end{aligned}$$

where $\{W(t) : t \geq 0\}$ denotes a Wiener process, and $(\sin u)/u := 1$ for $u = 0$. By applying (32.1) to $x = 0$, we obtain a formula due to Smirnov (1936, 1937, 1948), showing that, for each $u \in \mathbb{R}$ with $|\text{Re}(u)| < \frac{\pi}{2}$,

$$\mathbb{E}\Big(\exp\Big\{\frac{u^2}{2}\int_0^1 B^2(t)dt\Big\}\Big) = \Big\{\frac{\sin u}{u}\Big\}^{-1/2}, \qquad (32.2)$$

where $\{B(t) : 0 \leq t \leq 1\}$ denotes a Brownian bridge. On the other hand, by integrating (32.1) with respect to $\varphi(x)dx$, where $\varphi(x) = (2\pi)^{-1/2}e^{-x^2/2}$ is the

Gaussian $N(0,1)$ density, we obtain that, for each $u \in \mathbb{R}$ with $|\mathrm{Re}(u)| < \frac{\pi}{2}$,

$$\mathbb{E}\Big(\exp\Big\{ -\frac{u^2}{2} \int_0^1 W^2(t)dt \Big\}\Big) = \Big\{ \cos u \Big\}^{-1/2} \tag{32.3}$$

In this paper, we show that (32.1), (32.2) and (32.3) are particular cases of a more general formula holding for the *weighted Wiener process* $\{t^\beta W(t) : 0 < t \leq 1\}$, where $\beta > -1$ is an arbitrary constant.

The study of quadratic functionals of the form

$$\int_0^1 \psi^2(t) W^2(t) dt \quad \text{or} \quad \int_0^1 \psi^2(t) B^2(t) dt, \tag{32.4}$$

where $\psi(\cdot)$ is a general weight function, has received much attention in the literature [see, e.g., Anderson and Darling (1952), Darling (1957), Durbin (1973), Shorack and Wellner (1986), del Barrio et al. (2000), for surveys and references]. For a systematic treatment of the quadratic functionals in (32.4) with a general $\psi(\cdot)$, we refer to Donati-Martin and Yor (1991, 1992, 1997), Martynov (1975, 1977), Pitman and Yor (1981, 1982, 1986, 1999), Biane and Yor (1987), Abadir (1995), Shi (1996), Shi and Yor (1997), Ghomrasni (2003) and the references therein. Recently, Deheuvels and Martynov (2003) [see also Scott (1999)] have provided closed form expressions, in terms of *Bessel functions*, for the Laplace transforms of the quadratic functionals in (32.4) when $\psi(t) = t^\beta$ for $0 < t \leq 1$ (see, e.g., Corollary 32.1.1 in the sequel). The purpose of this paper is to show that the latter results follow from the general version of Lévy's formula (32.1), stated in Theorem 32.1.1 below.

We shall denote by $J_\nu(\cdot)$ the *Bessel function of the first kind and index* $\nu \in \mathbb{R}$ (refer to Watson (1952), and (32.14) in Section 32.2 for definitions). For each $\nu > -1$, we set $0 < z_{\nu,1} < z_{\nu,2} < \cdots$ for the positive zeros of $J_\nu(\cdot)$. Given this notation, our main result may be stated as follows. Set $\mathbf{C}^* = \mathbf{C} - \{0\}$.

Theorem 32.1.1 *For any $\beta > -1$, set $\nu = 1/(2(\beta+1))$. Then, for each $x \in \mathbb{R}$ and $u \in \mathbf{C}^*$ with $|\mathrm{Re}(u)| < z_{\nu,1}/(2\nu)$, we have*

$$\mathbb{E}\Big(\exp\Big\{ \frac{u^2}{2} \int_0^1 t^{2\beta} W^2(t) dt \Big\} \Big| W(1) = x \Big) \tag{32.5}$$
$$= \frac{(u\nu)^{\frac{\nu-1}{2}}}{\sqrt{\Gamma(\nu)}} \Big\{ \frac{J_\nu(2u\nu)}{u} \Big\}^{-1/2} \exp\Big\{ -\frac{x^2}{2} \Big(\frac{uJ_{\nu-1}(2u\nu)}{J_\nu(2u\nu)} - 1 \Big) \Big\}$$
$$= \frac{(u\nu)^{\frac{\nu}{2}}}{\sqrt{\Gamma(\nu+1)}} \Big\{ J_\nu(2u\nu) \Big\}^{-1/2} \exp\Big\{ \frac{x^2}{2} \Big(\frac{uJ_{\nu+1}(2u\nu)}{J_\nu(2u\nu)} \Big) \Big\}.$$

Remark 32.1.1 In view of the classical relation $\Gamma(\nu+1) = \nu\Gamma(\nu)$, the last equality in (32.5) is a direct consequence of the following well-known recursion relationship between Bessel functions [see, e.g., in Spanier and Oldham (1987, 53:5:1, p. 524)]. For any $\nu > 0$ and $x > 0$,

$$J_{\nu+1}(x) = \frac{2\nu}{x} J_\nu(x) - J_{\nu-1}(x). \tag{32.6}$$

Corollary 32.1.1 below provides some extensions of (32.3)–(32.4) which turn out to be direct consequences of Theorem 32.1.1.

Corollary 32.1.1 *Under the assumptions of Theorem 32.1.1, we have, for any $u \in \mathbf{C}^*$ with $|\mathrm{Re}(u)| < z_{\nu-1,1}/(2\nu)$,*

$$\mathbb{E}\Big(\exp\Big\{ \frac{u^2}{2} \int_0^1 t^{2\beta} W^2(t)dt \Big\}\Big) = \frac{(u\nu)^{\frac{\nu-1}{2}}}{\sqrt{\Gamma(\nu)}} \Big\{ J_{\nu-1}(2u\nu)\Big\}^{-1/2}, \tag{32.7}$$

and, for any $u \in \mathbf{C}$ with $|\mathrm{Re}(u)| < z_{\nu,1}/(2\nu)$,

$$\mathbb{E}\Big(\exp\Big\{ \frac{u^2}{2} \int_0^1 t^{2\beta} B^2(t)dt \Big\}\Big) = \frac{(u\nu)^{\frac{\nu-1}{2}}}{\sqrt{\Gamma(\nu)}} \Big\{ \frac{J_\nu(2u\nu)}{u}\Big\}^{-1/2}. \tag{32.8}$$

Remark 32.1.2 Some calculus shows that the right-hand sides of (32.5), (32.7) and (32.8) tend to 1 when $u \to 0$. For this reason, if we make use of the convention that each of these quantities is defined by 1 for $u = 0$, we may extend the validity of Theorem 32.1.1 and Corollary 32.1.1 to $u \in \mathbf{C}$.

Remark 32.1.3 We note that (32.7)–(32.8) are in agreement with Theorem 1.6 and (1.52)–(1.53) in Deheuvels and Martynov (2003).

The proofs of Theorem 32.1.1 and Corollary 32.1.1 are postponed until Section 32.2. In Section 32.3, we give an alternate proof of Theorem 32.1.1, based upon the results of Biane and Yor (1987). We conclude the present section by discussing the relationships of our results with other well-known facts in the literature. In the following example, we consider the particular case where $\beta = 0$.

Example 32.1.1 The Bessel function $J_\nu(\cdot)$ has a simple expression when either $\nu = -\frac{1}{2}$ or $\nu = \frac{1}{2}$. We obtain, namely [see, e.g., Watson (1952)] that

$$J_{-\frac{1}{2}}(x) = \Big\{\frac{2}{\pi x}\Big\}^{1/2} \cos x \quad \text{and} \quad J_{\frac{1}{2}}(x) = \Big\{\frac{2}{\pi x}\Big\}^{1/2} \sin x. \tag{32.9}$$

The corresponding positive zeros are given by

$$z_{-\frac{1}{2},k} = \Big\{k - \tfrac{1}{2}\Big\}\pi \quad \text{and} \quad z_{\frac{1}{2},k} = k\pi \quad \text{for} \quad k = 1, 2, \dots . \tag{32.10}$$

We have therefore, for $\beta = 0$, $\nu = \frac{1}{2}$, $\Gamma(\frac{1}{2}) = \sqrt{\pi}$, and

$$\mathbb{E}\Big(\exp\Big\{\frac{u^2}{2}\int_0^1 W^2(t)dt\Big\}\Big) = \Big\{\frac{\pi u}{2}\Big\}^{-1/4}\Big\{J_{-\frac{1}{2}}(u)\Big\}^{-1/2} \tag{32.11}$$
$$= \Big\{\Big\{\frac{\pi u}{2}\Big\}^{1/2}\Big\{\frac{2}{\pi u}\Big\}^{1/2}\cos u\Big\}^{-1/2} = \Big\{\cos u\Big\}^{-1/2} \text{ for } |\mathrm{Re}(u)| < \tfrac{\pi}{2},$$

which is (32.3). Likewise,

$$\mathbb{E}\Big(\exp\Big\{\frac{u^2}{2}\int_0^1 B^2(t)dt\Big\}\Big) = \Big\{\frac{\pi u}{2}\Big\}^{-1/4}\Big\{\frac{J_{\frac{1}{2}}(u)}{u}\Big\}^{-1/2} \tag{32.12}$$
$$= \Big\{\Big\{\frac{\pi u}{2}\Big\}^{1/2}\Big\{\frac{2}{\pi u}\Big\}^{1/2}\frac{\sin u}{u}\Big\}^{-1/2} = \Big\{\frac{\sin u}{u}\Big\}^{-1/2} \text{ for } |\mathrm{Re}(u)| < \pi,$$

which is (32.2).

32.2 Proofs

We start by recalling some basic facts about *Bessel functions* which will be needed in our proofs. We refer to Watson (1952), Bowman (1958), Korenev (2002) and §2 in Deheuvels and Martynov (2003) for further details and references. For each $\nu \in \mathbb{R}$, the second order homogeneous differential equation

$$x^2y'' + xy' + (x^2 - \nu^2)y = 0, \tag{32.13}$$

has a solution on $(0, \infty)$ given by the *Bessel function of the first kind and index* ν, denoted by $J_\nu(\cdot)$. For $\nu > -1$, this function is defined by

$$J_\nu(x) = (\tfrac{1}{2}x)^\nu \sum_{k=1}^{\infty} \frac{(-\frac{1}{4}x^2)^k}{\Gamma(\nu + k + 1)\Gamma(k + 1)} \quad \text{for} \quad x > 0. \tag{32.14}$$

Moreover [see, e.g., Watson (1952, Ch. XV, pp. 478–521), Korenev (2002, p. 96) and Kamke (1959, p. 440)], for each $\nu > -1$, the positive roots of $J_\nu(\cdot)$ are isolated and form an increasing sequence

$$0 < z_{\nu,1} < z_{\nu,2} < \cdots, \tag{32.15}$$

such that, as $k \to \infty$,

$$z_{\nu,k} = \{k + \tfrac{1}{2}(\nu - \tfrac{1}{2})\}\pi + o(1). \tag{32.16}$$

Moreover, for each specified $k \geq 1$, $z_{\nu,k}$ is an increasing function of $\nu > -1$. In particular, for each $\nu > 0$, we have

$$0 < z_{\nu-1,k} < z_{\nu,k}. \tag{32.17}$$

Given this notation, we have *Euler's formula* [see, e.g., Watson (1952, (2)-(3), p. 498)]

$$J_\nu(x) = \frac{(\frac{1}{2}x)^\nu}{\Gamma(\nu+1)} \prod_{k=1}^{\infty} \Big\{1 - \frac{x^2}{z_{\nu,k}^2}\Big\} \quad \text{for} \quad x > 0, \tag{32.18}$$

and the following recurrence relation between $J_\nu(\cdot)$ and $J_{\nu+1}(\cdot)$ [see, e.g., Korenev (2002, p. 43) and Bowman (1958, p. 109)]. For any $\nu > -1$,

$$\frac{J_{\nu+1}(x)}{xJ_\nu(x)} = -2\sum_{k=1}^{\infty} \Big\{\frac{1}{x^2 - z_{\nu,k}^2}\Big\} \quad \text{for} \quad x > 0,\ x \notin \{z_{\nu,k} : k \geq 1\}. \tag{32.19}$$

Our proofs will rely on the *Karhunen–Loeve expansion* of the weighted process $\{t^\beta W(t) : 0 < t \leq 1\}$ recently obtained for an arbitrary $\beta > -1$ by Deheuvels and Martynov (2003). The Karhunen–Loeve [KL] expansion of a general process defines its representation in term of the eigenvalues and eigenfunctions of the Hilbert–Schmidt operator associated to its covariance function [see, e.g., Kac and Siegert (1947), Kac (1951), Ash and Gardner (1975), Adler (1990) and the references therein].

We will make use of the following basic fact, due to Deheuvels and Martynov (2003) (see their Theorem 1.3).

Fact 32.2.1 Fix any $\beta > -1$, and set $\nu = 1/(2(1+\beta))$. Then, the Karhunen–Loeve decomposition of $\{t^\beta W(t) : 0 < t \leq 1\}$ is given by

$$t^\beta W(t) = \sum_{k=1}^{\infty} \sqrt{\lambda_k}\, \omega_k\, e_k(t), \tag{32.20}$$

where $\{\omega_k : k \geq 1\}$ is a sequence of independent $N(0,1)$ random variables, and, for $k = 1, 2, \ldots$ and $0 < t \leq 1$,

$$\lambda_k = \Big\{\frac{2\nu}{z_{\nu-1,k}}\Big\}^2 \quad \text{and} \quad e_k(t) = t^{\frac{1}{2\nu}-\frac{1}{2}} \Big\{\frac{J_\nu(z_{\nu-1,k}\, t^{\frac{1}{2\nu}})}{\sqrt{\nu}\, J_\nu(z_{\nu-1,k})}\Big\}. \tag{32.21}$$

In addition, the series (32.20) is almost surely uniformly convergent on each interval of the form $[\epsilon, 1]$ for $\epsilon > 0$.

Remark 32.2.1 Since (32.16) and (32.21) entail that

$$\sqrt{\lambda_k} = (1 + o(1))\Big\{\frac{\pi}{k(1+\beta)}\Big\} \quad \text{as} \quad k \to \infty, \tag{32.22}$$

an easy argument based on (32.22) establishes the last claim of Fact 32.2.1, relative to the almost sure uniform convergence of the series (32.20) on $[\epsilon, 1]$ for $\epsilon > 0$.

We infer from (32.20)–(32.21) that, almost surely,

$$\int_0^1 t^{2\beta} W^2(t)dt = \sum_{k=1}^{\infty} \lambda_k \, \omega_k^2 = \sum_{k=1}^{\infty} \Big\{ \frac{4\nu^2}{z_{\nu-1,k}^2} \Big\} \omega_k^2, \tag{32.23}$$

and

$$W(1) = \sum_{k=1}^{\infty} \sqrt{\lambda_k} \, \omega_k \, e_k(1) = 2\sqrt{\nu} \sum_{k=1}^{\infty} \Big\{ \frac{1}{z_{\nu-1,k}} \Big\} \omega_k. \tag{32.24}$$

Lemma 32.2.1 *Let $\omega \overset{d}{=} N(0,1)$ denote a standard normal $N(0,1)$ random variable. Then, for arbitrary $x, y \in \mathbf{C}$ such that $|\mathrm{Re}(x)| < 1$,*

$$\mathbb{E}\Big(\exp\Big\{ \frac{x}{2} \, \omega^2 + y\, \omega \Big\} \Big) = \frac{1}{\sqrt{1-x}} \exp\Big\{ \frac{y^2}{2(1-x)} \Big\}. \tag{32.25}$$

PROOF. Let $\omega \overset{d}{=} N(0,1)$. We have

$$\mathbb{E}\Big(\exp\Big\{ \frac{x}{2} \, \omega^2 + y\, \omega \Big\} \Big) = \frac{1}{\sqrt{2\pi}} \int_{\mathbb{R}} \exp\Big\{ -\tfrac{1}{2}(1-x)t^2 + yt \Big\} dt$$

$$= \frac{1}{\sqrt{1-x}} \exp\Big\{ \frac{y^2}{2(1-x)} \Big\},$$

as sought. ∎

Lemma 32.2.2 *Let $(Z, Y) \in \mathbb{R}^2$ be a random vector such that $Y \overset{d}{=} N(0,1)$ follows a standard normal distribution. Assume further that, on an open subset $\mathcal{D}_0$ of $\mathbf{C}^2$ such that $(\mathrm{i}\mathbb{R} \times \mathrm{i}\mathbb{R}) \subseteq \mathcal{D}_0$, we have, for all $(u, v) \in \mathcal{D}_0$,*

$$\mathbb{E}\Big(\exp\{uZ + vY\} \Big) = \Psi_1(u) \exp\Big\{ \frac{v^2}{2} \Psi_2(u) \Big\}, \tag{32.26}$$

where $\Psi_1(\cdot)$ and $\Psi_2(\cdot)$ are two analytic functions defined on an open subset $\mathcal{D}_1$ of $\mathbf{C}$ such that $\mathrm{i}\mathbb{R} \subseteq \mathcal{D}_1$. Then, we have, for each $y \in \mathbb{R}$ and $u \in \mathcal{D}_1$,

$$\mathbb{E}\Big(\exp\{uZ\} | Y = y \Big) \tag{32.27}$$

$$= \Psi_1(u) \Big\{ \Psi_2(u) \Big\}^{-1/2} \exp\Big\{ -\frac{y^2}{2} \Big(\frac{1}{\Psi_2(u)} - 1 \Big) \Big\}.$$

PROOF. Since the Laplace transform defines a one-to-one mapping, it is readily verified from (32.26) that, for each u, the function

$$y \to h(u, y) := \mathbb{E}(\exp\{uZ\} | Y = y),$$

is uniquely defined via the integral equation

$$\int_{-\infty}^{\infty} h(u,y) \times \exp\Big\{vy\Big\} \times \frac{e^{-y^2/2}}{\sqrt{2\pi}}\, dy \tag{32.28}$$
$$= \mathbb{E}\Big(\exp\{uZ+vY\}\Big) = \Psi_1(u)\exp\Big\{\frac{v^2}{2}\Psi_2(u)\Big\}.$$

By combining (32.28) with the observation that

$$\int_{-\infty}^{\infty}\Big\{\Psi_1(u)\Big\{\Psi_2(u)\Big\}^{-1/2}\exp\Big\{-\frac{y^2}{2}\Big(\frac{1}{\Psi_2(u)}-1\Big)\Big\}\Big\}$$
$$\times\exp\Big\{vy\Big\}\times\frac{e^{-y^2/2}}{\sqrt{2\pi}}\,dy = \frac{\Psi_1(u)}{\sqrt{2\pi\Psi_2(u)}}\int_{-\infty}^{\infty}\exp\Big\{vy-\frac{y^2}{2\Psi_2(u)}\Big\}dy$$
$$= \Psi_1(u)\exp\Big\{\frac{v^2}{2}\Psi_2(u)\Big\},$$

we obtain therefore (32.27). ■

First Proof of Theorem 32.1.1. Recall from (32.16) that $z_{\nu-1,1}/(2\nu) < z_{\nu,1}/(2\nu)$. By combining (32.23)–(32.24) with Fact 1, (32.15), (32.17) and (32.26), we readily obtain that, for each $u \in \mathbf{C}$ fulfilling $|\mathrm{Re}(u)| < z_{\nu-1,1}/(2\nu)$,

$$\mathbb{E}\Big(\exp\Big\{\frac{u^2}{2}\int_0^1 t^{2\beta}W^2(t)dt + yW(1)\Big\}\Big) \tag{32.29}$$
$$= \prod_{k=1}^{\infty}\mathbb{E}\Big(\exp\Big\{\frac{1}{2}\Big\{\frac{4u^2\nu^2}{z^2_{\nu-1,k}}\Big\}\omega_k^2 + \Big\{\frac{2y\sqrt{\nu}}{z_{\nu-1,k}}\Big\}\omega_k\Big\}\Big)$$
$$= \Big\{\prod_{k=1}^{\infty}\Big(1-\Big\{\frac{4u^2\nu^2}{z^2_{\nu-1,k}}\Big\}\Big)\Big\}^{-1/2}\exp\Big\{\sum_{k=1}^{\infty}\frac{\Big\{\frac{4y^2\nu}{z^2_{\nu-1,k}}\Big\}}{2\Big(1-\Big\{\frac{4u^2\nu^2}{z^2_{\nu-1,k}}\Big\}\Big)}\Big\},$$

which, by (32.18) and (32.19), equals

$$\frac{(u\nu)^{\frac{\nu-1}{2}}}{\sqrt{\Gamma(\nu)}}\Big\{J_{\nu-1}(2u\nu)\Big\}^{-1/2}\exp\Big\{-2y^2\nu\sum_{k=1}^{\infty}\frac{1}{4u^2\nu^2-z^2_{\nu-1,k}}\Big\} \tag{32.30}$$
$$= \frac{(u\nu)^{\frac{\nu-1}{2}}}{\sqrt{\Gamma(\nu)}}\Big\{J_{\nu-1}(2u\nu)\Big\}^{-1/2}\exp\Big\{\frac{y^2}{2}\Big(\frac{J_\nu(2u\nu)}{uJ_{\nu-1}(2u\nu)}\Big)\Big\}.$$

By combining (32.29)–(32.30) with (32.26)–(32.27), we see that

$$\mathbb{E}\Big(\exp\Big\{\frac{u^2}{2}\int_0^1 t^{2\beta}W^2(t)dt\Big\}\Big|W(1)=x\Big) \tag{32.31}$$
$$= \frac{(u\nu)^{\frac{\nu-1}{2}}}{\sqrt{\Gamma(\nu)}}\Big\{\frac{J_\nu(2u\nu)}{u}\Big\}^{-1/2}\exp\Big\{-\frac{x^2}{2}\Big(\frac{uJ_{\nu-1}(2u\nu)}{J_\nu(2u\nu)}-1\Big)\Big\},$$

which is (32.5). ■

Proof of Corollary 32.1.1. We note that (32.7) is a particular case of (32.29)–(32.30) corresponding to $y = 0$. We obtain (32.8) by setting $x = 0$ in (32.5), and making use of the observation that the distribution of $\{W(t) : 0 \leq t \leq 1\}$ conditional on $W(1) = 0$ is that of a Brownian bridge. ■

32.3 An Alternative Proof of Theorem 32.1.1 Based upon a Result of Biane and Yor (1987)

The formula (32.1) is closely related to the so-called stochastic area of the Brownian motion, formally defined by

$$A = \int_0^1 \Big\{W_1(t)dW_2(t) - W_2(t)dW_1(t)\Big\}, \tag{32.32}$$

where $W_1(\cdot)$ and $W_2(\cdot)$ are independent Wiener processes. Setting $Z(t) = W_1(t) + \mathrm{i}W_2(t)$, Lévy (1951) showed that, for any $z \in \mathbf{C}$,

$$\mathbb{E}\Big(\exp\Big\{uA\Big|Z(1) = z\Big\}\Big) = \Big\{\frac{\sin u}{u}\Big\}^{-1} \exp\Big\{-\frac{|z|^2}{2}\Big(u\cot u - 1\Big)\Big\}. \tag{32.33}$$

A classical conditioning argument [see, e.g., Donati-Martin and Yor (1991)] shows the identity

$$\begin{aligned}&\mathbb{E}\Big(\exp\Big\{uA\Big|Z(1) = z\Big\}\Big) \\ &= \mathbb{E}\Big(\exp\Big\{\frac{u^2}{2}\int_0^1 \Big\{W_1^2(t) + W_2^2(t)\Big\}dt\Big|Z(1) = z\Big\}\Big).\end{aligned} \tag{32.34}$$

The independence of $W_1(\cdot)$ and $W_2(\cdot)$, in combination with (32.33) and (32.34), taken with $z = x(1+\mathrm{i})$, readily implies (32.1). Making use of a Legendre polynomial decomposition, Biane and Yor (1987) have established in their Théorème 3 the following extension of (32.33). For each $p > -\frac{1}{2}$, setting $\nu = p + \frac{1}{2} > 0$ and

$$V_p(t) = \frac{1}{t^p}\int_0^t s^p dZ(s) = \frac{1}{t^p}\Big\{\int_0^t s^p dW_1(s) + \mathrm{i}\int_0^t s^p dW_2(s)\Big\}, \tag{32.35}$$

they showed that, for each $v \in \mathbb{R}$ [see, e.g., Biane and Yor (1987, p. 371)],

$$\begin{aligned}&\mathbb{E}\Big(\exp\Big\{\frac{v^2}{2}\int_0^1 |V_p(t)|^2 dt\Big|V_p(1) = z\Big\}\Big) \\ &= \frac{(\mathrm{i}v)^\nu}{2^\nu\Gamma(\nu+1)I_\nu(\mathrm{i}v)}\exp\Big\{-\frac{|z|^2}{2}\Big(\frac{\mathrm{i}vI_{\nu+1}(\mathrm{i}v)}{I_\nu(\mathrm{i}v)}\Big)\Big\},\end{aligned} \tag{32.36}$$

where $I_\nu(\cdot)$ denotes the *modified Bessel function* of index ν. Recalling the fact [see, e.g., Watson (1952)] that $I_\nu(iv) = i^\nu J_\nu(v)$, we may rewrite (32.36) into

$$\mathbb{E}\Big(\exp\Big\{\frac{v^2}{2}\int_0^1 |V_p(t)|^2 dt \Big| V_p(1) = z\Big\}\Big) \tag{32.37}$$
$$= \frac{(\frac{v}{2})^\nu}{\Gamma(\nu+1)J_\nu(v)} \exp\Big\{\frac{|z|^2}{2}\Big(\frac{vJ_{\nu+1}(v)}{J_\nu(v)}\Big)\Big\}.$$

It is clear that (32.37) must be related to (32.5). It is, in fact, possible to show, after some calculations, that both statements are equivalent. This, however, is not completely obvious, and we provide below the necessary arguments.

Lemma 32.3.1 *For any Wiener process $\{W(t) : t \geq 0\}$, we have the distributional identity between stochastic processes*

$$\int_0^t s^p dW(s) \stackrel{d}{=} \frac{W(t^{2p+1})}{\sqrt{2p+1}} \quad \textit{for} \quad t \geq 0. \tag{32.38}$$

PROOF. $Z_p(\cdot)$ is a centered Gaussian process with covariance function

$$\begin{aligned}\mathbb{E}(Z_p(s)Z_p(t)) &= \int_0^{s\wedge t} u^{2p} du = \frac{(s\wedge t)^{2p+1}}{2p+1} \\ &= \mathbb{E}\Big(\Big\{\frac{W(s^{2p+1})}{\sqrt{2p+1}}\Big\}\Big\{\frac{W(t^{2p+1})}{\sqrt{2p+1}}\Big\}\Big).\end{aligned}$$

This, in turn, readily yields (32.39). ■

SECOND PROOF OF THEOREM 32.1.1. As follows from (32.35) and Lemma 32.3.1, we have the distributional equality between stochastic processes

$$|V_p(t)|^2 \stackrel{d}{=} \frac{1}{t^{2p}(2p+1)}\Big\{W_1^2(t^{2p+1}) + W_2^2(t^{2p+1})\Big\}, \tag{32.39}$$

where $\{W_1(t) : t \geq 0\}$ and $\{W_2(t) : t \geq 0\}$ are independent (standard) Wiener processes. It follows from (32.39) that

$$\int_0^1 |V_p(t)|^2 dt \stackrel{d}{=} \frac{1}{2p+1}\Big\{\int_0^1 \frac{W_1^2(t^{2p+1})dt}{t^{2p}} + \int_0^1 \frac{W_2^2(t^{2p+1})dt}{t^{2p}}\Big\}. \tag{32.40}$$

Setting $t = s^{1/(2p+1)}$, $\beta = -2p/(2p+1)$ and $\nu = 1/(2(\beta+1)) = p + \frac{1}{2}$ in (32.40), we obtain in turn that

$$\begin{aligned}\int_0^1 |V_p(t)|^2 dt &\stackrel{d}{=} \frac{1}{(2p+1)^2}\Big\{\int_0^1 u^{-\frac{4p}{2p+1}} W_1^2(s)ds \\ &\qquad + \int_0^1 s^{-\frac{4p}{2p+1}} W_2^2(s)ds\Big\} \\ &= \frac{1}{4\nu^2}\Big\{\int_0^1 s^{2\beta}W_1^2(s)ds + \int_0^1 s^{2\beta}W_2^2(s)ds\Big\},\end{aligned} \tag{32.41}$$

whereas, via (32.35) and (32.38), on the same probability space,

$$V_p(1) \stackrel{d}{=} \frac{1}{\sqrt{2\nu}}\{W_1(1) + iW_2(1)\}. \tag{32.42}$$

Therefore, setting $z = \frac{1}{2\nu}\{x + iy\}$ and $v = 2u\nu$, we get, via (32.37),

$$\begin{aligned}
&\mathbb{E}\Big(\exp\Big\{\frac{(2u\nu)^2}{2}\int_0^1 |V_p(t)|^2 dt\Big| V_p(1) = \frac{1}{\sqrt{2\nu}}\{x+iy\}\Big\}\Big) \qquad (32.43)\\
&= \frac{(u\nu)^\nu}{\Gamma(\nu+1)J_\nu(2u\nu)}\exp\Big\{\frac{x^2+y^2}{4\nu}\Big(\frac{2u\nu J_{\nu+1}(2u\nu)}{J_\nu(2u\nu)}\Big)\Big\}\\
&= \frac{(u\nu)^\nu}{\Gamma(\nu+1)J_\nu(2u\nu)}\exp\Big\{\frac{x^2+y^2}{2}\Big(\frac{uJ_{\nu+1}(2u\nu)}{J_\nu(2u\nu)}\Big)\Big\}\\
&= \mathbb{E}\Big(\exp\Big\{\frac{u^2}{2}\int_0^1 u^{2\beta}W_1^2(u)dt\Big| W_1(1) = x\Big)\\
&\times \mathbb{E}\Big(\exp\Big\{\frac{u^2}{2}\int_0^1 u^{2\beta}W_2^2(u)dt\Big| W_2(1) = y\Big).
\end{aligned}$$

By setting $y = x$ in (32.43) readily obtain (32.5). ■

Remark 32.3.1 It is obvious from the above arguments that the conclusion of the Théorème 3 of Biane and Yor (1987) turns out to be equivalent to that of Theorem 32.1.1. Moreover, if $Z(t) = W_1(t) + iW_2(t)$ denotes a standard Wiener process in $\mathbf{C}$, both results are equivalent to the formula

$$\begin{aligned}
&\mathbb{E}\Big(\exp\Big\{\frac{u^2}{2}\int_0^1 t^{2\beta}|Z(t)|^2 dt\Big\}\Big| Z(1) = z\Big) \qquad (32.44)\\
&= \frac{(u\nu)^{\nu-1}}{\Gamma(\nu)}\Big\{\frac{u}{J_\nu(2u\nu)}\Big\}\exp\Big\{-\frac{|z|^2}{2}\Big(\frac{uJ_{\nu-1}(2u\nu)}{J_\nu(2u\nu)} - 1\Big)\Big\}\\
&= \frac{(u\nu)^\nu}{\Gamma(\nu+1)}\Big\{J_\nu(2u\nu)\Big\}^{-1}\exp\Big\{\frac{|z|^2}{2}\Big(\frac{uJ_{\nu+1}(2u\nu)}{J_\nu(2u\nu)}\Big)\Big\}.
\end{aligned}$$

Acknowledgments. We thank Marc Yor and Giovanni Peccati for stimulating comments on this problem, and in particular, for pointing out the connection of our results with that of Biane and Yor (1987), discussed in Section 32.3.

References

1. Abadir, K. M. (1995). The joint density of two functionals of Brownian motion, *Mathematical Methods of Statistics*, **4**, 449–462.

2. Adler, R. J. (1990). *An Introduction to Continuity, Extrema and Related Topics for General Gaussian Processes*, IMS Lecture Notes-Monograph Series **12**, Institute of Mathematical Statistics, Hayward, California.

3. Anderson, T. W. and Darling, D. A. (1952). Asymptotic theory of certain "goodness of fit" criteria based on stochastic processes, *Annals of Mathematical Statistics*, **23**, 193–212.

4. Ash, R. B. and Gardner, M. F. (1975). *Topics in Stochastic Processes,* Academic Press, New York.

5. del Barrio, E., Cuesta-Albertos, J. A. and Matran, C. (2000). Contributions of empirical and quantile processes to the asymptotic theory of goodness-of-fit tests, *Test*, **9**, 1–96.

6. Biane, Ph. and Yor, M. (1987). Variations sur une formule de Paul Lévy, *Ann. Inst. Henri Poincaré,* **23**, 359–371.

7. Bowman, F. (1958). *Introduction to Bessel Functions,* Dover, New York.

8. Darling, D. A. (1957). The Kolmogorov-Smirnov, Cramér-von Mises tests, *Annals of Mathematical Statistics*, **28**, 823–838.

9. Deheuvels, P. and Martynov, G. V. (2003). Karhunen-Loeve expansions for weighted Wiener processes and brownian bridges via Bessel functions, In *High Dimensional Probability* (Eds., M. B. Marcus and J. A. Wellner), Birkhäuser, Boston, Massachusetts (to appear).

10. Donati-Martin, C. and Yor, M. (1991). Fubini's theorem double Wiener integrals and the variance of the Brownian path, *Ann. Inst. Henri Poincaré,* **275**, 181–200.

11. Donati-Martin, C. and Yor, M. (1992). Extension d'une formule de Paul Lévy pour la variation quadratique d'un mouvement brownien plan, *Bull. Sci. Math., Ser. 2*, **116**, 353–382.

12. Donati-Martin, C. and Yor, M. (1997). Some Brownian functionals and their laws, *Annals of Probability*, **25**, 1011–1058.

13. Durbin, J. (1973). *Distribution Theory for Tests Based on the Sample Distribution Function,* Regional Conference Series in Applied Mathematics, **9**, SIAM, Philadelphia, Pennsylvania.

14. Gomrasni, R. (2003). On distributions associated with the generalized Lévy's stochastic area formula, *Studia Math. Sci. Hungar.* (to appear).

15. Kac, M. (1951). On some connections between probability theory and differential and integral equations, In *Proceedings of the Second Berkeley Symposium on Mathematical Statistics and Probability*, pp. 180–215.

16. Kac, M. and Siegert, A. J. F. (1947). An explicit representation of a stationary Gaussian process, *Annals of Mathematical Statistics*, **18**, 438–442.

17. Kamke, E. (1959). *Differentialgleichungen Lösungsmethoden und Lösungen,* Chelsea, New York.

18. Korenev, B. G. (2002). *Bessel Functions and their Applications,* Taylor & Francis, London, England.

19. Lévy, P. (1951). Wiener random functions and other Laplacian random functions, In *Proceedings of the 6th Berkeley Symposium on Probability Theory and Mathematical Statistics*, **2**, 171–186.

20. Lévy, P. (1953). La mesure de Hausdorff de la courbe du mouvement brownien, *Giorn. ist. Ital. Attuari.*, **16**, 1–37.

21. Martynov, G. V. (1975). Computation of the distribution function of quadratic forms in normal random variables, *Theory of Probability and its Applications*, **20**, 797–809.

22. Martynov, G. V. (1977). Generalization of the Smirnov formula for the quadratic forms distributions, *Theory of Probability and its Applications*, **22**, 614–620.

23. Pitman, J. W. and Yor, M. (1981). Bessel processes and infinitely divisible laws, In *Stochastic Integrals* (Ed., D. Williams), Lecture Notes in Mathematics, No. **1123**, pp. 285–370. Springer-Verlag, New York.

24. Pitman, J. W. and Yor, M. (1982). A decomposition of Bessel bridges, *Z. Wahrscheinlichkeit. verw. Gebiete.*, **59**, 425–457.

25. Pitman, J. W. and Yor, M. (1986). Some divergent integrals of Brownian motion, In *Analytic and Geometric Stochastics* (Ed., D. Kendall), Supplement to *Advances in Applied Probability*, 109–116.

26. Pitman, J. W. and Yor, M. (1999). The law of the maximum of a Bessel bridge, *Electronic Journal of Probability*, **4**, 1–35.

27. Scott, W. F. (1999). A weighted Cramér-von Mises statistic, with some applications to clinical trials, *Communications in Statistics—Theory and Methods*, **28**, 3001–3008.

28. Shi, Z. (1996). Small ball probabilities for a Wiener process under weighted sup-norms, with an application to the supremum of Bessel local times, *Journal of Theoretical Probability*, **9**, 915–929.

29. Shi, Z. and Yor, M. (1997). On an identity in law for the variance of the Brownian bridge, *Bulletin of the London Mathematical Society*, **29**, 103–108.

30. Shorack, G. R. and Wellner, J. A. (1986). *Empirical Processes with Applications to Statistics*, John Wiley & Sons, New York.

31. Smirnov, N. V. (1936). Sur la distribution de ω^2, *C. R. Acad. Sci. Paris.*, **202**, 449–452.

32. Smirnov, N. V. (1937). On the distribution of the ω^2 criterion, *Rec. Math. (Mat. Sbornik).*, **6**, 3–26.

33. Smirnov, N. V. (1948). Table for estimating the goodness of fit of empirical distributions, *Annals of Mathematical Statistics*, **19**, 279–281.

34. Spanier, J. and Oldham, K. B. (1987). *An Atlas of Functions*, Springer-Verlag, New York.

35. Watson, G. N. (1952). *A Treatise on the Theory of Bessel Functions*, Cambridge University Press, Cambridge, England.

33

Sur l'Inégalité de Concentration de Doeblin–Lévy, Rogozin–Kesten

Jean Bretagnolle

Université Paris-Sud, Orsay, France

Abstract: We sharpen the Concentration Inequality in the form given by Rogozin and Kesten. Our form is sharp for small and large concentrations. Moreover we give an explicit constant. The proof is based on Sperner's Lemma and explicit expression of extremal probabilities with given concentrations.

Keywords and phrases: Concentration

33.1 Introduction et résultat principal

Soient X une variable aléatoire réelle (v.a.r.) et I un intervalle compact (donc fermé) non vide. La *fonction de concentration* est la fonction de X et de I — en fait de la loi de X et de la longueur $\ell(I)$ de I — définie par

$$\gamma(X;I) = Sup_{x\in\mathbb{R}}\ P(X \in I + x)\cdot \tag{33.1}$$

Cette fonction est invariante par translation, sur X ou I, elle est bien adaptée aux problèmes de convergence essentielle. En fait, dans leur travail pionnier Doeblin et Lévy (1936), s'intéressaient plutôt à son inverse, la fonction de dispersion, qui représente la longueur minimum d'un intervalle devant contenir une probabilité donnée: Soit S_n la somme de n v.a.r. indépendantes, les X_i. *Grosso-modo* ils montraient que s'il faut à chaque X_i une longueur minimum ℓ pour porter la probabilité α, il faut à leur somme au minimum $C(\alpha,\beta)\ell\sqrt{n}$ pour porter la probabilité β. Compte-tenu de l'évidente inégalité $\gamma(X;2\ell(I)) \leq 2\gamma(X;\ell(I))$, leur résultat revenait en termes de Concentration à:

Si un même $\alpha < 1$ majore les $\gamma(X_i;I)$, alors $C(\alpha)/\sqrt{n}$ majore $\gamma(S_n;I)$.

Après eux, de nombreux auteurs ont majoré la concentration de la somme en fonction de celle des composants. Si on réduit leurs résultats au cas où l'on ne dispose que des concentrations sur I, je veux dire sans informations sur la concentration sur les sous-intervalles de I, les plus intéressants me semblent celui de Rogozin (1961)

$$\gamma(\Sigma_i X_i; I) \leq C[\Sigma_i(1 - \gamma(X_i; I)]^{-1/2} \tag{33.2}$$

pour C universelle, qui prend en compte à la fois les faibles concentrations (les γ proches de 1) et la disparité des composants, et celui de Kesten (1969), qui, pour prouver un Théorème Taubérien sur les domaines d'attraction des lois stables, eut besoin d'un résultat du côté des fortes concentrations et prouva que

$$\gamma(\Sigma_i X_i; I) \leq d \times C[\Sigma_i(1 - \gamma(X_i; I))\gamma(X_i; I)] \times [\Sigma_i(1 - \gamma(X_i; I)]^{-3/2} \tag{33.3}$$

où C est la constante de Rogozin et d une constante explicite. Pour illustrer ces résultats, prenons le cas particulier des variables *équidistribuées* en nombre n et de somme S_n. Suivant Doeblin et Lévy (1936), Rogozin (1961a,b) et Kesten (1969) on passe de Doeblin et Lévy (1936)

$$\gamma(S_n; I) \leq g(\gamma(X; I))/\sqrt{n}$$

à Rogozin (1961)

$$(S_n; I) \leq [C/\sqrt{1 - \gamma(X; I)}]/\sqrt{n}$$

puis à Kesten (1969)

$$\gamma(S_n; I) \leq [C'\gamma(X; I)/\sqrt{1 - \gamma(X; I)}]/\sqrt{n},$$

où g est une fonction finie et C, C' sont des constantes universelles.

A ma connaissance, aucune amélioration ultérieure n'évalue les constantes ou n'est satisfaisante pour l'expression de la concentration de la somme en fonction de celle de ses composants. Je reprends ici mon travail [*B*] ancien amélioré dans sa forme et dans l'expression pour les petites concentrations. (Ce travail est utilisé dans l'article de Rogozin (1987) où il prouve la maximalité de la concentration pour les distributions uniformes, cf. le Lemme 2 *infra*).

Si les X_i sont normales $\mathcal{N}(0, \sigma_i^2)$, de somme $\mathcal{N}(0, \sigma)$ avec $\sigma^2 = \Sigma_i \sigma_i^2$, si $\ell = \ell(I)$ est négligeable devant chaque σ_i, chaque concentration équivaut à $\ell/\sigma_i\sqrt{2\pi}$, et la forme

$$\gamma(\Sigma_i X_i; I) \leq C/\sqrt{\Sigma_i 1/\gamma(X_i; I)^2} \tag{33.4}$$

est naturelle quand les $\gamma(X_i, I)$ sont tous petits.

Si les X_i suivent les lois de Poisson de paramètres $\lambda_i \leq 1$ et si $I = \{0\}$, $\gamma(X_i; I) = exp(-\lambda_i)$. Si $\Lambda = \Sigma_i \lambda_i$ est grand, s'impose la forme

$$\gamma(\Sigma_i X_i; I) \leq C/\sqrt{\Sigma_i - Log(\gamma(X_i; I))}. \tag{33.5}$$

Enfin, si X_i est le résultat d'un jeu équilibré de pile ou face, la somme S_n suit la loi binomiale $Bi(n, 1/2)$. Pour $I = \{0\}$, $\gamma(X_i; I) = 1/2$, et pour n grand $\gamma(\Sigma_i X_i; I)$ est équivalent à $\sqrt{2/\pi n} = \sqrt{2/\pi}[\Sigma_i 1]^{-1/2}$. Assurer le raccordement entre ces différentes expressions et disposer de la meilleure constante suggère l'énoncé suivant, dont le travail ci-après propose une preuve:

Théorème 33.1.1 *Soit φ la fonction définie sur* $[0,1]$ *par*

$$\varphi(x) = 1/(2x)^2 \text{ si } 0 < x \le 1/2 \text{ et } \varphi(x) = -Log(x)/Log(2) \text{ sinon }. \quad (33.6)$$

Soit I un intervalle compact non vide et X_i une suite de variables aléatoires indépendantes. On a

$$\gamma(\Sigma_i X_i; I) \le \sqrt{2/\pi} \times [\Sigma_i \varphi(\gamma(X_i; I))]^{-1/2}. \quad (33.7)$$

Remarque: Dans le cas où I est réduit à un point, qu'on peut toujours supposer être 0, $\gamma(X, \{0\}) = 1$ caractérise les lois dégénérées, cependant que $\gamma(X, \{0\}) = 0$ caractérise les lois diffuses, auquel cas φ est infini: si l'un des composants est diffus, la somme l'est. Mais si $\ell(I) > 0$, $\gamma(X, I) > 0$.

On peut présenter le résultat, avec quelques hypothèses supplémentaires, sous une forme sous-additive plus parlante :

Corollaire 33.1.1 *Soit $G(X; I) = \varphi(\gamma(X; I))$. Soit X_i une suite de variables indépendantes. Si $Sup_i\ G(X_i; I) \ge 1$ ou bien si $\Sigma_i\ G(X_i; I) \ge 8/\pi$, alors*

$$G(\Sigma_i X_i; I) \ge (\pi/8)\Sigma_i G(X_i). \quad (33.8)$$

(La seconde hypothèse et (33.7) entraînent que $\gamma(\Sigma_i X_i; I) \le 1/2$. On verra plus loin (propriété (P1)) que la première entraîne également $\gamma(\Sigma_i X_i; I) \le 1/2$, et donc dans les deux cas $G(\Sigma_i X_i; I) = 1/(2\gamma(\Sigma X_i; I))^2$.)

Qualité de la majoration

Dans ce paragraphe, I est l'intervalle fermé $[0,1]$, β_o est défini par:

$$\beta_o = Sup\ \sqrt{\pi[\Sigma_i \varphi(\gamma(X_i; I))]/2} \times \gamma(\Sigma_i\ X_i; I).$$

Supremum pris sur toutes les sommes finies de variables aléatoires indépendantes Le Théorème assure que β_o est moindre que 1. Pour évaluer la qualité de la majoration, *minorons* en fonction de q ($0 < q < 1$) les quantités

$$\beta(q) := Sup_{S(q)}\ \sqrt{\pi[\Sigma_i \varphi(\gamma(X_i; I))]/2} \times \gamma(\Sigma_i X_i; I),$$

$\beta_-(q)$ associé à $S_-(q)$, enfin $\beta_+(q)$ associé à $S_+(q)$, où $S(q)$, resp $S_-(q)$, enfin $S_+(q)$ notent la famille des sommes finies de v.a.r. indépendantes X_i vérifiant toutes $\gamma(X_i; I) = q$, resp $\gamma(X_i; I) \leq q$, enfin $\gamma(X_i; I) \geq q$. β_o majore évidemment les trois fonctions ci-dessus.

Soient m l'entier positif et le p le réel vérifiant $mq + p = 1$ et $0 \leq p < q$. Soit $\xi(q)$ la loi sur $\mathbb{Z}$ définie par $\xi(q) = p\Sigma_{0\leq i\leq m}\delta_{2i} + (q-p)\Sigma_{1\leq i\leq m}\delta_{2i-1}$, dont $V(q)$ désignera la Variance. Pour $I = (0,1)$, on a $\gamma(\xi_q; I) = q$. Reparamétrant en posant $q - p = \Delta/m$, $(0 < \Delta \leq 1)$, on trouve $3V(q) = m^2 + 2m - \Delta(2m+1)$. D'après le Théorème de limite locale pour les variables entières (voir par exemple Petrov (1975, Chapitre VII, p. 187, Theorem 1)) $lim_n \sqrt{n} \times \gamma(S_n(\xi(q); I) = 2 \times \sqrt{1/(2\pi Var(\xi(q))}$. (On prendra garde que pour $p = 0$, le pas de la marche aléatoire S_n est 2, mais que tout translaté de I ne contient qu'un point du support de la loi, alors que sinon, le pas est 1 mais I contient *deux* points du support, de sorte que le résultat conserve la même forme). Il s'ensuit que $\beta(q) \geq \sqrt{\varphi(q)/V(q)}$.

I. *La majoration est atteinte:* $\beta_o = 1$.

(En effet, pour $q = 1/2$, $\varphi = V = 1$, soit $1 \leq \beta(1/2) \leq \beta_o \leq 1$).

II. *Pour les faibles concentrations* $(q \geq 1/2)$: $\beta_+(q) \geq 1/\sqrt{2Log(2)} > \cdot 849$.

(Pour $q \geq 1/2$, on observe que $\varphi(q)/V(q) = -Log(1-q)/[2(1-q)Log(2)]$ est monotone et a pour minimum $1/(2Log(2))$).

III. *Du côté des fortes concentrations* $(q \leq 1/2)$: $\beta_-(q) \geq \sqrt{3/4} > \cdot 866$.

(De ce côté, $\varphi(q)/V(q)$ n'est pas monotone de q mais il est facile de vérifier que $\beta_- \geq lim_{q\to 0}\sqrt{\varphi(q)/V(q)} = \sqrt{3/4}$).

IV. *Dans le cas où la somme ne comprend que deux termes*, l'évaluation n'est pas trop mauvaise: pour chaque m entier avec $m \geq 2$, si X_1,X_2 sont indépendantes et de loi $\xi(1/m)$, les concentrations de X_1,X_2,$X_1 + X_2$ sont égales à $1/m$. On a donc dans chacun de ces cas

$$\sqrt{\pi[\varphi(\gamma(X_1; I)) + \varphi(\gamma(X_2; I))]/2} \times \gamma(X_1 + X_2; I) = \sqrt{\pi/4} > \cdot 88.$$

33.2 Notations et premières réductions

Une propriété essentielle de la concentration est que *la convolution déconcentre*:

(P1) *Si X et Y sont indépendantes,* $\gamma(X+Y) \leq min(\gamma(X), \gamma(Y))$.

On dispose des résultats de continuité suivants:

La concentration est atteinte:
(comme I est fermé, il existe un x avec $P(X \in I + x) = \gamma(X; I)$.

La concentration est continue à droite:
(si $I_n \downarrow I$, $\gamma(X;I) = lim\ \gamma(X;I_n)$).

En ce qui concerne la *convergence en loi* on a seulement: si les S_n convergent en loi vers S, $limsup\ \gamma(S_n;I) \le \gamma(S;I)$ et, pour tout I' contenant *strictement* I, $\gamma(S;I) \le liminf\ \gamma(S_n;I')$.

Il suffit de démontrer (33.7) pour les sommes finies: En effet si les X_i sont en nombre infini, soit S_n la somme des n premiers. Si la série converge (vers S), S est la somme indépendante de S_n et de $S - S_n$, et par (P1) $\gamma(S;I) \le \gamma(S_n;I)$, ce qui suffit. En fait, Paul Lévy a démontré que la convergence essentielle d'une série équivaut à la finitude de la série de ses concentrations, pour tout I (non vide), et dans le cas infini, la formule a un sens si la série diverge mais converge essentiellement.

Il suffit de démontrer (33.7) quand toutes les concentrations sont strictement comprises entre 0 et 1: En effet, si l'un des $\gamma(X_i;I) = 0$, I est réduit à un point, la loi de X_i est diffuse, celle de $\Sigma_i X_i$ aussi et $\gamma(\Sigma X_i;I) = 0$.

A l'inverse, supposons que X_1 par exemple vérifie $\gamma(X_1) = 1$. Comme par (P1) $\gamma(\Sigma X_i) \le \gamma(\Sigma_{i>1} X_i)$ et comme d'autre part $\Sigma_i \varphi(\gamma(X_i)) = \Sigma_{i>1}\varphi(\gamma(X_i))$ puisque $\varphi(1) = 0$, on peut éliminer X_1 de l'assertion à démontrer.

Il suffit de démontrer (33.7) dans le cas où toutes les X_i sont à valeurs entières: Notons $Int(x)$ la partie entière du réel x. Soit $\varepsilon > 0$. A la variable réelle X et à l'intervalle I de longueur $\ell(I) > 0$, faisons correspondre la variable $\xi = Int(X/\varepsilon)$ et l'intervalle (de $\mathbb{Z}$) $J = [0, Int(\ell(I)/\varepsilon) + 1]$. On a $\gamma(X;I) \le \gamma(\xi;J)$. Par ailleurs quand ε tend vers 0 $\gamma(\xi;J)$ tend vers, $\gamma(X;I)$ et cela suffit. Si $\ell(I) = 0$, on prendra $J = \{0\}$.

Il suffit de démontrer (33.7) dans le cas où les X_i en nombre fini prennent leurs valeurs dans un même compact de $\mathbb{Z}$.

PREUVE. En effet, pour une suite S_n de variables $\mathbb{Z}$-valuées et convergeant en loi vers S, on a $\gamma(S;I) = lim\ \gamma(S_n;I)$. Soit $X_i^n = X_i 1_{|X_i|<n} + n 1_{|X_i|\ge n}$. Comme $P(X_i \in I + x)$ tend vers 0 quand $|x|$ tend vers l'infini, pour n assez grand, $\gamma(X_i^n;I) = \gamma(X_i;I)$. On peut réaliser cette égalité pour toutes les X_i en même temps au delà d'un même n, puisque les X_i sont en nombre fini. Prendre $S_n = \Sigma X_i^n$ et $S = \Sigma X_i$ donne le résultat. ■

Dans la suite, la notation $Z = \alpha X + (1-\alpha)Y$, où $0 \le \alpha \le 1$, signifie l'égalité *en loi* suivante: Pour tout borélien U,

$$P(\alpha X + (1-\alpha)Y \in U) = \alpha P(X \in U) + (1-\alpha)P(Y \in U).$$

(On peut transformer l'égalité en loi en égalité ordinaire en posant $Z = 1_A X + (1 - 1_A)Y$ où A est un ensemble de Probabilité α indépendant de (X, Y)).

Quelques propriétés de convexité

Soient $Z = \alpha X + (1-\alpha)Y$ (défini *supra*) et X' indépendantes de Z; on a:

(P2) $\gamma(Z;I) \leq \alpha\gamma(X;I) + (1-\alpha)\gamma(Y;I)$.

(P3) $\gamma(X'+Z;I) \leq \alpha\gamma(X'+X;I) + (1-\alpha)\gamma(X'+Y;I)$.

(P4) $\mathbf{C}(q;I)$ *est convexe*

où $\mathbf{C}(q;I)$ note l'ensemble des lois vérifiant $\gamma(X;I) \leq q$.

($P(Z \in I+x) = \alpha P(X \in I+x) + (1-\alpha)P(Y \in I+x) \leq \alpha\gamma(X;I) + (1-\alpha)\gamma(Y;I)$ prouve (P2); (P4) est une conséquence immédiate de (P2); pour (P3), on peut imposer X' indépendante de X et Y, et alors on peut écrire $X'+Z = \alpha(X'+X) + (1-\alpha)(X'+Y)$ au sens défini *supra*; (P3) se ramène alors à (P2).)

33.3 Présentation de la preuve

Désormais, les X_i sont indépendantes, en nombre fini, elles prennent leurs valeurs dans $\mathbb{Z}$ et leurs lois sont à support compact. On note I_o le *segment* de $\mathbb{Z}$ (c'est-à-dire un intervalle compact non vide de $\mathbb{Z}$) pour lequel on évalue les concentrations, qui sont notées $\gamma(X)$. On réserve la lettre I au segment courant translaté de I_o, autrement à l'élément courant de $\mathbf{J}$, qui note l'ensemble des $I_o + x$ où $x \in \mathbb{Z}$. Nous avons à montrer dans ce cadre la formule

$$\gamma(\Sigma_i X_i) \leq \sqrt{2/\pi} \times [\Sigma_i \varphi(\gamma(X_i))]^{-1/2} \tag{33.9}$$

où toutes les concentrations vérifient $0 < \gamma(X_i) < 1$.

Donnons maintenant quelques définitions et propriétés utiles pour la suite.

Segments et des segments stricts

Soit A une partie finie de $\mathbb{Z}$; on notera $|A|$ son cardinal. On peut l'énumérer dans l'ordre de $\mathbb{Z}$, $A = (s_1 < \cdots < s_{|A|})$. On appellera *segment de A* tout sous-ensemble B de A composé de points consécutifs (de A) soit $B = (s_i | i_o \leq i \leq i_1)$ avec $1 \leq i_o \leq i_1 \leq |A|$ (un segment est donc un compact non vide), et *segment strict* un segment comportant au moins deux points. (Attention: Si A n'est pas un segment de $\mathbb{Z}$, une partie B peut être un segment de A sans être un segment de $\mathbb{Z}$!).

Définition des lois modérées

Si $\gamma(X) \leq 1/2$, la loi X est dite *modérée*, sinon elle est dite *exceptionnelle*.

Décomposition de la concentration

A tout réel q tel que $0 < q \le 1$, on associe l'entier m et le reste p tels que $qm + p = 1$, avec $0 \le p < q$; on pose $\Delta = 1 - (m+1)p$. On a $0 < \Delta \le 1$ et $\Delta = 1$ si et seulement si $q = 1/m$. On définit $m(X)$ et $\Delta(X)$ comme les quantités associées à $q = \gamma(X)$. On a

(P5) $\qquad \gamma(X) = \Delta(X)/m(X) + (1-\Delta(X))/(m(X)+1),$

(P6) $\qquad$ *X est exceptionnelle si et seulement si $m(X) = 1$ et $0 < \Delta(X) < 1$.*

Définition des lois régulières et extrémales

1. *X est dite régulière si $1/\gamma(X) = m(X)$ (> 1) (et alors $\Delta(X) = 1$), et plus précisément m-régulière si $\gamma(X) = 1/m$.*
2. *X est dite m-extrémale si c'est un élément extrémal du convexe $\mathbf{C}(1/m)$.*
3. *X est dite extrémale dans son convexe si X est extrémale dans $\mathbf{C}(\gamma(X))$.*

Organisation de la preuve

On commence par décrire les lois extrémales de $\mathbf{C}(1/m)$ (elles sont bien sûr à support fini). Puis grâce au Lemme de Sperner (1928), on donne une première majoration de la concentration d'une somme dont tous les termes sont m-extrémaux. On prouve ensuite le résultat dans ce cas, puis on l'étend par convexité au cas où tous les composants sont modérés. On termine par une récurrence sur le nombre de composants irréguliers. Les preuves trop techniques ou trop longues sont rejetées en l'Appendice.

33.4 Preuve de l'assertion (33.7)

Etape 1: *Les lois m-extrémales*

Une suite x_j est dite **J**-*séparée* si pour I de **J**, $\Sigma_j 1_{x_j \in I} \le 1$.

Lemme 33.4.1 *Les lois m-extrémales s'écrivent $(1/m)\Sigma_{1\le j\le m}\delta_{x_j}$ où les x_j sont **J**-séparés.*

La preuve de ce lemme est en Appendice. Dans la suite, le support d'une telle loi sera noté $\Sigma = \{x_1, x_2, \cdot\cdot, x_m\}$, et les x_j seront supposés rangés dans l'ordre croissant.

Etape 2: *Le Lemme de Sperner*

Soit $m \geq 2$. Si $m = 2k+1$, on pose $T_m = (j | j \in \mathbb{Z};\ |j| \leq k)$. Si $m = 2k$, $T_m = (j | 2j \in 2\mathbb{Z}+1;\ |2j| \leq 2k-1)$, de sorte que T_m comporte m points et a 0 pour centre de symétrie. On note $(y_j | j = 1, 2, \cdot\cdot, m)$ les points de T_m rangés dans l'ordre croissant et $\xi^{(m)}$ est la loi uniforme sur T_m, soit $(1/m)\Sigma_{1\leq j\leq m}\delta_{y_j}$.

Soient, pour $1 \leq i \leq n$, des $m_i \geq 2$. Posons $\Omega = \Pi_{1\leq i\leq n} T_{m_i}$, de point courant $\omega = (\theta_i)$. On appelle *valuation de* ω le nombre $v(\omega) = \Sigma_i \theta_i$. Ordonnons Ω par l'ordre (partiel) $(\theta_i) \leq (\eta_i)$ si et seulement si pour tout i, $\theta_i \leq \eta_i$. Une *suite de Sperner* est une suite $\sigma = (\omega^{(1)}, \omega^{(2)}, \cdot\cdot\, \omega^{(\ell)})$ de points de Ω vérifiant:

a) pour tout $u \in [1, \ell-1]$, $\omega^{(u)} \leq \omega^{(u+1)}$ et $v(\omega^{(u+1)}) = v(\omega^{(u)}) + 1$.

b) $v(\omega^{(1)}) + v(\omega^{(\ell)}) = 0$.

(La propriété a) signifie qu'on passe de $\omega^{(u)}$ à $\omega^{(u+1)}$ en laissant toutes les coordonnées invariantes sauf une, qui croit de 1 exactement, et la propriété b) est une propriété de symétrie de la suite). Sperner (1928) a démontré le lemme suivant et son corollaire:

Lemme 33.4.2 *Il existe une partition de* Ω *en suites de Sperner.*

En conséquence des propriétés (a), (b), toute suite de Sperner contient un élément de valuation 0 ou 1/2 suivant la parité du nombre de m_i pairs.

Corollaire 33.4.1 *Le nombre de suites de Sperner est le nombre des points de* Ω *de valuation* 0 *ou de valuation* 1/2 *suivant la parité du nombre de* m_i *pairs.*

Dans la suite, on note $N((m_i))$ le nombre de suites de Sperner de Ω.

Interprétation du Lemme en termes de concentration

Soient les v.a.r Y_i, indépendantes de lois respectives $\xi^{(m_i)}$. Le vecteur (Y_i) est réparti uniformément sur Ω (chaque point de masse $1/\Pi m_i$). Soit V la somme des Y_i, dont les valeurs possibles appartiennent soit à $\mathbb{Z}$, soit à $\mathbb{Z} + 1/2$. Le point de masse maximum pour la loi de V est soit 0, soit 1/2, et sa Probabilité vaut $N((m_i))/\Pi m_i$. Cette valeur est aussi $\gamma(V, \{0\})$.

Etape 3: *Première majoration*

Soient des X_i m_i-extrémales en nombre fini n et S leur somme.

Lemme 33.4.3 $\gamma(S) \leq \gamma(\Sigma_i \xi^{(m_i)}, \{0\}) = N((m_i))/\Pi_i m_i$.

Preuve. Soit $\Omega' = \Pi\Sigma_i$ le produit des supports des lois des X_i. Pour chaque i, soit f_i la bijection croissante de T_{m_i} sur le support Σ_i de la loi de X_i, et F la bijection produit de Ω sur Ω'. Soit s l'application qui à un point ω' de Ω' fait correspondre la somme de ses coordonnées. Quand ω décrit une suite de Sperner, la *valuation image* $s(F(\omega))$ croît en vertu de la propriété a) des suites de Sperner, et décrit un ensemble **J**-séparé en vertu de la **J**-séparation de chaque support. Pour tout I de **J**, le nombre de points de Ω' de valuation prenant valeur dans I est donc majoré par $N((m_i))$, d'où le résultat. ■

Etape 4: *Preuve de* (33.7) *quand toutes les* X_i *sont m-extrémales*

Soit ξ une loi discrète. Dans ce paragraphe, $\mu(\xi) := Sup_x P(\xi = x)$ note la masse maximale de ξ, i.e. la concentration dans le cas où $I_o = \{0\}$, et $S_n(\xi)$ note la somme de n composants indépendants de loi ξ.

De même $\Sigma\xi_j$ notera la somme de v.a.r. indépendantes de lois respectives ξ_j.

Comme on l'a vu dans l'interprétation du Lemme de Sperner

$$N((m_i))/\Pi_i m_i = \mu(\Sigma_i \xi^{(m_i)}).$$

Soit $\gamma(m, n)$ la quantité $\mu(S_n(\xi^m)$ avec la convention $\gamma(m, 0) = 1$. Les propriétés bien connues de la loi binomiale donnent immédiatement

(P7) Pour $n = 2k - 1$ et $n = 2k$ avec $n > 0$, $\gamma(2, n) = \beta_k$ où $\beta_k = \binom{2k}{k} 2^{-2k}$.

(P8) $\beta_{k+1}/\beta_k = (2k + 1)/(2k + 2)$ et $lim_k \sqrt{\pi k}\beta_k = 1$.

Lemme 33.4.4 *a) Pour* $m \geq 2,\ n \geq 2$, *on a* $m\gamma(m, n) \leq 2\gamma(2, n)$.
b) Pour $m \geq 2,\ t \geq 2$, *on a* $|\int f_m|^t \leq [2/m] \times \sqrt{2/(\pi t)}$ *où* f_m *est la fonction caractéristique de la loi* $\xi^{(m)}$ *et où* $\int f$ *signifie* $\int_0^{2\pi} f(t)dt/2\pi$.

La preuve de ce lemme est en l'Appendice.

Lemme 33.4.5 *L'assertion (33.7) vaut si tous les composants sont m-extrémaux et en nombre fini.*

PREUVE. On a $\gamma(S) \leq \mu(\Sigma\xi_{m_i}) \leq \int \Pi_i |f_{m_i}| \leq \Pi_i [\int |f_{m_i}|^{t_i}]^{1/t_i}$ pour tout choix des t_i positifs avec $\Sigma 1/t_i = 1$, donc, en vertu du Lemme 33.4.4b), $\sqrt{\pi/8}\gamma(S)$ est majorée par $\Pi(m_i\sqrt{t_i})^{-1/t_i}$, si les t_i vérifient la condition supplémentaire $t_i \geq 2$. Si $sup_i m_i^2/\Sigma_j m_j^2 \leq 1/2$, le choix $t_i m_i^2 = \Sigma_j m_j^2$ donne le résultat, puisque $8/(\pi\Sigma m_i^2) = 2/(\pi\Sigma\varphi(q_i))$. Sinon, par exemple $m_1^2 > \Sigma_j m_j^2/2$. Par (P1) on a $\gamma(S) \leq 1/m_1$, moindre que $\sqrt{8/(\pi\Sigma m_i^2)}$ puisque $2 < 8/\pi$. ■

Etape 5: *Preuve de* **(33.7)** *quand toutes les X_i sont régulières*

Lemme 33.4.6 *Soit X une loi modérée non régulière et extrémale dans son convexe. On a alors $X = \Delta(X)Y + (1-\Delta(X))Z$ où Y (resp Z) sont $m(X)-$ (resp $(m(X)+1)-$)extrémales.*

(On remarquera que l'hypothèse entraîne $0 < \Delta(X) < 1$ mais que le résultat est trivialement vrai pour une loi régulière, qui se réduit alors au seul terme Y.)

Ce Lemme, qui est en quelque sorte une extension de la propriété (P5), est prouvé en Appendice.

Lemme 33.4.7 *L'assertion (33.7) vaut si tous les composants sont modérés et extrémaux dans leur convexe.*

PREUVE. Par récurrence sur le nombre n de composants non réguliers. Si $n = 0$, on applique le Lemme 33.4.5. Supposons le résultat vrai pour n. Si S a $n+1$ composants non réguliers, posons $S = X_1 + T$ où X_1 est l'un des non réguliers. On a, par le Lemme 33.4.6 et la propriété (P3): $\gamma(S) \leq \Delta(X_1)\gamma(Y+T) + (1-\Delta(X_1))\gamma(Z+T)$, soit, par l'hypothèse de récurrence, posant $\Phi = \Sigma_{i\geq 2}\varphi(\gamma(X_i))$, $\sqrt{\pi/2}\gamma(S) \leq \Delta(X_1)[\varphi(1/m(X))+\Phi]^{-1/2}+(1-\Delta(X))[\varphi(1/(m(X)+1))+\Phi]^{-1/2}$.

Or la fonction $x \to [\varphi(x)+\Phi]^{-1/2}$ est concave pour tout $\Phi \geq 0$, et par (P5) $\gamma(X) = \Delta(X)/m(X) + (1-\Delta(X))/(m(X)+1)$. Soit $\sqrt{\pi/2}\gamma(S) \leq [\varphi(\gamma(X)) + \Phi]^{-1/2}$ et le résultat est prouvé. ■

Etape 6: *Quelques résultats sur les lois exceptionnelles*

Soit X exceptionnelle de concentration $\gamma(X)$. On a $\Delta(X) = 2\gamma(X) - 1$, avec $0 < \Delta(X) < 1$. On pose $D(X) = Max(\Delta(X), 0)$, de sorte que

- X régulière équivaut à $D(X) = 0$.
- X exceptionnelle équivaut à $0 < D(X) = \Delta(X) < 1$, et alors $2\gamma(X) = 1 + D(X)$.
- dans tous les cas, $2\gamma(X) \leq 1 + D(X)$.
- (P1) donne $D(X+Y) \leq min(D(X), D(Y))$.

Lemme 33.4.8 *Si X est exceptionnelle et extrémale dans son convexe, alors $X = \Delta(X)Y + (1-\Delta(X))Z$ où Y est dégénérée et $Z \in \mathbf{C}(1/2)$.*

La preuve de ce lemme est en l'Appendice.

L'argument de convexité du Lemme 33.4.7 ne fonctionne pas car $x \to [\varphi(x) + \Phi]^{-1/2}$ n'est pas concave de x variant entre $1/2$ et 1 pour tous les Φ positifs. Etablissons une première majoration *ad hoc.*

Lemme 33.4.9 $D(\Sigma_i X_i) \leq \Pi_i D(X_i)$.

PREUVE. Rappelons que les X_i sont en nombre fini et à support dans un même compact. Etant donnée la structure de l'assertion, il suffit de la démontrer dans le cas où il n'y a que deux composants.

En utilisant le même argument que dans l'*étape finale* ci après, il suffit de prouver le résultat quand ces deux composants sont extrémaux dans leur compact. Notons D_i, Δ_i pour $D(X_i)$, $\Delta(X_i)$ et D pour $D(X_1 + X_2)$:

– si l'un ou l'autre des D_i est nul, (P1) donne le résultat.

– sinon, $D_i = \Delta_i$ et par le Lemme 8, $X_i = \Delta_i Y_i + (1 - \Delta_i) Z_i$ où $\gamma(Y_i) = 1$ et $\gamma(Z_i) = 1/2$. Notant Γ pour $\gamma(S)$, $\tilde{\Delta}_i$ pour $(1 - \Delta_i)$, on a donc par (P3) :

$$\Gamma \leq \Delta_1\Delta_2\gamma(Y_1 + Y_2) + \Delta_1\tilde{\Delta}_2\gamma(Y_1 + Z_2) + \tilde{\Delta}_1\Delta_2\gamma(Z_1 + Y_2) + \tilde{\Delta}_1\tilde{\Delta}_2\gamma(Z_1 + Z_2)$$

où dans le second membre les trois dernières concentrations sont moindres que 1/2 et la première vaut 1. Soit $2\Gamma - 1 \leq \Delta_1\Delta_2$ et le résultat. ■

Corollaire 33.4.2 *Soit $S = \Sigma_i X_i$. Si $\Pi_i D(X_i) \geq .038$, alors l'assertion (33.7) vaut.*

(Noter qu'alors tout les X_i sont exceptionnels.)

PREUVE. Posons $D_i = \Delta(X_i)$. Comme $0 < D_i < 1$, on a $0 < \sqrt{D_i} < (1+D_i)/2$, soit $Log(2)\varphi(\gamma(X_i)) \leq Log(D_i)/2$. Via le lemme précédent, il suffit donc de prouver que $-Log(x) \times (1 + x)^2 \leq 16Log(2)/\pi$ pour $1 \geq x \geq \cdot038$, ce qui est immédiat. ■

Lemme 33.4.10 *Soit $S = X + Y$. Si $D(X)D(Y) > 0$, alors l'assertion (33.7) vaut.*

(Ou: le résultat vaut si S n'a que deux composants, tous deux exceptionnels.)

PREUVE. Posons $D(X) = 1 - \xi$, $D(Y) = 1 - \eta$, $\gamma = \gamma(S)$, $\Phi = F(\varphi(X) + \varphi(Y))$ où $F(x) = \sqrt{2/\pi x}$. On a $\Phi^2 = 2Log(2)/[-\pi Log((1 - \xi/2)(1 - \eta/2))]$. Par ailleurs, $\gamma \leq (1 + D(S))/2 \leq (2 - \xi - \eta + \xi\eta)/2$ par le Lemme 33.4.9. Comme pour $u < 1/2$ on a $-Log(1 - u) \leq 2uLog(2)$, $F^2 \geq 2/[\pi(\xi + \eta)]$, il suffit de prouver que $Sup_{0<\xi,\eta<1}\ (2-\xi-\eta+\xi\eta)^2(\xi+\eta) \leq 8/\pi$. Or, comme $4\xi\eta \leq (\xi+\eta)^2$, $Sup_{0<\xi,\eta<1}\ (2 - \xi - \eta + \xi\eta)^2(\xi + \eta) \leq Sup_{0<s<2}\ s(2 - s + s^2/4) = 2 \leq 8/\pi$. ■

Lemme 33.4.11 *Soit $S = \Sigma X_i$, et X exceptionnelle indépendante de S. Soient les trois hypothèses:*

(h1) Le résultat vaut pour S.

(h2) Pour toute Y 2-extrémale, le résultat vaut pour $S + Y$.

(h3) $\gamma(S) \leq .5821$.

Sous ces hypothèses, l'assertion (33.7) vaut pour $S + X$.

La preuve de ce lemme est en l'Appendice.

Etape 7: *Preuve de* **(33.7)** *quand toutes les composants sont extrémaux dans leur convexe*

Soit ν (resp ν') le nombre de composants réguliers (resp exceptionnels) de S. Le résultat vaut si $\nu' = 0$ (Lemme 33.4.7). Il vaut si $\nu = 0$ et $\nu' = 1$, car pour $1/2 < q < 1$, $q < \sqrt{2/(\pi\varphi(q))}$, également si $\nu = 0$ et $\nu' = 2$ (Lemme 33.4.10).

Supposons le vrai pour $\nu' < n$, avec $n > 0$. Supposons $\nu' = n$, et soit X composant de concentration maximum, soit $q > 1/2$. Décomposons S comme $S = T + X$. Le résultat vaut pour T, puisque $\nu'(T) = n - 1$. Pour la même raison, il vaut aussi pour $T + Y$, où Y est la variable auxiliaire du Lemme 33.4.11.

(A) Si $\nu > 0$, T contient au moins un facteur régulier, donc d'après P1, $\gamma(T) \leq 1/2$. On applique le Lemme 33.4.10.

(B) Si $\nu = 0$ mais $\gamma(T) \leq .5821$, on applique encore le même Lemme.

(C) Sinon, S et T ne comportent que des composants exceptionnels, et on conclut par le Lemme 33.4.10 si $\nu' = 1$ ou 2.

(D) Sinon, $\nu' \geq 3$, $\gamma(T) > .5821$ donc $[\Pi_{1\leq i\leq\nu'}\Delta(X_i)]/\Delta(X) > 2 \times \cdot 5821 - 1 = \cdot 1642$ d'après le Lemme 33.4.9. Mais comme $\Delta(X) = Sup_i\Delta(X_i)$ et $\nu' \geq 3$, $D(S) = [\Pi_{1\leq i\leq\nu'}\Delta(X_i)] > (.1642)^{3/2} > .066$ et on conclut par le Corollaire 33.4.1.

Etape finale: *Preuve du résultat*

Soit $\mathbf{C}(q, K)$ le sous ensemble de $\mathbf{C}(q)$ dont les supports sont contenus dans le compact K. $\mathbf{C}(q, K)$ est compact et métrisable pour la convergence en loi (c'est un compact de $\mathbb{R}^K$), et toute $X \in \mathbf{C}(q, K)$ barycentre de points extrémaux. En utilisant la propriété de convexité (P3) le résultat est prouvé.

33.5 Appendice

I. *Démonstration des Lemmes 33.4.1, 33.4.6 et 33.4.8*

Comme n'apparaît qu'une loi dans ce paragraphe, on note $P(A)$ pour $P(X \in A)$. Soit $X \in \mathbf{C}(q)$ fixée dans la suite et S son support; on a $S = (s_1 < s_2 < \cdots < s_n)$ avec $n > m$ si $p > 0$ et $n \geq m$ sinon, et donc $n \geq 2$.

Soit $\mathbf{I}$ l'ensemble des I de $\mathbf{J}$ vérifiant $P(I) = q$. Comme $P(I_o + x)$ tend vers 0 quand x tend vers l'infini, il n'y a qu'un nombre fini de I vérifiant $P(I) > q/2$. En particulier

(a1) Le cardinal de $\mathbf{I}$ est fini.

(b1) Soit $q' := Sup_{I \in \mathbf{J} \setminus \mathbf{I}} P(I)$; alors $q' < q$.
Décrivons les traces de $\mathbf{I}$ sur S. Pour $I \in \mathbf{I}$, soit $K(I) = I \cap S$; $P(K) = q$, K est non vide, et, comme trace d'un segment de $\mathbb{Z}$, c'est un segment de S. Soit $\mathbf{K}$ l'ensemble -fini!- des $K(I)$ quand I varie dans $\mathbf{I}$. On pose $A = (\cup_{I \in \mathbf{I}} I)$, $T = \cup_{K \in \mathbf{K}} K = S \cap A$.

Lemme 33.5.1 *S'il existe μ, mesure non nulle de support compact M vérifiant:*
(h4) $M \subset S$, $\mu(S) = 0$ et $Sup_x |\mu(x)| \leq 1$,
(h5) $\mu(I) = 0$ pour tout $I \in \mathbf{C}$,
alors P n'est pas extrémale.

PREUVE. Soient P_1, P_2 les mesures $P + \varepsilon\mu, P - \varepsilon\mu$. Elles sont de support S, elles vérifient $P_i(\mathbb{Z}) = 1$, et pour tout $I \in \mathbf{I}$, $P_i(I) = P(I) = q$. Comme $Sup_{I \in \mathbf{J}} |\mu(I)| \leq |I_o|$, le cardinal de I_o, on a pour $I \in \mathbf{J} \setminus \mathbf{I}$, $P_i(I) \leq q' + |\varepsilon||I_o|$, moindre que q pour ε suffisamment petit. Enfin, soit $\rho = Inf_{x \in M} P(X = x)$. On a $0 < \rho$ et donc pour $|\varepsilon| \leq \rho$ P_1 et P_2 sont des mesures positives. Ce sont donc deux probabilités distinctes de $\mathbf{C}(q)$ et de demi-somme P. ■

On suppose désormais P extrémale. Voici trois conséquences du Lemme 33.5.1:

(A1) $S \setminus T$ comporte au plus un point, $\mathbf{I}$, $\mathbf{K}$ sont non vides et S est fini. (Car pour tout couple x, y de $S \setminus T$, $\mu = \delta_x - \delta_y$ satisfait (h4) et (h5), d'où la première partie; puis on utilise $n \geq 2$ et (a1)).

(A2) Soient $K \neq K'$ deux éléments non disjoints de $\mathbf{K}$. Alors $K \setminus K'$ et $K' \setminus K$ sont des segments *non vides* de S, et K,K' sont deux segments stricts de S. (Car $K \subset K'$ conduit à la contradiction $q = P(X \in K) < P(X \in K') = q$.)

(A3) Réciproquement, si $K \in \mathbf{K}$ est un segment strict de S (i.e. n'est pas réduit à un point), il existe alors $K' \in \mathbf{K}$ vérifiant avec K les hypothèses de (A2). (Sinon, soit x, y un couple de K; $\mu = \delta_x - \delta_y$ satisfait (h4) et (h5).)

Grâce à (A2), on peut énumérer $\mathbf{K} = (K_1 < K_2 < \cdots < K_{n'})$ où la notation $K < K'$ signifie que leurs extrémités gauche $\alpha_K, \alpha_{K'}$ vérifient $\alpha_K < \alpha_{K'}$. On appellera *suite stricte maximale* une sous-suite $(K_i, K_{i+1}, \cdot\cdot, K_{i+j})$ avec $j > 0$ telle qu'aucun des $K_{i+h} \cap K_{i+h+1}$ ne soit vide et telle qu'il n'en existe pas de plus grande. Un K n'appartenant à aucune suite stricte sera dit *isolé*.

Les *éléments isolés* se réduisent à un point (voir A3); leur ensemble représente une partie de T, notée $\mathbf{Is} = (t_1, \cdot\cdot, t_u)$, avec $P(\{t_j\}) = q$. $\mathbf{K} \setminus \mathbf{Is}$, se décompose en $\mathbf{St} = (\sigma_1, \cdots, \sigma_v)$, famille de suites strictes maximales. Noter que $\mathbf{Is}$ et $\mathbf{St}$ ne peuvent être tous deux vides.

Pour $\sigma \in \mathbf{St}$, soit $T_\sigma = \cup_{K \in \sigma} K$. T_σ est un segment strict de S et la famille composée de $\mathbf{Is}$, d'une part, $(T_\sigma | \sigma \in \mathbf{St})$ d'autre part, est une partition de T.

(A4) Soit $\sigma \in \mathbf{St}$. Comme tous les K de σ ont au moins deux points et comme ils sont totalement ordonnés, $|\sigma| < |T_\sigma|$; donc T_σ porte une mesure μ

non nulle vérifiant $Sup_x|\mu(x)| \leq 1$ et $\mu(K) = 0$ si $K \in \sigma$. Comme elle est portée par T_σ, elle vérifie que $\mu(K) = 0$ pour tout $K \in \mathbf{K}$, et μ vérifie h5. Elle vérifie aussi h4 si de plus $\mu(S) = 0$, c'est-à-dire $\mu(T_\sigma) = 0$, ce qu'on peut imposer s'il existe un $K \in \sigma$ de cardinal au moins trois, car alors $|\sigma| < |T_\sigma| - 1$, et alors P ne peut être extrémale.

(A5) Si tous les K de σ sont de cardinal 2, soit μ_σ la mesure chargeant alternativement les points successifs de T_σ de 1 et -1. Si $|T_\sigma| \in 2\mathbb{Z}$, $\mu_\sigma(T_\sigma) = 0$ et vérifie h4 et h5. P n'est pas extrémale.

Appelons exceptionnelle une suite σ vérifiant : les K de σ se composent de deux points *et* $|T_\sigma| \in 2\mathbb{Z} + 1$. Dans ces conditions μ_σ vérifie h5 et $\mu_\sigma(T_\sigma) = 1$. S'il existe *deux* suites strictes maximales et exceptionnelles σ et σ', il est immédiat que $\mu = \mu_\sigma - \mu_{\sigma'}$ vérifie les hypothèses du Lemme A et P n'est pas extrémale.

Conclusion. Si P est extrémale, deux cas peuvent se produire:

1. **St** est réduit à *une* suite maximale exceptionnelle, de support noté T_1.
2. **St** est vide.

Etude du cas 1. On a $T_1 = (s_0, s_2 \cdot \cdot, s_{2k})$ $(k > 0)$ avec: $P(s_{2j}) = \alpha$ pour $0 \leq j \leq k$ et $P(s_{2j-1}) = q - \alpha$ pour $1 \leq j \leq k$, avec $0 < \alpha < q$. Si $S \backslash T$ n'est pas vide, il se réduit à un point (cf. A1); soit $S \setminus T = \{\xi\}$, avec $0 < \rho = P(\{\xi\}) < q$ et P s'écrit

$$P = \rho\delta_\xi + \Sigma_{s \in \mathbf{Is}} q\delta_s + \alpha\delta_{s_0} + \Sigma_{1 \leq j \leq k}(\alpha\delta_{s_{2j}} + (q - \alpha)\delta_{s_{2j-1}}).$$

ξ n'est contenu dans aucun I de $\mathbf{I}$, donc ξ est **J**-séparé de **Is**, et $\xi < s_0$ ou $\xi > s_{2k}$ (car les s représentent tous les points du support compris entre s_0 et s_{2k}). Traitons le premier cas. Ou bien ξ est **J**-séparé de s_0 (cas 1a), ou non (cas 1b). La masse totale de la mesure P est 1, soit

$$1 = \rho + q|\mathbf{Is}| + \alpha + kq,$$

avec les conditions $0 < \rho, \alpha < q$ dans le cas 1a, et la condition supplémentaire $\rho + \alpha < q$ dans le cas 1b. Quoiqu'il en soit, ces conditions sur α et ρ sont ouvertes et P ne peut être extrémale. Autrement dit, $S \setminus T$ est vide, et

$$P = \Sigma_{s \in \mathbf{Is}} q\delta_s + \alpha\delta_{s_0} + \Sigma_{1 \leq j \leq k}(\alpha\delta_{s_{2j}} + (q - \alpha)\delta_{s_{2j-1}}).$$

On a alors $(k + |\mathbf{Is}|)q + \alpha = 1$, soit $m = k + |\mathbf{Is}|$ et $\alpha = p$.

De plus l'ensemble composé de **Is** et des s_{2j}, ainsi que celui composé de **Is** et des s_{2j-1} sont **J**-séparés (sinon il est aisé de construire un I de probabilité strictement supérieure à q) de sorte que si on pose

$$(m+1)P_1 = \Sigma_{s \in \mathbf{Is}}\delta_s + \delta_{s_0} + \Sigma_{1 \leq j \leq k}\delta_{s_{2j}}, \; mP_2 = \Sigma_{s \in \mathbf{Is}}\delta_s + \Sigma_{1 \leq j \leq k}\delta_{s_{2j}-1},$$

P_1 (resp P_2) est $(m+1)$ (resp m)-extrémale et on a la décomposition $P = (m+1)pP_1 + m(q-\alpha)P_2 = (1-\Delta(X))P_1 + \Delta(X)P_2$ (Lemme 33.4.5).

Etude du cas 2. qui se sépare en 2a: $S \setminus T$ vide et 2b: $S \setminus T = \{\xi\}$.

Comme précédemment **Is** (resp l'ensemble composé de **Is** et ξ) est **J**-séparé. Le premier donne

$$(*\text{cas } 2a) \qquad P = \Sigma_{s\in \mathbf{Is}} q\delta_s, \ |Is|q = 1, \ \text{soit } m = |Is| \text{ et } mq = 1, \ \text{donc } m > 1$$

et on trouve la décomposition du Lemme 33.4.1, le second donne

$$(*\text{cas } 2b) \qquad P = \rho\delta_\xi + \Sigma_{s\in \mathbf{Is}} q\delta_s, \ |Is|q + \rho = 1, \ \text{soit } m = |Is| \text{ et } \rho = p$$

et on se trouve dans la situation du Lemme 33.4.6. Comme on a épuisé toutes les mesures extrémales, les Lemmes 33.4.1 et 33.4.6 sont démontrés. Le Lemme 33.4.8 n'est que le cas particulier correspondant à $m = 1$.

II. *Preuve du Lemme 33.4.4*

Dans cette preuve, la concentration est la concentration ponctuelle, celle de la masse maximum. Les lois étant symétriques, le calcul Fourier donne

$$\gamma(m,n) \leq \int |f_m|^n,$$

pour tout entier n, avec égalité si n est pair.

PREUVE DE (A) POUR m PAIR. Posons $m = 2k$, avec $k > 1$, car il n'y a rien à prouver si $m = 2$. Soit P_m la probabilité definie par $mP_m = \Sigma_{0\leq i<m}\delta_i$. Soient X de loi P_m, b de loi P_2, et U de loi P_k. On a (en loi) $X = k \times b + U$ où b et U sont indépendantes.

Soient X_j (en nombre n) indépendantes de loi P_m, de somme S, S_1 (resp. S_2) celle des b_j associée (resp. celle des U_j associée).

La loi de U modulo k est de concentration $2/m$; aussi bien sur $\mathbb{Z}/k$ que sur $\mathbb{Z}$, la convolution déconcentre, la concentration d'une somme indépendante de telles variables est moindre que $2/m$, et donc pour tout r, $P(S_2 \in k\mathbb{Z} + r) \leq 2/m$. On en déduit immédiatement que $Sup_j P(S = j) \leq (2/m) Sup_u P(S_1 = u)$ (décomposer j comme $k \times v + r$). On notera que le résultat vaut quelle que soit la parité de n. ■

PREUVE DE (a) POUR m IMPAIR. Malheureusement, nous n'avons pu trouver de preuve combinatoire et nous proposons une preuve basée sur la fonction caractéristique. Soit $m = 2k+1$, $n = 2t$. On doit montrer que pour $t > 1$, $k > 0$, $m\gamma(m, 2t) \leq 2\gamma(2, 2t)$, soit

$$\int_0^\pi (sin(mx/2)/(msin(x/2)))^{2t} \times mdx \leq 2\pi\gamma(2,2t).$$

On écrit l'intégrale $C+D$, où $C = \int_0^{2\pi/m}$ et $D = \int_{2\pi/m}^{\pi}$. Majorant le dénominateur par 1, on a $D \leq 2m^{1-2t} \int_{\pi/m}^{\pi/2}(1/sin(y))^{2t}dy$. Comme $d(1/tg(y)) = -dy/(sin(y))^2$, comme $sin(y)$ est croissante dans l'intervalle, il vient

$$\text{pour } 2t > 2,\ D \leq 2m1 - 2t(sin(\pi/m)^{-2t+2}/tg(\pi/m).$$

Les monotonies de $sin(x)/x$ et de $tg(x)/x$ donnent finalement, puisque $m \geq 3$ et $t > 1$, $D/(2\pi\gamma(2,2t)) \leq (4/27)^{t-1}/(\pi^2\gamma(2,2t) \leq 32/(81\pi^2)$. Le changement de variables $y = mx$ montre que C est monotone de m et prend son maximum pour $m = 3$. Se restreignant désormais à ce cas, on obtient $C \leq I_{2t}$ où $I_n = 3\int_0^{2\pi/3}((1+2cosx)/3)^n dx$.

Soient les trois inéquations suivantes, où $n \geq 0$ et $t \geq 1$:

(i) $3(n+2)I_{n+2} = (2n+3)I_{n+1} + (n+1)I_n.$

(j) $(2n+1)/(2n+2) \leq I_{n+1}/I_n \leq (2n+2)/(2n+3).$

(k) $I_{2t+2}/I_{2t} \leq (2t+1)/(2t+2)$, soit $I_{2t+2}/I_{2t} \leq \gamma(2,2t+2)/\gamma(2,2t).$

On vérifie (i) en intégrant par parties. On a $I_o = 2\pi$, $I_1 = 2\pi/3 + \sqrt{3}$, $I_2 = 2\pi/9 + \sqrt{3}/6$. On vérifie (j) pour $n = 1$. Supposons (j) vraie pour n; par (i) et (j), $I_{n+2}/I_{n+1} \geq (2n+3)/(2n+4)$. De même $I_{n+2}/I_{n+1} \leq (6n^2+12n+5)/(6n^2+15n+6)$ dont on vérifie qu'il est moindre que $(2n+4)/(2n+5)$. (j) est donc vérifiée par récurrence. Revenant à (i) où on majore I_{n+1} en fonction de I_n, $3(n+2)I_{n+2} \leq (2n+2)I_n + (n+1)I_n$, qui donne (k) si $n = 2t.$, soit encore $I_{2t} \leq 2\gamma(2,2t)I_2$. Finalement, pour $t \geq 2$

$$(C+D)/2\pi\gamma(2,2t) \leq I_2/\pi + 32/(81\pi^2) = 2/3 + \sqrt{3}/(2\pi) + 32/(81\pi^2) \leq \cdot 983$$

et la preuve est terminée. ∎

Preuve de (b). Pour $t = 2k$, a) peut se récrire $m\int |f_m|^{2k} \leq 2\beta_k$. En appliquant Hölder entre $2k$ et $2k+2$, le résultat sera prouvé si l'on montre que, pour tout x avec $0 < x < 1$, pour tout $k \geq 1$, $\sqrt{\pi(2k+2x)}\beta_k^{1-x}\beta_{k+1}^x \leq \sqrt{2}$, soit, compte tenu de l'équation $\beta_{k+1}/\beta_k = (2k+1)/(2k+2)$, si l'on prouve que

$$A(k,x) := (\sqrt{\pi k}\beta_k) \times (\sqrt{1+x/k}((2k+1)/(2k+2))^x \leq 1.$$

Soit f le logarithme du deuxième facteur. Comme la dérivée seconde de $Log(1+y)$ est croissante, on a

$$f(x) \leq x \times (1/2k + log((2k+1)/(2k+2))) - x^2/(2k+2)^2$$

soit encore

$$4(k+1)^2 f(x) \leq 2x(1+1/k) - x^2 \leq 1 + 2/k$$

et donc

$$A(k,x) \leq B(k) := \sqrt{\pi k}\beta_k \times \exp(\psi(k))$$

avec

$$\psi(k) = (k+2)/(4k(k+1)^2).$$

Or $lim_{k\to\infty} B(k) = 1$. Il suffit donc de prouver que $r_k := (B(k+1)/B(k))^2 \geq 1$. Le rapport des premiers termes vaut $(1+u)$ avec $4k(k+1)u = 1$. $Log(1+u) \geq v = u - u^2/2$. Multipliant $v + 2\psi(k+1) - 2\psi(k)$ par $32(k(k+1)(k+2))^2$, on trouve $Log(r_k) \geq 32k^4 + 128k^3 + 111k^2 - 4k - 4$, expression positive pour $k \geq 1$. ■

III. *Preuve du Lemme 33.4.11*

Dans le cours de la preuve, on cherchera la plus *grande* constante γ_o telle que $S + X$ vérifie F dès lors que S vérifie les trois conditions où l'on substitue γ_o à $\cdot 5821$. Soit $\delta = \Delta(X)$, $\xi = \varphi(X) = 1 - Log(1+\delta)/Log(2)$, $\Phi = \Phi(S)$. Avec ces notations, $\Phi(S+X) = \Phi + \xi$. Par (P3), $\gamma(S+X) \leq \delta\gamma(S) + (1-\delta)\gamma(S+Y)$ où Y est 2-maximale. On a donc

$$\gamma(S+X) \leq \delta \min(\gamma_o, \sqrt{2/(\pi\Phi)}) + (1-\delta)\min(1/2, \sqrt{2/(\pi(\Phi+1))}).$$

Soit B le membre de droite de cette inégalité. Il suffit de prouver que $\sqrt{\pi/2}B/C \leq 1$, où $C = \sqrt{1/[\Phi+\xi]}$. Imposons à γ_o de vérifier la condition

(c1) $\quad a \leq b$ où $a = (8-\pi)/\pi$ et $b = 2/(\pi\gamma_o^2)$.

Si (c1) est vérifiée, $\sqrt{\pi/2}B$ vaut:

(I1) $\quad \sqrt{\pi/2}[\delta\gamma_o + (1-\delta)/2]$ si $\Phi \leq a$

(I2) $\quad \sqrt{\pi/2}\delta\gamma_o + (1-\delta)\sqrt{1/(\Phi+1)}$ si $a \leq \Phi \leq b$

(I3) $\quad \delta\sqrt{1/\Phi} + (1-\delta)\sqrt{1/(\Phi+1)}$ si $b \leq \Phi$.

(1) Dans le cas I1, le maximum de $\sqrt{\pi/2}B/C$ est obtenu pour $\Phi = a$.

(2) Dans le cas I2,

$$\sqrt{\pi/2}B/C = \sqrt{\pi/2}\gamma_o\delta\sqrt{(\Phi+\xi)} + (1-\delta)\sqrt{(\Phi+\xi)/(\Phi+1)}.$$

Comme $\sqrt{(\Phi+\xi)}$ et $\sqrt{(\Phi+\xi)/(\Phi+1)}$ sont croissantes puisque $0 < \xi < 1$, $\sqrt{\pi/2}B/C$ est croissante de Φ et prend son maximum pour $\Phi = b$.

(3) Dans le cas I3,

$$\sqrt{\pi/2}B/C = \delta\sqrt{(\Phi+\xi)/\Phi} + (1-\delta)\sqrt{(\Phi+\xi)/(\Phi+1)}.$$

On a

$$[2\sqrt{\Phi+\xi}(\sqrt{\Phi})^3(\sqrt{\Phi+1})^3] \times [\partial(\sqrt{\pi/2}B/C)/\partial\Phi] = T$$

avec

$$T = \xi\delta(\sqrt{\Phi})^3 - \sqrt{\Phi+1})^3) + (1-\xi-\delta)\sqrt{\Phi})^3.$$

Comme $1 - \xi - \delta = Log(1 + \delta)/Log(2)$ et $\sqrt{\Phi})^3 - \sqrt{\Phi + 1})^3$ sont négatifs, $\sqrt{\pi/2}B/C$ est décroissante de Φ et prend son maximum pour $\Phi = b$. Finalement il suffit de vérifier l'inégalité en $\Phi = b$ soit $\delta\sqrt{(b + \xi)/b} + (1 - \delta)\sqrt{(b + \xi)/(b + 1)} \leq 1$, soit encore

$$\text{(I4)} \qquad g(\delta) \leq 0 \text{ pour tout } \delta \text{ avec } 0 < \delta < 1,$$

où $g = \xi + b - b/[\delta(1 - \sqrt{b/(b + 1)}) + \sqrt{b/(b + 1)}]^2$ et $\xi = \xi(\delta) = 1 - Log(1 + \delta)/Log(2)$. Nous vérifierons (I4) sur $0 \leq \delta \leq 1$, avec $\xi(0) = 1$ et $\xi(1) = 0$. (I4) est vraie aux deux extrémités. Posant $x = \sqrt{b/(b + 1)}$, soit $b = x^2/(1 - x^2)$, puis $N(\delta) = (1 + \delta)Log(2))[\delta(1 - x) + x]^3$, enfin $M(\delta) = -[\delta(1 - x) + x]^3 + 2x^2Log(2)(1 + \delta)/(1 + x)$, il vient $N(\delta)(dg/d\delta) = M(\delta)$. Comme $N(\delta) \geq 0$, $dg/d\delta$ a le signe de la fonction $M(\delta)$, laquelle est visiblement concave pour $\delta \geq 0$. Si nous prouvons

$$\text{(I5)}\ -x^3 + 2x^2Log(2)/(1 + x) = M(0) < 0 < M(1) = -1 + 4x^2Log(2)/(1 + x),$$

nous prouvons du même coup que M a un zéro entre 0 et 1, donc (I4). Une calculette suffit pour montrer que $\gamma_o = \cdot5821$ vérifie (c1) et que le x associé vérifie (I5).

References

1. Bretagnolle, J. (1982). Sur l'inégalité de concentrattion de Doeblin–Levy–Kolmogorov–Rogozin–Kesten, *Preprint*, Université Paris-Nord, 1–13.

2. Doeblin W. et Lévy P. (1936). Sur les sommes de variables aléatoires indépendantes à dispersion bornée inférieurement, *Comptes Rendus Acad. Sci.*, **202**, 2027–2028.

3. Kesten H. (1969). A sharper form of the Doeblin–Lévy–Kolmogorov–Rogozin inequality for concentration functions, *Math. Scand.*, **25**, 133–144.

4. Petrov, V. V. (1975). *Sums of Independent Random Variables*, Springer-Verlag, New York.

5. Rogozin, B. A. (1961a). An estimate for concentration functions, *Teor. Veroyatn. Primen.*, **6**, 103–105 (traduction *Theory of Probability and its Applications*, **6**, 94–96).

6. Rogozin, B. A. (1961b). On the Increase of Dispersion of Sums of Independent Random Variables, *Teor. Veroyatn. Primen.*, **6**, 106–108 (translation *Theory of Probability and its Applications*, **6**, 97–99).

7. Rogozin, B. A. (1987). Inequalities for Concentration Functions of Convolutions of Arithmetic Distributions and Distributions with Bounded Densities, *Teor. Veroyatn. Primen.*, **32**, 325–329 (translation *Theory of Probability and its Applications*, **32**, 351–355).

8. Sperner, E. (1928). Ein Satz über Untermengen einer endlichen Menge *Math. Zeit. Z*, **27**, 544–548.

Index

MIX
Papier aus verantwortungsvollen Quellen
Paper from responsible sources
FSC® C105338

If you have any concerns about our products,
you can contact us on
ProductSafety@springernature.com

In case Publisher is established outside the EU,
the EU authorized representative is:
Springer Nature Customer Service Center GmbH
Europaplatz 3, 69115 Heidelberg, Germany

Printed by Libri Plureos GmbH
in Hamburg, Germany